Lie Algebras of Finite and Affine Type

Lie Algebras of Finite and Affine Type

R. W. CARTER

Mathematics Institute
University of Warwick

CAMBRIDGE
UNIVERSITY PRESS

University Printing House, Cambridge CB2 8BS, United Kingdom

Cambridge University Press is part of the University of Cambridge.

It furthers the University's mission by disseminating knowledge in the pursuit of education, learning and research at the highest international levels of excellence.

www.cambridge.org
Information on this title: www.cambridge.org/9780521851381

© Cambridge University Press 2005

This publication is in copyright. Subject to statutory exception and to the provisions of relevant collective licensing agreements, no reproduction of any part may take place without the written permission of Cambridge University Press.

First published 2005

A catalogue record for this publication is available from the British Library

ISBN 978-0-521-85138-1 Hardback

Digitally Printed in Korea at Kyobo Book Centre

Cambridge University Press has no responsibility for the persistence or accuracy of URLs for external or third-party internet websites referred to in this publication, and does not guarantee that any content on such websites is, or will remain, accurate or appropriate.

Dedicated to Sandy Green

Contents

	Preface	*page* xiii
1	**Basic concepts**	1
	1.1 Elementary properties of Lie algebras	1
	1.2 Representations and modules	5
	1.3 Abelian, nilpotent and soluble Lie algebras	7
2	**Representations of soluble and nilpotent Lie algebras**	11
	2.1 Representations of soluble Lie algebras	11
	2.2 Representations of nilpotent Lie algebras	14
3	**Cartan subalgebras**	23
	3.1 Existence of Cartan subalgebras	23
	3.2 Derivations and automorphisms	25
	3.3 Ideas from algebraic geometry	27
	3.4 Conjugacy of Cartan subalgebras	33
4	**The Cartan decomposition**	36
	4.1 Some properties of root spaces	36
	4.2 The Killing form	39
	4.3 The Cartan decomposition of a semisimple Lie algebra	45
	4.4 The Lie algebra $\mathfrak{sl}_n(\mathbb{C})$	52
5	**The root system and the Weyl group**	56
	5.1 Positive systems and fundamental systems of roots	56
	5.2 The Weyl group	59
	5.3 Generators and relations for the Weyl group	65

6	**The Cartan matrix and the Dynkin diagram**	69
	6.1 The Cartan matrix	69
	6.2 The Dynkin diagram	72
	6.3 Classification of Dynkin diagrams	74
	6.4 Classification of Cartan matrices	80
7	**The existence and uniqueness theorems**	88
	7.1 Some properties of structure constants	88
	7.2 The uniqueness theorem	93
	7.3 Some generators and relations in a simple Lie algebra	96
	7.4 The Lie algebras $L(A)$ and $\tilde{L}(A)$	98
	7.5 The existence theorem	105
8	**The simple Lie algebras**	121
	8.1 Lie algebras of type A_l	122
	8.2 Lie algebras of type D_l	124
	8.3 Lie algebras of type B_l	128
	8.4 Lie algebras of type C_l	132
	8.5 Lie algebras of type G_2	135
	8.6 Lie algebras of type F_4	138
	8.7 Lie algebras of types E_6, E_7, E_8	140
	8.8 Properties of long and short roots	145
9	**Some universal constructions**	152
	9.1 The universal enveloping algebra	152
	9.2 The Poincaré–Birkhoff–Witt basis theorem	155
	9.3 Free Lie algebras	160
	9.4 Lie algebras defined by generators and relations	163
	9.5 Graph automorphisms of simple Lie algebras	165
10	**Irreducible modules for semisimple Lie algebras**	176
	10.1 Verma modules	176
	10.2 Finite dimensional irreducible modules	186
	10.3 The finite dimensionality criterion	190
11	**Further properties of the universal enveloping algebra**	201
	11.1 Relations between the enveloping algebra and the symmetric algebra	201
	11.2 Invariant polynomial functions	207
	11.3 The structure of the ring of polynomial invariants	216
	11.4 The Killing isomorphisms	222

	11.5 The centre of the enveloping algebra	226
	11.6 The Casimir element	238
12	**Character and dimension formulae**	**241**
	12.1 Characters of L-modules	241
	12.2 Characters of Verma modules	244
	12.3 Chambers and roots	246
	12.4 Composition factors of Verma modules	255
	12.5 Weyl's character formula	258
	12.6 Complete reducibility	262
13	**Fundamental modules for simple Lie algebras**	**267**
	13.1 An alternative form of Weyl's dimension formula	267
	13.2 Fundamental modules for A_l	268
	13.3 Exterior powers of modules	270
	13.4 Fundamental modules for B_l and D_l	274
	13.5 Clifford algebras and spin modules	281
	13.6 Fundamental modules for C_l	292
	13.7 Contraction maps	295
	13.8 Fundamental modules for exceptional algebras	303
14	**Generalised Cartan matrices and Kac–Moody algebras**	**319**
	14.1 Realisations of a square matrix	319
	14.2 The Lie algebra $\tilde{L}(A)$ associated with a complex matrix	322
	14.3 The Kac–Moody algebra $L(A)$	331
15	**The classification of generalised Cartan matrices**	**336**
	15.1 A trichotomy for indecomposable GCMs	336
	15.2 Symmetrisable generalised Cartan matrices	344
	15.3 The classification of affine generalised Cartan matrices	351
16	**The invariant form, Weyl group and root system**	**360**
	16.1 The invariant bilinear form	360
	16.2 The Weyl group of a Kac–Moody algebra	371
	16.3 The roots of a Kac–Moody algebra	377

17	**Kac–Moody algebras of affine type**	386
	17.1 Properties of the affine Cartan matrix	386
	17.2 The roots of an affine Kac–Moody algebra	394
	17.3 The Weyl group of an affine Kac–Moody algebra	404
18	**Realisations of affine Kac–Moody algebras**	416
	18.1 Loop algebras and central extensions	416
	18.2 Realisations of untwisted affine Kac–Moody algebras	421
	18.3 Some graph automorphisms of affine algebras	426
	18.4 Realisations of twisted affine algebras	429
19	**Some representations of symmetrisable Kac–Moody algebras**	452
	19.1 The category $\mathcal{O}$ of $L(A)$-modules	452
	19.2 The generalised Casimir operator	459
	19.3 Kac' character formula	466
	19.4 Generators and relations for symmetrisable algebras	474
20	**Representations of affine Kac–Moody algebras**	484
	20.1 Macdonald's identities	484
	20.2 Specialisations of Macdonald's identities	491
	20.3 Irreducible modules for affine algebras	494
	20.4 The fundamental modules for $L(\tilde{A}_1)$	504
	20.5 The basic representation	508
21	**Borcherds Lie algebras**	519
	21.1 Definition and examples of Borcherds algebras	519
	21.2 Representations of Borcherds algebras	524
	21.3 The Monster Lie algebra	530

Appendix	540
Summary pages – explanation	540
Type A_l	543
Type B_l	545
Type C_l	547
Type D_l	549
Type E_6	551
Type E_7	553
Type E_8	555
Type F_4	557
Type G_2	559

Type $\tilde{A}_1$	561
Type $\tilde{A}'_1 = {}^2\tilde{A}_2$ (1st description)	563
(2nd description)	565
Type $\tilde{A}_l$	567
Type $\tilde{B}_l$	570
Type $\tilde{B}^t_l = {}^2\tilde{A}_{2l-1}$	573
Type $\tilde{C}_l$	576
Type $\tilde{C}^t_l = {}^2\tilde{D}_{l+1}$	579
Type $\tilde{C}'_l = {}^2\tilde{A}_{2l}$ (1st description)	582
(2nd description)	585
Type $\tilde{D}_4$	588
Type $\tilde{D}_l, \quad l \geq 5$	590
Type $\tilde{E}_6$	593
Type $\tilde{E}_7$	596
Type $\tilde{E}_8$	599
Type $\tilde{F}_4$	602
Type $\tilde{F}^t_4 = {}^2\tilde{E}_6$	604
Type $\tilde{G}_2$	606
Type $\tilde{G}^t_2 = {}^3\tilde{D}_4$	608
Notation	610
Bibliography of books on Lie algebras	619
Bibliography of articles on Kac–Moody algebras	621
Index	629

Preface

Lie algebras were originally introduced by S. Lie as algebraic structures used for the study of Lie groups. The tangent space of a Lie group at the identity element has the natural structure of a Lie algebra, called by Lie the infinitesimal group. However, Lie algebras also proved to be of interest in their own right. The finite dimensional simple Lie algebras over the complex field were investigated independently by E. Cartan and W. Killing and the classification of such algebras was achieved during the decade 1890–1900. Basic ideas on the structure and representation theory of these Lie algebras were also contributed at a later stage by H. Weyl. Since then the theory of finite dimensional simple Lie algebras has found many and varied applications both in mathematics and in mathematical physics, to the extent that it is now generally regarded as one of the classical branches of mathematics.

In 1967 V. G. Kac and R. V. Moody independently introduced the Lie algebras now known as Kac–Moody algebras. The finite dimensional simple Lie algebras are examples of Kac–Moody algebras; but the theory of Kac–Moody algebras is much broader, including many infinite dimensional examples. The Kac–Moody theory has developed rapidly since its introduction and has also turned out to have applications in many areas of mathematics, including among others group theory, combinatorics, modular forms, differential equations and invariant theory. It has also proved important in mathematical physics, where it has applications to statistical physics, conformal field theory and string theory. The representation theory of affine Kac–Moody algebras has been particularly useful in such applications.

In view of these applications it seems clear that the theory of Lie algebras, of both finite and affine types, will continue to occupy a central position in mathematics into the twenty-first century. This expectation provides the motivation for the present volume, which aims to give a mathematically rigorous development of those parts of the theory of Lie algebras most relevant to the understanding of the finite dimensional simple Lie algebras and the Kac–Moody algebras of affine type. A number of books on Lie algebras are confined to the finite dimensional theory, but this seemed too restrictive for the present volume in view of the many current applications of the Kac–Moody theory. On the other hand the Kac–Moody theory needs a prior knowledge of the finite dimensional theory, both to motivate it and to supply many technical details. For this reason I have included an account both of the Cartan–Killing–Weyl theory of finite dimensional simple Lie algebras and of the Kac–Moody theory, concentrating particularly on the Kac–Moody algebras of affine type. We work with Lie algebras over the complex field, although any algebraically closed field of characteristic zero would do equally well.

I was introduced to the theory of Lie algebras by an inspiring course of lectures given by Philip Hall at Cambridge University in the late 1950s. I have given a number of lecture courses on finite dimensional Lie algebras at Warwick University, and also two lecture courses on Kac–Moody algebras. The present book has developed as a considerably expanded version of the lecture notes of these courses. The main prerequisite for study of the book is a sound knowledge of linear algebra. I have in fact aimed to make this the sole prerequisite, and to explain from first principles any other techniques which are used in the development.

The most influential book on Kac–Moody algebras is the volume *Infinite-Dimensional Lie Algebras*, third edition (1990), by V. Kac. That formidable treatise contains a development of the Kac–Moody theory presupposing a knowledge of the finite dimensional theory, and includes information on several of the applications. The present volume will not rival Kac' account for experts on Kac–Moody algebras. About half of the theory covered in the 3rd edition of Kac' book has been included. However, for those new to the Kac–Moody theory, our account may be useful in providing a gentler introduction, making use of ideas from the finite dimensional theory developed earlier in the book.

The content of the book can be summarised as follows. The basic definitions of Lie algebras, their subalgebras and ideals, representations and modules, are given in Chapter 1. In Chapter 2 the standard results are proved on the representation theory of soluble and nilpotent Lie algebras. The results on representations of nilpotent Lie algebras are used extensively in the subsequent development. The key idea of a Cartan subalgebra is introduced in Chapter 3, where the existence and conjugacy of Cartan subalgebras are proved. We make use of some ideas from algebraic geometry to prove the conjugacy of Cartan subalgebras. In Chapter 4 the Killing form is introduced and used to describe the Cartan decomposition of a semisimple Lie algebra into root spaces with respect to a Cartan subalgebra. The well-known example of the special linear Lie algebra is used to illustrate the general ideas. In Chapter 5 the Weyl group is introduced and shown to be a Coxeter group. This leads on to the definition of the Cartan matrix and the Dynkin diagram. The possible Dynkin diagrams and Cartan matrices are classified in Chapter 6, and in Chapter 7 the existence and uniqueness of a semisimple Lie algebra with a given Cartan matrix are proved. In Chapter 8 the finite dimensional simple Lie algebras are discussed individually and their root systems determined.

Chapters 9 to 13 are concerned with the representation theory of finite dimensional semisimple Lie algebras. We begin in Chapter 9 with the introduction of the universal enveloping algebra, of free Lie algebras and of Lie algebras defined by generators and relations. The finite dimensional irreducible modules for semisimple Lie algebras are obtained in Chapter 10 as quotients of infinite dimensional Verma modules with dominant integral highest weight. In Chapter 11 the enveloping algebra is studied in more detail. Its centre is shown to be isomorphic to the algebra of polynomial functions on a Cartan subalgebra invariant under the Weyl group, and to the algebra of polynomial functions on the Lie algebra invariant under the adjoint group. This algebra is shown to be isomorphic to a polynomial algebra. The properties of the Casimir element of the centre of the enveloping algebra are also discussed. These are important in subsequent applications to representation theory. Characters of modules are introduced in Chapter 12, and Weyl's character formula for the irreducible modules is proved. The fundamental irreducible modules for the finite dimensional simple Lie algebras are discussed individually in Chapter 13. Their discussion involves exterior powers of modules, Clifford algebras and spin modules, and contraction maps.

This concludes the development of the structure and representation theory of the finite dimensional Lie algebras. This development has concentrated particularly on the properties necessary to obtain the classification of the simple Lie algebras and their finite dimensional irreducible modules. Among the significant results omitted from our account are Ado's theorem on the existence of a faithful finite dimensional module, the radical splitting theorem of Levi, the theorem of Malcev and Harish-Chandra on the conjugacy of complements to the radical, and the cohomology theory of Lie algebras.

The theory of Kac–Moody algebras is introduced in Chapter 14, where the Kac–Moody algebra associated to a generalised Cartan matrix is defined. In fact there are two slightly different definitions of a Kac–Moody algebra which have been used. There is a definition in terms of generators and relations which appears the more natural, but there is a different definition, given by Kac in his book, which is more convenient when one wishes to show that a given Lie algebra is a Kac–Moody algebra. I have used the latter definition, but have included a proof that, at least for symmetrisable generalised Cartan matrices, the two definitions are equivalent.

The trichotomy of indecomposable generalised Cartan matrices into those of finite, affine and indefinite types is obtained in Chapter 15. The Kac–Moody algebras of finite type turn out to be precisely the non-trivial finite dimensional simple Lie algebras, and a classification of those of affine type is given. The important special case of symmetrisable Kac–Moody algebras is also introduced. This class includes all those of finite and affine types, and some of those of indefinite type. In Chapter 16 it is shown that symmetrisable algebras have an invariant bilinear form, which plays a key role in the subsequent development. The Weyl group and root system of a Kac–Moody algebra are also discussed. The roots divide into real roots and imaginary roots, and a remarkable theorem of Kac is proved which characterises the set of positive imaginary roots. Kac–Moody algebras of affine type are singled out for more detailed discussion in Chapter 17. In Chapter 18 it is shown how some of them can be realised in terms of a central extension of a loop algebra of a finite dimensional simple Lie algebra, whereas the remainder can be obtained as fixed point subalgebras of these under a twisted graph automorphism.

Chapters 19 and 20 are devoted to the representation theory of Kac–Moody algebras. The representations considered are those from the category $\mathcal{O}$ introduced by Bernstein, Gelfand and Gelfand. In Chapter 19 the irreducible modules in this category are classified, and their characters are obtained in Kac' character formula, a generalisation to the Kac–Moody situation of Weyl's character formula. In Chapter 20 the representations of affine Kac–Moody algebras are discussed. The remarkable identities of I. G. Macdonald are obtained by specialising the denominator of Kac' character formula, interpreted in two different ways; one as an infinite sum and the other as an infinite product. The phenomenon of strings of weights with non-decreasing multiplicities is investigated inside an irreducible module for an affine algebra.

Many of the applications of the representation theory of affine Kac–Moody algebras use the theory of vertex operators. This theory lies beyond the scope of the present volume. However, we have introduced the idea of a vertex operator in Chapter 20 with the aim of encouraging the reader to explore the subject further.

A theory of generalised Kac–Moody algebras was introduced in 1988 by R. Borcherds. These Lie algebras were introduced as part of Borcherds' proof of the Conway–Norton conjectures on the representation theory of the Monster simple group. They are now frequently called Borcherds algebras. In Chapter 21 we have given an account of Borcherds algebras, including the definition and statements of the main results concerning their structure and representation theory, but detailed proofs are not given. Many of the results on Borcherds algebras are quite similar to those for Kac–Moody algebras, but there are examples of Borcherds algebras which are quite different from Kac–Moody algebras. The best known such example is the Monster Lie algebra, which we describe in the final section.

We conclude with an appendix containing one section for each of the algebras of finite and affine types, in which the most important pieces of information about the algebra concerned are collected.

I would like to express my thanks to Roger Astley of Cambridge University Press for his encouragement to complete the half finished manuscript of this book. This was eventually achieved after I had reached the status of Emeritus Professor, and therefore had more time to devote to it. I would also like to thank my colleague Bruce Westbury for the sustained interest he has shown in this work.

1
Basic concepts

1.1 Elementary properties of Lie algebras

A **Lie algebra** is a vector space L over a field k on which a multiplication

$$L \times L \to L$$
$$(x, y) \to [xy]$$

is defined satisfying the following axioms:

(i) $(x, y) \to [xy]$ is linear in x and in y;
(ii) $[xx] = 0$ for all $x \in L$;
(iii) $[[xy]z] + [[yz]x] + [[zx]y] = 0$ for all $x, y, z \in L$.

Axiom (iii) is called the **Jacobi identity**.

Proposition 1.1 $[yx] = -[xy]$ *for all* $x, y \in L$.

Proof. Since $[x+y, x+y] = 0$ we have $[xx] + [xy] + [yx] + [yy] = 0$. It follows that $[xy] + [yx] = 0$, that is $[yx] = -[xy]$. $\square$

Proposition 1.1 asserts that multiplication in a Lie algebra is anticommutative.

Now let H, K be subspaces of a Lie algebra L. Then $[HK]$ is defined as the subspace spanned by all products $[xy]$ with $x \in H$ and $y \in K$. Each element of $[HK]$ is a sum

$$[x_1 y_1] + \cdots + [x_r y_r]$$

with $x_i \in H$, $y_i \in K$.

Proposition 1.2 $[HK] = [KH]$ *for all subspaces* H, K *of* L.

Proof. Let $x \in H$, $y \in K$. Then $[xy] = [-y, x] \in [KH]$. This shows that $[HK] \subset [KH]$. Similarly we have $[KH] \subset [HK]$ and so we have equality. □

Proposition 1.2 asserts that multiplication of subspaces in a Lie algebra is commutative.

Example 1.3 Let A be an associative algebra over k. Thus we have a map

$$A \times A \to A$$
$$(x, y) \to xy$$

satisfying the associative law

$$(xy)z = x(yz) \quad \text{for all } x, y, z \in A.$$

Then A can be made into a Lie algebra by defining the Lie product $[xy]$ by

$$[xy] = xy - yx$$

We verify the Lie algebra axioms. Product $[xy]$ is clearly linear in x and in y. It is also clear that $[xx] = 0$. Finally we check the Jacobi identity. We have

$$[[xy]z] = (xy - yx)z - z(xy - yx)$$
$$= xyz - yxz - zxy + zyx.$$

We have similar expressions for $[[yz]x]$ and $[[zx]y]$. Hence

$$[[xy]z] + [[yz]x] + [[zx]y] = xyz - yxz - zxy + zyx + yzx - zyx - xyz$$
$$+ xzy + zxy - xzy - yzx + yxz = 0. \quad \square$$

The Lie algebra obtained from the associative algebra A in this way will be denoted by $[A]$.

Now let L be a Lie algebra over k. A subset H of L is called a **subalgebra** of L if H is a subspace of L and $[HH] \subset H$. Thus H is itself a Lie algebra under the same operations as L.

A subset I of L is called an **ideal** of L if I is a subspace of L and $[IL] \subset I$. We observe that the latter condition is equivalent to $[LI] \subset I$. Thus there is no distinction between left ideals and right ideals in the theory of Lie algebras. Every ideal is two-sided.

Proposition 1.4 (i) *If H, K are subalgebras of L so is $H \cap K$.*
(ii) *If H, K are ideals of L so is $H \cap K$.*

(iii) *If H is an ideal of L and K a subalgebra of L then $H+K$ is a subalgebra of L.*
(iv) *If H, K are ideals of L then $H+K$ is an ideal of L.*

Proof. (i) $H \cap K$ is a subspace of L and $[H \cap K, H \cap K] \subset [HH] \cap [KK] \subset H \cap K$. Thus $H \cap K$ is a subalgebra.
(ii) This time we have $[H \cap K, L] \subset [HL] \cap [KL] \subset H \cap K$. Thus $H \cap K$ is an ideal of L.
(iii) $H+K$ is a subspace of L. Also $[H+K, H+K] \subset [HH]+[HK]+[KH]+[KK] \subset H+K$, since $[HH] \subset H, [HK] \subset H, [KK] \subset K$. Thus $H+K$ is a subalgebra.
(iv) This time we have $[H+K, L] \subset [HL]+[KL] \subset H+K$. Thus $H+K$ is an ideal of L. $\square$

We next introduce the idea of a factor algebra. Let I be an ideal of a Lie algebra L. Then I is in particular a subspace of L and so we can form the factor space L/I whose elements are the cosets $I+x$ for $x \in L$. $I+x$ is the subset of L consisting of all elements $y+x$ for $y \in I$.

Proposition 1.5 *Let I be an ideal of L. Then the factor space L/I can be made into a Lie algebra by defining*

$$[I+x, I+y] = I+[xy] \quad \text{for all } x, y \in L.$$

Proof. We must first show that this definition is unambiguous, that is if $I+x = I+x'$ and $I+y = I+y'$ then $I+[xy] = I+[x'y']$.

Now $I+x = I+x'$ implies that $x = x'+i_1$ for some $i_1 \in I$. Similarly $I+y = I+y'$ implies $y = y'+i_2$ for some $i_2 \in I$. Thus

$$I+[xy] = I+[x'+i_1, y'+i_2]$$
$$= I+[i_1 y']+[x' i_2]+[i_1 i_2]+[x' y']$$
$$= I+[x' y']$$

since $[i_1 y'], [x' i_2], [i_1 i_2]$ all lie in I. Thus our multiplication is well defined. We also have

$$[I+x, I+x] = I+[xx] = I$$

and the Jacobi identity in L/I clearly follows from the Jacobi identity in L. $\square$

Now suppose we have two Lie algebras L_1, L_2 over k. A map $\theta: L_1 \to L_2$ is called a **homomorphism of Lie algebras** if θ is linear and

$$\theta[xy] = [\theta x, \theta y] \quad \text{for all } x, y \in L_1.$$

The map $\theta: L_1 \to L_2$ is called an **isomorphism of Lie algebras** if θ is a bijective homomorphism of Lie algebras. The Lie algebras L_1, L_2 are said to be **isomorphic** if there exists an isomorphism $\theta: L_1 \to L_2$.

Proposition 1.6 *Let $\theta: L_1 \to L_2$ be a homomorphism of Lie algebras. Then the image of θ is a subalgebra of L_2, the kernel of θ is an ideal of L_1 and $L_1/\ker \theta$ is isomorphic to $\operatorname{im} \theta$.*

Proof. $\operatorname{im} \theta$ is a subspace of L_2. Moreover for x, y in L_1 we have

$$[\theta(x), \theta(y)] = \theta[xy] \in \operatorname{im} \theta.$$

Hence $\operatorname{im} \theta$ is a subalgebra of L_2.

Now $\ker \theta$ is a subspace of L_1. Let $x \in \ker \theta$ and $y \in L_1$. Then

$$\theta[xy] = [\theta(x), \theta(y)] = [0, \theta(y)] = 0.$$

Hence $[xy] \in \ker \theta$ and so $\ker \theta$ is an ideal of L_1.

Now let $x, y \in L_1$. We consider when $\theta(x)$ is equal to $\theta(y)$. We have

$$\theta(x) = \theta(y) \Leftrightarrow \theta(x-y) = 0 \Leftrightarrow x - y \in \ker \theta$$

$$\Leftrightarrow \ker \theta + x = \ker \theta + y.$$

This shows that there is a bijective map $\theta(x) \to \ker \theta + x$ between $\operatorname{im} \theta$ and $L_1/\ker \theta$. We show this bijection is an isomorphism of Lie algebras. It is clearly linear. Moreover given $x, y, z \in L_1$ we have

$$[\theta(x), \theta(y)] = \theta(z) \Leftrightarrow \theta[xy] = \theta(z)$$

$$\Leftrightarrow \ker \theta + [xy] = \ker \theta + z$$

$$\Leftrightarrow [\ker \theta + x, \ker \theta + y] = \ker \theta + z.$$

Thus the bijection preserves Lie multiplication, so is an isomorphism of Lie algebras. □

Proposition 1.7 *Let I be an ideal of L and H a subalgebra of L. Then*

(i) *I is an ideal of $I + H$.*
(ii) *$I \cap H$ is an ideal of H.*
(iii) *$(I + H)/I$ is isomorphic to $H/(I \cap H)$.*

Proof. We recall from Proposition 1.4 that $I \cap H$ and $I + H$ are subalgebras. We have $[I, I+H] \subset [IL] \subset I$, thus I is an ideal of $I+H$. Also $[I \cap H, H] \subset [IH] \cap [HH] \subset I \cap H$, thus $I \cap H$ is an ideal of H.

Let $\theta : H \to (I+H)/I$ be defined by $\theta(x) = I + x$. This is clearly a linear map, and is also evidently a homomorphism of Lie algebras. It is surjective since each element of $(I+H)/I$ has form $I+x$ for some $x \in H$. Finally its kernel is the set of $x \in H$ for which $I + x = I$, that is $I \cap H$. Thus $(I+H)/I$ is isomorphic to $H/(I \cap H)$ by Proposition 1.6. $\square$

1.2 Representations and modules

Let $M_n(k)$ be the associative algebra of all $n \times n$ matrices over the field k and let $[M_n(k)]$ be the corresponding Lie algebra. This is often called the **general linear Lie algebra** of degree n over k and we write

$$\mathfrak{gl}_n(k) = [M_n(k)].$$

We have $\dim \mathfrak{gl}_n(k) = n^2$.

A **representation** of a Lie algebra L over k is a homomorphism of Lie algebras

$$\rho : L \to \mathfrak{gl}_n(k)$$

for some n, and ρ is called a representation of degree n. Two representations ρ, ρ' of degree n are called **equivalent** if there exists a non-singular $n \times n$ matrix T such that

$$\rho'(x) = T^{-1} \rho(x) T \qquad \text{forall } x \in L.$$

A **left L-module** is a vector space V over k together with a multiplication

$$L \times V \to V$$

$$(x, v) \to xv$$

satisfying the axioms:

(i) $(x, v) \to xv$ is linear in x and in v;
(ii) $[xy]v = x(yv) - y(xv)$ for all $x, y \in L$ and $v \in V$.

Suppose V is a finite dimensional L-module. Let $e_1, \ldots, e_n$ be a basis of V. Let

$$xe_j = \sum_i \rho_{ij}(x) e_i$$

with $\rho_{ij}(x) \in k$ and let $\rho(x) = (\rho_{ij}(x))$. Then ρ is a representation of L. For we have

$$[xy]e_j = x(ye_j) - y(xe_j)$$

$$= x\left(\sum_k \rho_{kj}(y)e_k\right) - y\left(\sum_k \rho_{kj}(x)e_k\right)$$

$$= \sum_k \rho_{kj}(y)xe_k - \sum_k \rho_{kj}(x)ye_k$$

$$= \sum_k \rho_{kj}(y)\left(\sum_i \rho_{ik}(x)e_i\right) - \sum_k \rho_{kj}(x)\left(\sum_i \rho_{ik}(y)e_i\right)$$

$$= \sum_i \left(\sum_k (\rho_{ik}(x)\rho_{kj}(y) - \rho_{ik}(y)\rho_{kj}(x))\right)e_i$$

$$= \sum_i (\rho(x)\rho(y) - \rho(y)\rho(x))_{ij} e_i.$$

Thus $\rho[xy] = \rho(x)\rho(y) - \rho(y)\rho(x) = [\rho(x), \rho(y)]$ and ρ is a representation of L.

Suppose now we take a second basis $f_1, \ldots, f_n$ of V. Let ρ' be the representation of L obtained from this basis. Then ρ' is equivalent to ρ. For there exists a non-singular $n \times n$ matrix T such that

$$f_j = \sum_i T_{ij}e_i.$$

Thus we have

$$xf_j = \sum_k T_{kj}xe_k = \sum_k T_{kj}\left(\sum_i \rho_{ik}(x)e_i\right) = \sum_i \left(\sum_k \rho_{ik}(x)T_{kj}\right)e_i.$$

On the other hand

$$xf_j = \sum_k \rho'_{kj}(x)f_k = \sum_k \rho'_{kj}(x)\left(\sum_i T_{ik}e_i\right) = \sum_i \left(\sum_k T_{ik}\rho'_{kj}(x)\right)e_i.$$

It follows that $\rho(x)T = T\rho'(x)$, that is $\rho'(x) = T^{-1}\rho(x)T$ for all $x \in L$. Hence the representation ρ' is equivalent to ρ. □

Example 1.8 L is itself a left L-module.

The left action of L on L is defined as $x \cdot y = [xy]$. Then we have

$$[[xy]z] = [x[yz]] - [y[xz]]$$

which is a consequence of the Jacobi identity. This shows that L is a left L-module. This is called the **adjoint module**. We define $\mathrm{ad}\, x : L \to L$ by

$$\mathrm{ad}\, x \cdot y = [xy] \qquad \text{for } x, y \in L.$$

Then we have

$$\mathrm{ad}[xy] = \mathrm{ad}\, x\, \mathrm{ad}\, y - \mathrm{ad}\, y\, \mathrm{ad}\, x. \qquad \square$$

Now let V be a left L-module, U be a subspace of V and H a subspace of L. We define HU to be the subspace of V spanned by all elements of the form xu for $x \in H$, $u \in U$.

A **submodule** of V is a subspace U of V such that $LU \subset U$. In particular V is a submodule of V and the zero subspace $O = \{0\}$ is a submodule of V. A **proper submodule** of V is a submodule distinct from V and O.

An L-module V is called **irreducible** if it has no proper submodules. V is called **completely reducible** if it is a direct sum of irreducible submodules. V is called **indecomposable** if V cannot be written as a direct sum of two proper submodules. Of course every irreducible L-module is indecomposable, but the converse need not be true.

We may also define right L-modules, but we shall mainly work with left L-modules, and L-modules will be assumed to be left L-modules unless otherwise stated.

1.3 Abelian, nilpotent and soluble Lie algebras

A Lie algebra L is **abelian** if $[LL] = O$. Thus $[xy] = 0$ for all $x, y \in L$ when L is abelian.

Given any Lie algebra L we define the powers of L by

$$L^1 = L, \qquad L^{n+1} = [L^n L] \qquad \text{for } n \geq 1.$$

Thus L is abelian if and only if $L^2 = O$.

Proposition 1.9 *L^n is an ideal of L. Also*

$$L = L^1 \supset L^2 \supset L^3 \supset \cdots.$$

Proof. We first observe that if I, J are ideals of L then $[IJ]$ is also an ideal of L. For let $x \in I$, $y \in J$, $z \in L$. Then

$$[[xy]z] = [x[yz]] - [y[xz]] \in [IJ].$$

It follows that L^n is an ideal of L for each $n > 0$. Thus we have
$$L^{n+1} = [L^n L] \subset L^n.$$
□

A Lie algebra L is called **nilpotent** if $L^n = O$ for some $n \geq 1$. Thus every abelian Lie algebra is nilpotent. It is clear that every subalgebra and every factor algebra of a nilpotent Lie algebra are nilpotent.

We now consider a different kind of powers of L. We define
$$L^{(0)} = L, \quad L^{(n+1)} = [L^{(n)}, L^{(n)}] \quad \text{for } n \geq 0.$$

Proposition 1.10 $L^{(n)}$ *is an ideal of L. Also*
$$L = L^{(0)} \supset L^{(1)} \supset L^{(2)} \supset \cdots .$$

Proof. $L^{(n)}$ is an ideal of L since the product of two ideals is an ideal. Also
$$L^{(n+1)} = [L^{(n)}, L^{(n)}] \subset L^{(n)}.$$
□

A Lie algebra L is called **soluble** if $L^{(n)} = O$ for some $n \geq 0$.

Proposition 1.11 *(a) $[L^m L^n] \subset L^{m+n}$ for all $m, n \geq 1$. (b) $L^{(n)} \subset L^{2^n}$ for all $n \geq 0$. (c) Every nilpotent Lie algebra is soluble.*

Proof. (a). We use induction on n. The result is clear if $n = 1$. Suppose it is true for $n = r$. Then
$$[L^m L^{r+1}] = [L^m [L^r L]] = [[L^r L] L^m]$$
$$\subset [[LL^m] L^r] + [[L^m L^r] L] \quad \text{by the Jacobi identity}$$
$$\subset [L^{m+1} L^r] + [[L^m L^r] L]$$
$$\subset L^{m+r+1} \quad \text{by inductive hypothesis.}$$

Thus the result holds for $n = r + 1$, so for all n.

(b). We again use induction on n. The result is clear if $n = 1$. Suppose it is true for $n = r$. Then
$$L^{(r+1)} = [L^{(r)} L^{(r)}] \subset [L^{2^r} L^{2^r}] \subset L^{2^{r+1}}$$

by (a). Thus the result holds for $n = r + 1$, so for all n.

(c). Suppose L is nilpotent. Then $L^{2^n} = O$ for n sufficiently large. Hence $L^{(n)} = O$ by (b) and so L is soluble. □

1.3 Abelian, nilpotent and soluble Lie algebras

It is clear that every subalgebra and every factor algebra of a soluble Lie algebra are soluble.

Proposition 1.12 *Suppose I is an ideal of L and both I and L/I are soluble. Then L is soluble.*

Proof. Since L/I is soluble we have $(L/I)^{(n)} = O$ for some n. This implies $L^{(n)} \subset I$. Since I is soluble we have $I^{(m)} = O$ for some m. Hence

$$L^{(n+m)} = (L^{(n)})^{(m)} \subset I^{(m)} = O$$

and so L is soluble. □

Proposition 1.13 *Every finite dimensional Lie algebra L contains a unique maximal soluble ideal R. Also L/R contains no non-zero soluble ideal.*

Proof. Let I, J be soluble ideals of L. Then $I+J$ is also an ideal of L and $(I+J)/I$ is isomorphic to $J/(I \cap J)$ by Proposition 1.7. Now J is soluble, thus $J/(I \cap J)$ is soluble and so $(I+J)/I$ is soluble. Since I is soluble we see that $I+J$ is soluble by Proposition 1.12. Thus the sum of two soluble ideals of L is a soluble ideal. It follows that L has a unique maximal soluble ideal R.

If I/R is a soluble ideal of L/R then I is a soluble ideal of L by Proposition 1.12. Hence $I = R$ and $I/R = O$. □

The ideal R is called the **soluble radical** of L. A Lie algebra L is called **semisimple** if $R = O$. Thus L is semisimple if and only if L has no non-zero soluble ideal.

L is called **simple** if L has no proper ideal, that is no ideal other than L and O.

Suppose L is a Lie algebra of dimension 1 over k. Then L has a basis $\{x\}$ with 1 element. Since $[xx] = 0$ we have $L^2 = O$. Thus L is abelian. We see that any two 1-dimensional Lie algebras over k are isomorphic. Of course any such Lie algebra is simple, because L has no proper subspaces. The 1-dimensional Lie algebra is called the **trivial simple Lie algebra**. A non-trivial simple Lie algebra is a simple Lie algebra L with dim $L > 1$.

Proposition 1.14 *Each non-trivial simple Lie algebra is semisimple.*

Proof. Suppose L is simple but not semisimple. Then the soluble radical R satisfies $R \neq O$. Since R is an ideal of L this implies $R = L$. Thus L is soluble.

Hence $L^{(n)} = O$ for some $n \geq 0$. This implies that $L^{(1)} \neq L$ since $L^{(1)} = L$ would imply $L^{(n)} = L$ for all n. Now $L^{(1)}$ is an ideal of L, hence $L^{(1)} = O$ since L is simple. Thus $[LL] = O$. But then every subspace of L is an ideal of L. Since L is simple L has no proper subspaces, so $\dim L = 1$. Thus the only simple Lie algebra which is not semisimple is the trivial simple Lie algebra. □

2
Representations of soluble and nilpotent Lie algebras

2.1 Representations of soluble Lie algebras

We shall now and subsequently take the base field k to be the field $\mathbb{C}$ of complex numbers. We shall also assume until further notice that L is a finite dimensional Lie algebra over $\mathbb{C}$, although at a later stage we shall also consider infinite dimensional Lie algebras.

We first consider 1-dimensional representations of a Lie algebra L. A 1-dimensional representation is a linear map $\rho : L \to \mathbb{C}$ such that $\rho[xy] = [\rho(x), \rho(y)]$ for all $x, y \in L$.

Lemma 2.1 *A linear map $\rho : L \to \mathbb{C}$ is a 1-dimensional representation of L if and only if ρ vanishes on L^2.*

Proof. Suppose ρ is a representation. Then for $x, y \in L$ we have

$$\rho[xy] = [\rho(x), \rho(y)] = \rho(x)\rho(y) - \rho(y)\rho(x) = 0.$$

Hence ρ vanishes on L^2.

Conversely suppose that ρ vanishes on L^2. Then

$$\rho[xy] = 0 = [\rho(x), \rho(y)]$$

and so ρ is a representation of L. □

We shall now prove a theorem of Lie which shows that any irreducible representation of a soluble Lie algebra is 1-dimensional.

Theorem 2.2 *(Lie's theorem). Let L be a soluble Lie algebra and V be a finite dimensional irreducible L-module. Then $\dim V = 1$.*

Proof. Since L is soluble we have $L^2 \ne L$. Let I be a subspace of L such that $I \supset L^2$ and $\dim I = \dim L - 1$. Then I is an ideal of L since

$$[IL] \subset [LL] = L^2 \subset I.$$

Thus I is an ideal of L of codimension 1.

We shall prove Lie's theorem by induction on $\dim L$. Suppose $\dim L = 1$ and V be an irreducible L-module. Let $L = \mathbb{C}x$ and v be an eigenvector of x in V. Then $\mathbb{C}v$ is an L-submodule of V. Since V is irreducible we have $V = \mathbb{C}v$ and $\dim V = 1$.

Now suppose $\dim L > 1$ and V is an irreducible L-module. We may regard V as an I-module. Then V contains an irreducible I-submodule W and we may assume $\dim W = 1$ by induction. Let w be a non-zero vector in W. Then

$$yw = \lambda(y)w \quad \text{for all } y \in I$$

where λ is the 1-dimensional representation of I given by W. Let

$$U = \{u \in V \,;\, yu = \lambda(y)u \quad \text{for all } y \in I\}.$$

Then we have

$$0 \ne W \subset U \subset V.$$

We shall show that U is an L-submodule of V. Let $u \in U$, $x \in L$. Then

$$y(xu) = x(yu) - [xy]u = \lambda(y)xu - \lambda([xy])u$$

since $[xy] \in I$. We shall show $\lambda([xy]) = 0$. Once we know this we have $xu \in U$ and so U is an L-submodule. Since V is irreducible we have $U = V$. Hence

$$yv = \lambda(y)v \quad \text{for all } v \in V, y \in I.$$

Since $\dim I = \dim L - 1$ we can write $L = I \oplus \mathbb{C}x$, a direct sum of subspaces. Let v be an eigenvector for x on V. Then $\mathbb{C}v$ is an L-submodule of V, being invariant under the action of both I and x. Since V is irreducible we have $V = \mathbb{C}v$ and so $\dim V = 1$.

In order to complete the proof we must show that $\lambda([xy]) = 0$ for all $x \in L$, $y \in I$. In fact it is sufficient to prove this for the element x chosen above such that $L = I \oplus \mathbb{C}x$.

Let u be any non-zero element of U. We write

$$v_0 = u, \quad v_1 = xu, \quad v_2 = x(xu), \ldots$$

We have $v_0, v_1, v_2, \ldots \in V$ and so there exists $p \ge 0$ such that $v_0, v_1, \ldots, v_p$ are linearly independent and v_{p+1} is a linear combination of these. Consider the subspace $\langle v_0, v_1, \ldots, v_p \rangle$ of V spanned by these vectors. This subspace

2.1 Representations of soluble Lie algebras

is invariant under the action of x. We consider the effect on this subspace of elements $y \in I$. We have

$$yv_0 = yu = \lambda(y)u = \lambda(y)v_0.$$

We shall show

$$yv_i = \lambda(y)v_i + \text{a linear combination of } v_0, \ldots, v_{i-1}.$$

This is true for $i=0$. Assuming it for v_{i-1} we have

$$\begin{aligned} yv_i &= y(xv_{i-1}) = x(yv_{i-1}) - [xy]v_{i-1} \\ &= x(\lambda(y)v_{i-1} + \text{a linear combination of } v_0, \ldots, v_{i-2}) \\ &\quad - (\text{a linear combination of } v_0, \ldots, v_{i-1}) \\ &= \lambda(y)v_i + \text{a linear combination of } v_0, \ldots, v_{i-1}. \end{aligned}$$

Thus the subspace $\langle v_0, v_1, \ldots, v_p \rangle$ is invariant under the action of y for all $y \in I$, as well as being invariant under x. Hence it is an L-submodule of V. Since V is irreducible we have

$$V = \langle v_0, v_1, \ldots, v_p \rangle.$$

Now $[xy] \in I$ and we see from the above description of the action of I that

$$\text{trace}_V[xy] = (p+1)\lambda([xy]).$$

Thus we have $(p+1)\lambda([xy]) = \text{trace}_V[xy] = \text{trace}_V xy - \text{trace}_V yx = 0$, since $\text{trace}_V xy = \text{trace}_V yx$. Hence $\lambda([xy]) = 0$ and the proof is complete. □

Corollary 2.3 *Let L be soluble and V be a finite dimensional L-module. Then a basis can be chosen for V with respect to which we obtain a matrix representation ρ of L of the form*

$$\rho(x) = \begin{pmatrix} * & & & & & \\ 0 & * & & & * & \\ 0 & & \cdot & & & \\ \cdot & & & 0 & & \cdot \\ \cdot & & & & & * \\ 0 & \cdot & \cdot & 0 & 0 & * \end{pmatrix} \quad \text{for all } x \in L.$$

Thus the matrices representing elements of L are all of triangular form.

Corollary 2.4 *Let L be a soluble Lie algebra with $\dim L = n$. Then L has a chain of ideals*

$$0 = I_0 \subset I_1 \subset \cdots \subset I_{n-1} \subset I_n = L$$

with $\dim I_r = r$.

Proof. We apply Theorem 2.2 to the adjoint L-module L. The submodules of L are the ideals of L. By taking a maximal chain of submodules we obtain ideals of L with the required property. $\square$

2.2 Representations of nilpotent Lie algebras

When L is a nilpotent Lie algebra we can obtain even stronger results about its representations. Moreover these results on representations of nilpotent Lie algebras play a crucial role in the understanding of semisimple Lie algebras, which we shall deal with subsequently. We begin by recalling results from linear algebra related to the Jordan canonical form. Any $n \times n$ matrix over $\mathbb{C}$ is similar to a diagonal sum of Jordan block matrixes of form

$$\begin{pmatrix} \lambda & 1 & & & & \\ & \lambda & 1 & & 0 & \\ & & \cdot & \cdot & & \\ & & & \cdot & \cdot & \\ & & & & \cdot & 1 \\ & 0 & & & \lambda & 1 \\ & & & & & \lambda \end{pmatrix}$$

In a similar way any linear transformation $\theta : V \to V$ on a finite dimensional vector space V over $\mathbb{C}$ gives rise to a decomposition of V as in the following proposition.

Proposition 2.5 *Let $\theta : V \to V$ be a linear map with characteristic polynomial*

$$\chi(t) = (t - \lambda_1)^{m_1} (t - \lambda_2)^{m_2} \ldots (t - \lambda_r)^{m_r}$$

where $\lambda_1, \ldots, \lambda_r$ are the distinct eigenvalues of θ and $m_1, \ldots, m_r$ are their multiplicities. Let V_i be the set of all $v \in V$ annihilated by some power of $\theta - \lambda_i 1$. Then we have

$$V = V_1 \oplus V_2 \oplus \cdots \oplus V_r.$$

Moreover $\dim V_i = m_i$, $\theta(V_i) \subset V_i$ and the characteristic polynomial of θ on V_i is $(t - \lambda_i)^{m_i}$.

2.2 Representations of nilpotent Lie algebras

Proof. Although this is a standard result from linear algebra we shall prove it in view of its importance for the theory of Lie algebras.

We begin by showing that V_i is equal to $W_i = \{v \in V \,;\, (\theta - \lambda_i 1)^{m_i} v = 0\}$. It is clear that $W_i \subset V_i$. So let $v \in V_i$. Then

$$(\theta - \lambda_i 1)^N v = 0 \quad \text{for some } N.$$

We may choose $N \geq m_i$. Also

$$\prod_{j=1}^{r} (\theta - \lambda_j 1)^{m_j} v = 0$$

for, by the Cayley–Hamilton theorem, θ satisfies its own characteristic equation. Now the polynomials

$$(t - \lambda_i)^N, \quad \prod_{j=1}^{r} (t - \lambda_j)^{m_j}$$

have highest common factor $(t - \lambda_i)^{m_i}$. Thus there exist polynomials $p(t), q(t) \in \mathbb{C}[t]$ such that

$$(t - \lambda_i)^{m_i} = p(t)(t - \lambda_i)^N + q(t) \prod_{j=1}^{r} (t - \lambda_j)^{m_j}.$$

Hence

$$(\theta - \lambda_i 1)^{m_i} v = p(\theta)(\theta - \lambda_i 1)^N v + q(\theta) \prod_{j=1}^{r} (\theta - \lambda_j 1)^{m_j} v = 0.$$

Thus $v \in W_i$ and $V_i = W_i$.

We next show that $V = V_1 \oplus \cdots \oplus V_r$. Let $f_i(t) = (t - \lambda_1)^{m_1} \ldots (t - \lambda_{i-1})^{m_{i-1}} (t - \lambda_{i+1})^{m_{i+1}} \ldots (t - \lambda_r)^{m_r}$. Then the polynomials $f_1(t), \ldots, f_r(t)$ have highest common factor 1. Thus there exist polynomials $p_1(t), \ldots, p_r(t) \in \mathbb{C}[t]$ with $\sum_i f_i(t) p_i(t) = 1$.

Let $v \in V$. Then $v = \sum_i f_i(\theta) p_i(\theta) v$. Let $v_i = f_i(\theta) p_i(\theta) v$. Then

$$(\theta - \lambda_i 1)^{m_i} v_i = \chi(\theta)(p_i(\theta) v) = 0$$

by the Cayley–Hamilton theorem. Thus $v_i \in V_i$. Hence $v = v_1 + \cdots + v_r$ with $v_i \in V_i$, and so $V = V_1 + \cdots + V_r$.

In order to show the sum is direct we must prove

$$V_i \cap (V_1 + \cdots + V_{i-1} + V_{i+1} + \cdots + V_r) = 0.$$

Now the polynomials $(t-\lambda_i)^{m_i}$ and $f_i(t)$ have highest common factor 1, thus there exist $p(t), q(t) \in \mathbb{C}[t]$ with

$$p(t)(t-\lambda_i)^{m_i} + q(t)f_i(t) = 1.$$

Let $v \in V_i \cap (V_1 + \cdots + V_{i-1} + V_{i+1} + \cdots + V_r)$. Since $v \in V_i$ we have

$$(\theta - \lambda_i 1)^{m_i} v = 0.$$

Since $v \in V_1 + \cdots + V_{i-1} + V_{i+1} + \cdots + V_r$ we have

$$f_i(\theta)v = 0.$$

Hence $v = p(\theta)(\theta - \lambda_i 1)^{m_i} v + q(\theta) f_i(\theta) v = 0$. Thus we have shown that

$$V = V_1 \oplus \cdots \oplus V_r.$$

We next observe that θ acts on each V_i. For let $v \in V_i$. Then

$$(\theta - \lambda_i 1)^{m_i} \theta v = \theta(\theta - \lambda_i 1)^{m_i} v = \theta(0) = 0,$$

thus $\theta v \in V_i$.

We next show that the only eigenvalue of $\theta : V_i \to V_i$ is λ_i. Suppose if possible that λ_j is an eigenvalue for some $j \neq i$ and let $v \in V_i$ be an eigenvector for λ_j. Then $v \neq 0$, $(\theta - \lambda_i 1)^{m_i} v = 0$ and $(\theta - \lambda_j 1) v = 0$. But the polynomials $(t - \lambda_i)^{m_i}$ and $t - \lambda_j$ have highest common factor 1 so there exist $p(t), q(t) \in \mathbb{C}[t]$ with

$$p(t)(t - \lambda_i)^{m_i} + q(t)(t - \lambda_j) = 1.$$

Hence $v = p(\theta)(\theta - \lambda_i 1)^{m_i} v + q(\theta)(\theta - \lambda_j 1) v = 0$, a contradiction. So all eigenvalues of $\theta_i : V_i \to V_i$ are equal to λ_i. It follows that $\dim V_i \le m_i$ since m_i is the multiplicity of eigenvalue λ_i on V. But

$$\dim V = \dim V_1 + \cdots + \dim V_r = m_1 + \cdots + m_r.$$

It follows that $\dim V_i = m_i$. Finally the characteristic polynomial of θ on V_i is $(t - \lambda_i)^{m_i}$. □

The subspace V_i is called the **generalised eigenspace** of V with eigenvalue λ_i. Thus the ordinary eigenspace of λ_i lies in the generalised eigenspace. It is not in general true that V is the direct sum of its eigenspaces with respect to its different eigenvalues, but Proposition 2.5 shows that this result is true if the eigenspaces are replaced by the generalised eigenspaces.

The relevance of the decomposition into generalised eigenspaces for the representations of nilpotent Lie algebras is shown by the following theorem.

2.2 Representations of nilpotent Lie algebras

Theorem 2.6 *Let L be a nilpotent Lie algebra and V be an L-module. Let $y \in L$ and $\rho(y) : V \to V$ be the map $v \to yv$. Then the generalised eigenspaces V_i of V associated with $\rho(y)$ are all submodules of V.*

Before proving this theorem we need a preliminary result.

Proposition 2.7 *Let L be a Lie algebra and V be an L-module. Let $v \in V$, $x, y \in L$ and $\alpha, \beta \in \mathbb{C}$. Then*

$$(\rho(y) - (\alpha + \beta)1)^n xv = \sum_{i=0}^{n} \binom{n}{i} ((\operatorname{ad} y - \beta 1)^i x) ((\rho(y) - \alpha 1)^{n-i} v).$$

Proof. We use induction on n. The result is clear when $n = 0$. We assume it for $n = r$. We write

$$x_i = (\operatorname{ad} y - \beta 1)^i x \in L.$$

Then we have

$$(\rho(y) - (\alpha + \beta)1)^{r+1} xv = (\rho(y) - (\alpha + \beta)1) \sum_{i=0}^{r} \binom{r}{i} \rho(x_i)(\rho(y) - \alpha 1)^{r-i} v.$$

Now

$$(\rho(y) - (\alpha + \beta)1) \rho(x_i) = \rho([yx_i]) + \rho(x_i)\rho(y) - (\alpha + \beta)\rho(x_i)$$
$$= \rho((\operatorname{ad} y - \beta 1)x_i) + \rho(x_i)(\rho(y) - \alpha 1)$$
$$= \rho(x_{i+1}) + \rho(x_i)(\rho(y) - \alpha 1).$$

Hence

$$(\rho(y) - (\alpha + \beta)1)^{r+1} xv$$
$$= \sum_{i=0}^{r} \binom{r}{i} \rho(x_{i+1}) (\rho(y) - \alpha 1)^{r-i} v + \sum_{i=0}^{r} \binom{r}{i} \rho(x_i) (\rho(y) - \alpha 1)^{r+1-i} v$$
$$= \sum_{i=0}^{r+1} \binom{r}{i-1} \rho(x_i)(\rho(y) - \alpha 1)^{r+1-i} v + \sum_{i=0}^{r+1} \binom{r}{i} \rho(x_i)(\rho(y) - \alpha 1)^{r+1-i} v$$

$$\left(\text{interpreting } \binom{r}{-1} = 0 \text{ and } \binom{r}{r+1} = 0 \right)$$

$$= \sum_{i=0}^{r+1} \binom{r+1}{i} ((\operatorname{ad} y - \beta 1)^i x) ((\rho(y) - \alpha 1)^{r+1-i} v).$$

This completes the induction. □

Proof of Theorem 2.6. Let $v \in V_i$, $x, y \in L$. Then

$$(\rho(y) - \lambda_i 1)^n xv = \sum_{j=0}^{n} \binom{n}{j} ((\operatorname{ad} y)^j x)((\rho(y) - \lambda_i 1)^{n-j} v)$$

by Proposition 2.7 with $\alpha = \lambda_i$, $\beta = 0$. Since $v \in V_i$, $(\rho(y) - \lambda_i 1)^{n-j} v = 0$ if $n - j$ is sufficiently large. Since L is nilpotent $(\operatorname{ad} y)^j x = 0$ if j is sufficiently large. Thus $(\rho(y) - \lambda_i 1)^n xv = 0$ if n is sufficiently large. Hence $xv \in V_i$ and so V_i is a submodule of V. □

Corollary 2.8 *Let L be a nilpotent Lie algebra and V a finite dimensional indecomposable L-module. Then a basis can be chosen for V with respect to which we obtain a matrix representation ρ of L of the form*

$$\rho(x) = \begin{pmatrix} \lambda(x) & & & * \\ & \cdot & & \\ & & \cdot & \\ 0 & & & \lambda(x) \end{pmatrix} \quad \text{for all } x \in L.$$

Proof. We can choose a basis as in Corollary 2.3 with respect to which each $\rho(x)$ is triangular. The generalised eigenspaces of V with respect to $\rho(x)$ are all submodules of V by Theorem 2.6 and V is their direct sum. Since V is indecomposable only one of the generalised eigenspaces is non-zero. Thus all the eigenvalues of $\rho(x)$ on V are equal. Let this eigenvalue be $\lambda(x)$. Then the diagonal entries of the triangular matrix $\rho(x)$ are all equal to $\lambda(x)$. □

We observe that the map $x \to \lambda(x)$ is a 1-dimensional representation of L, as it arises from a 1-dimensional submodule of V.

We have seen from Proposition 2.5 and Theorem 2.6 how to obtain a direct decomposition of V into submodules for any element $y \in L$. We may use this result to obtain a direct decomposition of V into submodules which does not depend on the choice of any particular element of L.

Theorem 2.9 *Let L be a nilpotent Lie algebra and V a finite dimensional L-module. For any 1-dimensional representation λ of L we define $V_\lambda = \{v \in V;$ for each $x \in L$ there exists $N(x)$ such that $(\rho(x) - \lambda(x)1)^{N(x)} v = 0\}$. Then*

$$V = \bigoplus_\lambda V_\lambda$$

and each V_λ is a submodule of L.

2.2 Representations of nilpotent Lie algebras

Proof. We first express V as a direct sum of indecomposable L-modules. Each of these defines a 1-dimensional representation λ of L as in Corollary 2.8. Let W_λ be the direct sum of all indecomposable components giving rise to λ. Then we have

$$V = \bigoplus_\lambda W_\lambda.$$

We shall show that $W_\lambda = V_\lambda$ and so that W_λ is independent of the decomposition chosen into indecomposable components. It is clear that $W_\lambda \subset V_\lambda$ by Corollary 2.8. Suppose if possible that $W_\lambda \neq V_\lambda$. Then there exists $v \in V_\lambda \cap \bigoplus_{\mu \neq \lambda} W_\mu$ with $v \neq 0$. We write $v = \sum_{\mu \in S} w_\mu$ with $w_\mu \in W_\mu$, where the set S is finite. Since $w_\mu \in W_\mu$ there exists N_μ such that $(\rho(x) - \mu(x)1)^{N_\mu} w_\mu = 0$. Hence

$$\prod_{\mu \in S} (\rho(x) - \mu(x)1)^{N_\mu} v = 0.$$

However, we also have $(\rho(x) - \lambda(x)1)^{N_\lambda} v = 0$.

We recall from Lemma 2.1 that the 1-dimensional representations of L are in bijective correspondence with linear maps $L/L^2 \to \mathbb{C}$. The vector space L/L^2 over $\mathbb{C}$ cannot be expressed as the union of finitely many proper subspaces. For each $\mu \in S$ the set of x satisfying $\lambda(x) = \mu(x)$ is a proper subspace. Thus there exists $x \in L$ such that $\lambda(x) \neq \mu(x)$ for all $\mu \in S$. Thus the polynomials

$$\prod_{\mu \in S} (t - \mu(x))^{N_\mu}, \quad (t - \lambda(x))^{N_\lambda}$$

are coprime. Thus there exist polynomials $a(t), b(t) \in \mathbb{C}[t]$ such that

$$a(t) \prod_{\mu \in S} (t - \mu(x))^{N_\mu} + b(t)(t - \lambda(x))^{N_\lambda} = 1.$$

Hence

$$a(\rho(x)) \prod_{\mu \in S} (\rho(x) - \mu(x)1)^{N_\mu} v + b(\rho(x))(\rho(x) - \lambda(x)1)^{N_\lambda} v = v.$$

The left-hand side of this expression is zero, as we have seen above. Thus $v = 0$, a contradiction. Hence $V_\lambda = W_\lambda$, $V = \bigoplus_\lambda V_\lambda$ and each V_λ is a submodule of V. □

A 1-dimensional representation λ of L is called a **weight** of V if $V_\lambda \neq 0$, and V_λ is called the **weight space** of λ. The decomposition $V = \bigoplus_\lambda V_\lambda$ is called the **weight space decomposition** of V. It follows from Corollary 2.8 that a basis can be chosen for V_λ with respect to which the matrix representation of L on V_λ has form

$$\rho(x) = \begin{pmatrix} \lambda(x) & & * \\ & \ddots & \\ 0 & & \lambda(x) \end{pmatrix} \quad \text{for each } x \in L.$$

We shall make frequent use of the weight space decomposition in subsequent chapters.

We next prove a theorem of Engel which gives a useful characterisation of nilpotent Lie algebras in terms of the adjoint representation.

Theorem 2.10 *(Engel's theorem). A Lie algebra L is nilpotent if and only if $\operatorname{ad} x : L \to L$ is nilpotent for each $x \in L$.*

Proof. Suppose L is nilpotent. Then $L^n = O$ for some n. Let $y \in L$. Then we have

$$\operatorname{ad} x \cdot y \in L^2, \quad (\operatorname{ad} x)^2 \cdot y \in L^3, \quad \ldots$$

and so $(\operatorname{ad} x)^{n-1} y = 0$ for each $y \in L$. Thus $(\operatorname{ad} x)^{n-1} = 0$ and so $\operatorname{ad} x$ is a nilpotent linear map.

Now suppose conversely that $\operatorname{ad} x$ is a nilpotent linear map for each $x \in L$. We wish to show L is nilpotent. We suppose if possible that this is false and let H be a maximal nilpotent subalgebra of L. Thus H is nilpotent but any subalgebra properly containing H is not nilpotent. We may regard L as an H-module. Then H is an H-submodule of L and we can find an H-submodule M of L containing H such that M/H is an irreducible H-module. We have

$$\dim(M/H) = 1 \qquad \text{by Theorem 2.2.}$$

Moreover the 1-dimensional representation of H afforded by M/H is the zero representation, as otherwise $\operatorname{ad} x$ would fail to be nilpotent for some $x \in H$. Hence we have $[HM] \subset H$. Now there exists $x \in M$ such that

$$M = H \oplus \mathbb{C}x.$$

We have

$$[MM] \subset [HH] + [Hx] \subset H.$$

Thus M is a subalgebra of L and H is an ideal of M.

2.2 Representations of nilpotent Lie algebras

We shall show that for each positive integer i there exists a positive integer $e(i)$ such that
$$M^{e(i)} \subset H^i$$
This is true for $i=1$ since $M^2 \subset H$. We prove it by induction on i. Assume that $M^{e(r)} \subset H^r$. Then
$$M^{e(r)+1} = [M^{e(r)}, H + \mathbb{C}x] \subset H^{r+1} + [M^{e(r)}, x].$$
Hence $M^{e(r)+1} \subset H^{r+1} + \text{ad}\, x \cdot M^{e(r)}$.

We shall show that
$$M^{e(r)+j} \subset H^{r+1} + (\text{ad}\, x)^j \cdot M^{e(r)}$$
for each positive integer j. This is true for $j=1$. Assuming it inductively for j we have
$$M^{e(r)+j+1} \subset [H^{r+1} + (\text{ad}\, x)^j \cdot M^{e(r)}, M]$$
$$\subset H^{r+1} + [(\text{ad}\, x)^j M^{e(r)}, H + \mathbb{C}x]$$
$$\subset H^{r+1} + (\text{ad}\, x)^{j+1} M^{e(r)}$$
since H^{r+1} is an ideal of M and $(\text{ad}\, x)^j M^{e(r)} \subset H^r$. Thus we have shown
$$M^{e(r)+j} \subset H^{r+1} + (\text{ad}\, x)^j M^{e(r)} \qquad \text{for all } j.$$
Now we know that $(\text{ad}\, x)^j = 0$ when j is sufficiently large. For such j we have
$$M^{e(r)+j} \subset H^{r+1}.$$
Thus we define $e(r+1) = e(r) + j$ and then $M^{e(r+1)} \subset H^{r+1}$ as required.

Now H is nilpotent so $H^i = O$ for i sufficiently large. For such i we have $M^{e(i)} = O$. Thus M is nilpotent. But this contradicts the maximality of H. Thus our initial assumption was incorrect and so L must be nilpotent. $\square$

Corollary 2.11 *A Lie algebra L is nilpotent if and only if L has a basis with respect to which the adjoint representation of L has form*

$$\rho(x) = \begin{pmatrix} 0 & & & & * \\ & 0 & & & \\ & & \cdot & & \\ & & & \cdot & \\ & & 0 & & \\ & & & 0 & \\ & & & & 0 \end{pmatrix} \quad \text{for all } x \in L.$$

Proof. Suppose L is nilpotent. Then L has a series of ideals

$$L \supset L^2 \supset L^3 \supset \cdots \supset L^r = O \qquad \text{for some } r.$$

We refine this series by choosing a sequence of subspaces between consecutive terms, each of codimension 1 in its predecessor. Such subspaces are automatically ideals of L since if $L^i \supset I \supset L^{i+1}$ we have

$$[IL] \subset [L^i L] = L^{i+1} \subset I.$$

Thus we have a chain of ideals

$$L = I_n \supset I_{n-1} \supset \cdots \supset I_1 \supset I_0 = O$$

with $\dim I_k = k$ and $[LI_k] \subset I_{k-1}$. By choosing a basis of L adapted to this chain of ideals the map $\operatorname{ad} x : L \to L$ is represented by a matrix $\rho(x)$ of zero-triangular form (i.e. triangular with zeros on the diagonal).

Conversely if L has a basis with respect to which $\operatorname{ad} x$ is represented by a zero-triangular matrix $\rho(x)$ for all $x \in L$, we have $\rho(x)$ nilpotent and so $\operatorname{ad} x$ is nilpotent. Thus L must be a nilpotent Lie algebra by Engel's theorem (Theorem 2.10). □

3
Cartan subalgebras

3.1 Existence of Cartan subalgebras

Let H be a subalgebra of a Lie algebra L. Let

$$N(H) = \{x \in L;\ [hx] \in H \quad \text{for all } h \in H\}.$$

$N(H)$ is called the **normaliser** of H.

Lemma 3.1 (i) $N(H)$ is a subalgebra of L.
(ii) H is an ideal of $N(H)$.
(iii) $N(H)$ is the largest subalgebra of L containing H as an ideal.

Proof. (i) Let $x, y \in N(H)$. Then

$$[h[xy]] = [[yh]x] + [[hx]y] \in H.$$

Hence $[xy] \in N(H)$ and $N(H)$ is a subalgebra.
(ii) This is clear from the definition of $N(H)$.
(iii) If H is an ideal of M then $[HM] \subset H$ so $M \subset N(H)$. $\square$

Definition *A subalgebra H of L is called a* **Cartan subalgebra** *if H is nilpotent and $H = N(H)$. Cartan subalgebras play a very important role in the theory of semisimple Lie algebras. Our aim in this section is to show that L contains a Cartan subalgebra.*

Let us take an element $x \in L$ and consider the linear map $\operatorname{ad} x : L \to L$. Let $L_{0,x}$ be the generalised eigenspace of $\operatorname{ad} x$ with eigenvalue 0. Thus $L_{0,x} = \{y \in L\ ;\ \text{there exists } n \text{ such that } (\operatorname{ad} x)^n y = 0\}$, and $L_{0,x}$ will be called the **null component** of L with respect to x.

An element $x \in L$ is called **regular** if $\dim L_{0,x}$ is as small as possible. The Lie algebra L certainly contains regular elements.

Theorem 3.2 *Let x be a regular element of L. Then the null component $L_{0,x}$ is a Cartan subalgebra of L.*

Proof. Let $H = L_{0,x}$. We must show that H is a subalgebra of L, that H is nilpotent, and that $H = N(H)$.

We first show that H is a subalgebra. Let $y, z \in H$. We must show that $[yz] \in H$. By Proposition 2.7 we have

$$(\mathrm{ad}\, x)^n [yz] = \sum_{i=0}^{n} \binom{n}{i} [(\mathrm{ad}\, x)^i y, (\mathrm{ad}\, x)^{n-i} z].$$

(We take $V = L$, $\alpha = \beta = 0$ in Proposition 2.7 to obtain this.) Since $y \in H$ we have

$$(\mathrm{ad}\, x)^i y = 0 \quad \text{if } i \text{ is sufficiently large.}$$

Since $z \in H$

$$(\mathrm{ad}\, x)^{n-i} z = 0 \quad \text{if } n - i \text{ is sufficiently large.}$$

Hence $(\mathrm{ad}\, x)^n [yz] = 0$ if n is sufficiently large. Thus $[yz] \in H$ and H is a subalgebra of L.

We next show that H is nilpotent. To do this we shall prove that all the matrices in the adjoint representation of H are nilpotent and use Engel's theorem (Theorem 2.10). Let $\dim H = l$ and $b_1, \ldots, b_l$ be a basis for H. Let

$$y = \lambda_1 b_1 + \cdots + \lambda_l b_l \in H \qquad \lambda_1, \ldots, \lambda_l \in \mathbb{C}.$$

Consider the linear map $\mathrm{ad}\, y : L \to L$. We have $\mathrm{ad}\, y : H \to H$ since H is a subalgebra and we obtain an induced map $\mathrm{ad}\, y : L/H \to L/H$.

Let $\chi(t)$ be the characteristic polynomial of $\mathrm{ad}\, y$ on L, $\chi_1(t)$ be its characteristic polynomial on H and $\chi_2(t)$ be its characteristic polynomial on L/H. Then we have

$$\chi(t) = \chi_1(t) \chi_2(t).$$

Since $\chi(t) = \det(t\mathbf{1} - \mathrm{ad}\, y)$ and y depends linearly on $\lambda_1, \ldots, \lambda_l$ we see that the coefficients of $\chi(t)$ are polynomial functions of $\lambda_1, \ldots, \lambda_l$. The same applies to $\chi_1(t)$ and $\chi_2(t)$. Let

$$\chi_2(t) = d_0 + d_1 t + d_2 t^2 + \cdots$$

where $d_0, d_1, d_2, \ldots$ are polynomial functions of $\lambda_1, \ldots, \lambda_l$. We claim that d_0 is not the zero polynomial. For in the special case when $y = x$ we know that

all eigenvalues of $\operatorname{ad} y$ on L/H are non-zero, so $\chi_2(t)$ has non-zero constant term. Let
$$\chi_1(t) = t^m(c_0 + c_1 t + c_2 t^2 + \cdots)$$
where $c_0, c_1, c_2, \ldots$ are polynomial functions of $\lambda_1, \ldots, \lambda_l$ and c_0 is not the zero polynomial. We have
$$m \le l = \deg \chi_1(t).$$
We then have
$$\chi(t) = t^m(c_0 d_0 + \text{terms involving positive powers of } t).$$
Now $c_0 d_0$ is not the zero polynomial so we can choose $\lambda_1, \ldots, \lambda_l \in \mathbb{C}$ to make $c_0 d_0$ non-zero. For such an element $y \in H$ we have
$$\dim L_{0,y} = m.$$
Since x is regular and $\dim L_{0,x} = l$ we have $m \ge l$. Since we also know $m \le l$ we have $m = l$. Now $\chi_1(t)$ has degree l and is divisible by t^l, hence
$$\chi_1(t) = t^l.$$
It follows by the Cayley–Hamilton theorem that $(\operatorname{ad} y)^l : H \to H$ is zero. Hence by Engel's theorem we deduce that H is nilpotent.

Finally we show that $H = N(H)$. It is certainly true that $H \subset N(H)$. So let $z \in N(H)$. Then $[xz] \in H$. Thus
$$(\operatorname{ad} x)^n[xz] = 0 \quad \text{for some } n.$$
But then $(\operatorname{ad} x)^{n+1} z = 0$ and so $z \in H$. Thus $H = N(H)$ and we have shown that H is a Cartan subalgebra of L. □

3.2 Derivations and automorphisms

A **derivation** of a Lie algebra L is a linear map $\delta : L \to L$ such that
$$\delta[xy] = [\delta x, y] + [x, \delta y] \quad \text{for all } x, y \in L.$$

Lemma 3.3 *Let $x \in L$. Then $\operatorname{ad} x$ is a derivation of L.*

Proof. $\operatorname{ad} x[yz] = [\operatorname{ad} x \cdot y, z] + [y, \operatorname{ad} x \cdot z]$ by the Jacobi identity. □

An **automorphism** of L is an isomorphism $\theta : L \to L$. The automorphisms of L form a group $\operatorname{Aut} L$ under composition.

Proposition 3.4 *Let δ be a nilpotent derivation of L. Then $\exp \delta$ is an automorphism of L.*

Proof. Since δ is nilpotent we have $\delta^n = 0$ for some n. Then we have

$$\exp \delta = \sum_{r=0}^{n-1} \frac{\delta^r}{r!}$$

The map $\exp \delta : L \to L$ is clearly linear. Let $x, y \in L$. Then

$$\delta[xy] = [\delta x, y] + [x, \delta y]$$

$$\delta^r[xy] = \sum_{i=0}^{r} \binom{r}{i} [\delta^i x, \delta^{r-i} y]$$

as is easily seen by induction on r. Hence

$$\exp \delta \cdot [xy] = \sum_{r \geq 0} \sum_{i=0}^{r} \frac{1}{r!} \binom{r}{i} [\delta^i x, \delta^{r-i} y] = \sum_{i \geq 0} \sum_{j \geq 0} \frac{1}{i! j!} [\delta^i x, \delta^j y]$$

$$= \left[\sum_{i \geq 0} \frac{1}{i!} \delta^i x, \sum_{j \geq 0} \frac{1}{j!} \delta^j y \right] = [\exp \delta \cdot x, \exp \delta \cdot y].$$

Thus $\exp \delta : L \to L$ is a homomorphism. Similarly $\exp(-\delta)$ is a homomorphism and we have $\exp \delta \; \exp(-\delta) = 1$. Thus $\exp \delta : L \to L$ is an automorphism. □

The subgroup of $\mathrm{Aut}\, L$ generated by all automorphisms $\exp \mathrm{ad}\, x$ for all $x \in L$ with $\mathrm{ad}\, x$ nilpotent is called the **group of inner automorphisms** $\mathrm{Inn}\, L$. Every element of $\mathrm{Inn}\, L$ has form

$$\exp \mathrm{ad}\, x_1 \cdot \exp \mathrm{ad}\, x_2 \cdots \cdots \exp \mathrm{ad}\, x_r$$

where $x_1, \ldots, x_r \in L$ and $\mathrm{ad}\, x_1, \ldots, \mathrm{ad}\, x_r$ are all nilpotent.

Lemma 3.5 *$\mathrm{Inn}\, L$ is a normal subgroup of $\mathrm{Aut}\, L$.*

Proof. Let $\theta \in \mathrm{Aut}\, L$. It is sufficient to show that $\theta(\exp \mathrm{ad}\, x)\theta^{-1} \in \mathrm{Inn}\, L$ for all $x \in L$ with $\mathrm{ad}\, x$ nilpotent. Now we have

$$\theta(\mathrm{ad}\, x)\theta^{-1} y = \theta[x, \theta^{-1} y] = [\theta x, y] = (\mathrm{ad}\, \theta x) \cdot y$$

for all $y \in L$. Hence

$$\theta(\mathrm{ad}\, x)\theta^{-1} = \mathrm{ad}\, \theta x.$$

It follows that

$$\theta(\exp \operatorname{ad} x)\theta^{-1} = \exp \operatorname{ad}(\theta x) \in \operatorname{Inn} L.$$

Thus $\operatorname{Inn} L$ is normal in $\operatorname{Aut} L$. □

Two subalgebras M_1, M_2 of L are called **conjugate** in L if there exists $\theta \in \operatorname{Inn} L$ such that $\theta(M_1) = M_2$.

We wish to show that any two Cartan subalgebras of L are conjugate in L. However, we first need some concepts from algebraic geometry.

3.3 Ideas from algebraic geometry

Let H be a nilpotent subalgebra of a Lie algebra L and regard L as an H-module. Then we obtain a decomposition

$$L = \bigoplus L_\lambda$$

as in Theorem 2.9, where

$$L_\lambda = \{x \in L;\ \text{for each } h \in H \text{ there exists } n \text{ such that } (\operatorname{ad} h - \lambda(h)1)^n x = 0\}.$$

Now H lies in L_0 by Corollary 2.11. We shall suppose that the nilpotent subalgebra H satisfies the condition $H = L_0$. Then there exist 1-dimensional representations $\lambda_1, \ldots, \lambda_r$ of H with $\lambda_1 \neq 0, \ldots, \lambda_r \neq 0$ and

$$L = H \oplus L_{\lambda_1} \oplus \cdots \oplus L_{\lambda_r}.$$

Given $x \in L$ we then have

$$x = x_0 + x_1 + \cdots + x_r$$

with $x_0 \in H$ and $x_i \in L_{\lambda_i}$ for $i = 1, \ldots, r$. We claim that $\operatorname{ad} x_i : L \to L$ is nilpotent when $i \neq 0$.

To see this let $\mu : H \to \mathbb{C}$ be a weight of the H-module L and let $y \in L_\mu$. Then by Proposition 2.7 we have

$$(\operatorname{ad} h - \mu(h)1 - \lambda_i(h)1)^n [x_i, y] = \sum_{j=0}^{n} \binom{n}{j} \left[(\operatorname{ad} h - \lambda_i(h)1)^j x_i, (\operatorname{ad} h - \mu(h)1)^{n-j} y \right].$$

Because $x_i \in L_{\lambda_i}$ then $(\operatorname{ad} h - \lambda_i(h)1)^j x_i = 0$ if j is sufficiently large. Since $y \in L_\mu$ then $(\operatorname{ad} h - \mu(h)1)^{n-j} y = 0$ if $n - j$ is sufficiently large. Thus

$$(\operatorname{ad} h - \mu(h)1 - \lambda_i(h)1)^n [x_i, y] = 0$$

if n is sufficiently large, and so $[x_i y] \in L_{\lambda_i + \mu}$. Thus we have

$$\operatorname{ad} x_i \cdot L_\mu \subset L_{\lambda_i + \mu}.$$

Since $\lambda_i \neq 0$ and there are only finitely many $\mu : H \to \mathbb{C}$ for which $L_\mu \neq 0$ we see that $(\operatorname{ad} x_i)^N = 0$ if N is sufficiently large. Thus $\operatorname{ad} x_i$ is nilpotent.

We deduce that $\exp \operatorname{ad} x_i \in \operatorname{Aut} L$ for $i \neq 0$. We now define a map $f : L \to L$ by

$$f(x) = \exp \operatorname{ad} x_1 \cdot \exp \operatorname{ad} x_2 \cdots \exp \operatorname{ad} x_r \cdot x_0.$$

We shall discuss some properties of this function f. We choose a basis $\{b_{ij}\}$ of L for $0 \le i \le r$ where for fixed i the elements b_{ij} form a basis of L_{λ_i} with respect to which the elements of H are represented by triangular matrices, as in Corollary 2.3. Here $\lambda_0 = 0$.

Lemma 3.6 $f : L \to L$ *is a polynomial function. Thus*

$$f\left(\sum \lambda_{ij} b_{ij}\right) = \sum \mu_{ij} b_{ij}$$

where each μ_{ij} is a polynomial in the λ_{kl}.

Proof. Each map $\operatorname{ad} x_i : L \to L$ is linear. Also we have

$$\exp \operatorname{ad} x_i = \sum_{k=0}^{N} \frac{(\operatorname{ad} x_i)^k}{k!} \quad \text{for some } N$$

since $\operatorname{ad} x_i$ is nilpotent. Thus $\exp \operatorname{ad} x_i : L \to L$ is a polynomial function.

The given map f is a composition of the linear map $x \to x_0$ with polynomial functions $\exp \operatorname{ad} x_i$ for $i > 0$, so is a polynomial function. □

We write $\mu_{ij} = f_{ij}(\lambda_{kl})$ where f_{ij} is a polynomial. We define the Jacobian matrix

$$J(f) = (\partial f_{ij} / \partial \lambda_{kl})$$

and the Jacobian determinant $\det J(f)$ of f. $\det J(f)$ is an element of the polynomial ring $\mathbb{C}[\lambda_{kl}]$.

Proposition 3.7 $\det J(f)$ *is not the zero polynomial.*

Proof. We shall show $\det J(f)$ is not the zero polynomial by showing that it is non-zero when evaluated at a carefully chosen element of H. So let $h \in H$ and consider $(\partial f_{ij} / \partial \lambda_{kl})_h$.

3.3 Ideas from algebraic geometry

First suppose $k \neq 0$. Then

$$\begin{aligned}
(\partial f/\partial \lambda_{kl})_h &= \lim_{t \to 0} \frac{f(h+tb_{kl}) - f(h)}{t} \\
&= \lim_{t \to 0} \frac{(\exp \operatorname{ad} tb_{kl})h - h}{t} \\
&= \lim_{t \to 0} \frac{h + t[b_{kl}, h] + \cdots - h}{t} \\
&= [b_{kl}, h] = -[h b_{kl}] \\
&= -\lambda_k(h) b_{kl} + \text{a linear combination of } b_{k1}, \ldots, b_{k\,l-1}.
\end{aligned}$$

Next suppose $k = 0$. Then

$$\begin{aligned}
(\partial f/\partial \lambda_{0l})_h &= \lim_{t \to 0} \frac{f(h+tb_{0l}) - f(h)}{t} \\
&= \lim_{t \to 0} \frac{h + tb_{0l} - h}{t} = b_{0l}.
\end{aligned}$$

Thus $J(f)_h$ is a block matrix of form

$$\begin{pmatrix}
\begin{array}{c|c|c|c}
\begin{matrix} 1 & & \\ & \ddots & \\ & & 1 \end{matrix} & O & O & \\
\hline
\begin{matrix} O \end{matrix} & \begin{matrix} -\lambda_1(h) & & * \\ & \ddots & \\ O & & -\lambda_1(h) \end{matrix} & O & \\
\hline
& & \begin{matrix} -\lambda_2(h) & & * \\ & \ddots & \\ O & & -\lambda_2(h) \end{matrix} & \\
\hline
& & & \ddots
\end{array}
\end{pmatrix}$$

(with row labels $k=0$, $k=1$, $k=2$, $\ldots$ on the left)

and so $(\det J(f))_h = \pm \prod_{i=1}^r \lambda_i(h)^{d_i}$ where $d_i = \dim L_{\lambda_i}$.

Now the linear maps $\lambda_i : H \to \mathbb{C}$ for $i = 1, \ldots, v$ are all non-zero. Thus we can find an element $h \in H$ with $\lambda_i(h) \neq 0$ for $i = 1, \ldots, v$. For such an element h we have $(\det J(f))_h \neq 0$. Hence $\det J(f)$ is not the zero polynomial. $\square$

Proposition 3.8 *The polynomial functions f_{ij} are algebraically independent.*

Proof. Suppose if possible that there is a non-zero polynomial $F(x_{ij}) \in \mathbb{C}[x_{ij}]$ such that $F(f_{ij}) = 0$. We choose such a polynomial F whose total degree in the variables x_{ij} is as small as possible. Then

$$\frac{\partial}{\partial \lambda_{kl}} F(f_{ij}) = 0$$

and so

$$\sum_{i,j} \frac{\partial F}{\partial f_{ij}} \frac{\partial f_{ij}}{\partial \lambda_{kl}} = 0.$$

Let v be the vector $(\partial F/\partial f_{ij})$. Then

$$vJ(f) = (0, \ldots, 0).$$

Since $\det J(f)$ is non-zero this implies that $v = (0, \ldots, 0)$, that is

$$\partial F/\partial f_{ij} = 0 \quad \text{for each } f_{ij}.$$

Now $\partial F/\partial x_{ij}$ is a polynomial in $\mathbb{C}[x_{ij}]$ of smaller total degree than F. By the choice of F $\partial F/\partial x_{ij}$ must be the zero polynomial. Hence F does not involve the variable x_{ij}. Since this is true for all x_{ij} F must be a constant. Since $F(f_{ij}) = 0$ this constant must be zero. Thus F is the zero polynomial and we have a contradiction. □

Let $B = \mathbb{C}[f_{ij}]$ be the polynomial ring in the f_{ij} and $A = \mathbb{C}[\lambda_{ij}]$ the polynomial ring in the λ_{ij}. We have a homomorphism $\theta : B \to A$ uniquely determined by

$$\theta(f_{ij}) = f_{ij}(\lambda_{kl}) \in A.$$

Proposition 3.9 *The homomorphism $\theta : B \to A$ is injective.*

Proof. Suppose $F \in B$ satisfies $\theta(F) = 0$. Then $F(f_{ij}) = 0$, regarded as a function of the λ_{kl}. Since the f_{ij} are algebraically independent this implies that $F = 0$. Thus θ is injective. □

Thus we may regard B as a subring of A. A and B are integral domains with a common identity element and A is finitely generated over B. We next prove a general result which applies to this situation.

Proposition 3.10 *Let A and B be integral domains such that $B \subset A$, A, B have a common identity element 1, and A is finitely generated over B. Let p be a non-zero element of A. Then there exists a non-zero element q of B such that any homomorphism $\phi : B \to \mathbb{C}$ with $\phi(q) \neq 0$ can be extended to a homomorphism $\psi : A \to \mathbb{C}$ with $\psi(p) \neq 0$.*

3.3 Ideas from algebraic geometry

Proof. We may assume that A is generated over B by a single element ζ. For then by iterating the process we can prove the result when A is finitely generated over B. Thus we assume that $A = B[\zeta]$ for some $\zeta \in A$.

Suppose first that ζ is transcendental over B. Given a non-zero element $p = p(\zeta) \in A$ we choose $q \in B$ to be one of the non-zero coefficients of $p(\zeta)$. Suppose we are given a homomorphism $\phi : B \to \mathbb{C}$ with $\phi(q) \neq 0$. We write $\phi(b) = \bar{b} \in \mathbb{C}$. By applying ϕ to the coefficients of $p(\zeta)$ we obtain $\bar{p}(\zeta) \in \mathbb{C}[\zeta]$. The element $\bar{p}(\zeta)$ is not the zero polynomial since $\phi(q) \neq 0$. We can find an element $\beta \in \mathbb{C}$ with $\bar{p}(\beta) \neq 0$. We now define a homomorphism $\psi : A \to \mathbb{C}$ by

$$\psi(g(\zeta)) = \bar{g}(\beta).$$

ψ is well defined since ζ is transcendental over B, and ψ is a homomorphism, being a composite of the homomorphisms

$$A = B[\zeta] \to \mathbb{C}[\zeta] \to \mathbb{C} \tag{3.1}$$

$$g(\zeta) \quad \to \bar{g}(\zeta) \to \bar{g}(\beta) \tag{3.2}$$

ψ clearly extends ϕ. Finally we have $\psi(p) = \bar{p}(\beta) \neq 0$.

Next suppose that ζ is algebraic over B. Then we can find $f(t) \in B[t]$ of minimal degree such that $f(\zeta) = 0$. We write

$$f(t) = b_0 t^n + b_1 t^{n-1} + \cdots + b_n \quad b_0 \neq 0.$$

Now let $g(t)$ be any polynomial in $B[t]$ satisfying $g(\zeta) = 0$. We divide $g(t)$ by $f(t)$ using the Euclidean algorithm. We are working over an integral domain B rather than over a field. However, provided we multiply $g(t)$ by a sufficiently high power of the leading coefficient b_0 of $f(t)$ we can carry out the Euclidean process over B. We thus obtain

$$b_0^k g(t) = u(t) f(t) + v(t)$$

where $u(t), v(t) \in B[t]$ and $\deg v(t) < \deg f(t)$. Thus

$$v(\zeta) = b_0^k g(\zeta) - u(\zeta) f(\zeta) = 0.$$

Since $\deg v(t) < \deg f(t)$ this implies that $v(t) = 0$. Hence

$$b_0^k g(t) = u(t) f(t).$$

Let p be the given non-zero element of A. The element p is algebraic over B since A is generated over B by the single algebraic element ζ. Thus there exists a polynomial $h(t) \in B[t]$ with non-zero constant term h_m such that $h(p) = 0$. We define the element $q \in B$ by $q = b_0 h_m$. Thus $q \neq 0$. We assume we are given a homomorphism $\phi : B \to \mathbb{C}$ with $\phi(q) \neq 0$. Then $\phi(b_0) \neq 0$ and

$\phi(h_m) \neq 0$. We write $\phi(b) = \bar{b} \in \mathbb{C}$. The polynomial $f(t) \in B[t]$ gives rise to a polynomial $\bar{f}(t) \in \mathbb{C}[t]$. We choose an element $\beta \in \mathbb{C}$ with $\bar{f}(\beta) = 0$. We note that

$$\bar{b}_0^k \bar{g}(\beta) = \bar{u}(\beta)\bar{f}(\beta) = 0,$$

hence $\bar{g}(\beta) = 0$ since $\bar{b}_0 = \phi(b_0) \neq 0$.

We now define a homomorphism

$$\psi : A \to \mathbb{C}$$

by $\psi(g(\zeta)) = \bar{g}(\beta)$. We note that the map ψ is well defined, since we have shown that $g(\zeta) = 0$ implies $\bar{g}(\beta) = 0$. The map ψ is a homomorphism since the maps

$$B[t] \to \mathbb{C}[t] \to \mathbb{C}$$
$$g(t) \to \bar{g}(t) \to \bar{g}(\beta)$$

are homomorphisms. The definition of ψ shows that ψ extends ϕ. Finally we show $\psi(p) \neq 0$. Since $h(p) = 0$ we have $\bar{h}(\psi(p)) = 0$. However, the constant term of $h(t)$ is $\phi(h_m)$, which is non-zero. Since $\bar{h}(t)$ has non-zero constant term and $h(\psi(p)) = 0$ we must have $\psi(p) \neq 0$. □

We now apply this result to our earlier situation. Let $d = \dim L$ and

$$f : \mathbb{C}^d \to \mathbb{C}^d$$

be the polynomial function

$$(\lambda_{ij}) \to (f_{ij}(\lambda_{kl})).$$

We write $V = \mathbb{C}^d$ and for each polynomial $p \in \mathbb{C}[x_{ij}]$ we write

$$V_p = \{v \in V \; ; \; p(v) \neq 0\}.$$

Corollary 3.11 *For each non-zero polynomial $p \in \mathbb{C}[x_{ij}]$ there exists a non-zero polynomial $q \in \mathbb{C}[x_{ij}]$ such that $f(V_p) \supset V_q$.*

Proof. We apply Proposition 3.10 to the integral domains $B \subset A$ discussed earlier. Thus A is the polynomial ring $\mathbb{C}[\lambda_{ij}]$ and B is the polynomial ring $\mathbb{C}[f_{ij}]$. We choose a non-zero polynomial $p \in A$. Then there exists a non-zero polynomial $q \in B$ such that any homomorphism $\phi : B \to \mathbb{C}$ with $\phi(q) \neq 0$ can be extended to a homomorphism $\psi : A \to \mathbb{C}$ with $\psi(p) \neq 0$. This means that given any $v \in V_q$ we have $v = f(w)$ for some $w \in V_p$. Hence $V_q \subset f(V_p)$ as required. □

3.4 Conjugacy of Cartan subalgebras

We showed in Theorem 3.2 that the null component $L_{0,x}$ of a regular element $x \in L$ is a Cartan subalgebra of L. We shall now show conversely that any Cartan subalgebra is the null component of some regular element. We shall then prove that, given two regular elements, their null components are conjugate in L.

Proposition 3.12 *Let H be a Cartan subalgebra of L. Then there exists a regular element $x \in L$ such that $H = L_{0,x}$.*

Proof. Since H is nilpotent we may regard L as an H-module and decompose L into weight spaces with respect to H as in Theorem 2.9. H lies in the zero weight space L_0 by Corollary 2.11. Since $H = N(H)$ we can show that $H = L_0$. For if $H \neq L_0$ the H-module L_0/H will have a 1-dimensional submodule M/H on which H acts with weight 0. Hence $[HM] \subset H$ and so $M \subset N(H)$. This contradicts $H = N(H)$. Thus we have $H = L_0$. Let

$$L = H \oplus L_{\lambda_1} \oplus \cdots \oplus L_{\lambda_r} \qquad \lambda_1, \ldots, \lambda_r \neq 0$$

be the weight space decomposition of L with respect to H. Let $x \in L$ and

$$x = x_0 + x_1 + \cdots + x_r$$

with $x_0 \in H$ and $x_i \in L_{\lambda_i}$ for $i \neq 0$. Then we can define a polynomial function $f : L \to L$ as in Section 3.3 with

$$f(x) = \exp \operatorname{ad} x_1 \cdot \exp \operatorname{ad} x_2 \cdots \cdot \exp \operatorname{ad} x_r \cdot x_0.$$

We define $p : L \to \mathbb{C}$ by

$$p(x) = \lambda_1(x_0) \lambda_2(x_0) \cdots \lambda_r(x_0).$$

Then p is a polynomial function on L. p is not the zero polynomial since we can find $x_0 \in H$ for which each $\lambda_i(x_0) \neq 0$ for $i = 1, \ldots, r$. Hence by Corollary 3.11 there exists a non-zero polynomial function $q : L \to \mathbb{C}$ such that $f(L_p) \supset L_q$.

We next consider the set R of regular elements of L. Let $y \in L$ and

$$\chi(y) = \det(t1 - \operatorname{ad} y) = t^n + \mu_1(y) t^{n-1} + \cdots + \mu_n(y)$$

be the characteristic polynomial of $\operatorname{ad} y$ on L. Then $\mu_1, \mu_2, \ldots, \mu_n$ are polynomial functions on L. There exists a unique integer k such that μ_{n-k} is not the zero polynomial but $\mu_{n-k+1}, \ldots, \mu_n$ are identically zero. The generalised eigenspace of $\operatorname{ad} y$ with eigenvalue 0 has dimension k if $\mu_{n-k}(y) \neq 0$

and dimension greater than k if $\mu_{n-k}(y)=0$. Thus y is regular if and only if $\mu_{n-k}(y) \neq 0$.

Now there exists $y \in L$ such that $y \in L_q \cap R$. For we may choose y with $(q\mu_{n-k})(y) \neq 0$. Since $L_q \subset f(L_p)$ we can find $x \in L_p$ such that $f(x)=y$. Thus we have

$$\exp \operatorname{ad} x_1 \cdot \exp \operatorname{ad} x_2 \cdots \exp \operatorname{ad} x_r \cdot x_0 = y.$$

Hence x_0, y are conjugate elements of L. Since y is regular, x_0 must also be regular. Since $x \in L_p$ we have

$$\lambda_1(x_0)\lambda_2(x_0) \cdots \lambda_r(x_0) \neq 0.$$

Now $x_0 \in H$ and H is nilpotent, hence $L_{0,x_0} \supset H$ by Corollary 2.11. On the other hand L_{0,x_0} cannot be larger than H since

$$\lambda_1(x_0) \neq 0, \ldots, \lambda_r(x_0) \neq 0.$$

Hence $H = L_{0,x_0}$ where x_0 is regular. □

Theorem 3.13 *Any two Cartan subalgebras of L are conjugate.*

Proof. Let H, H' be Cartan subalgebras of L. We regard L as an H-module and decompose L into weight spaces with respect to H. We have seen in the proof of Proposition 3.12 that $H = L_0$. Let the weight space decomposition be

$$L = H \oplus L_{\lambda_1} \oplus \cdots \oplus L_{\lambda_r} \qquad \lambda_1, \ldots, \lambda_r \neq 0.$$

For each $x \in L$ we have

$$x = x_0 + x_1 + \cdots + x_r$$

with $x_0 \in H$ and $x_i \in L_{\lambda_i}$ for $i \neq 0$.

Now for each $x_0 \in H$ we have $L_{0,x_0} \supset H$ and for some $x_0 \in H$ we have $L_{0,x_0} = H$ since H is a Cartan subalgebra. An element $x_0 \in H$ is regular if and only if $L_{0,x_0} = H$. This is equivalent to the condition

$$\lambda_1(x_0)\lambda_2(x_0) \cdots \lambda_r(x_0) \neq 0.$$

We now consider the polynomial function $f : L \to L$ defined by

$$f(x) = \exp \operatorname{ad} x_1 \cdot \exp \operatorname{ad} x_2 \cdots \exp \operatorname{ad} x_r \cdot x_0.$$

Let $p : L \to \mathbb{C}$ be the function given by

$$p(x) = \lambda_1(x_0)\lambda_2(x_0) \cdots \lambda_r(x_0)$$

where p is a polynomial function on L which is not identically zero, since $p(x)$ is non-zero when x_0 is a regular element of H. By Corollary 3.11 there exists a non-zero polynomial function $q : L \to \mathbb{C}$ such that $f(L_p) \supset L_q$.

We now start with the second Cartan subalgebra H'. We can define a corresponding function $f' : L \to L$ and a corresponding function $p' : L \to \mathbb{C}$. There exists a non-zero polynomial function $q' : L \to \mathbb{C}$ such that $f'(L_{p'}) \supset L_{q'}$.

Now $L_q \cap L_{q'} = \{x \in L \ ; \ (qq')(x) \neq 0\}$. Thus $L_q \cap L_{q'}$ is non-empty. We choose $z \in L_q \cap L_{q'}$. Thus $z \in f(L_p) \cap f'(L_{p'})$. Thus there exists $x \in L$ with $z = f(x)$ and $p(x) \neq 0$. Similarly there exists $x' \in L$ with $z = f'(x')$ and $p'(x) \neq 0$. Thus

$$z = \exp \operatorname{ad} x_1 \cdot \exp \operatorname{ad} x_2 \cdots \exp \operatorname{ad} x_r \cdot x_0$$

and so z is conjugate to x_0. Since $p(x) \neq 0$ x_0 is regular. Similarly z is conjugate to x_0' and x_0' is regular. Thus we have found regular elements $x_0 \in H$ and $x_0' \in H'$ such that x_0, x_0' are conjugate in L.

Now we have $H = L_{0,x_0}$ and $H' = L_{0,x_0'}$ since x_0, x_0' are regular. Thus an inner automorphism of L which transforms x_0 to x_0' will transform H to H'. Hence H, H' are conjugate in L. $\square$

The dimension of the Cartan subalgebras of L will be called the **rank** of L.

4
The Cartan decomposition

4.1 Some properties of root spaces

Let L be a Lie algebra and H be a Cartan subalgebra of L. We regard L as an H-module. Since H is nilpotent we have a weight space decomposition

$$L = \bigoplus_\lambda L_\lambda$$

as in Theorem 2.9, where

$$L_\lambda = \{x \in L \text{ ; for each } h \in H \text{ there exists } n \text{ such that } (\operatorname{ad} h - \lambda(h)1)^n x = 0\}.$$

Proposition 4.1 $L_0 = H$.

Proof. The algebra H is contained in L_0 by Corollary 2.11. Suppose if possible that $H \neq L_0$. Then L_0/H is an H-module, and this module contains a 1-dimensional submodule M/H on which H acts with weight 0. Hence $[HM] \subset H$ and so $M \subset N(H)$. This implies $H \neq N(H)$, a contradiction. □

The 1-dimensional representations λ of H such that $\lambda \neq 0$ and $L_\lambda \neq O$ are called the **roots** of L with respect to H. The set of roots of L with respect to H will be denoted by Φ. Thus we have

$$L = H \oplus \left(\bigoplus_{\alpha \in \Phi} L_\alpha \right)$$

This decomposition is called the **Cartan decomposition** of L with respect to H. L_α is called the **root space** of α.

Proposition 4.2 *Let λ, μ be 1-dimensional representations of H. Then*

$$[L_\lambda, L_\mu] \subset L_{\lambda+\mu}.$$

Proof. Let $y \in L_\lambda, z \in L_\mu$. We show that $[yz] \in L_{\lambda+\mu}$. Let $x \in H$. Then by Proposition 2.7 we have

$$(\operatorname{ad} x - \lambda(x)1 - \mu(x)1)^n[yz] = \sum_{i=0}^{n}\binom{n}{i}[(\operatorname{ad} x - \lambda(x)1)^i y, (\operatorname{ad} x - \mu(x)1)^{n-i}z].$$

Since $y \in L_\lambda$ $(\operatorname{ad} x - \lambda(x)1)^i y = 0$ if i is sufficiently large. Since $z \in L_\mu$ $(\operatorname{ad} x - \mu(x)1)^{n-i}z = 0$ if $n-i$ is sufficiently large. Hence

$$(\operatorname{ad} x - \lambda(x)1 - \mu(x)1)^n[yz] = 0$$

if n is sufficiently large. This shows that $[yz] \in L_{\lambda+\mu}$. □

Corollary 4.3 *Let $\alpha, \beta \in \Phi$ be roots of L with respect to H. Then*

$$[L_\alpha, L_\beta] \subset L_{\alpha+\beta} \quad \text{if } \alpha+\beta \in \Phi$$
$$[L_\alpha, L_\beta] \subset H \quad \text{if } \beta = -\alpha$$
$$[L_\alpha, L_\beta] = 0 \quad \text{if } \alpha+\beta \neq 0 \text{ and } \alpha+\beta \notin \Phi.$$

Proof. This follows from Proposition 4.2 and the fact that $L_0 = H$. □

Proposition 4.4 *Let $\alpha \in \Phi$ and consider the subspace $[L_\alpha L_{-\alpha}]$ of H. Given any $\beta \in \Phi$ there exists a number $r \in \mathbb{Q}$, depending on α and β, such that $\beta = r\alpha$ on $[L_\alpha L_{-\alpha}]$.*

Proof. If $-\alpha$ is not a weight of L with respect to H then $L_{-\alpha} = 0$ and there is nothing to prove. Thus we assume $-\alpha$ is a weight. Then $-\alpha \in \Phi$ since $\alpha \neq 0$.

We consider the functions $i\alpha + \beta : H \to \mathbb{C}$ where $i \in \mathbb{Z}$. Since Φ is finite there exist $p, q \in \mathbb{Z}$ with $p \geq 0, q \geq 0$ such that

$$-p\alpha + \beta, \ldots, \beta, \ldots, q\alpha + \beta$$

are all in Φ but $-(p+1)\alpha + \beta$, $(q+1)\alpha + \beta$ are not in Φ. If either $-(p+1)\alpha + \beta = 0$ or $(q+1)\alpha + \beta = 0$ the result is obvious. Thus we assume $-(p+1)\alpha + \beta \neq 0$, $(q+1)\alpha + \beta \neq 0$. Thus $-(p+1)\alpha + \beta, (q+1)\alpha + \beta$ are not weights of L with respect to H.

Let M be the subspace of L given by

$$M = L_{-p\alpha+\beta} \oplus \cdots \oplus L_{q\alpha+\beta}.$$

Let $y \in L_\alpha, z \in L_{-\alpha}$. Let $x = [yz] \in [L_\alpha L_{-\alpha}]$. Then we have

$\quad$ ad $y(M) \subset M \quad$ by Proposition 4.2, since $L_{(q+1)\alpha+\beta} = 0$

$\quad$ ad $z(M) \subset M \quad$ by Proposition 4.2, since $L_{-(p+1)\alpha+\beta} = 0$.

Thus
$$\text{ad } x(M) = (\text{ad } y \text{ ad } z - \text{ad } z \text{ ad } y)M \subset M.$$

We calculate the trace $\text{tr}_M \text{ad } x$. Since $x \in H$ each weight space $L_{i\alpha+\beta}$ is invariant under ad x. Thus
$$\text{tr}_M \text{ad } x = \sum_{i=-p}^{q} \text{tr}_{L_{i\alpha+\beta}} \text{ad } x.$$

Now ad x acts on the weight space $L_{i\alpha+\beta}$ by means of a matrix of form

$$\begin{pmatrix} (i\alpha+\beta)x & & & * \\ & \cdot & & \\ & & \cdot & \\ & & & \cdot \\ 0 & & & (i\alpha+\beta)x \end{pmatrix}$$

Thus $\text{tr}_{L_{i\alpha+\beta}} \text{ad } x = \dim L_{i\alpha+\beta} \cdot (i\alpha+\beta)(x)$. Thus
$$\text{tr}_M \text{ad } x = \sum_{i=-p}^{q} \dim L_{i\alpha+\beta}(i\alpha(x)+\beta(x))$$
$$= \left(\sum_{i=-p}^{q} i \dim L_{i\alpha+\beta} \right) \alpha(x) + \left(\sum_{i=-p}^{q} \dim L_{i\alpha+\beta} \right) \beta(x).$$

On the other hand we have
$$\text{tr}_M \text{ad } x = \text{tr}_M(\text{ad } y \text{ ad } z - \text{ad } z \text{ ad } y)$$
$$= \text{tr}_M(\text{ad } y \text{ ad } z) - \text{tr}_M(\text{ad } z \text{ ad } y) = 0.$$

Hence
$$\left(\sum_{i=-p}^{q} i \dim L_{i\alpha+\beta} \right) \alpha(x) + \left(\sum_{i=-p}^{q} \dim L_{i\alpha+\beta} \right) \beta(x) = 0.$$

Moreover $\dim L_{i\alpha+\beta} > 0$ for $-p \le i \le q$. Hence for $x \in [L_\alpha L_{-\alpha}]$ we have
$$\beta(x) = -\frac{(\sum_{i=-p}^{q} i \dim L_{i\alpha+\beta})}{(\sum_{i=-p}^{q} \dim L_{i\alpha+\beta})} \alpha(x).$$

Thus $\beta(x) = r\alpha(x)$ for some $r \in \mathbb{Q}$ independent of x. Hence $\beta = r\alpha$ on $[L_\alpha L_{-\alpha}]$. $\square$

4.2 The Killing form

In order to make further progress in understanding the Cartan decomposition of L we introduce a bilinear form on L called the **Killing form**. We define a map

$$L \times L \to \mathbb{C}$$
$$x, y \to \langle x, y \rangle$$

given by

$$\langle x, y \rangle = \operatorname{tr}(\operatorname{ad} x \operatorname{ad} y).$$

We have $\operatorname{ad} x : L \to L$, $\operatorname{ad} y : L \to L$ and $\operatorname{ad} x \operatorname{ad} y : L \to L$, so $\operatorname{tr}(\operatorname{ad} x \operatorname{ad} y) \in \mathbb{C}$.

Proposition 4.5 (i) $\langle x, y \rangle$ *is bilinear, i.e. linear in x and y.*
(ii) $\langle x, y \rangle$ *is symmetric, i.e.* $\langle y, x \rangle = \langle x, y \rangle$.
(iii) $\langle x, y \rangle$ *is invariant, i.e.*

$$\langle [xy], z \rangle = \langle x, [yz] \rangle \qquad \textit{for all } x, y, z \in L.$$

Proof. (i) is clear from the definition.
(ii) follows from the fact that $\operatorname{tr} AB = \operatorname{tr} BA$.
(iii) $\langle [xy], z \rangle = \operatorname{tr}(\operatorname{ad}[xy] \operatorname{ad} z) = \operatorname{tr}((\operatorname{ad} x \operatorname{ad} y - \operatorname{ad} y \operatorname{ad} x) \operatorname{ad} z)$

$$= \operatorname{tr}(\operatorname{ad} x \operatorname{ad} y \operatorname{ad} z) - \operatorname{tr}(\operatorname{ad} y \operatorname{ad} x \operatorname{ad} z)$$
$$= \operatorname{tr}(\operatorname{ad} x \operatorname{ad} y \operatorname{ad} z) - \operatorname{tr}(\operatorname{ad} x \operatorname{ad} z \operatorname{ad} y)$$
$$= \operatorname{tr}(\operatorname{ad} x \, (\operatorname{ad} y \operatorname{ad} z - \operatorname{ad} z \operatorname{ad} y)) = \operatorname{tr}(\operatorname{ad} x \operatorname{ad}[yz]) = \langle x, [yz] \rangle.$$
$\square$

Proposition 4.6 *Let I be an ideal of L and $x, y \in I$. Then*

$$\langle x, y \rangle_I = \langle x, y \rangle_L.$$

Thus the Killing form of L restricted to I is the Killing form of I.

Proof. We choose a basis of I and extend it to give a basis of L. With respect to this basis $\operatorname{ad} x : L \to L$ is represented by a matrix of form

$$\begin{pmatrix} A_1 & A_2 \\ O & O \end{pmatrix}$$

since $x \in I$, and similarly $\operatorname{ad} y : L \to L$ is represented by a matrix of form

$$\begin{pmatrix} B_1 & B_2 \\ O & O \end{pmatrix}$$

Thus $\operatorname{ad} x \operatorname{ad} y : L \to L$ is represented by the matrix

$$\begin{pmatrix} A_1 B_1 & A_1 B_2 \\ O & O \end{pmatrix}$$

Hence $\operatorname{tr}_L(\operatorname{ad} x \operatorname{ad} y) = \operatorname{tr} A_1 B_1 = \operatorname{tr}_I(\operatorname{ad} x \operatorname{ad} y)$ and so $\langle x, y \rangle_L = \langle x, y \rangle_I$ □

For any subspace M of L we define $M^\perp$ by

$$M^\perp = \{ x \in L \, ; \, \langle x, y \rangle = 0 \quad \text{for all } y \in M \}.$$

$M^\perp$ is also a subspace of L.

Lemma 4.7 *If I is an ideal of L then $I^\perp$ is also an ideal of L.*

Proof. Let $x \in I^\perp$, $y \in L$. We must show that $[xy] \in I^\perp$. So let $z \in I$. Then

$$\langle [xy], z \rangle = \langle x, [yz] \rangle = 0$$

since $[yz] \in I$ and $x \in I^\perp$. Thus $[xy] \in I^\perp$ and $I^\perp$ is an ideal of L. □

We see in particular that $L^\perp$ is an ideal of L. The Killing form of L is said to be **non-degenerate** if $L^\perp = O$. This is equivalent to the condition that if $\langle x, y \rangle = 0$ for all $y \in L$ then $x = 0$.

The Killing form of L is **identically zero** if $L^\perp = L$. This means that $\langle x, y \rangle = 0$ for all $x, y \in L$.

We now prove a deeper result on the Killing form which will be very useful subsequently.

Proposition 4.8 *Let L be a Lie algebra such that $L \neq 0$ and $L^2 = L$. Let H be a Cartan subalgebra of L. Then there exists $x \in H$ such that $\langle x, x \rangle \neq 0$.*

Proof. We consider the Cartan decomposition of L with respect to H. Let this be $L = \oplus L_\lambda$. Then we have

$$L^2 = [LL] = \left[\bigoplus_\lambda L_\lambda, \bigoplus_\mu L_\mu \right] = \sum_{\lambda, \mu} [L_\lambda L_\mu].$$

4.2 The Killing form

Now we have $[L_\lambda L_\mu] \subset L_{\lambda+\mu}$ by Proposition 4.2. Now $L_{\lambda+\mu} = O$ if $\lambda+\mu$ is not a weight. Thus each non-zero product $[L_\lambda L_\mu]$ lies in some weight space L_ν. We consider the zero weight space L_0. Since $L^2 = L$ we have

$$L_0 = \sum_\lambda [L_\lambda L_{-\lambda}]$$

summed over all weights λ such that $-\lambda$ is also a weight. Now $L_0 = H$ by Proposition 4.1, thus we have

$$H = [HH] + \sum_\alpha [L_\alpha L_{-\alpha}]$$

summed over all roots $\alpha \in \Phi$ such that $-\alpha$ is also a root.

Now L is not nilpotent since $L^2 = L$. H is nilpotent and so $H \neq L$. So there is at least one root $\beta \in \Phi$. β is a 1-dimensional representation of H and so vanishes on $[HH]$ since

$$\beta[xy] = \beta(x)\beta(y) - \beta(y)\beta(x) = 0 \qquad x, y \in H.$$

But β does not vanish on H since $\beta \neq 0$. So using the above decomposition of H we see that there is some root $\alpha \in \Phi$ such that $-\alpha \in \Phi$ and β does not vanish on $[L_\alpha L_{-\alpha}]$.

We choose $x \in [L_\alpha L_{-\alpha}]$ such that $\beta(x) \neq 0$. Then we have

$$\langle x, x \rangle = \mathrm{tr}(\mathrm{ad}\, x\, \mathrm{ad}\, x) = \sum_\lambda \dim L_\lambda (\lambda(x))^2$$

since $\mathrm{ad}\, x$ is represented on L_λ by a matrix of form

$$\begin{pmatrix} \lambda(x) & & * \\ & \ddots & \\ O & & \lambda(x) \end{pmatrix}$$

Now by Proposition 4.4 there exists $r_{\lambda,\alpha} \in \mathbb{Q}$ such that $\lambda(x) = r_{\lambda,\alpha}\alpha(x)$. Thus we have

$$\langle x, x \rangle = \left(\sum_\lambda \dim L_\lambda r_{\lambda,\alpha}^2 \right) \alpha(x)^2.$$

Now $\beta(x) = r_{\beta,\alpha}\alpha(x)$ and $\beta(x) \neq 0$. Thus $\alpha(x) \neq 0$ and $r_{\beta,\alpha} \neq 0$. It follows that $\langle x, x \rangle \neq 0$. □

We shall now obtain some important consequences of this result.

Theorem 4.9 *If the Killing form of L is identically zero then L is soluble.*

Proof. We use induction on the dimension of L. If $\dim L = 1$ then L is soluble. So suppose $\dim L > 1$. By Proposition 4.8 we have $L \ne L^2$. L^2 is an ideal of L so the Killing form of L^2 is the restriction of the Killing form of L, by Proposition 4.6. Thus the Killing form of L^2 is identically zero. By induction L^2 is soluble. Since L/L^2 is soluble it follows that L is soluble, by Proposition 1.12. $\square$

Theorem 4.10 *The Killing form of L is non-degenerate if and only if L is semisimple.*

Proof. Suppose first that the Killing form of L is degenerate. Then $L^\perp \ne O$. Now $L^\perp$ is an ideal of L by Lemma 4.7. Thus the Killing form of $L^\perp$ is the restriction of that of L by Proposition 4.6. Thus the Killing form of $L^\perp$ is identically zero. This implies that $L^\perp$ is soluble, by Theorem 4.9. Thus L has a non-zero soluble ideal, so L is not semisimple.

Now suppose conversely that L is not semisimple. Then the soluble radical R of L is non-zero. We consider the chain

$$R \supset R^{(1)} \supset R^{(2)} \supset \cdots \supset R^{(k-1)} \supset R^{(k)} = O$$

where as usual $R^{(i+1)} = [R^{(i)} R^{(i)}]$. The subspaces $R^{(i)}$ are all ideals of L since the product of two ideals is an ideal. Let $I = R^{(k-1)}$. Then I is a non-zero ideal of L such that $I^2 = O$.

We choose a basis of I and extend it to a basis of L. Let $x \in I$ and $y \in L$. With respect to this basis $\operatorname{ad} x$ is represented by a matrix of form

$$\begin{pmatrix} O & A \\ O & O \end{pmatrix}$$

since $I^2 = O$ and I is an ideal of L, $\operatorname{ad} y$ is represented by a matrix of form

$$\begin{pmatrix} B_1 & B_2 \\ O & B_3 \end{pmatrix}$$

and $\operatorname{ad} x \operatorname{ad} y$ is represented by the matrix

$$\begin{pmatrix} O & AB_3 \\ O & O \end{pmatrix}$$

Hence $\langle x, y \rangle = \operatorname{tr}(\operatorname{ad} x \operatorname{ad} y) = 0$. Since this holds for all $x \in I$ and $y \in L$ we have $I \subset L^\perp$. Thus $L^\perp \ne O$ and so the Killing form of L is degenerate. $\square$

We now define the **direct sum** of Lie algebras L_1, L_2. $L_1 \oplus L_2$ is the vector space of all pairs (x_1, x_2) with $x_1 \in L_1$, $x_2 \in L_2$ under the Lie multiplication given by

$$[(x_1, x_2)(y_1, y_2)] = ([x_1 y_1], [x_2 y_2]).$$

In this direct sum we define $I_1 = \{(x_1, 0) \; ; \; x_1 \in L_1\}$ and $I_2 = \{(0, x_2) \; ; \; x_2 \in L_2\}$. Then I_1 and I_2 are ideals of $L_1 \oplus L_2$ such that $I_1 \cap I_2 = O$ and $I_1 + I_2 = L_1 \oplus L_2$. Moreover I_1 is isomorphic to L_1 and I_2 is isomorphic to L_2.

Conversely let L be a Lie algebra containing two ideals I_1, I_2 such that $I_1 \cap I_2 = O$ and $I_1 + I_2 = L$. Then the Lie algebra $I_1 \oplus I_2$ is isomorphic to L under the isomorphism

$$\theta : I_1 \oplus I_2 \to L$$
$$(x_1, x_2) \to x_1 + x_2.$$

For θ is certainly an isomorphism of vector spaces. But θ also preserves Lie multiplication. To see this we first observe that

$$[I_1 I_2] \subset I_1 \cap I_2 = O.$$

Thus

$$[\theta(x_1, x_2), \theta(y_1, y_2)] = [x_1 + x_2, y_1 + y_2] = [x_1 y_1] + [x_2 y_2]$$
$$= \theta([x_1 y_1], [x_2 y_2]) = \theta[(x_1, x_2), (y_1, y_2)].$$

Thus if a Lie algebra has two complementary ideals I_1, I_2 the Lie algebra is isomorphic to $I_1 \oplus I_2$.

We may in a similar way consider direct sums $L_1 \oplus L_2 \oplus \cdots \oplus L_n$ of more than two Lie algebras.

Theorem 4.11 *A Lie algebra L is semisimple if and only if L is isomorphic to a direct sum of non-trivial simple Lie algebras.*

Proof. Suppose L is semisimple. If L is simple then L must be non-trivial since the trivial simple Lie algebra is not semisimple. Thus we suppose L is not simple. Let I be a minimal non-zero ideal of L. Then $I \ne O$ and $I \ne L$. Consider the subspace $I^\perp$ of L; $I^\perp$ is also an ideal of L by Lemma 4.7. Now the Killing form of L is non-degenerate by Theorem 4.10. Thus an element $x \in L$ lies in $I^\perp$ if and only if the coordinates of x with respect to a basis of L satisfy $\dim I$ homogeneous linear equations which are linearly independent. It follows that

$$\dim I^\perp = \dim L - \dim I.$$

Now consider the subspace $I\cap I^\perp$. This is an ideal of L. Thus the Killing form of $I\cap I^\perp$ is the restriction of the Killing form of L, by Proposition 4.6. Hence $I\cap I^\perp$ is soluble, by Theorem 4.9. Since L is semisimple we have $I\cap I^\perp = O$. Thus

$$\dim(I+I^\perp) = \dim I + \dim I^\perp - \dim(I\cap I^\perp)$$
$$= \dim I + \dim I^\perp = \dim L.$$

Hence $I + I^\perp = L$. Thus L is the direct sum of its ideals I and $I^\perp$. Hence L is isomorphic to the Lie algebra $I \oplus I^\perp$.

We shall now show that I is a simple Lie algebra. Let J be an ideal of I. Then we have

$$[JL] \subset [JI] + [JI^\perp] \subset [JI] \subset J$$

since $[JI^\perp] \subset [II^\perp] \subset I\cap I^\perp = O$. Thus J is an ideal of L contained in I. Since I is a minimal ideal of L we have $J = O$ or $J = I$. Thus I is simple.

We show next that $I^\perp$ is semisimple. Let J be a soluble ideal of $I^\perp$. Then

$$[JL] \subset [JI] + [JI^\perp] \subset [JI^\perp] \subset J$$

since $[JI] \subset [I^\perp I] \subset I\cap I^\perp = O$. Thus J is an ideal of L. Since L is semisimple and J is soluble we have $J = O$. Thus $I^\perp$ is semisimple.

Now we know $\dim I^\perp < \dim L$. By induction we may assume $I^\perp$ is a direct sum of simple non-trivial Lie algebras. Since $L = I \oplus I^\perp$ and I is simple and non-trivial, L is also a direct sum of simple non-trivial Lie algebras.

Conversely suppose that

$$L = L_1 \oplus \cdots \oplus L_r$$

where each L_i is a simple non-trivial Lie algebra. Each L_i is semisimple so has non-degenerate Killing form by Theorem 4.10. Now each L_i is an ideal of L. Moreover if $x_i \in L_i$, $x_j \in L_j$ and $i \ne j$ then $\langle x_i, x_j \rangle = 0$. For

$$\operatorname{ad} x_i \operatorname{ad} x_j \cdot y \in L_i \cap L_j = O \quad \text{for all } y \in L$$

thus $\langle x_i, x_j \rangle = \operatorname{tr}(\operatorname{ad} x_i \operatorname{ad} x_j) = 0$.

Now let $x = x_1 + \cdots + x_r \in L^\perp$ with $x_i \in L_i$. Let $y_i \in L_i$. Then we have

$$\langle x_i, y_i \rangle = \langle x, y_i \rangle = 0.$$

Since this holds for all $y_i \in L_i$ we have $x_i = 0$. This holds for all i, hence $x = 0$. Thus $L^\perp = O$ and the Killing form of L is non-degenerate. This implies that L is semisimple by Theorem 4.10. $\square$

4.3 The Cartan decomposition of a semisimple Lie algebra

When L is semisimple we can say much more about its Cartan decomposition than in the general case. We shall now investigate this Cartan decomposition in detail.

Let L be semisimple, H be a Cartan subalgebra of L, and $L = \bigoplus L_\lambda$ be the Cartan decomposition of L with respect to H. We recall from Proposition 4.1 that $L_0 = H$.

Proposition 4.12 L_λ and L_μ are orthogonal with respect to the Killing form, provided $\mu \neq -\lambda$.

Proof. Let $x \in L_\lambda$, $y \in L_\mu$. We assume $\lambda + \mu \neq 0$ and must show that $\langle x, y \rangle = 0$. Now for any weight space L_ν we have

$$\text{ad } x \text{ ad } y \, L_\nu \subset L_{\lambda + \mu + \nu} \quad \text{by Proposition 4.2.}$$

We choose a basis of L adapted to the Cartan decomposition. With respect to such a basis $\text{ad } x \text{ ad } y$ will be represented by a block matrix of form

$$\begin{pmatrix} 0 & & & & \\ & 0 & & & * \\ & & \cdot & & \\ & & & \cdot & \\ & & & & \cdot \\ & * & & & 0 \\ & & & & & 0 \end{pmatrix}$$

since $\lambda + \mu + \nu \neq \nu$. Hence we have

$$\langle x, y \rangle = \text{tr}(\text{ad } x \text{ ad } y) = 0. \qquad \square$$

Proposition 4.13 *If α is a root of L with respect to H then $-\alpha$ is also a root.*

Proof. We recall that α is a root if $\alpha \neq 0$ and $L_\alpha \neq O$. Suppose if possible that $-\alpha$ is not a root. Since $-\alpha \neq 0$ we have $L_{-\alpha} = O$. By Proposition 4.12 we see that L_α is orthogonal to all L_λ, hence $L_\alpha \subset L^\perp$. But since L is semisimple we have $L^\perp = O$ by Theorem 4.10. Thus $L_\alpha = O$, which contradicts the fact that α is a root. $\qquad \square$

Proposition 4.14 *The Killing form of L remains non-degenerate on restriction to H. Thus if $x \in H$ satisfies $\langle x, y \rangle = 0$ for all $y \in H$ then $x = 0$.*

Proof. Let $x \in H$ and suppose $\langle x, y \rangle = 0$ for all $y \in H$. We also have $\langle x, y \rangle = 0$ for all $y \in L_\alpha$ where $\alpha \neq 0$, by Proposition 4.12. Thus $\langle x, y \rangle = 0$ for all $y \in L$ and so $x \in L^\perp$. Since L is semisimple $L^\perp = O$, hence $x = 0$ as required. □

Note that the Killing form of L restricted to H does not coincide with the Killing form of H. The latter is degenerate since H is not semisimple.

Theorem 4.15 $[HH] = O$. *Thus the Cartan subalgebras of a semisimple Lie algebra are abelian.*

Proof. Let $x \in [HH]$ and $y \in H$. Then we have

$$\langle x, y \rangle = \operatorname{tr}(\operatorname{ad} x \, \operatorname{ad} y) = \sum_\lambda \dim L_\lambda \, \lambda(x)\lambda(y)$$

since $\operatorname{ad} x \, \operatorname{ad} y$ is represented on L_λ by a matrix of form

$$\begin{pmatrix} \lambda(x)\lambda(y) & & & * \\ & \cdot & & \\ & & \cdot & \\ O & & \cdot & \\ & & & \lambda(x)\lambda(y) \end{pmatrix}$$

However, λ is a 1-dimensional representation of H and $x \in [HH]$, hence $\lambda(x) = 0$. Thus $\langle x, y \rangle = 0$ for all $y \in H$. This implies $x = 0$ by Proposition 4.14. Thus $[HH] = O$. □

Let $H^* = \operatorname{Hom}(H, \mathbb{C})$ be the dual space of H. This is the vector space of all linear maps from H to $\mathbb{C}$. We have $\dim H^* = \dim H$.

We define a map $H \to H^*$ using the Killing form of L. Given $h \in H$ we define $h^* \in H^*$ by

$$h^*(x) = \langle h, x \rangle \quad \text{for all } x \in H.$$

Lemma 4.16 *The map $h \to h^*$ is an isomorphism of vector spaces between H and H^*.*

Proof. The map is certainly linear. Suppose $h \in H$ lies in the kernel. Then $\langle h, x \rangle = 0$ for all $x \in H$. This implies $h = 0$ by Proposition 4.14. Thus the kernel is O. Hence the image must be the whole of H^*, since $\dim H^* = \dim H$. Hence our map is bijective. □

Now we have a finite subset $\Phi \subset H^*$, the set of roots of L with respect to H. For each $\alpha \in \Phi$ there is a unique element $h'_\alpha \in H$ such that

$$\alpha(x) = \langle h'_\alpha, x \rangle \quad \text{for all } x \in H.$$

4.3 The Cartan decomposition of a semisimple Lie algebra

(The notation h_α might seem more natural, but this will be reserved for the coroot of α, to be discussed in Chapter 7.)

Proposition 4.17 *The vectors h'_α for $\alpha \in \Phi$ span H.*

Proof. Suppose if possible that the h'_α lie in a proper subspace of H. Then there exists an element $x \in H$ with $x \neq 0$ and $\langle h'_\alpha, x \rangle = 0$ for all $\alpha \in \Phi$. Thus $\alpha(x) = 0$ for all $\alpha \in \Phi$. Let $y \in H$. Then we have

$$\langle x, y \rangle = \text{tr } (\text{ad } x \text{ ad } y) = \sum_\lambda \dim L_\lambda \, \lambda(x)\lambda(y) = 0$$

since $\lambda(x) = 0$ for all weights λ. Thus $\langle x, y \rangle = 0$ for all $y \in H$. This implies $x = 0$ by Proposition 4.14, a contradiction. □

Proposition 4.18 $h'_\alpha \in [L_\alpha L_{-\alpha}]$ *for all $\alpha \in \Phi$.*

Proof. L_α is an H-module. Since all irreducible H-modules are 1-dimensional L_α contains a 1-dimensional H-submodule $\mathbb{C}e_\alpha$. We have $[xe_\alpha] = \alpha(x)e_\alpha$ for all $x \in H$.

Let $y \in L_{-\alpha}$. Then $[e_\alpha y] \in [L_\alpha L_{-\alpha}] \subset H$. We shall show that $[e_\alpha y] = \langle e_\alpha, y \rangle h'_\alpha$. In order to prove this we define

$$z = [e_\alpha y] - \langle e_\alpha, y \rangle h'_\alpha \in H.$$

Let $x \in H$. Then

$$\langle x, z \rangle = \langle x, [e_\alpha y] \rangle - \langle e_\alpha, y \rangle \langle x, h'_\alpha \rangle$$
$$= \langle [xe_\alpha], y \rangle - \langle e_\alpha, y \rangle \alpha(x)$$
$$= \alpha(x) \langle e_\alpha, y \rangle - \langle e_\alpha, y \rangle \alpha(x) = 0.$$

Thus $\langle x, z \rangle = 0$ for all $x \in H$, and it follows that $z = 0$. Hence

$$[e_\alpha y] = \langle e_\alpha, y \rangle h'_\alpha \quad \text{for all} \quad y \in L_{-\alpha}.$$

Now we can choose $y \in L_{-\alpha}$ such that $\langle e_\alpha, y \rangle \neq 0$. Otherwise e_α would be orthogonal to $L_{-\alpha}$, so orthogonal to the whole of L by Proposition 4.12. Then $e_\alpha \in L^\perp$. But $L^\perp = 0$ since L is semisimple. Thus $e_\alpha = 0$, a contradiction. Thus we can find $y \in L_{-\alpha}$ with $\langle e_\alpha, y \rangle \neq 0$. Then

$$h'_\alpha = \frac{1}{\langle e_\alpha, y \rangle} [e_\alpha y] \in [L_\alpha L_{-\alpha}]. \qquad \square$$

48 The Cartan decomposition

Proposition 4.19 $\langle h'_\alpha, h'_\alpha \rangle \neq 0$ *for all* $\alpha \in \Phi$.

Proof. We suppose that $\langle h'_\alpha, h'_\alpha \rangle = 0$ for some $\alpha \in \Phi$ and obtain a contradiction. Let β be any element of Φ. By Proposition 4.4 there is a number $r_{\beta,\alpha} \in \mathbb{Q}$ such that $\beta = r_{\beta,\alpha} \alpha$ when restricted to $[L_\alpha L_{-\alpha}]$. Since $h'_\alpha \in [L_\alpha L_{-\alpha}]$ by Proposition 4.18 we obtain

$$\beta(h'_\alpha) = r_{\beta,\alpha} \alpha(h'_\alpha)$$

that is $\langle h'_\beta, h'_\alpha \rangle = r_{\beta,\alpha} \langle h'_\alpha, h'_\alpha \rangle = 0$.

This holds for all $\beta \in \Phi$. But by Proposition 4.17 the elements h'_β for $\beta \in \Phi$ span H. Thus we have $\langle x, h'_\alpha \rangle = 0$ for all $x \in H$. This implies that $h'_\alpha = 0$ by Proposition 4.14. This in turn implies that $\alpha = 0$, which contradicts $\alpha \in \Phi$. $\square$

Having obtained a number of results on the Cartan decomposition of a semisimple Lie algebra, each depending on previous results, we are now able to obtain one of the most important properties of the Cartan decomposition.

Theorem 4.20 dim $L_\alpha = 1$ *for all* $\alpha \in \Phi$.

Proof. We choose a 1-dimensional H-submodule $\mathbb{C} e_\alpha$ of L_α as in Proposition 4.18 and, as in the proof of that proposition, we can find an element $e_{-\alpha} \in L_{-\alpha}$ with $[e_\alpha e_{-\alpha}] = h'_\alpha$.

We consider the subspace M of L given by

$$M = \mathbb{C} e_\alpha \oplus \mathbb{C} h'_\alpha \oplus L_{-\alpha} \oplus L_{-2\alpha} \oplus \cdots$$

There are only finitely many summands of M since Φ is finite and there are only finitely many non-negative integers r with $L_{-r\alpha} \neq O$.

We observe that ad $e_\alpha M \subset M$. For

$$[e_\alpha e_\alpha] = 0$$
$$[e_\alpha h'_\alpha] = -\alpha(h'_\alpha) e_\alpha$$
$$[e_\alpha y] = \langle e_\alpha, y \rangle h'_\alpha \quad \text{for all } y \in L_{-\alpha},$$

by the proof of Proposition 4.18, and

$$\text{ad } e_\alpha \cdot L_{-r\alpha} \subset L_{-(r-1)\alpha} \quad \text{for all } r \geq 2,$$

by Proposition 4.2.

4.3 The Cartan decomposition of a semisimple Lie algebra

Similarly we can show that ad $e_{-\alpha} M \subset M$. For we have

$$[e_{-\alpha} e_\alpha] = -h'_\alpha$$
$$[e_{-\alpha} h'_\alpha] = \alpha(h'_\alpha) e_{-\alpha}$$

and ad $e_{-\alpha} L_{-r\alpha} \subset L_{-(r+1)\alpha}$ for all $r \geq 1$.
Now $h'_\alpha = [e_\alpha e_{-\alpha}]$ and so

$$\text{ad } h'_\alpha = \text{ad } e_\alpha \text{ ad } e_{-\alpha} - \text{ad } e_{-\alpha} \text{ ad } e_\alpha.$$

Hence ad $h'_\alpha M \subset M$. We shall calculate the trace of ad h'_α on M in two different ways. On the one hand we have

$$\text{tr}_M (\text{ad } h'_\alpha) = \alpha(h'_\alpha) + \dim L_{-\alpha}(-\alpha(h'_\alpha)) + \dim L_{-2\alpha}(-2\alpha(h'_\alpha)) + \cdots$$
$$= \alpha(h'_\alpha)(1 - \dim L_{-\alpha} - 2 \dim L_{-2\alpha} - \cdots).$$

On the other hand we have

$$\text{tr}_M (\text{ad } h'_\alpha) = \text{tr}_M (\text{ad } e_\alpha \text{ ad } e_{-\alpha} - \text{ad } e_{-\alpha} \text{ ad } e_\alpha) = 0.$$

Thus

$$\alpha(h'_\alpha)(1 - \dim L_{-\alpha} - 2 \dim L_{-2\alpha} - \cdots) = 0.$$

Now $\alpha(h'_\alpha) = \langle h'_\alpha, h'_\alpha \rangle \neq 0$ by Proposition 4.19. Thus

$$1 - \dim L_{-\alpha} - 2 \dim L_{-2\alpha} - \cdots = 0.$$

This implies that $\dim L_{-\alpha} = 1$ and $\dim L_{-r\alpha} = 0$ for all $r \geq 2$. Now $\alpha \in \Phi$ if and only if $-\alpha \in \Phi$, by Proposition 4.13. Thus $\dim L_\alpha = 1$ for all $\alpha \in \Phi$. $\square$

Note that although all the root spaces L_α are 1-dimensional the space $H = L_0$ need not be 1-dimensional.

Proposition 4.21 *If $\alpha \in \Phi$ and $r\alpha \in \Phi$ where $r \in \mathbb{Z}$ then $r = 1$ or -1.*

Proof. This follows from the proof of Theorem 4.20, where we showed that, for all $\alpha \in \Phi$, $-r\alpha \notin \Phi$ for all $r \geq 2$. This, together with the fact that $r\alpha \in \Phi$ if and only if $-r\alpha \in \Phi$, gives the required result. $\square$

We shall now obtain some further properties of the set Φ of roots. Let $\alpha, \beta \in \Phi$ be such that $\beta \neq \alpha$ and $\beta \neq -\alpha$. Then β cannot be an integer multiple of α, by Proposition 4.21. There exist integers $p \geq 0, q \geq 0$ such that the elements

$$-p\alpha + \beta, \ldots, -\alpha + \beta, \beta, \alpha + \beta, \ldots, q\alpha + \beta$$

all lie in Φ, but $-(p+1)\alpha+\beta$ and $(q+1)\alpha+\beta$ do not lie in Φ. The set of roots

$$-p\alpha+\beta, \ldots, q\alpha+\beta$$

is called the **α-chain** of roots through β. Let M be the subspace of L defined by

$$M = L_{-p\alpha+\beta} \oplus \cdots \oplus L_{q\alpha+\beta}.$$

Then we have $\operatorname{ad} e_\alpha M \subset M$. This follows from the fact that $\operatorname{ad} e_\alpha L_{r\alpha+\beta} \subset L_{(r+1)\alpha+\beta}$ and $L_{(q+1)\alpha+\beta} = 0$ since $(q+1)\alpha+\beta \notin \Phi$ and $(q+1)\alpha+\beta \neq 0$. Similarly we see that $\operatorname{ad} e_{-\alpha} M \subset M$.

We assume that $[e_\alpha e_{-\alpha}] = h'_\alpha$, as in the proof of Theorem 4.20. Then we have

$$\operatorname{ad} h'_\alpha = \operatorname{ad} e_\alpha \operatorname{ad} e_{-\alpha} - \operatorname{ad} e_{-\alpha} \operatorname{ad} e_\alpha$$

and so $\operatorname{ad} h'_\alpha, M \subset M$. We calculate the trace of $\operatorname{ad} h'_\alpha$ on M in two different ways. We have

$$\operatorname{tr}_M (\operatorname{ad} h'_\alpha) = \sum_{r=-p}^{q} (r\alpha+\beta)(h'_\alpha)$$

since $\dim L_{r\alpha+\beta} = 1$. We also have

$$\operatorname{tr}_M (\operatorname{ad} h'_\alpha) = \operatorname{tr}_M (\operatorname{ad} e_\alpha \operatorname{ad} e_{-\alpha}) - \operatorname{tr}_M (\operatorname{ad} e_{-\alpha} \operatorname{ad} e_\alpha) = 0.$$

Thus

$$\sum_{r=-p}^{q} (r\alpha+\beta)(h'_\alpha) = 0,$$

that is

$$\left(\frac{q(q+1)}{2} - \frac{p(p+1)}{2} \right) \alpha(h'_\alpha) + (p+q+1)\beta(h'_\alpha) = 0.$$

Since $p+q+1 \neq 0$ we obtain

$$\frac{(q-p)}{2} \langle h'_\alpha, h'_\alpha \rangle + \langle h'_\alpha, h'_\beta \rangle = 0,$$

that is

$$2 \frac{\langle h'_\alpha, h'_\beta \rangle}{\langle h'_\alpha, h'_\alpha \rangle} = p - q$$

since $\langle h'_\alpha, h'_\alpha \rangle \neq 0$ by Proposition 4.19. Thus we have proved the following result.

4.3 The Cartan decomposition of a semisimple Lie algebra

Proposition 4.22 *Let α, β be roots such that $\beta \ne \alpha$ and $\beta \ne -\alpha$. Let*

$$-p\alpha + \beta, \ldots, \beta, \ldots, q\alpha + \beta$$

be the α-chain of roots through β. Then

$$2\frac{\langle h'_\alpha, h'_\beta \rangle}{\langle h'_\alpha, h'_\alpha \rangle} = p - q. \qquad \square$$

This result has some useful corollaries. The first gives a strengthening of the result of Proposition 4.21.

Proposition 4.23 *If $\alpha \in \Phi$ and $\zeta\alpha \in \Phi$ where $\zeta \in \mathbb{C}$, then $\zeta = 1$ or -1.*

Proof. Suppose if possible that $\zeta \ne \pm 1$. We put $\beta = \zeta\alpha$ and apply Proposition 4.22. This gives

$$2\zeta = 2\frac{\langle h'_\alpha, h'_\beta \rangle}{\langle h'_\alpha, h'_\alpha \rangle} = p - q.$$

Hence $2\zeta \in \mathbb{Z}$. If $\zeta \in \mathbb{Z}$ then $\zeta = \pm 1$ by Proposition 4.21. Hence $\zeta \notin \mathbb{Z}$. Then the α-chain of roots through β is

$$-\left(\frac{p+q}{2}\right)\alpha, \ldots, \beta = \left(\frac{p-q}{2}\right)\alpha, \ldots, \left(\frac{p+q}{2}\right)\alpha.$$

Now p, q are not both 0 since $\beta \ne 0$. So all the roots in the α-chain are odd multiples of $\tfrac{1}{2}\alpha$. Since the first and the last are negatives of one another and consecutive roots differ by α it is clear that $\tfrac{1}{2}\alpha$ lies in the chain. Hence $\tfrac{1}{2}\alpha \in \Phi$. Since $\alpha \in \Phi$ we have a contradiction to Proposition 4.21. Hence ζ must be 1 or -1. $\qquad \square$

Thus the only roots which are scalar multiples of a root α are α and $-\alpha$.

Proposition 4.24 $\langle h'_\alpha, h'_\beta \rangle \in \mathbb{Q}$ *for all $\alpha, \beta \in \Phi$.*

Proof. We know from the outset that $\langle h'_\alpha, h'_\beta \rangle \in \mathbb{C}$. Now we have

$$2\frac{\langle h'_\alpha, h'_\beta \rangle}{\langle h'_\alpha, h'_\alpha \rangle} \in \mathbb{Z} \qquad \text{by Proposition 4.22.}$$

Thus $\dfrac{\langle h'_\alpha, h'_\beta \rangle}{\langle h'_\alpha, h'_\alpha \rangle} \in \mathbb{Q}$. It will therefore be sufficient to show that $\langle h'_\alpha, h'_\alpha \rangle \in \mathbb{Q}$. Now we have

$$\langle h'_\alpha, h'_\alpha \rangle = \operatorname{tr}(\operatorname{ad} h'_\alpha \operatorname{ad} h'_\alpha) = \sum_{\beta \in \Phi} (\beta(h'_\alpha))^2 = \sum_{\beta \in \Phi} \langle h'_\alpha, h'_\beta \rangle^2.$$

If follows that

$$\frac{1}{\langle h'_\alpha, h'_\alpha \rangle} = \sum_{\beta \in \Phi} \left(\frac{\langle h'_\alpha, h'_\beta \rangle}{\langle h'_\alpha, h'_\alpha \rangle} \right)^2 \in \mathbb{Q}.$$

Hence $\langle h'_\alpha, h'_\alpha \rangle \in \mathbb{Q}$ and the result is proved. □

4.4 The Lie algebra $\mathfrak{sl}_n(\mathbb{C})$

We shall now illustrate the general results about the Cartan decomposition of a semisimple Lie algebra by considering in detail the Lie algebra $\mathfrak{sl}_n(\mathbb{C})$. The **special linear Lie algebra** $\mathfrak{sl}_n(\mathbb{C})$ is the Lie algebra of all $n \times n$ matrices of trace 0 under Lie multiplication $[AB] = AB - BA$. $\mathfrak{sl}_n(\mathbb{C})$ is a subalgebra of $\mathfrak{gl}_n(\mathbb{C}) = [M_n(\mathbb{C})]$. We have

$$\dim \mathfrak{gl}_n(\mathbb{C}) = n^2, \quad \dim \mathfrak{sl}_n(\mathbb{C}) = n^2 - 1.$$

We shall assume $n \geq 2$. Then $\mathfrak{sl}_n(\mathbb{C})$ has a basis

$$E_{11} - E_{22}, \quad E_{22} - E_{33}, \quad \ldots, \quad E_{n-1,n-1} - E_{nn}, \quad E_{ij} \quad i \neq j$$

where the E_{ij} are elementary matrices.

Theorem 4.25 $\mathfrak{sl}_n(\mathbb{C})$ *is a simple Lie algebra.*

Proof. We have $\mathfrak{gl}_n(\mathbb{C}) = \mathfrak{sl}_n(\mathbb{C}) \oplus \mathbb{C} I_n$. Now every ideal of $\mathfrak{sl}_n(\mathbb{C})$ is an ideal of $\mathfrak{gl}_n(\mathbb{C})$. For $[I, \mathfrak{sl}_n(\mathbb{C})] \subset I$ implies $[I, \mathfrak{gl}_n(\mathbb{C})] \subset I$ since $[x, I_n] = 0$ for all $x \in I$. It will therefore be sufficient to show that the only non-zero ideal of $\mathfrak{gl}_n(\mathbb{C})$ contained in $\mathfrak{sl}_n(\mathbb{C})$ is equal to $\mathfrak{sl}_n(\mathbb{C})$.

Let I be a non-zero ideal of $\mathfrak{gl}_n(\mathbb{C})$ contained in $\mathfrak{sl}_n(\mathbb{C})$. Let $x \in I$ with $x \neq 0$. Then

$$x = \sum x_{pq} E_{pq} \quad \text{with } x_{pq} \in \mathbb{C}.$$

Not all x_{pq} are zero.

Suppose first that there exist $i \neq j$ with $x_{ij} \neq 0$. Then

$$\left[E_{ii}, \sum x_{pq} E_{pq} \right] = \sum_q x_{iq} E_{iq} - \sum_p x_{pi} E_{pi} \in I.$$

Also

$$[[E_{ii}, x], E_{jj}] = x_{ij} E_{ij} + x_{ji} E_{ji} \in I.$$

Hence
$$[E_{ii} - E_{jj}, x_{ij}E_{ij} + x_{ji}E_{ji}] = 2x_{ij}E_{ij} - 2x_{ji}E_{ji} \in I.$$

Thus $4x_{ij}E_{ij} \in I$. Since $x_{ij} \neq 0$ we have $E_{ij} \in I$.

Now suppose that $x_{ij} = 0$ for all $i \neq j$. Then $x = \sum x_{pp} E_{pp}$. Since $\sum x_{pp} = 0$ and not all $x_{pp} = 0$ the x_{pp} are not all equal. Suppose $x_{ii} \neq x_{jj}$. Then
$$[x, E_{ij}] = (x_{ii} - x_{jj}) E_{ij} \in I$$
and so $E_{ij} \in I$.

Thus in either case there exist $i \neq j$ with $E_{ij} \in I$. Let $q \neq i, j$. Then
$$[E_{ij}, E_{jq}] = E_{iq} \in I.$$

Thus $E_{iq} \in I$ for all $q \neq i$. Now let $p \neq i, q$. Then
$$[E_{pi}, E_{iq}] = E_{pq} \in I.$$

Hence $E_{pq} \in I$ for all $p \neq q$. Also
$$[E_{pq}, E_{qp}] = E_{pp} - E_{qq} \in I \qquad \text{for all } p \neq q.$$

But the $E_{pp} - E_{qq}$ for $p \neq q$ and the E_{pq} for $p \neq q$ generate $\mathfrak{sl}_n(\mathbb{C})$. Thus $I = \mathfrak{sl}_n(\mathbb{C})$ and $\mathfrak{sl}_n(\mathbb{C})$ is simple. □

We next determine a Cartan subalgebra of $\mathfrak{sl}_n(\mathbb{C})$. We write $L = \mathfrak{sl}_n(\mathbb{C})$.

Proposition 4.26 *Let H be the set of diagonal matrices in L. Then $\dim H = n - 1$ and H is a Cartan subalgebra of L.*

Proof. The vector space of diagonal $n \times n$ matrices of trace 0 clearly has dimension $n - 1$. It is a subalgebra H of L with $[HH] = O$. Thus H is nilpotent. To show H is a Cartan subalgebra we must show $H = N(H)$.

Let $\sum_{i,j} \lambda_{ij} E_{ij}$ lie in $N(H)$. Suppose if possible that $\lambda_{ij} \neq 0$ for some $i \neq j$. We have
$$\left[\sum_k \mu_k E_{kk}, \sum_{i,j} \lambda_{ij} E_{ij}\right] \in H$$
for all $\sum_k \mu_k E_{kk} \in H$. The coefficient of E_{ij} in this matrix is $(\mu_i - \mu_j) \lambda_{ij}$. Thus if we choose (i, j) such that $i \neq j$ and $\lambda_{ij} \neq 0$ and choose $\sum_k \mu_k E_{kk} \in H$ with $\mu_i \neq \mu_j$ we obtain a contradiction. Hence $\lambda_{ij} = 0$ for all $i \neq j$. Thus $N(H) = H$ and H is a Cartan subalgebra of L. □

We next obtain the Cartan decomposition of L with respect to H.

Proposition 4.27 *Let H be the subalgebra of diagonal matrices in L. Then the Cartan decomposition of L with respect to H is*

$$L = H \oplus \sum_{i \ne j} \mathbb{C} E_{ij}.$$

Proof. This is certainly a decomposition of L into a direct sum of subspaces. To show it is a Cartan decomposition it is sufficient to verify that the 1-dimensional subspaces $\mathbb{C} E_{ij}$ for $i \ne j$ are H-submodules of L. Now we have

$$\left[\sum_{k=1}^{n} \lambda_k E_{kk}, E_{ij} \right] = (\lambda_i - \lambda_j) E_{ij}$$

and so $\mathbb{C} E_{ij}$ is indeed an H-submodule. $\square$

We next obtain the roots of L with respect to H.

Proposition 4.28 *The roots of L with respect to H are the functions $H \to \mathbb{C}$ given by*

$$\begin{pmatrix} \lambda_1 & & O \\ & \ddots & \\ & & \ddots \\ O & & \lambda_n \end{pmatrix} \to \lambda_i - \lambda_j \qquad i \ne j.$$

Proof. This follows from the Cartan decomposition given in Proposition 4.27. $\square$

We next calculate the value of the Killing form $\langle x, y \rangle$ when $x, y \in H$.

Proposition 4.29 *Let $x = \sum_{i=1}^{n} \lambda_i E_{ii}$, $y = \sum_{i=1}^{n} \mu_i E_{ii}$ lie in H. Then $\langle x, y \rangle = 2n \, \mathrm{tr}(xy)$.*

Proof. We have

$$\langle x, y \rangle = \mathrm{tr}(\mathrm{ad}\, x \, \mathrm{ad}\, y) = \sum_{\substack{i,j \\ i \ne j}} (\lambda_i - \lambda_j)(\mu_i - \mu_j)$$

since $\mathrm{ad}\, x \, \mathrm{ad}\, y \, E_{ij} = (\lambda_i - \lambda_j)(\mu_i - \mu_j) E_{ij}$ for $i \ne j$, and $\mathrm{ad}\, x \, \mathrm{ad}\, y \, H = O$.

Hence
$$\langle x, y \rangle = \sum_{i,j} (\lambda_i - \lambda_j)(\mu_i - \mu_j)$$
$$= \sum_{i,j} \lambda_i \mu_i + \sum_{i,j} \lambda_j \mu_j - \sum_{i,j} \lambda_i \mu_j - \sum_{i,j} \lambda_j \mu_i$$
$$= 2n \operatorname{tr}(xy) - \left(\sum_i \lambda_i\right)\left(\sum_j \mu_j\right) - \left(\sum_j \lambda_j\right)\left(\sum_i \mu_i\right)$$
$$= 2n \operatorname{tr}(xy), \quad \text{since } \sum_i \lambda_i = \sum_i \mu_i = 0. \qquad \square$$

We may use this knowledge of the Killing form of L restricted to H to determine the elements $h'_\alpha \in H$ corresponding to the roots $\alpha \in \Phi$.

Proposition 4.30 *Let $\alpha_{ij} \in \Phi$ satisfy*

$$\alpha_{ij}\begin{pmatrix} \lambda_1 & & O \\ & \ddots & \\ O & & \lambda_n \end{pmatrix} = \lambda_i - \lambda_j \quad i \neq j.$$

Then $h'_{\alpha_{ij}} = \dfrac{1}{2n}(E_{ii} - E_{jj})$.

Proof. Let $x = \sum_{k=1}^n \lambda_k E_{kk} \in H$. Then we have
$$\left\langle \frac{1}{2n}(E_{ii} - E_{jj}), x \right\rangle = 2n \operatorname{tr}\left(\frac{1}{2n}(E_{ii} - E_{jj}) x\right)$$
$$= \lambda_i - \lambda_j = \alpha_{ij}(x), \quad \text{by Proposition 4.29.}$$

However, $h'_{\alpha_{ij}} \in H$ is uniquely determined by the condition $\left\langle h'_{\alpha_{ij}}, x \right\rangle = \alpha_{ij}(x)$ for all $x \in H$. Hence $h'_{\alpha_{ij}} = \dfrac{1}{2n}(E_{ii} - E_{jj})$. $\qquad \square$

5
The root system and the Weyl group

5.1 Positive systems and fundamental systems of roots

As before, let L be a semisimple Lie algebra and H be a Cartan subalgebra. Let Φ be the set of roots of L with respect to H. We know by Proposition 4.17 that the elements h'_α, $\alpha \in \Phi$, span H. Thus we can find a subset which forms a basis of H. Let $h'_{\alpha_1}, \ldots, h'_{\alpha_l}$ form a basis of H.

Proposition 5.1 *Let $\alpha \in \Phi$. Then $h'_\alpha = \sum_{i=1}^l \mu_i h'_{\alpha_i}$ where each μ_i lies in $\mathbb{Q}$.*

Proof. We know that $h'_\alpha = \sum_{i=1}^l \mu_i h'_{\alpha_i}$ for uniquely determined elements $\mu_i \in \mathbb{C}$. Let $\langle h'_{\alpha_i}, h'_{\alpha_j} \rangle = \xi_{ij}$. Then $\xi_{ij} \in \mathbb{Q}$ by Proposition 4.24. We consider the system of equations:

$$\langle h'_\alpha, h'_{\alpha_1} \rangle = \mu_1 \xi_{11} + \mu_2 \xi_{21} + \cdots + \mu_l \xi_{l1}$$
$$\langle h'_\alpha, h'_{\alpha_2} \rangle = \mu_1 \xi_{12} + \mu_2 \xi_{22} + \cdots + \mu_l \xi_{l2}$$
$$\vdots$$
$$\langle h'_\alpha, h'_{\alpha_l} \rangle = \mu_1 \xi_{1l} + \mu_2 \xi_{2l} + \cdots + \mu_l \xi_{ll}.$$

This is a system of l equations in l variables $\mu_1, \ldots, \mu_l$. Now $\det(\xi_{ij}) \neq 0$ since the Killing form on L is non-degenerate on restriction to H, by Proposition 4.14. Thus we may solve this system of equations for $\mu_1, \ldots, \mu_l$ by Cramer's rule. Since $\langle h'_\alpha, h'_{\alpha_i} \rangle \in \mathbb{Q}$ and all $\xi_{ij} \in \mathbb{Q}$ we deduce that $\mu_i \in \mathbb{Q}$ for $i = 1, \ldots, l$. □

We denote by $H_\mathbb{Q}$ the set of all elements of form $\sum_{i=1}^l \mu_i h'_{\alpha_i}$ for $\mu_i \in \mathbb{Q}$ and $H_\mathbb{R}$ the set of all such elements with $\mu_i \in \mathbb{R}$. Proposition 5.1 shows that $H_\mathbb{Q}$ and $H_\mathbb{R}$ are independent of the choice of basis h'_{α_i}. Also $H_\mathbb{Q}$ is the set of all

5.1 Positive systems and fundamental systems of roots

rational linear combinations of the h'_α, $\alpha \in \Phi$, and $H_\mathbb{R}$ is the set of all real linear combinations of such elements.

We show next that the Killing form of L behaves in a favourable manner when restricted to $H_\mathbb{R}$.

Proposition 5.2 *Let $x \in H_\mathbb{R}$. Then $\langle x, x \rangle \in \mathbb{R}$ and $\langle x, x \rangle \geq 0$. If $\langle x, x \rangle = 0$ then $x = 0$.*

Proof. Let $x = \sum_{i=1}^{l} \mu_i h'_{\alpha_i}$. Then we have

$$\langle x, x \rangle = \sum_{i=1}^{l} \sum_{j=1}^{l} \mu_i \mu_j \langle h'_{\alpha_i}, h'_{\alpha_j} \rangle$$

$$= \sum_i \sum_j \mu_i \mu_j \operatorname{tr}\left(\operatorname{ad} h'_{\alpha_i} \operatorname{ad} h'_{\alpha_j}\right)$$

$$= \sum_i \sum_j \mu_i \mu_j \sum_{\lambda \in \Phi} \lambda\left(h'_{\alpha_i}\right) \lambda\left(h'_{\alpha_j}\right)$$

$$= \sum_{\lambda \in \Phi} \sum_i \sum_j \mu_i \mu_j \lambda\left(h'_{\alpha_i}\right) \lambda\left(h'_{\alpha_j}\right)$$

$$= \sum_{\lambda \in \Phi} \left(\sum_i \mu_i \lambda\left(h'_{\alpha_i}\right)\right)^2.$$

Now $\lambda\left(h'_{\alpha_i}\right) = \langle h'_\lambda, h'_{\alpha_i} \rangle \in \mathbb{Q}$ by Proposition 4.24. Thus we have $\langle x, x \rangle \in \mathbb{R}$, and also $\langle x, x \rangle \geq 0$.

Suppose that $\langle x, x \rangle = 0$. Then we have $\sum_i \mu_i \lambda\left(h'_{\alpha_i}\right) = 0$ for all $\lambda \in \Phi$. In particular $\sum_i \mu_i \alpha_j\left(h'_{\alpha_i}\right) = 0$ for $j = 1, \ldots, l$. This gives $\sum_i \mu_i \langle h'_{\alpha_i}, h'_{\alpha_j} \rangle = 0$, that is $\sum_i \mu_i \xi_{ij} = 0$. Since the matrix (ξ_{ij}) is non-singular we deduce that $\mu_i = 0$ for all i. Thus $x = 0$. □

This proposition shows that the Killing form restricted to $H_\mathbb{R}$ is a map $H_\mathbb{R} \times H_\mathbb{R} \to \mathbb{R}$ which is a symmetric positive definite bilinear form. The vector space $H_\mathbb{R}$ endowed with this positive definite form is a Euclidean space. This Euclidean space contains all vectors h'_α for $\alpha \in \Phi$.

We recall from Lemma 4.16 that we have an isomorphism $h \to h^*$ from H to H^* given by $h^*(x) = \langle h, x \rangle$. We define $H_\mathbb{R}^*$ to be the image of $H_\mathbb{R}$ under this isomorphism. $H_\mathbb{R}^*$ is the real subspace of H^* spanned by Φ. We may also define a symmetric positive definite bilinear form on $H_\mathbb{R}^*$ by

$$\langle h_1^*, h_2^* \rangle = \langle h_1, h_2 \rangle \in \mathbb{R}.$$

Thus $H_{\mathbb{R}}^*$ becomes a Euclidean space containing all the roots $\alpha \in \Phi$. We shall investigate the configuration formed by the roots in the Euclidean space $H_{\mathbb{R}}^*$. We shall, for the time being, write $V = H_{\mathbb{R}}^*$.

A **total ordering** on V is a relation $<$ on V satisfying the following axioms.

(i) $\lambda < \mu$ and $\mu < \nu$ implies $\lambda < \nu$.
(ii) For each pair of elements $\lambda, \mu \in V$ just one of the conditions $\lambda < \mu$, $\lambda = \mu, \mu < \lambda$ holds.
(iii) If $\lambda < \mu$ then $\lambda + \nu < \mu + \nu$.
(iv) If $\lambda < \mu$ and $\xi \in \mathbb{R}$ with $\xi > 0$ then $\xi\lambda < \xi\mu$, and if $\xi < 0$ then $\xi\mu < \xi\lambda$.

Every real vector space has such total orderings. If $v_1, \ldots, v_l$ are a basis of V and $\lambda = \sum \lambda_i v_i, \mu = \sum \mu_i v_i$ with $\lambda \neq \mu$ then we may define $\lambda < \mu$ if the first non-zero coefficient $\mu_i - \lambda_i$ is positive. This gives us a total ordering on V.

A **positive system** $\Phi^+ \subset \Phi$ is the set of all roots $\alpha \in \Phi$ satisfying $0 < \alpha$ for some total ordering on V. Given such a positive system Φ^+ we define the **fundamental system** $\Pi \subset \Phi^+$ as follows: $\alpha \in \Pi$ if and only if $\alpha \in \Phi^+$ and α cannot be expressed as the sum of two elements of Φ^+. Φ^- is the corresponding set of negative roots.

Proposition 5.3 *Every root in Φ^+ is a sum of roots in Π.*

Proof. Let $\alpha \in \Phi^+$. Then either $\alpha \in \Pi$ or $\alpha = \beta + \gamma$ where $\beta, \gamma \in \Phi^+$ and $\beta < \alpha, \gamma < \alpha$. We continue this process, which must eventually terminate since Φ^+ is finite. Thus α is a sum of elements of Π. □

Proposition 5.4 *Let $\alpha, \beta \in \Pi$ with $\alpha \neq \beta$. Then $\langle \alpha, \beta \rangle \leq 0$.*

Proof. We first observe that $\alpha - \beta \notin \Phi$. For if $\alpha - \beta \in \Phi$ we would have either $\alpha - \beta \in \Phi^+$ or $\beta - \alpha \in \Phi^+$. If $\alpha - \beta \in \Phi^+$ then $\alpha = (\alpha - \beta) + \beta$ which contradicts $\alpha \in \Pi$. If $\beta - \alpha \in \Phi^+$ then $\beta = (\beta - \alpha) + \alpha$ which contradicts $\beta \in \Pi$. Hence $\alpha - \beta \notin \Phi$. We now consider the α-chain of roots through β. This has form

$$\beta, \alpha + \beta, \ldots, q\alpha + \beta$$

since $-\alpha + \beta \notin \Phi$. By Proposition 4.22 we deduce

$$2\frac{\langle h'_\alpha, h'_\beta \rangle}{\langle h'_\alpha, h'_\alpha \rangle} = -q.$$

However, $\langle h'_\alpha, h'_\alpha \rangle > 0$, hence $\langle h'_\alpha, h'_\beta \rangle \leq 0$. It follows that $\langle \alpha, \beta \rangle \leq 0$. □

Thus any two distinct roots in the fundamental system Π are inclined at an obtuse angle.

5.2 The Weyl group

Our next result shows the importance of the concept of a fundamental system of roots.

Theorem 5.5 *A fundamental system Π forms a basis of $V = H_{\mathbb{R}}^*$.*

Proof. We first show that Π spans V. We know by Proposition 4.17 that Φ spans V. Since $\alpha \in \Phi$ if and only if $-\alpha \in \Phi$ we see that Φ^+ spans V. By Proposition 5.3 we deduce that Π spans V.

We show now that the set Π is linearly independent. Suppose this were false. Then there would exist a non-trivial linear combination of the roots $\alpha_i \in \Pi$ equal to zero. We take all the terms with positive coefficient to one side of this relation. Thus we have

$$\mu_{i_1}\alpha_{i_1} + \cdots + \mu_{i_r}\alpha_{i_r} = \mu_{j_1}\alpha_{j_1} + \cdots + \mu_{j_s}\alpha_{j_s}$$

where $\mu_{i_1}, \ldots, \mu_{i_r}, \mu_{j_1}, \ldots, \mu_{j_s} > 0$ and $\alpha_{i_1}, \ldots, \alpha_{i_r}, \alpha_{j_1}, \ldots, \alpha_{j_s}$ are distinct elements of Π. We write

$$v = \mu_{i_1}\alpha_{i_1} + \cdots + \mu_{i_r}\alpha_{i_r} = \mu_{j_1}\alpha_{j_1} + \cdots + \mu_{j_s}\alpha_{j_s}.$$

Then we have $\langle v, v \rangle = \langle \mu_{i_1}\alpha_{i_1} + \cdots + \mu_{i_r}\alpha_{i_r}, \mu_{j_1}\alpha_{j_1} + \cdots + \mu_{j_s}\alpha_{j_s} \rangle$. We deduce $\langle v, v \rangle \leq 0$ by Proposition 5.4. Since the form is positive definite this implies that $v = 0$. However, $0 < v$ since we have $0 < \alpha_i$ for all $\alpha_i \in \Pi$ and $\mu_i > 0$. This gives a contradiction. Thus Π is linearly independent. $\square$

We see in particular that $|\Pi| = l = \dim H$. Thus the number of roots in a fundamental system is equal to the rank of the Lie algebra L.

Corollary 5.6 *Let Π be a fundamental system of roots. Then each $\alpha \in \Phi$ can be expressed in the form $\alpha = \sum n_i \alpha_i$ where $\alpha_i \in \Pi$, $n_i \in \mathbb{Z}$ and either $n_i \geq 0$ for all i or $n_i \leq 0$ for all i.*

Proof. The roots $\alpha \in \Phi^+$ have all $n_i \geq 0$ and the roots $\alpha \in \Phi^-$ have all $n_i \leq 0$. $\square$

5.2 The Weyl group

Inside the root system Φ a positive system Φ^+ can be chosen in many different ways. However, we shall show that any two positive systems in Φ can be transformed into one another by an element of a certain finite group W which acts on Φ.

For each $\alpha \in \Phi$ we define a linear map $s_\alpha : V \to V$ by

$$s_\alpha(x) = x - 2\frac{\langle \alpha, x \rangle}{\langle \alpha, \alpha \rangle}\alpha \quad \text{for all } x \in V.$$

As before, $V = H_\mathbb{R}^*$. This map s_α satisfies

$$s_\alpha(\alpha) = -\alpha$$
$$s_\alpha(x) = x \quad \text{if } \langle \alpha, x \rangle = 0.$$

There is a unique linear map satisfying these conditions – the reflection in the hyperplane of V orthogonal to α. Thus s_α is this reflection.

The group W of all non-singular linear maps on V generated by the s_α for all $\alpha \in \Phi$ is called the **Weyl group**. This group plays an important role in the Lie theory. It is a group of isometries of V, that is we have

$$\langle wx, wy \rangle = \langle x, y \rangle \quad \text{for all } x, y \in V.$$

Proposition 5.7 *W permutes the roots. Thus if $\alpha \in \Phi$ and $w \in W$ then $w(\alpha) \in \Phi$.*

Proof. It is sufficient to show that $s_\alpha(\beta) \in \Phi$ for all $\alpha, \beta \in \Phi$ since the elements s_α generate W. If $\beta = \alpha$ or $-\alpha$ this is clear. Thus suppose $\beta \neq \pm\alpha$. Let the α-chain of roots through β be

$$-p\alpha + \beta, \ldots, \beta, \ldots, q\alpha + \beta.$$

Then we have

$$s_\alpha(\beta) = \beta - 2\frac{\langle \alpha, \beta \rangle}{\langle \alpha, \alpha \rangle}\alpha = \beta - (p-q)\alpha$$

by Proposition 4.22. Now $\beta - (p-q)\alpha$ is one of the roots in the α-chain through β. Thus $s_\alpha(\beta) \in \Phi$.

In fact we observe that s_α inverts the above α-chain of roots. In particular we have

$$s_\alpha(q\alpha + \beta) = -p\alpha + \beta, \quad s_\alpha(-p\alpha + \beta) = q\alpha + \beta. \quad \square$$

Proposition 5.8 *The Weyl group W is finite.*

Proof. W permutes Φ and Φ is finite. If two elements of W induce the same permutation of Φ they must be equal, since Φ spans V. Since there are only finitely many permutations of Φ, W must be finite. $\square$

5.2 The Weyl group

Now suppose that Φ^+ is a positive system in Φ and that Π is the corresponding fundamental system.

Lemma 5.9 *Let $\alpha \in \Pi$. If $\beta \in \Phi^+$ and $\beta \ne \alpha$ then $s_\alpha(\beta) \in \Phi^+$.*

Proof. We can express β in the form

$$\beta = \sum_i n_i \alpha_i \qquad \alpha_i \in \Pi, \quad n_i \in \mathbb{Z}, \quad n_i \ge 0$$

by Corollary 5.6. Since $\beta \ne \alpha$ there must be some $n_i \ne 0$ with $\alpha_i \ne \alpha$. We then consider

$$s_\alpha(\beta) = \beta - 2 \frac{\langle \alpha, \beta \rangle}{\langle \alpha, \alpha \rangle} \alpha$$

and express this as a linear combination of the elements of Π. The coefficient of α_i in $s_\alpha(\beta)$ remains n_i. Since $n_i > 0$ we deduce from Corollary 5.6 that $s_\alpha(\beta) \in \Phi^+$. $\square$

Theorem 5.10 *Let Φ_1^+, Φ_2^+ be two positive systems in Φ. Then there exists $w \in W$ such that $w(\Phi_1^+) = \Phi_2^+$.*

Proof. Let $m = |\Phi_1^+ \cap \Phi_2^-|$. We shall use induction on m. If $m = 0$ we have $\Phi_1^+ = \Phi_2^+$ and so $w = 1$ has the required property. Thus we may assume $m > 0$.

Let Π_1 be the fundamental system in Φ_1^+. We cannot have $\Pi_1 \subset \Phi_2^+$ as this would imply $\Phi_1^+ \subset \Phi_2^+$, contrary to $m > 0$. Thus there exists $\alpha \in \Pi_1 \cap \Phi_2^-$.

We consider $s_\alpha(\Phi_1^+)$. This is also a positive system in Φ. By Lemma 5.9 $s_\alpha(\Phi_1^+)$ contains all roots in Φ_1^+ except α, together with $-\alpha$. Thus we have

$$|s_\alpha(\Phi_1^+) \cap \Phi_2^-| = m - 1.$$

By induction there exists $w' \in W$ such that $w' s_\alpha(\Phi_1^+) = \Phi_2^+$. Let $w = w' s_\alpha$. Then $w(\Phi_1^+) = \Phi_2^+$ as required. $\square$

Corollary 5.11 *Let Π_1, Π_2 be two fundamental systems in Φ. Then there exists $w \in W$ such that $w(\Pi_1) = \Pi_2$.*

Proof. Let Φ_1^+, Φ_2^+ be positive systems containing Π_1, Π_2 respectively. Let $\Phi_2^+ = w(\Phi_1^+)$. Then $w(\Pi_1)$ is a fundamental system contained in Φ_2^+, so $w(\Pi_1) = \Pi_2$. $\square$

Proposition 5.12 *Let Π be a fundamental system in Φ. Then for each $\alpha \in \Phi$ there exist $\alpha_i \in \Pi$ and $w \in W$ with $\alpha = w(\alpha_i)$.*

Proof. Let Φ^+ be the positive system with fundamental system Π. First suppose $\alpha \in \Phi^+$. Then we have

$$\alpha = \sum_i n_i \alpha_i \qquad \alpha_i \in \Pi, \quad n_i \in \mathbb{Z}, \quad n_i \geq 0$$

by Corollary 5.6. We define the **height** of α by

$$\operatorname{ht} \alpha = \sum_i n_i.$$

We shall argue by induction on $\operatorname{ht} \alpha$. If $\operatorname{ht} \alpha = 1$ then $\alpha = \alpha_i$ for some i and $\alpha \in \Pi$. The result is obvious in this case. Thus suppose $\operatorname{ht} \alpha > 1$. Then we have $n_i > 0$ for at least two values of i by Proposition 4.21. Now

$$\langle \alpha, \alpha \rangle = \sum_i n_i \langle \alpha, \alpha_i \rangle.$$

Since $\langle \alpha, \alpha \rangle > 0$ and each $n_i \geq 0$ there exist $\alpha_i \in \Pi$ with $\langle \alpha, \alpha_i \rangle > 0$. Let $s_i(\alpha) = \beta$. Then $\beta \in \Phi$ and

$$\beta = \alpha - 2 \frac{\langle \alpha_i, \alpha \rangle}{\langle \alpha_i, \alpha_i \rangle} \alpha_i.$$

Since $\langle \alpha_i, \alpha \rangle > 0$ we see that $\operatorname{ht} \beta < \operatorname{ht} \alpha$. On the other hand $\beta \in \Phi^+$ since only one coefficient n_i is changed in passing from α to β, thus at least one coefficient remains positive in β. By Corollary 5.6 this is sufficient to show that $\beta \in \Phi^+$. By induction there exist $\alpha_j \in \Pi$ and $w' \in W$ such that $\beta = w'(\alpha_j)$. Then

$$\alpha = s_i(\beta) = s_i w'(\alpha_j)$$

as required.

Finally we suppose that $\alpha \in \Phi^-$. Then $\alpha = s_\alpha(-\alpha)$ and $-\alpha \in \Phi^+$. Thus $-\alpha = w'(\alpha_i)$ for some $w' \in W$, $\alpha_i \in \Pi$. Hence $\alpha = s_\alpha w'(\alpha_i)$ as required. □

Thus each root is the image of some fundamental root under an element of the Weyl group.

We show next that W is generated by the reflections corresponding to roots in a given fundamental system.

Theorem 5.13 *Let $\Pi = \{\alpha_1, \ldots, \alpha_l\}$ be a fundamental system in Φ. Then the corresponding fundamental reflections $s_{\alpha_1}, \ldots, s_{\alpha_l}$ generate W.*

Proof. Let W_0 be the subgroup of W generated by $s_{\alpha_1}, \ldots, s_{\alpha_l}$. Since the s_α generate W for all $\alpha \in \Phi$ it is sufficient to show that each s_α lies in W_0. We may assume $\alpha \in \Phi^+$ since $s_\alpha = s_{-\alpha}$. Now the proof of Proposition 5.12

shows that $\alpha = w(\alpha_i)$ for some $\alpha_i \in \Pi$ and some $w \in W_0$. We consider the element $w s_{\alpha_i} w^{-1} \in W_0$. We have

$$w s_{\alpha_i} w^{-1}(\alpha) = w s_{\alpha_i}(\alpha_i) = w(-\alpha_i) = -\alpha.$$

We shall also show $w s_{\alpha_i} w^{-1}(x) = x$ if $\langle \alpha, x \rangle = 0$. For $\langle \alpha, x \rangle = 0$ implies $\langle w^{-1}(\alpha), w^{-1}(x) \rangle = 0$, that is $\langle \alpha_i, w^{-1}(x) \rangle = 0$. This gives $s_{\alpha_i} w^{-1}(x) = w^{-1}(x)$, i.e. $w s_{\alpha_i} w^{-1}(x) = x$. Thus $w s_{\alpha_i} w^{-1}$ is the reflection in the hyperplane orthogonal to α, that is $w s_{\alpha_i} w^{-1} = s_\alpha$. This shows that $s_\alpha \in W_0$. Hence $W_0 = W$. □

We now wish to obtain further information about the way in which the Weyl group W is generated by a set of its fundamental reflections. As before we let $\Pi = \{\alpha_1, \ldots, \alpha_l\}$ be a fundamental system of roots and consider the corresponding set of fundamental reflections. For simplicity we write

$$s_1 = s_{\alpha_1}, \quad s_2 = s_{\alpha_2}, \quad \ldots, \quad s_l = s_{\alpha_l}.$$

Then each element of W can be expressed as a product of elements s_i. (We do not need to introduce inverses since $s_i^{-1} = s_i$.) For each $w \in W$ we define $l(w)$ to be the minimal value of m such that w can be expressed as a product of m fundamental reflections s_i. $l(w)$ is called the **length** of w. It is clear that $l(1) = 0$ and $l(s_i) = 1$. An expression of w as a product of fundamental reflections s_i with $l(w)$ terms is called a **reduced expression** for w.

We shall relate $l(w)$ to another integer $n(w)$. We recall that each element $w \in W$ permutes the elements of Φ. We define $n(w)$ to be the number of roots $\alpha \in \Phi^+$ for which $w(\alpha) \in \Phi^-$. Thus $n(w)$ is the number of positive roots made negative by w. We aim to show that $l(w) = n(w)$.

Proposition 5.14 $n(w) \le l(w)$ *for all* $w \in W$.

Proof. We shall first compare $n(w)$ with $n(w s_i)$. We recall from Lemma 5.9 that s_i transforms α_i to $-\alpha_i$ and all positive roots other than α_i to positive roots. It follows that

$$n(w s_i) = n(w) \pm 1.$$

In order to determine the sign we consider the effect of w and $w s_i$ on α_i. If $w(\alpha_i) \in \Phi^+$ then w transforms α_i to a positive root and $w s_i$ transforms α_i to a negative root. Hence $n(w s_i) = n(w) + 1$. On the other hand if $w(\alpha_i) \in \Phi^-$ then we get the reverse situation and $n(w s_i) = n(w) - 1$.

Now let us take a reduced expression

$$w = s_{i_1} s_{i_2} \ldots s_{i_r} \qquad r = l(w).$$

Then we have

$$n(w) \le n\left(s_{i_1} \ldots s_{i_{r-1}}\right) + 1 \le n\left(s_{i_1} \ldots s_{i_{r-2}}\right) + 2 \le \cdots \le r.$$

Thus $n(w) \le l(w)$ as required. □

In order to prove the converse result $l(w) \le n(w)$ we shall first prove a result called the **deletion condition** which is important in its own right.

Theorem 5.15 *Let $w = s_{i_1} \ldots s_{i_r}$ be any expression of $w \in W$ as a product of fundamental reflections. Suppose $n(w) < r$. Then there exist integers j, k with $1 \le j < k \le r$ such that*

$$w = s_{i_1} \ldots \hat{s}_{i_j} \ldots \hat{s}_{i_k} \ldots s_{i_r}$$

where ^ denotes omission.

Proof. We recall from the proof of Proposition 5.14 that, for all $w \in W$, $n(ws_i) = n(w) \pm 1$. Consider the given expression

$$w = s_{i_1} \ldots s_{i_r}.$$

Since $n(w) < r$ there exists k with $1 < k \le r$ such that

$$n\left(s_{i_1} \ldots s_{i_k}\right) = n\left(s_{i_1} \ldots s_{i_{k-1}}\right) - 1.$$

This implies $s_{i_1} \ldots s_{i_{k-1}}(\alpha_{i_k}) \in \Phi^-$ as in the proof of Proposition 5.14. Since $\alpha_{i_k} \in \Phi^+$ there exists j with $1 \le j < k$ such that

$$s_{i_{j+1}} \ldots s_{i_{k-1}}(\alpha_{i_k}) \in \Phi^+$$
$$s_{i_j} s_{i_{j+1}} \ldots s_{i_{k-1}}(\alpha_{i_k}) \in \Phi^-.$$

By Lemma 5.9 s_{i_j} transforms only one positive root into a negative root, namely α_{i_j}. Thus we have

$$s_{i_{j+1}} \ldots s_{i_{k-1}}(\alpha_{i_k}) = \alpha_{i_j}.$$

It follows that the reflections s_{i_k}, s_{i_j} associated with the roots $\alpha_{i_k}, \alpha_{i_j}$ are related by

$$s_{i_j} = s_{i_{j+1}} \ldots s_{i_{k-1}} s_{i_k} s_{i_{k-1}} \ldots s_{i_{j+1}}.$$

This implies
$$s_{i_j} s_{i_{j+1}} \ldots s_{i_{k-1}} = s_{i_{j+1}} \ldots s_{i_{k-1}} s_{i_k}.$$

Thus we have
$$s_{i_1} \ldots s_{i_r} = s_{i_1} \ldots s_{i_{j-1}} s_{i_{j+1}} \ldots s_{i_{k-1}} s_{i_{k+1}} \ldots s_{i_r}$$

and so $w = s_{i_1} \ldots \hat{s}_{i_j} \ldots \hat{s}_{i_k} \ldots s_{i_r}$ as required. $\square$

Corollary 5.16 $n(w) = l(w)$.

Proof. We know from Proposition 5.14 that $n(w) \le l(w)$. Suppose if possible that $n(w) < l(w)$. Let $w = s_{i_1} \ldots s_{i_r}$ be a reduced expression, thus $r = l(w)$. Since $n(w) < r$ we may apply Theorem 5.15 to show that w is a product of $r - 2$ fundamental reflections. This contradicts the definition of $l(w)$. $\square$

Thus the length of w is equal to the number of positive roots made negative by w.

Proposition 5.17 (a) *The maximal length of any element of W is $|\Phi^+|$.*
(b) *W has a unique element w_0 with $l(w_0) = |\Phi^+|$.*
(c) $w_0(\Phi^+) = \Phi^-$
(d) $w_0^2 = 1$.

Proof. Since $l(w) = n(w)$ we have $l(w) \le |\Phi^+|$. For each fundamental system $\Pi \subset \Phi$, $-\Pi$ is also a fundamental system, coming from the opposite total ordering. Thus by Corollary 5.11 there exists $w_0 \in W$ with $w_0(\Pi) = -\Pi$. Hence $w_0(\Phi^+) = \Phi^-$ and $n(w_0) = |\Phi^+|$. Thus $l(w_0) = |\Phi^+|$ also and w_0 is an element of W of maximal length.

Now let $w_0' \in W$ also have $l(w_0') = |\Phi^+|$. Then $n(w_0') = |\Phi^+|$ and so $w_0'(\Phi^+) = \Phi^-$. Let $w = (w_0')^{-1} w_0$. Then $w(\Phi^+) = \Phi^+$ and so $n(w) = 0$. Hence $l(w) = 0$ and so $w = 1$. Thus $w_0' = w_0$ and the element w_0 of maximal length is unique.

Finally we have $w_0^2(\Phi^+) = \Phi^+$ and so $n(w_0^2) = 0$. Hence $l(w_0^2) = 0$ and $w_0^2 = 1$. $\square$

5.3 Generators and relations for the Weyl group

In this section we shall give a description of the Weyl group W by means of generators and relations. Let the order of the element $s_i s_j \in W$ be m_{ij} when $i \ne j$.

Theorem 5.18 *W is isomorphic to the abstract group given by generators and relations:*

$$\langle s_1, \ldots, s_l;\ s_i^2 = 1,\ (s_i s_j)^{m_{ij}} = 1\ \text{ for } i \neq j \rangle.$$

A group defined by generators and relations of this form is called a **Coxeter group**. Thus the theorem asserts that the Weyl group is a Coxeter group.

Proof. Since W is generated by $s_1, \ldots, s_l$ and the relations $s_i^2 = 1$ and $(s_i s_j)^{m_{ij}} = 1$ hold in W it is sufficient to show that every relation

$$s_{i_1} \ldots s_{i_r} = 1$$

in W is a consequence of the defining relations. Now each s_i is a reflection, thus $\det s_i = -1$. Hence $\det(s_{i_1} \ldots s_{i_r}) = (-1)^r$. If $s_{i_1} \ldots s_{i_r} = 1$ we deduce that r must be even. Let $r = 2q$. We shall show that

$$s_{i_1} \ldots s_{i_{2q}} = 1$$

is a consequence of the defining relations, by induction on q. If $q = 1$ the relation is $s_{i_1} s_{i_2} = 1$, hence $s_{i_2} = s_{i_1}^{-1} = s_{i_1}$. Our relation is thus $s_{i_1}^2 = 1$, which is one of the defining relations.

We may therefore assume inductively that all relations in W of length less than $2q$ are consequences of the defining relations.

Now the given relation can be written

$$s_{i_1} \ldots s_{i_q} s_{i_{q+1}} = s_{i_{2q}} \ldots s_{i_{q+2}}.$$

Thus $l\left(s_{i_1} \ldots s_{i_q} s_{i_{q+1}}\right) < q+1$. Hence, by the deletion condition Theorem 5.15, we have

$$s_{i_1} \ldots s_{i_{q+1}} = s_{i_1} \ldots \hat{s}_{i_j} \ldots \hat{s}_{i_k} \ldots s_{i_{q+1}}$$

for certain j, k with $1 \leq j < k \leq q+1$. Now unless $j = 1$ and $k = q+1$ this is a consequence of a relation with fewer than $2q$ terms. It can therefore be deduced from the defining relations. The relation

$$s_{i_1} \ldots \hat{s}_{i_j} \ldots \hat{s}_{i_k} \ldots s_{i_{q+1}} = s_{i_{2q}} \ldots s_{i_{q+2}}$$

has $2q - 2$ terms, so is also a consequence of the defining relations. Thus the given relation

$$s_{i_1} \ldots s_{i_{q+1}} = s_{i_{2q}} \ldots s_{i_{q+2}}$$

will be a consequence of the defining relations, unless we have $j = 1$ and $k = q+1$.

5.3 Generators and relations for the Weyl group

We may therefore assume that $j=1$ and $k=q+1$. Thus we have

$$s_{i_1} \ldots s_{i_{q+1}} = s_{i_2} \ldots s_{i_q},$$

that is

$$s_{i_1} \ldots s_{i_q} = s_{i_2} \ldots s_{i_{q+1}}.$$

We now write the original relation

$$s_{i_1} \ldots s_{i_{2q}} = 1$$

in the alternative form

$$s_{i_2} \ldots s_{i_{2q}} s_{i_1} = 1.$$

In exactly the same way this relation will be a consequence of the defining relations unless

$$s_{i_2} \ldots s_{i_{q+1}} = s_{i_3} \ldots s_{i_{q+2}}.$$

If this relation is a consequence of the defining relations then the relation

$$s_{i_2} \ldots s_{i_{2q}} s_{i_1} = 1$$

will also be a consequence of the defining relations, by the above argument, and we are done.

Now $s_{i_2} \ldots s_{i_{q+1}} = s_{i_3} \ldots s_{i_{q+2}}$ is equivalent to

$$s_{i_3} s_{i_2} s_{i_3} \ldots s_{i_q} s_{i_{q+1}} s_{i_{q+2}} s_{i_{q+1}} \ldots s_{i_4} = 1$$

and this will be a consequence of the defining relations unless

$$s_{i_3} s_{i_2} s_{i_3} \ldots s_{i_q} = s_{i_2} s_{i_3} \ldots s_{i_q} s_{i_{q+1}}.$$

We may therefore assume this to be true. But we also have

$$s_{i_1} s_{i_2} s_{i_3} \ldots s_{i_q} = s_{i_2} s_{i_3} \ldots s_{i_q} s_{i_{q+1}}$$

and so $s_{i_1} = s_{i_3}$. Hence the given relation

$$s_{i_1} \ldots s_{i_{2q}} = 1$$

will be a consequence of the defining relations unless $s_{i_1} = s_{i_3}$.

However, the given relation can be written in the equivalent forms

$$s_{i_2} \ldots s_{i_{2q}} s_{i_1} = 1$$

$$s_{i_3} \ldots s_{i_{2q}} s_{i_1} s_{i_2} = 1$$

and so on. Thus this relation will be a consequence of the defining relations unless we have

$$s_{i_1} = s_{i_3} = s_{i_5} = \ldots = s_{i_{2q-1}}$$
$$s_{i_2} = s_{i_4} = s_{i_6} = \ldots = s_{i_{2q}}.$$

Thus we may assume that the given relation has form

$$s_{i_1} s_{i_2} s_{i_1} s_{i_2} \ldots s_{i_1} s_{i_2} = 1$$

that is $(s_{i_1} s_{i_2})^q = 1$. Now the order of $s_{i_1} s_{i_2}$ is $m_{i_1 i_2}$, hence $m_{i_1 i_2}$ divides q. Thus the relation $(s_{i_1} s_{i_2})^q = 1$ is a consequence of the defining relation $(s_{i_1} s_{i_2})^{m_{i_1 i_2}} = 1$. This completes the proof. □

This remarkable proof, due to R. Steinberg, shows that the Weyl group W is a finite Coxeter group.

6
The Cartan matrix and the Dynkin diagram

6.1 The Cartan matrix

We shall now investigate in more detail the geometry of the system of roots Φ in the vector space $V = H_\mathbb{R}^*$. We recall from Proposition 5.2 that V is a Euclidean space with respect to the scalar product $\langle , \rangle$. The roots Φ span V but are not linearly independent. Any fundamental system $\Pi \subset \Phi$ forms a basis of V.

We first consider the possible angles between pairs of roots $\alpha, \beta \in \Phi$ and the relative lengths of the roots α, β. The angles will be taken to satisfy $0 \le \theta \le \pi$.

Proposition 6.1 *Let $\alpha, \beta \in \Phi$ be such that $\beta \ne \pm \alpha$. Then*:

(i) *the angle between α, β is one of $\pi/6, \pi/4, \pi/3, \pi/2, 2\pi/3, 3\pi/4, 5\pi/6$*
(ii) *if α, β are inclined at $\pi/3$ or $2\pi/3$ then α, β have the same length*
(iii) *if α, β are inclined at $\pi/4$ or $3\pi/4$ then the ratio of their lengths is $\sqrt{2}$*
(iv) *if α, β are inclined at $\pi/6$ or $5\pi/6$ then the ratio of their lengths is $\sqrt{3}$.*

Proof. Let θ be the angle between α, β. Then we have

$$\langle \alpha, \beta \rangle = |\alpha| |\beta| \cos \theta$$

where $|\alpha| = \sqrt{\langle \alpha, \alpha \rangle}$. Hence

$$\cos^2 \theta = \frac{\langle \alpha, \beta \rangle^2}{\langle \alpha, \alpha \rangle \langle \beta, \beta \rangle} = \frac{\langle \alpha, \beta \rangle}{\langle \alpha, \alpha \rangle} \cdot \frac{\langle \beta, \alpha \rangle}{\langle \beta, \beta \rangle}$$

and so

$$4 \cos^2 \theta = 2 \frac{\langle \alpha, \beta \rangle}{\langle \alpha, \alpha \rangle} \cdot 2 \frac{\langle \beta, \alpha \rangle}{\langle \beta, \beta \rangle}.$$

Now we recall from Proposition 4.22 that $2\frac{\langle\alpha,\beta\rangle}{\langle\alpha,\alpha\rangle}$ and $2\frac{\langle\beta,\alpha\rangle}{\langle\beta,\beta\rangle}$ are integers. Hence $4\cos^2\theta \in \mathbb{Z}$. Since $0 \le 4\cos^2\theta \le 4$ and $\beta \ne \pm\alpha$ we have $4\cos^2\theta \in \{0, 1, 2, 3\}$. We consider in each case the possible factorisations of $4\cos^2\theta$ into the product of two integers.

First suppose $4\cos^2\theta = 0$. Then $\theta = \pi/2$.

Next suppose $4\cos^2\theta = 1$. Then $\cos\theta = \frac{1}{2}$ or $-\frac{1}{2}$, hence $\theta = \pi/3$ or $2\pi/3$. The possible factorisations of $4\cos^2\theta$ are

$$1 = 1 \cdot 1 \quad \text{or} \quad 1 = -1 \cdot -1.$$

In either case we have

$$2\frac{\langle\alpha,\beta\rangle}{\langle\alpha,\alpha\rangle} = 2\frac{\langle\beta,\alpha\rangle}{\langle\beta,\beta\rangle}$$

and so $\langle\alpha,\alpha\rangle = \langle\beta,\beta\rangle$ and α, β have the same length.

Next suppose $4\cos^2\theta = 2$. Then $\cos\theta = 1/\sqrt{2}$ or $-1/\sqrt{2}$, thus $\theta = \pi/4$ or $3\pi/4$. The possible factorisations of $4\cos^2\theta$ are

$$2 = 1 \cdot 2 \quad \text{or} \quad 2 = -1 \cdot -2.$$

In either case, by choosing α, β in a suitable order, we have

$$2\frac{\langle\beta,\alpha\rangle}{\langle\beta,\beta\rangle} = 2 \cdot 2\frac{\langle\alpha,\beta\rangle}{\langle\alpha,\alpha\rangle},$$

that is $\langle\alpha,\alpha\rangle = 2\langle\beta,\beta\rangle$ and $|\alpha| = \sqrt{2}|\beta|$. Thus the ratio of the lengths of α, β is $\sqrt{2}$.

Finally suppose that $4\cos^2\theta = 3$. Then $\cos\theta = \sqrt{3}/2$ or $-\sqrt{3}/2$, so $\theta = \pi/6$ or $5\pi/6$. The possible factorisations of $4\cos^2\theta$ are

$$3 = 1 \cdot 3 \quad \text{or} \quad 3 = -1 \cdot -3$$

In either case, by choosing α, β in a suitable order, we have

$$2\frac{\langle\beta,\alpha\rangle}{\langle\beta,\beta\rangle} = 3 \cdot 2\frac{\langle\alpha,\beta\rangle}{\langle\alpha,\alpha\rangle},$$

that is $\langle\alpha,\alpha\rangle = 3\langle\beta,\beta\rangle$ and $|\alpha| = \sqrt{3}|\beta|$. Thus the ratio of the lengths of α, β is $\sqrt{3}$.

This completes the proof. We do not obtain any information about the relative lengths of α, β in the case when $\theta = \pi/2$. □

Corollary 6.2 *Let Π be a fundamental system of roots and let $\alpha, \beta \in \Pi$ with $\beta \ne \alpha$. Then the angle between α, β is one of $\pi/2, 2\pi/3, 3\pi/4, 5\pi/6$.*

Proof. This follows from Proposition 6.1 together with the fact, proved in Proposition 5.4, that the angle θ between two distinct fundamental roots satisfies $\pi/2 \le \theta < \pi$. □

Let $\Pi = \{\alpha_1, \ldots, \alpha_l\}$ be a fundamental system. We incorporate the information about the angles between the α_i and their relative lengths in the form of a matrix. We define A_{ij} by

$$A_{ij} = 2\frac{\langle \alpha_i, \alpha_j \rangle}{\langle \alpha_i, \alpha_i \rangle} \qquad i, j = 1, \ldots, l.$$

Thus $A_{ij} \in \mathbb{Z}$. The $l \times l$ matrix $A = (A_{ij})$ is called the **Cartan matrix**.

Proposition 6.3 *The Cartan matrix A has the following properties.*

(i) $A_{ii} = 2$ *for all i.*
(ii) $A_{ij} \in \{0, -1, -2, -3\}$ *if $i \ne j$.*
(iii) *If $A_{ij} = -2$ or -3 then $A_{ji} = -1$.*
(iv) $A_{ij} = 0$ *if and only if $A_{ji} = 0$.*

Proof. Properties (i), (iv) are obvious and (ii), (iii) follow from the proof of Proposition 6.1. □

If we number the fundamental roots in Π in a different way we may well get a different Cartan matrix A. However, apart from this ambiguity of numbering, the Cartan matrix A is uniquely determined by the semisimple Lie algebra L.

Proposition 6.4 *The Cartan matrix of L depends only on the numbering of the fundamental roots. It is independent of the choice of Cartan subalgebra H and fundamental system Π.*

Proof. The independence of the choice of Cartan subalgebra follows from the conjugacy of Cartan subalgebras, proved in Theorem 3.13.

Let Π' be a second fundamental system. By Corollary 5.11 there exists $w \in W$ with $w(\Pi) = \Pi'$. Let $w(\alpha_i) = \alpha_i'$. Since w is an isometry of V we have

$$2\frac{\langle \alpha_i, \alpha_j \rangle}{\langle \alpha_i, \alpha_i \rangle} = 2\frac{\langle \alpha_i', \alpha_j' \rangle}{\langle \alpha_i', \alpha_i' \rangle}$$

Thus the Cartan matrices defined by Π and Π' with respect to these labellings are the same. □

The only possible 1×1 Cartan matrix is (2). We also see that any 2×2 Cartan matrix must be one of the following:

$$\begin{pmatrix} 2 & 0 \\ 0 & 2 \end{pmatrix} \begin{pmatrix} 2 & -1 \\ -1 & 2 \end{pmatrix} \begin{pmatrix} 2 & -1 \\ -2 & 2 \end{pmatrix} \begin{pmatrix} 2 & -2 \\ -1 & 2 \end{pmatrix} \begin{pmatrix} 2 & -1 \\ -3 & 2 \end{pmatrix} \begin{pmatrix} 2 & -3 \\ -1 & 2 \end{pmatrix}$$

The pair

$$\begin{pmatrix} 2 & -1 \\ -2 & 2 \end{pmatrix} \begin{pmatrix} 2 & -2 \\ -1 & 2 \end{pmatrix}$$

are obtained from one another by reversing the labelling 1, 2, and so are the pair

$$\begin{pmatrix} 2 & -1 \\ -3 & 2 \end{pmatrix} \begin{pmatrix} 2 & -3 \\ -1 & 2 \end{pmatrix}$$

6.2 The Dynkin diagram

In order to determine the possible $l \times l$ Cartan matrices for larger values of l it is useful to introduce a graph called the **Dynkin diagram**. The Dynkin diagram is determined by the Cartan matrix. It is a graph with vertices labelled $1, \ldots, l$. If $i \neq j$ the vertices i, j are joined by n_{ij} edges, where

$$n_{ij} = A_{ij} A_{ji}.$$

We see from Proposition 6.4 that the Dynkin diagram is uniquely determined by the semisimple Lie algebra L.

The Dynkin diagrams of the Cartan matrices of degrees 1 and 2 are as follows.

Cartan matrix	Dynkin diagram
(2)	○
$\begin{pmatrix} 2 & 0 \\ 0 & 2 \end{pmatrix}$	○ ○
$\begin{pmatrix} 2 & -1 \\ -1 & 2 \end{pmatrix}$	○—○
$\begin{pmatrix} 2 & -1 \\ -2 & 2 \end{pmatrix} \begin{pmatrix} 2 & -2 \\ -1 & 2 \end{pmatrix}$	○═○
$\begin{pmatrix} 2 & -1 \\ -3 & 2 \end{pmatrix} \begin{pmatrix} 2 & -3 \\ -1 & 2 \end{pmatrix}$	○≡○

6.2 The Dynkin diagram

Proposition 6.5 $n_{ij} \in \{0, 1, 2, 3\}$ *for all* $i \neq j$.

Proof. This follows from Proposition 6.3 and the fact that $n_{ij} = A_{ij} A_{ji}$. □

Thus the number of edges joining any two distinct vertices of the Dynkin diagram is either 0, 1, 2 or 3.

Now the Dynkin diagram need not be a connected graph. However, if it is disconnected it will split into connected components. If we number the vertices so that those in each connected component are numbered consecutively, the Cartan matrix will split into blocks of the form

$$A = \begin{pmatrix} * & O & O & O \\ \hline O & * & O & O \\ \hline O & O & * & O \\ \hline O & O & O & * \end{pmatrix}.$$

with one diagonal block for each connected component. This diagonal block will be the Cartan matrix for the given connected component. The set $\Pi = \{\alpha_1, \ldots, \alpha_l\}$ will be partitioned into subsets in a corresponding way, such that roots in different subsets are mutually orthogonal.

Now the set of graphs which can occur as Dynkin diagrams of semisimple Lie algebras turns out to be quite restricted. In order to determine the possible Dynkin diagrams it is useful to introduce a quadratic form $Q(x_1, \ldots, x_l)$ which is defined in terms of the Dynkin diagram. We define

$$Q(x_1, \ldots, x_l) = 2 \sum_{i=1}^{l} x_i^2 - \sum_{\substack{i,j=1 \\ i \neq j}}^{l} \sqrt{n_{ij}} \, x_i x_j.$$

We illustrate this definition in the cases $l = 1, 2$.

Dynkin diagram	Quadratic form
o	$2x_1^2$
o o	$2x_1^2 + 2x_2^2$
o—o	$2x_1^2 - 2x_1 x_2 + 2x_2^2$
o=o	$2x_1^2 - 2\sqrt{2} x_1 x_2 + 2x_2^2$
o≡o	$2x_1^2 - 2\sqrt{3} x_1 x_2 + 2x_2^2$

Proposition 6.6 *The quadratic form* $Q(x_1, \ldots, x_l)$ *is positive definite.*

Proof. We have, for $i \neq j$,

$$n_{ij} = A_{ij}A_{ji} = 2\frac{\langle \alpha_i, \alpha_j \rangle}{\langle \alpha_i, \alpha_i \rangle} \cdot 2\frac{\langle \alpha_j, \alpha_i \rangle}{\langle \alpha_j, \alpha_j \rangle}$$

hence $-\sqrt{n_{ij}} = 2\frac{\langle \alpha_i, \alpha_j \rangle}{|\alpha_i||\alpha_j|}$ since $\langle \alpha_i, \alpha_j \rangle \leq 0$. For $i = j$ we have $\frac{2\langle \alpha_i, \alpha_j \rangle}{|\alpha_i||\alpha_j|} = 2$.

Thus the quadratic form may be written

$$Q(x_1, \ldots, x_l) = \sum_{i,j=1}^{l} \frac{2\langle \alpha_i, \alpha_j \rangle}{|\alpha_i||\alpha_j|} x_i x_j = 2\left\langle \sum_{i=1}^{l} \frac{x_i \alpha_i}{|\alpha_i|}, \sum_{j=1}^{l} \frac{x_j \alpha_j}{|\alpha_j|} \right\rangle$$

$$= 2\langle y, y \rangle \quad \text{where } y = \sum_{i=1}^{l} \frac{x_i \alpha_i}{|\alpha_i|}.$$

Thus $Q(x_1, \ldots, x_l) \geq 0$. Moreover if $Q(x_1, \ldots, x_l) = 0$ then $y = 0$. Since $\alpha_1, \ldots, \alpha_l$ are linearly independent this implies that $x_i = 0$ for all i. Thus the quadratic form is positive definite. □

Now the connected components of the Dynkin diagram of any semisimple Lie algebra satisfy the following conditions:

(A) The graph is connected.
(B) Any pair of distinct vertices are joined by 0, 1, 2 or 3 edges.
(C) The corresponding quadratic form $Q(x_1, \ldots, x_l)$ is positive definite.

We shall approach the problem of finding the possible Dynkin diagrams by determining all graphs satisfying conditions (A), (B), (C). Having determined all such graphs we shall consider subsequently which ones occur as Dynkin diagrams.

6.3 Classification of Dynkin diagrams

The main result which we shall obtain in this section is as follows.

Theorem 6.7 *The graphs satisfying conditions (A), (B), (C) shown in Section 6.2 are just those in the following list.*

6.3 Classification of Dynkin diagrams

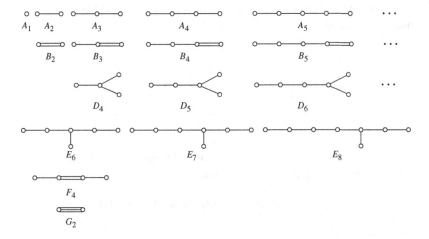

Proof. We shall show first that the graphs on this list satisfy conditions (A), (B), (C). It is obvious that they satisfy (A) and (B). We shall therefore concentrate on condition (C).

We recall from linear algebra that a quadratic form $\sum a_{ij} x_i x_j$ is positive definite if and only if all the leading minors of its symmetric matrix (a_{ij}) have positive determinant. This condition is

$$|a_{11}| > 0, \quad \begin{vmatrix} a_{11} & a_{12} \\ a_{21} & a_{22} \end{vmatrix} > 0, \ldots, \quad \det(a_{ij}) > 0.$$

Given a graph Γ with l vertices on the list in Theorem 6.7 we shall show that $Q(x_1, \ldots, x_l)$ is positive definite by induction on l. If $l = 1$ then $\Gamma = A_1$ and $Q(x_1) = 2x_1^2$ is positive definite. If $l = 2$ then Γ is A_2, B_2 or G_2. The symmetric matrix representing $Q(x_1, x_2)$ is then

$$\begin{pmatrix} 2 & -1 \\ -1 & 2 \end{pmatrix} \quad \begin{pmatrix} 2 & -\sqrt{2} \\ -\sqrt{2} & 2 \end{pmatrix} \quad \begin{pmatrix} 2 & -\sqrt{3} \\ -\sqrt{3} & 2 \end{pmatrix}$$
$$\quad A_2 \qquad\qquad B_2 \qquad\qquad G_2$$

In these cases the leading minors have positive determinant.

Now assume $l \geq 3$. Then inspection of the list of graphs in Theorem 6.7 shows that Γ contains at least one vertex which is joined to just one other vertex of Γ, and joined to it by a single edge. Let such a vertex be labelled l, and let the vertex it is joined to be labelled $l-1$. We write $\Gamma = \Gamma_l$, and the graph obtained from Γ_l by removing the vertex l by Γ_{l-1}, and the graph obtained from Γ_{l-1} by removing the vertex $l-1$ by Γ_{l-2}. Let $\det \Gamma_l$ be the determinant of the symmetric matrix representing the quadratic form $Q(x_1, \ldots, x_l)$ associated to Γ_l. We observe from the list of graphs that Γ_{l-1} and Γ_{l-2} also lie in the list. Moreover we have

$$\det \Gamma_l = \begin{vmatrix} & & & 0 \\ & & & \vdots \\ & & & 0 \\ & & 2 & -1 \\ 0 & \cdots & 0 & -1 & 2 \end{vmatrix} = 2 \det \Gamma_{l-1} - \det \Gamma_{l-2}$$

by expanding the determinant by its last row. This gives us an inductive way of calculating $\det \Gamma_l$. In particular we have

$$\det A_1 = 2, \quad \det A_2 = 3, \quad \det A_l = 2 \det A_{l-1} - \det A_{l-2}.$$

Thus $\det A_l = l + 1$.

$$\det A_1 = 2, \quad \det B_2 = 2, \quad \det B_3 = 2, \quad \det B_l = 2 \det B_{l-1} - \det B_{l-2}.$$

Thus $\det B_l = 2$.

$$\det A_3 = 4, \quad \det D_4 = 4, \quad \det D_5 = 4, \quad \det D_l = 2 \det D_{l-1} - \det D_{l-2}.$$

Thus $\det D_l = 4$.

$$\det E_6 = 2 \det D_5 - \det A_4 = 3$$
$$\det E_7 = 2 \det D_6 - \det A_5 = 2$$
$$\det E_8 = 2 \det D_7 - \det A_6 = 1$$
$$\det F_4 = 2 \det B_3 - \det A_2 = 1.$$

Thus we have shown that $\det \Gamma_l > 0$ for all Γ_l.

Now the leading minors of the symmetric matrix associated to Γ_l are the symmetric matrices associated to certain subgraphs of Γ_l. The numbering can be chosen so that all these subgraphs are connected. However, the list of graphs has the property that any connected subgraph of a graph on the list is also on the list. Thus the determinant of every leading minor of the given symmetric matrix is positive. Hence the quadratic form $Q(x_1, \ldots, x_l)$ associated to Γ_l is positive definite.

Thus we have shown that the graphs on our list satisfy conditions (A), (B), (C). We wish to prove the converse, i.e. that any graph satisfying conditions (A), (B), (C) is on our list. Before being able to prove this we shall need some lemmas.

6.3 Classification of Dynkin diagrams

Lemma 6.8 *For each of the graphs on the following list the corresponding quadratic form $Q(x_1, \ldots, x_l)$ has determinant 0.*

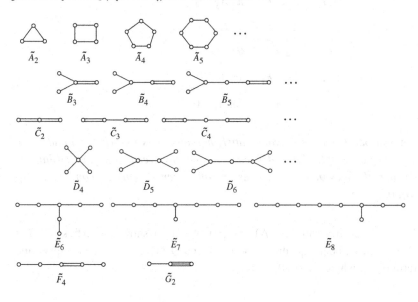

Proof. First consider the graphs $\Gamma = \tilde{A}_l$. Each row of the symmetric matrix of the given quadratic form contains one entry 2, two entries -1, and remaining entries 0. Thus the sum of the columns is zero and $\det \tilde{A}_l = 0$.

In all the other graphs Γ on the list we can find a vertex l joined to just one other vertex $l-1$. Moreover l is joined to $l-1$ by a single edge or a double edge. If there is a single edge we may use the formula

$$\det \Gamma_l = 2 \det \Gamma_{l-1} - \det \Gamma_{l-2}$$

as before. If there is a double edge we obtain instead

$$\det \Gamma_l = 2 \det \Gamma_{l-1} - 2 \det \Gamma_{l-2}.$$

We may use these formulae to calculate all the determinants inductively.

$$\det \tilde{B}_3 = 2 \det A_3 - 2 (\det A_1)^2 = 0$$

$$\det \tilde{B}_l = 2 \det D_l - 2 \det D_{l-1} = 0 \quad \text{for } l \geq 4$$

$$\det \tilde{C}_2 = 2 \det B_2 - 2 \det A_1 = 0$$

$$\det \tilde{C}_l = 2 \det B_l - 2 \det B_{l-1} = 0 \quad \text{for } l \geq 3$$

$$\det \tilde{D}_4 = 2 \det D_4 - (\det A_1)^3 = 0$$

$$\det \tilde{D}_l = 2 \det D_l - \det D_{l-2} \cdot \det A_1 = 0 \quad \text{for } l \geq 5$$

$$\det \tilde{E}_6 = 2 \det E_6 - \det A_5 = 0$$

$$\det \tilde{E}_7 = 2 \det E_7 - \det D_6 = 0$$

$$\det \tilde{E}_8 = 2 \det E_8 - \det E_7 = 0$$

$$\det \tilde{F}_4 = 2 \det F_4 - \det B_3 = 0$$

$$\det \tilde{G}_2 = 2 \det G_2 - \det A_1 = 0.$$

Lemma 6.9 *Let Γ be a graph satisfying conditions (A), (B), (C) and Γ' be a connected graph obtained from Γ by omitting vertices or decreasing the number of edges between vertices or both. Then Γ' satisfies conditions (A), (B), (C) also.*

Proof. Γ' clearly satisfies (A) and (B). We must show it satisfies (C). Let $Q(x_1, \ldots, x_l)$ be the quadratic form of Γ and $Q'(x_1, \ldots, x_m)$ be the quadratic form of Γ', where $m \leq l$. We have

$$Q(x_1, \ldots, x_l) = 2 \sum_{i=1}^{l} x_i^2 - \sum_{\substack{i,j=1 \\ i \neq j}}^{l} \sqrt{n_{ij}} x_i x_j$$

$$Q'(x_1, \ldots, x_m) = 2 \sum_{i=1}^{m} x_i^2 - \sum_{\substack{i,j=1 \\ i \neq j}}^{m} \sqrt{n'_{ij}} x_i x_j$$

where $n'_{ij} \leq n_{ij}$ for $i, j \in \{1, \ldots, m\}$. Suppose if possible that Q' is not positive definite. Then there exist $y_1, \ldots, y_m \in \mathbb{R}$, not all zero, with $Q'(y_1, \ldots, y_m) \leq 0$. Consider $Q(|y_1|, \ldots, |y_m|, 0, \ldots, 0)$. We have

$$Q(|y_1|, \ldots, |y_m|, 0, \ldots, 0) = 2 \sum_{i=1}^{m} |y_i|^2 - \sum_{\substack{i,j=1 \\ i \neq j}}^{m} \sqrt{n_{ij}} |y_i||y_j|$$

$$\leq 2 \sum_{i=1}^{m} y_i^2 - \sum_{\substack{i,j=1 \\ i \neq j}}^{m} \sqrt{n'_{ij}} |y_i||y_j|$$

$$\leq 2 \sum_{i=1}^{m} y_i^2 - \sum_{\substack{i,j=1 \\ i \neq j}}^{m} \sqrt{n'_{ij}} y_i y_j$$

$$= Q'(y_1, \ldots, y_m) \leq 0.$$

6.3 Classification of Dynkin diagrams

Hence $Q(|y_1|, \ldots, |y_m|, 0, \ldots, 0) \leq 0$ but $(|y_1|, \ldots, |y_m|, 0, \ldots, 0)$ is not the zero vector. This contradicts the fact that $Q(x_1, \ldots, x_l)$ is positive definite. Hence $Q'(x_1, \ldots, x_m)$ must be positive definite also. □

Having Lemmas 6.8 and 6.9 at our disposal we are now able to complete the proof of Theorem 6.7.

Let Γ be a graph satisfying conditions (A), (B), (C). Then, by Lemmas 6.8 and 6.9, Γ can have no subgraph of type $\tilde{A}_l, \tilde{B}_l, \tilde{C}_l, \tilde{D}_l, \tilde{E}_6, \tilde{E}_7, \tilde{E}_8, \tilde{F}_4$ or $\tilde{G}_2$. (By a subgraph of Γ we mean a graph obtainable from Γ by removing vertices, or removing edges, or both.) We shall use this information to show that Γ must be one of the graphs on the list in Theorem 6.7.

In the first place we see that Γ contains no cycles, otherwise it would contain a subgraph of type $\tilde{A}_l$ for some $l \geq 2$.

Suppose that Γ contains a triple edge. Then Γ must be the graph G_2, otherwise Γ would contain a subgraph $\tilde{G}_2$.

Thus we may assume that Γ contains no triple edge. Suppose Γ contains a double edge. Then Γ cannot have more than one double edge, otherwise it would contain a subgraph $\tilde{C}_l$ for some $l \geq 2$. Now Γ cannot contain a branch point in addition to a double edge, as otherwise it would contain a subgraph $\tilde{B}_l$ for some $l \geq 3$. Thus Γ is a chain containing just one double edge. If the double edge occurs at one end of the chain then $\Gamma = B_l$ for some $l \geq 2$. If not then we must have $\Gamma = F_4$, since otherwise Γ would contain a subgraph $\tilde{F}_4$.

Thus we may assume that Γ contains no double or triple edges. If Γ contains no branch point then $\Gamma = A_l$ for some $l \geq 1$. Thus we suppose that Γ contains at least one branch point. Now Γ cannot contain more than one branch point, as otherwise it would contain a subgraph $\tilde{D}_l$ for some $l \geq 5$. Thus Γ contains exactly one branch point. There must be exactly three branches emerging from this branch point, since otherwise Γ would contain a subgraph $\tilde{D}_4$. Let the number of vertices on the three branches be l_1, l_2, l_3 with $l_1 \geq l_2 \geq l_3$. Then the total number of vertices of Γ is $l = l_1 + l_2 + l_3 + 1$.

Now we must have $l_3 = 1$, as otherwise we have $l_i \geq 2$ for $i = 1, 2, 3$ and Γ contains a subgraph $\tilde{E}_6$. If $l_2 = 1$ then $\Gamma = D_l$ for some $l \geq 4$. Thus we may assume $l_2 \geq 2$. In fact we must have $l_2 = 2$, as otherwise we have $l_1 \geq 3, l_2 \geq 3$ and Γ contains a subgraph $\tilde{E}_7$. Thus we may assume $l_3 = 1, l_2 = 2$. We must have $l_1 \leq 4$ since otherwise Γ contains a subgraph $\tilde{E}_8$. Thus Γ has type E_6, E_7 or E_8.

Thus we have now determined all possibilities for Γ, and seen that Γ must be one of the graphs which appear on the list in Theorem 6.7. This completes the proof. □

Corollary 6.10 *Let Δ be the Dynkin diagram of a semisimple Lie algebra. Then each connected component of Δ must be one of the graphs*

$$A_l, \quad l \geq 1 \; ; \; B_l, \quad l \geq 2 \; ; \; D_l, \quad l \geq 4 \; ; \; E_6 \; ; \; E_7 \; ; \; E_8 \; ; \; F_4 \; ; \; G_2.$$

We shall consider later whether all these graphs actually occur as Dynkin diagrams.

6.4 Classification of Cartan matrices

We recall that the Dynkin diagram is determined by the Cartan matrix by the property

$$n_{ij} = A_{ij} A_{ji} \quad i \neq j.$$

However, the Cartan matrix is not always uniquely determined by the Dynkin diagram. If we know the integers $n_{ij} \in \{0, 1, 2, 3\}$ for all i, j with $i \neq j$ we consider to what extent the A_{ij} are determined. If $n_{ij} = 0$ then we must have $A_{ij} = 0$ and $A_{ji} = 0$ since $A_{ij} = 0$ if and only if $A_{ji} = 0$. If $n_{ij} = 1$ then we must have $A_{ij} = -1$ and $A_{ji} = -1$ since $A_{ij} \in \mathbb{Z}, A_{ji} \in \mathbb{Z}, A_{ij} \leq 0, A_{ji} \leq 0$. However, if $n_{ij} = 2$ there are two possibilities for the factorisation $n_{ij} = A_{ij} A_{ji}$. Either we have $2 = -1 \cdot -2$ or $2 = -2 \cdot -1$. Thus we have either $A_{ij} = -1, A_{ji} = -2$ or $A_{ij} = -2, A_{ji} = -1$. Similarly if $n_{ij} = 3$ we have either $A_{ij} = -1, A_{ji} = -3$ or $A_{ij} = -3, A_{ji} = -1$.

In the connected graphs in Corollary 6.10 the only ones which give rise to such an ambiguity are $B_l, \quad l \geq 2 \; ; \; F_4$ and G_2. In these graphs we shall place an arrow on the double or triple edges. The direction of the arrow is determined as follows. The arrow points from vertex i to vertex j if and only if $|\alpha_i| > |\alpha_j|$, that is $|A_{ji}| > |A_{ij}|$.

Thus in the situation

$$\underset{i \qquad j}{\circ\!\!=\!\!\!\Rightarrow\!\!=\!\!\circ}$$

we have $|\alpha_i| = \sqrt{2} |\alpha_j|$, $A_{ij} = -1$, $A_{ji} = -2$. In the situation

$$\underset{i \qquad j}{\circ\!\!\equiv\!\!\!\Rrightarrow\!\!\equiv\!\!\circ}$$

we have $|\alpha_i| = \sqrt{3} |\alpha_j|$, $A_{ij} = -1$, $A_{ji} = -3$. The arrow may thus be regarded as an inequality sign on the lengths of the fundamental roots at the vertices.

6.4 Classification of Cartan matrices

The set of possible connected Dynkin diagrams, including arrows, is shown on the following standard list.

6.11 Standard list of connected Dynkin diagrams

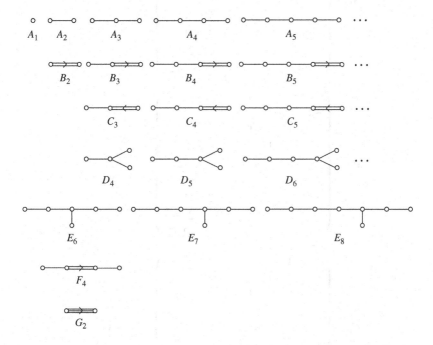

We note that, since the diagrams of types B_2, F_4, G_2 are symmetric, it does not matter in which direction the arrow is drawn in these cases.

The connected components of the Dynkin diagram of any semisimple Lie algebra must appear on this standard list.

We next obtain a standard list of corresponding Cartan matrices. We say that two Cartan matrices $(A_{ij}), (A'_{ij})$ are **equivalent** if they have the same degree l and there is a permutation σ of $1, \ldots, l$ such that

$$A'_{ij} = A_{\sigma(i)\sigma(j)}.$$

Equivalent Cartan matrices come from different labellings of the same Dynkin diagram. For each Dynkin diagram on the standard list 6.11 we choose a labelling and obtain a corresponding Cartan matrix which is uniquely determined. These Cartan matrices appear on the following list.

6.12 Standard list of indecomposable Cartan matrices

$$A_l = \begin{pmatrix} 2 & -1 & & & & & & \\ -1 & 2 & -1 & & & & & \\ & -1 & 2 & -1 & & & & \\ & & -1 & \cdot & \cdot & & & \\ & & & \cdot & \cdot & \cdot & & \\ & & & & \cdot & \cdot & -1 & \\ & & & & & -1 & 2 & -1 \\ & & & & & & -1 & 2 & -1 \\ & & & & & & & -1 & 2 \end{pmatrix} \qquad l \geq 1.$$

$$B_l = \begin{pmatrix} 2 & -1 & & & & & & \\ -1 & 2 & -1 & & & & & \\ & -1 & 2 & -1 & & & & \\ & & -1 & \cdot & \cdot & & & \\ & & & \cdot & \cdot & \cdot & & \\ & & & & \cdot & \cdot & -1 & \\ & & & & & -1 & 2 & -1 \\ & & & & & & -1 & 2 & -1 \\ & & & & & & & -2 & 2 \end{pmatrix} \qquad l \geq 2.$$

$$C_l = \begin{pmatrix} 2 & -1 & & & & & & \\ -1 & 2 & -1 & & & & & \\ & -1 & 2 & -1 & & & & \\ & & -1 & \cdot & \cdot & & & \\ & & & \cdot & \cdot & \cdot & & \\ & & & & \cdot & \cdot & -1 & \\ & & & & & -1 & 2 & -1 \\ & & & & & & -1 & 2 & -2 \\ & & & & & & & -1 & 2 \end{pmatrix} \qquad l \geq 3.$$

$$D_l = \begin{pmatrix} 2 & -1 & & & & & & \\ -1 & 2 & -1 & & & & & \\ & -1 & 2 & -1 & & & & \\ & & -1 & \cdot & \cdot & & & \\ & & & \cdot & \cdot & \cdot & & \\ & & & & \cdot & \cdot & -1 & \\ & & & & & -1 & 2 & -1 \\ & & & & & & -1 & 2 & -1 & -1 \\ & & & & & & & -1 & 2 & 0 \\ & & & & & & & -1 & 0 & 2 \end{pmatrix} \qquad l \geq 4.$$

6.4 Classification of Cartan matrices

$$E_6 = \begin{pmatrix} 2 & -1 & & & & \\ -1 & 2 & -1 & & & \\ & -1 & 2 & -1 & -1 & \\ & & -1 & 2 & & \\ & & -1 & & 2 & -1 \\ & & & & -1 & 2 \end{pmatrix}.$$

$$E_7 = \begin{pmatrix} 2 & -1 & & & & & \\ -1 & 2 & -1 & & & & \\ & -1 & 2 & -1 & & & \\ & & -1 & 2 & -1 & -1 & \\ & & & -1 & 2 & & \\ & & & -1 & & 2 & -1 \\ & & & & & -1 & 2 \end{pmatrix}.$$

$$E_8 = \begin{pmatrix} 2 & -1 & & & & & & \\ -1 & 2 & -1 & & & & & \\ & -1 & 2 & -1 & & & & \\ & & -1 & 2 & -1 & & & \\ & & & -1 & 2 & -1 & -1 & \\ & & & & -1 & 2 & & \\ & & & & -1 & & 2 & -1 \\ & & & & & & -1 & 2 \end{pmatrix}.$$

$$F_4 = \begin{pmatrix} 2 & -1 & & \\ -1 & 2 & -1 & \\ & -2 & 2 & -1 \\ & & -1 & 2 \end{pmatrix}.$$

$$G_2 = \begin{pmatrix} 2 & -1 \\ -3 & 2 \end{pmatrix}.$$

A Cartan matrix is called **indecomposable** if its Dynkin diagram is connected. Any Cartan matrix will determine a set of indecomposable Cartan matrices, unique up to equivalence, whose Dynkin diagrams are the connected components of the Dynkin diagram of the given Cartan matrix.

If A is the Cartan matrix of any semisimple Lie algebra, each indecomposable component of A will be equivalent to some Cartan matrix from the above standard list.

Proposition 6.13 *If a semisimple Lie algebra L has a connected Dynkin diagram then L is simple.*

Proof. Let $L = H \oplus \sum_{\alpha \in \Phi} L_\alpha$ be a Cartan decomposition giving rise to the Dynkin diagram Δ. Let I be a non-zero ideal of L. We shall show that $I = L$, thus proving that L is simple.

We first aim to prove that $I \cap H \neq O$. Suppose if possible that $I \cap H = O$. Let e_α be a non-zero element of L_α and choose a non-zero element $x \in I$ with

$$x = h + \sum_{\alpha \in \Phi} \mu_\alpha e_\alpha \quad h \in H, \mu_\alpha \in \mathbb{C}$$

such that the number of non-zero μ_α is as small as possible. Since $I \cap H = O$ there exists some $\mu_\beta \neq 0$. Then we have

$$[h'_\beta x] = \sum_{\alpha \in \Phi} \mu_\alpha [h'_\beta e_\alpha] = \sum_{\alpha \in \Phi} \mu_\alpha \alpha (h'_\beta) e_\alpha.$$

Now by Proposition 4.18 we can choose $e_\beta \in L_\beta$ and $e_{-\beta} \in L_{-\beta}$ such that $[e_\beta e_{-\beta}] = h'_\beta$. Thus $[[h'_\beta x] e_{-\beta}] = \sum_{\alpha \in \Phi} \mu_\alpha \alpha (h'_\beta) [e_\alpha e_{-\beta}] = \mu_\beta \beta (h'_\beta) h'_\beta + \sum_{\substack{\alpha \in \Phi \\ \alpha \neq \beta}} \mu_\alpha \alpha (h'_\beta) N_{\alpha, -\beta} e_{\alpha - \beta}$ where $[e_\alpha e_{-\beta}] = N_{\alpha, -\beta} e_{\alpha - \beta}$. Now we have $[[h'_\beta x] e_{-\beta}] \in I$ since $x \in I$ and $[[h'_\beta x] e_{-\beta}] \neq 0$ since $\mu_\beta \neq 0$ and $\beta (h'_\beta) = \langle h'_\beta, h'_\beta \rangle \neq 0$. Moreover the number of non-zero terms coming from the root spaces L_α is less for $[[h'_\beta x] e_{-\beta}]$ than it was for x. This contradicts the choice of x. We can therefore deduce that $I \cap H \neq O$.

The next step is to show that $I \supset H$. Suppose if possible this is not so. Then

$$O \neq I \cap H \neq H.$$

This implies that there exist $\alpha_i \in \Pi$ and $x \in I \cap H$ such that $\langle h'_{\alpha_i}, x \rangle \neq 0$. For if $I \cap H$ were orthogonal to each h'_{α_i} it would be orthogonal to the whole of H and would therefore be O. Then we have

$$[x e_{\alpha_i}] = \alpha_i(x) e_{\alpha_i} = \langle h'_{\alpha_i}, x \rangle e_{\alpha_i} \in I.$$

Since $\langle h'_{\alpha_i}, x \rangle \neq 0$ we deduce that $e_{\alpha_i} \in I$. Thus $[e_{\alpha_i} e_{-\alpha_i}] = h'_{\alpha_i} \in I$.

We can therefore divide the $\alpha_i \in \Pi$ into two classes, those with $h'_{\alpha_i} \in I$ and those with $h'_{\alpha_i} \notin I$. Both classes are non-empty. Furthermore if $h'_{\alpha_i} \in I$ and $h'_{\alpha_j} \notin I$ then $\langle h'_{\alpha_j}, h'_{\alpha_i} \rangle = 0$. This means that vertices i, j are not joined in the Dynkin diagram Δ, so Δ is disconnected. This is a contradiction, thus we deduce that $I \supset H$.

Finally we show that $I = L$. Let $\alpha \in \Phi$. Then we have

$$[h'_\alpha e_\alpha] = \alpha (h'_\alpha) e_\alpha = \langle h'_\alpha, h'_\alpha \rangle e_\alpha.$$

Since $h'_\alpha \in I$ we have $[h'_\alpha e_\alpha] \in I$, and since $\langle h'_\alpha, h'_\alpha \rangle \neq 0$ we deduce that $e_\alpha \in I$. This is true for all $\alpha \in \Phi$ and so $I = L$. Thus L is simple. $\square$

6.4 Classification of Cartan matrices

We next consider what happens when the Dynkin diagram of L is disconnected.

We first define an action of the Weyl group on H. The Weyl group was introduced in Section 5.2 as a group of non-singular linear transformations on the real vector space $H_\mathbb{R}^*$. This action can be extended by linearity to give an action of W on H^* by $\mathbb{C}$-linear transformations. We also define an action of W on H by $h \to wh$ where

$$\lambda(wh) = \left(w^{-1}\lambda\right)h \quad \text{for all } h \in H, \lambda \in H^*, w \in W.$$

There is a unique element $wh \in H$ satisfying this condition, and

$$w_1(w_2 h) = (w_1 w_2) h \quad \text{for all } w_1, w_2 \in W.$$

The actions of W on H^* and H are compatible with the isomorphism $H^* \to H$ given by $\lambda \to h'_\lambda$ where $\lambda(x) = \langle h'_\lambda, x \rangle$ for all $x \in H$. For suppose $w(\lambda) = \mu$ for $\lambda, \mu \in H^*$. Then

$$\langle w(h'_\lambda), x \rangle = \langle h'_\lambda, w^{-1}(x) \rangle = \lambda\left(w^{-1}(x)\right) = (w\lambda)x$$
$$= \mu(x) = \langle h'_\mu, x \rangle \quad \text{for all } x \in H.$$

Hence $w(\lambda) = \mu$ implies $w(h'_\lambda) = h'_\mu$.

Since we know that

$$s_\alpha(\lambda) = \lambda - 2\frac{\langle \alpha, \lambda \rangle}{\langle \alpha, \alpha \rangle}\alpha \quad \text{for } \alpha \in \Phi, \lambda \in H^*$$

it follows that

$$s_\alpha(x) = x - 2\frac{\langle h'_\alpha, x \rangle}{\langle h'_\alpha, h'_\alpha \rangle} h'_\alpha \quad \text{for } x \in H.$$

Proposition 6.14 *Let L be a semisimple Lie algebra whose Dynkin diagram Δ splits into connected components $\Delta_1, \ldots, \Delta_r$. Then we have*

$$L = L_1 \oplus \cdots \oplus L_r$$

a direct sum of Lie algebras, where L_i is a simple Lie algebra with Dynkin diagram Δ_i.

Proof. We have $\Delta = \Delta_1 \dot\cup \Delta_2 \dot\cup \cdots \dot\cup \Delta_r$. Let Π_i be the subset of Π corresponding to the vertices in Δ_i. Then we have

$$\Pi = \Pi_1 \dot\cup \Pi_2 \dot\cup \cdots \dot\cup \Pi_r.$$

Moreover we have $\langle \alpha, \beta \rangle = 0$ if $\alpha \in \Pi_i, \beta \in \Pi_j$ and $i \neq j$. Let H_i be the subspace of H spanned by the elements h'_α with $\alpha \in \Pi_i$. Then we have

$$H = H_1 \oplus H_2 \oplus \cdots \oplus H_r$$

where $\langle h, h' \rangle = 0$ if $h \in H_i$, $h' \in H_j$ and $i \neq j$.

Now let $\alpha \in \Pi_i$ and consider the fundamental reflection $s_\alpha \in W$. It is clear that s_α transforms H_i into itself and fixes each vector in H_j for all $j \neq i$. Thus we have

$$s_\alpha(H_j) = H_j \quad j = 1, \ldots, r.$$

Since the elements s_α generate the Weyl group W we deduce that

$$w(H_j) = H_j \quad j = 1, \ldots, r \quad w \in W.$$

Now for all $\alpha \in \Phi$ we have $h'_\alpha = w(h'_{\alpha_i})$ for some $\alpha_i \in \Pi$ and some $w \in W$, by Proposition 5.12 and the definition of the W-action on H. It follows that each h'_α, $\alpha \in \Phi$, lies in H_i for some i. Let Φ_i be the set of all $\alpha \in \Phi$ such that $h'_\alpha \in H_i$. Then we have

$$\Phi = \Phi_1 \dot\cup \Phi_2 \dot\cup \cdots \dot\cup \Phi_r.$$

We define L_i to be the subspace of L spanned by H_i and the e_α for all $\alpha \in \Phi_i$. We deduce from the Cartan decomposition of L that

$$L = L_1 \oplus L_2 \oplus \cdots \oplus L_r$$

a direct sum of vector spaces. In fact we can see that each L_i is a subalgebra of L. It is sufficient to verify that $[e_\alpha e_\beta] \in L_i$ if $\alpha, \beta \in \Phi_i$. If $\alpha + \beta \in \Phi$ then we have $\alpha + \beta \in \Phi_i$ since $h'_{\alpha+\beta} = h'_\alpha + h'_\beta \in H_i$. If $\alpha + \beta = 0$ then $[e_\alpha e_\beta]$ is a multiple of h'_α and so lies in H_i, thus in L_i. If $\alpha + \beta$ is non-zero but not a root then $[e_\alpha e_\beta] = 0$. In either case we have $[e_\alpha e_\beta] \in L_i$. Thus L_i is a subalgebra of L.

We show next that $[L_i L_j] = O$ if $i \neq j$. Let $\alpha \in \Phi_i$ and $\beta \in \Phi_j$. Then we have

$$[h'_\alpha e_\beta] = \beta(h'_\alpha) e_\beta = \langle h'_\beta, h'_\alpha \rangle e_\beta = 0$$

and similarly $[e_\alpha h'_\beta] = 0$. We also have $[e_\alpha e_\beta] = 0$. For $\alpha + \beta \notin \Phi$ since $h'_\alpha + h'_\beta$ does not lie in any subspace H_k of H. It follows that $[L_i L_j] = O$.

We now know that each L_i is an ideal of L, since

$$[L_i L] \subset \sum_j [L_i L_j] \subset [L_i L_i] \subset L_i.$$

6.4 Classification of Cartan matrices

This implies that
$$[x_1 + \cdots + x_r, y_1 + \cdots + y_r] = [x_1 y_1] + \cdots + [x_r y_r]$$
where $x_i, y_i \in L_i$. Hence
$$L = L_1 \oplus L_2 \oplus \cdots \oplus L_r$$
is a direct sum of Lie algebras.

Now each L_i is a semisimple Lie algebra. For let I be a soluble ideal of L_i. Since $[IL_j] = O$ for all $j \ne i$ the ideal I is an ideal of L. Since L is semisimple we have $I = O$. Hence L_i is semisimple.

We next observe that H_i is a Cartan subalgebra of L_i. The subalgebra H_i is abelian, hence nilpotent. Let $x \in L_i$ satisfy $x \in N(H_i)$. Then $[xh] \in H_i$ for all $h \in H_i$. We also have $[xh] = 0$ for all $h \in H_j$ with $j \ne i$. It follows that $[xh] \in H$ for all $h \in H$. Since H is a Cartan subalgebra of L we have $N(H) = H$. Hence $x \in H$. Thus $x \in H \cap L_i = H_i$. Thus H_i is a Cartan subalgebra of L_i.

We now consider the Cartan decomposition
$$L_i = H_i \oplus \sum_{\alpha \in \Phi_i} \mathbb{C} e_\alpha$$
of L_i with respect to H_i. We see that Φ_i is the root system of L_i with respect to H_i, that Π_i is a fundamental system of roots in Φ_i, and that Δ_i is the Dynkin diagram of L_i. Now Δ_i is connected. Thus the Lie algebra L_i must be simple, by Proposition 6.13. Thus we have obtained a decomposition of L as a direct sum of simple Lie algebras L_i, whose Dynkin diagrams are the connected components Δ_i of Δ. $\square$

Corollary 6.15 *A semisimple Lie algebra L has a connected Dynkin diagram if and only if L is simple.*

Proof. This follows from Propositions 6.13 and 6.14 $\square$

7
The existence and uniqueness theorems

We have seen that each non-trivial simple Lie algebra L has a Dynkin diagram Δ which appears on the standard list 6.11 of connected Dynkin diagrams. In the present chapter we shall consider the converse question. Given a Dynkin diagram Δ on the standard list, is there a simple Lie algebra L with Dynkin diagram Δ? If so, is L uniquely determined up to isomorphism? We shall show that both the existence and uniqueness properties hold. The proof of the uniqueness property is somewhat easier, and we shall prove this first. In order to do so we shall need some properties of the structure constants of the Lie algebra L.

7.1 Some properties of structure constants

Let L be a simple Lie algebra with Dynkin diagram Δ. Let H be a Cartan subalgebra of L and

$$L = H \oplus \sum_{\alpha \in \Phi} L_\alpha$$

be the Cartan decomposition of L with respect to H. We know from Theorem 4.20 that $\dim L_\alpha = 1$ for each $\alpha \in \Phi$. Let e_α be a non-zero element of L_α. Let Π be a fundamental system of roots in Φ. Then the elements h'_{α_i} for $\alpha_i \in \Pi$ form a basis for H. It will be convenient to choose a slightly different basis consisting of scalar multiples of the h'_{α_i}. We define $h_i \in H$ by

$$h_i = \frac{2 h'_{\alpha_i}}{\langle h'_{\alpha_i}, h'_{\alpha_i} \rangle}.$$

We note that $\alpha_i(h_i) = 2$. Then $\{h_i, \ i = 1, \ldots, l \ ; e_\alpha, \ \alpha \in \Phi\}$ is a basis of L. By Proposition 4.18 we know that $h'_\alpha \in [L_\alpha L_{-\alpha}]$ for all $\alpha \in \Phi$. Thus,

7.1 Some properties of structure constants

if we have already chosen the e_α for all $\alpha \in \Phi^+$, we may choose the $e_{-\alpha}$ uniquely for $\alpha \in \Phi^+$ to satisfy the condition

$$[e_\alpha e_{-\alpha}] = \frac{2h'_\alpha}{\langle h'_\alpha, h'_\alpha \rangle}.$$

(This relation will then be satisfied for $\alpha \in \Phi^-$ also.)

We define $h_\alpha \in H$ for each $\alpha \in \Phi$ by

$$h_\alpha = \frac{2h'_\alpha}{\langle h'_\alpha, h'_\alpha \rangle}.$$

The element h_α is called the **coroot** corresponding to the root α. In particular we have $h_i = h_{\alpha_i}$. We then have

$$[e_\alpha e_{-\alpha}] = h_\alpha \quad \text{for all } \alpha \in \Phi.$$

We next consider the product $[e_\alpha e_\beta]$ when $\alpha + \beta \neq 0$. We have $[L_\alpha L_\beta] = 0$ if $\alpha + \beta \neq 0$ and $\alpha + \beta \notin \Phi$. If $\alpha + \beta \in \Phi$ we have $[L_\alpha L_\beta] \subset L_{\alpha+\beta}$. We define $N_{\alpha,\beta} \in \mathbb{C}$ by the condition

$$[e_\alpha e_\beta] = N_{\alpha,\beta} e_{\alpha+\beta}.$$

The numbers $N_{\alpha,\beta}$ for $\alpha, \beta, \alpha + \beta \in \Phi$ will be called the **structure constants** of L. They clearly depend upon the choice of the elements $e_\alpha \in L_\alpha$.

We now consider the multiplication of the basis vectors $\{h_i, e_\alpha\}$ of L. We have

$$[h_i h_j] = 0$$
$$[h_i e_\alpha] = \alpha(h_i) e_\alpha$$
$$[e_\alpha e_{-\alpha}] = h_\alpha$$
$$[e_\alpha e_\beta] = N_{\alpha,\beta} e_{\alpha+\beta} \quad \text{if } \alpha, \beta, \alpha+\beta \in \Phi$$
$$[e_\alpha e_\beta] = 0 \quad \text{if } \alpha+\beta \neq 0 \text{ and } \alpha+\beta \notin \Phi.$$

In order to express $[e_\alpha e_{-\alpha}]$ as a linear combination of basis elements we may express h'_α as a linear combination of the h'_{α_i}, $\alpha_i \in \Pi$, and so also express h_α as a linear combination of the h_i.

We shall now derive some relations between the structure constants $N_{\alpha,\beta}$.

Proposition 7.1 *The structure constants $N_{\alpha,\beta}$ satisfy the following relations.*
(i) $N_{\beta,\alpha} = -N_{\alpha,\beta}$.
(ii) *If $\alpha, \beta, \gamma \in \Phi$ satisfy $\alpha + \beta + \gamma = 0$ then* $\dfrac{N_{\alpha,\beta}}{\langle \gamma, \gamma \rangle} = \dfrac{N_{\beta,\gamma}}{\langle \alpha, \alpha \rangle} = \dfrac{N_{\gamma,\alpha}}{\langle \beta, \beta \rangle}$.

90   *The existence and uniqueness theorems*

(iii) $N_{\alpha,\beta}N_{-\alpha,-\beta} = -(p+1)^2$ where the α-chain of roots through β is $-p\alpha+\beta, \ldots, \beta, \ldots, q\alpha+\beta$.

(iv) If $\alpha, \beta, \gamma, \delta \in \Phi$ satisfy $\alpha+\beta+\gamma+\delta = 0$ and no pair are negatives of one another, then

$$\frac{N_{\alpha,\beta}N_{\gamma,\delta}}{\langle \alpha+\beta, \alpha+\beta \rangle} + \frac{N_{\beta,\gamma}N_{\alpha,\delta}}{\langle \beta+\gamma, \beta+\gamma \rangle} + \frac{N_{\gamma,\alpha}N_{\beta,\delta}}{\langle \gamma+\alpha, \gamma+\alpha \rangle} = 0.$$

Proof. (i) This relation is clear.

(ii) Suppose $\alpha+\beta+\gamma = 0$. We consider the Jacobi identity

$$[[e_\alpha e_\beta] e_\gamma] + [[e_\beta e_\gamma] e_\alpha] + [[e_\gamma e_\alpha] e_\beta] = 0.$$

This gives

$$N_{\alpha,\beta}[e_{\alpha+\beta}e_{-(\alpha+\beta)}] + N_{\beta,\gamma}[e_{-\alpha}e_\alpha] + N_{\gamma,\alpha}[e_{-\beta}e_\beta] = 0$$

that is

$$2N_{\alpha,\beta}\frac{h'_{\alpha+\beta}}{\langle h'_{\alpha+\beta}, h'_{\alpha+\beta}\rangle} = 2N_{\beta,\gamma}\frac{h'_\alpha}{\langle h'_\alpha, h'_\alpha\rangle} + 2N_{\gamma,\alpha}\frac{h'_\beta}{\langle h'_\beta, h'_\beta\rangle}.$$

Now the roots α, β are linearly independent since, if they were not, $\alpha+\beta$ could not be a root. Thus h'_α, h'_β are linearly independent and $h'_{\alpha+\beta} = h'_\alpha + h'_\beta$. We deduce that

$$\frac{N_{\alpha,\beta}}{\langle h'_{\alpha+\beta}, h'_{\alpha+\beta}\rangle} = \frac{N_{\beta,\gamma}}{\langle h'_\alpha, h'_\alpha\rangle} = \frac{N_{\gamma,\alpha}}{\langle h'_\beta, h'_\beta\rangle}$$

that is

$$\frac{N_{\alpha,\beta}}{\langle \gamma, \gamma \rangle} = \frac{N_{\beta,\gamma}}{\langle \alpha, \alpha \rangle} = \frac{N_{\gamma,\alpha}}{\langle \beta, \beta \rangle}.$$

(iii) Now suppose $\alpha, \beta \in \Phi$ are linearly independent. We consider the Jacobi identity

$$[[e_\alpha e_{-\alpha}] e_\beta] + [[e_{-\alpha} e_\beta] e_\alpha] + [[e_\beta e_\alpha] e_{-\alpha}] = 0.$$

This gives

$$2\frac{[h'_\alpha e_\beta]}{\langle h'_\alpha, h'_\alpha \rangle} + N_{-\alpha,\beta}N_{-\alpha+\beta,\alpha}e_\beta + N_{\beta,\alpha}N_{\alpha+\beta,-\alpha}e_\beta = 0.$$

We deduce that

$$2\frac{\langle h'_\alpha, h'_\beta \rangle}{\langle h'_\alpha, h'_\alpha \rangle} + N_{-\alpha,\beta}N_{-\alpha+\beta,\alpha} + N_{\beta,\alpha}N_{\alpha+\beta,-\alpha} = 0.$$

7.1 Some properties of structure constants

Using relations (i) and (ii) this may be written

$$N_{\alpha,\beta}N_{-\alpha,-\beta}\frac{\langle\beta,\beta\rangle}{\langle\alpha+\beta,\alpha+\beta\rangle} - N_{\alpha,-\alpha+\beta}N_{-\alpha,\alpha-\beta}\frac{\langle-\alpha+\beta,-\alpha+\beta\rangle}{\langle\beta,\beta\rangle}$$

$$= 2\frac{\langle\alpha,\beta\rangle}{\langle\alpha,\alpha\rangle}.$$

(If $-\alpha+\beta$ is not a root $N_{-\alpha,\beta}$ is interpreted as 0 so the middle term disappears.) We now consider the α-chain of roots through β. Let it be

$$-p\alpha+\beta, \ldots, \beta, \ldots, q\alpha+\beta.$$

We apply the same formula to the pairs $(\alpha,\beta)(\alpha,-\alpha+\beta)\ldots(\alpha,-p\alpha+\beta)$ and obtain

$$N_{\alpha,\beta}N_{-\alpha,-\beta}\frac{\langle\beta,\beta\rangle}{\langle\alpha+\beta,\alpha+\beta\rangle} - N_{\alpha,-\alpha+\beta}N_{-\alpha,\alpha-\beta}\frac{\langle-\alpha+\beta,-\alpha+\beta\rangle}{\langle\beta,\beta\rangle} = 2\frac{\langle\alpha,\beta\rangle}{\langle\alpha,\alpha\rangle}$$

$$N_{\alpha,-\alpha+\beta}N_{-\alpha,\alpha-\beta}\frac{\langle-\alpha+\beta,-\alpha+\beta\rangle}{\langle\beta,\beta\rangle} - N_{\alpha,-2\alpha+\beta}N_{-\alpha,2\alpha-\beta}\frac{\langle-2\alpha+\beta,-2\alpha+\beta\rangle}{\langle-\alpha+\beta,-\alpha+\beta\rangle}$$

$$= 2\frac{\langle\alpha,-\alpha+\beta\rangle}{\langle\alpha,\alpha\rangle}$$

$$\vdots$$

$$N_{\alpha,-p\alpha+\beta}N_{-\alpha,p\alpha-\beta}\frac{\langle-p\alpha+\beta,-p\alpha+\beta\rangle}{\langle-(p-1)\alpha+\beta,-(p-1)\alpha+\beta\rangle} = \frac{2\langle\alpha,-p\alpha+\beta\rangle}{\langle\alpha,\alpha\rangle}.$$

(The last equation has only one term on the left since $-(p+1)\alpha+\beta$ is not a root.) Adding these equations we obtain

$$N_{\alpha,\beta}N_{-\alpha,-\beta}\frac{\langle\beta,\beta\rangle}{\langle\alpha+\beta,\alpha+\beta\rangle} = 2(p+1)\frac{\langle\alpha,\beta\rangle}{\langle\alpha,\alpha\rangle} - 2\frac{p(p+1)}{2}.$$

However, we know from Proposition 4.22 that $2\frac{\langle\alpha,\beta\rangle}{\langle\alpha,\alpha\rangle} = p-q$. Thus we have

$$N_{\alpha,\beta}N_{-\alpha,-\beta}\frac{\langle\beta,\beta\rangle}{\langle\alpha+\beta,\alpha+\beta\rangle} = -(p+1)q.$$

In order to obtain the required result $N_{\alpha,\beta}N_{-\alpha,-\beta} = -(p+1)^2$ we must show

$$\frac{\langle\alpha+\beta,\alpha+\beta\rangle}{\langle\beta,\beta\rangle} = \frac{p+1}{q}.$$

We recall from the proof of Proposition 6.1 that

$$2\frac{\langle\alpha,\beta\rangle}{\langle\alpha,\alpha\rangle}\cdot 2\frac{\langle\beta,\alpha\rangle}{\langle\beta,\beta\rangle}=4\cos^2\theta$$

where θ is the angle between α,β and hence that $2\frac{\langle\alpha,\beta\rangle}{\langle\alpha,\alpha\rangle}\in\{0,-1,-2,-3\}$. Also from Proposition 4.22 we know that $2\frac{\langle\alpha,\beta\rangle}{\langle\alpha,\alpha\rangle}=p-q$. If we choose β to be the initial root in its α-chain we have $p=0$ and hence $q\le 3$. This shows that each α-chain has at most four roots. Thus the possible positions of β in its α-chain are

β —— $\alpha+\beta$				$p=0$	$q=1$
β —— $\alpha+\beta$ —— $2\alpha+\beta$				$p=0$	$q=2$
$-\alpha+\beta$ —— β —— $\alpha+\beta$				$p=1$	$q=1$
β —— $\alpha+\beta$ —— $2\alpha+\beta$ —— $3\alpha+\beta$				$p=0$	$q=3$
$-\alpha+\beta$ —— β —— $\alpha+\beta$ —— $2\alpha+\beta$				$p=1$	$q=2$
$-2\alpha+\beta$ —— $-\alpha+\beta$ —— β —— $\alpha+\beta$				$p=2$	$q=1$

In the first case we have $\langle\alpha+\beta,\alpha+\beta\rangle=\langle\beta,\beta\rangle$ since $s_\alpha(\beta)=\alpha+\beta$. In the remaining cases the first and last roots in the α-chain are long roots and the remainder are short roots. The relative lengths are given in the proof of Proposition 6.1. We have

$$\frac{\langle\alpha+\beta,\alpha+\beta\rangle}{\langle\beta,\beta\rangle}=1,\tfrac{1}{2},2,\tfrac{1}{3},1,3$$

in the above six cases respectively. Thus in each case we have

$$\frac{\langle\alpha+\beta,\alpha+\beta\rangle}{\langle\beta,\beta\rangle}=\frac{p+1}{q}$$

and so $N_{\alpha,\beta}N_{-\alpha,-\beta}=-(p+1)^2$

(iv) Now suppose that $\alpha,\beta,\gamma,\delta\in\Phi$ satisfy $\alpha+\beta+\gamma+\delta=0$ with no pair equal and opposite. Consider the Jacobi identity

$$[[e_\alpha e_\beta]e_\gamma]+[[e_\beta e_\gamma]e_\alpha]+[[e_\gamma e_\alpha]e_\beta]=0.$$

This gives

$$N_{\alpha,\beta}N_{\alpha+\beta,\gamma} + N_{\beta,\gamma}N_{\beta+\gamma,\alpha} + N_{\gamma,\alpha}N_{\gamma+\alpha,\beta} = 0.$$

Using relations (ii) this gives

$$\frac{N_{\alpha,\beta}N_{\gamma,\delta}}{\langle \alpha+\beta, \alpha+\beta \rangle} + \frac{N_{\beta,\gamma}N_{\alpha,\delta}}{\langle \beta+\gamma, \beta+\gamma \rangle} + \frac{N_{\gamma,\alpha}N_{\beta,\delta}}{\langle \gamma+\alpha, \gamma+\alpha \rangle} = 0.$$

(As usual we interpret $N_{\theta,\phi}$ as 0 if $\theta + \phi$ is not a root.) □

Proposition 7.1 (iii) has a very useful corollary.

Corollary 7.2 *If* $\alpha, \beta, \alpha+\beta \in \Phi$ *then* $N_{\alpha,\beta} \neq 0$. *Thus* $[L_\alpha L_\beta] = L_{\alpha+\beta}$.

7.2 The uniqueness theorem

We shall now use the above relations between the structure constants to show that the Lie algebra L is uniquely determined up to isomorphism. A Dynkin diagram on the standard list 6.11 is given, and this determines uniquely a Cartan matrix $A = (A_{ij})$ on the standard list 6.12. Now the Cartan matrix determines the set Φ of roots as linear combinations of the fundamental roots $\Pi = \{\alpha_1, \ldots, \alpha_l\}$. For each root $\alpha \in \Phi$ has form $\alpha = w(\alpha_i)$ for some $\alpha_i \in \Pi$ and some $w \in W$, by Proposition 5.12. Moreover each element $w \in W$ is a product of elements $s_1, \ldots, s_l$ by Theorem 5.13. The actions of $s_1, \ldots, s_l$ on the fundamental roots $\alpha_1, \ldots, \alpha_l$ are given in terms of the Cartan matrix by

$$s_i(\alpha_j) = \alpha_j - A_{ij}\alpha_i.$$

Thus by applying the fundamental reflections successively to the fundamental roots we obtain all roots as linear combinations of the fundamental roots.

We next observe that all scalar products $\langle h'_\alpha, h'_\beta \rangle$ for $\alpha, \beta \in \Phi$ are determined by the Cartan matrix. By Proposition 4.22 $2\frac{\langle h'_\alpha, h'_\beta \rangle}{\langle h'_\alpha, h'_\alpha \rangle}$ is determined by the root system, hence by the Cartan matrix as shown above. Then $\langle h'_\alpha, h'_\alpha \rangle$ is determined by the formula

$$\frac{1}{\langle h'_\alpha, h'_\alpha \rangle} = \sum_{\beta \in \Phi} \left(\frac{\langle h'_\alpha, h'_\beta \rangle}{\langle h'_\alpha, h'_\alpha \rangle} \right)^2$$

of Proposition 4.24. Thus $\langle h'_\alpha, h'_\beta \rangle$ is also determined by the Cartan matrix. Thus we see that if the structure constants $N_{\alpha,\beta}$ are known the multiplication of basis elements

$$[h_i h_j] = 0$$
$$[h_i e_\alpha] = \alpha(h_i) e_\alpha$$
$$[e_\alpha e_{-\alpha}] = h_\alpha$$
$$[e_\alpha e_\beta] = N_{\alpha,\beta} e_{\alpha+\beta} \quad \text{if } \alpha, \beta, \alpha+\beta \in \Phi$$
$$[e_\alpha e_\beta] = 0 \quad \text{if } \alpha+\beta \neq 0 \text{ and } \alpha+\beta \notin \Phi$$

will be completely determined.

We shall show that for certain pairs (α, β) of roots the structure constants $N_{\alpha,\beta}$ can be chosen arbitrarily, and that the remaining structure constants are uniquely determined in terms of these by the relations of Proposition 7.1.

We choose a total ordering on the vector space $V = H^*_\mathbb{R}$ as in Section 5.1 giving rise to the positive system Φ^+ and fundamental system Π of roots. An ordered pair (α, β) of roots will be called **special** if $\alpha + \beta \in \Phi$ and $0 < \alpha < \beta$. The pair (α, β) will be called **extraspecial** if (α, β) is special and if, in addition, for all special pairs (γ, δ) such that $\alpha + \beta = \gamma + \delta$ we have $\alpha \leq \gamma$.

Lemma 7.3 *The structure constants $N_{\alpha,\beta}$ for extraspecial pairs (α, β) can be chosen as arbitrary non-zero elements of $\mathbb{C}$, by appropriate choice of the elements e_α.*

Proof. We choose the e_α for $\alpha \in \Phi^+$ in the order given by $<$. Suppose (α, β) is an extraspecial pair. Then we have

$$[e_\alpha e_\beta] = N_{\alpha,\beta} e_{\alpha+\beta}$$

and e_α, e_β have already been chosen. Moreover there is only one extraspecial pair with given sum $\alpha + \beta$. Thus $e_{\alpha+\beta}$ can be chosen to give any non-zero value of $N_{\alpha,\beta}$. □

Proposition 7.4 *All the structure constants $N_{\alpha,\beta}$ are determined by the structure constants for extraspecial pairs.*

Proof. We consider the set of all pairs of roots (α, β) such that $\alpha + \beta$ is a root. Let (α, β) be such a pair and let $\gamma = -\alpha - \beta$. Then the following 12 pairs of roots are of the given type.

$(\alpha, \beta) \ (\beta, \gamma) \ (\gamma, \alpha) \ (\beta, \alpha) \ (\gamma, \beta) \ (\alpha, \gamma)$

$(-\alpha, -\beta) \ (-\beta, -\gamma) \ (-\gamma, -\alpha) \ (-\beta, -\alpha) \ (-\gamma, -\beta) \ (-\alpha, -\gamma).$

Since $\alpha+\beta+\gamma=0$ either two or one of α, β, γ are positive. Thus either two of α, β, γ are positive or two of $-\alpha, -\beta, -\gamma$ are positive. By choosing two positive roots from α, β, γ or from $-\alpha, -\beta, -\gamma$ and by writing them in the appropriate order we obtain a special pair. Thus just one of the above 12 pairs of roots is a special pair.

Now the relations in Proposition 7.1 (i), (ii), (iii) enable us to express $N_{\beta,\alpha}, N_{\beta,\gamma}, N_{\gamma,\alpha}$ and $N_{-\alpha,-\beta}$ in terms of $N_{\alpha,\beta}$. Thus these relations enable us to express $N_{\theta,\phi}$ for all the 12 pairs (θ, ϕ) above in terms of $N_{\theta,\phi}$ for the special pair (θ, ϕ).

The next stage is to show that the $N_{\alpha,\beta}$ for all special pairs (α, β) are determined in terms of the $N_{\alpha,\beta}$ for extraspecial pairs. Suppose (α, β) is special but not extraspecial. Then there exists an extraspecial pair (γ, δ) such that $\alpha+\beta=\gamma+\delta$. Thus $\alpha+\beta+(-\gamma)+(-\delta)=0$ and no pair of $\alpha, \beta, -\gamma, -\delta$ are equal and opposite. By Proposition 7.1 (iv) we have

$$\frac{N_{\alpha,\beta}N_{-\gamma,-\delta}}{\langle\alpha+\beta, \alpha+\beta\rangle} + \frac{N_{\beta,-\gamma}N_{\alpha,-\delta}}{\langle\beta-\gamma, \beta-\gamma\rangle} + \frac{N_{-\gamma,\alpha}N_{\beta,-\delta}}{\langle-\gamma+\alpha, -\gamma+\alpha\rangle} = 0.$$

Now the roots $\alpha, \beta, \gamma, \delta$ are ordered by

$$0 < \gamma < \alpha < \beta < \delta.$$

Thus we may use relations (i), (ii), (iii) of Proposition 7.1 to express $N_{-\gamma,-\delta}$ in terms of $N_{\gamma,\delta}$; $N_{\beta,-\gamma}$ in terms of $N_{\gamma,\beta-\gamma}$; $N_{\alpha,-\delta}$ in terms of $N_{\alpha,\delta-\alpha}$; $N_{-\gamma,\alpha}$ in terms of $N_{\gamma,\alpha-\gamma}$; and $N_{\beta,-\delta}$ in terms of $N_{\beta,\delta-\beta}$. Thus $N_{\alpha,\beta}$ is expressed in terms of

$$N_{\gamma,\delta}, N_{\gamma,\beta-\gamma}, N_{\alpha,\delta-\alpha}, N_{\gamma,\alpha-\gamma}, N_{\beta,\delta-\beta}.$$

Now (γ, δ) is an extraspecial pair and $(\gamma, \beta-\gamma), (\alpha, \delta-\alpha), (\gamma, \alpha-\gamma)$ and $(\beta, \delta-\beta)$ are all pairs of positive roots whose sums are roots less than $\alpha+\beta=\gamma+\delta$ in the given ordering. We may therefore argue by induction on $\alpha+\beta$, using the given order, that $N_{\alpha,\beta}$ can be expressed in terms of $N_{\theta,\phi}$ for extraspecial pairs (θ, ϕ). □

We can now state our uniqueness theorem.

Theorem 7.5 *Any two simple Lie algebras with the same Cartan matrix are isomorphic.*

Proof. We choose the basis elements $\{h_i, e_\alpha\}$ of such a Lie algebra L such that $N_{\alpha,\beta}=1$ for all extraspecial pairs of roots (α, β). We may do this by Lemma 7.3. The remaining structure constants $N_{\alpha,\beta}$ are all then uniquely determined by Proposition 7.4. Thus the formulae expressing a Lie product of basis elements as a linear combination of basis elements are completely determined by the Cartan matrix. Thus the Lie algebra L is uniquely determined up to isomorphism.

7.3 Some generators and relations in a simple Lie algebra

We now turn to the question of the existence of a simple Lie algebra with Cartan matrix on the standard list 6.12. A proof of the existence theorem has been given by J. Tits (*IHES Publ. Math.* **31** (1966)) along the lines of the arguments used so far. The details are technically quite complicated, however, and so we prefer to give a different proof of the existence theorem.

Let L be a simple Lie algebra with Cartan matrix A. Let H be a Cartan subalgebra of L and

$$L = H \oplus \sum_{\alpha \in \Phi} L_\alpha$$

be the Cartan decomposition. As before we consider the elements $h_i \in H$ given by

$$h_i = \frac{2 h'_{\alpha_i}}{\langle h'_{\alpha_i}, h'_{\alpha_i} \rangle}$$

where $\Pi = \{\alpha_1, \ldots, \alpha_l\}$ is a fundamental system in Φ. As in Section 7.1 we can choose elements $e_i \in L_{\alpha_i}, f_i \in L_{-\alpha_i}$ such that $[e_i f_i] = h_i$.

We shall show that the elements $e_1, \ldots, e_l, h_1, \ldots, h_l, f_1, \ldots, f_l$ generate L. (Of course this is equivalent to saying that $e_1, \ldots, e_l, f_1, \ldots, f_l$ generate L, but it will be useful to include $h_1, \ldots, h_l$ in the generating set.)

Lemma 7.6 *If $\alpha \in \Phi^+$ and $\alpha \notin \Pi$ there exists $\alpha_i \in \Pi$ such that $\alpha - \alpha_i \in \Phi^+$. Thus every positive non-fundamental root is the sum of a fundamental root with a positive root.*

Proof. Suppose if possible that the result is false. Then $\alpha - \alpha_i$ is not a root and is non-zero for each i. (We can use Corollary 5.6 to see that $\alpha - \alpha_i$ cannot be a negative root.) Consider the α_i-chain of roots through α. This has form

$$\alpha, \alpha_i + \alpha, \ldots, q\alpha_i + \alpha.$$

7.3 Some generators and relations in a simple Lie algebra

By Proposition 4.22 we have

$$2\frac{\langle \alpha_i, \alpha \rangle}{\langle \alpha_i, \alpha_i \rangle} = -q.$$

This implies that $\langle \alpha_i, \alpha \rangle \le 0$. Now $\alpha \in \Phi^+$ has form $\alpha = \sum_i n_i \alpha_i$ with all $n_i \ge 0$. Thus

$$\langle \alpha, \alpha \rangle = \sum_i n_i \langle \alpha_i, \alpha \rangle \le 0.$$

This gives a contradiction, since we know $\langle \alpha, \alpha \rangle > 0$. □

Proposition 7.7 *The elements* $e_1, \ldots, e_l, h_1, \ldots, h_l, f_1, \ldots, f_l$ *generate* L.

Proof. Since $h_1, \ldots, h_l$ span H it will be sufficient to show that each L_α for $\alpha \in \Phi^+$ lies in the subalgebra generated by $e_1, \ldots, e_l$ and each L_α for $\alpha \in \Phi^-$ lies in the subalgebra generated by $f_1, \ldots, f_l$.

Let $\alpha \in \Phi^+$. If $\alpha = \alpha_i$ for some i we have $L_\alpha = \mathbb{C}e_i$. If $\alpha \notin \Pi$ we can write $\alpha = \alpha_i + \beta$ for some $\alpha_i \in \Pi$ and some $\beta \in \Phi^+$ by Lemma 7.6. We then have $[L_{\alpha_i} L_\beta] = L_\alpha$ by Corollary 7.2. Thus we may choose $e_\alpha = [e_i, e_\beta]$ for some $e_\beta \ne 0$ in L_β. By repeating this process we obtain

$$e_\alpha = [[e_{i_1} e_{i_2}] \cdots e_{i_k}]$$

for some sequence $i_1, \ldots, i_k$. Thus each L_α for $\alpha \in \Phi^+$ lies in the subalgebra generated by $e_1, \ldots, e_l$. Similarly each L_α for $\alpha \in \Phi^-$ lies in the subalgebra generated by $f_1, \ldots, f_l$. □

Proposition 7.8 *The generators* $e_1, \ldots, e_l, h_1, \ldots, h_l, f_1, \ldots, f_l$ *of* L *satisfy the following relations.*

(a) $[h_i h_j] = 0$
(b) $[h_i e_j] = A_{ij} e_j$
(c) $[h_i f_j] = -A_{ij} f_j$
(d) $[e_i f_i] = h_i$
(e) $[e_i f_j] = 0 \quad$ if $i \ne j$
(f) $[e_i [e_i \ldots [e_i e_j]]] = 0 \quad$ if $i \ne j$
$\quad \leftarrow 1 - A_{ij} \rightarrow$
(g) $[f_i [f_i \ldots [f_i f_j]]] = 0 \quad$ if $i \ne j$.
$\quad \leftarrow 1 - A_{ij} \rightarrow$

Note that in relations (f), (g) there are $1 - A_{ij}$ occurrences of e_i, f_i respectively. Since $A_{ij} \le 0$ for $i \ne j$ this number $1 - A_{ij}$ is a positive integer.

Proof. Relation (a) follows from $[HH]=0$. For relation (b), we have

$$[h_i e_j] = 2\frac{[h'_{\alpha_i} e_j]}{\langle h'_{\alpha_i}, h'_{\alpha_i}\rangle} = 2\frac{\alpha_j(h'_{\alpha_i})}{\langle h'_{\alpha_i}, h'_{\alpha_i}\rangle} e_j$$

$$= 2\frac{\langle h'_{\alpha_j}, h'_{\alpha_i}\rangle}{\langle h'_{\alpha_i}, h'_{\alpha_i}\rangle} e_j = 2\frac{\langle \alpha_i, \alpha_j\rangle}{\langle \alpha_i, \alpha_i\rangle} e_j = A_{ij} e_j.$$

Relation (c) is obtained similarly. Relation (d) holds by definition of f_i. Relation (e) holds because $[e_i f_j] \in L_{\alpha_i - \alpha_j}$ and $\alpha_i - \alpha_j$ is not a root when $i \neq j$, as follows from Corollary 5.6. In order to prove relation (f) we consider the α_i-chain of roots through α_j. Since $-\alpha_i + \alpha_j$ is not a root this chain has form

$$\alpha_j, \alpha_i + \alpha_j, \ldots, q\alpha_i + \alpha_j.$$

By Proposition 4.22 we have $A_{ij} = -q$. Thus $(1 - A_{ij})\alpha_i + \alpha_j$ is not a root. Since the element $[e_i[e_i \ldots [e_i e_j]]]$ lies in $L_{(1-A_{ij})\alpha_i + \alpha_j}$ this element must be 0. Relation (g) is obtained similarly. □

7.4 The Lie algebras $L(A)$ and $\tilde{L}(A)$

Let A be a Cartan matrix on the standard list 6.12. Motivated by Propositions 7.7 and 7.8 we shall construct a Lie algebra $L(A)$ which will be shown to be a finite dimensional simple Lie algebra with Cartan matrix A.

Suppose A is an $l \times l$ matrix. Let $\mathfrak{F}$ be the free associative algebra over $\mathbb{C}$ on the $3l$ generators $e_1, \ldots, e_l, h_1, \ldots, h_l, f_1, \ldots, f_l$. The set of all monomials in these generators form a basis for $\mathfrak{F}$. Let $[\mathfrak{F}]$ be the Lie algebra obtained from $\mathfrak{F}$ by redefining the multiplication in the usual way and let $\mathfrak{L}$ be the subalgebra of $[\mathfrak{F}]$ generated by the elements $e_1, \ldots, e_l, h_1, \ldots, h_l, f_1, \ldots, f_l$. Let J be the ideal of $\mathfrak{L}$ generated by the elements

$$[h_i h_j]$$
$$[h_i e_j] - A_{ij} e_j$$
$$[h_i f_j] + A_{ij} f_j$$
$$[e_i f_i] - h_i$$
$$[e_i f_j] \quad \text{for } i \neq j$$
$$[e_i[e_i \ldots [e_i e_j]]] \quad \text{for } i \neq j$$
$$[f_i[f_i \ldots [f_i f_j]]] \quad \text{for } i \neq j$$

7.4 The Lie algebras $L(A)$ and $\tilde{L}(A)$

where the number of occurrences of e_i, f_i respectively in the last two elements is $1 - A_{ij}$.

We define $L(A) = \mathfrak{L}/J$. We shall eventually be able to show that $L(A)$ is the Lie algebra we require to prove the existence theorem. This description of $L(A)$ by generators and relations is due to J. P. Serre.

In order to investigate the Lie algebra $L(A)$ it is convenient to define a second, larger, Lie algebra $\tilde{L}(A)$. Let $\tilde{J}$ be the ideal of $\mathfrak{L}$ generated by the elements

$$[h_i h_j]$$
$$[h_i e_j] - A_{ij} e_j$$
$$[h_i f_j] + A_{ij} f_j$$
$$[e_i f_i] - h_i$$
$$[e_i f_j] \quad \text{for } i \neq j.$$

Let $\tilde{L}(A) = \mathfrak{L}/\tilde{J}$. Since $\tilde{J} \subset J$ we have surjective Lie algebra homomorphisms

$$\mathfrak{L} \to \tilde{L}(A) \to L(A)$$

We shall investigate the properties of the Lie algebra $\tilde{L}(A)$. This is generated by the images of the generators of $\mathfrak{L}$ under the above homomorphism. These images will continue to be written $e_1, \ldots, e_l, h_1, \ldots, h_l, f_1, \ldots, f_l$. These elements satisfy the relations

$$[h_i h_j] = 0$$
$$[h_i e_j] = A_{ij} e_j$$
$$[h_i f_j] = -A_{ij} f_j$$
$$[e_i f_i] = h_i$$
$$[e_i f_j] = 0 \quad \text{for } i \neq j$$

Proposition 7.9 *Let $\mathfrak{F}^-$ be the free associative algebra over $\mathbb{C}$ with generators $f_1, \ldots, f_l$. Then $\mathfrak{F}^-$ may be made into an $\tilde{L}(A)$-module giving a representation $\rho : \tilde{L}(A) \to [\text{End } \mathfrak{F}^-]$ defined by:*

$$\rho(f_i) f_{i_1} \ldots f_{i_r} = f_i f_{i_1} \ldots f_{i_r}$$

$$\rho(h_i) f_{i_1} \ldots f_{i_r} = -\left(\sum_{k=1}^{r} A_{i i_k}\right) f_{i_1} \ldots f_{i_r}$$

$$\rho(e_i) f_{i_1} \ldots f_{i_r} = -\sum_{k=1}^{r} \delta_{i i_k} \left(\sum_{h=k+1}^{r} A_{i i_h}\right) f_{i_1} \ldots \hat{f}_{i_k} \ldots f_{i_r}$$

where as usual the symbol $\hat{f}_{i_k}$ means that f_{i_k} is omitted from the product.

Proof. Since the monomials $f_{i_1} \ldots f_{i_r}$ form a basis for $\mathfrak{F}^-$ the endomorphisms $\rho(f_i), \rho(h_i), \rho(e_i)$ are uniquely determined by the above formulae. Thus there is a unique homomorphism $\mathfrak{F} \to \text{End } \mathfrak{F}^-$ mapping e_i, h_i, f_i to $\rho(e_i), \rho(h_i), \rho(f_i)$ respectively. This induces a Lie algebra homomorphism $[\mathfrak{F}] \to [\text{End } \mathfrak{F}^-]$ and so, by restriction, a Lie algebra homomorphism $\mathfrak{L} \to [\text{End } \mathfrak{F}^-]$. In order to obtain a homomorphism $\tilde{L}(A) \to [\text{End } \mathfrak{F}^-]$ we must verify the following relations.

(a) $[\rho(h_i)\rho(h_j)] = 0$
(b) $[\rho(h_i)\rho(e_j)] = A_{ij}\rho(e_j)$
(c) $[\rho(h_i)\rho(f_j)] = -A_{ij}\rho(f_j)$
(d) $[\rho(e_i)\rho(f_i)] = \rho(h_i)$
(e) $[\rho(e_i)\rho(f_j)] = 0 \quad$ for $i \neq j$

Relation (a) is trivial since $\rho(h_i)$ multiplies each basis element of $\mathfrak{F}^-$ by a scalar.

To prove relation (b) we have

$$\rho(h_i)\rho(e_j) f_{i_1} \ldots f_{i_r} = -\sum_{k=1}^{r} \delta_{ji_k} \left(\sum_{h=k+1}^{r} A_{ji_h} \right) \left(-\sum_{g \neq k} A_{ii_g} \right) f_{i_1} \ldots \hat{f}_{i_k} \ldots f_{i_r}$$

$$\rho(e_j)\rho(h_i) f_{i_1} \ldots f_{i_r} = -\sum_{k=1}^{r} \delta_{ji_k} \left(\sum_{h=k+1}^{r} A_{ji_h} \right) \left(-\sum_{g=1}^{r} A_{ii_g} \right) f_{i_1} \ldots \hat{f}_{i_k} \ldots f_{i_r}$$

Thus

$$(\rho(h_i)\rho(e_j) - \rho(e_j)\rho(h_i)) f_{i_1} \ldots f_{i_r}$$

$$= A_{ij} \left(-\sum_{k=1}^{r} \delta_{ji_k} \left(\sum_{h=k+1}^{r} A_{ji_h} \right) \right) f_{i_1} \ldots \hat{f}_{i_k} \ldots f_{i_r}$$

$$= A_{ij}\rho(e_j) f_{i_1} \ldots f_{i_r}.$$

To prove relation (c) we have

$$\rho(h_i)\rho(f_j) f_{i_1} \ldots f_{i_r} = -\left(A_{ij} + \sum_{k=1}^{r} A_{ii_k} \right) f_j f_{i_1} \ldots f_{i_r}$$

$$\rho(f_j)\rho(h_i) f_{i_1} \ldots f_{i_r} = -\left(\sum_{k=1}^{r} A_{ii_k} \right) f_j f_{i_1} \ldots f_{i_r}.$$

Thus

$$(\rho(h_i)\rho(f_j) - \rho(f_j)\rho(h_i)) f_{i_1} \ldots f_{i_r} = -A_{ij} f_j f_{i_1} \ldots f_{i_r} = -A_{ij}\rho(f_j) f_{i_1} \ldots f_{i_r}.$$

7.4 The Lie algebras $L(A)$ and $\tilde{L}(A)$

We next consider relation (d). We have

$$\rho(e_i)\rho(f_i)f_{i_1}\ldots f_{i_r} = -\left(\sum_{h=1}^{r} A_{ii_h}\right) f_{i_1}\ldots f_{i_r}$$

$$-\sum_{k=1}^{r} \delta_{ii_k} \left(\sum_{h=k+1}^{r} A_{ii_h}\right) f_i f_{i_1}\ldots \hat{f}_{i_k}\ldots f_{i_r}$$

$$\rho(f_i)\rho(e_i)f_{i_1}\ldots f_{i_r} = -\sum_{k=1}^{r} \delta_{ii_k} \left(\sum_{h=k+1}^{r} A_{ii_h}\right) f_i f_{i_1}\ldots \hat{f}_{i_k}\ldots f_{i_r}.$$

Thus

$$(\rho(e_i)\rho(f_i)-\rho(f_i)\rho(e_i))f_{i_1}\ldots f_{i_r} = -\left(\sum_{h=1}^{r} A_{ii_h}\right) f_{i_1}\ldots f_{i_r} = \rho(h_i)f_{i_1}\ldots f_{i_r}.$$

Finally we consider relation (e). Suppose $i \neq j$. Then

$$\rho(e_i)\rho(f_j)f_{i_1}\ldots f_{i_r} = -\sum_{k=1}^{r} \delta_{ii_k} \left(\sum_{h=k+1}^{r} A_{ii_h}\right) f_j f_{i_1}\ldots \hat{f}_{i_k}\ldots f_{i_r}$$

$$= \rho(f_j)\rho(e_i)f_{i_1}\ldots f_{i_r}.$$

Thus all the relations are preserved and we have a homomorphism $\tilde{L}(A) \to$ [End $\mathfrak{F}^-$]. $\square$

We can deduce useful information about $\tilde{L}(A)$ from the existence of this homomorphism.

Proposition 7.10 *The elements $h_1, \ldots, h_l$ of $\tilde{L}(A)$ are linearly independent.*

Proof. We show that the elements $\rho(h_1), \ldots, \rho(h_l)$ of End $\mathfrak{F}^-$ are linearly independent. We have

$$\rho(h_i)f_j = -A_{ij}f_j.$$

Thus if $\sum \lambda_i \rho(h_i) = 0$ we would have $\sum_i \lambda_i A_{ij} = 0$ for all $j = 1, \ldots, l$. Since the Cartan matrix $A = (A_{ij})$ is non-singular this implies that $\lambda_i = 0$ for each i. Hence $\rho(h_1), \ldots, \rho(h_l)$ are linearly independent, and so $h_1, \ldots, h_l$ must be linearly independent also. $\square$

Let $\tilde{H}$ be the subspace of $\tilde{L}(A)$ spanned by $h_1, \ldots, h_l$. Then we have $\dim \tilde{H} = l$. Moreover $[\tilde{H}\tilde{H}] = O$, thus $\tilde{H}$ is an abelian subalgebra of $\tilde{L}(A)$. We consider the weight spaces of $\tilde{L}(A)$ with respect to $\tilde{H}$. We are no longer dealing with a finite dimensional $\tilde{H}$-module as in Theorem 2.9, but analogous ideas apply in our situation. Elements of $\operatorname{Hom}(\tilde{H}, \mathbb{C})$ will be called **weights**. For each weight $\mu : \tilde{H} \to \mathbb{C}$ we define the corresponding **weight space** $\tilde{L}(A)_\mu$ by

$$\tilde{L}(A)_\mu = \{x \in \tilde{L}(A) \, ; \, [hx] = \mu(h)x \quad \text{for all } h \in \tilde{H}\}.$$

Proposition 7.11 $\tilde{L}(A) = \bigoplus_\mu \tilde{L}(A)_\mu$. Thus $\tilde{L}(A)$ is the direct sum of its weight spaces.

Proof. We first show that $\tilde{L}(A) = \sum_\mu \tilde{L}(A)_\mu$. A vector which lies in a weight space will be called a weight vector. We observe that, if $x, y \in \tilde{L}(A)$ are weight vectors of weights λ, μ respectively, then $[xy]$ is a weight vector of weight $\lambda + \mu$. For we have

$$[h[xy]] = [[hx]y] + [x[hy]] = \lambda(h)[xy] + \mu(h)[xy]$$
$$= (\lambda + \mu)(h)[xy] \quad \text{for } h \in \tilde{H}.$$

Now $\tilde{L}(A)$ is generated by elements e_i, h_i, f_i. Let $\alpha_i \in \operatorname{Hom}(\tilde{H}, \mathbb{C})$ be defined by

$$\alpha_i(h_j) = A_{ji}.$$

Then e_i is a weight vector of weight α_i, f_i is a weight vector of weight $-\alpha_i$ and h_i is a weight vector of weight 0. Thus all Lie products of generators e_i, h_i, f_i are weight vectors. Since every element of $\tilde{L}(A)$ is a linear combination of such products we deduce that

$$\tilde{L}(A) = \sum_\mu \tilde{L}(A)_\mu.$$

We next show that this sum is direct. If this is not so we can find a non-zero vector $x \in \tilde{L}(A)_\mu$ such that $x = \sum_\nu x_\nu$ where $x_\nu \in \tilde{L}(A)_\nu$ and ν runs over a finite set of weights all distinct from μ. Since $x \in \tilde{L}(A)_\mu$ we have

$$(\operatorname{ad} h - \mu(h)1)x = 0.$$

Since $x = \sum_\nu x_\nu$ with $x_\nu \in \tilde{L}(A)_\nu$ we have

$$\prod_\nu (\operatorname{ad} h - \nu(h)1) \, x = 0.$$

Now we can find an element $h \in \tilde{H}$ such that $\mu(h) \neq \nu(h)$ for all such ν. For the elements satisfying $\mu(h) = \nu(h)$ for some fixed ν lie in a proper subspace of $\tilde{H}$, and the finite dimensional vector space $\tilde{H}$ over $\mathbb{C}$ cannot be expressed

7.4 The Lie algebras $L(A)$ and $\tilde{L}(A)$

as the union of a finite number of proper subspaces. Thus we choose $h \in \tilde{H}$ such that $\mu(h) \ne \nu(h)$ for all such ν. Then the polynomials

$$t - \mu(h), \quad \prod_\nu (t - \nu(h))$$

in $\mathbb{C}[t]$ are coprime. Thus there exist polynomials $a(t), b(t) \in \mathbb{C}[t]$ with

$$a(t)(t - \mu(h)) + b(t) \prod_\nu (t - \nu(h)) = 1.$$

If follows that

$$a(\operatorname{ad} h)(\operatorname{ad} h - \mu(h)1)x + b(\operatorname{ad} h)\prod_\nu (\operatorname{ad} h - \nu(h)1)x = x.$$

We deduce that $x = 0$, a contradiction. Thus the sum $\sum_\mu \tilde{L}(A)_\mu$ is direct. □

We next obtain information about the kind of weights μ which can occur, that is for which $\tilde{L}(A)_\mu \ne 0$. The weights $\alpha_1, \ldots, \alpha_l \in \operatorname{Hom}(\tilde{H}, \mathbb{C})$ are linearly independent since the Cartan matrix A is non-singular. Thus any weight has form $n_1 \alpha_1 + \cdots + n_l \alpha_l$ for $n_i \in \mathbb{C}$. We shall show that all weights μ which occur in $\tilde{L}(A)$ have this form with $n_i \in \mathbb{Z}$ and with either $n_i \ge 0$ for all i or $n_i \le 0$ for all i.

Let

$$Q = \{n_1 \alpha_1 + \cdots + n_l \alpha_l \ ; \ n_i \in \mathbb{Z}\}.$$
$$Q^+ = \{n_1 \alpha_1 + \cdots + n_l \alpha_l \ne 0 \ ; \ n_i \ge 0 \text{ for all } i\}$$
$$Q^- = \{n_1 \alpha_1 + \cdots + n_l \alpha_l \ne 0 \ ; \ n_i \le 0 \text{ for all } i\}.$$

Let

$$\tilde{L}(A)^+ = \sum_{\mu \in Q^+} \tilde{L}(A)_\mu$$
$$\tilde{L}(A)^- = \sum_{\mu \in Q^-} \tilde{L}(A)_\mu.$$

It follows from Proposition 7.11 that the sum $\tilde{L}(A)^- + \tilde{H} + \tilde{L}(A)^+$ is direct. We shall show that in fact

$$\tilde{L}(A) = \tilde{L}(A)^- \oplus \tilde{H} \oplus \tilde{L}(A)^+.$$

Let $\tilde{N}$ be the subalgebra of $\tilde{L}(A)$ generated by $e_1, \ldots, e_l$ and $\tilde{N}^-$ the subalgebra generated by $f_1, \ldots, f_l$. Since e_i has weight α_i and f_i has weight $-\alpha_i$ we have $\tilde{N} \subset \tilde{L}(A)^+$ and $\tilde{N}^- \subset \tilde{L}(A)^-$. Thus the sum $\tilde{N}^- + \tilde{H} + \tilde{N}$ is direct.

Proposition 7.12 (i) $\tilde{L}(A) = \tilde{N}^- \oplus \tilde{H} \oplus \tilde{N}$
(ii) $\tilde{N} = \tilde{L}(A)^+$, $\tilde{N}^- = \tilde{L}(A)^-$, $\tilde{H} = \tilde{L}(A)_0$
(iii) *Every non-zero weight of $\tilde{L}(A)$ lies in Q^+ or in Q^-.*

Proof. The relations $[h_i e_j] = A_{ij} e_j$ show that $[h_i, \tilde{N}] \subset \tilde{N}$ since the e_j generate $\tilde{N}$. Thus we have $[\tilde{H}, \tilde{N}] \subset \tilde{N}$. It follows that $\tilde{H} + \tilde{N}$ is a subalgebra of $\tilde{L}(A)$, since

$$[\tilde{H} + \tilde{N}, \tilde{H} + \tilde{N}] \subset [\tilde{H}\tilde{H}] + [\tilde{H}\tilde{N}] + [\tilde{N}\tilde{N}] \subset \tilde{H} + \tilde{N}.$$

Similarly $\tilde{N}^- + \tilde{H}$ is a subalgebra of $\tilde{L}(A)$. We now consider the subspace $\tilde{N}^- + \tilde{H} + \tilde{N}$. The relations $[e_i f_i] = h_i$ and $[e_i f_j] = 0$ if $i \neq j$ show that

$$[e_i, \tilde{N}^-] \subset \tilde{N}^- + \tilde{H}.$$

For this is true for the generators of $\tilde{N}^-$, and the relation

$$[e_i[xy]] = [[e_i x] y] + [x [e_i y]]$$

then shows it is true for all elements of $\tilde{N}^-$ since $\tilde{N}^- + \tilde{H}$ is a subalgebra. It follows that

$$[e_i, \tilde{N}^- + \tilde{H} + \tilde{N}] \subset \tilde{N}^- + \tilde{H} + \tilde{N}$$

since $\tilde{H} + \tilde{N}$ is a subalgebra. Similarly we have

$$[f_i, \tilde{N}^- + \tilde{H} + \tilde{N}] \subset \tilde{N}^- + \tilde{H} + \tilde{N}$$

and the relation

$$[h_i, \tilde{N}^- + \tilde{H} + \tilde{N}] \subset \tilde{N}^- + \tilde{H} + \tilde{N}$$

is clear. It follows that the set of all $x \in \tilde{L}(A)$ such that

$$[x, \tilde{N}^- + \tilde{H} + \tilde{N}] \subset \tilde{N}^- + \tilde{H} + \tilde{N}$$

contains e_i, h_i, f_i. However, the relation

$$[[xy]z] = [[xz]y] + [x[yz]]$$

for $z \in \tilde{N}^- + \tilde{H} + \tilde{N}$ shows that the set of such x is a subalgebra. This subalgebra must be the whole of $\tilde{L}(A)$. Thus $\tilde{N}^- + \tilde{H} + \tilde{N}$ is an ideal of $\tilde{L}(A)$. Since $\tilde{L}(A)$ is generated by e_i, h_i, f_i it follows that $\tilde{N}^- + \tilde{H} + \tilde{N} = \tilde{L}(A)$. We know that this sum is direct, so we have

$$\tilde{L}(A) = \tilde{N}^- \oplus \tilde{H} \oplus \tilde{N}.$$

Since $\tilde{N}^- \subset \tilde{L}(A)^-, \tilde{N} \subset \tilde{L}(A)^+$ and the sum $\tilde{L}(A)^- + \tilde{H} + \tilde{L}(A)^+$ is direct we deduce that $\tilde{N}^- = \tilde{L}(A)^-$ and $\tilde{N} = \tilde{L}(A)^+$. Since $\tilde{L}(A) = \tilde{L}(A)^- \oplus \tilde{H} \oplus \tilde{L}(A)^+$, $\tilde{H} \subset \tilde{L}(A)_0$, and the weights occurring in $\tilde{L}(A)^-$ and $\tilde{L}(A)^+$ are all non-zero, we deduce from Proposition 7.11 that $\tilde{H} = \tilde{L}(A)_0$. Thus all parts of the proposition have been proved. □

Proposition 7.13 $\dim \tilde{L}(A)_{\alpha_i} = 1$ and $\dim \tilde{L}(A)_{-\alpha_i} = 1$

Proof. We know that $e_i \in \tilde{L}(A)_{\alpha_i}$. Also the element $e_i \in \tilde{L}(A)$ is non-zero, since it induces a non-zero endomorphism $\rho(e_i)$ on the $\tilde{L}(A)$-module $\mathfrak{F}^-$ considered in Proposition 7.9. Hence $\dim \tilde{L}(A)_{\alpha_i} \geq 1$. On the other hand we have

$$\tilde{L}(A)_{\alpha_i} \subset \tilde{L}(A)^+ = \tilde{N}.$$

Now $\tilde{N}$ is generated by $e_1, \ldots, e_l$ so is spanned by monomials in these elements. All such monomials are weight vectors. The only monomial which has weight α_i is e_i, since the α_i are linearly independent. Thus we have $\dim \tilde{L}(A)_{\alpha_i} = 1$. The relation $\dim \tilde{L}(A)_{-\alpha_i} = 1$ is obtained similarly. □

7.5 The existence theorem

We now turn to a study of the Lie algebra $L(A)$, in order to show that it is a finite dimensional simple Lie algebra with Cartan matrix A. From the definitions of $L(A)$, $\tilde{L}(A)$ we see that $L(A)$ is isomorphic to $\tilde{L}(A)/I$ where I is the ideal of $\tilde{L}(A)$ generated by the elements

$$[e_i[e_i \ldots [e_i e_j]]]$$
$$[f_i[f_i \ldots [f_i f_j]]]$$

for all $i \neq j$. As usual we have $1 - A_{ij}$ factors e_i or f_i.

Proposition 7.14 (i) *Let I^+ be the ideal of $\tilde{N}$ generated by the elements $[e_i[e_i \ldots [e_i e_j]]]$ for all $i \neq j$. Then I^+ is an ideal of $\tilde{L}(A)$.*
(ii) *Let I^- be the ideal of $\tilde{N}^-$ generated by the elements $[f_i[f_i \ldots [f_i f_j]]]$ for all $i \neq j$. Then I^- is an ideal of $\tilde{L}(A)$.*
(iii) $I = I^+ \oplus I^-$.

Proof. We write $X_{ij} = [e_i[e_i \ldots [e_i e_j]]]$ and $Y_{ij} = [f_i[f_i \ldots [f_i f_j]]]$. Then I^+ is the set of all linear combinations of elements

$$[[X_{ij} e_{k_1}] \ldots e_{k_r}]$$

for all $i \neq j$ and all $k_1, \ldots, k_r$ in $\{1, \ldots, l\}$. For such linear combinations certainly lie in I^+, and form an ideal of $\tilde{N}$.

Now X_{ij} is a weight vector, being a Lie product of weight vectors e_i, e_j. Similarly $[[X_{ij} e_{k_1}] \ldots e_{k_r}]$ is a weight vector. It is therefore transformed by each of $h_1, \ldots, h_l$ into a scalar multiple of itself. In order to show that I^+ is an ideal of $\tilde{L}(A)$ it will therefore be sufficient to show

$$[f_k, [[X_{ij} e_{k_1}] \ldots e_{k_r}]] \in I^+$$

for all $i, j, k_1, \ldots, k_r, k$. We shall prove this by induction on r, beginning with $r = 0$. In the following lemma we shall show that $[f_k, X_{ij}] = 0$, thus beginning the induction. So let $r \geq 1$ and write $[[X_{ij} e_{k_1}] \ldots e_{k_{r-1}}] = y$. We assume $[f_k y] \in I^+$ by induction. Then

$$[f_k [y e_{k_r}]] = [[f_k y] e_{k_r}] + [y [f_k e_{k_r}]].$$

If $k_r \neq k$ then $[f_k e_{k_r}] = 0$ and so

$$[f_k [y e_{k_r}]] = [[f_k y] e_{k_r}] \in I^+.$$

If $k_r = k$ then

$$[f_k [y e_{k_r}]] = [[f_k y] e_{k_r}] + [h_k y] \in I^+.$$

This completes the induction. Thus I^+ is an ideal of $\tilde{L}(A)$. Similarly I^- is an ideal of $\tilde{L}(A)$. Hence $I^+ \oplus I^-$ is an ideal of $\tilde{L}(A)$ containing the elements X_{ij} and Y_{ij}. Moreover any ideal of $\tilde{L}(A)$ containing the X_{ij} and Y_{ij} must contain I^+ and I^-. Hence $I^+ \oplus I^- = I$. □

In order to complete the proof of Proposition 7.14 we need the following lemma.

Lemma 7.15 $[f_k, X_{ij}] = 0$ *for all* i, j, k *with* $i \neq j$.

Proof. If $k \notin \{i, j\}$ this relation is obvious since $[f_k e_i] = 0$ and $[f_k e_j] = 0$.
So suppose $k = j$. Then we have

$$[f_j [e_i e_j]] = [e_i [f_j e_j]] = [h_j e_i] = A_{ji} e_i$$

$$[f_j [e_i [e_i e_j]]] = [e_i [f_j [e_i e_j]]] = 0$$

$$[f_j [e_i [e_i \ldots [e_i e_j]]]] = 0 \quad \text{for } r \geq 2$$

7.5 The existence theorem

by induction on r. Hence $[f_j, X_{ij}] = 0$ if $1 - A_{ij} \geq 2$, that is $A_{ij} \leq -1$. If $A_{ij} = 0$ then $A_{ji} = 0$ and $[f_j, X_{ij}] = 0$ in this case also.

Finally suppose that $k = i$. In this case we shall show that, for $r \geq 1$,

$$[f_i[e_i[e_i \ldots \underbrace{[e_i e_j]]]]}_{\leftarrow r \rightarrow}] = -r(A_{ij} + r - 1)[e_i\underbrace{[e_i \ldots [e_i e_j]]]}_{\leftarrow r-1 \rightarrow}.$$

For $r = 1$ we have

$$[f_i[e_i e_j]] = [[f_i e_i] e_j] = -[h_i e_j] = -A_{ij} e_j.$$

For $r > 1$ we use induction. We have

$$[f_i[e_i\underbrace{[e_i \ldots [e_i e_j]]]]}_{\leftarrow r \rightarrow}]$$
$$= -[h_i[e_i\underbrace{[e_i \ldots [e_i e_j]]]]}_{\leftarrow r-1 \rightarrow}] - (r-1)(A_{ij} + r - 2)[e_i\underbrace{[e_i \ldots [e_i e_j]]]}_{\leftarrow r-1 \rightarrow}$$
$$= (-(2r - 2 + A_{ij}) - (r-1)(A_{ij} + r - 2))[e_i\underbrace{[e_i \ldots [e_i e_j]]]}_{\leftarrow r-1 \rightarrow}$$
$$= -r(A_{ij} + r - 1)[e_i\underbrace{[e_i \ldots [e_i e_j]]]}_{\leftarrow r-1 \rightarrow}$$

as required. We now put $r = 1 - A_{ij}$ and obtain $[f_i, X_{ij}] = 0$.

Corollary 7.16 $L(A) = N^- \oplus H \oplus N$ where H is isomorphic to $\tilde{H}$, N^- is isomorphic to $\tilde{N}^-/I^-$ and N is isomorphic to $\tilde{N}/I^+$.

Proof. This follows from the facts that $L(A)$ is isomorphic to $\tilde{L}(A)/I$, $\tilde{L}(A) = \tilde{N}^- \oplus \tilde{H} \oplus \tilde{N}$, and $I = I^+ \oplus I^-$. □

We shall continue to denote the generators of $L(A)$ by e_i, h_i, f_i. These are the images of the generators of $\mathfrak{L}$ under the natural homomorphism $\mathfrak{L} \to L(A)$.

Proposition 7.17 *The maps* $\mathrm{ad}\, e_i : L(A) \to L(A)$ *and* $\mathrm{ad}\, f_i : L(A) \to L(A)$ *are locally nilpotent.*

Proof. To show that $\mathrm{ad}\, e_i$ is locally nilpotent we must show that, for all $x \in L(A)$, there exists $n(x)$ such that $(\mathrm{ad}\, e_i)^{n(x)} x = 0$. Now if $\mathrm{ad}\, e_i$ acts locally nilpotently on x and y it also acts locally nilpotently on $[xy]$. For

$$(\mathrm{ad}\, e_i)^n [xy] = \sum_{r=0}^{n} \binom{n}{r} [(\mathrm{ad}\, e_i)^r x, (\mathrm{ad}\, e_i)^{n-r} y]$$

$(\mathrm{ad}\, e_i)^r x$ will be 0 if r is sufficiently large and $(\mathrm{ad}\, e_i)^{n-r} y$ will be 0 if $n - r$ is sufficiently large. Thus $(\mathrm{ad}\, e_i)^n [xy]$ will be 0 if n is sufficiently large.

It follows that the set of elements of $L(A)$ on which $\operatorname{ad} e_i$ acts locally nilpotently is a subalgebra. However, we have

$$\operatorname{ad} e_i \cdot e_i = 0$$

$$(\operatorname{ad} e_i)^{1-A_{ij}} e_j = 0 \quad \text{if } i \neq j$$

$$(\operatorname{ad} e_i)^2 h_j = 0 \quad \text{for all } j$$

$$(\operatorname{ad} e_i)^3 f_i = 0$$

$$\operatorname{ad} e_i \cdot f_j = 0 \quad \text{if } i \neq j.$$

Thus this subalgebra contains all the generators e_j, h_j, f_j of $L(A)$, so is the whole of $L(A)$.

We see similarly that $\operatorname{ad} f_i$ is locally nilpotent on $L(A)$. □

Now the proof of Proposition 3.4 shows that if $\delta : L \to L$ is a locally nilpotent derivation of a Lie algebra L then $\exp \delta$ is an automorphism of L. Thus $\exp \operatorname{ad} e_i$ and $\exp \operatorname{ad} f_i$ are automorphisms of $L(A)$. We define $\theta_i \in \operatorname{Aut} L(A)$ by

$$\theta_i = \exp \operatorname{ad} e_i \cdot \exp \operatorname{ad} (-f_i) \cdot \exp \operatorname{ad} e_i.$$

Proposition 7.18 (i) $\theta_i(H) = H$
(ii) $\theta_i(h) = s_i(h)$ for all $h \in H$ where $s_i : H \to H$ is the linear map given by $s_i(h_j) = h_j - A_{ji} h_i$.

Proof. We have

$$\exp \operatorname{ad} e_i \cdot h_j = (1 + \operatorname{ad} e_i) h_j = h_j - A_{ji} e_i$$

$$\exp \operatorname{ad} (-f_i) \cdot \exp \operatorname{ad} e_i \cdot h_j = \exp \operatorname{ad} (-f_i) \cdot (h_j - A_{ji} e_i)$$

$$= \left(1 - \operatorname{ad} f_i + \frac{(\operatorname{ad} f_i)^2}{2}\right) (h_j - A_{ji} e_i)$$

$$= h_j - A_{ji} e_i - A_{ji} f_i - A_{ji} h_i + A_{ji} f_i = h_j - A_{ji} h_i - A_{ji} e_i$$

$$\exp \operatorname{ad} e_i \cdot \exp \operatorname{ad} (-f_i) \cdot \exp \operatorname{ad} e_i \cdot h_j = \exp \operatorname{ad} e_i \left(h_j - A_{ji} h_i - A_{ji} e_i\right)$$

$$= (1 + \operatorname{ad} e_i) \left(h_j - A_{ji} h_i - A_{ji} e_i\right) = h_j - A_{ji} h_i - A_{ji} e_i - A_{ji} e_i + 2 A_{ji} e_i$$

$$= h_j - A_{ji} h_i.$$

□

7.5 The existence theorem

Now the action of s_i on H is precisely that of the fundamental reflection $s_i = s_{\alpha_i}$ defined in Section 6.4. We recall that

$$s_i(h) = h - 2\frac{\langle h'_{\alpha_i}, h \rangle}{\langle h'_{\alpha_i}, h'_{\alpha_i} \rangle} h'_{\alpha_i} \quad \text{for } h \in H$$

$$= h - \langle h'_{\alpha_i}, h \rangle h_i.$$

In particular

$$s_i(h_j) = h_j - \langle h'_{\alpha_i}, h_j \rangle h_i = h_j - 2\frac{\langle h'_{\alpha_i}, h'_{\alpha_j} \rangle}{\langle h'_{\alpha_j}, h'_{\alpha_j} \rangle} h_i = h_j - A_{ji} h_i.$$

Thus Proposition 7.18 shows that the automorphism θ_i of $L(A)$ induces the fundamental reflection s_i on H.

We now consider the decomposition of $L(A)$ into weight spaces with respect to H. This time the weights are elements of $\text{Hom}(H, \mathbb{C})$. For each weight $\mu : H \to \mathbb{C}$ we define the weight space $L(A)_\mu$ by

$$L(A)_\mu = \{x \in L(A) \; ; \; [hx] = \mu(h)x \quad \text{for all } h \in H\}.$$

Proposition 7.19 $L(A) = \bigoplus_\mu L(A)_\mu$.

Proof. The algebra $L(A)$ is the sum of its weight spaces, since its generators e_i, h_i, f_i are weight vectors. Moreover the sum of weight spaces is direct, just as in the proof of Proposition 7.11. □

It also follows from Proposition 7.12 and Corollary 7.16 that $L(A) = N^- \oplus H \oplus N$ where all weights coming from N are in Q^+ and all weights coming from N^- are in Q^-. We also have $H = L(A)_0$.

Proposition 7.20 $\dim L(A)_{\alpha_i} = 1$ and $\dim L(A)_{-\alpha_i} = 1$.

Proof. By Proposition 7.13 we certainly have $\dim L(A)_{\alpha_i} \leq 1$. However, the ideal I^+ of $\tilde{N}$ such that $\tilde{N}/I^+ \cong N$ has the property that I^+ is a sum of weight spaces, and all weights occurring in I^+ are sums of $\alpha_1, \ldots, \alpha_l$ involving at least two terms. This is clear from the proof of Proposition 7.14. Thus α_i is not a weight of I^+. Hence

$$\dim L(A)_{\alpha_i} = \dim \tilde{L}(A)_{\alpha_i} = 1.$$

One shows similarly that $\dim \tilde{L}(A)_{-\alpha_i} = 1$. □

Proposition 7.21 *The automorphism θ_i of $L(A)$ transforms $L(A)_\mu$ to $L(A)_{s_i\mu}$. Hence $\dim L(A)_\mu = \dim L(A)_{s_i\mu}$.*

Proof. Let $x \in L(A)_\mu$. Then $[hx] = \mu(h)x$ for all $h \in H$. We apply the automorphism θ_i. This fixes H by Proposition 7.18. We have

$$[\theta_i h, \theta_i x] = \mu(h) \theta_i x.$$

Hence

$$[h, \theta_i x] = \mu\left(\theta_i^{-1} h\right) \theta_i x = \mu\left(s_i^{-1} h\right) \theta_i x = (s_i \mu(h)) \theta_i x,$$

again by Proposition 7.18. Thus we have $\theta_i x \in L(A)_{s_i \mu}$. Hence

$$\theta_i \left(L(A)_\mu\right) \subset L(A)_{s_i \mu}.$$

Replacing θ_i by θ_i^{-1}, μ by $s_i \mu$ and recalling that $s_i^2 = 1$ we also obtain

$$\theta_i^{-1} \left(L(A)_{s_i \mu}\right) \subset L(A)_\mu.$$

Hence $\theta_i \left(L(A)_\mu\right) \supset L(A)_{s_i \mu}$ and we have $\theta_i \left(L(A)_\mu\right) = L(A)_{s_i \mu}$. □

We now define W to be the group of non-singular linear transformations of $H^* = \text{Hom}(H, \mathbb{C})$ generated by $s_1, \ldots, s_l$ and define Φ to be the set of elements $w(\alpha_i)$ for $w \in W$ and $i \in \{1, \ldots, l\}$. Then Φ is the root system determined by the given Cartan matrix A and W is its Weyl group.

Proposition 7.22 $\dim L(A)_\alpha = 1$ for all $\alpha \in \Phi$.

Proof. We have $\alpha = w(\alpha_i)$ for some i and some $w \in W$. Since W is generated by $s_1, \ldots, s_l$, w is a product of such elements. Thus it follows from Proposition 7.21 that

$$\dim L(A)_\alpha = \dim L(A)_{\alpha_i} = 1.$$ □

We aim to show that the Lie algebra $L(A)$ is finite dimensional. As a step in this direction we shall show that the Weyl group W is finite. W is isomorphic to the group of non-singular linear transformations of $H_\mathbb{R}$ generated by $s_1, \ldots, s_l$ where $H_\mathbb{R} = \mathbb{R} h_1 + \cdots + \mathbb{R} h_l$. We have $\dim H_\mathbb{R} = l$. We do not have the scalar product on $H_\mathbb{R}$ available from the Killing form, so we define a scalar product directly from the Cartan matrix A.

Proposition 7.23 *The Cartan matrix can be factorised as $A = DB$ where D is diagonal and B is symmetric. D is the diagonal matrix with entries $d_1, \ldots, d_l \in \{1, 2, 3\}$ defined as follows.*

If the Dynkin diagram has only single edges then all $d_i = 1$.

If the Dynkin diagram has a double edge then $d_i = 1$ if α_i is a long root and $d_i = 2$ if α_i is a short root.

7.5 The existence theorem

If the Dynkin diagram has a triple edge then $d_i = 1$ if α_i is a long root and $d_i = 3$ if α_i is a short root.

Proof. This may be checked from the standard list 6.12 of Cartan matrices. □

For example in type G_2 we have

$$\begin{pmatrix} 2 & -1 \\ -3 & 2 \end{pmatrix} = \begin{pmatrix} 1 & 0 \\ 0 & 3 \end{pmatrix} \begin{pmatrix} 2 & -1 \\ -1 & \frac{2}{3} \end{pmatrix}.$$

We now define a bilinear form on $H_{\mathbb{R}}$ by $\langle h_i, h_j \rangle = d_i d_j B_{ij}$. This form is symmetric since B is a symmetric matrix.

Proposition 7.24 *This scalar product is positive definite.*

Proof. We have $n_{ij} = A_{ij}A_{ji} = d_i d_j B_{ij}^2$, thus $-\sqrt{n_{ij}} = \sqrt{d_i}\sqrt{d_j}B_{ij}$. The matrix of our scalar product is

$$DBD = \begin{pmatrix} \sqrt{d_1} & & \\ & \ddots & \\ & & \sqrt{d_l} \end{pmatrix} \begin{pmatrix} 2 & & -\sqrt{n_{ij}} \\ & \ddots & \\ -\sqrt{n_{ij}} & & 2 \end{pmatrix} \begin{pmatrix} \sqrt{d_1} & & \\ & \ddots & \\ & & \sqrt{d_l} \end{pmatrix}.$$

This matrix is congruent to the matrix $\begin{pmatrix} 2 & & -\sqrt{n_{ij}} \\ & \ddots & \\ -\sqrt{n_{ij}} & & 2 \end{pmatrix}$ of the quadratic form $Q(x_1, \ldots, x_l)$ of Proposition 6.6, which is positive definite. Thus DBD is also positive definite. □

Proposition 7.25 *Our scalar product on $H_{\mathbb{R}}$ is invariant under W.*

Proof. We first observe that

$$\langle h_i, x \rangle = d_i \alpha_i(x) \quad \text{for all } x \in H_{\mathbb{R}}.$$

For $\langle h_i, h_j \rangle = d_i d_j B_{ij} = d_i A_{ji} = d_i \alpha_i(h_j)$. It is sufficient to show that $\langle s_i x, s_i y \rangle = \langle x, y \rangle$ for all $x, y \in H_{\mathbb{R}}$. We note that $s_i(x) = x - \alpha_i(x)h_i$ since $s_i(h_j) = h_j - A_{ji}h_i$. Hence

$$\langle s_i x, s_i y \rangle = \langle x - \alpha_i(x)h_i, y - \alpha_i(y)h_i \rangle$$
$$= \langle x, y \rangle - \alpha_i(x)\langle h_i, y \rangle - \alpha_i(y)\langle h_i, x \rangle + \alpha_i(x)\alpha_i(y)\langle h_i, h_i \rangle$$
$$= \langle x, y \rangle - d_i \alpha_i(x)\alpha_i(y) - d_i \alpha_i(x)\alpha_i(y) + 2d_i \alpha_i(x)\alpha_i(y) = \langle x, y \rangle. \quad □$$

Thus the Weyl group W acts as a group of isometries on the Euclidean space $H_\mathbb{R}$.

We define certain subsets of $H_\mathbb{R}$ as follows:

$$H_i = \{x \in H_\mathbb{R}; \langle h_i, x \rangle = 0\}$$
$$H_i^+ = \{x \in H_\mathbb{R}; \langle h_i, x \rangle > 0\}$$
$$H_i^- = \{x \in H_\mathbb{R}; \langle h_i, x \rangle < 0\}$$
$$C = H_1^+ \cap \cdots \cap H_l^+.$$

C is called the **fundamental chamber**.

Let W_{ij} be the subgroup of W generated by s_i, s_j, where $i \neq j$. $s_i s_j$ has finite order m_{ij} given in terms of the Cartan matrix by $2\cos(\pi/m_{ij}) = \sqrt{n_{ij}}$. Thus W_{ij} is a finite dihedral group.

Lemma 7.26 *Let $w \in W_{ij}$ with $i \neq j$. Then either $w\left(H_i^+ \cap H_j^+\right) \subset H_i^+$ or $w\left(H_i^+ \cap H_j^+\right) \subset H_i^-$ and $l(s_i w) = l(w) - 1$.*

Proof. Let U be the 2-dimensional subspace of $H_\mathbb{R}$ spanned by h_i, h_j and $U^\perp$ be the orthogonal subspace. Then $H_\mathbb{R} = U \oplus U^\perp$ and the elements of W_{ij} act trivially on $U^\perp$. It is therefore sufficient to prove the result in U. Let $\Gamma = U \cap H_i^+ \cap H_j^+$. We obtain a configuration of chambers in U as shown in Figure 7.1.

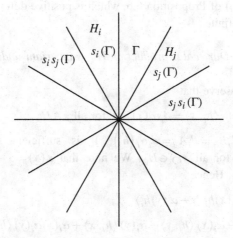

Figure 7.1 Configuration of chambers in U

7.5 The existence theorem

The chambers $\Gamma, s_j(\Gamma), s_j s_i(\Gamma), \ldots, \underset{\leftarrow m_{ij}-1 \rightarrow}{s_j s_i \ldots} (\Gamma)$ lie in H_i^+ and the chambers

$$s_i(\Gamma), s_i s_j(\Gamma), s_i s_j s_i(\Gamma), \ldots, \underset{\leftarrow m_{ij} \rightarrow}{s_i s_j \ldots} (\Gamma)$$

lie in H_i^-. The elements $s_i, s_i s_j, \ldots, \underset{\leftarrow m_{ij} \rightarrow}{s_i s_j \ldots}$ of W_{ij} all satisfy $l(s_i w) = l(w) - 1$. Thus for each $w \in W_{ij}$ we have either $w(H_i^+ \cap H_j^+) \subset H_i^+$ or $w(H_i^+ \cap H_j^+) \subset H_i^-$ and $l(s_i w) = l(w) - 1$. □

Proposition 7.27 (a) *Let $w \in W$. Then either $w(C) \subset H_i^+$ or $w(C) \subset H_i^-$ and $l(s_i w) = l(w) - 1$.*
(b) *Let $w \in W$ and $i \neq j$. Then there exists $w' \in W_{ij}$ such that $w(C) \subset w'(H_i^+ \cap H_j^+)$ and $l(w) = l(w') + l(w'^{-1} w)$.*

Note. Part (a) is the result we shall need. To prove it we must also prove part (b) at the same time.

Proof. We prove both statements together by induction on $l(w)$. If $l(w) = 0$ then $w = 1$ and (a), (b) are true. So suppose $l(w) > 0$. Then $w = s_j w'$ with $l(w') = l(w) - 1$ for some j. We prove (a).

First suppose $j = i$. By induction $w'(C) \subset H_i^+$ or $w'(C) \subset H_i^-$ and $l(s_i w') = l(w') - 1$. But $l(s_i w') = l(w') + 1$, so $w'(C) \subset H_i^+$. Then $w(C) \subset H_i^-$ and $l(s_i w) = l(w) - 1$.

Now suppose $j \neq i$. By induction there exists $w'' \in W_{ij}$ with $w'(C) \subset w''(H_i^+ \cap H_j^+)$ and $l(w') = l(w'') + l(w''^{-1} w')$. Thus $w(C) \subset s_j w''(H_i^+ \cap H_j^+)$. By Lemma 7.26 we have either $s_j w''(H_i^+ \cap H_j^+) \subset H_i^+$ or $s_j w''(H_i^+ \cap H_j^+) \subset H_i^-$ and $l(s_i s_j w'') = l(s_j w'') - 1$. In the first case $w(C) \subset H_i^+$. In the second case $w(C) \subset H_i^-$ and $l(s_i w) = l(s_i s_j w') = l(s_i s_j w'' w''^{-1} w') \leq l(s_i s_j w'') + l(w''^{-1} w') = l(s_j w'') - 1 + l(w''^{-1} w') \leq l(w'') + l(w''^{-1} w') = l(w') = l(w) - 1$. Thus $l(s_i w) = l(w) - 1$ and (a) is proved.

We now prove (b). If $w(C) \subset H_i^+ \cap H_j^+$ then (b) holds with $w' = 1$. Thus we may assume without loss of generality that $w(C) \not\subset H_i^+$. So by (a), which is now proved under the assumption of the inductive hypothesis, $w(C) \subset H_i^-$ and $l(s_i w) = l(w) - 1$. By induction there exists $w' \in W_{ij}$ such that $s_i w(C) \subset w'(H_i^+ \cap H_j^+)$ and $l(s_i w) = l(w') + l(w'^{-1} s_i w)$. Thus $w(C) \subset s_i w'(H_i^+ \cap H_j^+)$ and

$$l(w) = 1 + l(s_i w) = 1 + l(w') + l(w'^{-1} s_i w) \geq l(s_i w') + l((s_i w')^{-1} w) \geq l(w).$$

Thus we have equality throughout and $l(w) = l(s_i w') + l((s_i w')^{-1} w)$. Hence $s_i w' \in W_{ij}$ is the required element and (b) is proved. □

114   *The existence and uniqueness theorems*

Proposition 7.28 *If $w \in W$ satisfies $C \cap w(C)$ is non-empty, then $w = 1$.*

Proof. Suppose $w \neq 1$. Then $w = s_i w'$ with $l(w') = l(w) - 1$. By Proposition 7.27 (a) $w'(C) \subset H_i^+$. Thus $w(C) \subset H_i^-$. Hence

$$C \cap w(C) \subset H_i^+ \cap H_i^- = \emptyset.$$

So if $C \cap w(C)$ is non-empty, $w = 1$. □

Now the Euclidean space $H_\mathbb{R}$ has an orthonormal basis, and the isometries of $H_\mathbb{R}$ are represented by orthogonal matrices with respect to this basis. Thus $W \subset O_l$ where O_l is the group of $l \times l$ orthogonal matrices. $O_l \subset M_l$, the set of all $l \times l$ matrices over $\mathbb{R}$.

For any matrix $M = (m_{ij}) \in M_l$ we define $\|M\| = \sqrt{\sum_{i,j} m_{ij}^2}$ and for any column vector $v = (\lambda_1, \ldots, \lambda_l)^t \in \mathbb{R}^l$ we define $\|v\| = \sqrt{\sum_i \lambda_i^2}$.

Lemma 7.29 (a) *If $M \in O_l, v \in \mathbb{R}^l$ then $\|Mv\| = \|v\|$.*
(b) *If $M \in O_l, N \in M_l$ then $\|MN\| = \|N\|$.*
(c) *If $M \in M_l, v \in \mathbb{R}^l$ then $\|Mv\| \leq \|M\| \|v\|$.*

Proof. Straightforward. □

Proposition 7.30 (a) *W is finite.*
(b) *Φ is finite.*

Proof. Since $\Phi = W(\Pi)$ where $\Pi = \{\alpha_1, \ldots, \alpha_l\}$ it is clear that (a) implies (b).

Thus we show that W is finite. We consider the W-action on the Euclidean space $H_\mathbb{R}$. We give elements of $H_\mathbb{R}$ coordinates relative to our orthonormal basis. Let $v = (\lambda_1, \ldots, \lambda_l)^t \in C$. By definition of C there exists $r > 0$ such that $B_r(v) \subset C$ where

$$B_r(v) = \{x \in \mathbb{R}^l; \|x - v\| < r\}.$$

Let $w \in W$ with $w \neq 1$. Then $w(C) \cap C = \emptyset$ by Proposition 7.28. Thus $w(v) \notin C$ so

$$\|w(v) - v\| \geq r.$$

Hence

$$\|w - 1\| \|v\| \geq \|w(v) - v\| \geq r$$

so $\|w-1\| \geq \frac{r}{\|v\|}$. Put $\varepsilon = \frac{r}{\|v\|}$. Then $\|w-1\| \geq \varepsilon$ for all $w \neq 1$ in W. Now let $w, w' \in W$ have $w \neq w'$. Then

$$\|w-w'\| = \|w'(w'^{-1}w-1)\| = \|w'^{-1}w-1\| \geq \varepsilon$$

since $w' \in O_l$. Thus distinct elements of W are separated by a distance of at least ε. Since O_l, and hence W, is bounded it follows that W is finite. □

We now return to our Lie algebra $L(A)$. We know that $\dim L(A)_0 = l$ and $\dim L(A)_\alpha = 1$ for all $\alpha \in \Phi$.

If we can prove that $L(A)_\mu = O$ for all $\mu \in H^*$ with $\mu \notin \Phi \cup \{0\}$ we shall be able to deduce that $L(A)$ is finite dimensional.

Proposition 7.31 *Suppose $\mu \in H^*$ satisfies $\mu \neq 0$ and $\mu \notin \Phi$. Then $L(A)_\mu = O$.*

Proof. We assume that $\mu \neq 0$ and $L(A)_\mu \neq O$. Since $\dim L(A)_\mu \leq \dim \tilde{L}(A)_\mu$ we see by Proposition 7.12 (iii) that $\mu \in Q^+$ or $\mu \in Q^-$. In particular μ lies in the vector space $H_{\mathbb{R}}^*$ of real linear combinations of $\alpha_1, \ldots, \alpha_l$.

Suppose first that μ is a multiple of some root $\alpha \in \Phi$. Then $\mu = n\alpha$ or $-n\alpha$ with $n > 0$ and $\alpha \in \Phi^+$. Now $\alpha = w(\alpha_i)$ for some $w \in W$ and some $\alpha_i \in \Pi$ and we have $\dim L(A)_{n\alpha} = \dim L(A)_{n\alpha_i}$ by Proposition 7.21. Hence $L(A)_{n\alpha_i} \neq O$. Now N is generated by elements $e_1, \ldots, e_l$ and no non-zero Lie product of these can have weight $n\alpha_i$ unless $n = 1$. Thus $\mu = \alpha$ or $-\alpha$, that is $\mu \in \Phi$.

Secondly suppose μ is not a multiple of a root. Let

$$H_\mu = \{h \in H_{\mathbb{R}} \ ; \ \mu(h) = 0\}$$
$$H_\alpha = \{h \in H_{\mathbb{R}} \ ; \ \alpha(h) = 0\}.$$

Then H_μ is distinct from all the $H_\alpha, \alpha \in \Phi$. Since Φ is finite we can find $h \in H_\mu$ such that $h \notin H_\alpha$ for all $\alpha \in \Phi$. It follows that $w(h) \notin H_\alpha$ for all $\alpha \in \Phi$, since W permutes the H_α.

We claim there exists $w \in W$ such that $\alpha_i(w(h)) > 0$ for all $i = 1, \ldots, l$. To see this we define the height of an element of $H_{\mathbb{R}}$ by

$$\mathrm{ht}\left(\sum n_i h_i\right) = \sum n_i.$$

We choose an element $w \in W$ such that $\mathrm{ht}\, w(h)$ is maximum. This is certainly possible as W is finite. Then

$$s_i(w(h)) = w(h) - \alpha_i(w(h))h_i.$$

Since $\text{ht}\, s_i(w(h)) \leq \text{ht}\, w(h)$ we have $\alpha_i(w(h)) \geq 0$. However, $\alpha_i(w(h)) = 0$ would imply $w(h) \in H_{\alpha_i}$ which is impossible. Thus $\alpha_i(w(h)) > 0$ for each i and so $w(h) \in C$.

Now we have $(w(\mu))(w(h)) = \mu(h) = 0$. We write $w(\mu) = \sum_{i=1}^{l} m_i \alpha_i$. Then we have

$$\sum_{i=1}^{l} m_i \alpha_i(w(h)) = 0.$$

Since $\alpha_i(w(h)) > 0$ for each i we must have some $m_i > 0$ and some $m_j < 0$ in the sum. Thus $w(\mu) \notin Q^+$ and $w(\mu) \notin Q^-$. Hence $\tilde{L}(A)_{w(\mu)} = O$. By Proposition 7.21 we deduce $\tilde{L}(A)_{\mu} = O$, and so $L(A)_{\mu} = O$. This gives the required contradiction. □

Corollary 7.32 (i) $L(A) = H \oplus \sum_{\alpha \in \Phi} L(A)_\alpha$.
(ii) $\dim L(A) = l + |\Phi|$.

Proof. This is evident since $L(A)$ is the direct sum of its weight spaces. The 0-weight space is H and this has dimension l. The only non-zero weights are the elements of Φ and the corresponding weight spaces are 1-dimensional. Thus we have the required formula for the dimension of $L(A)$. □

We now know that $L(A)$ is a finite dimensional Lie algebra – indeed it has the dimension required for a simple Lie algebra with Cartan matrix A. We shall now be readily able to show that $L(A)$ has the required properties.

Proposition 7.33 *The Lie algebra $L(A)$ is semisimple.*

Proof. Let R be the soluble radical of $L(A)$ and consider the series

$$R = R^{(0)} \supset R^{(1)} \supset \cdots \supset R^{(n-1)} \supset R^{(n)} = O$$

where $R^{(i+1)} = [R^{(i)} R^{(i)}]$. We write $I = R^{(n-1)}$. We suppose if possible that $R \neq O$. Then I is a non-zero abelian ideal of L. Moreover I is invariant under all automorphisms of L.

Since $[HI] \subset I$ we may regard I as an H-module. We decompose it into its weight spaces. This weight space decomposition is

$$I = (H \cap I) \oplus \sum_{\alpha \in \Phi} (L_\alpha \cap I).$$

7.5 The existence theorem

For let $x \in I$ have $x = x_0 + \sum_{\alpha \in \Phi} x_\alpha$ where $x_0 \in H$ and $x_\alpha \in L_\alpha$. We show $x_0 \in I$ and each $x_\alpha \in I$. There exists $h \in H$ such that $\alpha(h) \neq 0$ and $\beta(h) \neq \alpha(h)$ for all $\beta \in \Phi$ with $\beta \neq \alpha$. Then

$$\text{ad } h \prod_{\substack{\beta \in \Phi \\ \beta \neq \alpha}} (\text{ad } h - \beta(h)1) \, x = \alpha(h) \prod_{\substack{\beta \in \Phi \\ \beta \neq \alpha}} (\alpha(h) - \beta(h)) \, x_\alpha$$

and this is an element of I. Hence $x_\alpha \in I$. Since this is true for each $\alpha \in \Phi$ we also have $x_0 \in I$. Hence

$$I = (H \cap I) \oplus \sum_{\alpha \in \Phi} (L_\alpha \cap I).$$

We claim that $L_\alpha \cap I = O$ for each $\alpha \in \Phi$. Otherwise we would have $L_\alpha \subset I$. Now $\alpha = w(\alpha_i)$ for some $w \in W$ and some $i = 1, \ldots, l$. By Proposition 7.21 we can find an automorphism of $L(A)$ which transforms L_α to L_{α_i}. Since I is invariant under all automorphism we would obtain $L_{\alpha_i} \subset I$. Hence $e_i \in I$. But then $[e_i f_i] = h_i \in I$ and we would have $[h_i e_i] = 2e_i \in I$, contradicting the fact that I is abelian. Hence $L_\alpha \cap I = O$ for all $\alpha \in \Phi$ and so $I \subset H$. Let $x \in I$. Then $[x e_i] = \alpha_i(x) e_i \in I$ hence $\alpha_i(x) = 0$. Since $\alpha_1, \ldots, \alpha_l$ are linearly independent we deduce that $x = 0$. Hence $I = O$, a contradiction. $\square$

Proposition 7.34 *H is a Cartan subalgebra of $L(A)$.*

Proof. Since H is abelian it is sufficient to show that $H = N(H)$. Let $x \in N(H)$. Then $x = h' + \sum_{\alpha \in \Phi} \lambda_\alpha e_\alpha$ for $h' \in H$, $e_\alpha \in L_\alpha$. Then for all $h \in H$ we have

$$[hx] = \sum_{\alpha \in \Phi} \lambda_\alpha \alpha(h) e_\alpha \in H.$$

However, we can find $h \in H$ such that $\alpha(h) \neq 0$ for all $\alpha \in \Phi$. We deduce that $\lambda_\alpha = 0$ for all $\alpha \in \Phi$, hence $x \in H$. $\square$

Proposition 7.35 *$L(A)$ is a simple Lie algebra with Cartan matrix A.*

Proof. $L(A) = H \oplus \sum_{\alpha \in \Phi} L(A)_\alpha$ is the Cartan decomposition of $L(A)$ with respect to H. Thus Φ is the root system of $L(A)$. The Cartan matrix $A' = (A'_{ij})$ of $L(A)$ is determined by the condition

$$s_i(\alpha_j) = \alpha_j - A'_{ij} \alpha_i.$$

However, we have
$$s_i(h_j) = h_j - A_{ji}h_i \quad \text{by Proposition 7.18.}$$
Using the facts that $(s_i\alpha_j)h_k = \alpha_j(s_ih_k)$ and $\alpha_j(h_k) = A_{kj}$ we deduce that
$$s_i(\alpha_j) = \alpha_j - A_{ij}\alpha_i.$$
Hence $A' = A$ and the Cartan matrix of $L(A)$ is A.

Since the Dynkin diagram of A is assumed connected, $L(A)$ must be a simple Lie algebra, by Proposition 6.13. □

Thus we have constructed, for each Cartan matrix on the standard list 6.12 a finite dimensional simple Lie algebra $L(A)$ with Cartan matrix A.

Theorem 7.36 *The finite dimensional non-trivial simple Lie algebras over $\mathbb{C}$ are*

$$\begin{array}{ll} A_l & l \geq 1 \\ B_l & l \geq 2 \\ C_l & l \geq 3 \\ D_l & l \geq 4 \\ E_6, E_7, E_8 & \\ F_4 & \\ G_2 & \end{array}$$

These Lie algebras are pairwise non-isomorphic.

Proof. For each Cartan matrix on the standard list 6.12 there is a corresponding finite dimensional simple Lie algebra, which by Theorem 7.5 is determined up to isomorphism. Simple Lie algebras with different Cartan matrices cannot be isomorphic since, by Proposition 6.4, the Cartan matrix on the standard list is uniquely determined by the Lie algebra. □

The description in Proposition 7.35 of the simple Lie algebras by generators and relations enables us to choose the root vectors e_α in a way which makes the structure constants $N_{\alpha,\beta}$ very simple.

Theorem 7.37 *It is possible to choose the root vectors e_α in the simple Lie algebra $L(A)$ in such a way that $N_{\alpha,\beta} = \pm(p+1)$ where $-p\alpha + \beta, \ldots, q\alpha + \beta$ are the α-chain of roots through β.*

7.5 The existence theorem

Proof. $L(A)$ is the Lie algebra generated by elements $e_1, \ldots, e_l$, $h_1, \ldots, h_l, f_1, \ldots f_l$ subject to relations

$$[h_i h_j] = 0$$
$$[h_i e_j] = A_{ij} e_j$$
$$[h_i f_j] = -A_{ij} f_j$$
$$[e_i f_i] = h_i$$
$$[e_i f_j] = 0 \quad \text{if } i \neq j$$
$$[e_i [e_i \ldots [e_i e_j]]] = 0 \quad \text{if } i \neq j$$
$$[f_i [f_i \ldots [f_i f_j]]] = 0 \quad \text{if } i \neq j$$

with $1 - A_{ij}$ occurrences of e_i, f_i respectively.

We now define

$$e_i' = -f_i, \quad h_i' = -h_i, \quad f_i' = -e_i.$$

It is straightforward to check that e_i', h_i', f_i' satisfy the above relations. Thus there is a homomorphism $\omega : L(A) \to L(A)$ satisfying $\omega(e_i) = -f_i$, $\omega(h_i) = -h_i$, $\omega(f_i) = -e_i$. Since $\omega^2 = 1$, ω is an automorphism of $L(A)$.

Let α be a positive root of $L(A)$. Then

$$[h_i e_\alpha] = \alpha(h_i) e_\alpha$$

and so

$$[-h_i, \theta(e_\alpha)] = \alpha(h_i) \theta(e_\alpha)$$

that is

$$[h_i, \theta(e_\alpha)] = -\alpha(h_i) \theta(e_\alpha)$$

whence $\theta(e_\alpha) \in L_{-\alpha}$. Let $\theta(e_\alpha) = \lambda e_{-\alpha}$. Then $\lambda \neq 0$ and we may choose $\mu \in \mathbb{C}$ with $\mu^2 = -\lambda^{-1}$. Then

$$\theta(\mu e_\alpha) = -\mu^{-1} e_{-\alpha}$$

and

$$[\mu e_\alpha, \mu^{-1} e_{-\alpha}] = [e_\alpha e_{-\alpha}] = \frac{2 h_\alpha'}{\langle h_\alpha', h_\alpha' \rangle}.$$

We now change our choice of the root vectors $e_\alpha \in L_\alpha$. For each positive root α we take μe_α as our root vector and for the corresponding negative root $-\alpha$ we take $\mu^{-1} e_{-\alpha}$ as the root vector. Changing the notation to call these new root vectors e_α, $e_{-\alpha}$ we retain the multiplication formulae of Section 7.1, except that the structure constants $N_{\alpha,\beta}$ may now be altered. We also now have $\omega(e_\alpha) = -e_{-\alpha}$. Now

$$[e_\alpha e_\beta] = N_{\alpha,\beta} e_{\alpha+\beta}$$

and so

$$[-e_{-\alpha}, -e_{-\beta}] = N_{\alpha,\beta}(-e_{-\alpha-\beta}).$$

This implies $N_{-\alpha,-\beta} = -N_{\alpha,\beta}$. By Proposition 7.1 (iii) we have $N_{\alpha,\beta} N_{-\alpha,-\beta} = -(p+1)^2$, where $-p\alpha+\beta, \ldots, q\alpha+\beta$ is the α-chain of roots through β. Hence $N_{\alpha,\beta}^2 = (p+1)^2$ and $N_{\alpha,\beta} = \pm(p+1)$. □

This result has important implications in the theory of Chevalley groups over arbitrary fields. (See, for example, R. W. Carter, *Simple Groups of Lie Type*, Wiley Classics Library, 1989.)

The signs in the formula $N_{\alpha,\beta} = \pm(p+1)$ can be chosen in various ways. By Lemma 7.3 and Proposition 7.4 the signs can be chosen arbitrarily for extraspecial pairs of roots (α, β) and are then determined for all other pairs (α, β).

8
The simple Lie algebras

Having obtained a classification of the finite dimensional simple Lie algebras over $\mathbb{C}$ we shall in the present chapter investigate them individually in order to obtain their dimensions and a description of their root systems. In the case of Lie algebras of type A_l, B_l, C_l or D_l we shall also give a description in terms of Lie algebras of matrices.

The strategy for obtaining this information will be as follows. Given a Cartan matrix A on the standard list 6.12 we shall describe a symmetric scalar product $\{,\}$ on an l-dimensional vector space V over $\mathbb{R}$ with basis $\alpha_1, \ldots, \alpha_l$ such that

$$2\frac{\{\alpha_i, \alpha_j\}}{\{\alpha_i, \alpha_i\}} = A_{ij} \qquad i, j = 1, \ldots, l.$$

We compare this scalar product with the Killing form $\langle \alpha_i, \alpha_j \rangle$ obtained when $\alpha_1, \ldots, \alpha_l$ are interpreted as a fundamental system of roots in the simple Lie algebra with Cartan matrix A. We claim there exists a constant κ such that

$$\langle \alpha_i, \alpha_j \rangle = \kappa \{\alpha_i, \alpha_j\} \qquad \text{for all } i, j.$$

In fact we can define κ by the equation

$$\langle \alpha_1, \alpha_1 \rangle = \kappa \{\alpha_1, \alpha_1\}.$$

Then we have

$$2\frac{\langle \alpha_i, \alpha_j \rangle}{\langle \alpha_i, \alpha_i \rangle} = 2\frac{\{\alpha_i, \alpha_j\}}{\{\alpha_i, \alpha_i\}} \qquad \text{for all } i, j$$

and since both scalar products are symmetric we deduce that

$$\frac{\langle \alpha_j, \alpha_j \rangle}{\langle \alpha_i, \alpha_i \rangle} = \frac{\{\alpha_j, \alpha_j\}}{\{\alpha_i, \alpha_i\}} \qquad \text{for all } i, j.$$

Putting $j=1$ we deduce

$$\langle \alpha_i, \alpha_i \rangle = \kappa \{\alpha_i, \alpha_i\} \quad \text{for all } i$$

and it follows that

$$\langle \alpha_i, \alpha_j \rangle = \kappa \{\alpha_i, \alpha_j\} \quad \text{for all } i, j.$$

Thus the symmetric scalar product $\{,\}$ is the same as the Killing form up to multiplication by the constant κ. In practise it will not be necessary to determine this constant.

We then consider the fundamental reflections $s_i : V \to V$ defined by

$$s_i(\alpha_j) = \alpha_j - A_{ij}\alpha_i.$$

The maps $s_1, \ldots, s_l$ generate the Weyl group W of transformations of V. The vectors in V of form $w(\alpha_i)$ for all $w \in W$ and all i will then give the full root system Φ. We shall then be able to obtain the dimension of the simple Lie algebra L by the formula

$$\dim L = l + |\Phi|.$$

8.1 Lie algebras of type A_l

It will be convenient to describe the vector space V as a subspace of a larger vector space $\tilde{V}$ of dimension $l+1$.

Let $\tilde{V}$ be a vector space over $\mathbb{R}$ with basis $\beta_1, \ldots, \beta_{l+1}$ and let the symmetric scalar product $\{,\}$ on $\tilde{V}$ be defined by

$$\{\beta_i, \beta_j\} = \delta_{ij} \quad i, j = 1, \ldots, l+1.$$

We define $\alpha_1, \ldots, \alpha_l$ by

$$\alpha_1 = \beta_1 - \beta_2, \quad \alpha_2 = \beta_2 - \beta_3, \quad \ldots, \quad \alpha_l = \beta_l - \beta_{l+1}.$$

Let V be the subspace of $\tilde{V}$ spanned by $\alpha_1, \ldots, \alpha_l$. Then we have $\dim V = l$. Our scalar product satisfies

$$\{\alpha_i, \alpha_i\} = 2, \quad \{\alpha_i, \alpha_{i+1}\} = -1, \quad \{\alpha_i, \alpha_j\} = 0 \quad \text{if } |i-j| > 1.$$

Hence

$$2\frac{\{\alpha_i, \alpha_j\}}{\{\alpha_i, \alpha_i\}} = A_{ij} \quad i, j = 1, \ldots, l$$

where $A = (A_{ij})$ is the Cartan matrix of type A_l.

8.1 Lie algebras of type A_l

We now consider the action of the fundamental reflections on V. We define linear maps $s_i : \tilde{V} \to \tilde{V}$ by

$$s_i(\beta_i) = \beta_{i+1}$$
$$s_i(\beta_{i+1}) = \beta_i$$
$$s_i(\beta_j) = \beta_j \quad j \neq i, i+1.$$

Then we have

$$s_i(\alpha_j) = \alpha_j - A_{ij}\alpha_i \quad i, j = 1, \ldots, l$$

and so s_i restricted to V is the ith fundamental reflection.

We consider the group of transformations of $\tilde{V}$ generated by $s_1, \ldots, s_l$. Since s_i acts on $\tilde{V}$ by permuting β_i, β_{i+1} and fixing the remaining β_j the group generated by the s_i is the group of all permutations of $\beta_1, \ldots, \beta_{l+1}$. This group leaves the subspace V invariant and induces on V the Weyl group W. Thus we have a surjective homomorphism

$$S_{l+1} \to W$$

whose kernel is trivial. Hence the Weyl group of type A_l is isomorphic to the symmetric group S_{l+1}.

The full root system Φ of type A_l is the set of vectors of form $w(\alpha_i)$ for all $w \in W$ and all i. This is the set

$$\Phi = \{\beta_i - \beta_j \, ; i \neq j \quad i, j = 1, \ldots, l+1\}.$$

Thus we have $|\Phi| = l(l+1)$ and $\dim L = l + |\Phi| = l(l+2)$.

We shall now show that $\mathfrak{sl}_{l+1}(\mathbb{C})$ is a simple Lie algebra of type A_l. We discussed this Lie algebra in Section 4.4. In particular we know from Proposition 4.26 that the subalgebra H of diagonal matrices in $L = \mathfrak{sl}_{l+1}(\mathbb{C})$ is a Cartan subalgebra. Moreover by Proposition 4.27

$$L = H \oplus \sum_{i \neq j} \mathbb{C} E_{ij}$$

is the Cartan decomposition of L with respect to H. By Theorem 4.25 L is a simple Lie algebra. By Proposition 4.28 the roots of L are the functions

$$\begin{pmatrix} \lambda_1 & & \\ & \cdot & \\ & & \cdot \\ & & & \lambda_{l+1} \end{pmatrix} \to \lambda_i - \lambda_j \quad i \neq j$$

and a system of fundamental roots is given by

$$\alpha_i \begin{pmatrix} \lambda_1 \\ \cdot \\ \cdot \\ \cdot \\ \lambda_{l+1} \end{pmatrix} = \lambda_i - \lambda_{i+1}.$$

We can now determine the Cartan matrix $A = (A_{ij})$ of L. We recall from Proposition 4.22 that the α_i-chain of roots through α_j when $i \neq j$ has the form

$$\alpha_j, \alpha_i + \alpha_j, \ldots, q\alpha_i + \alpha_j$$

where $q = -A_{ij}$. Since we know the roots we can determine the numbers q. We have $q = 1$ if $i = j - 1$ or $j + 1$ and $q = 0$ otherwise. Thus the Cartan matrix A is the same as the Cartan matrix of type A_l in the standard list 6.12. Thus we have proved:

Theorem 8.1 (i) *The simple Lie algebra of type A_l has dimension $l(l+2)$.*
(ii) *The Lie algebra $\mathfrak{sl}_{l+1}(\mathbb{C})$ of all $(l+1) \times (l+1)$ matrices of trace 0 is simple of type A_l.*

8.2 Lie algebras of type D_l

We recall that the Dynkin diagram of type D_l has form

Let V be a real vector space of dimension l and basis $\beta_1, \ldots, \beta_l$. Let the symmetric scalar product $\{,\}$ be defined by $\{\beta_i, \beta_j\} = \delta_{ij}$. We define $\alpha_1, \ldots, \alpha_l$ by

$$\alpha_1 = \beta_1 - \beta_2, \quad \alpha_2 = \beta_2 - \beta_3, \quad \ldots, \quad \alpha_{l-1} = \beta_{l-1} - \beta_l, \quad \alpha_l = \beta_{l-1} + \beta_l.$$

Then we have

$$\{\alpha_i, \alpha_i\} = 2 \quad \text{for all } i$$
$$\{\alpha_i, \alpha_{i+1}\} = -1 \quad \text{for } 1 \leq i \leq l-2$$
$$\{\alpha_i, \alpha_j\} = 0 \quad \text{for } i, j \in \{1, \ldots, l-1\} \text{ with } |i-j| > 1$$
$$\{\alpha_{l-2}, \alpha_l\} = -1$$
$$\{\alpha_i, \alpha_l\} = 0 \quad \text{for } i \neq l-2, l.$$

8.2 Lie algebras of type D_l

It follows that

$$2\frac{\{\alpha_i, \alpha_j\}}{\{\alpha_i, \alpha_i\}} = A_{ij} \quad \text{for all } i, j$$

and hence that the scalar product $\{,\}$ is a non-zero multiple of the Killing form.

We now consider the fundamental reflections s_i on V. For $1 \leq i \leq l-1$ we have

$$s_i(\beta_i) = \beta_{i+1}$$
$$s_i(\beta_{i+1}) = \beta_i$$
$$s_i(\beta_j) = \beta_j \quad \text{for } j \neq i, i+1.$$

For $i = l$ we have

$$s_l(\beta_{l-1}) = -\beta_l$$
$$s_l(\beta_l) = -\beta_{l-1}$$
$$s_l(\beta_j) = \beta_j \quad \text{for } j \neq l-1, l.$$

Thus the Weyl group W generated by $s_1, \ldots, s_l$ has form

$$w(\beta_i) = \pm\beta_{\sigma(i)} \quad w \in W$$

for some permutation σ of $1, \ldots, l$. Let $w(\beta_i) = \varepsilon_i \beta_{\sigma(i)}$. Then an even number of the signs ε_i are equal to -1. Conversely for any permutation σ of $1, \ldots, l$ and any set of signs ε_i with $\prod \varepsilon_i = 1$ there is an element $w \in W$ acting as above. It follows that the order of the Weyl group of type D_l is given by

$$|W| = 2^{l-1} l!.$$

We now consider the root system Φ. The elements of Φ have form $w(\alpha_i)$ for all $w \in W$ and all i. Since w acts on the β_i by a permutation combined with certain sign changes we obtain

$$\Phi = \{\pm\beta_i \pm \beta_j \; ; \; i \neq j \in \{1, \ldots, l\}\}.$$

All combinations of signs are possible. Hence $|\Phi| = 2l(l-1)$ and so

$$\dim L = l + |\Phi| = l(2l-1).$$

We now wish to describe L as a Lie algebra of matrices. We begin with a lemma which will be useful both for the type being considered and for certain other types also.

Lemma 8.2 *Let M be an $n \times n$ matrix over $\mathbb{C}$. Then the set of all $n \times n$ matrices X over $\mathbb{C}$ satisfying*

$$X^t M + M X = O$$

forms a Lie algebra under Lie multiplication of matrices.

Proof. The set of such matrices X is clearly closed under addition and scalar multiplication. Let X_1, X_2 be matrices satisfying the given condition. Thus we have

$$X_1^t M = -M X_1, \quad X_2^t M = -M X_2.$$

It follows that

$$[X_1 X_2]^t M = (X_1 X_2 - X_2 X_1)^t M = X_2^t X_1^t M - X_1^t X_2^t M$$
$$= -X_2^t M X_1 + X_1^t M X_2 = M X_2 X_1 - M X_1 X_2$$
$$= -M [X_1 X_2].$$

Thus the set of such matrices X forms a Lie algebra. □

We now consider the special case when M is the $2l \times 2l$ matrix

$$M = \begin{pmatrix} O & I_l \\ I_l & O \end{pmatrix}.$$

Then a $2l \times 2l$ matrix $X = \begin{pmatrix} X_{11} & X_{12} \\ X_{21} & X_{22} \end{pmatrix}$ satisfies $X^t M + M X = O$ if and only if $X_{22} = -X_{11}^t$ and X_{12}, X_{21} are skew-symmetric. Let L be the Lie algebra of all such matrices X and H be the set of diagonal matrices in L. The elements of H have form

$$h = \begin{pmatrix} \lambda_1 & & & & & & \\ & \ddots & & & & & \\ & & \lambda_l & & & & \\ & & & -\lambda_1 & & & \\ & & & & \ddots & & \\ & & & & & -\lambda_l \end{pmatrix}.$$

Let us number the rows and columns $1, \ldots, l, -1, \ldots, -l$. Then we have

$$L = H \oplus \sum_\alpha \mathbb{C} e_\alpha$$

8.2 Lie algebras of type D_l

where

$$e_\alpha = \begin{cases} E_{ij} - E_{-j,-i} \\ -E_{-i,-j} + E_{ji} \\ E_{i,-j} - E_{j,-i} \\ -E_{-i,j} + E_{-j,i} \end{cases} \text{ for } 0 < i < j,$$

that is for each pair i, j with $0 < i < j$ we have four vectors e_α as above. Moreover each of the 1-dimensional spaces $\mathbb{C}e_\alpha$ is a H-module, and we have:

$$[h, E_{ij} - E_{-j,-i}] = (\lambda_i - \lambda_j)(E_{ij} - E_{-j,-i})$$
$$[h, -E_{-i,-j} + E_{ji}] = (\lambda_j - \lambda_i)(-E_{-i,-j} + E_{ji})$$
$$[h, E_{i,-j} - E_{j,-i}] = (\lambda_i - \lambda_j)(E_{i,-j} - E_{j,-i})$$
$$[h, -E_{-i,j} + E_{-j,i}] = (-\lambda_i - \lambda_j)(-E_{-i,j} + E_{-j,i}).$$

We write $[he_\alpha] = \alpha(h)e_\alpha$ for all such e_α.

Now the argument of Proposition 7.34 shows that H is a Cartan subalgebra of L. The decomposition

$$L = H \oplus \sum_\alpha \mathbb{C}e_\alpha$$

is then the Cartan decomposition of L with respect to H.

We next verify that L is semisimple. Suppose not. Then L has a non-zero abelian ideal I. Since $[HI] \subset I$ we may regard I as a H-module and consider the decomposition of I into weight spaces with respect to H. This gives

$$I = (H \cap I) \oplus \sum_\alpha (\mathbb{C}e_\alpha \cap I)$$

just as in the proof of Proposition 7.33. Suppose if possible that $\mathbb{C}e_\alpha \cap I \neq 0$ for some α. Then we have $e_\alpha \in I$. We then define h_α by $h_\alpha = [e_\alpha e_{-\alpha}]$ and observe that $[h_\alpha e_\alpha] = 2e_\alpha$. Then $e_\alpha, h_\alpha \in I$ and we have a contradiction to the fact that I is abelian. Hence $\mathbb{C}e_\alpha \cap I = 0$ for all α and so $I \subset H$. Let $x \in I$. Then $[xe_\alpha] = \alpha(x)e_\alpha \in I$ so $\alpha(x) = 0$. This holds for all α and so $x = 0$. Thus $I = 0$, which gives a contradiction. Hence L is semisimple.

We now know that the functions $\alpha : H \to \mathbb{C}$ given above are the roots of L with respect to H. A system of fundamental roots is given by

$$\alpha_1(h) = \lambda_1 - \lambda_2$$
$$\alpha_2(h) = \lambda_2 - \lambda_3$$
$$\vdots$$
$$\alpha_{l-1}(h) = \lambda_{l-1} - \lambda_l$$
$$\alpha_l(h) = \lambda_{l-1} + \lambda_l$$

since all the other roots are integral combinations of these with coefficients all non-negative or all non-positive.

We now determine the Cartan matrix of L. Let the α_i-chain of roots through α_j for $i \neq j$ be

$$\alpha_j, \quad \alpha_i + \alpha_j, \quad \ldots, \quad q\alpha_i + \alpha_j.$$

Then $A_{ij} = -q$ by Proposition 4.22. Since we know the roots we can find the number q and hence A_{ij} for each $i \neq j$. This gives us the Cartan matrix $A = (A_{ij})$ of type D_l on the standard list 6.12.

Finally we note that since this Cartan matrix is indecomposable the Lie algebra L must be simple by Corollary 6.15. Thus we have proved the following result:

Theorem 8.3 (i) *The simple Lie algebra of type D_l has dimension $l(2l-1)$.*
(ii) *The Lie algebra of all $2l \times 2l$ matrices X satisfying $X^t M + MX = 0$ where*

$$M = \begin{pmatrix} O & I_l \\ I_l & O \end{pmatrix}$$

is simple of type D_l when $l \geq 4$.

8.3 Lie algebras of type B_l

We recall that the Dynkin diagram of type B_l has form

8.3 Lie algebras of type B_l

Let V be a real vector space of dimension l with basis $\beta_1, \ldots, \beta_l$. Let the scalar product $\{,\}$ on V be defined by $\{\beta_i, \beta_j\} = \delta_{ij}$. We define $\alpha_1, \ldots, \alpha_l \in V$ by

$$\alpha_1 = \beta_1 - \beta_2, \quad \alpha_2 = \beta_2 - \beta_3, \quad \ldots, \quad \alpha_{l-1} = \beta_{l-1} - \beta_l, \quad \alpha_l = \beta_l.$$

Then we have

$$\{\alpha_i, \alpha_i\} = 2 \quad \text{for } 1 \le i \le l-1$$

$$\{\alpha_l, \alpha_l\} = 1$$

$$\{\alpha_i, \alpha_{i+1}\} = -1 \quad \text{for } 1 \le i \le l-1$$

$$\{\alpha_i, \alpha_j\} = 0 \quad \text{if } |i-j| > 1.$$

It follows that

$$2 \frac{\{\alpha_i, \alpha_j\}}{\{\alpha_i, \alpha_i\}} = A_{ij} \quad \text{for all } i, j$$

where $A = (A_{ij})$ is the Cartan matrix of type B_l on the standard list 6.12. Thus the scalar product $\{,\}$ is a non-zero multiple of the Killing form.

We now consider the fundamental reflections s_i on V. We have, for $1 \le i \le l-1$,

$$s_i(\beta_i) = \beta_{i+1}$$
$$s_i(\beta_{i+1}) = \beta_i$$
$$s_i(\beta_j) = \beta_j \quad j \ne i, i+1.$$

For $i = l$ we have

$$s_l(\beta_l) = -\beta_l$$
$$s_l(\beta_i) = \beta_i \quad i \ne l.$$

Thus the Weyl group W generated by $s_1, \ldots, s_l$ consists of elements w of the form

$$w(\beta_i) = \pm \beta_{\sigma(i)}$$

for some permutation σ of $1, \ldots, l$. Let $w(\beta_i) = \varepsilon_i \beta_{\sigma(i)}$. Then, given any permutation σ of $1, \ldots, l$ and any set of signs $\varepsilon_i \in \{1, -1\}$ there is an element $w \in W$ such that $w(\beta_i) = \varepsilon_i \beta_{\sigma(i)}$ for all i. Thus the order of the Weyl group W of type B_l is

$$|W| = 2^l l!.$$

We now consider the root system Φ. The elements of Φ have form $w(\alpha_i)$ for all $w \in W$ and all i. Since w acts on the β_i by means of a permutation combined with sign changes we obtain

$$\Phi = \{\pm\beta_i \pm \beta_j \ i \neq j\,; \pm\beta_i\}.$$

All combinations of signs are possible. Thus we have $|\Phi| = 2l^2$ and so

$$\dim L = l + |\Phi| = l(2l+1).$$

We shall now describe L as a Lie algebra of matrices. We use Lemma 8.2 and this time we take the $(2l+1) \times (2l+1)$ matrix M given by

$$M = \begin{pmatrix} 2 & 0 & \cdots & 0 \\ 0 & O & & I_l \\ \vdots & & & \\ 0 & I_l & & O \end{pmatrix}.$$

Let L be the Lie algebra of all $(2l+1) \times (2l+1)$ matrices X satisfying the condition

$$X^t M + M X = O$$

We consider X as a block matrix

$$\begin{pmatrix} X_{00} & X_{01} & X_{02} \\ X_{10} & X_{11} & X_{12} \\ X_{20} & X_{21} & X_{22} \end{pmatrix} \begin{matrix} 1 \\ l \\ l \end{matrix}$$
$$\begin{matrix} 1 & l & l \end{matrix}$$

Then X satisfies $X^t M + MX = O$ if and only if $X_{22} = -X_{11}^t$, X_{12} and X_{21} are skew-symmetric, $X_{10} = -2X_{02}^t$, $X_{20} = -2X_{01}^t$ and $X_{00} = 0$.

Let H be the set of diagonal matrices in L. The elements of H have form

$$h = \begin{pmatrix} 0 & & & & & & \\ & \lambda_1 & & & & & \\ & & \ddots & & & & \\ & & & \lambda_l & & & \\ & & & & -\lambda_1 & & \\ & & & & & \ddots & \\ & & & & & & -\lambda_l \end{pmatrix}.$$

We number the rows and columns $0, 1, \ldots, l, -1, \ldots, -l$. Then we have

$$L = H \oplus \sum_\alpha \mathbb{C} e_\alpha$$

8.3 Lie algebras of type B_l

where

$$e_\alpha = \begin{cases} E_{ij} - E_{-j, -i} \\ -E_{-i, -j} + E_{ji} \\ E_{i, -j} - E_{j, -i} & \text{for } 0 < i < j \\ -E_{-i, j} + E_{-j, i} \\ 2E_{i0} - E_{0, -i} & \text{for } 0 < i \\ -2E_{-i0} + E_{0i} \end{cases}$$

Each of the 1-dimensional spaces $\mathbb{C}_{e_\alpha}$ is an H-module and we have

$$[h, E_{ij} - E_{-j, -i}] = (\lambda_i - \lambda_j)(E_{ij} - E_{-j, -i})$$
$$[h, -E_{-i, -j} + E_{ji}] = (\lambda_j - \lambda_i)(-E_{-i, -j} + E_{ji})$$
$$[h, E_{i, -j} - E_{j, -i}] = (\lambda_i + \lambda_j)(E_{i, -j} - E_{j, -i})$$
$$[h, -E_{-i, j} + E_{-j, i}] = (-\lambda_i - \lambda_j)(-E_{-i, j} + E_{-j, i})$$
$$[h, 2E_{i0} - E_{0, -i}] = \lambda_i (2E_{i0} - E_{0, -i})$$
$$[h, -2E_{-i, 0} + E_{0i}] = -\lambda_i (-2E_{-i, 0} + E_{0i}).$$

We write $[he_\alpha] = \alpha(h) e_\alpha$ for all such α.

We now show that H is a Cartan subalgebra of L; $L = H \oplus \sum_\alpha \mathbb{C} e_\alpha$ is the Cartan decomposition of L with respect to H, and L is semisimple. These facts can be proved in exactly the same way as that used in Section 8.2 for type D_l.

We now know that the functions $\alpha : H \to \mathbb{C}$ given above are the roots of L with respect to H. A system of fundamental roots is given by

$$\alpha_1(h) = \lambda_1 - \lambda_2$$
$$\alpha_2(h) = \lambda_2 - \lambda_3$$
$$\vdots$$
$$\alpha_{l-1}(h) = \lambda_{l-1} - \lambda_l$$
$$\alpha_l(h) = \lambda_l$$

since all the other roots are integral combinations of these with coefficients all non-negative or all non-positive.

We can now determine the Cartan integers A_{ij}. Let $\alpha_j, \alpha_i + \alpha_j, \ldots, q\alpha_i + \alpha_j$ be the α_i-chain of roots through α_j for $i \neq j$. By Proposition 4.22 we have $A_{ij} = -q$ and so the Cartan matrix $A = (A_{ij})$ can be determined. This turns out to be the Cartan matrix of type B_l on the standard list 6.12. Finally we observe that L must be a simple Lie algebra by Corollary 6.15, since its Cartan matrix is indecomposable. Thus we have

Theorem 8.4 (i) *The simple Lie algebra of type B_l has dimension $l(2l+1)$.*
(ii) *The Lie algebra of all $(2l+1) \times (2l+1)$ matrices X satisfying $X^t M + MX = O$ where*

$$M = \begin{pmatrix} 2 & 0 & \cdots & 0 \\ 0 & O & & I_l \\ \vdots & & & \\ 0 & I_l & & O \end{pmatrix}$$

is simple of type B_l when $l \geq 2$.

8.4 Lie algebras of type C_l

We recall that the Dynkin diagram of type C_l has form

$$\underset{1}{\circ} - \underset{2}{\circ} - \cdots - \underset{l-1}{\circ} \Leftarrow \underset{l}{\circ}$$

Let V be a real vector space of dimension l with basis $\beta_1, \ldots, \beta_l$. Let the scalar product $\{,\}$ on V be defined by $\{\beta_i, \beta_j\} = \delta_{ij}$. We define $\alpha_1, \ldots, \alpha_l \in V$ by

$$\alpha_1 = \beta_1 - \beta_2, \quad \alpha_2 = \beta_2 - \beta_3, \quad \ldots, \quad \alpha_{l-1} = \beta_{l-1} - \beta_l, \quad \alpha_l = 2\beta_l.$$

Then we have

$$\{\alpha_i, \alpha_i\} = 2 \qquad \text{for } 1 \leq i \leq l-1$$
$$\{\alpha_l, \alpha_l\} = 4$$
$$\{\alpha_i, \alpha_{i+1}\} = -1 \qquad \text{for } 1 \leq i \leq l-2$$
$$\{\alpha_{l-1}, \alpha_l\} = -2$$
$$\{\alpha_i, \alpha_j\} = 0 \qquad \text{for } |i-j| > 1.$$

It follows that

$$2 \frac{\{\alpha_i, \alpha_j\}}{\{\alpha_i, \alpha_i\}} = A_{ij} \qquad \text{for all } i, j$$

where $A = (A_{ij})$ is the Cartan matrix of type C_l on the standard list 6.12. Thus the scalar product $\{,\}$ is a non-zero multiple of the Killing form.

8.4 Lie algebras of type C_l

The fundamental reflections $s_1, \ldots, s_l$ act on $\beta_1, \ldots, \beta_l$ in exactly the same manner as in type B_l, considered in Section 8.3. Thus we have

$$|W| = 2^l l!$$

as in Section 8.3 and each $w \in W$ acts on the β_i by means of a permutation combined with sign changes. Both the permutation and the sign changes can be chosen arbitrarily. Thus we obtain the root system Φ as the set of all vectors of form $w(\alpha_i)$ for all $w \in W$ and all i. Thus

$$\Phi = \{\pm \beta_i \pm \beta_j \ \ i \neq j \ ; \pm 2\beta_i\}.$$

All combinations of signs are possible. Thus we have $|\Phi| = 2l^2$ and

$$\dim L = l + |\Phi| = l(2l+1).$$

We next describe L as a Lie algebra of matrices. Again we use Lemma 8.2. This time we take the $2l \times 2l$ matrix M given by

$$M = \begin{pmatrix} O & I_l \\ -I_l & O \end{pmatrix}.$$

Let L be the Lie algebra of all $2l \times 2l$ matrices satisfying the condition

$$X^t M + MX = O.$$

Let $X = \begin{pmatrix} X_{11} & X_{12} \\ X_{21} & X_{22} \end{pmatrix}$. Then X lies in L if and only if $X_{22} = -X_{11}^t$ and X_{12}, X_{21} are symmetric.

Let H be the set of diagonal matrices in L. The elements of H have form

$$h = \begin{pmatrix} \lambda_1 & & & & & & \\ & \ddots & & & & & \\ & & \lambda_l & & & & \\ & & & -\lambda_1 & & & \\ & & & & \ddots & \\ & & & & & -\lambda_l \end{pmatrix}.$$

We number the rows and columns $1, \ldots, l, -1, \ldots, -l$. Then we have

$$L = H \oplus \sum_\alpha \mathbb{C} e_\alpha$$

where

$$e_\alpha = \begin{cases} E_{ij} - E_{-j,\,-i} \\ -E_{-i,\,-j} + E_{ji} \\ E_{i,\,-j} + E_{j,\,-i} \\ E_{-i,\,j} + E_{-j,\,i} \\ E_{i,\,-i} \\ E_{-i,\,i} \end{cases} \begin{array}{l} 0 < i < j \\ \\ \\ \\ 0 < i. \end{array}$$

Each of the 1-dimensional spaces $\mathbb{C}e_\alpha$ is a H-module. We have

$$[h, E_{ij} - E_{-j,\,-i}] = (\lambda_i - \lambda_j)(E_{ij} - E_{-j,\,-i})$$
$$[h, -E_{-i,\,-j} + E_{ji}] = (\lambda_j - \lambda_i)(-E_{-i,\,-j} + E_{ji})$$
$$[h, E_{i,\,-j} + E_{j,\,-i}] = (\lambda_i + \lambda_j)(E_{i,\,-j} + E_{j,\,-i})$$
$$[h, E_{-i,\,j} + E_{-j,\,i}] = (-\lambda_i - \lambda_j)(E_{-i,\,j} + E_{-j,\,i})$$
$$[h, E_{i,\,-i}] = 2\lambda_i E_{i,\,-i}$$
$$[h, E_{-i,\,i}] = -2\lambda_i E_{-i,\,i}.$$

We write $[he_\alpha] = \alpha(h)e_\alpha$ for all such α.

We observe that H is a Cartan subalgebra of L, that $L = H \oplus \sum_\alpha \mathbb{C}e_\alpha$ is the Cartan decomposition of L with respect to H, and that the Lie algebra L is semisimple, using the same arguments as given in Section 8.2 for type D_l.

The functions $\alpha : H \to \mathbb{C}$ given above are the roots of L with respect to H. A system of fundamental roots is given by

$$\alpha_1(h) = \lambda_1 - \lambda_2$$
$$\alpha_2(h) = \lambda_2 - \lambda_3$$
$$\vdots$$
$$\alpha_{l-1}(h) = \lambda_{l-1} - \lambda_l$$
$$\alpha_l(h) = 2\lambda_l$$

since all the other roots are integral combinations of these with coefficients all non-negative or all non-positive.

We can now determine the Cartan integers A_{ij}. Let $\alpha_j, \alpha_i + \alpha_j, \ldots, q\alpha_i + \alpha_j$ be the α_i-chain of roots through α_j, for $i \neq j$. Then $A_{ij} = -q$. The Cartan matrix $A = (A_{ij})$ determined in this way turns out to be the Cartan matrix of type C_l on the standard list 6.12. Finally L is a simple Lie algebra, since its Cartan matrix is indecomposable. Thus we have

Theorem 8.5 (i) *The simple Lie algebra of type C_l has dimension $l(2l+1)$.*
(ii) *The Lie algebra of all $2l \times 2l$ matrices X satisfying $X^t M + MX = O$ where*

$$M = \begin{pmatrix} O & I_l \\ -I_l & O \end{pmatrix}$$

is simple of type C_l when $l \geq 3$. □

The Lie algebras of type A_l, B_l, C_l or D_l are called the **simple Lie algebras of classical type**. The remaining simple Lie algebras E_6, E_7, E_8, F_4, G_2 are called the **exceptional simple Lie algebras**. We now determine the dimensions and root systems of the exceptional Lie algebras.

8.5 Lie algebras of type G_2

The Dynkin diagram of type G_2 is

$$\underset{1}{\circ}\!\!\Rrightarrow\!\!\underset{2}{\circ}$$

and the corresponding Cartan matrix is

$$\begin{pmatrix} 2 & -1 \\ -3 & 2 \end{pmatrix}.$$

Let α_1, α_2 be the fundamental roots in a root system of type G_2. Then we have

$$s_1(\alpha_1) = -\alpha_1 \qquad s_2(\alpha_1) = \alpha_1 + 3\alpha_2$$
$$s_1(\alpha_2) = \alpha_1 + \alpha_2 \qquad s_2(\alpha_2) = -\alpha_2$$

and $W = \langle s_1, s_2 \rangle$. Thus each root in Φ is obtained from α_1 or α_2 by applying s_1, s_2 alternately. Now we have

$$\alpha_1 \underset{s_1}{\to} -\alpha_1 \underset{s_2}{\to} -\alpha_1 - 3\alpha_2 \underset{s_1}{\to} -2\alpha_1 - 3\alpha_2$$
$$\alpha_1 \underset{s_2}{\to} \alpha_1 + 3\alpha_2 \underset{s_1}{\to} 2\alpha_1 + 3\alpha_2$$
$$\alpha_2 \underset{s_1}{\to} \alpha_1 + \alpha_2 \underset{s_2}{\to} \alpha_1 + 2\alpha_2$$
$$\alpha_2 \underset{s_2}{\to} -\alpha_2 \underset{s_1}{\to} -\alpha_1 - \alpha_2 \underset{s_2}{\to} -\alpha_1 - 2\alpha_2$$

136   *The simple Lie algebras*

and

$$s_2(-2\alpha_1 - 3\alpha_2) = -2\alpha_1 - 3\alpha_2$$
$$s_2(2\alpha_1 + 3\alpha_2) = 2\alpha_1 + 3\alpha_2$$
$$s_1(\alpha_1 + 2\alpha_2) = \alpha_1 + 2\alpha_2$$
$$s_1(-\alpha_1 - 2\alpha_2) = -\alpha_1 - 2\alpha_2.$$

Thus all the vectors in the above sequences are roots, and we do not obtain new vectors by continuing the sequences further. Hence

$$\Phi = \{\alpha_1, \quad \alpha_2, \quad \alpha_1 + \alpha_2, \quad \alpha_1 + 2\alpha_2, \quad \alpha_1 + 3\alpha_2, \quad 2\alpha_1 + 3\alpha_2, \quad -\alpha_1,$$
$$-\alpha_2, -\alpha_1 - \alpha_2, \quad -\alpha_1 - 2\alpha_2, \quad -\alpha_1 - 3\alpha_2, \quad -2\alpha_1 - 3\alpha_2\}.$$

Thus we have $|\Phi| = 12$ and $\dim L = 14$. Hence we have proved

Theorem 8.6 *The simple Lie algebra of type G_2 has dimension* 14.

Figures 8.1, 8.2 and 8.3 compare the simple root systems of types A_2, B_2 and G_2.

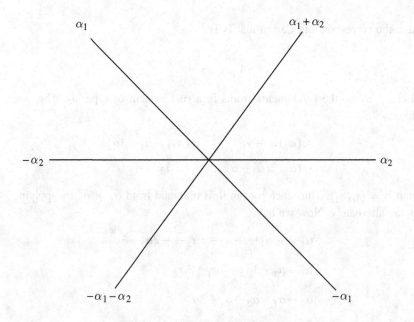

Figure 8.1   Simple root system of type A_2

8.5 Lie algebras of type G_2

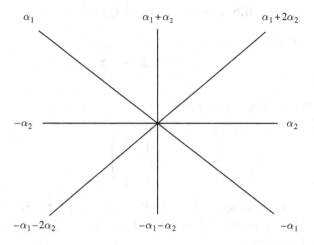

Figure 8.2   Simple root system of type B_2

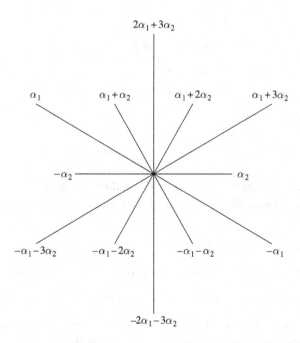

Figure 8.3   Simple root system of type G_2

8.6 Lie algebras of type F_4

The Dynkin diagram of type F_4 is

$$\underset{1}{\circ}\!\!-\!\!-\!\!-\!\!\underset{2}{\circ}\!\!\Rightarrow\!\!\underset{3}{\circ}\!\!-\!\!-\!\!-\!\!\underset{4}{\circ}$$

and the corresponding Cartan matrix is

$$\begin{pmatrix} 2 & -1 & 0 & 0 \\ -1 & 2 & -1 & 0 \\ 0 & -2 & 2 & -1 \\ 0 & 0 & -1 & 2 \end{pmatrix}$$

Let V be a real vector space with dim $V = 4$ and $\beta_1, \beta_2, \beta_3, \beta_4$ be a basis of V. Let the scalar product $\{,\}$ on V be defined by $\{\beta_i, \beta_j\} = \delta_{ij}$. We define $\alpha_1, \alpha_2, \alpha_3, \alpha_4 \in V$ by

$$\alpha_1 = \beta_1 - \beta_2 \quad \alpha_2 = \beta_2 - \beta_3 \quad \alpha_3 = \beta_3 \quad \alpha_4 = \tfrac{1}{2}(-\beta_1 - \beta_2 - \beta_3 + \beta_4).$$

Then we have

$$\{\alpha_1, \alpha_1\} = \{\alpha_2, \alpha_2\} = 2$$
$$\{\alpha_3, \alpha_3\} = \{\alpha_4, \alpha_4\} = 1$$
$$\{\alpha_1, \alpha_2\} = \{\alpha_2, \alpha_3\} = -1$$
$$\{\alpha_3, \alpha_4\} = -\tfrac{1}{2}$$
$$\{\alpha_i, \alpha_j\} = 0 \quad \text{if } |i-j| > 1.$$

It follows that

$$2\frac{\{\alpha_i, \alpha_j\}}{\{\alpha_i, \alpha_i\}} = A_{ij} \quad \text{for all } i, j.$$

Thus the scalar product $\{,\}$ is a non-zero multiple of the Killing form. We consider the action of the corresponding fundamental reflections s_1, s_2, s_3, s_4. We have

$$s_1(\beta_1) = \beta_2, \quad s_1(\beta_2) = \beta_1, \quad s_1(\beta_3) = \beta_3, \quad s_1(\beta_4) = \beta_4$$
$$s_2(\beta_1) = \beta_1, \quad s_2(\beta_2) = \beta_3, \quad s_2(\beta_3) = \beta_2, \quad s_2(\beta_4) = \beta_4$$
$$s_3(\beta_1) = \beta_1, \quad s_3(\beta_2) = \beta_2, \quad s_3(\beta_3) = -\beta_3, \quad s_3(\beta_4) = \beta_4.$$

We consider the subgroup $\langle s_1, s_2, s_3 \rangle$ of the Weyl group W generated by s_1, s_2, s_3. Elements in this subgroup all fix β_4 but act on $\beta_1, \beta_2, \beta_3$ by means of a permutation combined with sign changes. Thus $w(\beta_i) = \varepsilon_i \beta_{\sigma(i)}$ for $i = 1, 2, 3$.

8.6 Lie algebra of type F_4

Moreover each permutation σ and each choice of signs ε_i arise in this way. Applying the elements of this subgroup of W to $\alpha_1, \alpha_2, \alpha_3, \alpha_4$ we see that the vectors

$$\pm \beta_i \quad 1 \le i \le 3$$
$$\pm \beta_i \pm \beta_j \quad i \ne j \quad 1 \le i, j \le 3$$
$$\tfrac{1}{2}(\pm \beta_1 \pm \beta_2 \pm \beta_3 \pm \beta_4)$$

all lie in Φ. We next consider the action of s_4. We have

$$s_4(\beta_1) = \tfrac{1}{2}(\beta_1 - \beta_2 - \beta_3 + \beta_4)$$
$$s_4(\beta_2) = \tfrac{1}{2}(-\beta_1 + \beta_2 - \beta_3 + \beta_4)$$
$$s_4(\beta_3) = \tfrac{1}{2}(-\beta_1 - \beta_2 + \beta_3 + \beta_4)$$
$$s_4(\beta_4) = \tfrac{1}{2}(\beta_1 + \beta_2 + \beta_3 + \beta_4).$$

Since $s_4^2 = 1$ we have $s_4\left(\tfrac{1}{2}(\beta_1 + \beta_2 + \beta_3 + \beta_4)\right) = \beta_4$. Hence $\beta_4 \in \Phi$. We also have $s_4(\beta_1 + \beta_2) = -\beta_3 + \beta_4$. Hence $-\beta_3 + \beta_4 \in \Phi$. Thus, applying further elements of the subgroup $\langle s_1, s_2, s_3 \rangle$ we see that the vectors

$$\pm \beta_i \quad 1 \le i \le 4$$
$$\pm \beta_i \pm \beta_j \quad i \ne j \quad 1 \le i, j \le 4$$
$$\tfrac{1}{2}(\pm \beta_1 \pm \beta_2 \pm \beta_3 \pm \beta_4)$$

all lie in Φ, where the choice of signs is arbitrary.

We show this set of vectors is the whole of Φ. To do so it is sufficient to show that the set is invariant under s_1, s_2, s_3, s_4. The set is clearly invariant under s_1, s_2, s_3 because of the simple action of these reflections on $\beta_1, \beta_2, \beta_3, \beta_4$ described above. Thus it is sufficient to show the set is invariant under s_4. Now the action of s_4 given above shows that

$$s_4(\pm \beta_i) = \tfrac{1}{2}(\varepsilon_1 \beta_1 + \varepsilon_2 \beta_2 + \varepsilon_3 \beta_3 + \varepsilon_4 \beta_4)$$

where $\varepsilon_i \in \{1, -1\}$ and $\prod \varepsilon_i = 1$. Thus s_4 transforms vectors $\pm \beta_i$ into the given set, giving as images vectors $\tfrac{1}{2} \sum \varepsilon_i \beta_i$ with $\prod \varepsilon_i = 1$. Since there are eight such vectors they all appear as vectors $s_4(\pm \beta_i)$. Since $s_4^2 = 1$ we deduce that all vectors $\tfrac{1}{2} \sum \varepsilon_i \beta_i$ with $\prod \varepsilon_i = 1$ are transformed by s_4 into the given set.

The formulae for $s_4(\beta_i)$ also show that, for all $i \ne j$, $s_4(\pm\beta_i \pm \beta_j)$ has form $\pm\beta_k \pm \beta_l$ for certain $k \ne l$. Thus s_4 transforms vectors $\pm\beta_i \pm \beta_j$, $i \ne j$, into the given set.

It remains to show that s_4 transforms all vectors $\frac{1}{2}\sum \varepsilon_i \beta_i$ with $\prod \varepsilon_i = -1$ into the given set. We may clearly assume $\varepsilon_4 = 1$. There are four such vectors. One of them is α_4 and we have $s_4(\alpha_4) = -\alpha_4$. The other three are all orthogonal to α_4 and so are transformed into themselves by s_4.

Thus the given set of vectors is invariant under s_1, s_2, s_3, s_4 so is the whole of Φ. Thus we have

$$\Phi = \{\pm\beta_i \quad 1 \le i \le 4$$
$$\pm\beta_i \pm \beta_j \quad i \ne j \quad 1 \le i, j \le 4$$
$$\tfrac{1}{2}(\pm\beta_1 \pm \beta_2 \pm \beta_3 \pm \beta_4)\}.$$

In particular we have $|\Phi| = 48$, hence $\dim L = 52$. Thus we have

Theorem 8.7 *The simple Lie algebra of type F_4 has dimension 52.*

We observe that the roots of F_4 are of two different lengths. There are 24 short roots and 24 long roots. The short roots are $\pm\beta_i$ and $\frac{1}{2}(\pm\beta_1 \pm \beta_2 \pm \beta_3 \pm \beta_4)$. The long roots are $\pm\beta_i \pm \beta_j$.

8.7 Lie algebras of types E_6, E_7, E_8

We now consider the simple Lie algebra of type E_8. Its Dynkin diagram is

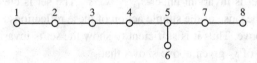

Let V be a real vector space with $\dim V = 8$ and with basis β_i $i = 1, \ldots, 8$. Let the scalar product $\{,\}$ on V be defined by $\{\beta_i, \beta_j\} = \delta_{ij}$. We wish to find a fundamental system of roots of type E_8 in V. We note that if the vertex 8 is removed from the Dynkin diagram we obtain a Dynkin diagram of type D_7. This indicates how the first seven vectors in the fundamental system should be chosen. The last one is chosen to be linearly independent of the others and

8.7 Lie algebras of types E_6, E_7, E_8

to satisfy the appropriate conditions relating to the scalar product. Thus we define $\alpha_1, \ldots, \alpha_8 \in V$ by:

$$\alpha_i = \beta_i - \beta_{i+1} \quad 1 \le i \le 6$$
$$\alpha_7 = \beta_6 + \beta_7$$
$$\alpha_8 = -\tfrac{1}{2}\sum_{i=1}^{8}\beta_i.$$

Then we have

$$\{\alpha_i, \alpha_i\} = 2 \quad \text{for } 1 \le i \le 8$$
$$\{\alpha_i, \alpha_{i+1}\} = -1 \quad \text{for } 1 \le i \le 5$$
$$\{\alpha_5, \alpha_7\} = -1$$
$$\{\alpha_7, \alpha_8\} = -1$$
$$\{\alpha_i, \alpha_j\} = 0 \quad \text{for all other pairs } i, j.$$

It follows that

$$2\frac{\{\alpha_i, \alpha_j\}}{\{\alpha_i, \alpha_i\}} = A_{ij}$$

where $A = (A_{ij})$ is the Cartan matrix of type E_8 on the standard list.

In order to obtain the remaining roots we consider the action of the fundamental reflections $s_1, \ldots, s_8$. We have

$$s_i(\beta_i) = \beta_{i+1}$$
$$s_i(\beta_{i+1}) = \beta_i$$
$$s_i(\beta_j) = \beta_j \quad \text{for } j \ne i, i+1$$

when $1 \le i \le 6$. Thus the subgroup of the Weyl group W generated by $s_1, \ldots, s_6$ will give all permutations of $\beta_1, \ldots, \beta_7$ and will fix β_8. The fundamental reflection s_7 acts by:

$$s_7(\beta_6) = -\beta_7$$
$$s_7(\beta_7) = -\beta_6$$
$$s_7(\beta_i) = \beta_i \quad i \ne 6, 7.$$

Thus the subgroup of W generated by $s_1, \ldots, s_7$ will act on $\beta_1, \ldots, \beta_7$ by permutations and sign changes, and will fix β_8. Moreover the number of sign changes will be even, and any permutation of $\beta_1, \ldots, \beta_7$ combined with any even number of sign changes will arise in this way.

It is then clear that the vectors

$$\pm\beta_i \pm \beta_j \quad 1 \le i, j \le 7 \quad i \ne j$$

$$\tfrac{1}{2}\left(\sum_{i=1}^{8} \varepsilon_i \beta_i\right) \quad \varepsilon_i = \pm 1, \quad \prod \varepsilon_i = 1$$

are all in the root system Φ. We also have

$$s_8(\beta_i) = \beta_i - 2\frac{\{\alpha_8, \beta_i\}}{\{\alpha_8, \alpha_8\}}\alpha_8 = \beta_i + \tfrac{1}{2}\alpha_8$$

for $1 \le i \le 8$. Thus

$$s_8(\beta_7 + \beta_8) = \tfrac{1}{2}(-\beta_1 - \beta_2 - \beta_3 - \beta_4 - \beta_5 - \beta_6 + \beta_7 + \beta_8) \in \Phi.$$

Since $s_8^2 = 1$ it follows that $\beta_7 + \beta_8 \in \Phi$. We then see that $\pm\beta_i \pm \beta_8 \in \Phi$ for all i with $1 \le i \le 7$. Thus the set of vectors

$$\pm\beta_i \pm \beta_j \quad 1 \le i, j \le 8 \quad i \ne j$$

$$\tfrac{1}{2}\left(\sum_{i=1}^{8} \varepsilon_i \beta_i\right) \quad \varepsilon_i = \pm 1, \quad \Pi \varepsilon_i = 1$$

lies in Φ. We shall show this is the full root system Φ. In order to do so we must verify that this set is invariant under $s_1, \ldots, s_8$. It is clearly invariant under $s_1, \ldots, s_7$ since these fix β_8 and act by permutations together with an even number of sign changes on $\beta_1, \ldots, \beta_7$. Thus it is sufficient to verify that this set is invariant under s_8. Now we have

$$s_8(\beta_i - \beta_j) = \beta_i - \beta_j \quad \text{for all } i \ne j.$$

Thus the set of vectors of form $\beta_i - \beta_j$, $i \ne j$, is invariant under s_8. Also

$$s_8(\beta_i + \beta_j) = \beta_i + \beta_j + \alpha_8 \quad \text{for all } i \ne j.$$

Thus s_8 transforms vectors of form $\beta_i + \beta_j$, $i \ne j$, into vectors $\tfrac{1}{2}(\sum \varepsilon_i \beta_i)$ with two ε_i equal to 1 and six equal to -1. Moreover all vectors $\tfrac{1}{2}(\sum \varepsilon_i \beta_i)$ with this property arise in this way. Similarly such vectors with six ε_i equal to 1 and two equal to -1 have the form $s_8(-\beta_i - \beta_j)$. Thus vectors of form $\beta_i + \beta_j$ or $-\beta_i - \beta_j$ with $i \ne j$ are transformed by s_8 into the given set, and so are vectors $\tfrac{1}{2}\sum \varepsilon_i \beta_i$ of type (2, 6) or (6, 2). The vectors of this form of type (0, 8) or (8, 0) are α_8 and $-\alpha_8$, which are transformed into one another by s_8. It remains to show that s_8 transforms vectors $\tfrac{1}{2}\sum \varepsilon_i \beta_i$ of type (4, 4) into the given set. However, since such vectors have four positive signs and four negative signs they are orthogonal to α_8, hence s_8 transforms each such

8.7 Lie algebras of types E_6, E_7, E_8

vector into itself. Thus the given set of vectors is invariant under $s_1, \ldots, s_8$ so is the full root system Φ.

There are $4 \cdot \binom{8}{2} = 112$ vectors of form $\pm \beta_i \pm \beta_j$ with $i \neq j$ and $2^7 = 128$ vectors of form $\frac{1}{2} \sum \varepsilon_i \beta_i$ with $\varepsilon_i = \pm 1$ and $\prod \varepsilon_i = 1$. Thus the total number of roots is

$$|\Phi| = 112 + 128 = 240.$$

Finally we have $\dim L = 8 + |\Phi| = 248$. Thus we have proved

Theorem 8.8 *The simple Lie algebra of type E_8 has dimension* 248.

We now turn to the simple Lie algebra of type E_7. Its Dynkin diagram is

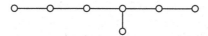

Thus the vectors $\alpha_2, \alpha_3, \alpha_4, \alpha_5, \alpha_6, \alpha_7, \alpha_8$ considered above form a fundamental root system of type E_7. In order to obtain the full root system we must transform these vectors repeatedly by $s_2, \ldots, s_8$ until no new vectors are obtained. Now the vectors $\alpha_2, \ldots, \alpha_8$ are all orthogonal to $\beta_1 - \beta_8$. Thus all their transforms by $s_2, \ldots, s_8$ will also be orthogonal to $\beta_1 - \beta_8$. These transforms are contained in the set of roots of E_8 obtained above.

Now the roots of E_8 orthogonal to $\beta_1 - \beta_8$ are:

$$\pm \beta_i \pm \beta_j, \quad 2 \leq i, j \leq 7, \quad i \neq j$$
$$\pm(\beta_1 + \beta_8)$$
$$\tfrac{1}{2} \sum \varepsilon_i \beta_i, \quad \varepsilon_i = \pm 1, \quad \prod \varepsilon_i = 1, \quad \varepsilon_1 = \varepsilon_8.$$

Thus the required root system of E_7 is contained in this set. We shall show it is the whole of this set.

We first consider the action of the subgroup of the Weyl group of E_7 generated by $s_2, s_3, s_4, s_5, s_6, s_7$. Elements of this subgroup fix β_1 and β_8 and act on $\beta_2, \beta_3, \beta_4, \beta_5, \beta_6, \beta_7$ by permutations combined with sign changes with an even number of negative signs. By applying elements of this subgroup to $\alpha_2, \ldots, \alpha_8$ we see that the vectors

$$\pm \beta_i \pm \beta_j, \quad 2 \leq i, j \leq 7, \quad i \neq j$$
$$\tfrac{1}{2} \sum \varepsilon_i \beta_i, \quad \varepsilon_i = \pm 1, \quad \prod \varepsilon_i = 1, \quad \varepsilon_1 = \varepsilon_8$$

are all roots of E_7. It remains to show that $\pm(\beta_1+\beta_8)$ are also roots of E_7. However,

$$s_8(\beta_1+\beta_8) = \beta_1+\beta_8+\alpha_8 = \tfrac{1}{2}(\beta_1-\beta_2-\beta_3-\beta_4-\beta_5-\beta_6-\beta_7+\beta_8)$$

is a root of E_7, thus so is $\beta_1+\beta_8$ and $-\beta_1-\beta_8$.

There are $4\binom{6}{2}=60$ roots of form $\pm\beta_i\pm\beta_j$, $2\le i,j\le 7$, $i\ne j$ and $2^6=64$ roots of form $\tfrac{1}{2}\sum\varepsilon_i\beta_i$ with $\varepsilon_i=\pm 1$, $\prod\varepsilon_i=1$ and $\varepsilon_1=\varepsilon_8$. Thus the number of roots of E_7 is given by

$$|\Phi|=60+2+64=126.$$

Also we have

$$\dim L = 7+|\Phi| = 133.$$

Thus we have shown:

Theorem 8.9 *The simple Lie algebra of type E_7 has dimension* 133.

Finally we consider the simple Lie algebra of type E_6. Its Dynkin diagram is

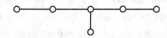

Thus the vectors $\alpha_3, \alpha_4, \alpha_5, \alpha_6, \alpha_7, \alpha_8$ considered above form a fundamental root system of type E_6. In order to obtain the full root system of E_6 we must transform these vectors successively by the fundamental reflections $s_3, s_4, s_5, s_6, s_7, s_8$.

Now the vectors $\alpha_3, \ldots, \alpha_8$ are all orthogonal to both $\beta_1-\beta_8$ and $\beta_2-\beta_8$. Thus the full root system of E_6 is orthogonal to $\beta_1-\beta_8$ and $\beta_2-\beta_8$.

Now the roots of E_8 orthogonal to both $\beta_1-\beta_8$ and $\beta_2-\beta_8$ are:

$$\pm\beta_i\pm\beta_j \quad 3\le i,j\le 7, \quad i\ne j$$

$$\tfrac{1}{2}\left(\sum_{i=1}^{8}\varepsilon_i\beta_i\right) \quad \varepsilon_i=\pm 1, \quad \prod\varepsilon_i=1, \quad \varepsilon_1=\varepsilon_2=\varepsilon_8.$$

Thus the required root system of E_6 is contained in this set. We shall show it is equal to this set of vectors.

Consider the action of the subgroup of the Weyl group of type E_6 generated by s_3, s_4, s_5, s_6, s_7. Elements of this subgroup fix $\beta_1, \beta_2, \beta_8$ and act on $\beta_3, \beta_4, \beta_5, \beta_6, \beta_7$ by permutations combined with sign changes with an even number of negative signs. By applying elements of this subgroup to $\alpha_3, \alpha_4, \alpha_5, \alpha_6, \alpha_7, \alpha_8$ we can obtain all vectors of form $\pm\beta_i\pm\beta_j$ with

8.8 Properties of long and short roots

Table 8.11 *The simple Lie algebras*

L		dim H	$\|\Phi\|$	dim L
A_l	$l \geq 1$	l	$l(l+1)$	$l(l+2)$
B_l	$l \geq 2$	l	$2l^2$	$l(2l+1)$
C_l	$l \geq 3$	l	$2l^2$	$l(2l+1)$
D_l	$l \geq 4$	l	$2l(l-1)$	$l(2l-1)$
E_6		6	72	78
E_7		7	126	133
E_8		8	240	248
F_4		4	48	52
G_2		2	12	14

$3 \leq i, j \leq 7$, $i \neq j$, and (up to sign) all vectors of form $\frac{1}{2} \sum \varepsilon_i e_i$ with $\varepsilon_i = \pm 1$, $\prod \varepsilon_i = 1$, $\varepsilon_1 = \varepsilon_2 = \varepsilon_8$. Hence the vectors in the above set are all roots of E_6. There are $\binom{5}{2} \cdot 4 = 40$ vectors of type $\pm \beta_i \pm \beta_j$ with $3 \leq i, j \leq 7, i \neq j$, and $2^5 = 32$ vectors of type $\frac{1}{2} \sum \varepsilon_i e_i$ with $\varepsilon_i = \pm 1$, $\prod \varepsilon_i = 1$ and $\varepsilon_1 = \varepsilon_i = \varepsilon_8$. Thus the total number of roots is

$$|\Phi| = 40 + 32 = 72$$

and we have

$$\dim L = 6 + |\Phi| = 78.$$

Thus:

Theorem 8.10 *The simple Lie algebra of type E_6 has dimension* 78.

We have now determined the dimensions of all the simple Lie algebras. We summarise the information we have obtained in Table 8.11. In this table L is a simple Lie algebra, H is a Cartan subalgebra and Φ the system of roots of L with respect to H.

8.8 Properties of long and short roots

Proposition 8.12 *In the simple Lie algebras of types A_l, D_l, E_6, E_7, E_8 all the roots have the same length. In the Lie algebras of types B_l, C_l, F_4, G_2 there are two possible lengths of roots. These are called the long roots and short roots.*

Proof. This is clear from the preceding results. □

Proposition 8.13 (i) *Let Φ be a root system of type B_l with fundamental system*

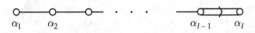

Then the long roots form a subsystem of type D_l with fundamental system

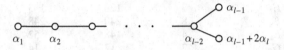

and the short roots form a subsystem of type $(A_1)^l$ with fundamental system

$$
\underset{\alpha_1+\cdots+\alpha_l}{\circ} \quad \underset{\alpha_2+\cdots+\alpha_l}{\circ} \quad \cdots \quad \underset{\alpha_{l-1}+\alpha_l}{\circ} \quad \underset{\alpha_l}{\circ}
$$

(ii) *Let Φ be a root system of type C_l with fundamental system*

Then the long roots form a subsystem of type $(A_1)^l$ with fundamental system

$$
\underset{2\alpha_1+\cdots+2\alpha_{l-1}+\alpha_l}{\circ} \quad \underset{2\alpha_2+\cdots+2\alpha_{l-1}+\alpha_l}{\circ} \quad \cdots \quad \underset{2\alpha_{l-1}+\alpha_l}{\circ} \quad \underset{\alpha_l}{\circ}
$$

and the short roots form a subsystem of type D_l with fundamental system

(iii) *Let Φ be a root system of type F_4 with fundamental system*

$$\underset{\alpha_1}{\circ}\!-\!\underset{\alpha_2}{\circ}\!\Rightarrow\!\underset{\alpha_3}{\circ}\!-\!\underset{\alpha_4}{\circ}$$

Then the long roots form a subsystem of type D_4 with fundamental system

$$
\begin{array}{c}
\alpha_2+2\alpha_3+2\alpha_4 \;\circ \\
 \diagdown \\
\circ\!\!-\!\!\!\!\begin{array}{c}\circ\\ \alpha_1\end{array}\!\!\!\!-\!\!\circ\;\alpha_2 \\
\diagup \\
\alpha_2+2\alpha_3\;\circ
\end{array}
$$

8.8 Properties of long and short roots

and the short roots form a subsystem of type D_4 with fundamental system

$$\alpha_1+\alpha_2+\alpha_3 \quad \alpha_2+\alpha_3 \quad \alpha_4 \quad \alpha_3$$

(iv) Let Φ be a root system of type G_2 with fundamental system

$$\alpha_1 \Rrightarrow \alpha_2$$

Then the long roots form a subsystem of type A_2 with fundamental system

$$\alpha_1 \quad\quad \alpha_1+3\alpha_2$$

and the short roots form a subsystem of type A_2 with fundamental system

$$\alpha_1+\alpha_2 \quad\quad \alpha_2$$

Proof. (i) We saw in Section 8.3 that the roots of type B_l have form $\pm\beta_i\pm\beta_j, i\neq j$, and $\pm\beta_i$. The former are the long roots and the latter the short roots. The long roots form a system of type D_l with fundamental system $\beta_1-\beta_2, \beta_2-\beta_3, \ldots, \beta_{l-1}-\beta_l, \beta_{l-1}+\beta_l$. These are $\alpha_1, \alpha_2, \ldots, \alpha_{l-1}, \alpha_{l-1}+2\alpha_l$ respectively. The short roots form a system of type $(A_1)^l$ with fundamental system $\beta_1, \ldots, \beta_l$. These are $\alpha_1+\cdots+\alpha_l, \alpha_2+\cdots+\alpha_l, \ldots, \alpha_l$ respectively.

(ii) We saw in Section 8.4 that the roots of type C_l have form $\pm\beta_i\pm\beta_j, i\neq j$, and $\pm 2\beta_i$. The former are the short roots and the latter the long roots. Thus the short roots form a subsystem of type D_l and the long roots a subsystem of type $(A_1)^l$.

(iii) We saw in Section 8.6 that the roots of type F_4 have form

$$\pm\beta_i\pm\beta_j \quad i\neq j$$
$$\pm\beta_i$$
$$\tfrac{1}{2}(\pm\beta_1\pm\beta_2\pm\beta_3\pm\beta_4).$$

Roots of the first type are long and those of the second and third types are short. The long roots form a subsystem of type D_4 with fundamental system $\beta_4-\beta_1, \beta_1-\beta_2, \beta_2-\beta_3, \beta_2+\beta_3$. These are $\alpha_2+2\alpha_3+2\alpha_4, \alpha_1, \alpha_2,$

$\alpha_2 + 2\alpha_3$ respectively. The short roots also form a subsystem of type D_4, with fundamental system $\beta_1, \beta_2, \beta_3, \frac{1}{2}(-\beta_1 - \beta_2 - \beta_3 + \beta_4)$. These are $\alpha_1 + \alpha_2 + \alpha_3, \alpha_2 + \alpha_3, \alpha_3, \alpha_4$ respectively.

(iv) The long and short roots of type G_2 are evident from Section 8.5. □

Let $\Pi = \{\alpha_i\}$ be a fundamental system of roots in a simple Lie algebra whose Dynkin diagram has a double or triple edge, and let Φ be the root system with fundamental system Π. Consider the simple Lie algebra whose Dynkin diagram is obtained from that above by reversing the direction of the arrow. Let $\Pi^v = \{\alpha_i^v\}$ be the corresponding fundamental system, labelled as before, and Φ^v be the root system with fundamental system Π^v.

System Φ^v is called the **dual root system** of Φ. The possible types of Φ, Φ^v are as shown.

Φ	Φ^v
B_l	C_l
C_l	B_l
F_4	F_4
G_2	G_2

We note that α_i is a short root in Π if and only if α_i^v is a long root in Π^v.

We suppose as usual that we have symmetric scalar products $\{,\}$ on $\mathbb{R}\Pi$ and $\mathbb{R}\Pi^v$ such that

$$2\frac{\{\alpha_i, \alpha_j\}}{\{\alpha_i, \alpha_i\}} = A_{ij}, \quad 2\frac{\{\alpha_i^v, \alpha_j^v\}}{\{\alpha_i^v, \alpha_i^v\}} = A_{ij}^v$$

for all i, j, where $A = (A_{ij})$, $A^v = (A_{ij}^v)$ are the Cartan matrices of Φ, Φ^v respectively.

We consider the free abelian groups $\mathbb{Z}\Pi, \mathbb{Z}\Pi^v$ generated by Π, Π^v. We define a homomorphism

$$\theta : \mathbb{Z}\Pi^v \to \mathbb{Z}\Pi$$

by

$$\theta(\alpha_i^v) = \begin{cases} p\alpha_i & \text{if } \alpha_i \text{ is a short root} \\ \alpha_i & \text{if } \alpha_i \text{ is a long root} \end{cases}$$

where p is the ratio of the squared lengths of the long and short roots. (Thus $p = 2$ in types B_l, C_l, F_4 and $p = 3$ in type G_2.)

Lemma 8.14 $2\dfrac{\{\theta(\alpha_i^\vee),\theta(\alpha_j^\vee)\}}{\{\theta(\alpha_i^\vee),\theta(\alpha_i^\vee)\}}=A_{ij}^\vee$ *for all i,j.*

Proof. We write $\theta(\alpha_i^\vee)=\xi_i\alpha_i$ where $\xi_i=p$ if α_i is short and $\xi_i=1$ if α_i is long. Then

$$2\frac{\{\theta(\alpha_i^\vee),\theta(\alpha_j^\vee)\}}{\{\theta(\alpha_i^\vee),\theta(\alpha_i^\vee)\}}=\xi_i^{-1}\xi_j 2\frac{\{\alpha_i,\alpha_j\}}{\{\alpha_i,\alpha_i\}}=\xi_i^{-1}\xi_j A_{ij}.$$

However, $\xi_i^{-1}\xi_j A_{ij}=A_{ij}^\vee$. This follows from the following observations.

If $A_{ij}\ne 0$ and α_i, α_j have the same length then we have $\xi_i=\xi_j$ and $A_{ij}=A_{ij}^\vee=-1$.

If $A_{ij}\ne 0$, α_i is long, α_j is short then $\xi_i=1$, $\xi_j=p$, $A_{ij}=-1$ and $A_{ij}^\vee=-p$.

If $A_{ij}\ne 0$, α_i is short, α_j is long then $\xi_i=p$, $\xi_j=1$, $A_{ij}=-p$, $A_{ij}^\vee=-1$.

Thus in all cases we have $\xi_i^{-1}\xi_j A_{ij}=A_{ij}^\vee$. $\square$

Lemma 8.15 *The diagram*

$$\begin{array}{ccc}\mathbb{Z}\Pi^\vee & \xrightarrow{\theta} & \mathbb{Z}\Pi \\ s_i^\vee \downarrow & & \downarrow s_i \\ \mathbb{Z}\Pi^\vee & \xrightarrow{\theta} & \mathbb{Z}\Pi\end{array}$$

commutes.

Proof. On the one hand $\theta s_i^\vee(\alpha_j^\vee)=\theta(\alpha_j^\vee-A_{ij}^\vee\alpha_i^\vee)=\theta(\alpha_j^\vee)-A_{ij}^\vee\theta(\alpha_i^\vee)$. On the other hand

$$s_i\theta(\alpha_j^\vee)=s_i(\xi_j\alpha_j)=\xi_j s_i(\alpha_j)=\xi_j(\alpha_j-A_{ij}\alpha_i)$$
$$=\xi_j\alpha_j-\xi_i^{-1}\xi_j A_{ij}(\xi_i\alpha_i)=\theta(\alpha_j^\vee)-A_{ij}^\vee\theta(\alpha_i^\vee).$$

Thus $\theta s_i^\vee=s_i\theta$. $\square$

Let $W, W^\vee$ be the Weyl groups of $\Phi, \Phi^\vee$. There is a natural isomorphism $W\cong W^\vee$ under which s_i corresponds to $s_i^\vee$, since the root lengths are irrelevant as far as the structure of the Weyl group is concerned. We shall use this isomorphism to identify $W^\vee$ with W. Then Lemma 8.15 shows that $\theta w=w\theta$ for all $w\in W$.

Proposition 8.16 *Given $\alpha^\vee\in\Phi^\vee$ there is a unique $\alpha\in\Phi$ such that*

$$\theta(\alpha^\vee)=\begin{cases}p\alpha & \text{if }\alpha\text{ is short}\\ \alpha & \text{if }\alpha\text{ is long.}\end{cases}$$

Proof. We have $\alpha^\vee = w(\alpha_i^\vee)$ for some $w \in W$, $\alpha_i^\vee \in \Pi^\vee$. Thus

$$\theta(\alpha^\vee) = \theta w(\alpha_i^\vee) = w\theta(\alpha_i^\vee) = \begin{cases} pw(\alpha_i) & \text{if } \alpha_i \text{ is short} \\ w(\alpha_i) & \text{if } \alpha_i \text{ is long.} \end{cases}$$

Let $\alpha = w(\alpha_i) \in \Phi$. Then α is uniquely determined since

$$w(\alpha_i^\vee) = w'(\alpha_j^\vee) \Rightarrow w'^{-1}w(\alpha_i^\vee) = \alpha_j^\vee \Rightarrow w'^{-1}w(\alpha_i) = \alpha_j \Rightarrow w(\alpha_i) = w'(\alpha_j).$$

Thus $\theta(\alpha^\vee) = \begin{cases} p\alpha & \text{if } \alpha \text{ is short} \\ \alpha & \text{if } \alpha \text{ is long.} \end{cases}$ $\square$

This proposition determines a bijection $\Phi^\vee \to \Phi$ under which $\alpha^\vee \to \alpha$. $\alpha^\vee$ is called the **dual root** of α. $\alpha^\vee$ is long if and only if α is short.

Proposition 8.17 *Let $\alpha \in \Phi$ satisfy $\alpha = \sum n_i \alpha_i$. Then α is a long root if and only if p divides n_i for all i for which α_i is a short root.*

Proof. Suppose α is long. Then $\theta(\alpha^\vee) = \alpha$ and so

$$\alpha^\vee = \sum_{\alpha_i \text{ long}} n_i \alpha_i^\vee + \sum_{\alpha_i \text{ short}} n_i p^{-1} \alpha_i^\vee.$$

Since $\alpha^\vee \in \sum_i \mathbb{Z}\alpha_i^\vee$ we deduce that p divides n_i whenever α_i is short.

Now suppose conversely that p divides n_i for all i for which α_i is short. Let $n_i = pm_i$ for such i. Suppose if possible that α is short. Then $\theta(\alpha^\vee) = p\alpha$. Thus

$$\alpha^\vee = \sum_{\alpha_i \text{ long}} pn_i \alpha_i^\vee + \sum_{\alpha_i \text{ short}} n_i \alpha_i^\vee$$

$$= p\left(\sum_{\alpha_i \text{ long}} n_i \alpha_i^\vee + \sum_{\alpha_i \text{ short}} m_i \alpha_i^\vee \right).$$

This gives $\alpha^\vee \in \sum p\mathbb{Z}\alpha_i^\vee$ which is impossible. Thus α is a long root. $\square$

Proposition 8.18 (i) *The abelian group generated by the short roots in Φ is $\sum_{i=1}^l \mathbb{Z}\alpha_i$.*
(ii) *The abelian group generated by the long roots in Φ is $\sum_{\alpha_i \text{ long}} \mathbb{Z}\alpha_i + \sum_{\alpha_i \text{ short}} p\mathbb{Z}\alpha_i$.*

8.8 Properties of long and short roots

Proof. By considering the fundamental system of the subsystem of short roots described in Proposition 8.13 it is clear that the abelian group generated by the short roots contains $\alpha_1, \ldots, \alpha_l$ so is $\sum \mathbb{Z}\alpha_i$.

The abelian group generated by the long roots lies in $\sum_{\alpha_i \text{ long}} \mathbb{Z}\alpha_i + \sum_{\alpha_i \text{ short}} p\mathbb{Z}\alpha_i$ by Proposition 8.17. However, this group contains α_i for α_i long and $p\alpha_i$ for α_i short, again by Proposition 8.17, so must be $\sum_{\alpha_i \text{ long}} \mathbb{Z}\alpha_i + \sum_{\alpha_i \text{ short}} p\mathbb{Z}\alpha_i$. □

9
Some universal constructions

9.1 The universal enveloping algebra

Let L be a Lie algebra over $\mathbb{C}$. We shall show in this section how to construct an associative algebra $\mathfrak{U}(L)$, the universal enveloping algebra of L, such that the representation theory of $\mathfrak{U}(L)$ is the same as the representation theory of the Lie algebra L. Even if L is finite dimensional its enveloping algebra $\mathfrak{U}(L)$ will be infinite dimensional.

We begin by forming the tensor powers of L. We define T^0 to be the 1-dimensional vector space $\mathbb{C}1$, $T^1 = L$, $T^2 = L \otimes_\mathbb{C} L$ and, in general,

$$T^n = L \otimes_\mathbb{C} L \otimes_\mathbb{C} \cdots \otimes_\mathbb{C} L \qquad (n \text{ factors})$$

T^n is a vector space over $\mathbb{C}$ of dimension $(\dim L)^n$.

We next form the tensor algebra $T = T(L)$ of L. We define T as the direct sum of vector spaces

$$T = T^0 \oplus T^1 \oplus T^2 \oplus \cdots.$$

Thus elements of T are finite sums of elements, each of which lies in some T^n. We may define a bilinear map

$$T^m \times T^n \to T^{m+n}$$

satisfying

$$(x_1 \otimes \cdots \otimes x_m) \cdot (y_1 \otimes \cdots \otimes y_n) = x_1 \otimes \cdots \otimes x_m \otimes y_1 \otimes \cdots \otimes y_n$$

for $x_i, y_j \in L$ and then extend this map by linearity to give a multiplication map

$$T \times T \to T.$$

In this way T becomes an associative algebra called the **tensor algebra** of L. The element $1 \in T^0$ is the identity element of T.

9.1 The universal enveloping algebra

Let J be the 2-sided ideal of T generated by all elements of the form

$$x \otimes y - y \otimes x - [xy] \quad \text{for } x, y \in L.$$

J is in particular a subspace of T. Let $\mathfrak{U}(L) = T/J$. Then $\mathfrak{U}(L)$ is an associative algebra over $\mathbb{C}$ called the **universal enveloping algebra** of L.

Example 9.1 Let L be an n-dimensional abelian Lie algebra over $\mathbb{C}$. Then L has basis $x_1, \ldots, x_n$ and we have $[x_i x_j] = 0$ for all i, j. Thus J is the 2-sided ideal of T generated by all elements of the form $x \otimes y - y \otimes x$ for $x, y \in L$. Thus $\mathfrak{U}(L)$ is a commutative algebra, and is generated as an algebra by the identity 1 and the elements $x_1, \ldots, x_n$. In fact $\mathfrak{U}(L)$ is isomorphic to the polynomial algebra $\mathbb{C}[x_1, \ldots, x_n]$. $\square$

In general we have linear maps

$$L \to T^1 \to T \to \mathfrak{U}(L)$$

and we denote by $\sigma : L \to \mathfrak{U}(L)$ the composite linear map. We now show that $\mathfrak{U}(L)$ has a certain universal property which justifies its name.

Proposition 9.2 Let A be any associative algebra with 1 over $\mathbb{C}$ and $[A]$ the corresponding Lie algebra. Then given any Lie algebra homomorphism $\theta : L \to [A]$ there exists a unique associative algebra homomorphism $\phi : \mathfrak{U}(L) \to A$ such that $\phi \circ \sigma = \theta$.

Note Associative algebra homomorphisms will be understood to be homomorphisms of associative algebras with identity in this chapter. Thus the homomorphism will map identity to identity.

Proof. We first observe that the linear map $\theta : L \to A$ can be extended to an associative algebra homomorphism from T to A. If $x_i, i \in I$, are a basis for L then the set of all monomials $x_{i_1} \ldots x_{i_r}$ for $i_1, \ldots, i_r \in I$ form a basis for T. The case $r = 0$ gives the identity element. The map

$$x_{i_1} \ldots x_{i_r} \to \theta(x_{i_1}) \ldots \theta(x_{i_r})$$

can then be extended by linearity to give an associative algebra homomorphism from T to A. Let this map be $\theta' : T \to A$. Let $x, y \in L$. Then we have

$$\theta'(x \otimes y - y \otimes x - [xy])$$
$$= \theta(x)\theta(y) - \theta(y)\theta(x) - \theta[xy]$$
$$= [\theta(x), \theta(y)] - \theta[xy] = 0$$

since $\theta : L \to [A]$ is a Lie algebra homomorphism. Thus all the generators of the 2-sided ideal J of T lie in the kernel of θ'. Since the kernel is a 2-sided ideal, J lies in the kernel of θ'. This shows there is an induced homomorphism

$$\phi : T/J \to A$$

such that the diagram

commutes. When we restrict the domain to T^1 we deduce that $\phi \circ \sigma = \theta$. This proves the existence of a homomorphism $\phi : \mathfrak{U}(L) \to A$ of the required type.

We now prove the uniqueness of ϕ. Let $\phi' : \mathfrak{U}(L) \to A$ be another such homomorphism. Now T is generated by T^1 as an associative algebra with 1. Thus its factor algebra $\mathfrak{U}(L)$ is generated by $\sigma(L)$, which is the image of T^1 in $\mathfrak{U}(L)$. Let $x \in L$. Then

$$\phi'(\sigma(x)) = \theta(x) = \phi(\sigma(x)).$$

Thus ϕ, ϕ' agree on $\sigma(x)$ for all $x \in L$. Since $\sigma(L)$ generates $\mathfrak{U}(L)$ it follows that ϕ, ϕ' agree on $\mathfrak{U}(L)$, so $\phi' = \phi$. □

Using this universal property we can relate representations of the Lie algebra L to representations of the associative algebra $\mathfrak{U}(L)$. If V is a vector space over $\mathbb{C}$ the set End V of all linear maps of V into itself forms an associative algebra with 1, and the corresponding Lie algebra is $[\text{End } V]$. A representation of L is a Lie algebra homomorphism $L \to [\text{End } V]$ and a representation of $\mathfrak{U}(L)$ is an associative algebra homomorphism $\mathfrak{U}(L) \to \text{End } V$.

Proposition 9.3 *There is a bijective correspondence between representations $\theta : L \to [\text{End } V]$ and representations $\phi : \mathfrak{U}(L) \to \text{End } V$. Corresponding representations are related by the condition*

$$\phi(\sigma(x)) = \theta(x) \quad \text{for all } x \in L.$$

Proof. Let $\theta : L \to [\text{End } V]$ be a representation of L. Then by Proposition 9.2 there exists a unique associative algebra homomorphism $\phi : \mathfrak{U}(L) \to \text{End } V$ such that $\phi \circ \sigma = \theta$.

Conversely, given an associative algebra homomorphism $\phi : \mathfrak{U}(L) \to \text{End } V$ we wish to define a corresponding Lie algebra homomorphism

9.2 The Poincaré–Birkhoff–Witt basis theorem

$\theta : L \to [\text{End } V]$. Now we have a linear map $\sigma : L \to \mathfrak{U}(L)$. Since $\mathfrak{U}(L) = T/J$ and $x \otimes y - y \otimes x - [xy] \in J$ for all $x, y \in L$ we see that

$$\sigma(x)\sigma(y) - \sigma(y)\sigma(x) - \sigma[xy] = 0$$

for $x, y \in L$. This gives

$$[\sigma(x), \sigma(y)] = \sigma[xy]$$

and so $\sigma : L \to [\mathfrak{U}(L)]$ is a Lie algebra homomorphism. We now define $\theta : L \to [\text{End } V]$ by $\theta = \phi \circ \sigma$. Then θ is a Lie homomorphism of the required type.

It is clear from the definitions that the maps $\theta \to \phi$ and $\phi \to \theta$ are inverse to one another. □

We shall find this result very useful in the subsequent development, when we shall obtain information about representations of finite dimensional Lie algebras by considering the representation theory of the corresponding universal enveloping algebra.

9.2 The Poincaré–Birkhoff–Witt basis theorem

We shall now describe how to obtain a basis for the universal enveloping algebra $\mathfrak{U}(L)$.

Theorem 9.4 (*Poincaré–Birkhoff–Witt*). *Let L be a Lie algebra with basis $\{x_i ; i \in I\}$. Let $<$ be a total order on the index set I. Let $\sigma : L \to \mathfrak{U}(L)$ be the natural linear map from L into its enveloping algebra. Let $\sigma(x_i) = y_i$. Then the elements*

$$y_{i_1}^{r_1} \ldots y_{i_n}^{r_n}$$

for all $n \geq 0$, all $r_i \geq 0$, and all $i_1, \ldots, i_n \in I$ with $i_1 < i_2 < \cdots < i_n$ form a basis for $\mathfrak{U}(L)$.

Proof. (a) We first show that the above elements $y_{i_1}^{r_1} \ldots y_{i_n}^{r_n}$ span $\mathfrak{U}(L)$. We know that the elements of form $x_{j_1} \otimes \ldots \otimes x_{j_k}$ for all k and all $j_1, \ldots, j_k \in I$ span T. By applying the natural homomorphism $T \to \mathfrak{U}(L)$ it follows that the elements of form $y_{j_1} \ldots y_{j_k}$ span $\mathfrak{U}(L)$. It is therefore sufficient to show that every product $y_{j_1} \ldots y_{j_k}$ is a linear combination of the given elements of form $y_{i_1}^{r_1} \ldots y_{i_n}^{r_n}$. We shall prove this by induction on k. It is obvious if $k = 1$. For arbitrary k it is clear when $j_1 \leq \cdots \leq j_k$. If this is not so we may use relations of form

$$y_i y_j = y_j y_i + \sigma[x_i x_j].$$

We note that $[x_i x_j]$ is a linear combination of elements x_t for $t \in I$ and so $\sigma[x_i x_j]$ is a linear combination of y_t for $t \in I$. Thus we may interchange the order of two consecutive terms y_i, y_j in a monomial of degree k provided we introduce a certain linear combination of monomials of degree less than k. By performing such interchanges a finite number of times we may express the terms y_i in the monomial with the i in the given order $<$ on I. Thus

$$y_{j_1} \ldots y_{j_k} = y_{i_1}^{r_1} \ldots y_{i_n}^{r_n} + \text{a linear combination of monomials of degree less than } k$$

where $r_1 + \cdots + r_n = k$ and $i_1 < \cdots < i_n$. By induction we may assume that all monomials of degree less than k are expressible as linear combinations of monomials with terms in the given order $<$. The required result then follows.

(b) We now show that the given monomials of form $y_{i_1}^{r_1} \ldots y_{i_n}^{r_n}$ are linearly independent. This is not so easy to see, and we shall prove it by an indirect argument. We introduce the polynomial ring $R = \mathbb{C}[z_i \; ; \; i \in I]$ and shall make use of the following lemma.

Lemma 9.5 *There exists a linear map* $\theta : T \to R$ *satisfying the conditions*

$$\theta(x_{i_1} \otimes \cdots \otimes x_{i_n}) = z_{i_1} \ldots z_{i_n} \quad \text{if } i_1 \leq \cdots \leq i_n$$

$$\theta\left(x_{i_1} \otimes \cdots \otimes x_{i_k} \otimes x_{i_{k+1}} \otimes \cdots \otimes x_{i_n} - x_{i_1} \otimes \cdots \otimes x_{i_{k+1}} \otimes x_{i_k} \otimes \cdots \otimes x_{i_n}\right)$$
$$= \theta\left(x_{i_1} \otimes \cdots \otimes [x_{i_k} x_{i_{k+1}}] \otimes \cdots \otimes x_{i_n}\right) \quad \text{for all } i_1, \ldots, i_n \text{ and all } k$$

with $1 \leq k < n$.

Proof. We define the index of the monomial $x_{i_1} \otimes \cdots \otimes x_{i_n}$ to be the number of pairs (r, s) with $1 \leq r < s \leq n$ satisfying $i_r > i_s$. Thus the monomials of index 0 are those whose terms appear in their natural order. Let $T^{n,j}$ be the subspace of T^n spanned by all monomials $x_{i_1} \otimes \cdots \otimes x_{i_n}$ of index at most j. Thus

$$T^{n,0} \subset T^{n,1} \subset \cdots \subset T^n.$$

We define $\theta : T^0 \to R$ by $\theta(1) = 1$. Suppose inductively that $\theta : T^0 \oplus \cdots \oplus T^{n-1} \to R$ has already been defined satisfying the required conditions. We shall show that θ can be extended to $\theta : T^0 \oplus \cdots \oplus T^n \to R$. We define $\theta : T^{n,0} \to R$ by

$$\theta(x_{i_1} \otimes \cdots \otimes x_{i_n}) = z_{i_1} \ldots z_{i_n}$$

if the monomial $x_{i_1} \otimes \cdots \otimes x_{i_n}$ has index 0. We suppose $\theta : T^{n,i} \to R$ has already been defined, thus giving a linear map from $T^0 \oplus \cdots \oplus T^{n-1} \oplus T^{n,i}$ to

9.2 The Poincaré–Birkhoff–Witt basis theorem

R satisfying the required conditions. We wish to define $\theta : T^{n,i+1} \to R$. Thus suppose the monomial $x_{i_1} \otimes \cdots \otimes x_{i_n}$ has index $i+1$. Then there exists k with $1 \leq k < n$ such that $x_{i_1} \otimes \cdots \otimes x_{i_{k-1}} \otimes x_{i_{k+1}} \otimes x_{i_k} \otimes x_{i_{k+2}} \otimes \cdots \otimes x_{i_n}$ has index i. We then wish to define $\theta\left(x_{i_1} \otimes \cdots \otimes x_{i_n}\right)$ by the formula

$$\theta\left(x_{i_1} \otimes \cdots \otimes x_{i_k} \otimes x_{i_{k+1}} \otimes \cdots \otimes x_{i_n}\right)$$
$$= \theta\left(x_{i_1} \otimes \cdots \otimes x_{i_{k+1}} \otimes x_{i_k} \otimes \cdots \otimes x_{i_n}\right)$$
$$+ \theta\left(x_{i_1} \otimes \cdots \otimes \left[x_{i_k} x_{i_{k+1}}\right] \otimes \cdots \otimes x_{i_n}\right)$$

noting that the terms on the right-hand side have already been defined. However, there may be more than one possible choice of k and we must check that if we choose a different one the linear map $\theta : T^{n,i+1} \to R$ will still be the same. So suppose k' also satisfies $1 \leq k' < n$. We may without loss of generality assume that $k < k'$.

We suppose first that $k+1 < k'$. Let $x_{i_k} = a$, $x_{i_{k+1}} = b$, $x_{i_{k'}} = c$, $x_{i_{k'+1}} = d$. Then the definition using the integer k gives

$$\theta(\cdots \otimes a \otimes b \otimes \cdots \otimes c \otimes d \otimes \cdots)$$
$$= \theta(\cdots \otimes b \otimes a \otimes \cdots \otimes c \otimes d \otimes \cdots)$$
$$+ \theta(\cdots \otimes [ab] \otimes \cdots \otimes c \otimes d \otimes \cdots)$$
$$= \theta(\cdots \otimes b \otimes a \otimes \cdots \otimes d \otimes c \otimes \cdots)$$
$$+ \theta(\cdots \otimes b \otimes a \otimes \cdots \otimes [cd] \otimes \cdots)$$
$$+ \theta(\cdots \otimes [ab] \otimes \cdots \otimes d \otimes c \otimes \cdots)$$
$$+ \theta(\cdots \otimes [ab] \otimes \cdots \otimes [cd] \otimes \cdots)$$

using the inductive assumptions.

The second definition using the integer k' gives

$$\theta(\cdots \otimes a \otimes b \otimes \cdots \otimes c \otimes d \otimes \cdots)$$
$$= \theta(\cdots \otimes a \otimes b \otimes \cdots \otimes d \otimes c \otimes \cdots)$$
$$+ \theta(\cdots \otimes a \otimes b \otimes \cdots \otimes [cd] \otimes \cdots)$$
$$= \theta(\cdots \otimes b \otimes a \otimes \cdots \otimes d \otimes c \otimes \cdots)$$
$$+ \theta(\cdots \otimes [ab] \otimes \cdots \otimes d \otimes c \otimes \cdots)$$
$$+ \theta(\cdots \otimes b \otimes a \otimes \cdots \otimes [cd] \otimes \cdots)$$
$$+ \theta(\cdots \otimes [ab] \otimes \cdots \otimes [cd] \otimes \cdots)$$

using the inductive assumptions. These two expressions using integers k, k' are the same.

Now suppose that $k' = k+1$. Let $x_{i_k} = a$, $x_{i_{k+1}} = b$, $x_{i_{k+2}} = c$. We compare the two ways of calculating $\theta(\cdots \otimes a \otimes b \otimes c \otimes \cdots)$. The first method, using the integer k, gives

$$\theta(\cdots \otimes a \otimes b \otimes c \otimes \cdots)$$
$$= \theta(\cdots \otimes b \otimes a \otimes c \otimes \cdots) + \theta(\cdots \otimes [ab] \otimes c \otimes \cdots)$$
$$= \theta(\cdots \otimes b \otimes c \otimes a \otimes \cdots) + \theta(\cdots \otimes b \otimes [ac] \otimes \cdots)$$
$$+ \theta(\cdots \otimes c \otimes [ab] \otimes \cdots) + \theta(\cdots \otimes [[ab]c] \otimes \cdots)$$
$$= \theta(\cdots \otimes c \otimes b \otimes a \otimes \cdots) + \theta(\cdots \otimes [bc] \otimes a \otimes \cdots)$$
$$+ \theta(\cdots \otimes b \otimes [ac] \otimes \cdots) + \theta(\cdots \otimes c \otimes [ab] \otimes \cdots)$$
$$+ \theta(\cdots \otimes [[ab]c] \otimes \cdots)$$
$$= \theta(\cdots \otimes c \otimes b \otimes a \otimes \cdots) + \theta(\cdots \otimes a \otimes [bc] \otimes \cdots)$$
$$+ \theta(\cdots \otimes b \otimes [ac] \otimes \cdots) + \theta(\cdots \otimes c \otimes [ab] \otimes \cdots)$$
$$+ \theta(\cdots \otimes [[ab]c] \otimes \cdots) + \theta(\cdots \otimes [[bc]a] \otimes \cdots)$$

using the inductive assumptions. The second method, using the integer $k' = k+1$, gives

$$\theta(\cdots \otimes a \otimes b \otimes c \otimes \cdots)$$
$$= \theta(\cdots \otimes a \otimes c \otimes b \otimes \cdots) + \theta(\cdots \otimes a \otimes [bc] \otimes \cdots)$$
$$= \theta(\cdots \otimes c \otimes a \otimes b \otimes \cdots) + \theta(\cdots \otimes [ac] \otimes b \otimes \cdots)$$
$$+ \theta(\cdots \otimes a \otimes [bc] \otimes \cdots)$$
$$= \theta(\cdots \otimes c \otimes b \otimes a \otimes \cdots) + \theta(\cdots \otimes c \otimes [ab] \otimes \cdots)$$
$$+ \theta(\cdots \otimes b \otimes [ac] \otimes \cdots) + \theta(\cdots \otimes [[ac]b] \otimes \cdots)$$
$$+ \theta(\cdots \otimes a \otimes [bc] \otimes \cdots)$$

again using the inductive assumptions. Comparing the two expressions obtained we see that they are equal since

$$[[ac]b] = [[ab]c] + [[bc]a].$$

Thus $\theta : T^{n,i+1} \to R$ is now defined and this gives

$$\theta : T^0 \oplus \cdots \oplus T^{n-1} \oplus T^{n,i+1} \to R.$$

9.2 The Poincaré–Birkhoff–Witt basis theorem

Since $T^n = T^{n,r}$ for r sufficiently large we have

$$\theta \ : \ T^0 \oplus \cdots \oplus T^n \to R.$$

Since $T = T^0 \oplus T^1 \oplus T^2 \oplus \cdots$ we have defined $\theta \ : \ T \to R$ satisfying the required conditions. □

We now return to part (b) of the proof of Theorem 9.4. We have $\mathfrak{U}(L) = T/J$ and the elements

$$x_{i_1} \otimes \cdots \otimes x_{i_k} \otimes x_{i_{k+1}} \otimes \cdots \otimes x_{i_n} - x_{i_1} \otimes \cdots \otimes x_{i_{k+1}} \otimes x_{i_k} \otimes \cdots \otimes x_{i_n}$$

$$- x_{i_1} \otimes \cdots \otimes \left[x_{i_k} x_{i_{k+1}} \right] \otimes \cdots \otimes x_{i_n}$$

$$= x_{i_1} \ldots x_{i_{k-1}} \left(x_{i_k} \otimes x_{i_{k+1}} - x_{i_{k+1}} \otimes x_{i_k} - \left[x_{i_k} x_{i_{k+1}} \right] \right) x_{i_{k+2}} \ldots x_{i_n}$$

all lie in J. In fact the definition of J shows that each element of J is a linear combination of such elements. Thus the linear map $\theta \ : \ T \to R$ of Lemma 9.5 annihilates all elements of J, and so induces a linear map $\bar{\theta} \ : \ T/J \to R$, that is $\bar{\theta} \ : \ \mathfrak{U}(L) \to R$. Now the monomial $y_{i_1}^{r_1} \ldots y_{i_n}^{r_n} \in \mathfrak{U}(L)$ for $i_1 < \cdots < i_n$ is mapped by $\bar{\theta}$ to $z_{i_1}^{r_1} \ldots z_{i_n}^{r_n} \in R$. Since the elements $z_{i_1}^{r_1} \ldots z_{i_n}^{r_n}$ are linearly independent in the polynomial ring R it follows that the elements $y_{i_1}^{r_1} \ldots y_{i_n}^{r_n}$ given in the statement of Theorem 9.4 must be linearly independent in $\mathfrak{U}(L)$. This completes the proof. □

We now deduce some consequences of the Poincaré–Birkhoff–Witt basis theorem. (We shall subsequently call it the **PBW basis theorem**.)

Corollary 9.6 *The map* $\sigma \ : \ L \to \mathfrak{U}(L)$ *is injective.*

Proof. The elements $x_i, i \in I$, form a basis for L and $\sigma(x_i) = y_i$. By the PBW basis theorem the elements $y_i, i \in I$, are linearly independent. Thus the kernel of σ is zero. □

Corollary 9.7 *The subspace $\sigma(L)$ is a Lie subalgebra of $[\mathfrak{U}(L)]$ isomorphic to L. Thus σ identifies L with a Lie subalgebra of $[\mathfrak{U}(L)]$.*

Proof. By Corollary 9.6 we know that $\sigma \ : \ L \to \sigma(L)$ is bijective. The elements $y_i, i \in I$, form a basis of $\sigma(L)$ and we have

$$y_i y_j - y_j y_i = \sigma \left[x_i x_j \right].$$

It follows that $\left[y_i y_j \right] \in \sigma(L)$ and so $\sigma(L)$ is a Lie subalgebra of $[\mathfrak{U}(L)]$. □

It is often convenient to consider L as a subspace of $\mathfrak{U}(L)$ without mentioning the map σ explicitly.

Corollary 9.8 $\mathfrak{U}(L)$ *has no zero-divisors.*

Proof. Let $a, b \in \mathfrak{U}(L)$ have $a \neq 0, b \neq 0$. Then we have

$$a = \sum \lambda_{i_1,\ldots,i_n,r_1,\ldots,r_n}\, y_{i_1}^{r_1} \ldots y_{i_n}^{r_n}$$
$$b = \sum \mu_{i_1,\ldots,i_n,r_1,\ldots,r_n}\, y_{i_1}^{r_1} \ldots y_{i_n}^{r_n}.$$

We write

$$a = f(y_i) + \text{a sum of terms of smaller degree}$$

where $f(y_i)$ is the sum of all terms $\lambda_{i_1,\ldots,i_n,r_1,\ldots,r_n}\, y_{i_1}^{r_1} \ldots y_{i_n}^{r_n}$ of maximal total degree $r = r_1 + \cdots + r_n$. Similarly we have

$$b = g(y_i) + \text{a sum of terms of smaller degree}.$$

Now we have

$$y_i y_j = y_j y_i + \text{a sum of terms of degree 1}$$

and so

$$f(y_i) g(y_i) = (fg)(y_i) + \text{a sum of terms of smaller degree}.$$

Hence

$$ab = (fg)(y_i) + \text{a sum of terms of smaller degree}.$$

Now f is not the zero polynomial since $a \neq 0$ and g is not the zero polynomial since $b \neq 0$. Thus fg is not the zero polynomial. The PBW basis theorem then implies that $ab \neq 0$. $\square$

9.3 Free Lie algebras

It is well known how to define groups by generators and relations. One first constructs the free group on the given set of generators and then forms the factor group with respect to the smallest normal subgroup containing the elements specified by the given relations. We shall show that something similar can be done in the theory of Lie algebras. We first introduce the idea of the free Lie algebra $FL(X)$ on a set X.

9.3 Free Lie algebras

Let $X = \{x_i, i \in I\}$ be a set of elements parametrised by an index set I. We first define the free associative algebra $F(X)$ on the set X. $F(X)$ is the set of all finite sums of the form

$$\sum_{k \geq 0} \sum_{i_1, \ldots, i_k \in I} \lambda_{i_1, \ldots, i_k} x_{i_1} \ldots x_{i_k}$$

with $\lambda_{i_1, \ldots, i_k} \in \mathbb{C}$, summed over all non-negative integers k and all ordered k-tuples $i_1, \ldots, i_k$ from I (repetitions being allowed). When $k = 0$ the product $x_{i_1} \ldots x_{i_k}$ is the empty product, and is written as 1. The operations of addition, multiplication and scalar multiplication are defined in an obvious way and make $F(X)$ into an associative algebra over $\mathbb{C}$ with identity 1.

Let $[F(X)]$ be the Lie algebra obtained from the associative algebra $F(X)$ in the usual manner. X is a subset of $[F(X)]$. We define $FL(X)$ to be the intersection of all the Lie subalgebras of $[F(X)]$ containing X, i.e. the Lie subalgebra of $[F(X)]$ generated by X. $FL(X)$ is called **the free Lie algebra** on the set X. It is clear that X is contained in $FL(X)$ so we have an injective map $i : X \to FL(X)$.

In order to justify its name, we show that the free Lie algebra $FL(X)$ has the following universal property.

Proposition 9.9 *Let $\theta : X \to L$ be any map from the set X into a Lie algebra L. Then there is a unique homomorphism $\phi : FL(X) \to L$ such that $\phi \circ i = \theta$.*

Proof. Consider the maps $X \xrightarrow{\theta} L \xrightarrow{\sigma} \mathfrak{U}(L)$. Let $\theta' : X \to \mathfrak{U}(L)$ be given by $\theta' = \sigma \circ \theta$. The map θ' from X into $\mathfrak{U}(L)$ can be extended uniquely (in an obvious way) to an associative algebra homomorphism $\phi' : F(X) \to \mathfrak{U}(L)$. The same map gives a Lie algebra homomorphism $\phi' : [F(X)] \to [\mathfrak{U}(L)]$. Now we have $\phi'(X) \subset \sigma(L)$ and we know from Corollary 9.7 that $\sigma(L)$ is a Lie subalgebra of $[\mathfrak{U}(L)]$ isomorphic to L. The set of elements of $[F(X)]$ mapped by ϕ' into $\sigma(L)$ is therefore a Lie subalgebra of $[F(X)]$ containing X, and this contains $FL(X)$. Hence we have $\phi' : FL(X) \to \sigma(L)$. We define $\phi : FL(X) \to L$ by $\phi = \sigma^{-1} \circ \phi'$. We check that $\phi \circ i = \theta$. For if $x \in X$ we have

$$\phi \circ i(x) = \sigma^{-1} \phi' i(x) = \sigma^{-1} \phi'(x) = \sigma^{-1} \theta'(x) = \theta(x).$$

Thus we have a homomorphism ϕ of the required type. Finally we show that ϕ is unique. Let $\bar{\phi} : FL(X) \to L$ be another such homomorphism. Then we have

$$\phi i(x) = \theta(x) = \bar{\phi} i(x) \qquad \text{for all } x \in X.$$

Thus ϕ agrees with $\bar{\phi}$ on X. Now the set of elements of $FL(X)$ for which ϕ agrees with $\bar{\phi}$ is a Lie subalgebra of $FL(X)$ containing X. Since X generates $FL(X)$ as a Lie algebra we deduce that ϕ agrees with $\bar{\phi}$ on $FL(X)$. □

We next identify the universal enveloping algebra of the free Lie algebra $FL(X)$. This turns out to be isomorphic to the free associative algebra $F(X)$.

Proposition 9.10 *The universal enveloping algebra* $\mathfrak{U}(FL(X))$ *is isomorphic to* $F(X)$.

Proof. We have an inclusion map $\sigma : FL(X) \to F(X)$. We shall show that the universal property of enveloping algebras given in Proposition 9.2 is satisfied by $F(X)$. Thus we shall show that if A is any associative algebra with 1 over $\mathbb{C}$ and if $\theta : FL(X) \to [A]$ is any Lie algebra homomorphism then there exists a unique associative algebra homomorphism $\phi : F(X) \to A$ such that $\phi \circ \sigma = \theta$.

Now the Lie homomorphism $\theta : FL(X) \to [A]$ restricts to a map $\theta : X \to A$. This map can be extended to a unique associative algebra homomorphism $\phi : F(X) \to A$. This same map gives a Lie algebra homomorphism $\phi : [F(X)] \to [A]$. By restriction we obtain a Lie algebra homomorphism $\phi : FL(X) \to [A]$. However, ϕ agrees with θ on X and X generates $FL(X)$ as a Lie algebra. Hence ϕ agrees with θ on $FL(X)$. It follows that $\phi \circ \sigma = \theta$ as required. Thus there exists an algebra homomorphism ϕ of the required kind. On the other hand ϕ is clearly unique since $FL(X)$ contains X and therefore generates the associative algebra $F(X)$.

Thus $F(X)$ satisfies the above universal property. Of course $\mathfrak{U}(FL(X))$ satisfies it also. This implies that $\mathfrak{U}(FL(X))$ is isomorphic to $F(X)$. For suppose we are given a Lie algebra L and two associative algebras $\mathfrak{U}, \mathfrak{U}'$ with maps $\sigma : L \to \mathfrak{U}, \sigma' : L \to \mathfrak{U}'$ both satisfying the universal property. Then we obtain unique algebra homomorphisms $\phi : \mathfrak{U} \to \mathfrak{U}'$ and $\phi' : \mathfrak{U}' \to \mathfrak{U}$ such that $\sigma' = \phi \circ \sigma$ and $\sigma = \phi' \circ \sigma'$.

It follows that

$$\phi'\phi\sigma(x) = \sigma(x), \quad \phi\phi'\sigma'(x) = \sigma'(x)$$

for all $x \in L$. Now $\sigma(L)$ generates $\mathfrak{U}$ and $\sigma'(L)$ generates $\mathfrak{U}'$ as associative algebras, by the uniqueness of ϕ and ϕ'. It follows that

$$\phi'\phi = \mathrm{Id}_{\mathfrak{U}}, \quad \phi\phi' = \mathrm{Id}_{\mathfrak{U}'}$$

and so ϕ, ϕ' are inverse isomorphisms between $\mathfrak{U}$ and $\mathfrak{U}'$. $\square$

9.4 Lie algebras defined by generators and relations

Let $X = \{x_i, i \in I\}$ be a given set. A **Lie monomial** in the elements of X is a finite product of elements of X bracketed by Lie brackets in any manner. For example

$$[[[x_3[x_1x_2]]x_3][x_2[x_1x_1]]]$$

is a Lie monomial on the set $X = \{x_1, x_2, x_3\}$. A **Lie word** in the elements on X is a finite linear combination of Lie monomials on X with coefficients in $\mathbb{C}$. For example

$$3[[[x_3[x_1x_2]]x_3][x_2[x_1x_1]]] + 2[x_2[[x_1x_2][x_3x_2]]]$$

is a Lie word on the set $X = \{x_1, x_2, x_3\}$.

Let $R = \{w_j, j \in J\}$ be a set of Lie words in the elements of X. We shall define a Lie algebra $L(X; R)$ called the Lie algebra generated by X subject to relations R.

Now the elements of X all lie in the free Lie algebra $FL(X)$ and all the Lie words w_j also lie in $FL(X)$. Of course different Lie words can give the same element of $FL(X)$ because of relations such as $[x_ix_i] = 0$ and the Jacobi identity. Let $\langle R \rangle$ be the ideal of $FL(X)$ generated by R. Thus $\langle R \rangle$ is the intersection of all ideals of $FL(X)$ containing R. We define $L(X; R)$ by

$$L(X; R) = FL(X)/\langle R \rangle.$$

Lemma 9.11 *Let R, R' be sets of Lie words in X such that $R' \subset R$. Then $L(X; R)$ is isomorphic to a factor algebra of $L(X; R')$.*

Proof. Since $R' \subset R$ we have

$$\langle R' \rangle \subset \langle R \rangle \subset FL(X).$$

It follows that

$$L(X;R) = \frac{FL(X)}{\langle R \rangle} \cong \frac{FL(X)}{\langle R' \rangle} / \frac{\langle R \rangle}{\langle R' \rangle} = \frac{L(X;R')}{I}$$

where $I = \langle R \rangle / \langle R' \rangle$. □

Example 9.12 Let A be a Cartan matrix on the standard list 6.12. In Section 7.4 we defined a Lie algebra $L(A)$ associated with A, and $L(A)$ was subsequently shown in Proposition 7.35 to be a simple Lie algebra. In fact all the finite dimensional non-trivial simple Lie algebras over $\mathbb{C}$ have form $L(A)$, as A runs over all Cartan matrices on the standard list. The definition of $L(A)$ given in Section 7.4 shows that $L(A)$ can conveniently be described in terms of generators and relations. In fact we have

$$L(A) \cong L(X;R)$$

where $X = \{e_1, \ldots, e_l, h_1, \ldots, h_l, f_1, \ldots, f_l\}$ and R is the set of Lie words in X given by

$$[h_i h_j]$$
$$[h_i e_j] - A_{ij} e_j$$
$$[h_i f_j] + A_{ij} f_j$$
$$[e_i f_i] - h_i$$
$$[e_i f_j] \quad \text{for } i \neq j$$
$$[e_i [e_i [\ldots [e_i e_j]]]] \quad \text{for } i \neq j$$
$$[f_i [f_i [\ldots [f_i f_j]]]] \quad \text{for } i \neq j$$

where the number of occurrences of e_i, f_i respectively in the last two words is $1 - A_{ij}$.

Example 9.13 Again let A be a Cartan matrix on the standard list 6.12. In Section 7.4 we also defined a certain Lie algebra $\tilde{L}(A)$ depending on A which contains $L(A)$ as a factor algebra. The algebra $\tilde{L}(A)$ is infinite dimensional. It can also conveniently be described by generators and relations. In fact we have

$$\tilde{L}(A) \cong L(X;R')$$

where $X = \{e_1, \ldots, e_l, h_1, \ldots, h_l, f_1, \ldots, f_l\}$ and R' is the set of Lie words on X given by

9.5 Graph automorphisms of simple Lie algebras

$$[h_i h_j]$$
$$[h_i e_j] - A_{ij} e_j$$
$$[h_i f_j] + A_{ij} f_j$$
$$[e_i f_i] - h_i$$
$$[e_i f_j] \quad \text{for } i \neq j.$$

We observe that R' is a proper subset of the set R of relations in Example 9.12. This explains why $L(A)$ is isomorphic to a factor algebra of $\tilde{L}(A)$, as in Lemma 9.11.

9.5 Graph automorphisms of simple Lie algebras

Let A be a Cartan matrix on the standard list 6.12 and σ be a permutation of $\{1, \ldots, l\}$ such that $A_{\sigma(i)\sigma(j)} = A_{ij}$ for all i, j. Let $L(A)$ be the simple Lie algebra associated with A. $L(A)$ can be generated by $e_1, \ldots, e_l, h_1, \ldots, h_l, f_1, \ldots, f_l$. We define a permutation of this generating set by

$$e_i \to e_{\sigma(i)} \quad f_i \to f_{\sigma(i)} \quad h_i \to h_{\sigma(i)}.$$

Under this permutation of the generators each of the defining relations of $L(A)$ in Example 9.12 is transformed into a defining relation. Let

$$L(A) = L(X\ ;\ R) = \frac{FL(X)}{\langle R \rangle}.$$

The given permutation of X extends to a Lie algebra homomorphism of $FL(X)$ into itself, and this homomorphism maps the ideal $\langle R \rangle$ into itself. It therefore induces a Lie algebra homomorphism of $L(X; R)$ into itself. Since the permutation of X is invertible, so is this Lie algebra homomorphism. It is thus an isomorphism of $L(X; R)$ into itself, that is an automorphism of $L(A)$. This automorphism is called a graph automorphism of $L(A)$ and will also be denoted by σ. The possible non-trivial graph automorphisms can be described in terms of the action of σ on the Dynkin diagram of $L(A)$. These possibilities are listed below.

Type A_{2k}

$$\sigma(i) = 2k + 1 - i$$

Type A_{2k-1}

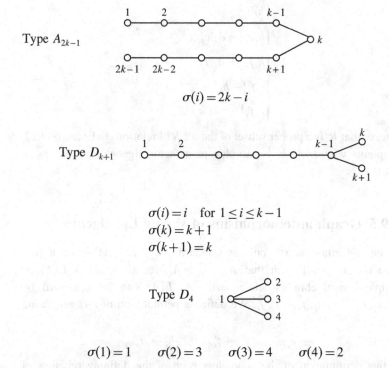

$\sigma(i) = 2k - i$

Type D_{k+1}

$\sigma(i) = i$ for $1 \leq i \leq k-1$
$\sigma(k) = k+1$
$\sigma(k+1) = k$

Type D_4

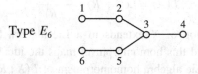

$\sigma(1) = 1 \quad \sigma(2) = 3 \quad \sigma(3) = 4 \quad \sigma(4) = 2$

(The inverse of σ is also a graph automorphism, which can be obtained from σ by renumbering the vertices.)

Type E_6

$\sigma(1) = 6 \quad \sigma(2) = 5 \quad \sigma(3) = 3 \quad \sigma(4) = 4 \quad \sigma(5) = 2 \quad \sigma(6) = 1$

Our main aim in the present section is to determine the fixed point subalgebra

$$L(A)^\sigma = \{x \in L(A)\,;\, \sigma(x) = x\}.$$

We begin by considering the action of σ on $V = H_{\mathbb{R}}^*$ given by $\sigma(\alpha_i) = \alpha_{\sigma(i)}$ and extending by linearity. Let $V^1 = \{v \in V\,;\, \sigma(v) = v\}$. For each orbit J of σ on $\{1, \ldots, l\}$ we define $\alpha_J = \frac{1}{|J|} \sum_{j \in J} \alpha_j$. Then $\alpha_J \in V^1$ and the α_J form a basis of V^1 as J runs over the σ-orbits on $\{1, \ldots, l\}$. α_J is simply the projection of α_j on to the subspace V^1 of the Euclidean space V. We see from the above diagrams that the orbits J have the following possible types.

9.5 Graph automorphisms of simple Lie algebras

(a) $|J|=1$ and $J=\{j\}$ with $\sigma(j)=j$.
(b) $|J|=2$ and $J=\{j\bar{j}\}$ where $\sigma(j)=\bar{j}, \sigma(\bar{j})=j$ and $\alpha_j+\alpha_{\bar{j}} \notin \Phi$.
(c) $|J|=3$ and $J=\{j\bar{j}\bar{\bar{j}}\}$ where $\sigma(j)=\bar{j}, \sigma(\bar{j})=\bar{\bar{j}}, \sigma(\bar{\bar{j}})=j$ and $\alpha_j+\alpha_{\bar{j}}, \alpha_j+\alpha_{\bar{\bar{j}}}, \alpha_{\bar{j}}+\alpha_{\bar{\bar{j}}}$ do not lie in Φ.
(d) $|J|=2$ and $J=\{j\bar{j}\}$ where $\sigma(j)=\bar{j}, \sigma(\bar{j})=j$ and $\alpha_j+\alpha_{\bar{j}} \in \Phi$.

These four will be called orbits of types A_1, $A_1 \times A_1$, $A_1 \times A_1 \times A_1$ and A_2 respectively.

We next consider the possible pairs J, K of distinct orbits.

Lemma 9.14 *The vectors α_J, α_K for distinct σ-orbits J, K form a fundamental system of roots of rank 2. The type of this rank 2 system is as follows.*

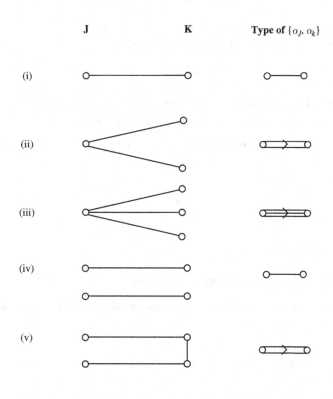

(vi) If no node in J is joined to any node in K then the type of $\{\alpha_j, \alpha_k\}$ is $A_1 \times A_1$.

Proof. This is straightforward. Suppose for example we have case (v) with roots numbered

```
1 o──────o 2
4 o──────o 3
```

Then
$$\alpha_J = \frac{\alpha_1 + \alpha_4}{2}, \quad \alpha_K = \frac{\alpha_2 + \alpha_3}{2}.$$

We have
$$\langle \alpha_J, \alpha_J \rangle = \tfrac{1}{2} \langle \alpha_1, \alpha_1 \rangle$$
$$\langle \alpha_K, \alpha_K \rangle = \tfrac{1}{4} \langle \alpha_1, \alpha_1 \rangle$$
$$\langle \alpha_J, \alpha_K \rangle = -\tfrac{1}{4} \langle \alpha_1, \alpha_1 \rangle.$$

Thus $\langle \alpha_J, \alpha_J \rangle = 2 \langle \alpha_K, \alpha_K \rangle$ and $2 \langle \alpha_J, \alpha_K \rangle / \langle \alpha_J, \alpha_J \rangle = -1$. Hence we have a fundamental system with diagram

$$\alpha_J \Longrightarrow \alpha_K$$

Corollary 9.15 *Let Π^1 be the set of vectors α_J for all σ-orbits J on $\{1, \ldots, l\}$. Then Π^1 is a fundamental system of roots of the following type:*

Type Π	Order of σ	Type Π^1
A_{2k}	2	B_k
A_{2k-1}	2	C_k
D_{k+1}	2	B_k
D_4	3	G_2
E_6	2	F_4

Proof. This follows immediately from Lemma 9.14. □

The relationship between Π and Π^1 may be illustrated in the following diagrams.

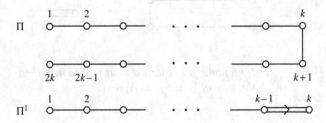

9.5 Graph automorphisms of simple Lie algebras

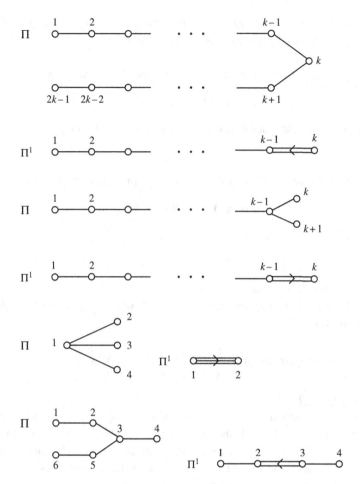

Now let Φ^1 be the root system in V^1 with fundamental system Π^1. Let W^1 be the Weyl group of Φ^1. Then $\Phi^1 = W^1(\Pi^1)$. Let A^1 be the Cartan matrix of Φ^1.

Proposition 9.16 *Let I, J be distinct σ-orbits on $\{1, \ldots, l\}$. Then*

$$A^1_{IJ} = \begin{cases} \sum_{i \in I} A_{ij} & \text{for any } j \in J, \text{ if } I \text{ has type } A_1, A_1 \times A_1 \text{ or } A_1 \times A_1 \times A_1 \\ 2\sum_{i \in I} A_{ij} & \text{for any } j \in J, \text{ if } I \text{ has type } A_2. \end{cases}$$

Proof. This follows from Lemma 9.14. □

Proposition 9.17 *Let $W^\sigma = \{w \in W \; ; \; w\sigma = \sigma w \text{ on } V\}$. Then there is an isomorphism $W^1 \to W^\sigma$ under which the fundamental reflection $s_J \in W^1$*

corresponding to α_J maps to $(w_0)_J \in W^\sigma$, the element of maximal length in the Weyl subgroup W_J of W generated by the s_i for $i \in J$.

Proof. We first observe that W^σ acts on V^1. For let $w \in W^\sigma$, $v \in V^1$. Then

$$\sigma(wv) = w(\sigma v) = wv, \quad \text{thus } wv \in V^1.$$

Secondly we note that $(w_0)_J \in W^\sigma$ for each σ-orbit J. For $j \in J$ we have

$$\sigma s_j \sigma^{-1} = s_{\sigma(j)}$$

thus $\sigma W_J \sigma^{-1} = W_J$. Since σ preserves the sign of each root we have

$$\sigma (w_0)_J \sigma^{-1} (\Phi_J^+) = \Phi_J^-$$

and hence

$$\sigma (w_0)_J \sigma^{-1} = (w_0)_J$$

by Proposition 5.17. Thus $(w_0)_J \in W^\sigma$.

Thirdly we note that the element $(w_0)_J \in W^\sigma$, when restricted to V^1, coincides with s_J. For

$$(w_0)_J (\alpha_J) = (w_0)_J \left(\frac{1}{|J|} \sum_{j \in J} \alpha_j \right) = -\frac{1}{|J|} \sum_{j \in J} \alpha_j = -\alpha_J$$

since $(w_0)_J (\Phi_J^+) = \Phi_J^-$.

Also if $v \in V^1$ satisfies $\langle \alpha_J, v \rangle = 0$ then it satisfies $\langle \alpha_j, v \rangle = 0$ for all $j \in J$. It follows that $(w_0)_J (v) = v$. Thus $(w_0)_J$ coincides with s_J on restriction to V^1.

We next show that the elements $(w_0)_J$ for all σ-orbits J generate W^σ. Let $w \in W^\sigma$ satisfy $w \neq 1$. Then there exists a fundamental root α_j with $w(\alpha_j) \in \Phi^-$. Let J be the σ-orbit containing j. Then

$$w\sigma(\alpha_j) = \sigma w(\alpha_j) \in \Phi^-$$

since σ preserves the sign of each root. Thus $w(\alpha_i) \in \Phi^-$ for all $i \in J$. Now $(w_0)_J$ changes the signs of all roots in Φ_J but of none in $\Phi - \Phi_J$. Hence

$$l(w(w_0)_J) = l(w) - l((w_0)_J) < l(w).$$

We assume by induction on $l(w)$ that $w(w_0)_J$ lies in the subgroup generated by the $(w_0)_I$ for all σ-orbits I. It follows that w has the same property. Hence the $(w_0)_I$ generate W^σ.

We may now define a homomorphism $W^\sigma \to W^1$, by restricting the action of $w \in W^\sigma$ from V to V^1. Since W^σ is generated by the elements $(w_0)_J$ and $(w_0)_J$ restricted to V^1 is s_J, the image of the homomorphism is generated by the s_J and so is W^1. Finally we show our map is injective. Suppose $w \in W^\sigma$ and $w \neq 1$. Then there exists a σ-orbit J such that $w(\alpha_i) \in \Phi^-$ for all $i \in J$. Thus $w(\alpha_J) \neq \alpha_J$ and so w acts non-trivially on V^1. Thus our map $W^\sigma \to W^1$ is an isomorphism under which $(w_0)_J \in W^\sigma$ corresponds to $s_J \in W^1$. □

We next consider the relation between the root systems Φ and Φ^1. For each $\alpha \in \Phi$ we denote by α^1 its projection into V^1.

Proposition 9.18 (a) *For each $\alpha \in \Phi$, α^1 is a positive multiple of a root in Φ^1.*
(b) *Let $\sim$ be the equivalence relation on Φ given by $\alpha \sim \beta$ if and only if α^1 is a positive multiple of β^1. Then the equivalence classes are the subsets of Φ of form $w(\Phi_J^+)$ where J is a σ-orbit in $\{1, \ldots, l\}$ and $w \in W^\sigma$.*
(c) *There is a bijection between equivalence classes on Φ and roots in Φ^1 given by $w(\Phi_J^+) \leftrightarrow w^1(\alpha_J)$ where w^1 is the restriction of w to V^1.*

Proof. We first show that each $\alpha \in \Phi$ lies in $w(\Phi_J^+)$ for some σ-orbit J and some $w \in W^\sigma$. Consider the element $w_0 \in W$ of maximal length. w_0 transforms each positive root to a negative root. Since σ does not change the sign of any root we have

$$\sigma w_0 \sigma^{-1}(\Phi^+) = \Phi^-.$$

Since $\sigma w_0 \sigma^{-1} \in W$ it follows that $\sigma w_0 \sigma^{-1} = w_0$, that is $w_0 \in W^1$. By Proposition 9.17 the elements $(w_0)_J$ for all σ-orbits J generate W^σ and so

$$w_0 = (w_0)_{J_1} \ldots (w_0)_{J_r}$$

for some $J_1, \ldots, J_r$. Let $\alpha \in \Phi^+$. Then $w_0(\alpha) \in \Phi^-$. Thus there exists i such that

$$(w_0)_{J_{i+1}} \ldots (w_0)_{J_r}(\alpha) \in \Phi^+$$
$$(w_0)_{J_i}(w_0)_{J_{i+1}} \ldots (w_0)_{J_r}(\alpha) \in \Phi^-.$$

Since the only positive roots made negative by $(w_0)_{J_i}$ are those in $\Phi_{J_i}^+$ we have

$$(w_0)_{J_{i+1}} \ldots (w_0)_{J_r}(\alpha) \in \Phi_{J_i}^+,$$

that is $\alpha \in (w_0)_{J_r} \ldots (w_0)_{J_{i+1}}(\Phi_{J_i}^+)$ and $-\alpha \in (w_0)_{J_r} \ldots (w_0)_{J_{i+1}}(w_0)_{J_i}(\Phi_{J_i}^+)$. Hence each root in Φ lies in $w(\Phi_J^+)$ for some σ-orbit J and some $w \in W^\sigma$.

Now consider the projection α^1 for $\alpha \in \Phi_J^+$. If J has type A_1, $A_1 \times A_1$ or $A_1 \times A_1 \times A_1$ then $\Phi_J^+ = \Pi_J$ and so $\alpha^1 = \alpha_J$ for $\alpha \in \Phi_J^+$. If J has type A_2, however, then $\Pi_J = \{\alpha_j, \alpha_{\bar{j}}\}$ and $\Phi_J^+ = \{\alpha_j, \alpha_{\bar{j}}, \alpha_j + \alpha_{\bar{j}}\}$. We have

$$\alpha^1 = \begin{cases} \alpha_J & \text{when } \alpha = \alpha_j \text{ or } \alpha_{\bar{j}} \\ 2\alpha_J & \text{when } \alpha = \alpha_j + \alpha_{\bar{j}}. \end{cases}$$

Thus α^1 is a positive multiple of α_J when $\alpha \in \Phi_J^+$. Hence for $\alpha \in w\left(\Phi_J^+\right)$ with $w \in W^\sigma$ we see that α^1 is a positive multiple of $w(\alpha_J) \in \Phi^1$.

Now consider the equivalence relation on Φ defined in (b). The elements of each set $w\left(\Phi_J^+\right)$ for $w \in W^\sigma$ lie in an equivalence class. Suppose $w\left(\Phi_J^+\right)$, $w'\left(\Phi_K^+\right)$ lie in the same equivalence class for σ-orbits J, K and $w, w' \in W^\sigma$. Then

$$w(\alpha_J) = w'(\alpha_K) \in \Phi^1.$$

Hence $w'^{-1}w(\alpha_J) = \alpha_K$.

Consider the root $w'^{-1}w(\alpha_j) \in \Phi$ for $j \in J$. This root has the property that

$$\left(w'^{-1}w(\alpha_j)\right)^1 = \alpha_K.$$

Since K is a σ-orbit this implies that $w'^{-1}w(\alpha_j)$ is a non-negative combination of the α_k for $k \in K$. Hence

$$w'^{-1}w(\Pi_J) \subset \Phi_K^+$$

and so $w'^{-1}w\left(\Phi_J^+\right) \subset \Phi_K^+$. By symmetry we also have

$$w'w^{-1}\left(\Phi_K^+\right) \subset \Phi_J^+.$$

Hence we have equality, that is

$$w\left(\Phi_J^+\right) = w'\left(\Phi_K^+\right).$$

Hence the equivalence classes are the subsets of Φ of form $w\left(\Phi_J^+\right)$.

Now any root in Φ^1 has form $w(\alpha_J)$ for some $w \in W^1$ and some σ-orbit J. The set of roots $\alpha \in \Phi$ such that α^1 is a positive multiple of $w(\alpha_J)$ is $w\left(\Phi_J^+\right)$, as shown above. Thus

$$w\left(\Phi_J^+\right) \leftrightarrow w(\alpha_J)$$

is a bijective correspondence between equivalence classes on Φ and elements of Φ^1. □

Theorem 9.19 *Let σ be a graph automorphism of the simple Lie algebra $L(A)$. Then the subalgebra $L(A)^\sigma$ is isomorphic to the simple Lie algebra $L(A^1)$.*

9.5 Graph automorphisms of simple Lie algebras

Proof. For each σ-orbit J on $\{1,\ldots,l\}$ we define elements e_J, h_J, f_J of $L(A)^\sigma$ by

$$e_J = \sum_{j \in J} e_j \qquad f_J = \sum_{j \in J} f_j \qquad h_J = \sum_{j \in J} h_j$$

if J has type A_1, $A_1 \times A_1$ or $A_1 \times A_1 \times A_1$ and

$$e_J = \sqrt{2} \sum_{j \in J} e_j, \qquad f_J = \sqrt{2} \sum_{j \in J} f_j, \qquad h_J = 2 \sum_{j \in J} h_j$$

if J has type A_2. Then we have

$$[e_I f_I] = h_I$$
$$[e_I f_J] = 0 \qquad \text{if } I \neq J.$$
$$[h_I h_J] = 0$$

We consider $[h_I e_J]$. If I, J have type A_1, $A_1 \times A_1$ or $A_1 \times A_1 \times A_1$ we have

$$[h_I e_J] = \left[\sum_{i \in I} h_i, \sum_{j \in J} e_j \right] = \sum_j \left(\sum_i A_{ij} e_j \right) = A_{IJ}^1 e_J.$$

We also have $[h_I e_J] = A_{IJ}^1 e_J$ if one or both of I, J has type A_2. Similarly

$$[h_I f_J] = -A_{IJ}^1 f_J \qquad \text{for all } I, J.$$

We also check the relation

$$[e_I[e_I[\ldots[e_I e_J]]]] = 0 \qquad \text{for } I \neq J$$

where there are $1 - A_{IJ}^1$ factors e_I. This follows from the following observations, which can be checked from Lemma 9.14.

If $A_{IJ}^1 = 0$ then $[e_i e_j] = 0$ for all $i \in I, j \in J$.
If $A_{IJ}^1 = -1$ then $[e_i[e_{i'} e_j]] = 0$ for all $i, i' \in I, j \in J$.
If $A_{IJ}^1 = -2$ then $[e_i[e_{i'}[e_{i''} e_j]]] = 0$ for all $i, i', i'' \in I, j \in J$.
If $A_{IJ}^1 = -3$ then $[e_i[e_{i'}[e_{i''}[e_{i'''} e_j]]]] = 0$ for all $i, i', i'', i''' \in I, j \in J$.

Similarly we obtain the relation

$$[f_I[f_I[\ldots[f_I f_J]]]] = 0 \qquad \text{for } I \neq J$$

with $1 - A_{IJ}^1$ factors f_I.

We now consider the generators and defining relations for the simple Lie algebra $L(A^1)$ given in Example 9.12. All these relations are satisfied by the elements e_J, f_J, h_J of $L(A)^\sigma$. Thus there is a homomorphism

$$L(A^1) \to L(A)^\sigma$$

under which the generators of $L(A^1)$ map to the elements e_J, f_J, h_J of $L(A)^\sigma$. Since $L(A^1)$ is simple this homomorphism is injective. We show it is also surjective and that the map is therefore an isomorphism. It will be sufficient to show that

$$\dim L(A)^\sigma = \dim L(A^1).$$

We consider the decomposition of Φ into equivalence classes given in Proposition 9.18. For each equivalence class S let

$$L_S = \bigoplus_{\alpha \in S} L_\alpha.$$

Then $\sigma(L_S) = L_S$ and

$$L = H \oplus \sum_S L_S$$

$$L^\sigma = H^\sigma \oplus \sum_S (L_S)^\sigma.$$

Now $\dim (L_S)^\sigma \leq 1$ for each equivalence class S. This is clear if S has type $A_1, A_1 \times A_1$ or $A_1 \times A_1 \times A_1$. Suppose then that S has type A_2. Then $S = \{\alpha, \beta, \alpha + \beta\}$. We have

$$\sigma(e_\alpha) = \lambda e_\beta, \qquad \sigma(e_\beta) = \lambda^{-1} e_\alpha$$

for some $\lambda \in \mathbb{C}$. Hence

$$\sigma[e_\alpha e_\beta] = [e_\beta e_\alpha] = -[e_\alpha e_\beta].$$

Thus $(L_S)^\sigma$ consists of all multiples of $e_\alpha + \lambda e_\beta$ and $\dim (L_S)^\sigma = 1$. It follows that

$$\dim L^\sigma \leq \dim H^\sigma + \text{no. of equivalence classes } S$$
$$= \dim H^1 + |\Phi^1|$$
$$= \dim L(A^1).$$

Hence $\dim L^\sigma \leq \dim L(A^1)$.

This shows that the homomorphism $L(A^1) \to L^\sigma$ is surjective and hence is an isomorphism. We note in particular that $\dim (L_S)^\sigma = 1$ for each equivalence class S. □

9.5 Graph automorphisms of simple Lie algebras

Thus we have shown that $L(A)^\sigma$ is isomorphic to $L(A^1)$. To be specific we have:

$$L(A_{2k})^\sigma \cong L(B_k)$$
$$L(A_{2k-1})^\sigma \cong L(C_k)$$
$$L(D_{k+1})^\sigma \cong L(B_k)$$
$$L(D_4)^\sigma \cong L(G_2)$$
$$L(E_6)^\sigma \cong L(F_4)$$

where σ is a graph automorphism of order 2, 2, 2, 3, 2 respectively.

10
Irreducible modules for semisimple Lie algebras

In the present chapter we shall determine the finite dimensional irreducible modules for a semisimple Lie algebra over $\mathbb{C}$. We begin by investigating certain important modules for such algebras known as Verma modules.

10.1 Verma modules

We begin with a lemma on universal enveloping algebras. Let L be a finite dimensional Lie algebra over $\mathbb{C}$ and K a subalgebra of L.

Lemma 10.1 *There exists a unique algebra homomorphism* $\theta : \mathfrak{U}(K) \to \mathfrak{U}(L)$ *such that the diagram*

$$\begin{array}{ccc} K & \xrightarrow{\sigma_K} & \mathfrak{U}(K) \\ {}_i\downarrow & & \downarrow_\theta \\ L & \xrightarrow{\sigma_L} & \mathfrak{U}(L) \end{array}$$

commutes, where i is the embedding of K in L and σ_K, σ_L are the embeddings of K, L in $\mathfrak{U}(K)$, $\mathfrak{U}(L)$ respectively.

Also θ is injective.

Proof. Let $x \in K$. Then we must have

$$\theta(\sigma_K(x)) = \sigma_L(i(x)).$$

Thus $\theta(\sigma_K(x))$ is uniquely determined. Since $\mathfrak{U}(K)$ is generated by $\sigma_K(K)$ as algebra with 1 we see that θ is uniquely determined.

We now show that θ exists. We recall that

$$\mathfrak{U}(L) = T(L)/J(L), \quad \mathfrak{U}(K) = T(K)/J(K)$$

10.1 Verma modules

where $J(L)$ is the 2-sided ideal of $T(L)$ generated by the elements

$$x \otimes y - y \otimes x - [xy] \quad \text{for } x, y \in L.$$

Now the map $i : K \to L$ induces an algebra homomorphism $i : T(K) \to T(L)$ and we have

$$i(x \otimes y - y \otimes x - [xy]) = i(x) \otimes i(y) - i(y) \otimes i(x) - [i(x)i(y)]$$

for all $x, y \in K$. This shows that

$$i(J(K)) \subset J(L).$$

Thus there is an algebra homomorphism $\theta : \mathfrak{U}(K) \to \mathfrak{U}(L)$ such that the required diagram commutes.

Finally we show that θ is injective. This follows from the PBW basis theorem 9.4. Let $x_1, \ldots, x_r$ be a basis of K. Suppose if possible there exists $u \in \mathfrak{U}(K)$ such that $u \neq 0$ and $\theta(u) = 0$. Then u is a non-zero linear combination of monomials $x_1^{e_1} \ldots x_r^{e_r}$. However, since $x_1, \ldots, x_r$ can be chosen as part of a basis of L, the PBW basis theorem for L shows that such a combination of monomials cannot be zero in L. Hence $\theta(u) \neq 0$, a contradiction. Thus θ is injective. $\square$

This lemma shows that $\mathfrak{U}(K)$ may be regarded in a natural way as a subalgebra of $\mathfrak{U}(L)$.

We now suppose that L is a finite dimensional semisimple Lie algebra over $\mathbb{C}$. Let H be a Cartan subalgebra of L and

$$L = H \oplus \sum_{\alpha \in \Phi} L_\alpha$$

be the Cartan decomposition of L with respect to H. Let Φ^+ be the positive system of roots in Φ. Then we have a triangular decomposition

$$L = N^- \oplus H \oplus N$$

where $N^- = \bigoplus_{\alpha \in \Phi^-} L_\alpha$, $N = \bigoplus_{\alpha \in \Phi^+} L_\alpha$. We recall that H, N, N^- are all subalgebras of L.

Let $B = H \oplus N$.

Lemma 10.2 (i) *B is a subalgebra of L.*
(ii) *N is an ideal of B.*
(iii) *B/N is isomorphic to H.*

Proof. (i) We have
$$[BB] = [H+N, H+N] \subset H+N = B$$
since H, N are subalgebras and $[HN] \subset N$.
(ii) $[NB] = [N, N+H] \subset N$.
(iii) $B/N = (H+N)/N \cong H/H \cap N \cong H$
since $H \cap N = 0$, using Proposition 1.7. □

Definition 10.3 *Let* $\lambda \in H^*$, *i.e.* λ *be a linear map from* H *to* $\mathbb{C}$. *We recall that* L *has a basis*
$$\{e_\alpha, \alpha \in \Phi \; ; \; h_i, i = 1, \ldots, l\}.$$
We define
$$K_\lambda = \sum_{\alpha \in \Phi^+} \mathfrak{U}(L) e_\alpha + \sum_{i=1}^{l} \mathfrak{U}(L)(h_i - \lambda(h_i)).$$
Thus K_λ *is the left ideal of* $\mathfrak{U}(L)$ *generated by the elements* $e_\alpha, \alpha \in \Phi^+$, *and* $h_i - \lambda(h_i)$ *for* $i = 1, \ldots, l$.

(We are as usual here embedding L *in* $\mathfrak{U}(L)$.*) We also define*
$$M(\lambda) = \mathfrak{U}(L)/K_\lambda.$$
$M(\lambda)$ is a left $\mathfrak{U}(L)$-module called the **Verma module** determined by λ. It is our aim in this section to describe some of the properties of $M(\lambda)$.

We note that the elements $e_\alpha, \alpha \in \Phi^+$, and $h_i - \lambda(h_i)$ for $i = 1, \ldots, l$ all lie in $\mathfrak{U}(B)$. We define
$$K'_\lambda = \sum_{\alpha \in \Phi^+} \mathfrak{U}(B) e_\alpha + \sum_{i=1}^{l} \mathfrak{U}(B)(h_i - \lambda(h_i))$$
to be the left ideal of $\mathfrak{U}(B)$ generated by these elements. Let
$$\Phi^+ = \{\beta_1, \ldots, \beta_N\}.$$
Then the set
$$h_1, \ldots, h_l, \quad e_{\beta_1}, \ldots, e_{\beta_N}$$
is a basis of B. It follows from the PBW basis theorem that the elements
$$h_1^{s_1} \ldots h_l^{s_l} \quad e_{\beta_1}^{t_1} \ldots e_{\beta_N}^{t_N} \qquad s_i \geq 0 \quad t_i \geq 0$$
form a basis of $\mathfrak{U}(B)$.

10.1 Verma modules

Proposition 10.4 (i) $\dim \mathfrak{U}(B)/K'_\lambda = 1$.

(ii) *The elements*
$$(h_1 - \lambda(h_1))^{s_1} \ldots (h_l - \lambda(h_l))^{s_l} e_{\beta_1}^{t_1} \ldots e_{\beta_N}^{t_N}$$
with $s_i \geq 0, t_i \geq 0$, excluding the element with $s_i = t_i = 0$ for all i, form a basis for K'_λ.

Proof. It is not difficult to see that the elements
$$(h_1 - \lambda(h_1))^{s_1} \ldots (h_l - \lambda(h_l))^{s_l} e_{\beta_1}^{t_1} \ldots e_{\beta_N}^{t_N}$$
with $s_i \geq 0, t_i \geq 0$ also form a basis for $\mathfrak{U}(B)$. This can be seen, for example, by defining a partial ordering on the basis elements $h_1^{s_1} \ldots h_l^{s_l} e_{\beta_1}^{t_1} \ldots e_{\beta_N}^{t_N}$. We say that $h_1^{s'_1} \ldots h_l^{s'_l} e_{\beta_1}^{t'_1} \ldots e_{\beta_N}^{t'_N}$ is lower than $h_1^{s_1} \ldots h_l^{s_l} e_{\beta_1}^{t_1} \ldots e_{\beta_N}^{t_N}$ if $s'_1 \leq s_1, \ldots, s'_l \leq s_l$, $t'_1 = t_1, \ldots, t'_N = t_N$. Then there are only a finite number of basis elements lower than a given one, and the element
$$(h_1 - \lambda(h_1))^{s_1} \ldots (h_l - \lambda(h_l))^{s_l} e_{\beta_1}^{t_1} \ldots e_{\beta_N}^{t_N}$$
is the sum of $h_1^{s_1} \ldots h_l^{s_l} e_{\beta_1}^{t_1} \ldots e_{\beta_N}^{t_N}$ with a linear combination of strictly lower basis elements in the partial order. An induction argument on the partial order will then show that the elements
$$(h_1 - \lambda(h_1))^{s_1} \ldots (h_l - \lambda(h_l))^{s_l} e_{\beta_1}^{t_1} \ldots e_{\beta_N}^{t_N}$$
$s_i \geq 0, t_i \geq 0$ span $\mathfrak{U}(B)$ and are linearly independent. Now all these elements clearly lie in K'_λ, with the exception of the element with $s_i = 0, t_i = 0$ for all i. This is the unit element 1. However, 1 does not lie in K'_λ, as the following argument shows.

Consider the representation λ of H mapping h_i to $\lambda(h_i)$ for $i = 1, \ldots, l$. Since B/N is isomorphic to H there is a 1-dimensional representation λ of B with N in the kernel agreeing with the above representation on $B/N \cong H$. This in turn gives a 1-dimensional representation ρ of $\mathfrak{U}(B)$ under which

$$e_\alpha \to 0 \quad \alpha \in \Phi^+$$
$$h_i \to \lambda(h_i) \quad i = 1, \ldots, l$$
$$1 \to 1$$

Now $\ker \rho$ is a 2-sided ideal of $\mathfrak{U}(B)$ containing $e_\alpha, \alpha \in \Phi^+$ and $h_i - \lambda(h_i)$ so containing K'_λ. Thus we have

$$K'_\lambda \subset \ker \rho \quad 1 \notin \ker \rho$$

hence $1 \notin K'_\lambda$.

It follows that the elements

$$(h_1 - \lambda(h_1))^{s_1} \ldots (h_l - \lambda(h_l))^{s_l} e_{\beta_1}^{t_1} \ldots e_{\beta_N}^{t_N}$$

for $s_i \geq 0$, $t_i \geq 0$, excluding 1, form a basis of K_λ' and that

$$\dim \mathfrak{U}(B)/K_\lambda' = 1.$$

Proposition 10.5 $K_\lambda \cap \mathfrak{U}(N^-) = O.$

Proof. We are here regarding $\mathfrak{U}(N^-)$ as a subalgebra of $\mathfrak{U}(L)$ as in Lemma 10.1. Now we have

$$L = N^- \oplus B.$$

Regarding $\mathfrak{U}(B)$ as a subalgebra of $\mathfrak{U}(L)$ also we assert that

$$\mathfrak{U}(L) = \mathfrak{U}(N^-) \mathfrak{U}(B),$$

i.e. each element of $\mathfrak{U}(L)$ is a finite sum $\sum x_i y_i$ with $x_i \in \mathfrak{U}(N^-)$, $y_i \in \mathfrak{U}(B)$. This follows from the PBW basis theorem, choosing bases for N^- and for B and combining them to give a basis of L. We then have

$$K_\lambda = \sum_{\alpha \in \Phi^+} \mathfrak{U}(L) e_\alpha + \sum_{i=1}^{l} \mathfrak{U}(L)(h_i - \lambda(h_i))$$

$$= \sum_{\alpha \in \Phi^+} \mathfrak{U}(N^-) \mathfrak{U}(B) e_\alpha + \sum_{i=1}^{l} \mathfrak{U}(N^-) \mathfrak{U}(B)(h_i - \lambda(h_i))$$

$$= \mathfrak{U}(N^-) \left(\sum_{\alpha \in \Phi^+} \mathfrak{U}(B) e_\alpha + \sum_{i=1}^{l} \mathfrak{U}(B)(h_i - \lambda(h_i)) \right) = \mathfrak{U}(N^-) K_\lambda'.$$

It follows that each element of K_λ is a linear combination of terms of form

$$f_{\beta_1}^{r_1} \ldots f_{\beta_N}^{r_N} (h_1 - \lambda(h_1))^{s_1} \ldots (h_l - \lambda(h_l))^{s_l} e_{\beta_1}^{t_1} \ldots e_{\beta_N}^{t_N}$$

where $f_\alpha = e_{-\alpha}$, $\alpha \in \Phi^+$, and $r_i \geq 0$, $s_i \geq 0$, $t_i \geq 0$ with $(s_1, \ldots s_l, t_1, \ldots t_N) \neq (0, \ldots 0, 0 \ldots 0)$. No non-zero element of $\mathfrak{U}(N^-)$ can be a linear combination of such terms by the PBW basis theorem. Hence

$$K_\lambda \cap \mathfrak{U}(N^-) = O. \qquad \square$$

Let $m_\lambda \in M(\lambda)$ be defined by $m_\lambda = 1 + K_\lambda$. Thus 1 maps to m_λ under the natural homomorphism

$$\mathfrak{U}(L) \to \mathfrak{U}(L)/K_\lambda = M(\lambda).$$

10.1 Verma modules

Theorem 10.6 (i) *Each element of $M(\lambda)$ is uniquely expressible in the form um_λ for some $u \in \mathfrak{U}(N^-)$.*
(ii) *The elements $f_{\beta_1}^{r_1} \ldots f_{\beta_N}^{r_N} m_\lambda$ for all $r_i \geq 0$ form a basis for $M(\lambda)$.*

Proof. Each element of $\mathfrak{U}(L)$ has form $u \cdot 1$ for some $u \in \mathfrak{U}(L)$. Thus each element of $M(\lambda) = \mathfrak{U}(L)/K_\lambda$ has form um_λ for some $u \in \mathfrak{U}(L)$.

Now $f_{\beta_1}, \ldots, f_{\beta_N}, h_1, \ldots, h_l, e_{\beta_1}, \ldots, e_{\beta_N}$ are a basis of L so the elements

$$f_{\beta_1}^{r_1} \ldots f_{\beta_N}^{r_N} \quad h_1^{s_1} \ldots h_l^{s_l} \quad e_{\beta_1}^{t_1} \ldots e_{\beta_N}^{t_N}$$

$r_i \geq 0$, $s_i \geq 0$, $t_i \geq 0$ form a basis of $\mathfrak{U}(L)$ by the PBW basis theorem. Thus u is a linear combination of such elements, and um_λ is a linear combination of elements

$$f_{\beta_1}^{r_1} \ldots f_{\beta_N}^{r_N} \quad h_1^{s_1} \ldots h_l^{s_l} \quad e_{\beta_1}^{t_1} \ldots e_{\beta_N}^{t_N} m_\lambda.$$

Now this element is 0 if any t_i is positive. Suppose then that all $t_i = 0$. Then

$$h_1^{s_1} \ldots h_l^{s_l} m_\lambda = \gamma m_\lambda \quad \text{for some } \gamma \in \mathbb{C}$$

since $h_i m_\lambda = \lambda(h_i) m_\lambda$. Thus um_λ is a linear combination of elements of form

$$f_{\beta_1}^{r_1} \ldots f_{\beta_N}^{r_N} m_\lambda.$$

Thus elements of this form for $r_i \geq 0$ span $M(\lambda)$. They are also linearly independent. For if we have

$$\sum_{r_1, \ldots, r_N} \xi_{r_1, \ldots, r_N} f_{\beta_1}^{r_1} \ldots f_{\beta_N}^{r_N} m_\lambda = 0$$

with $\xi_{r_1, \ldots, r_N} \in \mathbb{C}$ then it follows that

$$\sum_{r_1, \ldots, r_N} \xi_{r_1, \ldots, r_N} f_{\beta_1}^{r_1} \ldots f_{\beta_N}^{r_N} \in K_\lambda \cap \mathfrak{U}(N^-).$$

Hence this element is 0 by Proposition 10.5. Thus each $\xi_{r_1, \ldots, r_N} = 0$ by the PBW basis theorem for $\mathfrak{U}(N^-)$.

Thus the elements $f_{\beta_1}^{r_1} \ldots f_{\beta_N}^{r_N} m_\lambda$ for $r_1 \geq 0, \ldots, r_N \geq 0$ form a basis for $M(\lambda)$. It follows that each element of $M(\lambda)$ is uniquely expressible in the form um_λ for $u \in \mathfrak{U}(N^-)$. □

We now regard $M(\lambda)$ as an H-module. For each 1-dimensional representation μ of H we define

$$M(\lambda)_\mu = \{ m \in M(\lambda) \, ; \, xm = \mu(x)m \quad \text{for all } x \in H \}.$$

$M(\lambda)_\mu$ is a subspace of $M(\lambda)$.

Theorem 10.7 (i) $M(\lambda) = \bigoplus_{\mu \in H^*} M(\lambda)_\mu$.
(ii) $M(\lambda)_\mu \neq 0$ if and only if $\lambda - \mu$ is a sum of positive roots.
(iii) $\dim M(\lambda)_\mu = \mathfrak{P}(\lambda - \mu)$, the number of ways of expressing $\lambda - \mu$ as a sum of positive roots.
$\mathfrak{P}(\lambda - \mu)$ is the number of vectors $(r_1, \ldots, r_N)$ with $r_i \in \mathbb{Z}$, $r_i \geq 0$ such that
$$\lambda - \mu = r_1 \beta_1 + \cdots + r_N \beta_N.$$

Proof. We know from Theorem 10.6 that the elements $f_{\beta_1}^{r_1} \ldots f_{\beta_N}^{r_N} m_\lambda$ with $r_i \geq 0$ form a basis for $M(\lambda)$. We show that
$$x f_{\beta_1}^{r_1} \ldots f_{\beta_N}^{r_N} m_\lambda = (\lambda - r_1 \beta_1 - \cdots - r_N \beta_N)(x) f_{\beta_1}^{r_1} \ldots f_{\beta_N}^{r_N} m_\lambda \quad \text{for all } x \in H.$$

We prove this by induction on $r_1 + \cdots + r_N$, the result being clear if all $r_i = 0$. So suppose not all r_i are 0 and let i be the least integer with $r_i > 0$. Then we have
$$x f_{\beta_i}^{r_i} \ldots f_{\beta_N}^{r_N} m_\lambda = f_{\beta_i} x f_{\beta_i}^{r_i - 1} \ldots f_{\beta_N}^{r_N} m_\lambda - \beta_i(x) f_{\beta_i}^{r_i} \ldots f_{\beta_N}^{r_N} m_\lambda.$$
It follows that
$$x f_{\beta_i}^{r_i} \ldots f_{\beta_N}^{r_N} m_\lambda = (\lambda - r_i \beta_i - \cdots - r_N \beta_N)(x) f_{\beta_i}^{r_i} \ldots f_{\beta_N}^{r_N} m_\lambda$$
as required.

This implies that $f_{\beta_1}^{r_1} \ldots f_{\beta_N}^{r_N} m_\lambda \in M(\lambda)_\mu$ where $\mu = \lambda - r_1 \beta_1 - \cdots - r_N \beta_N$. Since these elements form a basis of $M(\lambda)$ we see that
$$M(\lambda) = \sum_\mu M(\lambda)_\mu.$$

We now show this sum is direct. To see this we must show that if a finite sum $\sum v_\mu$ is 0 with $v_\mu \in M(\lambda)_\mu$, then each v_μ is 0.

It is sufficient to show that
$$M(\lambda)_\mu \cap (M(\lambda)_{\mu_1} + \cdots + M(\lambda)_{\mu_k}) = 0$$
where the elements $\mu, \mu_1, \ldots, \mu_k \in H^*$ are all distinct.

Let v lie in this intersection. Then we have
$$v = v_{\mu_1} + \cdots + v_{\mu_k}$$
where $v \in M(\lambda)_\mu$, $v_{\mu_i} \in M(\lambda)_{\mu_i}$. Thus
$$(x - \mu(x)) v = 0$$
$$(x - \mu_i(x)) v_{\mu_i} = 0$$

10.1 Verma modules

for all $x \in H$. Hence

$$(x - \mu_1(x)) \ldots (x - \mu_k(x))(v_{\mu_1} + \cdots + v_{\mu_k}) = 0,$$

that is $(x - \mu_1(x)) \ldots (x - \mu_k(x)) v = 0$. Since the vector space H over $\mathbb{C}$ cannot be expressed as the union of finitely many proper subspaces we can find $x \in H$ such that

$$\mu(x) \neq \mu_1(x), \quad \mu(x) \neq \mu_2(x), \quad \ldots, \quad \mu(x) \neq \mu_k(x).$$

Thus the polynomials

$$t - \mu(x), \quad (t - \mu_1(x))(t - \mu_2(x)) \ldots (t - \mu_k(x))$$

in $\mathbb{C}[t]$ for this element x are coprime. Thus there exist polynomials $p(t), q(t) \in \mathbb{C}[t]$ such that

$$p(t)(t - \mu(x)) + q(t)(t - \mu_1(x)) \ldots (t - \mu_k(x)) = 1.$$

Thus we have

$$p(x)(x - \mu(x)) + q(x)(x - \mu_1(x)) \ldots (x - \mu_k(x)) = 1.$$

It follows that

$$p(x)(x - \mu(x)) v + q(x)(x - \mu_1(x)) \ldots (x - \mu_k(x)) v = v.$$

The above conditions show that the left-hand side is zero, hence we have $v = 0$. Thus

$$M(\lambda) = \bigoplus_\mu M(\lambda)_\mu.$$

Let $\Lambda = \{\mu \in H^* \, ; \, \lambda - \mu \text{ is a sum of positive roots}\}$. For each $\lambda \in \Lambda$ let N_μ be the subspace of $M(\lambda)$ spanned by the basis vectors

$$f_{\beta_1}^{r_1} \ldots f_{\beta_N}^{r_N} m_\lambda$$

with $\lambda - r_1 \beta_1 - \cdots - r_N \beta_N = \mu$. Since these vectors for all such μ form a basis of $M(\lambda)$ we have

$$M(\lambda) = \bigoplus_{\mu \in \Lambda} N_\mu.$$

On the other hand we know that

$$N_\mu \subset M(\lambda)_\mu$$

and $M(\lambda) = \bigoplus_{\mu \in H^*} M(\lambda)_\mu$. It follows that $M(\lambda)_\mu = N_\mu$ for all $\mu \in \Lambda$, and that $M(\lambda)_\mu = O$ for all $\mu \in H^*$ with $\mu \notin \Lambda$.

Thus we have $\dim M(\lambda)_\mu = \dim N_\mu =$ the number of vectors $(r_1, \ldots, r_N)$ with $r_i \in \mathbb{Z}$, $r_i \geq 0$ such that

$$\lambda - r_1\beta_1 - \cdots - r_N\beta_N = \mu.$$

This gives $\dim M(\lambda)_\mu = \mathfrak{P}(\lambda - \mu)$ as required. □

Definition 10.8 $\mu \in H^*$ is called a **weight** of $M(\lambda)$ if $M(\lambda)_\mu \neq O$, and $M(\lambda)_\mu$ is called the **weight space** of $M(\lambda)$ with weight μ.

We note that since $M(\lambda) = \oplus_\mu M(\lambda)_\mu$ an element m of $M(\lambda)$ satisfying the condition that, for all $x \in H$, $(x - \mu(x))^k m = 0$ for some $k > 0$ can have no non-zero component in any $M(\lambda)_\nu$ for $\nu \neq \mu$, and must therefore lie in $M(\lambda)_\mu$. Thus we have

$$M(\lambda)_\mu = \{ m \in M(\lambda) \; ; \text{ for each } x \in H \text{ there exists } k \text{ such that}$$
$$(x - \mu(x))^k m = 0\}.$$

This shows that our definitions of weight and weight space here in the context of H-modules are compatible with the definitions in Chapter 2 in the context of representations of nilpotent Lie algebras.

Theorem 10.7 asserts that a Verma module is the direct sum of its weight spaces. There are infinitely many weights, but each weight space is finite dimensional.

We now proceed to another very important property of Verma modules.

Theorem 10.9 $M(\lambda)$ has a unique maximal submodule.

Proof. Let V be a $\mathfrak{U}(L)$-submodule of $M(\lambda)$ with $V \neq M(\lambda)$. Let $v \in V$. By Theorem 10.7 we have

$$v = \sum_i v_{\mu_i} \qquad v_{\mu_i} \in M(\lambda)_{\mu_i}$$

summed over a finite set of distinct weights μ_i. We aim to show that each v_{μ_i} lies in V also. We have

$$xv_{\mu_i} = \mu_i(x)v_{\mu_i} \qquad x \in H.$$

Hence

$$\prod_{\substack{j \\ j \neq i}} (x - \mu_j(x))\, v = \prod_{\substack{j \\ j \neq i}} (x - \mu_j(x))\, v_{\mu_i} = \prod_{\substack{j \\ j \neq i}} (\mu_i(x) - \mu_j(x))\, v_{\mu_i}.$$

Since H is not the union of finitely many proper subspaces we can find $x \in H$ with $\mu_i(x) \ne \mu_j(x)$ for all $j \ne i$. For such an x we have

$$\prod_{\substack{j \\ j \ne i}} (\mu_i(x) - \mu_j(x)) \, v_{\mu_i} \in V$$

and

$$\prod_{\substack{j \\ j \ne i}} (\mu_i(x) - \mu_j(x)) \ne 0.$$

It follows that $v_{\mu_i} \in V$.

We now define $V_\mu = V \cap M(\lambda)_\mu$. We have shown that $V = \sum_\mu V_\mu$. Since we know that $M(\lambda) = \bigoplus_\mu M(\lambda)_\mu$ it follows that the sum in V must be direct, that is

$$V = \bigoplus_\mu V_\mu.$$

Thus every submodule V of $M(\lambda)$ is also the direct sum of its weight spaces. Now $V_\lambda = O$. For if $V_\lambda \ne O$ then $V_\lambda = M(\lambda)_\lambda$ since $\dim M(\lambda)_\lambda = 1$. This would imply that $m_\lambda \in V$. But then

$$M(\lambda) = \mathfrak{U}(L)m_\lambda \subset V$$

so $V = M(\lambda)$, a contradiction. Thus $V_\lambda = O$ and we have

$$V = \bigoplus_{\substack{\mu \\ \mu \ne \lambda}} V_\mu \subset \sum_{\substack{\mu \\ \mu \ne \lambda}} M(\lambda)_\mu.$$

Thus every proper submodule V of $M(\lambda)$ lies in the subspace $\sum_{\substack{\mu \\ \mu \ne \lambda}} M(\lambda)_\mu$ of codimension 1 in $M(\lambda)$. Let $J(\lambda)$ be the sum of all the proper submodules of $M(\lambda)$. $J(\lambda)$ lies in the above subspace of codimension 1, so is a proper submodule of $M(\lambda)$. Thus $J(\lambda)$ is the unique maximal submodule of $M(\lambda)$, since it contains all proper submodules of $M(\lambda)$. □

Definition 10.10 *Let $\lambda, \mu \in H^*$. In view of Theorem 10.7 it is natural to make the following definition.*

We say that $\lambda \succ \mu$ if $\lambda - \mu$ is a sum of positive roots. This is a partial order on H^*.

Theorem 10.7 shows that the weights of $M(\lambda)$ are precisely the $\mu \in H^*$ with $\mu \prec \lambda$. Thus λ is the highest weight of $M(\lambda)$ with respect to this partial order.

*$M(\lambda)$ is called the **Verma module with highest weight** λ.*

186   *Irreducible modules for semisimple Lie algebras*

We also define $L(\lambda) = M(\lambda)/J(\lambda)$. Since $J(\lambda)$ is a maximal submodule of $M(\lambda)$, $L(\lambda)$ is an irreducible $\mathfrak{U}(L)$-module. In subsequent sections of this chapter we shall determine under what circumstances the irreducible module $L(\lambda)$ is finite dimensional. We note that λ is a weight of $L(\lambda)$, since $J(\lambda)_\lambda = O$. Thus $\dim L(\lambda)_\lambda = 1$ and λ is the highest weight of $L(\lambda)$.

10.2 Finite dimensional irreducible modules

Now let V be any finite dimensional irreducible L-module where, as usual in this chapter, L is a finite dimensional semisimple Lie algebra over $\mathbb{C}$. Let H be a Cartan subalgebra of L and

$$\{e_\alpha, \alpha \in \Phi^+; h_i, i = 1, \ldots, l\,; f_\alpha, \alpha \in \Phi^+\}$$

be a basis of L adapted to H. We may regard V as an H-module. Now H is abelian, so in particular nilpotent, thus we may apply the representation theory of nilpotent Lie algebras developed in Chapter 2. By Theorem 2.9 we have

$$V = \bigoplus_\lambda V_\lambda$$

where $V_\lambda = \{v \in V\,;\ \text{for each } x \in H \text{ there exists } k \text{ such that } (x - \lambda(x))^k v = 0\}$. We also know from Chapter 2 that each non-zero V_λ contains a non-zero vector v such that

$$xv = \lambda(x)v \qquad \text{for all } x \in H.$$

We shall show that in our present situation the weight spaces V_λ can be defined more simply.

Proposition 10.11 *Let $W_\lambda = \{v \in V\,;\ xv = \lambda(x)v \text{ for all } x \in H\}$. Then $W_\lambda = V_\lambda$.*

Proof. It is clear that $W_\lambda \subset V_\lambda$ and that $W_\lambda \neq O$ whenever $V_\lambda \neq O$. Let $W = \sum_\lambda W_\lambda$. Since $V = \bigoplus_\lambda V_\lambda$ and $W_\lambda \subset V_\lambda$ we see that $W = \bigoplus_\lambda W_\lambda$. We shall show that W is a submodule of V. To see this it is sufficient to show that $h_i w, e_\alpha w, f_\alpha w$ lie in W for all $w \in W_\lambda$, all $i = 1, \ldots, l$ and all $\alpha \in \Phi^+$. Now we have

$$h_i w = \lambda(h_i) w \in W$$

$$x(e_\alpha w) = e_\alpha(xw) + \alpha(x) e_\alpha w$$
$$= \lambda(x) e_\alpha w + \alpha(x) e_\alpha w$$
$$= (\lambda + \alpha)(x) e_\alpha w.$$

10.2 Finite dimensional irreducible modules

Hence $e_\alpha w \in W_{\lambda+\alpha} \subset W$. Similarly we have $f_\alpha w \in W_{\lambda-\alpha} \subset W$. Thus W is a $\mathfrak{U}(L)$-submodule of V. Since $W \neq O$ and V is irreducible we have $W = V$. It follows that $W_\lambda = V_\lambda$ for each $\lambda \in H^*$. $\square$

Thus the irreducible module V is the direct sum of its weight spaces V_λ and V_λ is the set of all $v \in V$ such that $xv = \lambda(x)v$ for all $x \in H$.

We now consider the set of all weights λ for V, that is the set of all $\lambda \in H^*$ for which $V_\lambda \neq O$. This is a finite set, so will contain at least one weight maximal in the partial order $\succ$ defined in Definition 10.10. Let λ be such a weight of V. If $\mu \succ \lambda$ and $\mu \neq \lambda$ then μ is not a weight of V.

We may choose $v_\lambda \in V_\lambda$ with $v_\lambda \neq 0$.

Proposition 10.12 (i) $xv_\lambda = \lambda(x)v_\lambda$ for all $x \in H$.
(ii) $e_\alpha v_\lambda = 0$ for all $\alpha \in \Phi^+$.
(iii) $V = \mathfrak{U}(N^-)v_\lambda$
(iv) λ is the highest weight of V.

Proof. Condition (i) is clear. We have

$$x(e_\alpha v_\lambda) = e_\alpha(xv_\lambda) + \alpha(x)e_\alpha v_\lambda$$

for all $x \in H$. Now if $e_\alpha v_\lambda \neq 0$ this implies that $\lambda + \alpha$ is a weight of V. But $\lambda + \alpha \succ \lambda$ so this cannot be the case. Hence $e_\alpha v_\lambda = 0$ for $\alpha \in \Phi^+$.

Now $V = \mathfrak{U}(L)v_\lambda$ since $v_\lambda \neq 0$ and V is an irreducible $\mathfrak{U}(L)$-module. Thus each element of V is a linear combination of elements of the form

$$f_{\beta_1}^{r_1} \ldots f_{\beta_N}^{r_N} h_1^{s_1} \ldots h_l^{s_l} e_{\beta_1}^{t_1} \ldots e_{\beta_N}^{t_N} v_\lambda.$$

This element is 0 unless all t_i are 0. In that case it is a scalar multiple of

$$f_{\beta_1}^{r_1} \ldots f_{\beta_N}^{r_N} v_\lambda.$$

Hence $V = \mathfrak{U}(N^-)v_\lambda$.

Finally we have

$$x\left(f_{\beta_1}^{r_1} \ldots f_{\beta_N}^{r_N} v_\lambda\right) = (\lambda - r_1\beta_1 - \cdots - r_N\beta_N)(x) f_{\beta_1}^{r_1} \ldots f_{\beta_N}^{r_N} v_\lambda$$

as in the proof of Theorem 10.7 ; thus all weights of V have form

$$\mu = \lambda - r_1\beta_1 - \cdots - r_N\beta_N.$$

Thus $\lambda \succ \mu$ for all weights μ of V. $\square$

It follows from Proposition 10.12 that the set of weights of V has a unique maximal element λ with respect to the partial order $\succ$.

We now compare the finite dimensional module V with the Verma module $M(\lambda)$.

Proposition 10.13 *There exists a surjective homomorphism* $\theta : M(\lambda) \to V$ *of* $\mathfrak{U}(L)$-*modules such that* $\theta(m_\lambda) = v_\lambda$.

Proof. We recall from Theorem 10.6 that each element of $M(\lambda)$ is uniquely expressible in the form um_λ with $u \in \mathfrak{U}(N^-)$. We define a linear map

$$\theta : M(\lambda) \to V$$

by $\theta(um_\lambda) = uv_\lambda$ $u \in \mathfrak{U}(N^-)$. Then θ is surjective by Proposition 10.12 (iii). We must check that θ is a homomorphism of $\mathfrak{U}(L)$-modules. Thus we must show

$$\theta(yum_\lambda) = yuv_\lambda \quad \text{for all } y \in \mathfrak{U}(L).$$

By the PBW basis theorem we know that the element yu of $\mathfrak{U}(L)$ can be written as a finite sum

$$yu = \sum_i a_i b_i c_i$$

where $a_i \in \mathfrak{U}(N^-)$, $b_i \in \mathfrak{U}(H)$, $c_i \in \mathfrak{U}(N)$. Thus

$$yum_\lambda = \sum_i a_i b_i c_i m_\lambda.$$

Now $b_i c_i m_\lambda = \xi_i m_\lambda$ for some $\xi_i \in \mathbb{C}$. Hence

$$yum_\lambda = \left(\sum_i \xi_i a_i \right) m_\lambda.$$

Since $\sum_i \xi_i a_i \in \mathfrak{U}(N^-)$ we have

$$\theta(yum_\lambda) = \theta\left(\left(\sum_i \xi_i a_i \right) m_\lambda \right) = \left(\sum_i \xi_i a_i \right) v_\lambda.$$

On the other hand we have

$$yuv_\lambda = \left(\sum_i a_i b_i c_i \right) v_\lambda = \left(\sum_i \xi_i a_i \right) v_\lambda$$

since $b_i c_i v_\lambda = \xi_i v_\lambda$. Hence $\theta(yum_\lambda) = yuv_\lambda$ for all $y \in \mathfrak{U}(L)$. Thus θ is a homomorphism of $\mathfrak{U}(L)$-modules. $\square$

10.2 Finite dimensional irreducible modules

Corollary 10.14 *V is isomorphic to $L(\lambda)$.*

Proof. Since V is irreducible the kernel of θ must be a maximal submodule of $M(\lambda)$. But $M(\lambda)$ has a unique maximal submodule $J(\lambda)$, by Theorem 10.9. Thus $\ker \theta = J(\lambda)$. Hence V is isomorphic to $M(\lambda)/J(\lambda) = L(\lambda)$. $\square$

Thus we have seen that every finite dimensional irreducible L-module is isomorphic to one of the irreducible modules $L(\lambda)$ obtained as irreducible quotients of Verma modules. However, we shall see that by no means all the $L(\lambda)$ are finite dimensional.

Proposition 10.15 *Suppose $L(\lambda)$ is finite dimensional. Then $\lambda(h_i)$ is a non-negative integer for each $i = 1, \ldots, l$.*

Proof. Let v_λ be a highest weight vector of $L(\lambda)$, that is $v_\lambda \in L(\lambda)_\lambda$ and $v_\lambda \neq 0$. As in Section 7.1 we shall choose elements $e_i \in L_{\alpha_i}, f_i \in L_{-\alpha_i}$ such that $[e_i f_i] = h_i$. We consider the sequence of elements

$$v_\lambda, \quad f_i v_\lambda, \quad f_i^2 v_\lambda, \quad \ldots$$

of $L(\lambda)$. We have

$$x\left(f_i^k v_\lambda\right) = (\lambda - k\alpha_i)(x)\left(f_i^k v_\lambda\right)$$

for all $x \in H$. Thus we have

$$v_\lambda \in L(\lambda)_\lambda, \quad f_i v_\lambda \in L(\lambda)_{\lambda - \alpha_i}, \quad f_i^2 v_\lambda \in L(\lambda)_{\lambda - 2\alpha_i}$$

and so on. Now $L(\lambda)$, being finite dimensional, has only finitely many distinct weights. Thus there exists $p \in \mathbb{Z}, p \geq 0$ such that

$$f_i^k v_\lambda \neq 0 \quad \text{for } k \leq p$$
$$f_i^{p+1} v_\lambda = 0.$$

Let $M = \mathbb{C} v_\lambda + \mathbb{C} f_i v_\lambda + \cdots + \mathbb{C} f_i^p v_\lambda$. This sum is direct since $v_\lambda, f_i v_\lambda, \ldots, f_i^p v_\lambda$ all lie in different weight spaces. We show that M is a submodule with respect to the subalgebra $\langle e_i, h_i, f_i \rangle$ of L. It is clear from the definitions that $h_i M \subset M$ and $f_i M \subset M$. We shall show that $e_i M \subset M$ also.

We verify that $e_i f_i^k v_\lambda \in M$ by induction on k. If $k = 0$ we have $e_i v_\lambda = 0$. If $k > 0$ we have

$$e_i f_i^k v_\lambda = f_i e_i f_i^{k-1} v_\lambda + h_i f_i^{k-1} v_\lambda.$$

Now $e_i f_i^{k-1} v_\lambda \in M$ by induction, hence $e_i f_i^k v_\lambda \in M$ also. Thus M is an $\langle e_i, h_i, f_i \rangle$-submodule. We consider the trace of h_i on M. This can be calculated in two ways. On the one hand we have

$$\text{trace}_M h_i = \text{trace}_M [e_i f_i] = \text{trace}_M (e_i f_i - f_i e_i) = 0.$$

On the other hand we have

$$\text{trace}_M h_i = \lambda(h_i) + (\lambda - \alpha_i)(h_i) + \cdots + (\lambda - p\alpha_i)(h_i)$$
$$= (p+1)\lambda(h_i) - p(p+1)$$

since $\alpha_i(h_i) = 2$. Hence

$$\text{trace}_M h_i = (p+1)\lambda(h_i) - p(p+1).$$

It follows that

$$(p+1)(\lambda(h_i) - p) = 0,$$

that is $\lambda(h_i) = p$. Thus $\lambda(h_i) \in \mathbb{Z}$ and $\lambda(h_i) \geq 0$. $\square$

The condition $\lambda(h_i) \in \mathbb{Z}$, $\lambda(h_i) \geq 0$ for all $i = 1, \ldots, l$ is therefore necessary for $L(\lambda)$ to be finite dimensional. In the next section we shall show that this condition is also sufficient.

10.3 The finite dimensionality criterion

We consider the set of $\lambda \in H^*$ such that $\lambda(h_i) \in \mathbb{Z}$, $\lambda(h_i) \geq 0$ for $i = 1, \ldots, l$.

Definition 10.16 Let $\omega_i \in H^*$ be the element satisfying $\omega_i(h_i) = 1$, $\omega_i(h_j) = 0$ if $j \neq i$. The elements $\omega_1, \ldots, \omega_l \in H^*$ are called the **fundamental weights**.

We note that $\omega_1, \ldots, \omega_l$ are linearly independent, since this is true of $h_1, \ldots, h_l \in H$. Thus $\omega_1, \ldots, \omega_l$ form a basis of H^*. Let $X = \{n_1 \omega_1 + \cdots + n_l \omega_l \; ; \; n_1, \ldots, n_l \in \mathbb{Z}\}$. X is a free abelian subgroup of H^* with basis the set of fundamental weights and is called the **lattice of integral weights** or, briefly, the **weight lattice**. It is clear that an element $\lambda \in H^*$ lies in X if and only if $\lambda(h_i) \in \mathbb{Z}$ for $i = 1, \ldots, l$. Let $X^+ = \{n_1 \omega_1 + \cdots + n_l \omega_l \; ; \; n_i \in \mathbb{Z}, n_i \geq 0 \text{ for } i = 1, \ldots, l\}$. X^+ is called the set of **dominant integral weights**.

An element $\lambda \in H^*$ lies in X^+ if and only if $\lambda(h_i) \in \mathbb{Z}$ and $\lambda(h_i) \geq 0$ for $i = 1, \ldots, l$.

We have seen, therefore, that if $L(\lambda)$ is finite dimensional then λ is a dominant integral weight, and wish to prove the converse.

10.3 The finite dimensionality criterion

We shall first explore the connection between the fundamental weights $\omega_1, \ldots, \omega_l$ and the fundamental roots $\alpha_1, \ldots, \alpha_l$.

Proposition 10.17 $\alpha_i = \sum_j A_{ji} \omega_j$. *Thus the matrix expressing the fundamental roots as linear combinations of the fundamental weights is the transpose of the Cartan matrix.*

Proof. Since $\omega_1, \ldots, \omega_l$ are a basis for H^* there exist $c_{ij} \in \mathbb{C}$ such that

$$\alpha_i = \sum_j c_{ij} \omega_j.$$

Then we have

$$\alpha_i(h_j) = c_{ij}.$$

Hence

$$c_{ij} = \alpha_i(h_j) = \alpha_i\left(\frac{2h'_{\alpha_j}}{\langle h'_{\alpha_j}, h'_{\alpha_j}\rangle}\right) = \left\langle h'_{\alpha_i}, \frac{2h'_{\alpha_j}}{\langle h'_{\alpha_j}, h'_{\alpha_j}\rangle}\right\rangle = 2\frac{\langle h'_{\alpha_i}, h'_{\alpha_j}\rangle}{\langle h'_{\alpha_j}, h'_{\alpha_j}\rangle} = A_{ji}.$$

Thus we obtain $\alpha_i = \sum_j A_{ji} \omega_j$. □

In particular we note that all the fundamental roots are integral combinations of the fundamental weights, so lie in the weight lattice X. However, it is not true that the fundamental weights are, in general, integral combinations of the fundamental roots. We have

$$\omega_i = \sum_j (A^{-1})_{ji} \alpha_j.$$

For example, when L has type A_1 we have

$$\alpha_1 = 2\omega_1, \qquad \omega_1 = \tfrac{1}{2}\alpha_1.$$

When L has type A_2 we have

$$\alpha_1 = 2\omega_1 - \omega_2$$
$$\alpha_2 = -\omega_1 + 2\omega_2$$

and so

$$\omega_1 = \tfrac{2}{3}\alpha_1 + \tfrac{1}{3}\alpha_2$$
$$\omega_2 = \tfrac{1}{3}\alpha_1 + \tfrac{2}{3}\alpha_2.$$

We show that in general the coefficients expressing the ω_i in terms of the α_j are non-negative rational numbers.

Proposition 10.18 (i) $\langle \omega_i, \omega_j \rangle \geq 0$ *for all* i, j.
(ii) ω_i *is a non-negative rational combination of* $\alpha_1, \ldots, \alpha_l$.
(iii) *The coefficients of the inverse* A^{-1} *of the Cartan matrix are non-negative rational numbers.*

Proof. We shall show that condition (i) implies the others, and prove (i) in a subsequent lemma. Since the coefficients of A are integers the coefficients of A^{-1} are rational numbers. We show they are all non-negative.

We know that $\omega_i(h_j) = \delta_{ij}$. This condition is equivalent to

$$\left\langle \omega_i, \frac{2\alpha_j}{\langle \alpha_j, \alpha_j \rangle} \right\rangle = \delta_{ij}$$

where $\langle , \rangle$ is now the Killing form on H^* as defined in Section 5.1. Thus we have

$$\langle \omega_i, \alpha_j \rangle = \delta_{ij} \frac{\langle \alpha_j, \alpha_j \rangle}{2}.$$

Now let $\omega_i = \sum_j c_{ij} \alpha_j$. Then we have

$$\langle \omega_i, \omega_j \rangle = c_{ij} \langle \alpha_j, \omega_j \rangle = c_{ij} \frac{\langle \alpha_j, \alpha_j \rangle}{2}.$$

Thus

$$c_{ij} = 2 \frac{\langle \omega_i, \omega_j \rangle}{\langle \alpha_j, \alpha_j \rangle} \geq 0,$$

since $\langle \omega_i, \omega_j \rangle \geq 0$ and $\langle \alpha_j, \alpha_j \rangle > 0$.

We must now show that $\langle \omega_i, \omega_j \rangle \geq 0$. This will follow from the fact that

$$\left\langle \omega_i, \frac{2\alpha_j}{\langle \alpha_j, \alpha_j \rangle} \right\rangle = \delta_{ij}$$

and the fact that

$$\left\langle \frac{2\alpha_i}{\langle \alpha_i, \alpha_i \rangle}, \frac{2\alpha_j}{\langle \alpha_j, \alpha_j \rangle} \right\rangle \leq 0 \qquad \text{if } i \neq j$$

by Proposition 5.4. The following lemma on Euclidean spaces will give us what we need.

10.3 The finite dimensionality criterion

Lemma 10.19 *Let V be an n-dimensional Euclidean space with a basis $v_1, \ldots, v_n$ satisfying $\langle v_i, v_j \rangle \leq 0$ for all $i \neq j$. Let $w_1, \ldots, w_n$ be the dual basis of V uniquely determined by the conditions $\langle v_i, w_j \rangle = \delta_{ij}$. Then $\langle w_i, w_j \rangle \geq 0$ for all i, j.*

Proof. We use induction on n. If $n = 1$ there is nothing to prove. So assume $n > 1$ and let U be the $(n-1)$-dimensional subspace of V spanned by $v_1, \ldots, v_{n-1}$. Let $w'_1, \ldots, w'_{n-1}$ be the dual basis of $v_1, \ldots, v_{n-1}$ in U. Thus we have

$$\langle v_i, w'_j \rangle = \delta_{ij} \qquad i, j = 1, \ldots, n-1.$$

Let $U^\perp = \{v \in V \ ; \ \langle v, u \rangle = 0 \text{ for all } u \in U\}$. Then $\dim U^\perp = 1$ and $U^\perp$ is the subspace of V spanned by w_n. We see also that $w_i - w'_i \in U^\perp$ for $i = 1, \ldots, n-1$. Thus we have

$$w_i = w'_i + \lambda_i w_n \qquad \text{for some } \lambda_i \in \mathbb{R},$$

for $i = 1, \ldots, n-1$. Taking the scalar product with v_n we have

$$0 = \langle v_n, w'_i \rangle + \lambda_i$$

hence $\lambda_i = -\langle v_n, w'_i \rangle$. We wish to determine the sign of λ_i. By induction we know $\langle w'_i, w'_j \rangle \geq 0$ for $i, j = 1, \ldots, n-1$. This implies that w'_i is a non-negative combination of $v_1, \ldots, v_{n-1}$. Since

$$\langle v_n, v_1 \rangle \leq 0, \ldots \langle v_n, v_{n-1} \rangle \leq 0$$

we see that $\langle v_n, w'_i \rangle \leq 0$ and so $\lambda_i \geq 0$. Hence for $i, j = 1, \ldots, n-1$

$$\langle w_i, w_j \rangle = \langle w'_i + \lambda_i w_n, w'_j + \lambda_j w_n \rangle$$
$$= \langle w'_i, w'_j \rangle + \lambda_i \lambda_j \langle w_n, w_n \rangle \geq 0$$

since $\langle w'_i, w'_j \rangle \geq 0$, $\lambda_i \geq 0$, $\lambda_j \geq 0$, $\langle w_n, w_n \rangle > 0$. It remains to show that $\langle w_i, w_n \rangle \geq 0$ for $i = 1, \ldots, n-1$. We have

$$\langle w_i, w_n \rangle = \langle w'_i, w_n \rangle + \lambda_i \langle w_n, w_n \rangle \geq 0$$

since $\langle w'_i, w_n \rangle = 0$, $\lambda_i \geq 0$, $\langle w_n, w_n \rangle > 0$. □

By applying this lemma in the case where $v_i = \frac{2\alpha_i}{\langle \alpha_i, \alpha_i \rangle}$, $w_i = \omega_i$, we deduce that $\langle \omega_i, \omega_j \rangle \geq 0$ and so Proposition 10.18 is proved. □

We now turn to the main theorem of the present section.

Theorem 10.20 *Suppose $\lambda \in H^*$ is dominant and integral, that is $\lambda \in X^+$. Then the irreducible L-module $L(\lambda)$ is finite dimensional.*

Proof. We have $L(\lambda) = M(\lambda)/J(\lambda)$. We know from Theorem 10.7 that $M(\lambda)$ is the direct sum of its weight spaces and from Theorem 10.9 that any submodule of $M(\lambda)$ is also the direct sum of its weight spaces. This applies in particular to $J(\lambda)$. It follows that $L(\lambda) = M(\lambda)/J(\lambda)$ is also the direct sum of its weight spaces. In fact the same proof as given in Theorem 10.9 shows that any H-submodule of $L(\lambda)$ is the direct sum of its weight spaces.

Let v_λ be a highest weight vector of $L(\lambda)$. Thus $v_\lambda \in L(\lambda)_\lambda$ and $v_\lambda \neq 0$. We consider the sequence of elements

$$v_\lambda, \quad f_i v_\lambda, \quad f_i^2 v_\lambda, \quad \ldots$$

We wish to show that terms in this sequence eventually become zero. In fact we show

$$f_i^{k_i} v_\lambda = 0 \quad \text{where} \quad k_i = \lambda(h_i) + 1.$$

Let m_λ be a highest weight vector of the Verma module $M(\lambda)$ such that $m_\lambda + J(\lambda) = v_\lambda$. We consider the submodule $\mathfrak{U}(L) f_i^{k_i} m_\lambda$ of $M(\lambda)$. As usual we choose elements $e_i \in L_{\alpha_i}, f_i \in L_{-\alpha_i}$ such that $[e_i f_i] = h_i$. We have

$$e_i f_i = f_i e_i + h_i$$
$$e_i f_i^2 = f_i e_i f_i + h_i f_i = f_i^2 e_i + 2 f_i h_i - 2 f_i$$
$$= f_i^2 e_i + 2 f_i (h_i - 1)$$

and inductively we obtain

$$e_i f_i^n = f_i^n e_i + n f_i^{n-1}(h_i - (n-1)).$$

Thus we have

$$e_i f_i^{k_i} m_\lambda = f_i^{k_i} e_i m_\lambda + k_i f_i^{k_i - 1}(h_i - (k_i - 1)) m_\lambda = 0$$

since $e_i m_\lambda = 0$ and $h_i m_\lambda = \lambda(h_i) m_\lambda = (k_i - 1) m_\lambda$. Also if $j \neq i$ then

$$e_j f_i^{k_i} m_\lambda = f_i^{k_i} e_j m_\lambda = 0.$$

Thus $e_j f_i^{k_i} m_\lambda = 0$ for all $j = 1, \ldots, l$. It follows that $e_\alpha f_i^{k_i} m_\lambda = 0$ for all $\alpha \in \Phi^+$, since $e_1, \ldots, e_l$ generate N, by Proposition 7.7. We also know that

$$h_j f_i^{k_i} m_\lambda = (\lambda - k_i \alpha_i)(h_j) f_i^{k_i} m_\lambda$$

since $f_i^{k_i} m_\lambda$ is a weight vector with weight $\lambda - k_i \alpha_i$.

10.3 The finite dimensionality criterion

We now consider an arbitrary basis vector of $\mathfrak{U}(L)$ applied to $f_i^{k_i} m_\lambda$:

$$f_{\beta_1}^{r_1} \cdots f_{\beta_N}^{r_N} h_1^{s_1} \cdots h_l^{s_l} e_{\beta_1}^{t_1} \cdots e_{\beta_N}^{t_N} \left(f_i^{k_i} m_\lambda \right)$$

is zero unless all $t_i = 0$, in which case it will be a scalar multiple of

$$f_{\beta_1}^{r_1} \cdots f_{\beta_N}^{r_N} f_i^{k_i} m_\lambda.$$

This shows that

$$\mathfrak{U}(L) f_i^{k_i} m_\lambda = \mathfrak{U}(N^-) f_i^{k_i} m_\lambda.$$

Now $\mathfrak{U}(N^-) f_i^{k_i}$ is a proper subspace of $\mathfrak{U}(N^-)$ since $k_i = \lambda(h_i) + 1 > 0$. It follows from Theorem 10.6 that $\mathfrak{U}(N^-) f_i^{k_i} m_\lambda$ is a proper subspace of $M(\lambda)$. Hence $\mathfrak{U}(L) f_i^{k_i} m_\lambda$ is a proper submodule of $M(\lambda)$. It therefore lies in the unique maximal submodule $J(\lambda)$ of $M(\lambda)$. Hence $f_i^{k_i} m_\lambda \in J(\lambda)$ and this implies $f_i^{k_i} v_\lambda = 0$.

Now let K be the finite dimensional subspace of $L(\lambda)$ given by

$$K = \mathbb{C} v_\lambda + \mathbb{C} f_i v_\lambda + \cdots + \mathbb{C} f_i^{k_i-1} v_\lambda.$$

We clearly have $HK \subset K$ since each $f_i^n v_\lambda$ is a weight vector. We have $f_i K \subset K$ since $f_i^{k_i} v_\lambda = 0$. We also have $e_i K \subset K$ since

$$\begin{aligned} e_i f_i^n v_\lambda &= f_i^n e_i v_\lambda + n f_i^{n-1} (h_i - (n-1)) v_\lambda \\ &= n (\lambda(h_i) - (n-1)) f_i^{n-1} v_\lambda. \end{aligned}$$

Thus K is a submodule of $L(\lambda)$ for the subalgebra $\langle e_i, H, f_i \rangle$ of L of dimension $l+2$. We shall consider non-zero finite dimensional $\langle e_i, H, f_i \rangle$-submodules of $L(\lambda)$. K is such a submodule. If U is any finite dimensional $\langle e_i, H, f_i \rangle$-submodule of $L(\lambda)$ we claim that LU is also. For LU is finite dimensional and we have, for $u \in U$, $z \in L$, $y \in \langle e_i, H, f_i \rangle$

$$y(zu) = z(yu) + [yz]u \in LU$$

since $yu \in U$ and $[yz] \in L$.

Let V be the sum of all finite dimensional $\langle e_i, H, f_i \rangle$-submodules of $L(\lambda)$. Then $V \neq O$ since V contains K. V is an L-submodule of $L(\lambda)$, since if U is a finite dimensional $\langle e_i, H, f_i \rangle$-submodule of $L(\lambda)$ so is LU. Since $L(\lambda)$ is an irreducible L-module we see that $V = L(\lambda)$. Thus $L(\lambda)$ is a sum of finite dimensional $\langle e_i, H, f_i \rangle$-submodules.

Now each such finite dimensional $\langle e_i, H, f_i \rangle$-submodule of $L(\lambda)$ is the direct sum of its weight spaces, as observed above. Thus we may choose a basis for it consisting of weight vectors, that is vectors spanning 1-dimensional H-modules. Hence we can find a basis of $L(\lambda)$ consisting of weight vectors, each of which lies in some finite dimensional $\langle e_i, H, f_i \rangle$-submodule of $L(\lambda)$. This fact will give useful information about the set of weights of $L(\lambda)$.

Let Λ be the set of all weights of $L(\lambda)$. Thus $\mu \in \Lambda$ if and only if $L(\lambda)_\mu \ne O$. Of course all weights of $L(\lambda)$ are weights of $M(\lambda)$ so have form

$$\lambda - r_1 \beta_1 - \cdots - r_N \beta_N$$

by Theorem 10.7. In particular $\Lambda \subset X$, since $\lambda \in X$ and each $\beta_i \in X$ by Theorem 10.7. Let μ be any element of Λ. Then there is a weight vector $v_\mu \in L(\lambda)$ for μ such that v_μ lies in a finite dimensional $\langle e_i, H, f_i \rangle$-submodule U of $L(\lambda)$.

We consider the vectors

$$\ldots, f_i^2 v_\mu, f_i v_\mu, v_\mu, e_i v_\mu, e_i^2 v_\mu, \ldots$$

These vectors all lie in U and have weights

$$\ldots, \mu - 2\alpha_i, \mu - \alpha_i, \mu, \mu + \alpha_i, \mu + 2\alpha_i, \ldots$$

Since $\dim U$ is finite U has only finitely many weights so there exist $p, q \ge 0$ such that

$$f_i^n v_\mu \ne 0 \quad \text{for } 0 \le n \le p, \qquad f_i^{p+1} v_\mu = 0$$
$$e_i^n v_\mu \ne 0 \quad \text{for } 0 \le n \le q, \qquad e_i^{q+1} v_\mu = 0.$$

Let $V = \mathbb{C} f_i^p v_\mu + \cdots + \mathbb{C} f_i v_\mu + \mathbb{C} v_\mu + \mathbb{C} e_i v_\mu + \cdots + \mathbb{C} e_i^q v_\mu$. Then V is a $\langle e_i, H, f_i \rangle$-submodule of $L(\lambda)$. This follows readily from the relations

$$e_i f_i^n = f_i^n e_i + n f_i^{n-1} (h_i - (n-1))$$
$$f_i e_i^n = e_i^n f_i - n e_i^{n-1} (h_i + (n-1))$$

and the fact that $f_i^{p+1} v_\mu = 0$, $e_i^{q+1} v_\mu = 0$. We consider the trace of h_i on V. On the one hand we have

$$\text{trace}_V h_i = (\mu(h_i) - p\alpha_i(h_i)) + \cdots + \mu(h_i) + \cdots + (\mu(h_i) + q\alpha_i(h_i))$$
$$= (p+q+1)\mu(h_i) + \left(\frac{q(q+1)}{2} - \frac{p(p+1)}{2}\right) \alpha_i(h_i)$$
$$= (p+q+1)\mu(h_i) + (q-p)(p+q+1)$$

since $\alpha_i(h_i) = 2$. On the other hand we have

$$\text{trace}_V h_i = \text{trace}_V [e_i f_i] = \text{trace}_V (e_i f_i - f_i e_i) = 0.$$

Hence $\mu(h_i) = p - q$.

We may make an important interpretation of this result in terms of the Weyl group W. We recall from Section 5.2 that W is the group of linear transformations of $H_{\mathbb{R}}^*$ generated by the reflections s_α with respect to the roots $\alpha \in \Phi$. We write $s_i = s_{\alpha_i}$ and recall from Theorem 5.13 that W is generated by $s_1, \ldots, s_l$. We have

$$s_i(\mu) = \mu - 2\frac{\langle \alpha_i, \mu \rangle}{\langle \alpha_i, \alpha_i \rangle}\alpha_i = \mu - \mu(h_i)\alpha_i$$

since

$$\mu(h_i) = \mu\left(\frac{2h'_{\alpha_i}}{\langle h'_{\alpha_i}, h'_{\alpha_i} \rangle}\right) = \left\langle \mu, \frac{2\alpha_i}{\langle \alpha_i, \alpha_i \rangle}\right\rangle = 2\frac{\langle \alpha_i, \mu \rangle}{\langle \alpha_i, \alpha_i \rangle}.$$

Choosing μ as above, where $\mu(h_i) = p - q$, we have

$$s_i(\mu) = \mu - (p-q)\alpha_i = \mu + (q-p)\alpha_i.$$

Now $\mu + (q-p)\alpha_i$ is one of the weights in the list

$$\mu - p\alpha_i, \ldots, \mu - \alpha_i, \mu, \mu + \alpha_i, \ldots, \mu + q\alpha_i$$

of weights of V. Thus we have shown that if μ is any weight of V then $s_i(\mu)$ is a weight of $L(\lambda)$ also. Since $s_1, \ldots, s_l$ generate W it follows that for any $\mu \in \Lambda$ and any $w \in W$ we have $w(\mu) \in \Lambda$ also. Thus the set of weights Λ of $L(\mu)$ is invariant under the Weyl group. We recall also from Proposition 5.8 that W is finite.

We now claim that for each $\mu \in \Lambda$ there exists $w \in W$ such that $w(\mu) \in X^+$. To see this we consider the finite set of weights $\{w(\mu) \; ; \; w \in W\}$ and pick one maximal in the partial order $\succ$ on H^*. Let ν be such a weight. Then

$$s_i(\nu) = \nu - \nu(h_i)\alpha_i.$$

We know that $\nu \in X$ since $\Lambda \subset X$, hence $\nu(h_i) \in \mathbb{Z}$. If $\nu(h_i) < 0$ we would have $s_i(\nu) \succ \nu$, a contradiction to the choice of ν. Hence $\nu(h_i) \geq 0$. This holds for all $i = 1, \ldots, l$ and so $\nu \in X^+$. Thus each weight in Λ has a W-transform which lies in X^+.

We shall now concentrate on the set $\Lambda \cap X^+$. For any weight $\nu \in \Lambda \cap X^+$ we have $\nu \prec \lambda$. We express λ and ν in terms of the fundamental roots α_j. Since $\lambda, \nu \in X^+$ these weights are non-negative integral combinations of the fundamental weights $\omega_1, \ldots, \omega_l$. By Proposition 10.18 they are therefore non-negative rational combinations of the fundamental roots $\alpha_1, \ldots, \alpha_l$. Thus we have

$$\lambda = \sum_{i=1}^{l} q_i \alpha_i \qquad q_i \in \mathbb{Q} \quad q_i \geq 0$$

$$\nu = \sum_{i=1}^{l} q'_i \alpha_i \qquad q'_i \in \mathbb{Q} \quad q'_i \geq 0.$$

The condition $\lambda \succ \nu$ means simply that $q_i - q'_i$ is a non-negative integer for each $i = 1, \ldots, l$. Now given q_i there are only finitely many q'_i such that $q'_i \geq 0$ and $q_i - q'_i$ is a non-negative integer. Thus given $\lambda \in X^+$ there are only finitely many $\nu \in X^+$ such that $\nu \prec \lambda$. Thus $\Lambda \cap X^+$ is finite. Since every element of Λ can be transformed by an element of W into one of $\Lambda \cap X^+$ and since W is finite we see that Λ is finite. Thus $L(\lambda)$ has only finitely many weights. However, each weight space $L(\lambda)_\mu$ of $L(\lambda)$ is finite dimensional, since

$$\dim L(\lambda)_\mu \leq \dim M(\lambda)_\mu$$

and $\dim M(\lambda)_\mu$ is finite by Theorem 10.7. Thus we have

$$L(\lambda) = \bigoplus_\mu L(\lambda)_\mu$$

with finitely many summands, each finite dimensional. Hence $L(\lambda)$ is finite dimensional. □

We conclude by summarising the main ideas in this somewhat lengthy proof. In order to show that $L(\lambda)$ is finite dimensional it is sufficient to show that $L(\lambda)$ has only finitely many weights, since each weight space is known to be finite dimensional. This can be proved if the set of weights is known to be invariant under the Weyl group, since each weight will be W-equivalent to one in X^+, and there are only finitely many elements of X^+ lower than λ in the partial ordering. It is therefore necessary to show that, for any weight μ of $L(\lambda)$, $s_i(\mu)$ is a weight also. This can be shown provided we know that any weight μ comes from a weight vector lying in a finite dimensional $\langle e_i, H, f_i \rangle$-submodule of $L(\lambda)$. We therefore have to show that $L(\lambda)$ is the sum of its finite dimensional $\langle e_i, H, f_i \rangle$-submodules. This comes from the irreducibility of $L(\lambda)$ provided $L(\lambda)$ has a non-zero finite dimensional $\langle e_i, H, f_i \rangle$-submodule. The existence of such a submodule K is proved above.

We have now completed the determination of the finite dimensional irreducible L-modules where L is a finite dimensional semisimple Lie algebra over $\mathbb{C}$.

10.3 The finite dimensionality criterion

Theorem 10.21 *Let L be a finite dimensional semisimple Lie algebra over $\mathbb{C}$. Then the finite dimensional irreducible L-modules are the modules $L(\lambda)$ for $\lambda \in X^+$. These modules are pairwise non-isomorphic.*

Proof. The fact that any finite dimensional irreducible L-module is isomorphic to $L(\lambda)$ for some λ is proved in Corollary 10.14. The fact that λ must lie in X^+ is proved in Proposition 10.15. The fact that $L(\lambda)$ is finite dimensional when $\lambda \in X^+$ is proved in Theorem 10.20. The fact that the $L(\lambda)$ are pairwise non-isomorphic follows from the fact that λ is the highest weight of $L(\lambda)$. Thus if $\lambda \neq \mu$, $L(\lambda)$ and $L(\mu)$ have different highest weights so cannot be isomorphic. □

A property of $L(\lambda)$ which will be very useful subsequently is given by the following proposition.

Proposition 10.22 *Let $\lambda \in X^+$ and $w \in W$. Then*

$$\dim L(\lambda)_\mu = \dim L(\lambda)_{w(\mu)}.$$

Proof. Since W is generated by the fundamental reflections $s_1, \ldots, s_l$ it is sufficient to show that

$$\dim L(\lambda)_\mu = \dim L(\lambda)_{s_i(\mu)}.$$

We recall from Section 7.5 that there is an automorphism θ_i of L such that $\theta_i(H) = H$ and $\theta_i(h) = s_i(h)$ for all $h \in H$. We define an L-module $\bar{L}(\lambda)$ which is the same space $L(\lambda)$ as before but with a different L-action. For $\bar{v} \in \bar{L}(\lambda)$ we have

$$x\bar{v} = \overline{\theta_i(x)v}$$

where v is the corresponding element of $L(\lambda)$. It is clear that this action makes $\bar{L}(\lambda)$ into an L-module.

Now let $v \in L(\lambda)_\mu$. For $x \in H$ we have

$$x\bar{v} = \overline{\theta_i(x)v} = \overline{s_i(x)v} = \overline{\mu(s_i(x))v}$$
$$= (s_i(\mu))(x)\bar{v}.$$

Thus $\bar{v} \in \bar{L}(\lambda)_{s_i(\mu)}$. A similar argument shows that if $\bar{v} \in \bar{L}(\lambda)_{s_i(\mu)}$ then $v \in L(\lambda)_\mu$. Hence

$$\dim \bar{L}(\lambda)_{s_i(\mu)} = \dim L(\lambda)_\mu.$$

Now $\bar{L}(\lambda)$ is an irreducible L-module, since $L(\lambda)$ is irreducible. For if $\bar{M}$ were a submodule of $\bar{L}(\lambda)$ the corresponding subspace M would be a submodule of $L(\lambda)$. Let Λ be the set of weights of $L(\lambda)$. Then we have seen that $s_i(\Lambda)$ is the set of weights of $\bar{L}(\lambda)$. But we showed in the proof of Theorem 10.20 that $w(\Lambda) = \Lambda$ for all $w \in W$. Hence the set of weights of $\bar{L}(\lambda)$ is also Λ. In particular the highest weight of $\bar{L}(\lambda)$ is λ. Thus $\bar{L}(\lambda)$ is a finite dimensional irreducible L-module with highest weight λ. Hence $\bar{L}(\lambda)$ is isomorphic to $L(\lambda)$ by Theorem 10.21. Thus we have

$$\dim L(\lambda)_\mu = \dim \bar{L}(\lambda)_{s_i(\mu)} = \dim L(\lambda)_{s_i(\mu)}.$$

Since each $w \in W$ is a product of elements s_i we deduce that

$$\dim L(\lambda)_\mu = \dim L(\lambda)_{w(\mu)}$$

as required. □

11
Further properties of the universal enveloping algebra

11.1 Relations between the enveloping algebra and the symmetric algebra

Let L be any finite dimensional Lie algebra over $\mathbb{C}$. Let T be the tensor algebra of L. We recall that the enveloping algebra $\mathfrak{U}(L)$ is defined by

$$\mathfrak{U}(L) = T/J$$

where J is the 2-sided ideal of T generated by all elements of the form

$$x \otimes y - y \otimes x - [xy]$$

for $x, y \in L$. The symmetric algebra $S(L)$ is defined by

$$S(L) = T/I$$

where I is the 2-sided ideal of T generated by all elements of the form

$$x \otimes y - y \otimes x$$

for $x, y \in L$.

$S(L)$ is isomorphic, as $\mathbb{C}$-algebra, to the polynomial ring $\mathbb{C}[z_1, \ldots, z_n]$ where $n = \dim L$. We have

$$S(L) = \bigoplus_k S^k(L)$$

where $S^k(L) = (T^k + I)/I$.

$S^k(L)$ is the set of homogeneous elements of $S(L)$ of degree k. In particular we have an isomorphism

$$L = T^1 \to S^1(L)$$

thus L can be regarded as a subspace of $S(L)$.

If $x_1, \ldots, x_n$ are a basis of L then the elements

$$x_1^{r_1} \ldots x_n^{r_n} \qquad r_1, \ldots, r_n \geq 0$$

form a basis of $S(L)$.

We now explain how $S(L)$ can be regarded as a left L-module. In the first place L is an L-module under the adjoint action. Then T may be made into an L-module by means of the action

$$y\left(x_{i_1} \otimes \cdots \otimes x_{i_k}\right) = [yx_{i_1}] \otimes x_{i_2} \otimes \cdots \otimes x_{i_k} + \cdots + x_{i_1} \otimes \cdots \otimes x_{i_{k-1}} \otimes [yx_{i_k}].$$

The ideal I of T is then a submodule, and so $S(L) = T/I$ can be given the structure of a left L-module. We have

$$y\left(x_{i_1} \ldots x_{i_k}\right) = [yx_{i_1}] x_{i_2} \ldots x_{i_k} + \cdots + x_{i_1} \ldots x_{i_{k-1}} [yx_{i_k}]$$

where $y \in L$ and the x_{i_α} are basis vectors of L. We note that each $S^k(L)$ is an L-submodule of $S(L)$.

Similarly $\mathfrak{U}(L) = T/J$ can be made into a left L-module. For the ideal J of T is also a submodule since, for $a, b \in L$, we have

$$y(a \otimes b - b \otimes a - [ab]) = [ya] \otimes b + a \otimes [yb] - [yb] \otimes a - b \otimes [ya] - [y[ab]]$$
$$= [ya] \otimes b - b \otimes [ya] - [[ya]b] + a \otimes [yb]$$
$$- [yb] \otimes a - a[yb]$$

since

$$[y[ab]] = [[ya]b] + [a[yb]].$$

We shall find it useful to compare the enveloping algebra $\mathfrak{U}(L)$ with the symmetric algebra $S(L)$. We first compare their $\mathbb{C}$-algebra structures. Of course they need not be isomorphic as $\mathbb{C}$-algebras since $S(L)$ is commutative whereas $\mathfrak{U}(L)$ is in general non-commutative. However, there is a relation between these two algebras: it is the relation between a filtered algebra and the corresponding graded algebra.

A **filtered algebra** is an associative algebra A with a chain of subspaces

$$A_0 \subset A_1 \subset A_2 \subset \cdots$$

such that $\cup_i A_i = A$ and $A_i A_j \subset A_{i+j}$.

A **graded algebra** is an associative algebra A with a decomposition

$$A = A_0 \oplus A_1 \oplus A_2 \oplus \cdots$$

into a direct sum of subspaces such that $A_i A_j \subset A_{i+j}$ for all i, j.

11.1 Relations between enveloping and symmetric algebra

Given any filtered algebra we may obtain a corresponding graded algebra as follows. Let $A = \bigcup_i A_i$ be a filtered algebra. We define vector spaces $B_0, B_1, B_2, \ldots$ by

$$B_0 = A_0, \quad B_1 = A_1/A_0, \quad B_2 = A_2/A_1, \quad \ldots$$

and define the vector space B by

$$B = B_0 \oplus B_1 \oplus B_2 \oplus \cdots.$$

We define a multiplication on B to make it into a graded algebra. It is sufficient to define xy when $x \in B_i$, $y \in B_j$ and to extend this multiplication by linearity. Thus let $x \in A_i/A_{i-1}$, $y \in A_j/A_{j-1}$. Let $x = A_{i-1} + a_i$, $y = A_{j-1} + a_j$. Then, for any pair of elements $u \in A_{i-1}$, $v \in A_{j-1}$ we have

$$(u + a_i)(v + a_j) = uv + ua_j + a_i v + a_i a_j \in A_{i+j-1} + a_i a_j.$$

Thus the coset in A_{i+j}/A_{i+j-1} containing the product of any element in x with any element in y is the same. Thus we may without ambiguity define $xy \in B_{i+j}$ by

$$xy = A_{i+j-1} + a_i a_j.$$

It is readily checked that this multiplication when extended by linearity makes B into a graded algebra. B is called the **associated graded algebra** of the filtered algebra A.

We may regard $\mathfrak{U}(L)$ as a filtered algebra as follows. Let $\mathfrak{U}_i(L)$ be the subspace of $\mathfrak{U}(L)$ generated by all products $a_1 a_2 \ldots a_j$ for $j \le i$, where $a_k \in L$. We also define $\mathfrak{U}_0(L) = \mathbb{C}1$. Then we have

$$\bigcup_i \mathfrak{U}_i(L) = \mathfrak{U}(L)$$

and

$$\mathfrak{U}_0(L) \subset \mathfrak{U}_1(L) \subset \mathfrak{U}_2(L) \subset \cdots.$$

Moreover $\mathfrak{U}_i(L)\mathfrak{U}_j(L) \subset \mathfrak{U}_{i+j}(L)$. Thus $\mathfrak{U}(L)$ is a filtered algebra. We consider its associated graded algebra.

Proposition 11.1 *The associated graded algebra of the filtered algebra $\mathfrak{U}(L)$ is isomorphic to $S(L)$.*

Proof. Let $B = B_0 \oplus B_1 \oplus B_2 \oplus \cdots$ be the associated graded algebra of $\mathfrak{U}(L)$. We first observe that B is a commutative algebra. B is generated as an algebra by 1 and B_1, and $B_1 = \mathfrak{U}_1(L)/\mathfrak{U}_0(L)$. The natural map

$$L \to \mathfrak{U}_1(L)/\mathfrak{U}_0(L)$$

is an isomorphism of vector spaces. For elements $x, y \in L$ we have

$$xy - yx = [xy] \quad \text{in} \quad \mathfrak{U}(L).$$

Thus

$$(\mathfrak{U}_0(L) + x)(\mathfrak{U}_0(L) + y) \equiv (\mathfrak{U}_0(L) + y)(\mathfrak{U}_0(L) + x) \mod \mathfrak{U}_1(L).$$

Hence any two elements of $B_1 = \mathfrak{U}_1(L)/\mathfrak{U}_0(L)$ commute in B, where their product lies in $\mathfrak{U}_2(L)/\mathfrak{U}_1(L)$. It follows that B is a commutative algebra.

We now compare B with the symmetric algebra $S(L)$. Let $x_1, \ldots, x_n$ be a basis of L. Then it follows from the PBW basis theorem that the elements

$$x_1^{r_1} \ldots x_n^{r_n} \qquad r_1 + \cdots + r_n \leq i$$

form a basis of $\mathfrak{U}_i(L)$. Moreover the elements $\mathfrak{U}_{i-1}(L) + x_1^{r_1} \ldots x_n^{r_n}$ with $r_1 + \cdots + r_n = i$ form a basis for $\mathfrak{U}_i(L)/\mathfrak{U}_{i-1}(L) = B_i$. Now we have

$$\left(\mathfrak{U}_{i-1}(L) + x_1^{r_1} \ldots x_n^{r_n}\right)\left(\mathfrak{U}_{j-1}(L) + x_1^{s_1} \ldots x_n^{s_n}\right) = \mathfrak{U}_{i+j-1}(L) + x_1^{r_1} \ldots x_n^{r_n} x_1^{s_1} \ldots x_n^{s_n}.$$

This is equal to

$$\mathfrak{U}_{i+j-1}(L) + x_1^{r_1+s_1} \ldots x_n^{r_n+s_n}$$

since multiplication in B is commutative. This shows that the linear map $S(L) \to B$ defined by

$$x_1^{r_1} \ldots x_n^{r_n} \to \mathfrak{U}_{i-1}(L) + x_1^{r_1} \ldots x_n^{r_n} \qquad \sum r_k = i$$

extends to an isomorphism of algebras. Thus the associated graded algebra of $\mathfrak{U}(L)$ is isomorphic to $S(L)$. $\square$

We now wish to compare the enveloping algebra $\mathfrak{U}(L)$ and the symmetric algebra $S(L)$ as left L-modules. We shall show that they are isomorphic as L-modules. In order to do so we shall first find a complement to $\mathfrak{U}_{i-1}(L)$ in $\mathfrak{U}_i(L)$.

We have $T^i = L \otimes \cdots \otimes L$ (i factors). The symmetric group S_i operates on T^i by

$$\sigma(y_1 \otimes \cdots \otimes y_i) = y_{\sigma^{-1}(1)} \otimes \cdots \otimes y_{\sigma^{-1}(i)}$$

11.1 Relations between enveloping and symmetric algebra

and extending by linearity. A tensor in T^i is called symmetric if it is fixed by all $\sigma \in S_i$. The natural map $T \to \mathfrak{U}(L)$ induces a map $T^i \to \mathfrak{U}_i(L)$. Let $\mathfrak{U}^i(L)$ be the image under this map of the space of symmetric tensors in T^i.

Proposition 11.2 (i) $\mathfrak{U}_i(L) = \mathfrak{U}_{i-1}(L) \oplus \mathfrak{U}^i(L)$.
(ii) *These spaces are all L-submodules of* $\mathfrak{U}(L)$.

Proof. We first show that

$$\mathfrak{U}_i(L) = \mathfrak{U}_{i-1}(L) + \mathfrak{U}^i(L).$$

Let $x_1^{r_1} \ldots x_n^{r_n}$ be a basis element of $\mathfrak{U}_i(L)$ with $r_1 + \cdots + r_n = i$. For each $\sigma \in S_i$ we define $\sigma(x_1^{r_1} \ldots x_n^{r_n})$ to be the element obtained from $x_1^{r_1} \ldots x_n^{r_n}$ by permuting the factors by the permutation σ. Since multiplication in the graded algebra of $\mathfrak{U}(L)$ is commutative we have

$$x_1^{r_1} \ldots x_n^{r_n} = \frac{1}{i!} \sum_{\sigma \in S_i} \sigma(x_1^{r_1} \ldots x_n^{r_n}) + u$$

where $u \in \mathfrak{U}_{i-1}(L)$. Since the sum lies in $\mathfrak{U}^i(L)$ we have

$$\mathfrak{U}_i(L) = \mathfrak{U}_{i-1}(L) + \mathfrak{U}^i(L).$$

We next show that $\mathfrak{U}_{i-1}(L) \cap \mathfrak{U}^i(L) = O$. Any element of $\mathfrak{U}^i(L)$ has the form

$$\sum_{\substack{r_1, \ldots, r_n \\ r_1 + \cdots + r_n = i}} \lambda_{r_1, \ldots, r_n} \sum_{\sigma \in S_i} \sigma(x_1^{r_1} \ldots x_n^{r_n}).$$

We express this element as a linear combination of basis elements of $\mathfrak{U}(L)$. We obtain

$$\sum_{\substack{r_1, \ldots, r_n \\ r_1 + \cdots + r_n = i}} \lambda_{r_1, \ldots, r_n} \sum_{\sigma \in S_i} \sigma(x_1^{r_1} \ldots x_n^{r_n}) = i! \sum_{\substack{r_1, \ldots, r_n \\ r_1 + \cdots + r_n = i}} \lambda_{r_1, \ldots, r_n} x_1^{r_1} \ldots x_n^{r_n} + u$$

where $u \in \mathfrak{U}_{i-1}(L)$, since multiplication in the graded algebra of $\mathfrak{U}(L)$ is commutative. This element can only lie in $\mathfrak{U}_{i-1}(L)$ if each $\lambda_{r_1}, \ldots, \lambda_{r_n}$ is 0. Thus $\mathfrak{U}_{i-1}(L) \cap \mathfrak{U}^i(L) = O$. Hence we have

$$\mathfrak{U}_i(L) = \mathfrak{U}_{i-1}(L) \oplus \mathfrak{U}^i(L).$$

Finally these subspaces are all L-submodules. The subspaces $\mathfrak{U}_i(L)$ and $\mathfrak{U}_{i-1}(L)$ are evidently submodules by the definition of the L-action. $\mathfrak{U}^i(L)$ is an L-submodule since the L-action commutes with the S_i-action on T^i. $\square$

Let T^i_{sym} be the subspace of symmetric tensors in T^i.

Proposition 11.3 *There is a commutative diagram of vector space isomorphisms*

$$T^i_{\text{sym}}(L) \begin{array}{c} \xrightarrow{\alpha} \mathfrak{U}^i(L) \xrightarrow{\gamma} \\ \searrow_{\beta} S^i(L) \nearrow_{\delta} \end{array} \mathfrak{U}_i(L)/\mathfrak{U}_{i-1}(L)$$

where α is induced by the map $T(L) \to \mathfrak{U}(L)$, β is induced by $T(L) \to S(L)$, γ is induced by $\mathfrak{U}_i(L) \to \mathfrak{U}_i(L)/\mathfrak{U}_{i-1}(L)$ and δ is the map of Proposition 11.1.

Example

$$x_1 \otimes x_2 + x_2 \otimes x_1 \begin{array}{c} \xrightarrow{\alpha} x_1 x_2 + x_2 x_1 \xrightarrow{\gamma} \\ \searrow_{\beta} 2x_1 x_2 \nearrow_{\delta} \end{array} \begin{array}{c} \mathfrak{U}_1 + x_1 x_2 + x_2 x_1 \\ \mathfrak{U}_1 + 2 x_1 x_2 \end{array}$$

Proof. It is sufficient to show that $\gamma\alpha(t) = \delta\beta(t)$ where $t = \sum_{\sigma \in S_i} y_{\sigma^{-1}(1)} \otimes \cdots \otimes y_{\sigma^{-1}(i)}$ and $y_k \in L$. We have

$$\alpha(t) = \sum_\sigma y_{\sigma^{-1}(1)} \cdots y_{\sigma^{-1}(i)}$$

$$\gamma\alpha(t) = \mathfrak{U}_{i-1}(L) + \sum_\sigma y_{\sigma^{-1}(1)} \cdots y_{\sigma^{-1}(i)}$$

$$\beta(t) = \sum_\sigma y_{\sigma^{-1}(1)} \cdots y_{\sigma^{-1}(i)}$$

$$\delta\beta(t) = \mathfrak{U}_{i-1}(L) + \sum_\sigma y_{\sigma^{-1}(1)} \cdots y_{\sigma^{-1}(i)}$$

since the difference between $y_{\sigma^{-1}(1)} \cdots y_{\sigma^{-1}(i)}$ and the corresponding element in canonical form lies in $\mathfrak{U}_{i-1}(L)$. Hence $\gamma\alpha(t) = \delta\beta(t)$. □

We now define $\theta : S^i(L) \to \mathfrak{U}^i(L)$ by $\theta = \gamma^{-1}\delta$, and extend this map by linearity to give $\theta : S(L) \to \mathfrak{U}(L)$. θ is called the operation of **symmetrisation**. We have

$$\theta(y_1 y_2 \cdots y_i) = \frac{1}{i!} \sum_{\sigma \in S_i} y_{\sigma^{-1}(1)} \cdots y_{\sigma^{-1}(i)}.$$

Proposition 11.4 $\theta : S(L) \to \mathfrak{U}(L)$ *is an isomorphism of L-modules.*

Proof. We know that θ is an isomorphism of vector spaces and so must show that

$$x \cdot \theta(P) = \theta(x \cdot P) \quad \text{for all } x \in L, P \in S(L).$$

A derivation of an associative algebra A is a linear map $D : A \to A$ such that
$$D(ab) = D(a)b + aD(b)$$
for all $a, b \in A$. It follows from the definition of the L-action that the maps
$$S(L) \to S(L) \qquad \mathfrak{U}(L) \to \mathfrak{U}(L)$$
$$P \to x \cdot P \qquad u \to x \cdot u$$
for $x \in L$ are derivations. Now L may be identified with a subspace of $S(L)$ and the map $P \to x \cdot P$ when restricted to L is $\operatorname{ad} x$. Similarly L may be identified with a subspace of $\mathfrak{U}(L)$ and the map $u \to x \cdot u$ when restricted to L is again $\operatorname{ad} x$. Now $S(L)$ is generated as an algebra by L and 1. We have $D(1) = 0$ for any derivation of $S(L)$. Thus there is a unique derivation of $S(L)$ extending $\operatorname{ad} x$ on L. Similarly $u \to x \cdot u$ is the unique derivation of $\mathfrak{U}(L)$ extending $\operatorname{ad} x$ on L.

Let $D : \mathfrak{U}(L) \to \mathfrak{U}(L)$ be this derivation. D transforms $\mathfrak{U}^i(L)$ into $\mathfrak{U}^i(L)$ for each i. Using the isomorphism γ of Proposition 11.3, D determines a map
$$\bigoplus_i \frac{\mathfrak{U}_i(L)}{\mathfrak{U}_{i-1}(L)} \to \bigoplus_i \frac{\mathfrak{U}_i(L)}{\mathfrak{U}_{i-1}(L)}$$
which is still a derivation. Using the isomorphism δ of Proposition 11.3 we obtain a map $S(L) \to S(L)$ that is still a derivation and which acts as $\operatorname{ad} x$ on L. Thus it is the map $P \to x \cdot P$. Hence for $P \in S(L)$ we have
$$\theta^{-1}(x \cdot \theta(P)) = x \cdot P.$$
Thus $x \cdot \theta(P) = \theta(x \cdot P)$ as required. □

Note The L-action on $\mathfrak{U}(L)$ considered here may be described simply by
$$x \cdot u = xu - ux \qquad x \in L, u \in \mathfrak{U}(L)$$
For this is a derivation of $\mathfrak{U}(L)$ which extends $\operatorname{ad} x : L \to L$. □

11.2 Invariant polynomial functions

Let $G = \operatorname{Inn} L$ be the group of inner automorphisms of the Lie algebra L. We recall from Section 3.2 that G is generated by automorphisms of the form $\exp \operatorname{ad} x$ for elements $x \in L$ such that $\operatorname{ad} x$ is nilpotent. We define an action of G on L^* by
$$(gf)x = f(g^{-1}x) \qquad g \in G, f \in L^*, x \in L.$$

The tensor algebra

$$T(L^*) = \bigoplus_{k \geq 0} \underbrace{(L^* \otimes \cdots \otimes L^*)}_{k \text{ factors}}$$

may then be made into a G-module satisfying

$$g(f_1 \otimes \cdots \otimes f_k) = gf_1 \otimes \cdots \otimes gf_k$$

for $g \in G$, $f_i \in L^*$. Let I be the 2-sided ideal of $T(L^*)$ generated by all elements of form

$$f \otimes g - g \otimes f$$

for $f, g \in L^*$. Then I is a G-submodule of $T(L^*)$. Let

$$S(L^*) = T(L^*)/I.$$

Then $S(L^*)$ may also be made into a G-module. $S(L^*)$ is the symmetric algebra on L^*. The algebra $S(L^*)$ may be identified with the algebra of polynomial functions on L. The element $I + f_1 \otimes \cdots \otimes f_k$ of $S(L^*)$ gives rise to the polynomial function $f_1 f_2 \ldots f_k$ on L. We define $P(L) = S(L^*)$ and $P^m(L) = S^m(L^*)$. This is the image of $T^m(L^*)$ under the natural homomorphism $T(L^*) \to S(L^*)$. $P^k(L)$ is the space of homogeneous polynomial functions of degree k on L. In particular $P^1(L)$ may be identified with L^*. Each subspace $P^k(L)$ is clearly a G-submodule of $P(L)$.

We now prove some lemmas which will help in understanding the action of G on $P(L)$.

Lemma 11.5 *The linear map* $\alpha : T^m(L^*) \to (T^m(L))^*$ *uniquely determined by*

$$(\alpha(f_1 \otimes \cdots \otimes f_m))(x_1 \otimes \cdots \otimes x_m) = f_1(x_1) f_2(x_2) \ldots f_m(x_m)$$

is an isomorphism of G-modules. Here $x_1, \ldots, x_m$ lie in L and $f_1, \ldots, f_m$ in L^*.

Proof. The linear map α is clearly injective. Since $T^m(L^*)$ and $(T^m(L))^*$ have the same dimension, α must also be surjective. Thus α is an isomorphism of vector spaces. We must also show that

$$\alpha(\gamma \cdot f_1 \otimes \cdots \otimes f_m) = \gamma(\alpha(f_1 \otimes \cdots \otimes f_m))$$

for all $\gamma \in G$. Now we have

$$\gamma(f_1 \otimes \cdots \otimes f_m) = \gamma f_1 \otimes \cdots \otimes \gamma f_m.$$

11.2 Invariant polynomial functions

Thus

$$\alpha(\gamma \cdot f_1 \otimes \cdots \otimes f_m)(x_1 \otimes \cdots \otimes x_m) = (\gamma f_1)(x_1) \ldots (\gamma f_m)(x_m)$$
$$= f_1(\gamma^{-1} x_1) \ldots f_m(\gamma^{-1} x_m).$$

On the other hand

$$\gamma(\alpha(f_1 \otimes \cdots \otimes f_m))(x_1 \otimes \cdots \otimes x_m) = \alpha(f_1 \otimes \cdots \otimes f_m)(\gamma^{-1} x_1 \otimes \cdots \otimes \gamma^{-1} x_m)$$
$$= f_1(\gamma^{-1} x_1) \ldots f_m(\gamma^{-1} x_m).$$

This gives the required equality. □

Lemma 11.6 *Consider the maps*

$$T^m(L)^* \xrightarrow{\alpha^{-1}} T^m(L^*) \xrightarrow{\beta} S^m(L^*)$$

and let $\theta : (T^m(L))^* \longrightarrow S^m(L^*)$ *be given by* $\theta = \beta \alpha^{-1}$. *Thus* θ *is a homomorphism of G-modules. Then we have* $(\theta f)x = f(x \otimes \cdots \otimes x)$ *with m factors, for* $x \in L$.

Proof. It is sufficient to prove this when f has the form

$$f(x_1 \otimes \cdots \otimes x_m) = f_1(x_1) \ldots f_m(x_m)$$

that is when $\alpha^{-1} f = f_1 \otimes \cdots \otimes f_m$. In this case we have

$$(\theta f)x = f_1(x) f_2(x) \ldots f_m(x) = f(x \otimes \cdots \otimes x). \quad \square$$

An element $f \in (T^m(L))^*$ is called symmetric if

$$f(x_1 \otimes \cdots \otimes x_m) = f(x_{\sigma(1)} \otimes \cdots \otimes x_{\sigma(m)})$$

for all $x_1, \ldots, x_m \in L$ and all $\sigma \in S_m$. The set of symmetric elements of $(T^m(L))^*$ will be denoted by $(T^m(L))^*_{\text{sym}}$.

An element of $T^m(L^*)$ is called symmetric if it is invariant under the linear maps which transform $f_1 \otimes \cdots \otimes f_m$ to $f_{\sigma(1)} \otimes \cdots \otimes f_{\sigma(m)}$ for all $\sigma \in S_m$.

The set of symmetric elements of $T^m(L^*)$ will be denoted by $T^m(L^*)_{\text{sym}}$.

Lemma 11.7 *The subspaces* $T^m(L)^*_{\text{sym}}$ *and* $T^m(L^*)_{\text{sym}}$ *are G-submodules. Moreover the maps* α^{-1}, β *give isomorphisms*

$$T^m(L)^*_{\text{sym}} \xrightarrow{\alpha^{-1}} T^m(L^*)_{\text{sym}} \xrightarrow{\beta} S^m(L^*).$$

Proof. The subspaces are G-submodules since the G-action commutes with the S_m-action on $T^m(L)^*$ and $T^m(L^*)$. The map α transforms $T^m(L^*)_{\text{sym}}$ into $T^m(L)^*_{\text{sym}}$ and, since α is an isomorphism and these two spaces have the same dimension, we have

$$\alpha\left(T^m(L^*)_{\text{sym}}\right) = T^m(L)^*_{\text{sym}}.$$

Again the spaces $T^m(L^*)_{\text{sym}}$ and $S^m(L^*)$ have the same dimension and the map

$$\beta : T^m(L^*)_{\text{sym}} \longrightarrow S^m(L^*)$$

is surjective, since it transforms

$$\frac{1}{m!} \sum_{\sigma \in S_m} f_{\sigma(1)} \otimes \cdots \otimes f_{\sigma(m)}$$

into $f_1 f_2 \ldots f_m$. Thus this map is also an isomorphism. $\square$

The G-module isomorphism

$$P^m(L) \xrightarrow{\beta^{-1}\alpha} T^m(L)^*_{\text{sym}}$$

is useful in determining the G-action on $P^m(L)$, since it is often easier to calculate the action on the linear functions in $T^m(L)^*_{\text{sym}}$ than on the polynomial functions in $P^m(L)$.

We shall now assume that the Lie algebra L is semisimple. The group G of inner automorphisms is called the **adjoint group** of L. A polynomial function $P \in P(L)$ is called **invariant** if $\gamma(P) = P$ for all $\gamma \in G$. The set of invariant polynomial functions on L is denoted by $P(L)^G$. This is clearly a subalgebra of $P(L)$. We shall investigate the algebra of invariant polynomial functions on L by relating it to the algebra of polynomial functions on a Cartan subalgebra of L invariant under the Weyl group.

Let H be a Cartan subalgebra of L and $P(H) = S(H^*)$ be the algebra of polynomial functions on H. Let W be the Weyl group of L. Then we know that both H and H^* are W-modules. (We recall from Section 5.2 that an action of W was defined on the real subspace $H^*_{\mathbb{R}}$ of H^*, and this gives rise to a W-action on H^* by linearity.) The W-actions on H and H^* are related by

$$(wf)h = f(w^{-1}h) \qquad w \in W, f \in H^*, h \in H.$$

There is then a W-action on $T(H^*)$ satisfying

$$w(f_1 \otimes \cdots \otimes f_m) = wf_1 \otimes \cdots \otimes wf_m.$$

11.2 Invariant polynomial functions

This in turn induces a W-action on $S(H^*) = T(H^*)/I$ since I is a W-submodule of $T(H^*)$. Thus $P(H) = S(H^*)$ may be regarded as a W-module.

A polynomial function $P \in P(H)$ is called W-**invariant** if $w(P) = P$ for all $w \in W$. The set of W-invariant polynomial functions on H will be denoted by $P(H)^W$. This is a subalgebra of $P(H)$.

Now we have an algebra homomorphism

$$\psi : P(L) \to P(H)$$

given by restriction from L to H. We consider the image of $P(L)^G$ under this restriction map. We show first that this image lies in the subalgebra $P(H)^W$.

Proposition 11.8 $\psi\left(P(L)^G\right) \subset P(H)^W$.

Proof. We use the element $\theta_i \in G$ given by

$$\theta_i = \exp \operatorname{ad} e_i \cdot \exp \operatorname{ad}(-f_i) \cdot \exp \operatorname{ad} e_i.$$

We recall from Proposition 7.18 that $\theta_i(H) = H$ and that $\theta_i(h) = s_i(h)$ for all $h \in H$, where $s_i \in W$ is a fundamental reflection. Thus θ_i acts on H in the same way as s_i. It follows that θ_i and s_i also act in the same way on H^*, and on $S(H^*) = P(H)$.

Let $P \in P(L)^G$. Then $\theta_i(P) = P$. We have $\psi(P) \in P(H)$ and so $s_i(\psi(P)) = \psi(P)$. However, the Weyl group W is generated by its fundamental reflections $s_1, \ldots, s_l$, by Theorem 5.13. Thus we have

$$w(\psi(P)) = \psi(P) \qquad \text{for all } w \in W$$

and so $\psi(P) \in P(H)^W$. □

Proposition 11.9 *The map* $\psi : P(L)^G \to P(H)^W$ *is injective.*

Proof. Let R be the set of regular elements of L. We recall from the proof of Proposition 3.12 that there is a polynomial function $F \in P(L)$ such that $x \in R$ if and only if $F(x) \neq 0$. We also recall from Theorem 3.2 that every regular element lies in some Cartan subalgebra and from Theorem 3.13 that any two Cartan subalgebras are conjugate. Thus given any regular element $x \in R$ there exists $\gamma \in G$ such that $\gamma(x) \in H$.

Now suppose $P \in P(L)^G$ satisfies $\psi(P) = O$. Let x be a regular element and let $\gamma \in G$ be such that $\gamma(x) \in H$. Then

$$\psi(P)(\gamma(x)) = 0,$$

that is $P(\gamma(x)) = 0$. Hence $(\gamma^{-1}P)(x) = 0$. Since $P \in P(L)^G$ we have $\gamma^{-1}P = P$ and we may deduce that

$$P(x) = 0.$$

Thus P annihilates all regular elements of L. Hence $P(x) = 0$ whenever $F(x) \neq 0$. By the principle of irrelevance of algebraic inequalities we have

$$P(x) = 0 \quad \text{for all } x \in L,$$

that is $P = O$. □

Finally we show that the map ψ is also surjective.

Theorem 11.10 $\psi : P(L)^G \to P(H)^W$ *is surjective, and is therefore an isomorphism of algebras.*

Proof. We make use of ideas from the representation theory of L. Let $\lambda \in H^*$ be a dominant integral weight and $L(\lambda)$ be the finite dimensional irreducible L-module with highest weight λ. We can choose a basis of $L(\lambda)$ with respect to which $L(\lambda)$ decomposes into a direct sum of 1-dimensional H-modules. Let ρ be the representation of L afforded by this basis. Consider the function $P : L \to \mathbb{C}$ given by

$$P(x) = \operatorname{tr}((\rho(x))^m) \quad x \in L.$$

We claim that $P \in P^m(L)$. For let $b_1, \ldots, b_n$ be a basis of L and let

$$x = \xi_1 b_1 + \cdots + \xi_n b_n \quad \xi_i \in \mathbb{C}.$$

Then we have

$$\rho(x) = \sum_i \xi_i \rho(b_i)$$

$$(\rho(x))^m = \sum_{i_1, \ldots, i_m} \xi_{i_1} \ldots \xi_{i_m} \rho(b_{i_1}) \ldots \rho(b_{i_m})$$

$$\operatorname{trace}(\rho(x))^m = \sum_{i_1, \ldots, i_m} \operatorname{tr}(\rho(b_{i_1}) \ldots \rho(b_{i_m})) \xi_{i_1} \ldots \xi_{i_m}.$$

This is evidently a polynomial function on L which is homogeneous of degree m. Thus $P \in P^m(L)$.

We wish to show that P is an invariant polynomial function, that is $P \in (P^m(L))^G$. We shall make use of the isomorphism

$$P^m(L) \to T^m(L)^*_{\text{sym}}$$

11.2 Invariant polynomial functions

obtained in Lemma 11.7. The element $f \in T^m(L)^*_{\text{sym}}$ corresponding to $P \in P^m(L)$ is given by

$$f(x_1 \otimes \cdots \otimes x_m) = \frac{1}{m!} \sum_{\sigma \in S_m} \operatorname{tr}\left(\rho\left(x_{\sigma(1)}\right) \cdots \rho\left(x_{\sigma(m)}\right)\right).$$

For f certainly lies in $T^m(L)^*_{\text{sym}}$ and

$$f(x \otimes \cdots \otimes x) = \operatorname{tr}(\rho(x)^m).$$

Lemma 11.6 now shows that f corresponds to P.

We recall that $T^m(L)$ may be regarded as an L-module under the action

$$x \cdot (x_1 \otimes \cdots \otimes x_m) = \sum_i x_1 \otimes \cdots \otimes [xx_i] \otimes \cdots \otimes x_m.$$

Its dual space $T^m(L)^*$ then becomes an L-module under the action

$$(xf)(x_1 \otimes \cdots \otimes x_m) = -f(x(x_1 \otimes \cdots \otimes x_m))$$

for $x \in L$, $f \in T^m(L)^*$.

We now consider xf where $f \in T^m(L)^*_{\text{sym}}$ is the function defined above. We have

$$(xf)(x_1 \otimes \cdots \otimes x_m) = -\sum_i f(x_1 \otimes \cdots \otimes [xx_i] \otimes \cdots \otimes x_m)$$

$$= -\frac{1}{m!} \sum_{\sigma \in S_m} \sum_i \operatorname{tr}\left(\rho\left(x_{\sigma(1)}\right) \cdots \rho\left(\left[xx_{\sigma(i)}\right]\right) \cdots \rho\left(x_{\sigma(m)}\right)\right)$$

$$= -\frac{1}{m!} \sum_{\sigma \in S_m} \sum_i \operatorname{tr}\left(\rho\left(x_{\sigma(1)}\right) \cdots \rho(x)\rho\left(x_{\sigma(i)}\right) \cdots \rho\left(x_{\sigma(m)}\right)\right)$$

$$+ \frac{1}{m!} \sum_{\sigma \in S_m} \sum_i \operatorname{tr}\left(\rho\left(x_{\sigma(1)}\right) \cdots \rho\left(x_{\sigma(i)}\right) \rho(x) \cdots \rho\left(x_{\sigma(m)}\right)\right).$$

All the terms in these expressions cancel except those for which $\rho(x)$ occurs at the beginning or the end of the product. Thus we have

$$(xf)(x_1 \otimes \cdots \otimes x_m) = \frac{1}{m!} \sum_{\sigma \in S_m} \left(\operatorname{tr}\left(\rho\left(x_{\sigma(1)}\right) \cdots \rho\left(x_{\sigma(m)}\right) \rho(x)\right)\right.$$

$$\left. - \operatorname{tr}\left(\rho(x)\rho\left(x_{\sigma(1)}\right) \cdots \rho\left(x_{\sigma(m)}\right)\right)\right)$$

$$= 0 \quad \text{since} \quad \operatorname{tr}(AB) = \operatorname{tr}(BA).$$

Thus $xf = 0$ for all $x \in L$.

We now compare the L-action on $T^m(L)^*_{\text{sym}}$ with the G-action.

Let x be an element of L such that $\operatorname{ad} x$ is nilpotent. Then $\exp \operatorname{ad} x \in G$ and G is generated by all such elements. Let

$$\tau(x) : T^m(L)^*_{\text{sym}} \to T^m(L)^*_{\text{sym}}$$

be the linear map given by

$$\tau(x)f' = xf'.$$

Then we have

$$\tau(x)f(x_1 \otimes \cdots \otimes x_m) = \sum_i f(x_1 \otimes \cdots \otimes \operatorname{ad}(-x) \cdot x_i \otimes \cdots \otimes x_m).$$

Thus

$$\left(\frac{\tau(x)^k}{k!}f\right)(x_1 \otimes \cdots \otimes x_m) = \sum_{\substack{i_1,\ldots,i_m \\ i_1+\cdots+i_m=k}} f\left(\frac{(\operatorname{ad}-x)^{i_1}}{i_1!}x_1 \otimes \cdots \otimes \frac{(\operatorname{ad}-x)^{i_m}}{i_m!}x_m\right).$$

Since $\operatorname{ad} x$ is nilpotent the right-hand side is 0 for k sufficiently large. Hence

$$(\exp \tau(x) \cdot f)(x_1 \otimes \cdots \otimes x_m) = \sum_{i_1,\ldots,i_m} f\left(\frac{(\operatorname{ad}-x)^{i_1}}{i_1!}x_1 \otimes \cdots \otimes \frac{(\operatorname{ad}-x)^{i_m}}{i_m!}x_m\right)$$

$$= f(\exp \operatorname{ad} -x \cdot x_1 \otimes \cdots \otimes \exp \operatorname{ad} -x \cdot x_m)$$

$$= (\exp \operatorname{ad} -x \cdot f)(x_1 \otimes \cdots \otimes x_m).$$

Thus we see that

$$\exp \operatorname{ad} -x \cdot f = \exp \tau(x) \cdot f.$$

Now we have shown that $xf = 0$, hence $\tau(x)f = 0$. Thus $\exp \tau(x) \cdot f = f$. It follows that

$$\exp \operatorname{ad} -x \cdot f = f.$$

Since this holds for all $x \in L$ with $\operatorname{ad} x$ nilpotent we deduce that

$$f \in \left(T^m(L)^*_{\text{sym}}\right)^G.$$

By Lemma 11.7 it follows that $P \in P^m(L)^G$.

11.2 Invariant polynomial functions

The restriction $\psi(P)$ therefore lies in $P^m(H)^W$. Let $\lambda_1, \lambda_2, \ldots, \lambda_k$ be the weights of $L(\lambda)$ with $\lambda_1 = \lambda$. Then we have

$$\rho(x) = \begin{pmatrix} \lambda_1(x) & & O \\ & \cdot & \\ & \cdot & \\ O & & \lambda_k(x) \end{pmatrix} \quad x \in H$$

$$\rho(x)^m = \begin{pmatrix} \lambda_1(x)^m & & O \\ & \cdot & \\ & \cdot & \\ O & & \lambda_k(x)^m \end{pmatrix}$$

$$\operatorname{tr} \rho(x)^m = \lambda_1^m(x) + \cdots + \lambda_k^m(x).$$

Hence $\psi(P) = \lambda_1^m + \cdots + \lambda_k^m$.

We shall show that polynomial functions of this kind span $P^m(H)^W$. In the first place we know that H^* is spanned by the lattice X of integral weights. It follows that $P^m(H)$ is spanned by the set of monomials of degree m in the integral weights. However, it is well known that the process of polarisation can be used to express such a monomial as a linear combination of mth powers. (For example the formula

$$\lambda_1 \lambda_2 = \tfrac{1}{2}(\lambda_1 + \lambda_2)^2 - \tfrac{1}{2}\lambda_1^2 - \tfrac{1}{2}\lambda_2^2$$

expresses the monomial $\lambda_1 \lambda_2$ as a linear combination of squares.) Thus the elements λ^m for $\lambda \in X$ span $P^m(H)$. It follows that every W-invariant element of $P^m(H)$ is a linear combination of elements of form

$$\sum_{w \in W} w(\lambda)^m \quad \lambda \in X.$$

Since each W-orbit of integral weights contains a dominant integral weight we see that elements of form

$$\sum_{w \in W} w(\lambda)^m \quad \lambda \in X^+$$

span $P^m(H)$.

Now we have $\psi(P) = \lambda_1^m + \cdots + \lambda_k^m$ where $\lambda_1 = \lambda$. λ appears with multiplicity 1 in the set $\{\lambda_1, \ldots, \lambda_k\}$ and each $w(\lambda)$ also appears in this set. Moreover this set is W-invariant, so is a union of W-orbits.

It follows from these facts that $\psi(P) = \sum_{w \in W} w(\lambda)^m + $ a linear combination of terms $\sum_{w \in W} w(\mu)^m$ for $\mu \in X^+$ with $\mu \prec \lambda$. There are only finitely many weights $\mu \in X^+$ with $\mu \prec \lambda$. Therefore we may invert these equations and express $\sum_{w \in W} w(\lambda)^m$ as a linear combination of functions of the form $\psi(P)$ coming from representations with highest weight $\mu \prec \lambda$. Thus $P^m(H)^W$ is spanned by functions of the form $\psi(P)$. Hence $P^m(H)^W$ lies in the image of ψ. Since this is true for all m the image of ψ must be the whole of $P(H)^W$. Thus ψ is surjective.

We therefore have an isomorphism of algebras

$$\psi : P(L)^G \to P(H)^W.$$
□

11.3 The structure of the ring of polynomial invariants

In this section we shall prove a theorem of Chevalley which shows that the ring $P(H)^W$ of W-invariant polynomials on H is isomorphic to a polynomial ring in l variables over $\mathbb{C}$.

We write $I = P(H)^W$ and define $\theta : P(H) \to P(H)$ to be the operation of averaging over W. Thus

$$\theta(P) = \frac{1}{|W|} \sum_{w \in W} w(P).$$

It is clear that $\theta(P(H)) = I$, that θ acts as the identity on I, and that $\theta^2 = \theta$, i.e. θ is idempotent.

Let $P(H)^+$ be the set of polynomial functions with constant term 0, and let $I^+ = I \cap P(H)^+$. Let $P(H)I^+$ be the ideal of $P(H)$ generated by I^+. The elements of $P(H)I^+$ have form

$$P_1 J_1 + \cdots + P_k J_k$$

with $P_i \in P(H), J_i \in I^+$.

Lemma 11.11 *Suppose $J_1, \ldots, J_k$ are elements of I such that J_1 does not lie in the ideal of I generated by $J_2, \ldots, J_k$. Let $P_1, P_2, \ldots, P_k \in P(H)$ be homogeneous polynomials such that*

$$P_1 J_1 + P_2 J_2 + \cdots + P_k J_k = 0.$$

Then $P_1 \in P(H)I^+$.

Proof. We shall show that J_1 does not lie in the ideal of $P(H)$ generated by $J_2, \ldots, J_k$. Suppose this were false. Then we have

$$J_1 = Q_2 J_2 + \cdots + Q_k J_k \quad \text{with } Q_i \in P(H).$$

11.3 The structure of the ring of polynomial invariants

Applying $w \in W$ we obtain

$$J_1 = w(Q_2) J_2 + \cdots + w(Q_k) J_k$$

and therefore

$$J_1 = \theta(Q_2) J_2 + \cdots + \theta(Q_k) J_k.$$

However, $\theta(Q_i) \in I$ and so J_1 lies in the ideal of I generated by $J_2, \ldots, J_k$. This gives the required contradiction.

We now show that $P_1 \in P(H)I^+$ by induction on the degree of the homogeneous polynomial P_1.

If $\deg P_1 = 0$ then P_1 is constant. Since $P_1 J_1 + \cdots + P_k J_k = O$ and J_1 is not in the ideal of $P(H)$ generated by $J_2, \ldots, J_k$ this implies that $P_1 = O$. Thus $P_1 \in P(H)I^+$ in this case.

Now suppose $\deg P_1 > 0$. We recall that W is generated by its fundamental reflections $s_1, \ldots, s_l$. In the W-action on H each s_j has a fixed point set which is a hyperplane in H given by an equation $H_j = O$ where $H_j \in P(H)$ is a homogeneous polynomial of degree 1. We have

$$(s_j(P_i)) x = P_i(s_j(x)) = P_i(x)$$

where $H_j(x) = 0$. Thus the polynomial $s_j(P_i) - P_i$ vanishes at all $x \in H$ for which H_j vanishes. It follows that

$$s_j(P_i) - P_i = H_j \bar{P}_i$$

for some $\bar{P}_i \in P(H)$.

Since P_i is homogeneous, $s_j(P_i)$ is also homogeneous of the same degree, hence $s_j(P_i) - P_i$ is homogeneous. Thus $\bar{P}_i$ is also homogeneous with $\deg \bar{P}_i < \deg P_i$.

Now the relation

$$P_1 J_1 + \cdots + P_k J_k = O$$

implies

$$s_j(P_1) J_1 + \cdots + s_j(P_k) J_k = O$$

and so

$$H_j (\bar{P}_1 J_1 + \cdots + \bar{P}_k J_k) = O.$$

Since H_j is not the zero polynomial this implies that

$$\bar{P}_1 J_1 + \cdots + \bar{P}_k J_k = O.$$

Since $\deg \bar{P}_1 < \deg P_1$ we may deduce by induction that $\bar{P}_1 \in P(H)I^+$. Hence $s_j(P_1) - P_1 \in P(H)I^+$ also.

Now $P(H)I^+$ is a W-submodule of $P(H)$, thus $P(H)/P(H)I^+$ is also a W-module. We have

$$s_j(P_1) \equiv P_1 \mod P(H)I^+$$

and since W is generated by $s_1, \ldots, s_l$ it follows that

$$w(P_1) \equiv P_1 \mod P(H)I^+$$

for all $w \in W$. Hence

$$\theta(P_1) \equiv P_1 \mod P(H)I^+.$$

Now P_1 is a homogeneous polynomial of positive degree, therefore $\theta(P_1) \in I^+$. In particular $\theta(P_1) \in P(H)I^+$ and so $P_1 \in P(H)I^+$ as required. □

Now the ideal $P(H)I^+$ of $P(H)$ is generated by the homogeneous elements of I of positive degree. By Hilbert's basis theorem there is a finite subset of this generating set which generates $P(H)I^+$. Let $I_1, \ldots, I_n$ be a set of homogeneous polynomials in I such that $I_1, \ldots, I_n$ generates $P(H)I^+$ but no proper subset generates $P(H)I^+$.

Proposition 11.12 *The polynomials $I_1, \ldots, I_n$ are algebraically independent.*

Proof. Suppose the result is false. Then there is a non-zero polynomial P in n variables such that

$$P(I_1, \ldots, I_n) = 0.$$

We may assume, by comparing terms of a given degree, that all monomials in $I_1, \ldots, I_n$ which occur in P have the same degree d in $x_1, \ldots, x_l$. Let $P_i = \partial P/\partial I_i$. Then

$$P_i(I_1, \ldots, I_n) \qquad i = 1, \ldots, n$$

are elements of I and not all the P_i are zero.

Let J be the ideal of I generated by $P_1, P_2, \ldots, P_n$. We may choose the notation so that $P_1, \ldots, P_m$ but no proper subset generate J as an ideal in I. Thus there exist polynomials $Q_{i,j} \in I$ such that

$$P_i = \sum_{j=1}^{m} Q_{i,j} P_j \qquad i = m+1, \ldots, n.$$

11.3 The structure of the ring of polynomial invariants

Now each P_i is homogeneous in $x_1, \ldots, x_l$ of degree $d - \deg I_i$. Thus, by comparing terms of the same degree in $x_1, \ldots, x_l$ on both sides, we may assume that each $Q_{i,j}$ is homogeneous of degree $\deg P_i - \deg P_j$.

Now $P(I_1, \ldots, I_n) = O$ thus $\partial P/\partial x_k = 0$ for $k = 1, \ldots, l$. Hence

$$\sum_{i=1}^{n} \frac{\partial P}{\partial I_i} \frac{\partial I_i}{\partial x_k} = 0,$$

that is

$$\sum_{i=1}^{n} P_i \partial I_i / \partial x_k = 0.$$

It follows that

$$\sum_{i=1}^{m} P_i \partial I_i / \partial x_k + \sum_{i=m+1}^{n} \sum_{j=1}^{m} Q_{i,j} P_j \partial I_i / \partial x_k = 0$$

that is

$$\sum_{i=1}^{m} P_i \left(\partial I_i / \partial x_k + \sum_{j=m+1}^{n} Q_{j,i} \partial I_j / \partial x_k \right) = 0.$$

We now apply Lemma 11.11. $P_1, \ldots, P_m$ are in I and P_1 is not in the ideal of I generated by $P_2, \ldots, P_m$. Each of the polynomials

$$\partial I_i / \partial x_k + \sum_{j=m+1}^{n} Q_{j,i} \partial I_j / \partial x_k \qquad i = 1, \ldots, m$$

is homogeneous in $x_1, \ldots, x_l$ of degree $\deg I_i - 1$. For

$$\deg Q_{j,i} = \deg P_j - \deg P_i = \deg I_i - \deg I_j.$$

It follows from Lemma 11.11 that

$$\partial I_1 / \partial x_k + \sum_{j=m+1}^{n} Q_{j,1} \partial I_j / \partial x_k \in P(H) I^+.$$

We now multiply this polynomial by x_k and sum over $k = 1, \ldots, l$. For a homogeneous polynomial I_j in $x_1, \ldots, x_l$ we have, by Euler's formula,

$$\sum_{k=1}^{l} x_k \frac{\partial I_j}{\partial x_k} = \deg I_j \cdot I_j.$$

Thus we have

$$\deg I_1 \cdot I_1 + \sum_{j=m+1}^{n} \deg I_j \cdot Q_{j,1} I_j = \sum_{i=1}^{n} I_i R_i$$

where each $R_i \in P(H)^+$. We note that all the terms on the left-hand side are homogeneous polynomials of degree $\deg I_1$. Comparing terms of this degree on the two sides we obtain

$$\deg I_1 \cdot I_1 + \sum_{j=m+1}^{n} \deg I_j \cdot Q_{j,1} I_j = \sum_i I_i R_i$$

where the sum on the right extends over a subset of $1, \ldots, n$ not including $i = 1$, since $I_1 R_1$ has degree greater than $\deg I_1$. It follows that I_1 is in the ideal of $P(H)$ generated by $I_2, \ldots, I_n$. However, this contradicts the definition of $I_1, \ldots, I_n$. Thus the proposition is proved. □

Proposition 11.13 *Every element of I is a polynomial in $I_1, \ldots, I_n$.*

Proof. It is sufficient to prove this for homogeneous polynomials in I. Let $J \in I$ be homogeneous. We use induction on $\deg J$, the result being clear if $\deg J = 0$. Suppose $\deg J > 0$. Then $J \in I^+$ and in particular $J \in P(H)I^+$. Thus we have

$$J = P_1 I_1 + \cdots + P_n I_n$$

for certain polynomials $P_1, \ldots, P_n \in P(H)$. Since $J, I_1, \ldots, I_n$ are all homogeneous we may clearly assume that each P_i is homogeneous also, with

$$\deg P_i = \deg J - \deg I_i.$$

Then we have

$$J = \theta(P_1) I_1 + \cdots + \theta(P_n) I_n.$$

$\theta(P_1), \ldots, \theta(P_n)$ are homogeneous polynomials in I of degree less than $\deg J$. Thus they are polynomials in $I_1, \ldots, I_n$ by induction, and so J is also. □

Corollary 11.14 *The algebra $P(H)^W = \mathbb{C}[I_1, \ldots, I_n]$ is isomorphic to the polynomial ring in n generators over $\mathbb{C}$.*

Proof. This follows from Propositions 11.12 and 11.13. □

The set $I_1, \ldots, I_n$ is called a set of **basic polynomial invariants** of W. We now determine the number of invariants in a basic set.

Proposition 11.15 *The number n of invariants in a basic set is equal to the dimension l of H.*

11.3 The structure of the ring of polynomial invariants

Proof. Let $K = \mathbb{C}(x_1, \ldots, x_l)$ be the field of rational functions in $x_1, \ldots, x_l$ over $\mathbb{C}$. Also let $k = \mathbb{C}(I_1, \ldots, I_n)$ be the field of rational functions in $I_1, \ldots, I_n$ over $\mathbb{C}$. Then we have inclusions

$$\mathbb{C} \subset k \subset K.$$

Since $x_1, \ldots, x_l$ are algebraically independent over $\mathbb{C}$ the transcendence degree of K over $\mathbb{C}$ is given by

$$\operatorname{tr deg} K/\mathbb{C} = l.$$

Since $I_1, \ldots, I_n$ are algebraically independent over $\mathbb{C}$, by Proposition 11.12, the transcendence degree of k over $\mathbb{C}$ is given by

$$\operatorname{tr deg} k/\mathbb{C} = n.$$

Since we have

$$\operatorname{tr deg} K/\mathbb{C} = \operatorname{tr deg} k/\mathbb{C} + \operatorname{tr deg} K/k$$

we shall consider $\operatorname{tr deg} K/k$. Now K is generated over k by $x_1, \ldots, x_l$. However, each x_i is an algebraic element over k. For the polynomial

$$\prod_{w \in W} (t - w(x_i))$$

has x_i as a root, and its coefficients are the elementary symmetric functions in the $w(x_i)$ as w runs over W. These coefficients are W-invariants and therefore lie in I. In particular this polynomial lies in $k[t]$ and so x_i is algebraic over k. Thus K is generated by a finite number of algebraic elements over k and so

$$\operatorname{tr deg} K/k = 0.$$

It follows that

$$\operatorname{tr deg} K/\mathbb{C} = \operatorname{tr deg} k/\mathbb{C},$$

that is $n = l$. □

Now the set $I_1, \ldots, I_l$ of basic polynomial invariants of W is not uniquely determined. We show, however, that the degrees of these polynomials are uniquely determined.

Proposition 11.16 *Let $I_1, \ldots, I_l$ and $I'_1, \ldots, I'_l$ be two sets of basic polynomial invariants of W in $P(H)$. Then we may arrange the numbering so that*

$$\deg I_i = \deg I'_i \quad \text{for } i = 1, \ldots, l.$$

Proof. Each of $I'_1, \ldots, I'_l$ is expressible as a polynomial in $I_1, \ldots, I_l$ and conversely. Consider the matrices

$$\left(\partial I_i / \partial I'_j\right) \quad \left(\partial I'_i / \partial I_j\right).$$

These are inverse matrices, thus the determinant

$$\det \left(\partial I_i / \partial I'_j\right)$$

is non-zero. It follows that for some permutation σ of $1, \ldots, l$

$$\prod_{i=1}^{l} \frac{\partial I_i}{\partial I'_{\sigma(i)}} \neq 0.$$

By renumbering $I'_1, \ldots, I'_l$ if necessary we may assume σ is the identity. Thus

$$\prod_{i=1}^{l} \frac{\partial I_i}{\partial I'_i} \neq 0$$

and so $\partial I_i / \partial I'_i \neq 0$ for each i. This means that I_i, as a polynomial in $I'_1, \ldots, I'_l$, involves I'_i and so

$$\deg I_i \geq \deg I'_i.$$

This implies that

$$\sum_{i=1}^{l} \deg I_i \geq \sum_{i=1}^{l} \deg I'_i.$$

By symmetry we must have equality. This implies

$$\deg I_i = \deg I'_i \quad \text{for each } i. \qquad \square$$

We summarise the results of this section in the following theorem, due to C. Chevalley.

Theorem 11.17 (a) *The algebra $P(H)^W$ of W-invariant polynomials on H is isomorphic to a polynomial ring in l variables over $\mathbb{C}$.*
(b) *$P(H)^W$ may be generated as a polynomial ring by l homogeneous invariant polynomials $I_1, \ldots, I_l$.*
(c) *The degrees $d_1, \ldots, d_l$ of $I_1, \ldots, I_l$ are independent of the system of generators chosen.*

11.4 The Killing isomorphisms

In the preceding sections we have investigated the algebras $P(L)^G$ and $P(H)^W$ of invariant polynomial functions on L and H respectively. Assuming again that the Lie algebra L is semisimple we show now how to relate these algebras

11.4 The Killing isomorphisms

to algebras $S(L)^G$ and $S(H)^W$ of invariants on the symmetric algebras of L and H.

The action of G on the Lie algebra L may be extended to a G-action on $T(L)$ satisfying

$$\gamma(x_1 \otimes \cdots \otimes x_m) = \gamma x_1 \otimes \cdots \otimes \gamma x_m \qquad \gamma \in G.$$

We then obtain an induced action on $S(L) = T(L)/I$ since I is a G-submodule. $S(L)^G$ is the subalgebra of all G-invariant elements of $S(L)$. We shall relate this to $P(L)^G$ by means of the Killing form.

We recall from Theorem 4.10 that the Killing form on the semisimple Lie algebra L is non-degenerate. This implies that the linear map $L \to L^*$ given by $x \to x^*$ where $x^*(y) = \langle x, y \rangle$ is bijective. We wish to show that this is an isomorphism of G-modules.

Proposition 11.18 *Let $\gamma \in G$ and $x, y \in L$. Then $\langle \gamma x, \gamma y \rangle = \langle x, y \rangle$. Thus the adjoint group preserves the Killing form.*

Proof. Since G is generated by elements $\exp \operatorname{ad} z$ where $z \in L$ is such that $\operatorname{ad} z$ is nilpotent, it is sufficient to show that

$$\langle \exp \operatorname{ad} z \cdot x, \exp \operatorname{ad} z \cdot y \rangle = \langle x, y \rangle.$$

We recall from Proposition 4.5 that

$$\langle [xz], y \rangle = \langle x, [zy] \rangle.$$

Thus $\langle \operatorname{ad} z \cdot x, y \rangle = \langle x, \operatorname{ad} -z \cdot y \rangle$. Iterating we obtain

$$\langle (\operatorname{ad} z)^i x, y \rangle = \langle x, (\operatorname{ad} -z)^i y \rangle.$$

Now we have

$$\exp \operatorname{ad} z = 1 + \operatorname{ad} z + \frac{(\operatorname{ad} z)^2}{2!} + \cdots + \frac{(\operatorname{ad} z)^k}{k!}$$

for some k, since $\operatorname{ad} z$ is nilpotent. Hence

$$\langle \exp \operatorname{ad} z \cdot x, y \rangle = \langle x, \exp \operatorname{ad} -z \cdot y \rangle$$

and so

$$\langle \exp \operatorname{ad} z \cdot x, \exp \operatorname{ad} z \cdot y \rangle = \langle x, y \rangle. \qquad \square$$

Corollary 11.19 *The Killing map $L \to L^*$ is an isomorphism of G-modules.*

Proof. We must show that $(\gamma x)^* = \gamma x^*$ for all $\gamma \in G, x \in L$. We have

$$(\gamma x)^*(y) = \langle \gamma x, y \rangle = \langle x, \gamma^{-1} y \rangle = x^* \left(\gamma^{-1} y \right) = (\gamma x^*)(y).$$

Thus $(\gamma x)^* = \gamma x^*$ as required. □

The Killing map $L \to L^*$ induces an isomorphism $T(L) \to T(L^*)$ and then an isomorphism $S(L) \to S(L^*)$ in an obvious way. This is again an isomorphism of G-modules. There is therefore an isomorphism between $S(L)^G$ and $S(L^*)^G$. We recall that $S(L^*) = P(L)$ and so obtain a Killing isomorphism of algebras $S(L) \to P(L)$ which induces a Killing isomorphism $S(L)^G \to P(L)^G$ between the subalgebras of invariants.

We now consider the action of the Weyl group W on the Cartan subalgebra H of L. We recall from Proposition 4.14 that the Killing form of L remains non-degenerate on restriction to H. Thus the map $H \to H^*$ given by $x \to x^*$ where $x^*(y) = \langle x, y \rangle$ for all $y \in H$ is bijective.

Proposition 11.20 *The Killing map $H \to H^*$ is an isomorphism of W-modules.*

Proof. We have

$$(wh)^* x = \langle wh, x \rangle = \langle h, w^{-1} x \rangle = h^* \left(w^{-1} x \right) = (wh^*) x \quad \text{for all } x \in H.$$

Hence $(wh)^* = wh^*$ as required. □

The Killing isomorphism $H \to H^*$ induces an isomorphism $T(H) \to T(H^*)$ and then an isomorphism $S(H) \to S(H^*)$. This is again an isomorphism of W-modules. Since $S(H^*) = P(H)$ we obtain a Killing isomorphism of algebras $S(H) \to P(H)$ which induces an isomorphism $S(H)^W \to P(H)^W$ between the subalgebras of invariants.

We now consider the relation between $S(L)$ and $S(H)$. We recall that L may be identified with a subspace of $S(L)$ and that L has a triangular decomposition

$$L = N^- \oplus H \oplus N.$$

Let K be the ideal of $S(L)$ generated by N and N^-. Then we have $S(L)/K$ isomorphic to $S(H)$. Let $\eta : S(L) \to S(H)$ be the natural homomorphism given in this way.

11.4 The Killing isomorphisms

Proposition 11.21 *We have a commutative diagram of algebra homomorphisms*

$$\begin{array}{ccc} S(L) & \xrightarrow{\alpha} & P(L) \\ \eta \downarrow & & \downarrow \psi \\ S(H) & \xrightarrow{\beta} & P(H) \end{array}$$

where α, β are the Killing isomorphisms, ψ is restriction from $P(L)$ to $P(H)$, and η is projection from $S(L)$ to $S(H)$.

Proof. We must show $\psi\alpha(Q) = \beta\eta(Q)$ for all $Q \in S(L)$. It is sufficient to prove this when

$$Q = f_{\beta_1}^{r_1} \ldots f_{\beta_N}^{r_N} h_1^{s_1} \ldots h_l^{s_l} e_{\beta_1}^{t_1} \ldots e_{\beta_N}^{t_N}$$

where $\Phi^+ = \{\beta_1, \ldots, \beta_N\}$.

If $r_i = 0$ and $t_i = 0$ for each i then $\eta(Q) = Q$. Moreover $\beta(Q) = \psi\alpha(Q)$. Thus the diagram commutes.

If not all the r_i and t_i are 0 then $\eta(Q) = 0$. Thus $\beta\eta(Q) = 0$. We have

$$\alpha(Q) = \alpha\left(f_{\beta_1}\right)^{r_1} \ldots \alpha\left(f_{\beta_N}\right)^{r_N} \alpha(h_1)^{s_1} \ldots \alpha(h_l)^{s_l} \alpha\left(e_{\beta_1}\right)^{t_1} \ldots \alpha\left(e_{\beta_N}\right)^{t_N}.$$

Therefore, for $x \in H$ we have

$$(\alpha Q)x = \langle f_{\beta_1}, x\rangle^{r_1} \ldots \langle f_{\beta_N}, x\rangle^{r_N} \langle h_1, x\rangle^{s_1} \ldots \langle h_l, x\rangle^{s_l} \langle e_{\beta_1}, x\rangle^{t_1} \ldots \langle e_{\beta_N}, x\rangle^{t_N}.$$

This is 0 since $\langle N^-, H\rangle = 0$ and $\langle N, H\rangle = 0$, and some r_i or t_i is non-zero. Thus the diagram commutes in this case also. □

Corollary 11.22 *We have a commutative diagram of algebra isomorphisms*

$$\begin{array}{ccc} S(L)^G & \xrightarrow{\alpha} & P(L)^G \\ \eta \downarrow & & \downarrow \psi \\ S(H)^W & \xrightarrow{\beta} & P(H)^W \end{array}$$

Proof. We have seen that the Killing isomorphisms α, β map $S(L)^G$ to $P(L)^G$ and $S(H)^W$ to $P(H)^W$, respectively. We also know from Theorem 11.10 that $\psi : P(L)^G \to P(H)^W$ is an isomorphism of algebras. Thus η acts on $S(L)^G$ in the same way as $\beta^{-1}\psi\alpha$. Hence $\eta : S(L)^G \to S(H)^W$ is an algebra isomorphism.

We note by Theorem 11.17 that the four algebras $S(L)^G, P(L)^G, S(H)^W, P(H)^W$ are all isomorphic to the polynomial algebra $\mathbb{C}[z_1, \ldots, z_l]$.

11.5 The centre of the enveloping algebra

The centre $Z(L)$ of $\mathfrak{U}(L)$ is defined by

$$Z(L) = \{z \in \mathfrak{U}(L)\; ; \; zu = uz \quad \text{for all } u \in \mathfrak{U}(L)\}.$$

Proposition 11.23 *The centre $Z(L)$ acts on each Verma module $M(\lambda)$ by scalar multiplications.*

Proof. Let m_λ be the highest weight vector of $M(\lambda)$. Let $z \in Z(L)$ and $h \in H$. Then

$$h(zm_\lambda) = z(hm_\lambda) = \lambda(h)zm_\lambda.$$

Thus $zm_\lambda \in M(\lambda)_\lambda$. Now the λ-weight space of $M(\lambda)$ is 1-dimensional – in fact $M(\lambda)_\lambda = \mathbb{C}m_\lambda$. Hence

$$zm_\lambda = \xi m_\lambda \quad \text{for some } \xi \in \mathbb{C}.$$

Now let $u \in \mathfrak{U}(L)$. Then we have

$$z(um_\lambda) = u(zm_\lambda) = \xi um_\lambda.$$

Since $M(\lambda) = \mathfrak{U}(L)m_\lambda$ we see that z acts on $M(\lambda)$ as scalar multiplication by ξ. □

We write $\chi_\lambda(z) = \xi$. Thus $\chi_\lambda : Z(L) \to \mathbb{C}$ is a 1-dimensional representation of $Z(L)$. χ_λ is called the **central character** of $M(\lambda)$. We shall show how to determine this central character.

We consider $\mathfrak{U}(L)$ as an L-module, as described in Section 11.1. The L-action on $\mathfrak{U}(L)$ is given by

$$x \cdot u = xu - ux \qquad x \in L, u \in \mathfrak{U}(L).$$

$\mathfrak{U}(L)$ has basis

$$f_{\beta_1}^{r_1} \ldots f_{\beta_N}^{r_N} \; h_1^{s_1} \ldots h_l^{s_l} \; e_{\beta_1}^{t_1} \ldots e_{\beta_N}^{t_N}$$

where $\Phi^+ = \{\beta_1, \ldots, \beta_N\}$. If $x \in H$ we have

$$x \cdot f_{\beta_1}^{r_1} \ldots f_{\beta_N}^{r_N} \; h_1^{s_1} \ldots h_l^{s_l} \; e_{\beta_1}^{t_1} \ldots e_{\beta_N}^{t_N} = (-r_1\beta_1 - \cdots - r_N\beta_N$$
$$+ t_1\beta_1 + \cdots + t_N\beta_N)(x) f_{\beta_1}^{r_1} \ldots f_{\beta_N}^{r_N} \; h_1^{s_1} \ldots h_l^{s_l} \; e_{\beta_1}^{t_1} \ldots e_{\beta_N}^{t_N}.$$

Thus $f_{\beta_1}^{r_1} \ldots f_{\beta_N}^{r_N} \; h_1^{s_1} \ldots h_l^{s_l} \; e_{\beta_1}^{t_1} \ldots e_{\beta_N}^{t_N}$ is a weight vector with weight $(t_1 - r_1)\beta_1 + \cdots + (t_N - r_N)\beta_N$.

11.5 The centre of the enveloping algebra

We consider the zero weight space $\mathfrak{U}(L)_0$. This has basis $f_{\beta_1}^{r_1}\ldots f_{\beta_N}^{r_N} h_1^{s_1}\ldots h_l^{s_l}\, e_{\beta_1}^{t_1}\ldots e_{\beta_N}^{t_N}$ where $(t_1-r_1)\beta_1+\cdots+(t_N-r_N)\beta_N=0$. We have

$$\mathfrak{U}(L)_0=\{u\in\mathfrak{U}(L)\,;\,xu-ux=0 \quad \text{for all } x\in H\}$$

thus $\mathfrak{U}(L)_0$ is a subalgebra of $\mathfrak{U}(L)$. It is clear that $Z(L)\subset\mathfrak{U}(L)_0$.

Proposition 11.24 (i) $\mathfrak{U}(L)N\cap\mathfrak{U}(L)_0=N^-\mathfrak{U}(L)\cap\mathfrak{U}(L)_0=K$.
(ii) *The subspace K of* (i) *is a 2-sided ideal of* $\mathfrak{U}(L)_0$.
(iii) $\mathfrak{U}(L)_0=K\oplus\mathfrak{U}(H)$.

Proof. (i) $\mathfrak{U}(L)N$ is spanned by the basis vectors of $\mathfrak{U}(L)$ with some $t_i>0$. $N^-\mathfrak{U}(L)$ is spanned by the basis vectors with some $r_i>0$. $\mathfrak{U}(L)N\cap\mathfrak{U}(L)_0$ is spanned by the basis vectors of $\mathfrak{U}(L)$ with $\sum t_i\beta_i=\sum r_i\beta_i$ and some $t_i>0$. $N^-\mathfrak{U}(L)\cap\mathfrak{U}(L)_0$ is spanned by the basis vectors of $\mathfrak{U}(L)$ with $\sum t_i\beta_i=\sum r_i\beta_i$ and some $r_i>0$. These are clearly equal.
(ii) $\mathfrak{U}(L)N\cap\mathfrak{U}(L)_0$ is clearly a left ideal of $\mathfrak{U}(L)_0$ and $N^-\mathfrak{U}(L)\cap\mathfrak{U}(L)_0$ is a right ideal of $\mathfrak{U}(L)_0$. Thus K is a 2-sided ideal of $\mathfrak{U}(L)_0$.
(iii) $\mathfrak{U}(H)$ is spanned by the basis vectors with all $r_i=0$ and all $t_i=0$. This shows that $\mathfrak{U}(L)_0$ is the direct sum of its subspaces K and $\mathfrak{U}(H)$. □

Let $\phi:\mathfrak{U}(L)_0\to\mathfrak{U}(H)$ be the projection map obtained from the decomposition

$$\mathfrak{U}(L)_0=K\oplus\mathfrak{U}(H).$$

Since K is a 2-sided ideal of $\mathfrak{U}(L)_0$, ϕ is a homomorphism of algebras. ϕ is called the **Harish-Chandra homomorphism**.

We can now determine the central character χ_λ. The weight $\lambda\in H^*$ determines a 1-dimensional representation of $\mathfrak{U}(H)$, also denoted by λ.

Theorem 11.25 *The central character* $\chi_\lambda:Z(L)\to\mathbb{C}$ *is given by* $\chi_\lambda(z)=\lambda(\phi(z))$ *where ϕ is the Harish-Chandra homomorphism.*

Proof. We have

$$\mathfrak{U}(L)_0=(\mathfrak{U}(L)N\cap\mathfrak{U}(L)_0)\oplus\mathfrak{U}(H)$$

and $Z(L)\subset\mathfrak{U}(L)_0$. Let $z\in Z(L)$. Then we can write

$$z=u_1n_1+\cdots+u_kn_k+\phi(z)$$

where $u_i \in \mathfrak{U}(L)$ and $n_i \in N$. Thus

$$zm_\lambda = (u_1 n_1 + \cdots + u_k n_k + \phi(z)) m_\lambda$$
$$= \lambda(\phi(z)) m_\lambda$$

since $Nm_\lambda = O$ and $\phi(z) m_\lambda = \lambda(\phi(z)) m_\lambda$. Thus $\chi_\lambda(z) = \lambda(\phi(z))$. □

We have seen that the Harish-Chandra homomorphism maps $Z(L)$ into $\mathfrak{U}(H)$. Since the Lie algebra H is abelian we have $\mathfrak{U}(H) = S(H)$. We shall show that by combining the Harish-Chandra homomorphism with a 'twisting homomorphism' we get a homomorphism from $Z(L)$ into $S(H)$ with very favourable properties. The twisting homomorphism $\tau : S(H) \to S(H)$ is defined as follows. We recall that $S(H)$ is a polynomial algebra over $\mathbb{C}$ with generators $h_1, \ldots, h_l$. Thus there is a unique algebra homomorphism

$$\tau : S(H) \to S(H)$$

such that $\tau(h_i) = h_i - 1$. τ is in fact an automorphism of algebras. Its inverse is given by $\tau^{-1}(h_i) = h_i + 1$.

Let $\rho \in X$ be the element of the weight lattice given by

$$\rho = \omega_1 + \cdots + \omega_l.$$

Thus ρ is the sum of the fundamental weights. We recall from Section 10.3 that

$$\omega_i(h_i) = 1 \qquad w_i(h_j) = 0 \qquad \text{if } j \neq i.$$

Thus $\rho(h_i) = 1$ for each $i = 1, \ldots, l$.

Now any element $\lambda \in H^*$ extends to a 1-dimensional representation of $S(H)$. $\lambda - \rho$ is also a 1-dimensional representation of $S(H)$. We have

$$\lambda\tau(h_i) = \lambda(h_i - 1) = (\lambda - \rho) h_i.$$

Since $\lambda\tau$ and $\lambda - \rho$ are 1-dimensional representations of $S(H)$ and the h_i generate $S(H)$ we have

$$\lambda\tau(Q) = (\lambda - \rho)(Q) \qquad \text{for all } Q \in S(H).$$

The homomorphism

$$\tau\phi : Z(L) \to S(H)$$

is called the **twisted Harish-Chandra homomorphism**. We wish to show that the image of $Z(L)$ under the twisted Harish-Chandra homomorphism lies in $S(H)^W$. To do so we first need a result on Verma modules.

11.5 The centre of the enveloping algebra

Proposition 11.26 *Let $\lambda \in H^*$ and $M(\lambda)$ be the corresponding Verma module with highest weight vector m_λ. Suppose $(\lambda+\rho)(h_i) \in \mathbb{Z}$ and $(\lambda+\rho)(h_i) > 0$ for some i. Let*

$$v = f_i^{(\lambda+\rho)(h_i)} m_\lambda.$$

Then the submodule of $M(\lambda)$ generated by v is isomorphic to $M(\mu)$ where

$$\mu + \rho = s_i(\lambda+\rho).$$

Proof. We recall from Theorem 10.6 that there is an isomorphism of $\mathfrak{U}(N^-)$-modules between $\mathfrak{U}(N^-)$ and $M(\lambda)$ given by $u \to um_\lambda$. Since $f_i^{(\lambda+\rho)(h_i)} \neq 0$ in $\mathfrak{U}(N^-)$ we see that $v \neq 0$ in $M(\lambda)$. Since $m_\lambda \in M(\lambda)_\lambda$ we have $v \in M(\lambda)_\mu$ where

$$\mu = \lambda - (\lambda+\rho)(h_i)\alpha_i.$$

Thus we have

$$\mu + \rho = (\lambda+\rho) - (\lambda+\rho)(h_i)\alpha_i = s_i(\lambda+\rho).$$

We shall show that $Nv = 0$. It is sufficient to show that $e_j v = 0$ for $j = 1, \ldots, l$. If $j \neq i$ we have

$$e_j v = e_j f_i^{(\lambda+\rho)(h_i)} m_\lambda = f_i^{(\lambda+\rho)(h_i)} e_j m_\lambda = 0.$$

If $j = i$ we have

$$e_i v = e_i f_i^{(\lambda+\rho)(h_i)} m_\lambda$$
$$= f_i^{(\lambda+\rho)(h_i)} e_i + (\lambda+\rho)(h_i) f_i^{(\lambda+\rho)(h_i)-1}(h_i - (\lambda+\rho)(h_i)-1))m_\lambda$$
$$= (\lambda+\rho)(h_i) f_i^{(\lambda+\rho)(h_i)-1}(\lambda(h_i) - \lambda(h_i) - 1 + 1) m_\lambda = 0.$$

Thus $Nv = 0$.

Let V be the submodule of $M(\lambda)$ generated by v. Since $Nv = 0$ and $h_i v = \mu(h_i) v$ for $i = 1, \ldots, l$, there is a surjective homomorphism of $\mathfrak{U}(L)$-modules from $M(\mu)$ into V given by

$$um_\mu \to uv \quad u \in \mathfrak{U}(N^-).$$

(See Proposition 10.13.) We consider the kernel of this homomorphism. Let $u \in \mathfrak{U}(N^-)$ be such that $uv = 0$. Then

$$u f_i^{(\lambda+\rho)(h_i)} m_\lambda = 0.$$

Since $u f_i^{(\lambda+\rho)(h_i)} \in \mathfrak{U}(N^-)$ this implies that $u f_i^{(\lambda+\rho)(h_i)} = 0$. Since $f_i^{(\lambda+\rho)(h_i)} \neq 0$ and $\mathfrak{U}(L)$ has no zero-divisors we have $u = 0$. Thus our homomorphism is an isomorphism and so V is isomorphic to $M(\mu)$. □

Proposition 11.27 *The twisted Harish-Chandra homomorphism $\tau\phi$ maps $Z(L)$ into $S(H)^W$.*

Proof. We must show that $\tau\phi(z) \in S(H)^W$ for all $z \in Z(L)$. Since W is generated by $s_1, \ldots, s_l$ it will be sufficient to show that

$$s_i(\tau\phi(z)) = \tau\phi(z).$$

Since $S(H) = P(H^*)$ it will be sufficient to show these elements take the same value for all $\lambda \in H^*$, i.e. that

$$\lambda(s_i(\tau\phi(z))) = \lambda(\tau\phi(z)) \qquad \text{for all } \lambda \in H^*.$$

In fact it will be sufficient to prove this for elements of H^* of the form $\lambda + \rho$ where $\lambda \in X^+$ is dominant and integral. For such weights form a dense subset of H^* in the Zariski topology, for which the closed sets are the algebraic sets.

Thus suppose $\lambda \in X^+$. Then we have

$$(\lambda + \rho)(\tau(\phi(z))) = \lambda(\phi(z)) = \chi_\lambda(z)$$

using Theorem 11.25 and the definition of τ. Similarly we have

$$(\lambda + \rho)(s_i(\tau(\phi(z)))) = (\mu + \rho)(\tau(\phi(z))) = \mu(\phi(z)) = \chi_\mu(z)$$

where $s_i(\lambda + \rho) = \mu + \rho$.

We now apply Proposition 11.26. Since $\lambda \in X^+$ we have $\lambda(h_i) \geq 0$, so $(\lambda + \rho)(h_i) > 0$. Thus the Verma module $M(\lambda)$ contains a submodule isomorphic to $M(\mu)$. Now $z \in Z(L)$ acts on $M(\lambda)$ as scalar multiplication by $\chi_\lambda(z)$ and on $M(\mu)$ as scalar multiplication by $\chi_\mu(z)$. Since $M(\mu)$ is isomorphic to a submodule of $M(\lambda)$ we must have

$$\chi_\lambda(z) = \chi_\mu(z).$$

Thus

$$(\lambda + \rho)(\tau(\phi(z))) = (\lambda + \rho)(s_i(\tau(\phi(z))))$$

and hence

$$\tau(\phi(z)) = s_i(\tau(\phi(z))).$$

Thus $\tau\phi(z) \in S(H)^W$ as required. □

In fact we shall show that the twisted Harish-Chandra map

$$\tau\phi : Z(L) \to S(H)^W$$

is an isomorphism of algebras.

11.5 The centre of the enveloping algebra

To see this we first recall the operation $\theta : S(L) \to \mathfrak{U}(L)$ of symmetrisation which was shown in Proposition 11.4 to be an isomorphism of L-modules. Now the adjoint group G acts on both $S(L)$ and $\mathfrak{U}(L)$. For the G-action on L can be extended to a G-action on $T(L)$ as described in Section 11.4 and these induce G-actions on the quotients $S(L)$ and $\mathfrak{U}(L)$. Suppose $x \in L$ is such that $\operatorname{ad} x$ is nilpotent. Then $\exp \operatorname{ad} x \in G$. Let x induce the linear maps $\alpha(x)$ on $S(L)$ and $\beta(x)$ on $\mathfrak{U}(L)$. The definition of the G-actions then shows that $\exp \operatorname{ad} x$ acts as $\exp \alpha(x)$ on $S(L)$ and as $\exp \beta(x)$ on $\mathfrak{U}(L)$. Since θ is an isomorphism of L-modules we have

$$\theta \alpha(x) = \beta(x) \theta.$$

It follows that

$$\theta \frac{\alpha(x)^i}{i!} = \frac{\beta(x)^i}{i!} \theta \quad \text{for all } i,$$

and therefore that

$$\theta \exp \alpha(x) = \exp \beta(x) \theta.$$

(Note that both $\alpha(x)$ and $\beta(x)$ are nilpotent.) Since G is generated by such elements $\exp \operatorname{ad} x$ it follows that θ is an isomorphism of G-modules. We deduce that θ restricts to an isomorphism between $S(L)^G$ and $\mathfrak{U}(L)^G$.

Proposition 11.28 $\mathfrak{U}(L)^G = Z(L)$.

Proof. We first note that $Z(L) \subset \mathfrak{U}(L)^G$. Let $z \in Z(L)$. Let $x \in L$ be such that $\operatorname{ad} x$ is nilpotent. Thus $\exp \operatorname{ad} x \in G$. Since $z \in Z(L)$ we have

$$x \cdot z = xz - zx = 0.$$

Hence $\beta(x)z = 0$. Thus

$$\exp \operatorname{ad} x \cdot z = \exp \beta(x) \cdot z = \left(1 + \beta(x) + \frac{\beta(x)^2}{2!} + \cdots \right) z = z.$$

Thus z is invariant under $\exp \operatorname{ad} x$. Since such elements generate G we have $z \in \mathfrak{U}(L)^G$.

Conversely we show that $\mathfrak{U}(L)^G \subset Z(L)$. Let $u \in \mathfrak{U}(L)^G$. Then $\exp \operatorname{ad} x \cdot u = u$ for all $x \in L$ with $\operatorname{ad} x$ nilpotent. Suppose $(\operatorname{ad} x)^t \neq 0$ but $(\operatorname{ad} x)^{t+1} = 0$. We choose elements $\xi_1, \ldots, \xi_{t+1} \in \mathbb{C}$ which are all distinct. Then $\operatorname{ad}(\xi_i x)$ is also nilpotent and

$$\exp \operatorname{ad}(\xi_i x) = 1 + \operatorname{ad}(\xi_i x) + \cdots + \frac{1}{t!}(\operatorname{ad}(\xi_i x))^t$$

$$= 1 + \xi_i(\operatorname{ad} x) + \cdots + \frac{\xi_i^t}{t!}(\operatorname{ad} x)^t.$$

Now the determinant

$$\begin{vmatrix} 1 & \xi_1 & \frac{\xi_1^2}{2!} & \cdots & \frac{\xi_1^t}{t!} \\ 1 & \xi_2 & \frac{\xi_2^2}{2!} & \cdots & \frac{\xi_2^t}{t!} \\ \cdot & \cdot & \cdot & \cdots & \cdot \\ \cdot & \cdot & \cdot & \cdots & \cdot \\ \cdot & \cdot & \cdot & \cdots & \cdot \\ 1 & \xi_{t+1} & \frac{\xi_{t+1}^2}{2!} & \cdots & \frac{\xi_{t+1}^t}{t!} \end{vmatrix} = \frac{1}{2!3!\ldots t!}\prod_{i<j}(\xi_i - \xi_j)$$

is non-zero. Thus the vector $(0, 1, 0, \ldots, 0)$ is a linear combination of the rows of the determinant. Thus there exist $\eta_1, \ldots, \eta_{t+1} \in \mathbb{C}$ such that

$$\text{ad } x = \eta_1 \exp \text{ ad}(\xi_1 x) + \cdots + \eta_{t+1} \exp \text{ ad}(\xi_{t+1} x).$$

So $\text{ad } x \cdot u = (\eta_1 + \cdots + \eta_{t+1})u$. Since ad x acts nilpotently on u it follows that $\eta_1 + \cdots + \eta_{t+1} = 0$ and that ad $x \cdot u = 0$. This means that $xu - ux = 0$. This holds for all $x \in L$ with ad x nilpotent, in particular for $x = e_i$ and $x = f_i$. However, $e_1, \ldots, e_l, f_1, \ldots, f_l$ generate $\mathfrak{U}(L)$, together with 1. It follows that $xu - ux = 0$ for all $x \in \mathfrak{U}(L)$, that is $u \in Z(L)$. □

Thus the operation θ of symmetrisation gives an isomorphism of vector spaces

$$\theta : S(L)^G \to Z(L).$$

Now we also have an isomorphism of algebras

$$\eta : S(L)^G \to S(H)^W$$

given in Corollary 11.22. Combining these maps we obtain an isomorphism of vector spaces

$$\eta\theta^{-1} : Z(L) \to S(H)^W.$$

Thus we have two maps $\eta\theta^{-1}$ and $\tau\phi$ from $Z(L)$ into $S(H)^W$. The first is an isomorphism of vector spaces and the second a homomorphism of algebras. We shall compare these maps, using the structure of $Z(L)$ and $S(H)$ as filtered algebras.

We recall from Section 11.1 that $\mathfrak{U}(L)$ may be regarded as a filtered algebra with filtration

$$\mathfrak{U}_0(L) \subset \mathfrak{U}_1(L) \subset \mathfrak{U}_2(L) \subset \cdots.$$

11.5 The centre of the enveloping algebra

We define $Z_i(L) = Z(L) \cap \mathfrak{U}_i(L)$. This makes $Z(L)$ into a filtered algebra. $S(H)$ also has a natural structure as a filtered algebra, where $S_i(H)$ is the subspace of $S(H)$ generated by all products $a_1 a_2 \ldots a_j$, $j \le i$, where $a_k \in H$. We also define

$$\left(S(H)^W \right)_i = S(H)^W \cap S_i(H).$$

This makes $S(H)^W$ into a filtered algebra.

We shall make use of the following lemma on filtered and graded algebras.

Lemma 11.29 *Let* $A = \bigcup_{i \ge 0} A_i$ *and* $B = \bigcup_{i \ge 0} B_i$ *be filtered algebras with*

$$A_0 \subset A_1 \subset A_2 \subset \cdots$$

and

$$B_0 \subset B_1 \subset B_2 \subset \cdots.$$

Let

$$\operatorname{gr} A = A_0 \oplus A_1/A_0 \oplus A_2/A_1 \oplus \cdots$$

and

$$\operatorname{gr} B = B_0 \oplus B_1/B_0 \oplus B_2/B_1 \oplus \cdots$$

be the corresponding graded algebras. Let $\alpha : A \to B$ *be a linear map such that* $\alpha(A_i) \subset B_i$ *for each* i. *Then:*

(a) *There is a linear map* $\operatorname{gr} \alpha : \operatorname{gr} A \to \operatorname{gr} B$ *satisfying* $\operatorname{gr} \alpha (A_{i-1} + a_i) = B_{i-1} + \alpha(a_i)$ *for* $a_i \in A_i$.
(b) *If* $\alpha(A_i) = B_i$ *for each* i *and* α *is bijective then* $\operatorname{gr} \alpha$ *is bijective.*
(c) *If* $\operatorname{gr} \alpha$ *is bijective then* α *is bijective.*

Proof. (a) We must show that $\operatorname{gr} \alpha : A_i/A_{i-1} \to B_i/B_{i-1}$ is well defined. Suppose $A_{i-1} + a_i = A_{i-1} + a'_i$ where $a_i, a'_i \in A_i$. Then $a_i - a'_i \in A_{i-1}$, so $\alpha(a_i - a'_i) \in B_{i-1}$. Thus $B_{i-1} + \alpha(a_i) = B_{i-1} + \alpha(a'_i)$ and so $\operatorname{gr} \alpha$ is well defined.
(b) Suppose now that $\alpha(A_i) = B_i$ for each i and that α is bijective. Then the induced map $\operatorname{gr} \alpha : A_i/A_{i-1} \to B_i/B_{i-1}$ is bijective. It follows that $\operatorname{gr} \alpha : \operatorname{gr} A \to \operatorname{gr} B$ is bijective.
(c) Suppose conversely that $\operatorname{gr} \alpha : \operatorname{gr} A \to \operatorname{gr} B$ is bijective. This implies that

$$\operatorname{gr} \alpha : A_i/A_{i-1} \to B_i/B_{i-1}$$

is bijective for each i. We show first that α is surjective. B_0 lies in the image of α since $\alpha : A_0 \to B_0$ agrees with $\operatorname{gr} \alpha : A_0 \to B_0$. Assume by induction that B_{i-1} lies in the image of α. Let $b_i \in B_i$. Then there exists $a_i \in A_i$ such that

$$B_{i-1} + \alpha(a_i) = B_{i-1} + b_i.$$

Thus $b_i - \alpha(a_i) \in B_{i-1}$. Hence $b_i - \alpha(a_i)$ lies in the image of α, thus b_i does also. Thus α is surjective.

Now let $a \in \ker \alpha$. If $a \in A_0$ then $a = 0$ since α agrees with $\operatorname{gr} \alpha$ on A_0. Otherwise there exists $i > 0$ such that $a \in A_i$ but $a \notin A_{i-1}$. But then $A_{i-1} + a \neq 0$ whereas $\operatorname{gr} \alpha (A_{i-1} + a) = 0$, a contradiction. Hence $\ker \alpha = O$ and so α is bijective. □

Theorem 11.30 *The twisted Harish-Chandra map $\tau\phi$ gives an isomorphism of algebras $Z(L) \to S(H)^W$.*

Proof. We have maps $\tau\phi : Z(L) \to S(H)^W$ and $\eta\theta^{-1} : Z(L) \to S(H)^W$. Those induce maps

$$\operatorname{gr}(\tau\phi) : \operatorname{gr} Z(L) \to \operatorname{gr} S(H)^W$$

$$\operatorname{gr}\left(\eta\theta^{-1}\right) : \operatorname{gr} Z(L) \to \operatorname{gr} S(H)^W.$$

We shall show that $\operatorname{gr}(\tau\phi) = \operatorname{gr}\left(\eta\theta^{-1}\right)$. Let $z \in Z(L)$. Then there exists d such that $z \in Z_d(L)$ but $z \notin Z_{d-1}(L)$. Then z has the form

$$z = \sum_{\sum r_i + \sum s_i + \sum t_i \leq d} \xi(\underline{r}, \underline{s}, \underline{t}) f_{\beta_1}^{r_1} \ldots f_{\beta_N}^{r_N} h_1^{s_1} \ldots h_l^{s_l} e_{\beta_1}^{t_1} \ldots e_{\beta_N}^{t_N}$$

where $\Phi^+ = \{\beta_1, \ldots, \beta_N\}$ and $\xi(\underline{r}, \underline{s}, \underline{t}) \in \mathbb{C}$. Then

$$\phi(z) = \sum_{\sum s_i \leq d} \xi(\underline{0}, \underline{s}, \underline{0}) h_1^{s_1} \ldots h_l^{s_l}$$

$$\tau\phi(z) = \sum_{\sum s_i \leq d} \xi(\underline{0}, \underline{s}, \underline{0}) (h_1 - 1)^{s_1} \ldots (h_l - 1)^{s_l}$$

$$\theta^{-1}(z) \equiv \sum \xi(\underline{r}, \underline{s}, \underline{t}) f_{\beta_1}^{r_1} \ldots f_{\beta_N}^{r_N} h_1^{s_1} \ldots h_l^{s_l} e_{\beta_1}^{t_1} \ldots e_{\beta_N}^{t_N} \mod S_{d-1}(L)$$

$$\eta\theta^{-1}(z) \equiv \sum \xi(\underline{0}, \underline{s}, \underline{0}) h_1^{s_1} \ldots h_l^{s_l} \mod S_{d-1}(H).$$

Now it is apparent that

$$\tau\phi(z) \equiv \phi(z) \mod S_{d-1}(H)$$

hence

$$\tau\phi(z) \equiv \eta\theta^{-1}(z) \mod S_{d-1}(H).$$

11.5 The centre of the enveloping algebra

Since $\tau\phi(z)$ and $\eta\theta^{-1}(z)$ both lie in $S(H)^W$ they satisfy $\tau\phi(z) \equiv \eta\theta^{-1}(z)$ mod $(S(H)^W)_{d-1}$. Thus $\operatorname{gr}(\tau\phi) = \operatorname{gr}(\eta\theta^{-1})$.

Now the maps $\theta^{-1} : Z(L) \to S(L)^G$ and $\eta : S(L)^G \to S(H)^W$ satisfy

$$\theta^{-1}(Z_d(L)) = (S(L)^G)_d$$
$$\eta(S(L)^G)_d = (S(H)^W)_d.$$

Thus we have

$$\eta\theta^{-1}(Z_d(L)) = (S(H)^W)_d.$$

We may now apply Lemma 11.29. The map

$$\eta\theta^{-1} : Z(L) \to S(H)^W$$

is bijective and satisfies

$$\eta\theta^{-1}(Z_d(L)) = (S(H)^W)_d$$

for each d. Hence

$$\operatorname{gr}(\eta\theta^{-1}) : \operatorname{gr} Z(L) \to \operatorname{gr} S(H)^W$$

is bijective. This is turn implies that

$$\tau\phi : Z(L) \to S(H)^W$$

is bijective. Since $\tau\phi$ is known to be a homomorphism of algebras, it must therefore be an algebra isomorphism. $\square$

We can deduce from this theorem a necessary and sufficient condition for two central characters χ_λ, χ_μ to be equal.

Theorem 11.31 *Let $\lambda, \mu \in H^*$. Then $\chi_\lambda = \chi_\mu$ if and only if $\mu + \rho = w(\lambda + \rho)$ for some $w \in W$.*

Proof. Suppose first that $\mu + \rho = w(\lambda + \rho)$. Then, for $z \in Z(L)$, we have

$$\chi_\mu(z) = \mu(\phi(z)) = (w(\lambda + \rho) - \rho)(\phi(z))$$
$$= w(\lambda + \rho)(\tau(\phi(z))) = (\lambda + \rho)(w^{-1}\tau(\phi(z))).$$

Now $\tau\phi(z) \in S(H)^W$ and so is fixed by w^{-1}. Hence

$$\chi_\mu(z) = (\lambda + \rho)(\tau(\phi(z))) = \lambda(\phi(z)) = \chi_\lambda(z),$$

by Theorem 11.25. Hence $\chi_\mu = \chi_\lambda$.

Suppose conversely that $\mu+\rho \ne w(\lambda+\rho)$ for all $w \in W$. Then the finite sets $W(\lambda+\rho)$ and $W(\mu+\rho)$ do not intersect. Therefore there exists a polynomial function $Q \in P(H^*)$ such that Q takes values 1 on $W(\lambda+\rho)$ and values 0 on $W(\mu+\rho)$. We have

$$Q \in S(H) = P(H^*).$$

By replacing Q by $\dfrac{1}{|W|} \sum_{w \in W} w(Q)$ we may assume Q lies in $S(H)^W$.

We now make use of the isomorphism $\tau\phi : Z(L) \to S(H)^W$. There exists $z \in Z(L)$ such that $\tau\phi(z) = Q$. Thus we have

$$\chi_\lambda(z) = \lambda(\phi(z)) = (\lambda+\rho)(\tau\phi(z)) = (\lambda+\rho)Q = 1$$
$$\chi_\mu(z) = \mu(\phi(z)) = (\mu+\rho)(\tau\phi(z)) = (\mu+\rho)Q = 0.$$

Hence $\chi_\lambda \ne \chi_\mu$. □

A second deduction from Theorem 11.30 is the following important result.

Theorem 11.32 *The centre $Z(L)$ of $\mathfrak{U}(L)$ is isomorphic to the polynomial ring over $\mathbb{C}$ in l variables, where L is semisimple and $l = \mathrm{rank}\, L$.*

Proof. This follows from Theorem 11.30, Corollary 11.22 and Theorem 11.17. □

As an example we consider the Lie algebra L of type A_1. The algebra L has a basis f, h, e with

$$[he] = 2e, \quad [hf] = -2f, \quad [ef] = h.$$

The algebras

$$S(H)^W, P(H)^W, S(L)^G, P(L)^G, Z(L)$$

are all isomorphic to the polynomial ring over $\mathbb{C}$ in one variable. We find a generator of each of these algebras.

We have $W = \langle s \rangle$ where $s(h) = -h$. Thus $S(H)^W$ is the polynomial algebra generated by h^2.

We now consider the isomorphism $S(L)^G \to S(H)^W$ given by projection. The element of $S(L)^G$ mapping to h^2 is homogeneous of degree 2 in e, h, f

11.5 The centre of the enveloping algebra

and has weight 0. It must therefore have form $h^2 + \xi fe$ for some $\xi \in \mathbb{C}$. We determine the constant ξ. We have

$$\text{ad } e \cdot h = -2e, \quad \text{ad } e \cdot f = h, \quad \text{ad } e \cdot e = 0.$$

Thus

$$(\exp \text{ad } e)h = h - 2e$$
$$(\exp \text{ad } e)f = f + h - e$$
$$(\exp \text{ad } e)e = e$$
$$(\exp \text{ad } e)\left(h^2 + \xi fe\right) = (h-2e)^2 + \xi(f+h-e)e$$
$$= h^2 + \xi fe + (\xi - 4)he + (4-\xi)e^2.$$

Thus $\exp \text{ad } e$ fixes $h^2 + \xi fe$ if and only if $\xi = 4$. Hence $S(L)^G$ is the polynomial ring generated by $h^2 + 4fe$.

Next we consider the Killing isomorphism $L \to L^*$. L^* has basis f^*, h^*, e^* dual to f, h, e, that is $y^*(x) = 1$ if $y = x$ and $y^*(x) = 0$ if $y \ne x$. Now the Killing form satisfies

$$\langle h, h \rangle = 8, \quad \langle f, e \rangle = 4, \quad \langle h, f \rangle = 0, \langle h, e \rangle = 0, \quad \langle e, e \rangle = 0, \quad \langle f, f \rangle = 0.$$

Thus under the Killing isomorphism $L \to L^*$ we have $e \to 4f^*, h \to 8h^*, f \to 4e^*$. This induces a map $S(L) \to P(L)$ under which $h^2 + 4fe$ maps to $64\left(h^{*2} + f^*e^*\right)$. Thus $P(L)^G$ is the polynomial ring generated by $h^{*2} + f^*e^*$.

We also have a map $S(L)^G \to Z(G)$ given by symmetrisation. Under this map $h^2 + 4fe$ is transformed into

$$h^2 + 2fe + 2ef = h^2 + 2h + 4fe.$$

Thus $Z(L)$ is the polynomial ring generated by $h^2 + 2h + 4fe$. We also note that the element of $Z(L)$ mapping to $h^2 \in S(H)^W$ under the twisted Harish-Chandra homomorphism is $h^2 + 2h + 1 + 4fe$.

Thus we have:

$$S(H)^W = \mathbb{C}\left[h^2\right]$$
$$P(H)^W = \mathbb{C}\left[h^{*2}\right]$$
$$S(L)^G = \mathbb{C}\left[h^2 + 4fe\right]$$
$$P(L)^G = \mathbb{C}\left[h^{*2} + f^*e^*\right]$$
$$Z(L) = \mathbb{C}\left[h^2 + 2h + 4fe\right]$$

11.6 The Casimir element

We now introduce an element of the centre $Z(L)$ of $\mathfrak{U}(L)$ which has useful properties. Let $x_1, \ldots, x_n$ be a basis of L. Since the Killing form of L is non-degenerate by Theorem 4.10 there is a unique dual basis $y_1, \ldots, y_n$ of L satisfying

$$\langle x_i, y_j \rangle = \delta_{ij}.$$

Let $c \in \mathfrak{U}(L)$ be defined by

$$c = \sum_{i=1}^{n} x_i y_i.$$

Proposition 11.33 *The element c is independent of the choice of basis $x_1, \ldots, x_n$ of L.*

Proof. Suppose $x'_1, \ldots, x'_n$ are a second basis of L and $y'_1, \ldots, y'_n$ are the dual basis. Let

$$x'_i = \sum_j \sigma_{ij} x_j \qquad y'_i = \sum_j \tau_{ij} y_j.$$

Then we have

$$\langle x'_i, y'_j \rangle = \left\langle \sum_k \sigma_{ik} x_k, \sum_l \tau_{jl} y_l \right\rangle = \sum_{k,l} \sigma_{ik} \tau_{jl} \langle x_k, y_l \rangle = \sum_k \sigma_{ik} \tau_{jk}.$$

Hence if $\sigma = (\sigma_{ij})$, $\tau = (\tau_{ij})$ we have $\sigma \tau^t = I$. We then have

$$\sum_i x'_i y'_i = \sum_i \left(\sum_j \sigma_{ij} x_j \right) \left(\sum_k \tau_{ik} y_k \right) = \sum_{j,k} \left(\sum_i \sigma_{ij} \tau_{ik} \right) x_j y_k.$$

Now $\sigma^t \tau = I$ so $\sum_i \sigma_{ij} \tau_{ik} = \delta_{jk}$. Hence $\sum_i x'_i y'_i = \sum_i x_i y_i$. □

Definition c is called the **Casimir element** of $\mathfrak{U}(L)$.

Proposition 11.34 *c lies in the centre $Z(L)$ of $\mathfrak{U}(L)$.*

Proof. It is sufficient to show that $cx = xc$ for all $x \in L$. We have

$$cx = \sum_i x_i y_i x = \sum_i x_i (xy_i + [y_i, x])$$

$$= \sum_i ((xx_i + [x_i, x]) y_i + x_i [y_i, x])$$

$$= xc + \sum_i ([x_i, x] y_i + x_i [y_i, x]).$$

11.6 The Casimir element

Let $[x_i x] = \sum_j \alpha_{ij} x_j$ and $[y_i x] = \sum_j \beta_{ij} y_j$. Since $\langle [x_i x], y_j \rangle = \langle x_i, [xy_j] \rangle$ we have $\alpha_{ij} = -\beta_{ji}$. It follows that

$$\sum_i ([x_i x] y_i + x_i [y_i x]) = \sum_i \sum_j \alpha_{ij} x_j y_i + \sum_i \sum_j \beta_{ij} x_i y_j$$
$$= \sum_{i,j} (\alpha_{ij} + \beta_{ji}) x_j y_i = 0.$$

Thus $cx = xc$ and so $c \in Z(L)$. $\square$

We now recall from Proposition 4.18 that for each $e_\alpha \in L_\alpha$ we can find $f_\alpha \in L_{-\alpha}$ such that $[e_\alpha f_\alpha] = h'_\alpha$, and that we then have $\langle e_\alpha, f_\alpha \rangle = 1$. Since the Killing form of L remains non-degenerate on H we may choose a basis $h'_1, \ldots, h'_l$ of H and there will be a dual basis $h''_1, \ldots, h''_l$ satisfying

$$\langle h'_i, h''_j \rangle = \delta_{ij}.$$

Then $h'_1, \ldots, h'_l, e_\alpha \ (\alpha \in \Phi^+), f_\alpha \ (\alpha \in \Phi^+)$ are a basis of L and its dual basis is

$$h''_1, \ldots, h''_l, \quad f_\alpha (\alpha \in \Phi^+), \quad e_\alpha (\alpha \in \Phi^+).$$

Using this pair of dual bases we have

$$c = h'_1 h''_1 + \cdots + h'_l h''_l + \sum_{\alpha \in \Phi^+} e_\alpha f_\alpha + \sum_{\alpha \in \Phi^+} f_\alpha e_\alpha.$$

Thus we obtain:

Proposition 11.35 *The Casimir element of $Z(L)$ is given by*

$$c = \sum_{i=1}^{l} h'_i h''_i + \sum_{\alpha \in \Phi^+} h'_\alpha + 2 \sum_{\alpha \in \Phi^+} f_\alpha e_\alpha$$

where $h'_1, \ldots, h'_l; h''_1, \ldots, h''_l$ are any pair of dual bases of H. $\square$

The properties of the Casimir element will be useful as we explore further the representation theory of L.

Proposition 11.36 *Let $c \in Z(L)$ be the Casimir element. Then*

$$\chi_\lambda(c) = \langle \lambda + \rho, \lambda + \rho \rangle - \langle \rho, \rho \rangle.$$

Thus c acts on the Verma module $M(\lambda)$ as scalar multiplication by $\langle \lambda + \rho, \lambda + \rho \rangle - \langle \rho, \rho \rangle$.

Proof. We consider the action of c on the highest weight vector m_λ of $M(\lambda)$. By Proposition 11.35 we have

$$cm_\lambda = \left(\sum_{i=1}^{l} h'_i h''_i + \sum_{\alpha \in \Phi^+} h'_\alpha + 2 \sum_{\alpha \in \Phi^+} f_\alpha e_\alpha \right) m_\lambda$$

$$= \left(\sum_{i=1}^{l} \lambda(h'_i) \lambda(h''_i) + \sum_{\alpha \in \Phi^+} \lambda(h'_\alpha) \right) m_\lambda$$

Now $\sum_{\alpha \in \Phi^+} \lambda(h'_\alpha) = \sum_{\alpha \in \Phi^+} \langle \lambda, \alpha \rangle = \langle \lambda, \sum_{\alpha \in \Phi^+} \alpha \rangle = 2\langle \lambda, \rho \rangle$.

Let $h'_\lambda \in H$ be the element corresponding to $\lambda \in H^*$ under the isomorphism defined by the Killing form. Thus

$$\lambda(h'_i) = \langle h'_\lambda, h'_i \rangle$$

$$\lambda(h''_i) = \langle h'_\lambda, h''_i \rangle.$$

We express h'_λ in terms of the dual bases $h'_1, \ldots, h'_l$ and $h''_1, \ldots, h''_l$ of H. Let

$$h'_\lambda = a_1 h'_1 + \cdots + a_l h'_l$$

$$h'_\lambda = b_1 h''_1 + \cdots + b_l h''_l.$$

Since $\langle h'_i, h''_j \rangle = \delta_{ij}$ we have

$$\langle h'_\lambda, h'_\lambda \rangle = a_1 b_1 + \cdots + a_l b_l$$

$$\langle h'_\lambda, h'_i \rangle = b_i \qquad \langle h'_\lambda, h''_i \rangle = a_i.$$

It follows that

$$\sum_{i=1}^{l} \lambda(h'_i) \lambda(h''_i) = \sum_{i=1}^{l} \langle h'_\lambda, h'_i \rangle \langle h'_\lambda, h''_i \rangle = \langle h'_\lambda, h'_\lambda \rangle = \langle \lambda, \lambda \rangle.$$

Hence

$$cm_\lambda = (\langle \lambda, \lambda \rangle + 2\langle \lambda, \rho \rangle) m_\lambda$$

$$= (\langle \lambda + \rho, \lambda + \rho \rangle - \langle \rho, \rho \rangle) m_\lambda.$$

Thus the value of the central character χ_λ at c is given by

$$\chi_\lambda(c) = \langle \lambda + \rho, \lambda + \rho \rangle - \langle \rho, \rho \rangle. \qquad \square$$

12
Character and dimension formulae

12.1 Characters of L-modules

Let V be an L-module where L is semisimple. We say that V admits a character if V is the direct sum of its weight spaces and each weight space of V is finite dimensional. Thus we have

$$V = \bigoplus_{\lambda \in H^*} V_\lambda \quad \dim V_\lambda \quad \text{finite}$$

where $V_\lambda = \{v \in V \; ; \; hv = \lambda(h)v \text{ for all } h \in H\}$. The character of V is then the function $\operatorname{ch} V : H^* \to \mathbb{Z}$ given by

$$(\operatorname{ch} V)(\lambda) = \dim V_\lambda.$$

We see that if V admits a character then the structure of V as an H-module is determined by $\operatorname{ch} V$.

In this chapter we shall obtain formulae for the characters of the Verma modules $M(\lambda)$ for $\lambda \in H^*$ and for the finite dimensional irreducible modules $L(\lambda)$ for $\lambda \in X^+$.

We first identify a certain ring of functions $H^* \to \mathbb{Z}$ in which it will be convenient to work. Given a function $f : H^* \to \mathbb{Z}$ we define $\operatorname{Supp} f$, the support of f, to be the set of $\lambda \in H^*$ for which $f(\lambda) \neq 0$. For example the support of the function $\operatorname{ch} M(\lambda)$ is the set of all $\mu \in H^*$ which have form

$$\mu = \lambda - n_1 \alpha_1 - \cdots - n_l \alpha_l \quad n_i \in \mathbb{Z}, \quad n_i \geq 0.$$

This follows from Theorem 10.7. We define

$$S(\lambda) = \operatorname{Supp}(\operatorname{ch} M(\lambda)).$$

Definition $\mathfrak{R}$ denotes the set of all functions $f : H^* \to \mathbb{Z}$ such that there exists a finite set $\lambda_1, \ldots, \lambda_k \in H^*$ with

$$\operatorname{Supp} f \subset S(\lambda_1) \cup \cdots \cup S(\lambda_k).$$ □

It is clear that $\operatorname{ch} M(\lambda)$ for $\lambda \in H^*$ and $\operatorname{ch} L(\lambda)$ for $\lambda \in X^+$ lie in $\mathfrak{R}$. It is also clear that if $f, g \in \mathfrak{R}$ then $f + g \in \mathfrak{R}$, since

$$\operatorname{Supp}(f+g) \subset \operatorname{Supp} f \cup \operatorname{Supp} g.$$

Thus $\mathfrak{R}$ is an additive group. We can also define a product on $\mathfrak{R}$ which makes it into a ring. Given $f, g \in \mathfrak{R}$ we define $fg : H^* \to \mathbb{Z}$ by

$$(fg)(\lambda) = \sum_{\substack{\mu,\nu \in H^* \\ \mu+\nu=\lambda}} f(\mu)g(\nu).$$

We note that the sum is finite, so that fg is well defined. For we may assume $\mu \in \operatorname{Supp} f$ and $\nu \in \operatorname{Supp} g$. Suppose

$$\operatorname{Supp} f \subset S(\mu_1) \cup \cdots \cup S(\mu_h)$$
$$\operatorname{Supp} g \subset S(\nu_1) \cup \cdots \cup S(\nu_h).$$

If $\mu \in S(\mu_i)$ and $\nu \in S(\nu_j)$ we have

$$\mu = \mu_i - m_1 \alpha_1 - \cdots - m_l \alpha_l \qquad m_k \in \mathbb{Z}, \quad m_k \geq 0$$
$$\nu = \nu_j - n_1 \alpha_1 - \cdots - n_l \alpha_l \qquad n_k \in \mathbb{Z}, \quad n_k \geq 0.$$

Since $\mu + \nu = \lambda$ we have

$$\lambda = (\mu_i + \nu_j) - r_1 \alpha_1 - \cdots - r_l \alpha_l \qquad r_k \in \mathbb{Z}, r_k \geq 0$$

where $r_k = m_k + n_k$. However, given i, j and λ the non-negative integers m_k, n_k with $m_k + n_k = r_k$ can be chosen in only finitely many ways, thus our sum is finite. Also we see that $\operatorname{Supp}(fg) \subset \bigcup_{i,j} S(\mu_i + \nu_j)$, hence $fg \in \mathfrak{R}$. It is also readily checked that $(fg)h = f(gh)$, thus $\mathfrak{R}$ becomes a ring.

For each $\lambda \in H^*$ we define $e_\lambda : H^* \to \mathbb{Z}$ by $e_\lambda(\lambda) = 1$, $e_\lambda(\mu) = 0$ if $\mu \neq \lambda$. Thus e_λ is the characteristic function of λ. All such characteristic functions lie in $\mathfrak{R}$. In fact if f is any function in $\mathfrak{R}$ it is convenient to write

$$f = \sum_{\lambda \in H^*} f(\lambda) e_\lambda$$

even though the sum may be infinite.

We note that $e_\lambda e_\mu = e_{\lambda + \mu}$.

12.1 Characters of L-modules

Lemma 12.1 *Suppose that the L-module V admits a character and let U be a submodule of V. Then both U and V/U admit a character, and*

$$\operatorname{ch} U + \operatorname{ch} \frac{V}{U} = \operatorname{ch} V.$$

Proof. We have $V = \bigoplus_\lambda V_\lambda$. Also $U_\lambda = U \cap V_\lambda$. Thus the sum $\sum_\lambda U_\lambda$ is direct. Moreover $U = \sum_\lambda U_\lambda$ since if $u \in U$ and $u = \sum u_\lambda$ with $u_\lambda \in V_\lambda$ then $u_\lambda \in U$, as in the proof of Theorem 10.9. Hence we have

$$U = \bigoplus_\lambda U_\lambda$$

with $U_\lambda \subset V_\lambda$, so U admits a character. We also have

$$V/U = \bigoplus_\lambda (V_\lambda/U_\lambda)$$

and V_λ/U_λ can be identified with the λ-weight space $(V/U)_\lambda$. Thus V/U admits a character. Finally we have

$$(\operatorname{ch} U)(\lambda) + (\operatorname{ch}(V/U))(\lambda) = \dim U_\lambda + \dim (V_\lambda/U_\lambda) = \dim V_\lambda.$$

Thus $\operatorname{ch} U + \operatorname{ch}(V/U) = \operatorname{ch} V$. $\square$

Lemma 12.2 *Suppose V_1, V_2 are L-modules which both admit characters such that $\operatorname{ch} V_1$ and $\operatorname{ch} V_2$ lie in $\mathfrak{R}$. Then $V_1 \otimes V_2$ admits a character and $\operatorname{ch}(V_1 \otimes V_2) = \operatorname{ch} V_1 \operatorname{ch} V_2$.*

Proof. Since V_1, V_2 admit characters we have $V_1 = \bigoplus_\lambda (V_1)_\lambda$ and $V_2 = \bigoplus_\mu (V_2)_\mu$. Hence

$$V_1 \otimes V_2 = \bigoplus_{\lambda,\mu} \left((V_1)_\lambda \otimes (V_2)_\mu \right).$$

$V_1 \otimes V_2$ may be made into an L-module by means of the action

$$x(v_1 \otimes v_2) = xv_1 \otimes v_2 + v_1 \otimes xv_2$$

extended by linearity. In particular, if $x \in H$, $v_1 \in (V_1)_\lambda$ and $v_2 \in (V_2)_\mu$ we have

$$x(v_1 \otimes v_2) = (\lambda(x) + \mu(x)) v_1 \otimes v_2.$$

Thus $(V_1)_\lambda \otimes (V_2)_\mu \subset (V_1 \otimes V_2)_{\lambda+\mu}$. It follows that

$$V_1 \otimes V_2 = \oplus (V_1 \otimes V_2)_\nu$$

where $(V_1 \otimes V_2)_\nu = \sum_{\lambda+\mu=\nu} ((V_1)_\lambda \otimes (V_2)_\mu)$. Thus $V_1 \otimes V_2$ admits a character. Moreover we have

$$(\operatorname{ch}(V_1 \otimes V_2))(\nu) = \dim(V_1 \otimes V_2)_\nu = \sum_{\substack{\lambda,\mu \\ \lambda+\mu=\nu}} \dim(V_1)_\lambda \dim(V_2)_\mu$$

$$= \sum_{\substack{\lambda,\mu \\ \lambda+\mu=\nu}} (\operatorname{ch} V_1)(\lambda)(\operatorname{ch} V_2)(\mu) = (\operatorname{ch} V_1 \operatorname{ch} V_2)(\nu).$$

Thus $\operatorname{ch}(V_1 \otimes V_2) = \operatorname{ch} V_1 \operatorname{ch} V_2$ as required. $\square$

12.2 Characters of Verma modules

We now consider the character of the Verma module $M(\lambda)$ where $\lambda \in H^*$. We recall from Theorem 10.7 that

$$(\operatorname{ch} M(\lambda))(\mu) = \mathfrak{P}(\lambda - \mu)$$

where $\mathfrak{P}(\lambda - \mu)$ is the number of ways of expressing $\lambda - \mu$ as a sum of positive roots. Thus we have

$$\operatorname{ch} M(\lambda) = \sum_{\mu \in H^*} \mathfrak{P}(\lambda - \mu) e_\mu = \sum_{\nu \in H^*} \mathfrak{P}(\nu) e_{\lambda - \nu}$$

$$= \sum_{\nu \in H^*} \mathfrak{P}(\nu) e_\lambda e_{-\nu} = e_\lambda \sum_{\nu \in H^*} \mathfrak{P}(\nu) e_{-\nu}.$$

We write $\Gamma = \sum_{\nu \in H^*} \mathfrak{P}(\nu) e_{-\nu}$. We have $\Gamma \in \mathfrak{R}$ since $\operatorname{Supp} \Gamma \subset S(0)$. Then we have

$$\operatorname{ch} M(\lambda) = e_\lambda \Gamma.$$

Lemma 12.3 Γ *has an inverse in the ring* $\mathfrak{R}$ *given by*

$$\Gamma^{-1} = \prod_{\alpha \in \Phi^+} (1 - e_{-\alpha}).$$

Proof. Let $\Phi^+ = \{\beta_1, \ldots, \beta_N\}$. Then $\mathfrak{P}(\nu) \neq 0$ if and only if there exist non-negative integers $r_1, \ldots, r_N$ such that $\nu = r_1 \beta_1 + \cdots + r_N \beta_N$. In fact $\mathfrak{P}(\nu)$ is the number of such sets $(r_1, \ldots, r_N)$. Thus we have

$$\Gamma = \sum_\nu \mathfrak{P}(\nu) e_{-\nu} = \sum_{r_1, \ldots, r_N \geq 0} e_{-r_1 \beta_1 - \cdots - r_N \beta_N}$$

$$= \sum_{r_1, \ldots, r_N \geq 0} e_{-\beta_1}^{r_1} \cdots e_{-\beta_N}^{r_N} = \prod_{i=1}^{N} \left(\sum_{r_i \geq 0} e_{-\beta_i}^{r_i} \right).$$

12.2 Characters of Verma modules

This factorisation of Γ in $\mathfrak{R}$ gives us the required result. For the element
$$1 + e_{-\beta_i} + e_{-\beta_i}^2 + \cdots \quad \text{of } \mathfrak{R}$$
has an inverse $1 - e_{-\beta_i} \in \mathfrak{R}$. Thus Γ has an inverse
$$\Gamma^{-1} = \prod_{i=1}^{N} (1 - e_{-\beta_i}) = \prod_{\alpha \in \Phi^+} (1 - e_{-\alpha}). \qquad \square$$

This gives us a useful formula for the character of the Verma module $M(\lambda)$.

Proposition 12.4 $\operatorname{ch} M(\lambda) = e_{\lambda+\rho} / \Delta$ where $\Delta = e_\rho \prod_{\alpha \in \Phi^+} (1 - e_{-\alpha})$.

Proof. We have
$$\operatorname{ch} M(\lambda) = e_\lambda \Gamma = \frac{e_\lambda e_\rho}{\Delta} = \frac{e_{\lambda+\rho}}{\Delta}$$
by Lemma 12.3. $\qquad \square$

The denominator Δ is an element of $\mathfrak{R}$ which can be expressed in a number of alternative ways.

We recall that $\rho \in X$ was defined by
$$\rho = \omega_1 + \cdots + \omega_l$$
i.e. ρ is the sum of the fundamental weights. This element can also be expressed simply in terms of the roots.

Proposition 12.5 $\rho = \frac{1}{2} \sum_{\alpha \in \Phi^+} \alpha$. Thus ρ is one half the sum of the positive roots.

Proof. Let $\rho' = \frac{1}{2} \sum_{\alpha \in \Phi^+} \alpha$. We can express ρ' as a linear combination of the fundamental weights. Let
$$\rho' = \sum_{i=1}^{l} c_i \omega_i \quad \text{with } c_i \in \mathbb{Q}.$$
Now the fundamental reflection $s_i \in W$ transforms α_i to $-\alpha_i$ and transforms every other positive root to a positive root, by Lemma 5.9. Thus we have
$$s_i(\rho') = \rho' - \alpha_i.$$
On the other hand we have
$$2 \frac{\langle \alpha_j, \omega_i \rangle}{\langle \alpha_j, \alpha_j \rangle} = \delta_{ij}$$

by Proposition 10.18. This shows that $s_j(\omega_i) = \omega_i$ if $i \ne j$ and $s_j(\omega_j) = \omega_j - \alpha_j$. Thus we have

$$s_i(\rho') = \rho' - c_i \alpha_i.$$

Comparing this with the above formula for $s_i(\rho')$ we deduce that $c_i = 1$. Hence $\rho' = \rho$ as required. $\square$

Corollary 12.6 $\Delta = e_{-\rho} \prod_{\alpha \in \Phi^+} (e_\alpha - 1)$

Proof. We have

$$\Delta = e_\rho \prod_{\alpha \in \Phi^+} (1 - e_{-\alpha}) = e_\rho \prod_{\alpha \in \Phi^+} e_{-\alpha}(e_\alpha - 1) = e_\rho \left(\prod_{\alpha \in \Phi^+} e_{-\alpha} \right) \prod_{\alpha \in \Phi^+} (e_\alpha - 1)$$

$$= e_\rho e_{-2\rho} \prod_{\alpha \in \Phi^+} (e_\alpha - 1) = e_{-\rho} \prod_{\alpha \in \Phi^+} (e_\alpha - 1).$$

There is a further useful expression for the denominator Δ. Before proving it we shall need some information about the geometry of the action of the Weyl group W on the Euclidean space $V = H_{\mathbb{R}}^*$.

12.3 Chambers and roots

We recall that the Weyl group is a finite group of isometries of the Euclidean space V generated by the reflections s_α for $\alpha \in \Phi$. We have

$$s_\alpha(v) = v - 2 \frac{\langle \alpha, v \rangle}{\langle \alpha, \alpha \rangle} \alpha \qquad v \in V.$$

Let

$$L_\alpha = \{v \in V \,;\, s_\alpha(v) = v\}$$
$$= \{v \in V \,;\, \langle \alpha, v \rangle = 0\}.$$

L_α is the reflecting hyperplane orthogonal to the root α. We consider the complement

$$V - \bigcup_{\alpha \in \Phi} L_\alpha$$

of the set of reflecting hyperplanes. This is an open subset of V. The connected components of this set are called the **chambers** of V. Two points of

12.3 Chambers and roots

$V - \bigcup_{\alpha \in \Phi} L_\alpha$ lie in the same chamber if and only if they lie on the same side of each reflecting hyperplane.

Let C be a chamber in V and $\delta(C)$ be the boundary of C. Then the hyperplanes L_α such that $L_\alpha \cap \delta(C)$ is not contained in any proper subspace of L_α are called the bounding hyperplanes, or walls, of C.

Now let $\Pi = \{\alpha_1, \ldots, \alpha_l\}$ be a fundamental system of roots. Then the set

$$C = \{v \in V; \langle \alpha_i, v \rangle > 0 \quad \text{for } i = 1, \ldots, l\}$$

is a chamber of V. For if α is any positive root we have $\langle \alpha, v \rangle > 0$ for all $v \in C$. Thus all elements of C lie on the same side of each reflecting hyperplane L_α. Thus C lies in $V - \bigcup_{\alpha \in \Phi} L_\alpha$ and C is connected. Moreover any subset of $V - \bigcup_{\alpha \in \Phi} L_\alpha$ larger than C would contain an element v with $\langle \alpha_i, v \rangle < 0$ for some i, and so would be disconnected. C is called the **fundamental chamber** corresponding to the fundamental system Π. The bounding hyperplanes of C are $L_{\alpha_1}, \ldots, L_{\alpha_l}$. For $L_{\alpha_i} \cap \delta(C)$ consists of all $v \in V$ such that $\langle \alpha_i, v \rangle = 0$ but $\langle \alpha_j, v \rangle \geq 0$ for $j \neq i$. Since $\alpha_1, \ldots, \alpha_l$ are linearly independent $L_{\alpha_i} \cap \delta(C)$ is not contained in any proper subspace of L_{α_i}. On the other hand let α be a positive root which is not fundamental. Then $\alpha = \sum_{i=1}^{l} n_i \alpha_i$ with each $n_i \geq 0$ and at least two $n_i > 0$. If $v \in L_\alpha \cap \delta(C)$ then

$$\sum n_i \langle \alpha_i, v \rangle = 0$$

and so $\langle \alpha_i, v \rangle = 0$ whenever $n_i > 0$. Thus $L_\alpha \cap \delta(C)$ lies in a proper subspace of L_α. Hence the bounding hyperplanes of C are $L_{\alpha_1}, \ldots, L_{\alpha_l}$. In fact the set $\Pi = \{\alpha_1, \ldots, \alpha_l\}$ of fundamental roots may be characterised as the roots orthogonal to the bounding hyperplanes of C which point into C, that is such that α_i lies on the same side of L_{α_i} as C.

Now the Weyl group acts on V in a way which permutes the roots. It therefore permutes the reflecting hyperplanes L_α, and so acts on $V - \bigcup_{\alpha \in \Phi} L_\alpha$. Since W is a group of isometries of V, W permutes the connected components of $V - \bigcup_{\alpha \in \Phi} L_\alpha$. Thus the Weyl group W acts on the set of chambers of V.

Proposition 12.7 (i) *Given any two chambers C, C' of V there is a unique element $w \in W$ such that $w(C) = C'$.*
(ii) *The number of chambers of V is equal to the order of the Weyl group.*
(iii) *If C is a chamber in V its closure $\bar{C}$ contains just one element from each W-orbit on V.*

Proof. Let Π be a fundamental system of roots and C be the chamber defined by $v \in C$ if and only if $\langle \alpha_i, v \rangle > 0$ for $i = 1, \ldots, l$. Let C' be any chamber and let $v \in C'$. We recall from Section 5.1 that Π is associated with a total ordering $>$ on V. We consider the set of transforms $w(v)$ for $w \in W$ and let v' be the one which is greatest in the above total ordering. Then we have

$$s_{\alpha_i}(v') = v' - 2\frac{\langle \alpha_i, v' \rangle}{\langle \alpha_i, \alpha_i \rangle} \alpha_i \qquad \alpha_i \in \Pi$$

and since $s_{\alpha_i}(v') \leq v'$ we must have $\langle \alpha_i, v' \rangle \geq 0$. This holds for all $i = 1, \ldots, l$, thus $v' \in C$. Now let $v' = w(v)$. Since $v \in C'$ we have $v' \in w(C')$. Thus $w(C')$ is a chamber which intersects $\bar{C}$. However, the only chamber intersecting $\bar{C}$ is C. Thus $w(C') = C$. Hence any chamber C' is in the same W-orbit as C. Thus W acts transitively on the set of chambers. It follows that any chamber is associated to some fundamental system of roots in the manner described above.

Now suppose $w(C) = C$. Then we have $w(\Pi) = \Pi$ where Π is the fundamental system determined by C, i.e. the set of roots orthogonal to the walls of C and pointing into C. It follows that $w(\Phi^+) = \Phi^+$, so w makes every positive root positive. Hence $n(w) = 0$. It follows from Corollary 5.16 that $l(w) = 0$, i.e. $w = 1$. Thus W acts simply transitively on the set of chambers.

It is a consequence of this that the number of chambers of V is equal to $|W|$.

We now consider the closure $\bar{C}$ of a chamber C. Since each vector lies in the closure of some chamber and W acts transitively on the chambers each orbit of W on V intersects $\bar{C}$. We must also show that if $v_1, v_2 \in \bar{C}$ and $w(v_1) = v_2$ then $v_1 = v_2$. We prove this by induction on $l(w)$. It is clear when $l(w) = 0$, i.e. $w = 1$. Thus we assume $l(w) > 0$. Then $n(w) > 0$ so there exists $\alpha_i \in \Pi$ with $w(\alpha_i) < 0$. Thus

$$0 \leq \langle v_1, \alpha_i \rangle = \langle v_2, w(\alpha_i) \rangle \leq 0.$$

Hence $\langle v_1, \alpha_i \rangle = 0$ and $s_{\alpha_i}(v_1) = v_1$. But now $ws_{\alpha_i}(v_1) = v_2$. The only positive root made negative by s_{α_i} is α_i. Thus the positive roots made negative by w and ws_{α_i} are the same, apart from α_i, which is made negative by w and positive by ws_{α_i}. Thus

$$n(w) = n(ws_{\alpha_i}) + 1$$

and so

$$l(ws_{\alpha_i}) = l(w) - 1$$

by Corollary 5.16. We can then deduce that $v_1 = v_2$ by induction, as required. □

We shall now suppose that Π is a fixed fundamental system of roots and C is the corresponding fundamental chamber.

12.3 Chambers and roots

Proposition 12.8 (i) $v \in C$ if and only if $v = \sum_{i=1}^{l} n_i \omega_i$ with $n_i > 0$ for all i.
(ii) $v \in \bar{C}$ if and only if $v = \sum_{i=1}^{l} n_i \omega_i$ with $n_i \geq 0$ for all i.

Proof. Since $\omega_1, \ldots, \omega_l$ are a basis of V we can write $v = \sum n_i \omega_i$ for each $v \in V$. Now $v \in C$ if and only if $\langle \alpha_i, v \rangle > 0$ for $i = 1, \ldots, l$. We recall from the definition of the fundamental weights $\omega_1, \ldots, \omega_l$ that

$$\langle \alpha_i, \omega_j \rangle = 0 \quad \text{if } i \neq j$$
$$\langle \alpha_i, \omega_i \rangle = 2 \langle \alpha_i, \alpha_i \rangle.$$

Thus we have $\langle \alpha_i, v \rangle = 2 n_i \langle \alpha_i, \alpha_i \rangle$. In particular $\langle \alpha_i, v \rangle > 0$ if and only if $n_i > 0$. Similarly $\langle \alpha_i, v \rangle \geq 0$ if and only if $n_i \geq 0$. The required result follows.
□

We show in Figures 12.1, 12.2 and 12.3 the chambers for the 2-dimensional root systems A_2, B_2 and G_2.

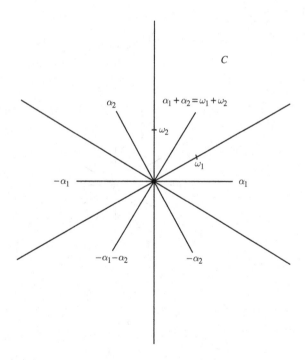

Figure 12.1 Two-dimensional root system type A_2

250   *Character and dimension formulae*

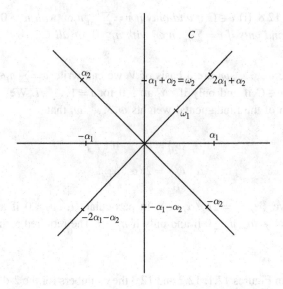

Figure 12.2   Two-dimensional root system type B_2

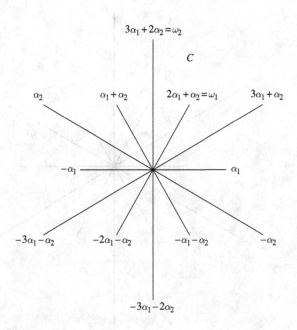

Figure 12.3   Two-dimensional root system type G_2

12.3 Chambers and roots

Proposition 12.9 *Suppose Φ is the root system of a simple Lie algebra and let C be the fundamental chamber.*

(i) *Suppose all roots in Φ have the same length. Then there exists a unique root $\theta_l = \sum_{i=1}^{l} a_i \alpha_i$ in $\bar{C}$. This root satisfies the condition that for any root $\alpha = \sum_{i=1}^{l} k_i \alpha_i$ we have $k_i \leq a_i$.*

(ii) *Now suppose there are two root lengths. Then there are just two roots*

$$\theta_l = \sum_{i=1}^{l} a_i \alpha_i, \qquad \theta_s = \sum_{i=1}^{l} c_i \alpha_i$$

in $\bar{C}$. θ_l is a long root and θ_s is a short root. θ_l satisfies the condition that for any root $\alpha = \sum_{i=1}^{l} k_i \alpha_i$ we have $k_i \leq a_i$. (In particular $c_i \leq a_i$.)

*θ_l is called the **highest root** and θ_s the **highest short root**.*

Proof. By Proposition 12.7 $\bar{C}$ contains just one root in each W-orbit on Φ. Now two roots lie in the same W-orbit if and only if they have the same length. For roots in the same orbit obviously have the same length; but any root is in the same orbit as a fundamental root, and any two fundamental roots of the same length can be joined in the Dynkin diagram by a sequence of fundamental roots all of this length. Two fundamental roots of the same length joined in the Dynkin diagram obviously lie in the same W-orbit.

Thus in case (i) $\bar{C}$ contains a unique root θ_l and in case (ii) $\bar{C}$ contains one long root θ_l and one short root θ_s.

We now introduce a partial order $\triangleright$ on the set Φ^+ of positive roots. Given

$$\alpha = \sum_{i=1}^{l} m_i \alpha_i, \qquad \beta = \sum_{i=1}^{l} n_i \alpha_i$$

in Φ^+ we write $\alpha \triangleright \beta$ if $m_i \geq n_i$ for each i. We consider maximal elements of Φ^+ with respect to this partial order. Let α be maximal. Then $\langle \alpha, \alpha_i \rangle \geq 0$ for each i, as otherwise $\alpha + \alpha_i$ would be a root higher than α. We also have $\langle \alpha, \alpha_i \rangle > 0$ for some i. Let $\alpha = \sum_{i=1}^{l} m_i \alpha_i$. We show that each $m_i > 0$. Suppose this is not so. Then there exist i, i' with $m_i \neq 0$, $m_{i'} = 0$ and $\langle \alpha_i, \alpha_{i'} \rangle < 0$. But then $\langle \alpha, \alpha_{i'} \rangle = \sum_{j=1}^{l} m_j \langle \alpha_j, \alpha_{i'} \rangle < 0$, a contradiction. Hence each $m_i > 0$.

We now show that α is the unique maximal element of Φ^+ with respect to $\triangleright$. Suppose if possible that β is also maximal and $\beta \neq \alpha$. Then $\alpha + \beta \notin \Phi$. Also $\alpha - \beta \notin \Phi$, as $\alpha - \beta \notin \Phi$ would imply $\alpha \triangleright \beta$ or $\beta \triangleright \alpha$. Hence $\langle \alpha, \beta \rangle = 0$ by Proposition 4.22. But

$$\langle \alpha, \beta \rangle = \sum_{i=1}^{l} m_i \langle \alpha_i, \beta \rangle > 0$$

since each $m_i > 0$, each $\langle \alpha_i, \beta \rangle \geq 0$, and some $\langle \alpha_i, \beta \rangle > 0$. Thus we have a contradiction. Hence α is the unique maximal element of Φ^+ with respect to $\triangleright$.

Now $\alpha \in \bar{C}$ since $\langle \alpha, \alpha_i \rangle \geq 0$ for each i. Thus $\alpha = \theta_l$ or θ_s. We wish to show $\alpha = \theta_l$. To do so we show that if $\alpha' \in \Phi \cap \bar{C}$ then $\langle \alpha', \alpha' \rangle \leq \langle \alpha, \alpha \rangle$. By the maximality of α, $\alpha - \alpha'$ is a non-negative combination of $\alpha_1, \ldots, \alpha_l$ and so $\langle \alpha - \alpha', x \rangle \geq 0$ for all $x \in \bar{C}$. In particular we have $\langle \alpha - \alpha', \alpha \rangle \geq 0$ and $\langle \alpha - \alpha', \alpha' \rangle \geq 0$. Hence

$$\langle \alpha, \alpha \rangle \geq \langle \alpha, \alpha' \rangle \geq \langle \alpha', \alpha' \rangle.$$

It follows that $\alpha = \theta_l$. Thus θ_l is the unique maximal element of Φ^+ with respect to $\triangleright$ and the result is proved. □

Definition *The number $h = 1 + \mathrm{ht}\,\theta_l$ is called the **Coxeter number** of L. It is known to be equal to the order of the element $s_1 s_2 \ldots s_l \in W$, and also to $|\Phi|/|\Pi|$. (See, for example, Bourbaki, Groupes et algèbres de Lie, Chapters 4, 5, 6.)* □

In order to prove Weyl's denominator formula we shall need some properties of the transforms $w(\rho)$.

Proposition 12.10 (i) $w(\rho) = \rho - \sum_{\alpha \in \Omega} \alpha$ *for some subset Ω of Φ^+.*
(ii) *Given any subset Ω of Φ^+ the vector $\rho - \sum_{\alpha \in \Omega} \alpha$ either lies in one of the reflecting hyperplanes L_α or has the form $w(\rho)$ for some $w \in W$.*
(iii) *If $\rho - \sum_{\alpha \in \Omega} \alpha$ lies in the fundamental chamber then Ω is empty.*

Proof. We know from Proposition 12.5 that $\rho = \frac{1}{2} \sum_{\alpha \in \Phi^+} \alpha$. Let $w \in W$. Then w permutes the roots and so

$$w(\rho) = \frac{1}{2} \sum_{\alpha \in \Phi^+} (\pm \alpha) = \rho - \sum_{\alpha \in \Omega} \alpha$$

where Ω is the set of positive roots made negative by w^{-1}.

Now suppose Ω is any subset of Φ^+. Suppose $\rho - \sum_{\alpha \in \Omega} \alpha$ lies in the fundamental chamber. We write $v = \sum_{\alpha \in \Omega} \alpha$. Then $(\rho - v)(h_i) > 0$ for each $i = 1, \ldots, l$. Moreover $(\rho - v)(h_i) \in \mathbb{Z}$ and $\rho(h_i) = 1$, since $\omega_i(h_i) = 1$ and $\omega_j(h_i) = 0$ if $j \neq i$. It follows that $v(h_i) \leq 0$, that is that $\langle v, \alpha_i \rangle \leq 0$ for $i = 1, \ldots, l$. However, v is a sum of positive roots so has form $v = \sum_{i=1}^{l} n_i \alpha_i$ where $n_i \geq 0$ for each i. Hence

$$\langle v, v \rangle = \sum_{i=1}^{l} n_i \langle v, \alpha_i \rangle \leq 0.$$

It follows that $v = 0$ and so Ω is empty.

12.3 Chambers and roots

Finally we must show that $\rho - \sum_{\alpha \in \Omega} \alpha$ either lies in a reflecting hyperplane or is a W-transform of ρ. Suppose it does not lie in any reflecting hyperplane. Then it lies in a chamber. Thus there exists $w \in W$ such that

$$w\left(\rho - \sum_{\alpha \in \Omega} \alpha\right)$$

lies in the fundamental chamber. However, $\rho - \sum_{\alpha \in \Omega} \alpha$ has the form $\frac{1}{2} \sum_{\alpha \in \Phi^+} (\pm \alpha)$, so $w(\rho - \sum_{\alpha \in \Omega} \alpha)$ also has form $\frac{1}{2} \sum_{\alpha \in \Phi^+} (\pm \alpha)$ since w permutes the roots. Hence

$$w\left(\rho - \sum_{\alpha \in \Omega} \alpha\right) = \rho - \sum_{\alpha \in \Omega'} \alpha$$

for some subset Ω' of Φ^+. Since this vector lies in the fundamental chamber, Ω' must be empty. Hence

$$w\left(\rho - \sum_{\alpha \in \Omega} \alpha\right) = \rho$$

and so $\rho - \sum_{\alpha \in \Omega} \alpha$ is a W-transform of ρ. □

We can now prove Weyl's denominator formula.

Theorem 12.11 (*Weyl's denominator formula*).

$$e_\rho \prod_{\alpha \in \Phi^+} (1 - e_{-\alpha}) = \sum_{w \in W} \varepsilon(w) e_{w(\rho)}$$

where $\varepsilon(w) = (-1)^{l(w)}$.

Proof. Let $\mathbb{Z}[H^*]$ be the set of functions $f : H^* \to \mathbb{Z}$ of finite support. This is the set of finite $\mathbb{Z}$-combinations of the characteristic functions e_λ. Weyl's denominator formula is an identity in $\mathbb{Z}[H^*]$. There is a natural action of W on $\mathbb{Z}[H^*]$ given by

$$(wf)\lambda = f\left(w^{-1}\lambda\right).$$

We define a map $\theta : \mathbb{Z}[H^*] \to \mathbb{Z}[H^*]$ by

$$\theta(f) = \sum_{w \in W} \varepsilon(w) wf.$$

It is clear that, for $w' \in W$, $\theta w' = \varepsilon(w') \theta$, hence

$$\theta\left(\varepsilon(w') w'\right) = \theta$$

and
$$\theta^2 = |W|\theta.$$

We now consider the effect of a fundamental reflection s_i on
$$\Delta = e_\rho \prod_{\alpha \in \Phi^+} (1 - e_{-\alpha}).$$

We have
$$s_i\left(e_\rho \prod_{\alpha \in \Phi^+} (1 - e_{-\alpha})\right) = e_{s_i(\rho)} \prod_{\alpha \in \Phi^+} (1 - e_{-s_i(\alpha)}).$$

Now $s_i(\rho) = \rho - \alpha_i$ by the proof of Proposition 12.5. Also s_i transforms every positive root to a positive root, except for α_i. Hence we have

$$s_i\left(e_\rho \prod_{\alpha \in \Phi^+} (1 - e_{-\alpha})\right) = e_{\rho - \alpha_i} \left(\prod_{\substack{\alpha \in \Phi^+ \\ \alpha \neq \alpha_i}} (1 - e_{-\alpha})\right)(1 - e_{\alpha_i})$$

$$= e_\rho \left(\prod_{\substack{\alpha \in \Phi^+ \\ \alpha \neq \alpha_i}} (1 - e_{-\alpha})\right)(e_{-\alpha_i} - 1)$$

$$= -e_\rho \prod_{\alpha \in \Phi^+} (1 - e_{-\alpha}).$$

Thus $s_i(\Delta) = -\Delta$. It follows that $w(\Delta) = \varepsilon(w)\Delta$ for all $w \in W$. Hence $\theta(\Delta) = |W|\Delta$.

We also have
$$\Delta = e_\rho \prod_{\alpha \in \Phi^+} (1 - e_{-\alpha})$$
$$= e_\rho \sum_{\Omega \subset \Phi^+} (-1)^{|\Omega|} e_{-\sum_{\alpha \in \Omega} \alpha}$$
$$= \sum_{\Omega \subset \Phi^+} (-1)^{|\Omega|} e_{\rho - \sum_{\alpha \in \Omega} \alpha}.$$

Now $\rho - \sum_{\alpha \in \Omega} \alpha$ either is of form $w(\rho)$ for some $w \in W$ or lies in some reflecting hyperplane, by Proposition 12.10. If v lies in a reflecting hyperplane then $\theta(v) = 0$ since the terms in $\theta(v)$ cancel out in pairs. For if $v \in L_\alpha$ then
$$\varepsilon(ww_\alpha) ww_\alpha v = -\varepsilon(w) wv.$$

Thus we have
$$\theta(\Delta) = \theta \sum_{w \in W} \varepsilon(w) e_{w\rho}$$

since if $\rho - \sum_{\alpha \in \Omega} \alpha = w(\rho)$ then $|\Omega| = l(w)$ by the proof of Proposition 12.10. Thus

$$\theta(\Delta) = \theta\left(\theta\left(e_\rho\right)\right) = \theta^2\left(e_\rho\right) = |W|\theta\left(e_\rho\right).$$

But $\theta(\Delta) = |W|\Delta$ as shown above. It follows that

$$\Delta = \theta\left(e_\rho\right) = \sum_{w \in W} \varepsilon(w) e_{w\rho}. \qquad \square$$

Corollary 12.12 $\operatorname{ch} M(\lambda) = \dfrac{e_{\lambda+\rho}}{\sum_{w \in W} \varepsilon(w) e_{w\rho}}.$

Proof. This follows from Proposition 12.4 and Theorem 12.11. $\qquad \square$

12.4 Composition factors of Verma modules

We shall show in this section that each Verma module $M(\lambda)$ has a composition series of finite length and that all its composition factors are irreducible modules of the form $L(\mu)$ where $\mu = w(\lambda + \rho) - \rho$ for some $w \in W$. It will be convenient to define

$$w \cdot \lambda = w(\lambda + \rho) - \rho.$$

We shall use these results in the following section to prove Weyl's character formula for the finite dimensional irreducible modules $L(\lambda)$.

We begin with a lemma on filtered algebras and their corresponding graded algebras. We recall the definitions as given in Section 11.1.

Lemma 12.13 *Let* $A = \bigcup_i A_i$ *be a filtered algebra with*

$$A_0 \subset A_1 \subset A_2 \subset \cdots$$

and let $B = B_0 \oplus B_1 \oplus B_2 \oplus \cdots$ *be the corresponding graded algebra.*

(i) *if I is a left ideal of A then* $\operatorname{gr} I = \bigoplus_i \dfrac{A_{i-1} + (A_i \cap I)}{A_{i-1}}$ *is a left ideal of B.*
(ii) *If $I_1 \subset I_2$ then $\operatorname{gr} I_1 \subset \operatorname{gr} I_2$.*
(iii) *If $I_1 \subset I_2$ and $\operatorname{gr} I_1 = \operatorname{gr} I_2$ then $I_1 = I_2$.*
(iv) *If B satisfies the maximal condition on left ideals so does A.*

Proof. We recall that $B = \bigoplus_i B_i$ where $B_i = A_i/A_{i-1}$. If $x \in A_i \cap I$, $y \in A_j$ then we have

$$(A_{j-1} + y)(A_{i-1} + x) = A_{i+j-1} + yx$$

where $yx \in A_{i+j} \cap I$. Thus $A_{i+j-1} + yx \in \operatorname{gr} I$. It follows that $\operatorname{gr} I$ is a left ideal of B.

It is clear from the definition that if $I_1 \subset I_2$ then $\operatorname{gr} I_1 \subset \operatorname{gr} I_2$.

We now suppose that $I_1 \subset I_2$ and $\operatorname{gr} I_1 = \operatorname{gr} I_2$. Then

$$A_{i-1} + (A_i \cap I_1) = A_{i-1} + (A_i \cap I_2)$$

for each i. Thus we have

$$A_i \cap I_2 = (A_i \cap I_1) + (A_{i-1} \cap I_2).$$

We shall show that $A_i \cap I_1 = A_i \cap I_2$ by induction on i. We know $A_0 \cap I_1 = A_0 \cap I_2$ since $\operatorname{gr} I_1 = \operatorname{gr} I_2$.

Assume inductively that $A_{i-1} \cap I_1 = A_{i-1} \cap I_2$. Then we have

$$A_i \cap I_2 = (A_i \cap I_1) + (A_{i-1} \cap I_1) = A_i \cap I_1.$$

Thus $A_i \cap I_2 = A_i \cap I_1$ for all i. Since $A = \cup_i A_i$ it follows that $I_1 = I_2$.

Now suppose that

$$I_1 \subset I_2 \subset I_3 \subset \cdots$$

is a chain of left ideals of A. Then

$$\operatorname{gr} I_1 \subset \operatorname{gr} I_2 \subset \operatorname{gr} I_3 \subset \cdots$$

is a chain of left ideals of B. Assume that B satisfies the maximal condition on left ideals. Then we have $\operatorname{gr} I_i = \operatorname{gr} I_j$ for all i, j sufficiently large. It follows that $I_i = I_j$ for all i, j sufficiently large. Hence A satisfies the maximal condition on left ideals. □

Proposition 12.14 $\mathfrak{U}(L)$ *satisfies the maximal condition on left ideals.*

Proof. $\mathfrak{U}(L)$ is a filtered algebra whose graded algebra is the symmetric algebra $S(L)$. However, $S(L)$ is isomorphic to the polynomial ring $\mathbb{C}[z_1, \ldots, z_n]$ where $n = \dim L$, so satisfies the maximal condition on (left) ideals, by Hilbert's basis theorem. Thus $\mathfrak{U}(L)$ satisfies the maximal condition on left ideals, by Lemma 12.13. □

Corollary 12.15 *The Verma module* $M(\lambda)$ *satisfies the maximal condition on submodules.*

12.4 Composition factors of Verma modules

Proof. The left ideals of $\mathfrak{U}(L)$ are the same as the $\mathfrak{U}(L)$-submodules. Thus $\mathfrak{U}(L)$ satisfies the maximal condition on submodules. We recall that

$$M(\lambda) = \mathfrak{U}(L)/K_\lambda$$

where K_λ is a submodule of $\mathfrak{U}(L)$. It follows that $M(\lambda)$ satisfies the maximal condition on submodules. □

Theorem 12.16 *The Verma module $M(\lambda)$ has a finite composition series*

$$M(\lambda) = N_0 \supset N_1 \supset N_2 \supset \cdots \supset N_r = O$$

where each N_i is a submodule of $M(\lambda)$ and N_{i+1} is a maximal submodule of N_i. Moreover N_i/N_{i+1} is isomorphic to $L(w \cdot \lambda)$ for some $w \in W$.

Proof. Since $M(\lambda)$ satisfies the maximal condition on submodules, every submodule of $M(\lambda)$ has a maximal submodule. Thus we have a descending series

$$M(\lambda) = N_0 \supset N_1 \supset N_2 \supset \cdots$$

of submodules, in which N_{i+1} is a maximal submodule of N_i. We wish to show that this series reaches O after finitely many steps.

Now $M(\lambda)$ is the direct sum of its weight spaces by Theorem 10.7. Thus every submodule of $M(\lambda)$ is also the direct sum of its weight spaces, by the proof of Theorem 10.9. It follows that each quotient N_i/N_{i+1} is the direct sum of its weight spaces. Moreover each weight μ of N_i/N_{i+1} is a weight of $M(\lambda)$ so satisfies $\mu \prec \lambda$ with respect to the natural partial order on weights. Thus we can choose a weight μ of N_i/N_{i+1} which is maximal in this partial order among the set of possible weights. Let v be a non-zero vector in N_i/N_{i+1} of weight μ. Then we have $e_i v = 0$ and $hv = \mu(h)v$ for all $h \in H$. Thus we have

$$\mathfrak{U}(L)v = \mathfrak{U}(N^-)v.$$

However, N_i/N_{i+1} is an irreducible $\mathfrak{U}(L)$-module, thus $\mathfrak{U}(L)v = N_i/N_{i+1}$. Thus we have a homomorphism $M(\mu) \to N_i/N_{i+1}$ given by $um_\mu \to uv$ for all $u \in \mathfrak{U}(N^-)$ as in Proposition 10.13. This homomorphism is surjective and its kernel is the unique maximal submodule of $M(\mu)$, since N_i/N_{i+1} is irreducible. It follows that N_i/N_{i+1} is isomorphic to $L(\mu)$, the unique irreducible quotient of $M(\mu)$.

We now consider the action of the centre $Z(L)$ of $\mathfrak{U}(L)$. $Z(L)$ acts on $M(\lambda)$ by scalar multiplications. The element $z \in Z(L)$ acts on $M(\lambda)$ by scalar multiplication by $\chi_\lambda(z)$, as in Section 11.5. Hence z acts on each submodule N_i and each quotient N_i/N_{i+1} as scalar multiplication by $\chi_\lambda(z)$. However,

z acts on $M(\mu)$ as scalar multiplication by $\chi_\mu(z)$, and so also on its quotient $L(\mu)$. Since N_i/N_{i+1} is isomorphic to $L(\mu)$ we deduce that $\chi_\lambda(z) = \chi_\mu(z)$ for all $z \in Z(L)$. Hence $\chi_\lambda = \chi_\mu$. It follows from Theorem 11.31 that $\mu + \rho = w(\lambda + \rho)$ for some $w \in W$. This is equivalent to $\mu = w \cdot \lambda$ for some $w \in W$.

Now W is finite and so there are only finitely many possible composition factors of $M(\lambda)$, up to isomorphism. Also each weight space of $M(\lambda)$ is finite dimensional. Thus $L(\mu)$, which contains μ as a weight, can appear as a composition factor with multiplicity at most the dimension of the μ weight space $M(\lambda)_\mu$. It follows that the series

$$M(\lambda) = N_0 \supset N_1 \supset N_2 \supset \cdots$$

must reach O after at most $\sum_{w \in W} \dim M(\lambda)_{w \cdot \lambda}$ steps. Thus $M(\lambda)$ has a finite composition series and each composition factor has form $L(w \cdot \lambda)$ for some $w \in W$. $\square$

12.5 Weyl's character formula

We now find a formula for the characters of the finite dimensional irreducible modules $L(\lambda)$ where $\lambda \in X^+$.

Theorem 12.17 (*Weyl's character formula*). *Let $\lambda \in X^+$. Then*

$$\operatorname{ch} L(\lambda) = \frac{\sum_{w \in W} \varepsilon(w) e_{w(\lambda+\rho)}}{\sum_{w \in W} \varepsilon(w) e_{w(\rho)}}.$$

(*This is an equality in the ring $\mathfrak{R}$ of Section 12.1 since the denominator*

$$\Delta = \sum_{w \in W} \varepsilon(w) e_{w(\rho)}$$

is an invertible element of $\mathfrak{R}$.)

Proof. Since λ is a dominant integral weight we have $\lambda(h_i) \geq 0$ for $i = 1, \ldots, l$. Hence $(\lambda + \rho)(h_i) = \lambda(h_i) + 1 > 0$ for $i = 1, \ldots, l$. Thus $\lambda + \rho$ lies in the fundamental chamber C. Hence $w(\lambda + \rho)$ lies in the chamber $w(C)$. It follows from Proposition 12.7 that the weights $w(\lambda + \rho)$ for $w \in W$ are all distinct.

Now the highest weight of the Verma module $M(w \cdot \lambda)$ is $w \cdot \lambda = w(\lambda + \rho) - \rho$. Thus the characters $\operatorname{ch} M(w \cdot \lambda) \in \mathfrak{R}$ are linearly independent as w runs over W. Similarly the characters $\operatorname{ch} L(w \cdot \lambda)$ are linearly independent for $w \in W$.

12.5 Weyl's character formula

Now $M(w \cdot \lambda)$ has a finite composition series with composition factors of form $L(y \cdot \lambda)$ for $y \in W$, by Theorem 12.16. Moreover, since $y \cdot \lambda$ is a weight of $L(y \cdot \lambda)$ and $w \cdot \lambda$ is the highest weight of $M(w \cdot \lambda)$ we have $y \cdot \lambda \prec w \cdot \lambda$ whenever $L(y \cdot \lambda)$ occurs as a composition factor of $M(w \cdot \lambda)$. Moreover $w \cdot \lambda$ occurs as a weight of $M(w \cdot \lambda)$ with multiplicity 1, thus $L(w \cdot \lambda)$ appears as a composition factor of $M(w \cdot \lambda)$ with multiplicity 1. We therefore have

$$\operatorname{ch} M(w \cdot \lambda) = \sum_{y \in W} a_{wy} \operatorname{ch} L(y \cdot \lambda)$$

where $a_{wy} \in \mathbb{Z}$, $a_{wy} \geq 0$, $a_{ww} = 1$, and $a_{wy} \neq 0$ only if $y \cdot \lambda \prec w \cdot \lambda$. If we write the elements of W in an order compatible with the partial order $y \cdot \lambda \prec w \cdot \lambda$ we see that the integers a_{wy} form a triangular $|W| \times |W|$ matrix with entries 1 on the diagonal. The determinant of this matrix is 1. Thus we may invert the above equations to obtain

$$\operatorname{ch} L(w \cdot \lambda) = \sum_{y \in W} b_{wy} \operatorname{ch} M(y \cdot \lambda)$$

where $b_{wy} \in \mathbb{Z}$ and $b_{ww} = 1$. (The b_{wy} will no longer be non-negative.) In particular we have

$$\operatorname{ch} L(\lambda) = \sum_{y \in W} c_y \operatorname{ch} M(y \cdot \lambda)$$

where $c_y = b_{1y}$. By Proposition 12.4 this gives

$$\operatorname{ch} L(\lambda) = \frac{\sum_{y \in W} c_y e_{y(\lambda + \rho)}}{\Delta}$$

where $c_1 = 1$. We wish to determine the remaining coefficients c_y.

We recall from Proposition 10.22 that

$$\dim L(\lambda)_\mu = \dim L(\lambda)_{w(\mu)}$$

for all $w \in W$. Thus we have

$$w(\operatorname{ch} L(\lambda)) = \operatorname{ch} L(\lambda) \qquad \text{for all } w \in W.$$

On the other hand we have

$$s_i(\Delta) = -\Delta$$

by Theorem 12.11 and thus

$$w(\Delta) = \varepsilon(w) \Delta.$$

It follows that
$$w\left(\sum_{y\in W} c_y e_{y(\lambda+\rho)}\right) = \varepsilon(w)\left(\sum_{y\in W} c_y e_{y(\lambda+\rho)}\right).$$
Thus
$$\sum_{y\in W} c_y e_{wy(\lambda+\rho)} = \sum_{y\in W} \varepsilon(w) c_y e_{y(\lambda+\rho)}$$
since $we_\lambda = e_{w\lambda}$. This is equivalent to
$$\sum_{y\in W} c_{w^{-1}y} e_{y(\lambda+\rho)} = \sum_{y\in W} \varepsilon(w) c_y e_{y(\lambda+\rho)}.$$
Since the functions $e_{y(\lambda+\rho)}$ for $y \in W$ are linearly independent we deduce that
$$c_{w^{-1}y} = \varepsilon(w) c_y.$$
In particular we have $c_{w^{-1}} = \varepsilon(w)$, thus $c_w = \varepsilon(w^{-1}) = \varepsilon(w)$. It follows that
$$\operatorname{ch} L(\lambda) = \frac{\sum_{w\in W} \varepsilon(w) e_{w(\lambda+\rho)}}{\Delta}$$
as required. $\square$

We note that in the special case $\lambda = 0$ we have $\operatorname{ch} L(0) = e_0$ as $L(0)$ is the trivial 1-dimensional representation of L. Thus we have
$$\sum_{w\in W} \varepsilon(w) e_{w(\rho)} = \Delta e_0 = \Delta.$$
This gives an alternative proof of Weyl's denominator formula, Theorem 12.11.

We also note that while the character $\operatorname{ch} L(\lambda)$ is invariant under the Weyl group both the numerator and the denominator in Weyl's character formula are alternating functions under the Weyl group, i.e. satisfy $w(a) = \varepsilon(w)a$.

We may deduce from Weyl's character formula a formula due to Kostant for the dimension of the weight space $L(\lambda)_\mu$ of $L(\lambda)$.

Theorem 12.18 (*Kostant's multiplicity formula*). *Let $\lambda \in X^+$ and $\mu \in X$. Then*
$$\dim L(\lambda)_\mu = \sum_{w\in W} \varepsilon(w) \mathfrak{P}(w(\lambda+\rho) - (\mu+\rho))$$
where $\mathfrak{P}$ is the partition function defined in Theorem 10.7.

Proof. We have $\operatorname{ch} L(\lambda) = \sum_\mu \dim L(\lambda)_\mu e_\mu$. Moreover we know that
$$\Delta^{-1} = e_{-\rho} \Gamma = e_{-\rho} \sum_\nu \mathfrak{P}(\nu) e_{-\nu}$$

12.5 Weyl's character formula

by Lemma 12.3. Thus Weyl's character formula gives the identity

$$\sum_\mu \dim L(\lambda)_\mu e_\mu = \left(\sum_{w\in W} \varepsilon(w) e_{w(\lambda+\rho)}\right) e_{-\rho} \sum_\nu \mathfrak{P}(\nu) e_{-\nu}$$

$$= \sum_{w\in W}\sum_\nu \varepsilon(w)\mathfrak{P}(\nu) e_{w(\lambda+\rho)-\rho-\nu}.$$

We compare the coefficients of e_μ on both sides. This gives

$$\dim L(\lambda)_\mu = \sum_{w\in W} \varepsilon(w)\mathfrak{P}(w(\lambda+\rho)-(\mu+\rho)). \qquad \square$$

We can also derive from Weyl's character formula a formula for the dimension of $L(\lambda)$.

Theorem 12.19 (*Weyl's dimension formula*). *Let $\lambda \in X^+$. Then*

$$\dim L(\lambda) = \frac{\prod_{\alpha \in \Phi^+} \langle \lambda+\rho, \alpha\rangle}{\prod_{\alpha\in\Phi^+}\langle\rho,\alpha\rangle}.$$

Proof. Let $\mathfrak{R}_0$ be the subring of $\mathfrak{R}$ consisting of all finite sums $\sum_{\mu\in X} n_\mu e_\mu$ with $n_\mu \in \mathbb{Z}$. Then the character formula

$$\Delta \operatorname{ch} L(\lambda) = \sum_{w\in W}\varepsilon(w) e_{w(\lambda+\rho)}$$

may be regarded as an identity in $\mathfrak{R}_0$. Let $A = \mathbb{R}[[t]]$ be the ring of formal power series in the variable t with real coefficients. Then for each weight $\xi \in X$ we have a ring homomorphism

$$\theta_\xi : \mathfrak{R}_0 \to A$$

given by

$$\theta_\xi(e_\mu) = \exp(\langle \xi,\mu\rangle t) = 1 + \langle\xi,\mu\rangle t + \frac{1}{2!}\langle\xi,\mu\rangle^2 t^2 + \cdots.$$

Consider $\theta_\xi\left(\sum_{w\in W}\varepsilon(w) e_{w\mu}\right)$. We have

$$\theta_\xi\left(\sum_{w\in W}\varepsilon(w)e_{w\mu}\right) = \sum_{w\in W}\varepsilon(w)\exp(\langle\xi,w\mu\rangle t) = \sum_{w\in W}\varepsilon(w)\exp(\langle\mu,w^{-1}\xi\rangle t)$$

$$= \sum_{w\in W}\varepsilon(w)\exp(\langle\mu,w\xi\rangle t) = \theta_\mu\left(\sum_{w\in W}\varepsilon(w)e_{w\xi}\right).$$

In particular we have

$$\theta_\rho \left(\sum_{w \in W} \varepsilon(w) e_{w(\lambda+\rho)} \right) = \theta_{\lambda+\rho} \left(\sum_{w \in W} \varepsilon(w) e_{w\rho} \right)$$

$$= \theta_{\lambda+\rho} \left(e_{-\rho} \prod_{\alpha \in \Phi^+} (e_\alpha - 1) \right)$$

$$= \exp(\langle \lambda+\rho, -\rho \rangle t) \prod_{\alpha \in \Phi^+} (\exp\langle \lambda+\rho, \alpha \rangle t - 1)$$

$$= \exp(\langle \lambda+\rho, -\rho \rangle t) \prod_{\alpha \in \Phi^+} (\langle \lambda+\rho, \alpha \rangle t + \cdots)$$

$$= t^N \left(\prod_{\alpha \in \Phi^+} \langle \lambda+\rho, \alpha \rangle + \cdots \right) \qquad \text{where } N = |\Phi^+|.$$

By putting $\lambda = 0$ we obtain

$$\theta_\rho \left(\sum_{w \in W} \varepsilon(w) e_{w\rho} \right) = t^N \left(\prod_{\alpha \in \Phi^+} \langle \rho, \alpha \rangle + \cdots \right).$$

Thus by applying θ_ρ to Weyl's character formula we obtain

$$t^N \left(\prod_{\alpha \in \Phi^+} \langle \rho, \alpha \rangle + \cdots \right) \sum_\mu \dim L(\lambda)_\mu \exp(\langle \rho, \mu \rangle t) = t^N \left(\prod_{\alpha \in \Phi^+} \langle \lambda+\rho, \alpha \rangle + \cdots \right).$$

By cancelling t^N and then taking the constant term we obtain

$$\left(\prod_{\alpha \in \Phi^+} \langle \rho, \alpha \rangle \right) \dim L(\lambda) = \prod_{\alpha \in \Phi^+} \langle \lambda+\rho, \alpha \rangle. \qquad \square$$

12.6 Complete reducibility

We have now attained a good understanding of the finite dimensional irreducible modules for a semisimple Lie algebra L. We now consider arbitrary finite dimensional L-modules. Each of these turns out to be a direct sum of irreducible L-modules.

Theorem 12.20 *Let L be a semisimple Lie algebra and V a finite dimensional L-module. Then V is completely reducible.*

Proof. We shall prove this result in a number of steps. If V is itself irreducible there is nothing to prove. Thus we suppose U is a proper submodule of V. It

12.6 Complete reducibility

will be sufficient to show that U has a complementary submodule U' in V, that is a submodule such that $V = U \oplus U'$.

(a) Suppose $\dim V = 2$, $\dim U = 1$. Then U and V/U are 1-dimensional L-modules. Since for $x, y \in L$, $u \in U$ we have

$$[xy]u = x(yu) - y(xu)$$

and since the actions of x and y on the 1-dimensional module commute we have

$$[xy]u = 0.$$

Thus $[LL]$ acts as 0 on U. Since L is semisimple we have $[LL] = L$. Thus U gives the trivial 1-dimensional representation $L(0)$. Similarly V/U is isomorphic to $L(0)$.

Now let $v \in V$. Then

$$[xy]v = x(yv) - y(xv).$$

Since L annihilates V/U we have $xv \in U$ and $yv \in U$. Since L annihilates U we have $x(yv) = 0$ and $y(xv) = 0$. Hence $[xy]v = 0$. This shows that $[LL]$ annihilates V, i.e. L annihilates V. But then any complementary subspace U' of U is a submodule of V.

(b) Suppose U is irreducible, $\dim U > 1$, and $\dim(V/U) = 1$.

Then U is isomorphic to $L(\lambda)$ for some $\lambda \in X^+$ with $\lambda \neq 0$. We consider the action of the Casimir element c on V. We recall from Proposition 11.36 that c acts on the irreducible module $L(\lambda)$ as scalar multiplication by $\langle \lambda + \rho, \lambda + \rho \rangle - \langle \rho, \rho \rangle$. In particular c acts on $L(0)$ as zero, and c acts on $L(\lambda)$ for $\lambda \in X^+$, $\lambda \neq 0$, as multiplication by a positive scalar. For then $\langle \lambda, \lambda \rangle > 0$ and $\langle \lambda, \rho \rangle \geq 0$ since $\lambda \in X^+$. Thus c has one eigenvalue 0 on V and $\dim V - 1$ eigenvalues $\chi_\lambda(c) = \langle \lambda + \rho, \lambda + \rho \rangle - \langle \rho, \rho \rangle > 0$. Let U' be the eigenspace of c on V with eigenvalue 0. Then we have

$$V = U \oplus U'.$$

Moreover U' is a submodule of V. For let $x \in L$, $u' \in U'$. Then

$$c(xu') = x(cu') = 0$$

since c lies in the centre $Z(L)$ of $\mathfrak{U}(L)$. Thus $xu' \in U'$ and U' is the required submodule of V.

(c) Suppose $\dim(V/U) = 1$ but U is not irreducible. We prove the existence of the required complementary submodule U' by induction on $\dim U$. Let $\bar{U}$ be a proper submodule of U. Then by induction we have

$$V/\bar{U} = U/\bar{U} \oplus \bar{V}/\bar{U}$$

for some submodule $\bar{V}$ of V containing $\bar{U}$. We have $\dim(\bar{V}/\bar{U}) = 1$ and $\dim \bar{U} < \dim U$. Thus we may apply induction again and conclude that there exists a submodule U' such that

$$\bar{V} = \bar{U} \oplus U'.$$

But then we have $V = U \oplus U'$ as required. □

(d) We now consider the general case when U is any proper submodule of V. We consider the set $\text{Hom}(V, U)$ of all linear maps from V to U. We can make this set into an L-module as follows. If $x \in L$ and $\theta \in \text{Hom}(V, U)$ we define $x\theta \in \text{Hom}(V, U)$ by

$$(x\theta)v = x(\theta(v)) - \theta(xv) \in U.$$

Then we have

$$(y(x\theta))v = y((x\theta)v) - (x\theta)(yv)$$
$$= y(x(\theta v)) - y(\theta(xv)) - x(\theta(yv)) + \theta(x(yv)).$$

Similarly

$$(x(y\theta))v = x(y(\theta v)) - x(\theta(yv)) - y(\theta(xv)) + \theta(y(xv)).$$

Thus

$$(x(y\theta) - y(x\theta))v = x(y(\theta v)) - y(x(\theta v)) + \theta(y(xv)) - \theta(x(yv))$$
$$= [xy](\theta v) - \theta([xy]v)$$
$$= ([xy]\theta)v.$$

Thus $\text{Hom}(V, U)$ is an L-module. Let S be the subspace of $\text{Hom}(V, U)$ of maps θ such that $\theta|_U$ is a scalar multiplication. Then S is a submodule of $\text{Hom}(V, U)$. For suppose $\theta \in S$, $x \in L$. Then for $u \in U$ we have

$$(x\theta)u = x(\theta u) - \theta(xu) = \xi x u - \xi x u = 0$$

where θ acts on U as multiplication by ξ. Thus S is a submodule of $\text{Hom}(V, U)$. Moreover let T be the subspace of S of maps θ such that $\theta|_U$ is zero. Then T is a submodule of S and $\dim(S/T) = 1$.

We know then from the earlier parts of the proof that there is a submodule T' of S such that $S = T \oplus T'$. We have $\dim T' = 1$. Suppose T' is

12.6 Complete reducibility

spanned by the non-zero element $f : V \to U$. We may choose f so that $f(u) = u$ for all $u \in U$. We have $xf = 0$ for all $x \in L$ since $\dim T' = 1$. Thus

$$(xf)v = x(fv) - f(xv) = 0$$

for all $v \in V$, that is

$$x(fv) = f(xv) \qquad \text{for all } x \in L, v \in V.$$

This shows that $f : V \to U$ is a homomorphism of L-modules. Let U' be the kernel of f. Then U' is a submodule of V. We have $U \cap U' = O$ since f acts as the identity on U, and

$$\dim V = \dim U + \dim U'$$

since f is surjective. Hence we have

$$V = U \oplus U'$$

and U' is the required complementary submodule. □

Note The crucial step in the above proof of complete reducibility is the fact that the Casimir element c acts on the irreducible module $L(\lambda)$ for $\lambda \in X^+$, $\lambda \neq 0$, as multiplication by a positive scalar.

The theorem of complete reducibility shows that every finite dimensional L-module is a direct sum of irreducible L-modules each isomorphic to $L(\lambda)$ for some $\lambda \in X^+$.

In particular the tensor product $L(\lambda) \otimes L(\mu)$ is a direct sum of irreducible modules $L(\nu)$ where $\lambda, \mu, \nu \in X^+$. It is natural to try to determine the multiplicity with which $L(\nu)$ occurs as a direct summand of $L(\lambda) \otimes L(\mu)$. This multiplicity is given in a formula of Steinberg.

Theorem 12.21 (*Steinberg's multiplicity formula*). *Let $\lambda, \mu \in X^+$ and*

$$L(\lambda) \otimes L(\mu) = \sum_{\nu \in X^+} c_{\lambda \mu \nu} L(\nu).$$

Then

$$c_{\lambda \mu \nu} = \sum_{w, w' \in W} \varepsilon(w) \varepsilon(w') \mathfrak{P}\left(w(\lambda + \rho) + w'(\mu + \rho) - (\nu + 2\rho) \right).$$

Proof. We have

$$\operatorname{ch} L(\lambda) \operatorname{ch} L(\mu) = \sum_{\nu \in X^+} c_{\lambda \mu \nu} \operatorname{ch} L(\nu)$$

by Lemma 12.2. We multiply both sides of this equation by the Weyl denominator Δ. By Weyl's character formula, Theorem 12.17, we have

$$\left(\sum_{w\in W} \varepsilon(w)e_{w(\lambda+\rho)}\right)\left(\sum_{\xi\in X} \dim L(\mu)_\xi e_\xi\right) = \sum_{\nu\in X^+} c_{\lambda\mu\nu}\left(\sum_{w\in W} \varepsilon(w)e_{w(\nu+\rho)}\right).$$

Thus $\sum_{w\in W}\sum_{\xi\in X} \varepsilon(w)\dim L(\mu)_\xi e_{w(\lambda+\rho)+\xi} = \sum_{\nu\in X^+}\sum_{w\in W} c_{\lambda\mu\nu}\varepsilon(w)e_{w(\nu+\rho)}$. Now $\nu\in X^+$, thus $\nu+\rho$ lies in the fundamental chamber C. Thus $w(\nu+\rho)$ lies in the chamber $w(C)$. Thus the elements $w(\nu+\rho)$ are all distinct as w, ν vary, and so the elements $e_{w(\nu+\rho)}$ are linearly independent. We may therefore compare the coefficients of $e_{w(\nu+\rho)}$ on both sides of the above equation. In fact we compare the coefficients of $e_{\nu+\rho}$ on both sides. This gives

$$c_{\lambda\mu\nu} = \sum_{\substack{w\in W\ \xi\in X \\ w(\lambda+\rho)+\xi=\nu+\rho}} \varepsilon(w)\dim L(\mu)_\xi$$

$$= \sum_{w\in W} \varepsilon(w)\dim L(\mu)_{\nu+\rho-w(\lambda+\rho)}.$$

We now use Kostant's multiplicity formula, Theorem 12.18. This gives

$$\dim L(\mu)_{\nu+\rho-w(\lambda+\rho)} = \sum_{w'\in W} \varepsilon(w')\mathfrak{P}\left(w'(\mu+\rho)+w(\lambda+\rho)-(\nu+2\rho)\right).$$

Thus we obtain

$$c_{\lambda\mu\nu} = \sum_{w,w'\in W} \varepsilon(w)\varepsilon(w')\mathfrak{P}\left(w(\lambda+\rho)+w'(\mu+\rho)-(\nu+2\rho)\right).$$

13

Fundamental modules for simple Lie algebras

13.1 An alternative form of Weyl's dimension formula

Let L be a finite dimensional simple Lie algebra. The irreducible L-modules $L(\omega_i)$ whose highest weights are the fundamental weights $\omega_1, \ldots, \omega_l$ are called the fundamental modules. In this chapter we shall determine the dimensions of the fundamental modules for the various simple Lie algebras. We shall first derive an alternative form of the Weyl dimension formula which will be useful in this respect.

Theorem 13.1 *Let $\lambda = \sum_{i=1}^{l} m_i \omega_i$ be a dominant integral weight. Then*

$$\dim L(\lambda) = \prod_{\alpha \in \Phi^+} d_\alpha$$

where $\alpha = \sum_{i=1}^{l} k_i \alpha_i$ and

$$d_\alpha = \frac{\sum_i (m_i + 1) k_i w_i}{\sum_i k_i w_i}.$$

Here the integer w_i, called the weight of α_i, is defined by

$$\langle \alpha_i, \alpha_i \rangle = w_i \langle \alpha_{i_0}, \alpha_{i_0} \rangle$$

where α_{i_0} is a short fundamental root. Thus $w_i \in \{1, 2, 3\}$ for each i.

Proof. We know from Theorem 12.19 that

$$\dim L(\lambda) = \prod_{\alpha \in \Phi^+} d_\alpha$$

where $d_\alpha = \dfrac{\langle \lambda + \rho, \alpha \rangle}{\langle \rho, \alpha \rangle}$.

Since $\lambda = \sum m_i \omega_i$, $\rho = \sum \omega_i$, $\alpha = \sum k_i \alpha_i$ we have
$$d_\alpha = \frac{\langle \sum (m_i+1)\omega_i, \sum k_i \alpha_i \rangle}{\langle \sum \omega_i, \sum k_i \alpha_i \rangle}.$$

Now we know from the proof of Proposition 10.18 that $\langle \omega_i, \alpha_j \rangle = 0$ if $i \ne j$ and $\langle \omega_i, \alpha_i \rangle = \frac{1}{2} \langle \alpha_i, \alpha_i \rangle$. Thus

$$d_\alpha = \frac{\sum_i (m_i+1) k_i \cdot \frac{1}{2} \langle \alpha_i, \alpha_i \rangle}{\sum_i k_i \cdot \frac{1}{2} \langle \alpha_i, \alpha_i \rangle}$$
$$= \frac{\sum_i (m_i+1) k_i w_i}{\sum_i k_i w_i}. \qquad \square$$

13.2 Fundamental modules for A_l

The fundamental weights $\omega_1, \ldots, \omega_l$ for a simple Lie algebra of type A_l will be numbered according to the vertices of the Dynkin diagram as shown.

$$\underset{1}{\circ} \!\!-\!\! \underset{2}{\circ} \!\!-\!\! \cdots \!\!-\!\! \underset{l-1}{\circ} \!\!-\!\! \underset{l}{\circ}$$

We shall use Theorem 13.1 to calculate $\dim L(\omega_j)$ for $j \in \{1, \ldots, l\}$. We have $\dim L(\omega_j) = \prod_{\alpha \in \Phi^+} d_\alpha$ where

$$d_\alpha = \frac{\sum_i (m_i+1) k_i}{\sum_i k_i}.$$

(All weights w_i are equal to 1.) Now $m_j = 1$ and $m_i = 0$ if $i \ne j$. Thus if α does not involve the fundamental root α_j we have $d_\alpha = 1$.

So suppose α does involve α_j. Then $\alpha = \alpha_i + \cdots + \alpha_j + \cdots + \alpha_k$ for some i with $1 \le i \le j$ and some k with $j \le k \le l$. For such a root α we have

$$d_\alpha = \frac{k-i+2}{k-i+1}.$$

Thus

$$\dim L(\omega_j) = \prod_{\substack{i \\ 1 \le i \le j}} \prod_{\substack{k \\ j \le k \le l}} \frac{k-i+2}{k-i+1}$$
$$= \prod_{\substack{k \\ j \le k \le l}} \frac{(k+1)k \ldots (k-j+2)}{k(k-1) \ldots (k-j+1)} = \prod_{\substack{k \\ j \le k \le l}} \frac{k+1}{k-j+1}$$
$$= \frac{(j+1)(j+2) \ldots (l+1)}{1.2. \ldots .(l+1-j)} = \frac{(l+1)!}{j!(l+1-j)!} = \binom{l+1}{j}.$$

Thus we have shown

Proposition 13.2 *The dimensions of the fundamental modules for the simple Lie algebra of type A_l are*

$$l+1 \quad \binom{l+1}{2} \quad \binom{l+1}{3} \quad \cdots \quad \binom{l+1}{3} \quad \binom{l+1}{2} \quad l+1$$

$\square$

These fundamental modules may be described in terms of exterior powers of the natural A_l-module of dimension $l+1$. We recall from Theorem 8.1 that A_l is isomorphic to the Lie algebra $\mathfrak{sl}_{l+1}(\mathbb{C})$ of all $(l+1) \times (l+1)$ matrices of trace 0. The identity map gives an $(l+1)$-dimensional representation of A_l called the natural representation. Its weights are the maps $\mu_1, \mu_2, \ldots, \mu_{l+1}$ given by

$$\begin{pmatrix} \lambda_1 & & & \\ & \lambda_2 & & \\ & & \ddots & \\ & & & \lambda_{l+1} \end{pmatrix} \xrightarrow{\mu_i} \lambda_i.$$

Then we have

$$\mu_1 - \mu_2 = \alpha_1$$
$$\vdots$$
$$\mu_l - \mu_{l+1} = \alpha_l$$
$$\mu_1 + \cdots + \mu_{l+1} = 0.$$

On the other hand we have $\alpha_i = \sum_j A_{ji} \omega_j$ by Proposition 10.17. Hence

$$\alpha_1 = 2\omega_1 - \omega_2$$
$$\alpha_2 = -\omega_1 + 2\omega_2 - \omega_3$$
$$\vdots$$
$$\alpha_{l-1} = -\omega_{l-2} + 2\omega_{l-1} - \omega_l$$
$$\alpha_l = \phantom{-\omega_{l-2} + 2\omega_{l-1}} -\omega_{l-1} + 2\omega_l.$$

Eliminating $\alpha_1, \ldots, \alpha_l$ we obtain

$$\mu_1 = \omega_1$$
$$\mu_2 = -\omega_1 + \omega_2$$
$$\vdots$$
$$\mu_l = -\omega_{l-1} + \omega_l$$
$$\mu_{l+1} = -\omega_l.$$

Now the weights of the natural module satisfy $\mu_1 \succ \mu_2 \succ \cdots \succ \mu_{l+1}$ since $\mu_i - \mu_{i+1} = \alpha_i$. Thus the highest weight of the natural module is $\mu_1 = \omega_1$. It follows that $L(\omega_1)$ is one of the irreducible direct summands of the natural module V. Since

$$\dim V = \dim L(\omega_1) = l + 1$$

it follows that $V = L(\omega_1)$. Thus we have shown

Proposition 13.3 *The natural A_l-module is an irreducible module with highest weight ω_1.*

To obtain the remaining fundamental A_l-modules we introduce exterior powers of modules.

13.3 Exterior powers of modules

Let V be a finite dimensional module for a Lie algebra L. Let

$$T(V) = T^0(V) \oplus T^1(V) \oplus T^2(V) \oplus \cdots$$

be the tensor algebra of V, where

$$T^n(V) = V \otimes \cdots \otimes V \quad (n \text{ factors}).$$

$T(V)$ may be made into an associative algebra in which

$$(x_1 \otimes \cdots \otimes x_m)(y_1 \otimes \cdots \otimes y_n) = x_1 \otimes \cdots \otimes x_m \otimes y_1 \otimes \cdots \otimes y_n$$

for $x_1, \ldots, x_m, y_1, \ldots, y_n \in V$.

$T(V)$ may also be given the structure of an L-module satisfying

$$x(x_1 \otimes \cdots \otimes x_m) = \sum_{i=1}^{m} x_1 \otimes \cdots \otimes x_{i-1} \otimes xx_i \otimes x_{i+1} \otimes \cdots \otimes x_m$$

for all $x \in L$.

13.3 Exterior powers of modules

Let J be the 2-sided ideal of $T(V)$ generated by the elements $v \otimes v$ for all $v \in V$.

Definition 13.4 $\Lambda(V) = T(V)/J$ is called the **exterior algebra** of V.

Let $v, v' \in V$. Then

$$(v+v') \otimes (v+v') = v \otimes v + v' \otimes v' + v \otimes v' + v' \otimes v.$$

Hence

$$v \otimes v' + v' \otimes v \in J.$$

Now let $v_1, \ldots, v_n$ be a basis of V. Then J is the 2-sided ideal of $T(V)$ generated by all elements of form

$$v_i \otimes v_i \quad i = 1, \ldots, n$$
$$v_i \otimes v_j + v_j \otimes v_i \quad i < j.$$

It follows from this that

$$J = \bigoplus_{k \geq 0} (T^k(V) \cap J)$$

and that $T^0(V) \cap J = 0$, $T^1(V) \cap J = 0$. Hence

$$\Lambda(V) = \Lambda^0 V \oplus \Lambda^1 V \oplus \Lambda^2 V \oplus \cdots$$

where $\Lambda^k V = T^k(V)/T^k(V) \cap J$. In particular we have

$$\Lambda^0 V \cong T^0(V) = \mathbb{C}1$$
$$\Lambda^1 V \cong T^1(V) = V.$$

Thus we may identify the subspace $\Lambda^0 V \oplus \Lambda^1 V$ of $\Lambda(V)$ with $\mathbb{C}1 \oplus V$.

Let $\sigma : T(V) \to \Lambda(V) = T(V)/J$ be the natural homomorphism. We define

$$\sigma(v \otimes v') = v \wedge v' \quad \text{for } v, v' \in V.$$

Then every element of $\Lambda(V)$ is a linear combination of elements

$$v_{i_1} \wedge \cdots \wedge v_{i_k} \quad i_1, \ldots, i_k \in \{1, \ldots, n\}.$$

The relations defining J may be written

$$v_i \wedge v_i = 0 \quad i \in \{1, \ldots, n\}$$
$$v_j \wedge v_i = -(v_i \wedge v_j) \quad i < j.$$

By applying these relations we see that each element of $\Lambda(V)$ is a linear combination of elements

$$v_{i_1} \wedge \cdots \wedge v_{i_k} \quad \text{for } i_1 < \cdots < i_k$$

and that the relations cannot be used further. Thus we have shown:

Proposition 13.5 (i) $\Lambda(V) = \Lambda^0 V \oplus \Lambda^1 V \oplus \cdots \oplus \Lambda^n V$
(ii) $\dim \Lambda^k V = \binom{n}{k}$
(iii) $\dim \Lambda(V) = 2^n$
(iv) *The elements $v_{i_1} \wedge \cdots \wedge v_{i_k}$ for subsets $\{i_1, \ldots, i_k\} \subset \{1, \ldots, n\}$ with $i_1 < \cdots < i_k$ form a basis of $\Lambda(V)$.* □

We now show that $\Lambda(V)$ has the structure of an L-module. We recall that $T(V)$ is an L-module and that its ideal J is generated by $v \otimes v$ for all $v \in V$. For $x \in L$ we have

$$x(v \otimes v) = xv \otimes v + v \otimes xv.$$

Since the right-hand side lies in J we see that J is a submodule of $T(V)$. Thus $\Lambda(V) = T(V)/J$ can be made into an L-module in the natural way. Each exterior power $\Lambda^k V$ is a submodule.

Proposition 13.6 *Let V be a finite dimensional module for the simple Lie algebra L. Then the weights of $\Lambda^k V$ are all sums of k distinct weights of V.*

Proof. Let H be a Cartan subalgebra of L. We consider V as an H-module. V is a direct sum of 1-dimensional H-submodules. Let $v_1, \ldots, v_n$ be a basis of V adapted to this decomposition. Let $\lambda_1, \ldots, \lambda_n \in H^*$ be the corresponding weights. Then

$$xv_i = \lambda_i(x)v_i \quad \text{for } x \in H.$$

Now $\Lambda^k V$ has basis $v_{i_1} \wedge \cdots \wedge v_{i_k}$ for all $i_1 < \cdots < i_k$. We have

$$x(v_{i_1} \wedge \cdots \wedge v_{i_k}) = \sum_{r=1}^{k} v_{i_1} \wedge \cdots \wedge xv_{i_r} \wedge \cdots \wedge v_{i_k}$$

$$= (\lambda_{i_1}(x) + \cdots + \lambda_{i_k}(x)) v_{i_1} \wedge \cdots \wedge v_{i_k} \quad x \in H.$$

Thus $v_{i_1} \wedge \cdots \wedge v_{i_k}$ is a weight vector with weight $\lambda_{i_1} + \cdots + \lambda_{i_k}$. Thus the weights of $\Lambda^k V$ are sums of k distinct weights of V.

13.3 Exterior powers of modules

Theorem 13.7 *Let V be the natural module for the simple Lie algebra A_l. Then the fundamental modules for A_l are*

$$\Lambda^1 V, \Lambda^2 V, \ldots, \Lambda^l V.$$

Proof. We have seen in Proposition 13.3 that $\Lambda^1 V = V$ is the fundamental module with highest weight ω_1. The weights of V are the maps $\mu_1, \ldots, \mu_{l+1}$ given by

$$\mu_i : \begin{pmatrix} \lambda_1 & & & \\ & \lambda_2 & & \\ & & \ddots & \\ & & & \lambda_{l+1} \end{pmatrix} \to \lambda_i.$$

Since $\mu_i - \mu_{i+1} = \alpha_i$ we have $\mu_i \succ \mu_{i+1}$. Thus the weights are ordered by

$$\mu_1 \succ \cdots \succ \mu_{l+1}.$$

Now $\mu_1 = \omega_1$, $\mu_i - \mu_{i+1} = \alpha_i$ and $\alpha_i = \sum_j A_{ji} \omega_j$ by Proposition 10.17. Hence

$$\alpha_1 = 2\omega_1 - \omega_2$$
$$\alpha_i = -\omega_{i-1} + 2\omega_i - \omega_{i+1} \quad \text{for } 2 \le i \le l-1$$
$$\alpha_l = -\omega_{l-1} + 2\omega_l.$$

It follows that

$$\mu_1 = \omega_1$$
$$\mu_2 = -\omega_1 + \omega_2$$
$$\mu_3 = -\omega_2 + \omega_3$$
$$\vdots$$
$$\mu_l = -\omega_{l-1} + \omega_l$$
$$\mu_{l+1} = -\omega_l.$$

By Proposition 13.6 the highest weight of $\Lambda^k V$ is $\mu_1 + \cdots + \mu_k = \omega_k$, for $1 \le k \le l$. Thus $\Lambda^k V$ contains the irreducible module $L(\omega_k)$ as one of its irreducible direct summands. However,

$$\dim L(\omega_k) = \dim \Lambda^k V = \binom{l+1}{k}$$

by Proposition 13.2. Hence $L(\omega_k) = \Lambda^k V$. $\square$

13.4 Fundamental modules for B_l and D_l

The fundamental weights $\omega_1, \ldots, \omega_l$ for a simple Lie algebra of type B_l or D_l will be numbered according to the labelling of the Dynkin diagrams:

$$\underset{1}{\circ}\!\!-\!\!\underset{2}{\circ}\!\!-\!\!\underset{3}{\circ}\!\!-\cdots-\underset{l-1}{\circ}\!\Rightarrow\!\underset{l}{\circ} \quad B_l$$

$$\underset{1}{\circ}\!\!-\!\!\underset{2}{\circ}\!\!-\!\!\underset{3}{\circ}\!\!-\cdots-\underset{l-2}{\circ}\!\!<\!\!{\underset{l}{\circ}\atop\overset{l-1}{\circ}} \quad D_l$$

We again use Theorem 13.1 to calculate $\dim L(\omega_j)$.

We suppose first that we have an algebra of type B_l. We know from Section 8.3 that the roots have the following form. Let

$$h = \begin{pmatrix} 0 & & & & & & \\ & \lambda_1 & & & & & \\ & & \ddots & & & & \\ & & & \lambda_l & & & \\ & & & & -\lambda_1 & & \\ & & & & & \ddots & \\ & & & & & & -\lambda_l \end{pmatrix}.$$

Then the fundamental roots are

$$\alpha_i(h) = \lambda_i - \lambda_{i+1} \quad \text{for } 1 \le i \le l-1$$
$$\alpha_l(h) = \lambda_l$$

The full set of positive roots is given by

$$h \to \lambda_i - \lambda_j \quad \text{for } i < j$$
$$h \to \lambda_i + \lambda_j \quad \text{for } i < j$$
$$h \to \lambda_i$$

where $i, j \in \{1, \ldots, l\}$. These positive roots can be expressed as combinations of fundamental roots as follows:

$$\alpha_i + \cdots + \alpha_{j-1} \quad \text{for } i < j$$
$$\alpha_i + \cdots + \alpha_{j-1} + 2\alpha_j + \cdots + 2\alpha_l \quad \text{for } i < j$$
$$\alpha_i + \cdots + \alpha_l.$$

13.4 Fundamental modules for B_l and D_l

The first two families are long roots and the third family are short roots. Thus the weights w_i are given by

$$w_1 = \cdots = w_{l-1} = 2 \quad w_l = 1.$$

According to Theorem 13.1 we have

$$\dim L(\omega_j) = \prod_{\alpha \in \Phi^+} d_\alpha$$

where $\alpha = \sum k_i \alpha_i$ and

$$d_\alpha = \frac{\sum_{i=1}^{l} k_i w_i + k_j w_j}{\sum_{i=1}^{l} k_i w_i}.$$

We have $d_\alpha = 1$ if α does not involve α_j. We first suppose $j \in \{1, \ldots, l-1\}$. Then the positive roots involving j are:

$$\alpha_i + \cdots + \alpha_j + \cdots + \alpha_k \quad 1 \le i \le j,\ j \le k \le l-1$$

$$\alpha_i + \cdots + \alpha_j + \cdots + \alpha_l \quad 1 \le i \le j$$

$$\alpha_i + \cdots + \alpha_j + \cdots + \alpha_{k-1} + 2\alpha_k + \cdots + 2\alpha_l \quad 1 \le i \le j,\ j \le k-1 < l$$

$$\alpha_i + \cdots + \alpha_{k-1} + 2\alpha_k + \cdots + 2\alpha_j + \cdots + 2\alpha_l \quad 1 \le i < k,\ k \le j \le l-1.$$

The values of d_α in these four cases are

$$\frac{k-i+2}{k-i+1},\ \frac{2l-2i+3}{2l-2i+1},\ \frac{2l-k-i+2}{2l-k-i+1},\ \frac{2l-k-i+3}{2l-k-i+1}$$

respectively. The product of all possible d_α in these four cases is

$$\frac{(j+1)(j+2)\cdots l}{1 \cdot 2 \cdots l-j},\ \frac{2l+1}{2l-2j+1},\ \frac{(2l-j)(2l-j-1)\cdots(l+1)}{(2l-2j)(2l-2j-1)\cdots(l+1-j)},$$

$$\frac{2l(2l-1)(2l-2)\cdots(2l+2-j)}{(2l-j)(2l-j-1)\cdots(2l-2j+3)(2l-2j+2)}$$

respectively. Finally the total product $\prod_{\alpha \in \Phi^+} d_\alpha$ is $\binom{2l+1}{j}$. We now take $j = l$. Then the positive roots involving l are

$$\alpha_i + \cdots + \alpha_l \quad 1 \le i \le l$$

$$\alpha_i + \cdots + \alpha_{k-1} + 2\alpha_k + \cdots + 2\alpha_l \quad 1 \le i < k \le l.$$

The values of d_α in these cases are $\frac{2l-2i+2}{2l-2i+1}$, $\frac{2l-i-j+2}{2l-i-j+1}$ respectively. The product of all possible d_α in the two cases is

$$\frac{2l(2l-2)(2l-4)\cdots 2}{(2l-1)(2l-3)\cdots 3\cdot 1}, \quad \frac{(2l-1)(2l-2)\cdots(l+1)}{(2l-2)(2l-4)\cdots 2}$$

respectively. Finally the total product $\prod_{\alpha\in\Phi^+} d_\alpha$ is 2^l.

Thus we have shown:

Proposition 13.8 *The dimensions of the fundamental modules for the simple Lie algebra of type B_l are*

$$2l+1 \quad \binom{2l+1}{2} \quad \binom{2l+1}{3} \quad \cdots \quad \binom{2l+1}{l-1} \quad 2^l$$
$$\circ\!\!-\!\!\circ\!\!-\!\!\circ\!\!-\!\!\circ\ \cdots\ \circ\!\!\Rightarrow\!\!\circ$$

The dimensions of the modules $L(\omega_j)$ for $1\le j\le l-1$ suggest that these modules are exterior powers of the $(2l+1)$-dimensional natural module. This is indeed the case.

Theorem 13.9 *Let V be the $(2l+1)$-dimensional natural module for the simple Lie algebra B_l (described in Section 8.3). Then the fundamental module $L(\omega_j)$ is isomorphic to $\Lambda^j V$ for $1\le j\le l-1$.*

Proof. Let

$$h = \begin{pmatrix} 0 & & & & & \\ & \lambda_1 & & & & \\ & & \ddots & & & \\ & & & \lambda_l & & \\ & & & & -\lambda_1 & \\ & & & & & \ddots \\ & & & & & & -\lambda_l \end{pmatrix}.$$

Then the weights of V are $0, \mu_1, \ldots, \mu_l, -\mu_1, \ldots, -\mu_l$ where $\mu_i(h) = \lambda_i$. Since $\mu_i - \mu_{i+1} = \alpha_i$ for $1\le i\le l-1$ and $\mu_l = \alpha_l$ we have

$$\mu_1 \succ \mu_2 \succ \cdots \succ \mu_l \succ 0.$$

13.4 Fundamental modules for B_l and D_l

Thus the highest weight of $\Lambda^j V$ for $1 \le j \le l$ is $\mu_1 + \mu_2 + \cdots + \mu_j$. Expressing the μs in terms of the αs gives

$$\mu_1 = \alpha_1 + \cdots + \alpha_l$$
$$\mu_2 = \alpha_2 + \cdots + \alpha_l$$
$$\vdots$$
$$\mu_l = \alpha_l.$$

We also have $\alpha_i = \sum A_{ji} \omega_j$, which in type B_l gives

$$\alpha_1 = 2\omega_1 - \omega_2$$
$$\alpha_i = -\omega_{i-1} + 2\omega_i - \omega_{i+1} \qquad 2 \le i \le l-2$$
$$\alpha_{l-1} = -\omega_{l-2} + 2\omega_{l-1} - 2\omega_l$$
$$\alpha_l = \qquad\qquad -\omega_{l-1} + 2\omega_l.$$

It follows that

$$\mu_1 = \omega_1$$
$$\mu_2 = -\omega_1 + \omega_2$$
$$\vdots$$
$$\mu_{l-1} = -\omega_{l-2} + \omega_{l-1}$$
$$\mu_l = -\omega_{l-1} + 2\omega_l.$$

Hence $\mu_1 + \cdots + \mu_j = \omega_j$ for $1 \le j \le l-1$

$$\mu_1 + \cdots + \mu_l = 2\omega_l.$$

Thus the highest weight of $\Lambda^j V$ is ω_j for $1 \le j \le l-1$. Since

$$\dim L(\omega_j) = \dim \Lambda^j V = \binom{2l+1}{j} \qquad j \le l-1$$

we deduce that $L(\omega_j)$ is isomorphic to $\Lambda^j V$. This argument fails when $j = l$ since the highest weight of $\Lambda^l V$ is $2\omega_l$ rather than ω_l. We shall see subsequently how to find the remaining fundamental module $L(\omega_l)$. □

We now consider the simple Lie algebra of type D_l. This algebra was described in Section 8.2. Its roots have the following form. Let

$$h = \begin{pmatrix} \lambda_1 & & & & & \\ & \ddots & & & & \\ & & \lambda_l & & & \\ & & & -\lambda_1 & & \\ & & & & \ddots & \\ & & & & & -\lambda_l \end{pmatrix}.$$

Then the fundamental roots are

$$\alpha_i(h) = \lambda_i - \lambda_{i+1} \quad \text{for } 1 \le i \le l-1$$
$$\alpha_l(h) = \lambda_{l-1} + \lambda_l.$$

The full set of positive roots is given by

$$h \to \lambda_i - \lambda_j \quad i < j$$
$$h \to \lambda_i + \lambda_j \quad i < j.$$

These are expressed as combinations of the fundamental roots by

$$\alpha_i + \cdots + \alpha_{j-1} \quad \text{for } 1 \le i < j \le l$$
$$\alpha_i + \cdots + \alpha_{j-1} + 2\alpha_j + \cdots + 2\alpha_{l-2} + \alpha_{l-1} + \alpha_l \quad \text{for } 1 \le i < j \le l-1$$
$$\alpha_i + \cdots + \alpha_{l-2} + \alpha_l \quad \text{for } 1 \le i \le l-2.$$

We take a fixed j with $1 \le j \le l-2$ and consider $\dim L(\omega_j)$. By Theorem 13.1 this is given by

$$\dim L(\omega_j) = \prod_{\alpha \in \Phi^+} d_\alpha$$

where $\alpha = \sum k_i \alpha_i$ and

$$d_\alpha = \frac{\sum_{i=1}^l k_i + k_j}{\sum_{i=1}^l k_i}.$$

13.4 Fundamental modules for B_l and D_l

(All weights w_i are equal to 1 in type D_l.) As usual $d_\alpha = 1$ if α does not involve α_j. The positive roots involving α_j are

$\alpha_i + \cdots + \alpha_j + \cdots + \alpha_k \qquad 1 \le i \le j, \ j \le k \le l-1$

$\alpha_i + \cdots + \alpha_j + \cdots + \alpha_k + 2\alpha_{k+1} + \cdots + 2\alpha_{l-2} + \alpha_{l-1} + \alpha_l \qquad 1 \le i \le j, \ j \le k \le l-2$

$\alpha_i + \cdots + \alpha_k + 2\alpha_{k+1} + \cdots + 2\alpha_j + \cdots + 2\alpha_{l-2} + \alpha_{l-1} + \alpha_l \qquad 1 \le i \le k < j$

$\alpha_i + \cdots + \alpha_j + \cdots + \alpha_{l-2} + \alpha_l \qquad 1 \le i \le j.$

The values of d_α in these four cases are

$$\frac{k-i+2}{k-i+1}, \quad \frac{2l-i-k}{2l-i-k-1}, \quad \frac{2l-i-k+1}{2l-i-k-1}, \quad \frac{l-i+1}{l-i}$$

respectively. The product of all possible d_α in these four cases is

$$\frac{(j+1)(j+2)\cdots l}{1\cdot 2 \cdots (l-j)}, \quad \frac{(2l-j-1)(2l-j-2)\cdots(l+1)}{(2l-2j-1)(2l-2j-2)\cdots(l+1-j)},$$

$$\frac{(2l-1)(2l-2)\cdots(2l-j+1)}{(2l-j-1)(2l-j-2)\cdots(2l-2j+1)}, \quad \frac{l}{l-j}$$

respectively. Finally the total product is $\binom{2l}{j}$.

We next suppose $j = l-1$. The positive roots involving α_{l-1} are

$\alpha_i + \cdots + \alpha_{l-1} \qquad 1 \le i \le l-1$

$\alpha_i + \cdots + \alpha_{k-1} + 2\alpha_k + \cdots + 2\alpha_{l-2} + \alpha_{l-1} + \alpha_l \qquad 1 \le i < k \le l-1.$

The values of d_α in these two cases are $\frac{l-i+1}{l-i}$, $\frac{2l-i-k+1}{2l-i-k}$ respectively. The product of all possible d_α in those two cases is l, $2^{l-1}/l$ respectively, and so the total product is 2^{l-1}.

Finally suppose $j = l$. The positive roots involving α_l are

α_l

$\alpha_i + \cdots + \alpha_{l-2} + \alpha_l \qquad 1 \le i \le l-2$

$\alpha_i + \cdots + \alpha_{k-1} + 2\alpha_k + \cdots + 2\alpha_{l-2} + \alpha_{l-1} + \alpha_l \qquad 1 \le i < k \le l-1.$

The values of d_α in these three cases are 2, $\frac{l-i+1}{l-i}$, $\frac{2l-i-k+1}{2l-i-k}$ respectively. The product of all possible d_α in these three cases is 2, $l/2$, $2^{l-1}/l$ respectively. Thus the total product is 2^{l-1}.

Thus we have shown

Proposition 13.10 *The dimensions of the fundamental modules for the simple Lie algebra of type D_l are*

$$2l \quad \binom{2l}{2} \quad \binom{2l}{3} \quad \cdots \quad \binom{2l}{l-2} \quad 2^{l-1} \quad 2^{l-1}$$

$\square$

Again the dimensions of these modules for $1 \le j \le l-2$ suggest that they are given by exterior powers of the natural module.

Theorem 13.11 *Let V be the $2l$-dimensional natural module for the simple Lie algebra D_l (described in Section 8.2). Then the fundamental module $L(\omega_j)$ is isomorphic to $\Lambda^j V$ for $1 \le j \le l-2$.*

Proof. Let

$$h = \begin{pmatrix} \lambda_1 & & & & & \\ & \ddots & & & & \\ & & \lambda_l & & & \\ & & & -\lambda_1 & & \\ & & & & \ddots & \\ & & & & & -\lambda_l \end{pmatrix}.$$

Then the weights of V are $\mu_1, \ldots, \mu_l, -\mu_1, \ldots, -\mu_l$ where $\mu_i(h) = \lambda_i$. We have

$$\mu_1 - \mu_2 = \alpha_1$$
$$\vdots$$
$$\mu_{l-1} - \mu_l = \alpha_{l-1}$$
$$\mu_{l-1} + \mu_l = \alpha_l$$

and

$$\alpha_1 = 2\omega_1 - \omega_2$$
$$\alpha_2 = -\omega_1 + 2\omega_2 - \omega_3$$
$$\vdots$$
$$\alpha_{l-3} = -\omega_{l-4} + 2\omega_{l-3} - \omega_{l-2}$$

$$\alpha_{l-2} = -\omega_{l-3} + 2\omega_{l-2} - \omega_{l-1} - \omega_l$$
$$\alpha_{l-1} = -\omega_{l-2} + 2\omega_{l-1}$$
$$\alpha_l = -\omega_{l-2} + 2\omega_l$$

using the Cartan matrix of type D_l. It follows that

$$\mu_1 = \omega_1$$
$$\mu_2 = -\omega_1 + \omega_2$$
$$\vdots$$
$$\mu_{l-2} = -\omega_{l-3} + \omega_{l-2}$$
$$\mu_{l-1} = -\omega_{l-2} + \omega_{l-1} + \omega_l$$
$$\mu_l = -\omega_{l-1} + \omega_l.$$

Since $\mu_1 \succ \mu_2 \succ \cdots \succ \mu_{l-1} \succ \mu_l$ the highest weight of $\Lambda^j V$ for $1 \le j \le l-2$ is $\mu_1 + \cdots + \mu_j$. Also we have $\mu_1 + \cdots + \mu_j = \omega_j$ for $j \le l-2$. Thus the highest weight of $\Lambda^j V$ is ω_j when $j \le l-2$. Since

$$\dim L(\omega_j) = \dim \Lambda^j V = \binom{2l}{j}, \quad j \le l-2$$

we deduce that $L(\omega_j)$ is isomorphic to $\Lambda^j V$ for $j \le l-2$. □

13.5 Clifford algebras and spin modules

There remain one fundamental module for B_l of dimension 2^l and two fundamental modules for D_l of dimension 2^{l-1} which cannot be obtained as exterior powers of the natural module. These are called spin modules and give rise to spin representations of B_l and D_l. We shall now show how these modules may be obtained in terms of the Clifford algebra.

Let V be a vector space of dimension n over $\mathbb{C}$ and suppose we are given a symmetric bilinear map $V \times V \to \mathbb{C}$ under which the pair v, v' maps to $(v, v') \in \mathbb{C}$. Thus we have

$$(v', v) = (v, v').$$

Let $T(V)$ be the tensor algebra of V and J be the two-sided ideal of $T(V)$ generated by elements

$$v \otimes v - (v, v)1 \quad \text{for all } v \in V.$$

Since

$$(v+v')\otimes(v+v')-(v+v',v+v')1 = (v\otimes v-(v,v)1)+(v'\otimes v'-(v',v')1)$$
$$+(v\otimes v'+v'\otimes v-2(v,v')1)$$

we see that

$$v\otimes v'+v'\otimes v-2(v,v')1 \in J \quad \text{for all } v, v' \in V.$$

Let $C(V) = T(V)/J$. Then $C(V)$ is an associative algebra called the **Clifford algebra** of V.

Now let $v_1, \ldots, v_n$ be a basis of V. Then the elements

$$v_i \otimes v_i - (v_i, v_i)1$$
$$v_i \otimes v_j + v_j \otimes v_i - 2(v_i, v_j)1 \quad i < j$$

lie in J and it is evident that these elements generate J as a 2-sided ideal. We observe also that

$$(\mathbb{C}1 \oplus V) \cap J = (T^0(V) \oplus T^1(V)) \cap J = 0$$

and so the natural map $T(V) \to C(V)$ is injective when restricted to $\mathbb{C}1 \oplus V$. We shall regard $\mathbb{C}1 \oplus V$ as a subspace of $C(V)$. Thus $C(V)$ is generated, as associative algebra with 1, by elements $v_1, \ldots, v_n$ subject to relations

$$v_i v_i = (v_i, v_i)1$$
$$v_j v_i = -v_i v_j + 2(v_i, v_j)1 \quad i < j.$$

By using these relations any polynomial in $v_1, \ldots, v_n$ can be written as a polynomial in which each monomial has form $v_{i_1} v_{i_2} \ldots v_{i_k}$ where $i_1 < i_2 < \cdots < i_k$ and $0 \le k \le n$. Moreover an element of $C(V)$ in this standard form cannot be simplified further by the use of the above relations. Thus we have shown

Proposition 13.12 (i) $\dim C(V) = 2^n$.
(ii) *The elements $v_{i_1} v_{i_2} \ldots v_{i_k}$ for $i_1 < i_2 < \cdots < i_k$ with $0 \le k \le n$ form a basis for $C(V)$. (The empty product is 1.)* □

We note that all generators of J lie in $T^0 V \oplus T^2 V$. We define $T(V)^+, T(V)^-$ by

$$T(V)^+ = \bigoplus_{i \text{ even}} T^i V$$

$$T(V)^- = \bigoplus_{i \text{ odd}} T^i V.$$

13.5 Clifford algebras and spin modules

Then we have

$$T(V) = T(V)^+ \oplus T(V)^-$$
$$J = (J \cap T(V)^+) \oplus (J \cap T(V)^-).$$

This follows from the fact that J is generated by elements of $T(V)^+$. Hence

$$C(V) \cong \frac{T(V)^+}{J \cap T(V)^+} \oplus \frac{T(V)^-}{J \cap T(V)^-}.$$

We write $C(V)^+ = \frac{T(V)^+}{J \cap T(V^+)}$ and $C(V)^- = \frac{T(V)^-}{J \cap T(V)^-}$. Then

$$C(V) = C(V)^+ \oplus C(V)^-.$$

In terms of our basis for $C(V)$, $C(V)^+$ has basis $v_{i_1} v_{i_2} \ldots v_{i_k}$ for $i_1 < i_2 < \cdots < i_k$ with k even and $C(V)^-$ has basis consisting of these elements with k odd. Thus

$$\dim C(V)^+ = \dim C(V)^- = 2^{n-1}.$$

Now the associative algebra $C(V)$ can be made into a Lie algebra $[C(V)]$ in the usual way by defining $[xy] = xy - yx$. Let L be the subspace of $[C(V)]$ spanned by the elements $[vv']$ for all $v, v' \in V$. Then L can be spanned by elements $[v_i v_j]$ for $i < j$, and since

$$[v_i v_j] = 2v_i v_j - 2(v_i, v_j) 1$$

these elements are linearly independent. Thus $\dim L = n(n-1)/2$. We shall show that L is a Lie subalgebra of $[C(V)]$.

Lemma 13.13 (i) *Let $x, y, z \in V$. Then $[[xy]z] = 4(y, z)x - 4(x, z)y$.*
(ii) *Let $x, y, z, w \in V$. Then*

$$[[xy], [zw]] = 4(y, z)[xw] - 4(y, w)[xz] + 4(x, w)[yz] - 4(x, z)[yw].$$

Proof. (i) $[[xy]z] = (xy - yx)z - z(xy - yx)$

$$= xyz - yxz - zxy + zyx$$
$$= -xzy + 2(y, z)x + yzx - 2(x, z)y$$
$$\quad + xzy - 2(x, z)y - yzx + 2(y, z)x$$
$$= 4(y, z)x - 4(x, z)y.$$

(ii) $[[xy],[zw]] = [xy]zw - [xy]wz - zw[xy] + wz[xy]$
$$= [[xy]z]w + z[xy]w - [[xy]w]z - w[xy]z - zw[xy] + wz[xy]$$
$$= [[xy]z]w - [[xy]w]z + z[[xy]w] - w[[xy]z]$$
$$= [[[xy]z]w] - [[[xy]w]z]$$
$$= [4(y,z)x - 4(x,z)y, w] - [4(y,w)x - 4(x,w)y, z]$$
$$= 4(y,z)[xw] - 4(x,z)[yw] - 4(y,w)[xz] + 4(x,w)[yz]. \quad \square$$

Corollary 13.14 *L is a Lie subalgebra of $[C(V)]$.* $\square$

Now $C(V)$ is a $[C(V)]$-module giving the adjoint representation so is in particular an L-module. Lemma 13.13 (i) shows that its subspace V is an L-submodule.

Proposition 13.15 *Suppose the symmetrix scalar product on V is non-degenerate. Then V is a faithful L-module.*

Proof. Let $x \in L$ and suppose $[xv] = 0$ for all $v \in V$. We must show $x = 0$. Let $x = \sum_{i<j} c_{ij}[v_i v_j]$. We may define a skew-symmetrix $n \times n$ matrix $C = (c_{ij})$ by $c_{ii} = 0$ and $c_{ji} = -c_{ij}$ for $i < j$. We have

$$\sum_{i<j} c_{ij}[[v_i v_j]v] = 0.$$

By Lemma 13.13 (i) we have

$$4\sum_{i<j} c_{ij}((v_j, v)v_i - (v_i, v)v_j) = 0.$$

The coefficient of v_i in this expression is

$$4\left(\sum_{\substack{j \\ j>i}} c_{ij}(v_j, v) - \sum_{\substack{j \\ j<i}} c_{ji}(v_j, v)\right) = 4\sum_{j=1}^{n} c_{ij}(v_j, v).$$

It follows that

$$\sum_{j=1}^{n} c_{ij}(v_j, v) = 0$$

for all i and all $v \in V$. Let $(v_j, v_k) = m_{jk}$. Then we have

$$\sum_{j=1}^{n} c_{ij} m_{jk} = 0 \quad \text{for all } i, k$$

that is $CM = O$ where $M = (m_{jk})$. If the scalar product on V is non-degenerate then M is a non-singular matrix. Then $C = O$ and so $x = 0$. □

Lemma 13.16 *Let $x \in L$ and $v, v' \in V$. Then*

$$([xv], v') + (v, [xv']) = 0.$$

Proof. It is sufficient to prove this when $x = [yz]$ for $y, z \in V$. Now

$$([[yz]v], v') = 4(z, v)(y, v') - 4(y, v)(z, v')$$
$$(v, [[yz]v']) = 4(z, v')(v, y) - 4(y, v')(v, z)$$

by Lemma 13.13 (i). The result follows. □

Thus we have a Lie algebra L of dimension $n(n-1)/2$, an L-module V of dimension n, and a symmetric bilinear scalar product on V invariant under L in the sense of Lemma 13.16.

We now consider some special cases of the above situation. First let V be a vector space with $\dim V = 2l+1$ and let $v_0, v_1, \ldots, v_l, v_{-1}, \ldots, v_{-l}$ be a basis of V. Consider the symmetric bilinear scalar product $V \times V \to \mathbb{C}$ determined by

$$(v_0, v_0) = 2$$
$$(v_i, v_{-i}) = 1 \quad i = 1, \ldots, l$$

and all other scalar products of basis elements 0. The matrix of this scalar product is

$$\begin{pmatrix} 2 & 0 & \cdots & 0 \\ 0 & O & & I_l \\ \vdots & & & \\ 0 & I_l & & O \end{pmatrix}$$

The condition

$$([xv], v') + (v, [xv']) = 0$$

of Lemma 13.16 tells us that if the element $x \in L$ is represented by the matrix X on V then

$$X^t M + MX = O.$$

Now L has dimension $l(2l+1)$ and, by Proposition 13.15, acts faithfully on V. However, the Lie algebra of all $(2l+1) \times (2l+1)$ matrices X satisfying $X^t M + MX = O$ is the simple Lie algebra B_l (cf. Section 8.3) and has dimension $l(2l+1)$. Since L is contained in this set of matrixes we must have $L = B_l$.

We aim to find the spin module for B_l inside the Clifford algebra $C(V)$. Recall that V has basis

$$v_0, v_1, \ldots, v_l, v_{-1}, \ldots, v_{-l}.$$

We define $u_i = v_0 v_i$ and $u_{-i} = v_0 v_{-i}$ for $i = 1, \ldots, l$. Let U be the subspace of $C(V)$ spanned by elements

$$u_{-j_1} u_{-j_2} \ldots u_{-j_t} u_1 u_2 \ldots u_l$$

for all subsets $j_1 < j_2 < \cdots < j_t$ of $\{1, \ldots, l\}$ with $0 \leq t \leq l$. Bearing in mind the natural basis of $C(V)$ we see that these elements are linearly independent. Thus $\dim U = 2^l$. We have $U \subset C(V)^+$.

Lemma 13.17 $C(V)^+ U \subset U$. Thus U is a left ideal of $C(V)^+$.

Proof. We first observe that $C(V)^+$ is generated by the elements u_i and u_{-i} for $i = 1, \ldots, l$. This follows from the fact that v_0 anticommutes with v_i and v_{-i} for all $i = 1, \ldots, l$.

We note next that $u_i^2 = 0$ and $u_{-i}^2 = 0$ for $i = 1, \ldots, l$, that, u_i, u_j anticommute and u_{-i}, u_{-j} anticommute when $i \neq j$ and both lie in $\{1, \ldots, l\}$, that u_i, u_{-j} anticommute for all $i, j \in \{1, \ldots, l\}$, and that

$$u_i u_{-i} + u_{-i} u_i = -4 \cdot 1.$$

For $u_i u_{-i} + u_{-i} u_i = v_0 v_i v_0 v_{-i} + v_0 v_{-i} v_0 v_i = -v_0^2 (v_i v_{-i} + v_{-i} v_i) = -2 \cdot 2(v_i, v_{-i}) 1 = -4 \cdot 1$.

It follows from these relations that

$$u_i \cdot u_{-j_1} \ldots u_{-j_t} u_1 \ldots u_l$$

$$= \begin{cases} \pm 4 u_{-j_1} \ldots \hat{u}_{-i} \ldots u_{-j_t} u_1 \ldots u_l & \text{if } i \in \{j_1, \ldots, j_t\} \\ 0 & \text{otherwise} \end{cases}$$

13.5 Clifford algebras and spin modules

where $\hat{u}_{-i}$ means the term u_{-i} is omitted.

$$u_{-i} \cdot u_{-j_1} \ldots u_{-j_t} u_1 \ldots u_l = \begin{cases} 0 & \text{if } i \in \{j_1, \ldots, j_t\} \\ \pm u_{-j_1} \ldots u_{-i} \ldots u_{-j_t} u_1 \ldots u_l & \text{if } i \notin \{j_1, \ldots, j_t\}. \end{cases}$$

This shows that $u_i U \subset U$ and $u_{-i} U \subset U$, so $C(V)^+ U \subset U$. $\square$

This lemma shows that we may regard U as a $C(V)^+$-module under left multiplication. U is therefore a $[C(V)^+]$-module under the same action. Since L is a Lie subalgebra of $[C(V)^+]$ we may regard U as an L-module under left multiplication.

Warning note Whereas the action of L on U is given by left multiplication the action of L on $C(V)$ considered earlier in this section was given by Lie multiplication.

We consider the weights of the L-module U. In order to do this we identify the diagonal Cartan subalgebra H of L.

Lemma 13.18 *Under the above isomorphism $L \cong B_l$ the element $[v_i v_{-i}] \in L$ corresponds to the diagonal matrix*

$$\begin{pmatrix} 0 & & & & & & & & \\ & \ddots & & & & & & & \\ & & 0 & & & & & & \\ & & & 4 & & & & & \\ & & & & 0 & & & & \\ & & & & & \ddots & & & \\ & & & & & & 0 & & \\ & & & & & & & -4 & \\ & & & & & & & & 0 \\ & & & & & & & & & \ddots \\ & & & & & & & & & & 0 \end{pmatrix} \begin{matrix} \\ \\ \\ i \\ \\ \\ -i \\ \\ \\ \\ \end{matrix}$$

Proof. The matrix representation of L comes from the L-module V with basis $v_0, v_1, \ldots, v_l, v_{-1}, \ldots, v_{-l}$. Now

$$[[v_i v_{-i}], v_0] = 0$$
$$[[v_i v_{-i}], v_j] = 4(v_{-i}, v_j) v_i - 4(v_i, v_j) v_{-i} = \delta_{ij} \cdot 4v_i$$
$$[[v_i v_{-i}], v_{-j}] = 4(v_{-i}, v_{-j}) v_i - 4(v_i, v_{-j}) v_{-i} = -\delta_{ij} \cdot 4v_{-i}$$

by Lemma 13.13 (i). $\square$

Fundamental modules for simple Lie algebras

It follows from this lemma that the element

$$h = \sum_{i=1}^{l} \frac{\lambda_i}{4[v_i v_{-i}]}$$

is represented by the diagonal matrix

$$\begin{pmatrix} 0 & & & & & & \\ & \lambda_1 & & & & & \\ & & \ddots & & & & \\ & & & \lambda_l & & & \\ & & & & -\lambda_1 & & \\ & & & & & \ddots & \\ & & & & & & -\lambda_l \end{pmatrix}$$

We recall from Section 8.3 that such matrices form a Cartan subalgebra H of L.

We consider the action of h on the L-module U. We have

$$[v_i v_{-i}] = v_i v_{-i} - v_{-i} v_i = \tfrac{1}{2}(v_i v_0 v_0 v_{-i} - v_{-i} v_0 v_0 v_i)$$
$$= -\tfrac{1}{2} u_i u_{-i} + \tfrac{1}{2} u_{-i} u_i.$$

Thus

$$[v_i v_{-i}] u_{-j_1} \ldots u_{-j_t} u_1 \ldots u_l = -\tfrac{1}{2} u_i u_{-i} \cdot u_{-j_1} \ldots u_{-j_t} u_1 \ldots u_l$$
$$+ \tfrac{1}{2} u_{-i} u_i \cdot u_{-j_1} \ldots u_{-j_t} u_1 \ldots u_l$$
$$= \begin{cases} -2 u_{-j_1} \ldots u_{-j_t} u_1 \ldots u_l & \text{if } i \in \{j_1, \ldots, j_t\} \\ 2 u_{-j_1} \ldots u_{-j_t} u_1 \ldots u_l & \text{if } i \notin \{j_1, \ldots, j_t\}. \end{cases}$$

Thus

$$h u_{-j_1} \ldots u_{-j_t} u_1 \ldots u_l = \tfrac{1}{2} \left(\sum_{i=1}^{l} \varepsilon_i \lambda_i \right) u_{-j_1} \ldots u_{-j_t} u_1 \ldots u_l$$

where $\varepsilon_i = \begin{cases} -1 & \text{if } i \in \{j_1, \ldots, j_t\} \\ 1 & \text{if } i \notin \{j_1, \ldots, j_t\} \end{cases}$. Let $\mu_i \in H^*$ be given by $\mu_i(h) = \lambda_i$. Then the weights of L coming from the L-module U are

$$\tfrac{1}{2} \sum_{i=1}^{l} \varepsilon_i \mu_i$$

13.5 Clifford algebras and spin modules

for all possible choices of the signs $\varepsilon_i = \pm 1$. In particular the highest weight is $\frac{1}{2} \sum_{i=1}^{l} \mu_i$.

We recall from the proof of Theorem 13.9 that $\omega_l = \frac{1}{2}(\mu_1 + \cdots + \mu_l)$. Thus U has highest weight ω_l. It follows that U contains the spin module $L(\omega_l)$ as an irreducible direct summand. But

$$\dim U = \dim L(\omega_l) = 2^l.$$

Thus we have proved

Theorem 13.19 *Let L be the simple Lie algebra of type B_l. Then the L-module U constructed as above in the Clifford algebra is the spin module $L(\omega_l)$ of dimension 2^l.* □

We now consider a second special case. This time let V be a vector space with $\dim V = 2l$ and let $v_1, \ldots, v_l, v_{-1}, \ldots, v_{-l}$ be a basis of V. Consider the symmetric bilinear scalar product $V \times V \to \mathbb{C}$ determined by

$$(v_i, v_{-i}) = 1 \qquad i = 1, \ldots, l$$

and all other scalar products of basis elements are 0. The matrix of this scalar product is

$$M = \begin{pmatrix} O & I_l \\ I_l & O \end{pmatrix}.$$

The condition of Lemma 13.16 implies that if $x \in L$ is represented by the matrix X with respect to this basis of V then

$$X^t M + M X = O.$$

Now L has dimension $l(2l-1)$ and acts faithfully on V. The Lie algebra of all $2l \times 2l$ matrices X satisfying $X^t M + M X = O$ is the simple Lie algebra D_l, by Section 8.2. Since $\dim D_l = l(2l-1)$ we have $L = D_l$.

We again aim to find the two spin modules for D_l inside the Clifford algebra $C(V)$. Let U be the subspace of $C(V)$ spanned by all elements of form

$$v_{-j_1} v_{-j_2} \cdots v_{-j_t} v_1 \cdots v_l$$

for all subsets $j_1 < \cdots < j_t$ of $\{1, \ldots, l\}$. These elements are linearly independent, so form a basis for U. We have $\dim U = 2^l$.

Lemma 13.20 $C(V)U \subset U$. *Thus U is a left ideal of $C(V)$.*

Proof. $C(V)$ is generated by elements v_i, v_{-i} and we have

$$v_i \cdot v_{-j_1} \ldots v_{-j_t} v_1 \ldots v_l = \begin{cases} \pm 2 v_{-j_1} \ldots \hat{v}_{-i} \ldots v_{-j_t} v_1 \ldots v_l & \text{if } i \in \{j_1, \ldots, j_t\} \\ 0 & \text{if } i \notin \{j_1, \ldots, j_t\} \end{cases}$$

$$v_{-i} \cdot v_{-j_1} \ldots v_{-j_t} v_1 \ldots v_l = \begin{cases} 0 & \text{if } i \in \{j_1, \ldots, j_t\} \\ \pm v_{-j_1} \ldots v_{-i} \ldots v_{-j_t} v_1 \ldots v_l & \text{if } i \notin \{j_1, \ldots, j_t\} \end{cases}.$$

Thus $v_i U \subset U$ and $v_{-i} U \subset U$, so $C(V) U \subset U$. $\square$

Let $U^+ = U \cap C(V)^+$ and $U^- = U \cap C(V)^-$. Then we have

$$C(V)^+ U^+ \subset U \cap C(V)^+ = U^+$$

$$C(V)^+ U^- \subset U \cap C(V)^- = U^-.$$

Since $L \subset C(V)^+$ it follows that

$$LU^+ \subset U^+, \quad LU^- \subset U^-.$$

Thus U^+ and U^- are L-modules under left multiplication, with

$$\dim U^+ = \dim U^- = 2^{l-1}.$$

We shall show that these are the two spin modules for L.

We consider the weights of the L-modules U^+, U^- by identifying the diagonal Cartan subalgebra H of L.

Lemma 13.21 *Under the above isomorphism $L \cong D_l$ the element $[v_i v_{-i}] \in L$ corresponds to the diagonal matrix*

$$\begin{pmatrix} 0 & & & & & & & \\ & \ddots & & & & & & \\ & & 0 & & & & & \\ & & & 4 & & & & \\ i & & & & 0 & & & \\ & & & & & \ddots & & \\ & & & & & & 0 & \\ -i & & & & & & & -4 \\ & & & & & & & & 0 \\ & & & & & & & & & \ddots \\ & & & & & & & & & & 0 \end{pmatrix}$$

13.5 Clifford algebras and spin modules

Proof. The proof is the same as that for Lemma 13.18 with the first row and column omitted. □

Thus the element

$$h = \sum_{i=1}^{l} \frac{\lambda_i}{4[v_i v_{-i}]}$$

is represented by the diagonal matrix

$$\begin{pmatrix} \lambda_1 & & & & & \\ & \ddots & & & & \\ & & \lambda_l & & & \\ & & & -\lambda_1 & & \\ & & & & \ddots & \\ & & & & & -\lambda_l \end{pmatrix}$$

We consider the action of h on the L-modules U^+ and U^-. We have

$$[v_i v_{-i}] v_{-j_1} \ldots v_{-j_t} v_1 \ldots v_l$$
$$= v_i v_{-i} v_{-j_1} \ldots v_{-j_t} v_1 \ldots v_l - v_{-i} v_i v_{-j_1} \ldots v_{-j_t} v_1 \ldots v_l$$
$$= \begin{cases} -2 v_{-j_1} \ldots v_{-j_t} v_1 \ldots v_l & \text{if } i \in \{j_1, \ldots, j_t\} \\ 2 v_{-j_1} \ldots v_{-j_t} v_1 \ldots v_l & \text{if } i \notin \{j_1, \ldots, j_t\} \end{cases}$$

since $v_{-i} v_i + v_i v_{-i} = 21$. Thus

$$h v_{-j_1} \ldots v_{-j_t} v_1 \ldots v_l = \tfrac{1}{2} \left(\sum_{i=1}^{l} \varepsilon_i \lambda_i \right) v_{-j_1} \ldots v_{-j_t} v_1 \ldots v_l$$

where $\varepsilon_i = \begin{cases} -1 & \text{if } i \in \{j_1, \ldots, j_t\} \\ 1 & \text{if } i \notin \{j_1, \ldots, j_t\} \end{cases}$. As before let $\mu_i \in H^*$ be defined by

$$\mu_i \begin{pmatrix} \lambda_1 & & & & & \\ & \ddots & & & & \\ & & \lambda_l & & & \\ & & & -\lambda_1 & & \\ & & & & \ddots & \\ & & & & & -\lambda_l \end{pmatrix} = \lambda_i.$$

Then the weights of the L-module U are

$$\tfrac{1}{2}\sum_{i=1}^{l}\varepsilon_i\mu_i$$

for all possible choices of the signs $\varepsilon_i = \pm 1$.

If l is even, the basis elements with t even lie in U^+ and those with t odd in U^-. Thus the weights of U^+ have an even number of ε_i negative and those of U^- have an odd number negative. Since $\mu_1 \succ \mu_2 \succ \cdots \succ \mu_l$ the highest weight of U^+ is $\tfrac{1}{2}\sum_{i=1}^{l}\mu_i$ and that of U^- is $\tfrac{1}{2}\left(\sum_{i=1}^{l-1}\mu_i - \mu_l\right)$.

If l is odd we have the reverse situation in which $\tfrac{1}{2}\sum_{i=1}^{l}\mu_i$ is the highest weight of U^- and $\tfrac{1}{2}\left(\sum_{i=1}^{l-1}\mu_i - \mu_l\right)$ is the highest weight of U^+.

Now by the proof of Theorem 13.11 we have

$$\tfrac{1}{2}(\mu_1 + \cdots + \mu_{l-1} + \mu_l) = \omega_l$$
$$\tfrac{1}{2}(\mu_1 + \cdots + \mu_{l-1} - \mu_l) = \omega_{l-1}.$$

Thus we have proved

Theorem 13.22 *Let L be the simple Lie algebra of type D_l. Then the L-modules U^+, U^- are the spin modules of dimension 2^{l-1}. If l is even we have $U^+ = L(\omega_l)$, $U^- = L(\omega_{l-1})$. If l is odd we have $U^+ = L(\omega_{l-1})$, $U^- = L(\omega_l)$.*

13.6 Fundamental modules for C_l

The fundamental weights $\omega_1, \ldots, \omega_l$ for a simple Lie algebra of type C_l will be numbered according to the labelling of the Dynkin diagram

$$\underset{1}{\circ}\!\!-\!\!\underset{2}{\circ}\!\!-\!\!\underset{3}{\circ}\cdots\underset{l-1}{\circ}\!\!\Leftarrow\!\!\underset{l}{\circ}$$

As before we shall use Theorem 13.1 to calculate $\dim L(\omega_j)$. We knows from Section 8.4 that the roots of C_l have the following form. Let

$$h = \begin{pmatrix} \lambda_1 & & & & & \\ & \ddots & & & & \\ & & \lambda_l & & & \\ & & & -\lambda_1 & & \\ & & & & \ddots & \\ & & & & & -\lambda_l \end{pmatrix}.$$

13.6 Fundamental modules for C_l

Then the fundamental roots are

$$\alpha_i(h) = \lambda_i - \lambda_{i+1} \quad \text{for } 1 \leq i \leq l-1$$
$$\alpha_l(h) = 2\lambda_l.$$

The full set of positive roots is given by

$$h \to \lambda_i - \lambda_j \quad \text{for } i < j$$
$$h \to \lambda_i + \lambda_j \quad \text{for } i < j$$
$$h \to 2\lambda_i$$

where $i, j \in \{1, \ldots, l\}$. These positive roots can be expressed as combinations of fundamental roots as follows:

$$\alpha_i + \cdots + \alpha_{j-1} \quad 1 \leq i < j \leq l$$
$$\alpha_i + \cdots + \alpha_{j-1} + 2\alpha_j + \cdots + 2\alpha_{l-1} + \alpha_l \quad 1 \leq i < j \leq l$$
$$2\alpha_i + \cdots + 2\alpha_{l-1} + \alpha_l \quad 1 \leq i \leq l.$$

The first two families are short roots and the third family are long roots. The weights w_i are given by

$$w_1 = \cdots = w_{l-1} = 1 \quad w_l = 2.$$

According to Theorem 13.1 we have

$$\dim L(\omega_j) = \prod_{\alpha \in \Phi^+} d_\alpha$$

where $\alpha = \sum k_i \alpha_i$ and

$$d_\alpha = \frac{\sum_{i=1}^l k_i w_i + k_j w_j}{\sum_{i=1}^l k_i w_i}.$$

We have $d_\alpha = 1$ if α does not involve α_j.

We first suppose that $j \in \{1, \ldots, l-1\}$. Then the positive roots involving j are:

$$\alpha_i + \cdots + \alpha_j + \cdots + \alpha_{k-1} \quad 1 \leq i \leq j < k \leq l$$
$$\alpha_i + \cdots + \alpha_j + \cdots + \alpha_{k-1} + 2\alpha_k + \cdots + 2\alpha_{l-1} + \alpha_l \quad 1 \leq i \leq j < k \leq l$$
$$\alpha_i + \cdots + \alpha_{k-1} + 2\alpha_k + \cdots + 2\alpha_j + \cdots + 2\alpha_{l-1} + \alpha_l \quad 1 \leq i < k \leq j$$
$$2\alpha_i + \cdots + 2\alpha_j + \cdots + 2\alpha_{l-1} + \alpha_l \quad 1 \leq i \leq j.$$

The values of d_α in these four cases are

$$\frac{k-i+1}{k-i}, \quad \frac{2l-i-k+3}{2l-i-k+2}, \quad \frac{2l-i-k+4}{2l-i-k+2}, \quad \frac{l-i+2}{l-i+1}$$

respectively. The product of all possible d_α in these four cases is

$$\frac{(j+1)(j+2)\cdots l}{1\cdot 2\cdots l-j}, \quad \frac{(2l-j+1)(2l-j)\cdots(l+2)}{(2l-2j+1)(2l-2j)\cdots(l-j+2)},$$

$$\frac{(2l+1)2l(2l-1)\cdots(2l-j+2)}{(2l-j+2)(2l-j+1)\cdots(2l-2j+3)}, \quad \frac{l+1}{l-j+1}$$

respectively. The total product $\prod_{\alpha \in \Phi^+} d_\alpha$ is

$$\frac{(2l)!}{(2l-j+2)!\,j!}(2l+1)(2l-2j+2).$$

This expression may be written in a more suggestive form by using the identity

$$\binom{2l}{j}-\binom{2l}{j-2}=\frac{(2l)!}{(2l-j+2)!\,j!}(2l+1)(2l-2j+2).$$

Thus $\dim L(\omega_j) = \binom{2l}{j}-\binom{2l}{j-2}$ for $1 \le j \le l-1$.

We now suppose that $j=l$. The positive roots involving α_l are

$$2\alpha_i + \cdots + 2\alpha_{l-1} + \alpha_l \quad 1 \le i \le l$$
$$\alpha_i + \cdots + \alpha_{j-1} + 2\alpha_j + \cdots + 2\alpha_{l-1} + \alpha_l \quad 1 \le i < j \le l.$$

The first family are long roots and the second short roots. The values of $1d_\alpha$ in these two cases are

$$\frac{l-i+2}{l-i+1}, \quad \frac{2l-i-j+4}{2l-i-j+2}$$

respectively. The product of all possible d_α in these cases is

$$l+1, \quad \frac{(2l+1)(2l)\cdots(l+2)}{(l+2)(l+1)\cdots 3}$$

respectively, and the total product $\prod_{\alpha \in \Phi^+} d_\alpha$ is $\frac{(2l+1)!2}{(l+2)!\,l!}$.

By using the identity

$$\binom{2l}{l} - \binom{2l}{l-2} = \frac{(2l+1)!2}{(l+2)!l!}$$

we see that

$$\dim L(\omega_l) = \binom{2l}{l} - \binom{2l}{l-2}.$$

Thus we have shown

Proposition 13.23 *The dimensions of the fundamental modules for the simple Lie algebra of type C_l are*

$$2l \quad \binom{2l}{2}-1 \quad \binom{2l}{3}-2l \quad \cdots \quad \binom{2l}{l-1}-\binom{2l}{l-3} \quad \binom{2l}{l}-\binom{2l}{l-2}$$

13.7 Contraction maps

We shall now identify the fundamental modules whose dimensions we have obtained. We begin with $L(\omega_1)$.

Proposition 13.24 *The natural $2l$-dimensional C_l-module is isomorplic to $L(\omega_1)$.*

Proof. Let V be the natural C_l-module. Let

$$h = \begin{pmatrix} \lambda_1 & & & & & \\ & \ddots & & & & \\ & & \lambda_l & & & \\ & & & -\lambda_1 & & \\ & & & & \ddots & \\ & & & & & -\lambda_l \end{pmatrix}.$$

Then the weights of V are $\mu_1, \ldots, \mu_l, -\mu_1, \ldots, -\mu_l$ where $\mu_i(h) = \lambda_i$. Since

$$\mu_i - \mu_{i+1} = \alpha_i \quad 1 \le i \le l-1$$
$$2\mu_l = \alpha_l$$

we have

$$\mu_1 \succ \mu_2 \succ \cdots \succ \mu_l \succ 0.$$

Thus the highest weight of V is μ_1. We have

$$\mu_1 = \alpha_1 + \cdots + \alpha_{l-1} + \tfrac{1}{2}\alpha_l$$
$$\mu_2 = \alpha_2 + \cdots + \alpha_{l-1} + \tfrac{1}{2}\alpha_l$$
$$\vdots$$
$$\mu_l = \tfrac{1}{2}\alpha_l.$$

We also have $\alpha_i = \sum_j A_{ji}\omega_j$ which in type C_l gives

$$\alpha_1 = 2\omega_1 - \omega_2$$
$$\alpha_2 = -\omega_1 + 2\omega_2 - \omega_3$$
$$\vdots$$
$$\alpha_{l-1} = -\omega_{l-2} + 2\omega_{l-1} - \omega_l$$
$$\alpha_l = -2\omega_{l-1} + 2\omega_l.$$

It follows that

$$\mu_1 = \omega_1$$
$$\mu_2 = -\omega_1 + \omega_2$$
$$\vdots$$
$$\mu_l = -\omega_{l-1} + \omega_l.$$

Thus V is a C_l-module with highest weight ω_1. It therefore contains $L(\omega_1)$ as an irreducible component. However,

$$\dim V = \dim L(\omega_1) = 2l$$

thus V is irreducible and isomorphic to $L(\omega_1)$. □

We now consider the fundamental modules $L(\omega_j)$ for $j \geq 2$. We have

$$\dim L(\omega_j) = \binom{2l}{j} - \binom{2l}{j-2}.$$

This suggests that we should look for $L(\omega_j)$ as a submodule of the exterior power $\Lambda^j V$. The key idea is to find a homomorphism of C_l-modules from $\Lambda^j V$ into $\Lambda^{j-2} V$, called a contraction map.

13.7 Contraction maps

Proposition 13.25 *Let $v, v' \to (v, v')$ be the skew-symmetric bilinear map $V \times V \to \mathbb{C}$ given by the matrix*

$$M = \begin{pmatrix} O & I_l \\ -I_l & O \end{pmatrix}.$$

Then there is a unique homomorphism of C_l-modules

$$\theta : \Lambda^j V \to \Lambda^{j-2} V$$

satisfying the condition

$$\theta(u_1 \wedge \cdots \wedge u_j) = \sum_{r<s} (-1)^{r+s-1} (u_r, u_s) u_1 \wedge \cdots \wedge \hat{u}_r \wedge \cdots \wedge$$

$$\hat{u}_s \wedge \cdots \wedge u_j \text{ for all } u_1, \ldots, u_j \in V. \quad (\dagger)$$

Here as usual the notation $\hat{u}_r, \hat{u}_s$ means that those terms are omitted.

Proof. It is clear that if such a map θ exists it will be unique. To prove the existence let $v_1, \ldots, v_{2l}$ be a basis of V. Then there is a unique linear map θ satisfying

$$\theta\left(v_{i_1} \wedge \cdots \wedge v_{i_j}\right) = \sum_{r<s} (-1)^{r+s-1} \left(v_{i_r}, v_{i_s}\right) v_{i_1} \wedge \cdots \wedge \hat{v}_{i_r} \wedge \cdots \wedge \hat{v}_{i_s} \wedge \cdots \wedge v_{i_j}$$

for all $i_1, \ldots, i_j \in \{1, \ldots, 2l\}$ with $i_1 < \cdots < i_j$. We show this map has the required properties. Since both sides of equation (†) are linear in $u_1, \ldots, u_j$ it will be sufficient to prove it when each u_k is one of the basis elements of V. If the same basis element appears twice both sides of (†) are 0. Thus we may assume the basis elements are all distinct. They may not occur in increasing order, thus we must show that the above formula defining θ remains valid if the factors $v_{i_1}, \ldots, v_{i_j}$ are permuted. In fact it is sufficient to see this if we transpose two consecutive terms $v_{i_k}, v_{i_{k+1}}$. When we carry out such a transposition the expression $v_{i_1} \wedge \cdots \wedge v_{i_j}$ changes in sign. We show that each term

$$(-1)^{r+s-1} \left(v_{i_r}, v_{i_s}\right) v_{i_1} \wedge \cdots \wedge \hat{v}_{i_r} \wedge \cdots \wedge \hat{v}_{i_s} \wedge \cdots \wedge v_{i_j}$$

changes in sign also. If neither of r, s lie in $\{k, k+1\}$ the term

$$v_{i_1} \wedge \cdots \wedge \hat{v}_{i_r} \wedge \cdots \wedge \hat{v}_{i_s} \wedge \cdots \wedge \hat{v}_{i_j}$$

will change in sign when we make the transposition. If just one of r, s lies in $\{k, k+1\}$ the term $(-1)^{r+s-1}$ will change in sign when the transposition is made. Finally if $r=k, s=k+1$ the term $\left(v_{i_r}, v_{i_s}\right)$ changes in sign, since the bilinear map is skew-symmetric. This shows that the linear map θ we have defined satisfies (†).

It remains to show that θ is a homomorphism of C_l-modules. Let x lie in the Lie algebra C_l. Then

$$x\theta(u_1 \wedge \cdots \wedge u_j) = x\sum_{r<s}(-1)^{r+s-1}(u_r, u_s)u_1 \wedge \cdots \wedge \hat{u}_r \wedge \cdots \wedge \hat{u}_s \wedge \cdots \wedge u_j$$

$$= \sum_{r<s} \sum_{\substack{k \\ k\neq r, k\neq s}} (-1)^{r+s-1}(u_r, u_s)u_1 \wedge \cdots \wedge xu_k \wedge \cdots \wedge \hat{u}_r \wedge \cdots$$

$$\wedge \hat{u}_s \wedge \cdots \wedge u_j.$$

On the other hand we have

$$\theta x(u_1 \wedge \cdots \wedge u_j)$$
$$= \theta \sum_k u_1 \wedge \cdots \wedge xu_k \wedge \cdots \wedge u_j = x\theta(u_1 \wedge \cdots \wedge u_j)$$
$$+ \sum_k \sum_{\substack{s \\ k<s}} (-1)^{k+s-1}(xu_k, u_s)u_1 \wedge \cdots \wedge \hat{u}_k \wedge \cdots \wedge \hat{u}_s \wedge \cdots \wedge u_j$$
$$+ \sum_k \sum_{\substack{r \\ r<k}} (-1)^{r+k-1}(u_r, xu_k)u_1 \wedge \cdots \wedge \hat{u}_r \wedge \cdots \wedge \hat{u}_k \wedge \cdots \wedge u_j.$$

Renaming the suffixes we see that the last two sums cancel since

$$(xu_k, u_s) + (u_k, xu_s) = 0.$$

This condition is equivalent to

$$X^t M + MX = O$$

where X is the matrix representing x on V, and we recall from Section 8.4 that the simple Lie algebra C_l satisfies this condition. It follows that

$$\theta x(u_1 \wedge \cdots \wedge u_j) = x\theta(u_1 \wedge \cdots \wedge u_j)$$

and so θ is a homomorphism of C_l-modules. □

This homomorphism $\theta : \Lambda^j V \to \Lambda^{j-2} V$ will be called a **contraction map**.

Now the weights of $\Lambda^j V$ are sums of j distinct weights of V. By the proof of Proposition 13.24 the weights of V are

$$\omega_1 \succ -\omega_1 + \omega_2 \succ -\omega_2 + \omega_3 \succ \cdots \succ -\omega_{l-1} + \omega_l$$
$$\succ \omega_{l-1} - \omega_l \succ \cdots \succ \omega_2 - \omega_3 \succ \omega_1 - \omega_2 \succ -\omega_1.$$

Thus if $j \leq l$ the highest weight of $\Lambda^j V$ is ω_j. Similarly the highest weight of $\Lambda^{j-2} V$ is ω_{j-2}. Since $\omega_j \succ \omega_{j-2}$ we see that ω_j is not a weight of $\Lambda^{j-2} V$. Since ω_j is the highest weight of $\Lambda^j V$ the module $L(\omega_j)$ must be an irreducible direct summand of $\Lambda^j V$. On the other hand $L(\omega_j)$ cannot be a submodule of $\Lambda^{j-2}V$, as ω_j is not a weight of this module. Thus $L(\omega_j)$ must lie in the kernel of the contraction map θ.

We shall show subsequently that when $j \le l$ the contraction map $\theta : \Lambda^j V \to \Lambda^{j-2}V$ is surjective. It will follow that

$$\dim(\ker \theta) = \binom{2l}{j} - \binom{2l}{j-2} = \dim L(\omega_j)$$

and therefore that $L(\omega_j) = \ker \theta$. This will identify the irreducible module $L(\omega_j)$ as the submodule of $\Lambda^j V$ which is the kernel of the contraction map θ.

Let $v_1, \ldots, v_l, v_{-1}, \ldots, v_{-l}$ be the natural basis of V with respect to which the skew-symmetric bilinear form is given by

$$(v_i, v_{-i}) = 1 \quad 1 \le i \le l$$
$$(v_{-i}, v_i) = -1$$

and all other scalar products zero. Let W be the subspace of V spanned by $v_1, \ldots, v_l$ and W^- the subspace spanned by $v_{-1}, \ldots, v_{-l}$. Then W, W^- are isotropic subspaces of V, i.e. the skew-symmetric form restricted to W and W^- is identically zero. Also we have $V = W \oplus W^-$. It follows that

$$\Lambda^j V = \bigoplus_{a+b=j} \left(\Lambda^a W \otimes \Lambda^b W^- \right).$$

The contraction map $\theta : \Lambda^j V \to \Lambda^{j-2}V$ satisfies

$$\theta \left(\Lambda^a W \otimes \Lambda^b W^- \right) \subset \Lambda^{a-1} W \otimes \Lambda^{b-1} W^-$$

since a basis element in W has a non-zero scalar product only with a basis element in W^-. Thus in order to show that $\theta : \Lambda^j V \to \Lambda^{j-2} V$ is surjective for $j \le l$ it will be sufficient to show that

$$\theta : \Lambda^a W \otimes \Lambda^b W^- \to \Lambda^{a-1} W \otimes \Lambda^{b-1} W^-$$

is surjective whenever $a + b \le l$.

For each subset $I \subset \{1, \ldots, l\}$ we define $v_I = v_{i_1} \wedge \cdots \wedge v_{i_k}$ where $I = \{i_1, \ldots, i_k\}$ with $i_1 < \cdots < i_k$. We also define $v_{-I} = v_{-i_1} \wedge \cdots \wedge v_{-i_k}$. Then any basis element of $\Lambda^{a-1} W \otimes \Lambda^{b-1} W^-$ can be written in the form

$$\pm (v_X \wedge v_T) \otimes (v_{-T} \wedge v_{-Y})$$

for some subsets T, X, Y of $\{1, \ldots, l\}$ with

$$T \cap X = \phi, \quad T \cap Y = \phi, \quad X \cap Y = \phi, \quad |X| + |T| = a-1, \quad |Y| + |T| = b-1.$$

We write $|T|=r$. Since $a+b\leq l$ we have $|X|+|Y|+2r+2\leq l$, that is $l-|X|-|Y|\geq 2r+2$. Thus it is possible to choose a subset S of $\{1,\ldots,l\}$ such that
$$|S|=2r+1, \quad S\cap X=\phi, \quad S\cap Y=\phi, \quad S\supset T.$$
We can now describe an element of $\Lambda^a W\otimes \Lambda^b W^-$ which maps under θ to a non-zero multiple of $(v_X\wedge v_T)\otimes(v_{-T}\wedge v_{-Y})$.

Proposition 13.26 *Suppose subsets T,X,Y,S of $\{1,\ldots,l\}$ are chosen as above, and let $\theta: \Lambda^j V \to \Lambda^{j-2} V$ be the contraction map. Then*

$$\theta\left(\sum_{i=0}^{r}(-1)^i i!(r-i)! \sum_{\substack{U\\U<S\\|U|=r+1\\|U\cap T|=i}} (v_X\wedge v_U)\otimes(v_{-U}\wedge v_{-Y})\right)$$
$$=(r+1)!\,(v_X\wedge v_T)\otimes(v_{-T}\wedge v_{-Y}).$$

Consequently the map
$$\theta: \Lambda^a W\otimes \Lambda^b W^- \to \Lambda^{a-1} W\otimes \Lambda^{b-1} W^-$$
is surjective when $a+b\leq l$.

Proof. We note that $S\supset T$, $|S|=2r+1$, $|T|=r$ and that we are summing over all subsets U of S with $|U|=r+1$ and $|U\cap T|=i$. Since $|X|+|T|=a-1$ and $|Y|+|T|=b-1$ we have $|X|+|U|=a$ and $|Y|+|U|=b$. Thus the left-hand side lies in $\Lambda^a W\otimes \Lambda^b W^-$.

By definition of θ we have
$$\theta(v_u\otimes v_{-u}) = (-1)^r \sum_{\substack{R\\R\subset U\\|R|=r}} v_R\otimes v_{-R}$$
where the right-hand side involves a sum over all r-element subsets R of U. Thus

$$\theta\left(\sum_{\substack{U\\|U\cap T|=i}} v_U\otimes v_{-U}\right) = (-1)^r \sum_{\substack{U\\|U|=r+1\\|U\cap T|=i}} \left(\sum_{\substack{R\\|R|=r\\R\subset U}} v_R\otimes v_{-R}\right)$$

$$= (-1)^r \sum_{\substack{R\\|R|=r}} \left(\sum_{\substack{U\\R\subset U\\|U|=r+1\\|U\cap T|=i}} 1\right) v_R\otimes v_{-R}.$$

13.7 Contraction maps

Since $|U \cap T| = i$ and R is obtained from U by omitting one element we have $|R \cap T| = i$ or $|R \cap T| = i - 1$. We split the sum according to those two possibilities. Thus

$$\theta \left(\sum_{\substack{U \\ |U \cap T| = i}} v_U \otimes v_{-U} \right)$$

$$= (-1)^r \sum_{\substack{R \\ |R| = r \\ |R \cap T| = i}} \left(\sum_{\substack{U \\ R \subset U \\ |U| = r+1 \\ |U \cap T| = i}} 1 \right) v_R \otimes v_{-R} + (-1)^r \sum_{\substack{R \\ |R| = r \\ |R \cap T| = i-1}} \left(\sum_{\substack{U \\ R \subset U \\ |U| = r+1 \\ |U \cap T| = i}} 1 \right) v_R \otimes v_{-R}$$

$$= (-1)^r \sum_{\substack{R \\ |R| = r \\ |R \cap T| = i}} (i+1) v_R \otimes v_{-R} + (-1)^r \sum_{\substack{R \\ |R| = r \\ |R \cap T| = i-1}} (r+1-i) v_R \otimes v_{-R}$$

since in the first case the additional element of U can be chosen in $i+1$ ways and in the second case in $r+1-i$ ways. Thus

$$\theta \left(\sum_{i=0}^{r} (-1)^i i! (r-i)! \sum_{\substack{U \\ |U \cap T| = i}} v_U \otimes v_{-U} \right)$$

$$= (-1)^r \sum_{i=0}^{r} (-1)^i i! (r-i)! \sum_{\substack{R \\ |R| = r \\ |R \cap T| = i}} (i+1) v_R \otimes v_{-R}$$

$$+ (-1)^r \sum_{i=0}^{r} (-1)^i i! (r-i)! \sum_{\substack{R \\ |R| = r \\ |R \cap T| = i-1}} (r+1-i) v_R \otimes v_{-R}.$$

We rename the variable i in the second sum to give

$$(-1)^r \sum_{i=0}^{r} (-1)^i i! (r-i)! \sum_{\substack{R \\ |R| = r \\ |R \cap T| = i}} (i+1) v_R \otimes v_{-R}$$

$$+ (-1)^r \sum_{i=-1}^{r-1} (-1)^{i+1} (i+1)! (r-i-1)! \sum_{\substack{R \\ |R| = r \\ |R \cap T| = i}} (r-i) v_R \otimes v_{-R}$$

$$= (-1)^r \sum_{i=0}^{r-1}(-1)^i(i+1)!(r-i)! \sum_{\substack{R \\ |R|=r \\ |R\cap T|=i}} (1-1)v_R \otimes v_{-R}$$

$$+ (r+1)! \sum_{\substack{R \\ |R|=r \\ |R\cap T|=r}} v_R \otimes v_{-R}$$

$$= (r+1)! v_T \otimes v_{-T}.$$

We now consider

$$\theta\left(\sum_{i=0}^{r}(-1)^i i!(r-i)! \sum_{\substack{U \\ U\subset S \\ |U|=r+1 \\ |U\cap T|=i}} (v_X \wedge v_U) \otimes (v_{-U} \wedge v_{-Y})\right).$$

Since the v_i for $i \in X$ and the v_{-i} for $i \in Y$ have scalar product 0 with all factors in the above product they are not involved in any contraction. Thus

$$\theta\left(\sum_{i=0}^{r}(-1)^i i!(r-i)! \sum_{\substack{U \\ |U\cap T|=i}} (v_X \wedge v_U) \otimes (v_{-U} \wedge v_{-Y})\right)$$

$$= v_X \wedge \theta\left(\sum_{i=0}^{r}(-1)^i i!(r-i)! \sum_{\substack{U \\ |U\cap T|=i}} v_U \otimes v_{-U}\right) \wedge v_{-Y}$$

$$= v_X \wedge ((r+1)! v_T \otimes v_{-T}) \wedge v_{-Y}$$

$$= (r+1)!(v_X \wedge v_T) \otimes (v_{-T} \wedge v_{-Y}). \qquad \square$$

Corollary 13.27 *The contraction map* $\theta : \Lambda^j V \to \Lambda^{j-2} V$ *is surjective when* $j \le l$.

The surjectivity of θ enables us to identify the fundamental modules $L(\omega_j)$.

Theorem 13.28 *The fundamental modules* $L(\omega_j)$ *for the simple Lie algebra* C_l *are given as follows.*

(a) $L(\omega_1)$ *is the natural $2l$-dimensional C_l-module V.*
(b) *For $2 \le j \le l$, $L(\omega_j)$ is the submodule of $\Lambda^j V$ given by the kernel of the contraction map* $\theta : \Lambda^j V \to \Lambda^{j-2} V$.

13.8 Fundamental modules for exceptional algebras

Proof. (a) was shown in Proposition 13.24. We also pointed out earlier that $L(\omega_j)$ is a submodule of $\Lambda^j V$ contained in the kernel of θ. Since $\theta : \Lambda^j V \to \Lambda^{j-2} V$ is surjective for $j \le l$ we have

$$\dim \ker \theta = \binom{2l}{j} - \binom{2l}{j-2}$$

and this is equal to $\dim L(\omega_j)$ by Proposition 13.23. It follows that $L(\omega_j) = \ker \theta$. □

13.8 Fundamental modules for exceptional algebras

By applying Theorem 13.1 to the exceptional simple Lie algebras and making use of the information about their root systems available in Sections 8.5, 8.6 and 8.7 we can show that the dimensions of the fundamental modules for these algebras are as shown. We omit the details.

G_2 14   7

F_4 52 $\binom{52}{2}-52$ $\binom{26}{2}-52$ 26

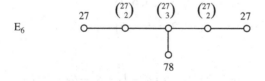

E_6: 27, $\binom{27}{2}$, $\binom{27}{3}$, $\binom{27}{2}$, 27, 78

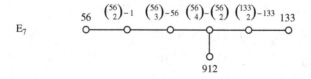

E_7: 56, $\binom{56}{2}-1$, $\binom{56}{3}-56$, $\binom{56}{4}-\binom{56}{2}$, $\binom{133}{2}-133$, 133, 912

E_8: 248, 30 380, 2 450 240, 146 325 270, 6 899 079 264, 6 696 000, 3875, 147 250

We shall show in each case how to obtain the fundamental module of smallest dimension. We begin by obtaining a 27-dimensional fundamental module for E_6.

Proposition 13.29 (a) *The number of positive roots of E_7 not in E_6 is 27.*
(b) *The subspace V of E_7 spanned by vectors e_α for such roots is a 27-dimensional fundamental E_6-module.*

Proof. We recall from Section 8.7 that the fundamental roots of E_7 are given by

$$\beta_2-\beta_3 \quad \beta_3-\beta_4 \quad \beta_4-\beta_5 \quad \beta_5-\beta_6 \quad \beta_6+\beta_7 \quad -\tfrac{1}{2}\sum_{i=1}^{8}\beta_i$$

$$\beta_6-\beta_7$$

and that the full set of roots of E_7 is

$$\pm\beta_i\pm\beta_j \quad i\neq j \quad i,j\in\{2,3,4,5,6,7\}$$
$$\pm(\beta_1+\beta_8)$$
$$\tfrac{1}{2}\sum\varepsilon_i\beta_i \quad \varepsilon_i\in\{1,-1\}, \quad \prod\varepsilon_i=1, \quad \varepsilon_1=\varepsilon_8.$$

The positive roots are

$$\beta_i-\beta_j \quad i\neq j \quad i,j\in\{2,3,4,5,6,7\}$$
$$\beta_i+\beta_j \quad i\neq j \quad i,j\in\{2,3,4,5,6,7\}$$
$$-\beta_1-\beta_8$$
$$\tfrac{1}{2}\sum\varepsilon_i\beta_i \quad \varepsilon_i\in\{1,-1\}, \quad \prod\varepsilon_i=1, \quad \varepsilon_1=\varepsilon_8=-1.$$

The positive roots of E_7 which are not roots of E_6 are

$$\beta_2-\beta_j \quad j\in\{3,4,5,6,7\}$$
$$\beta_2+\beta_j \quad j\in\{3,4,5,6,7\}$$
$$-\beta_1-\beta_8$$
$$\tfrac{1}{2}\sum\varepsilon_i\beta_i \quad \prod\varepsilon_i=1, \quad \varepsilon_1=\varepsilon_8=-1, \quad \varepsilon_2=1.$$

The number of such roots is 27.

Now let V be the subspace of E_7 spanned by the root vectors e_α for such roots α. Then dim $V=27$.

13.8 Fundamental modules for exceptional algebras

Now E_7 may be regarded as an E_7-module giving the adjoint representation. In particular E_7 may be regarded as an E_6-module. We observe that V is an E_6-submodule. To see this it is sufficient to show that $[e_\alpha e_\beta] \in V$ for all $\alpha \in \Phi(E_6), \beta \in \Phi^+(E_7) - \Phi^+(E_6)$. We have

$$[e_\alpha e_\beta] = \begin{cases} N_{\alpha,\beta} e_{\alpha+\beta} & \text{if } \alpha+\beta \in \Phi(E_7) \\ 0 & \text{otherwise.} \end{cases}$$

Suppose $\alpha+\beta \in \Phi(E_7)$. Since β is not a root of E_6, β will involve the fundamental root of E_7 not in E_6, and since β is positive this fundamental root will have positive coefficient in β. It will therefore have positive coefficient in $\alpha+\beta$, and so $\alpha+\beta \in \Phi^+(E_7)$. We claim that $\alpha+\beta \notin \Phi(E_6)$. Suppose to the contrary that $\alpha+\beta \in \Phi(E_6)$. Then $-\alpha \in \Phi(E_6)$ and

$$[e_{\alpha+\beta} e_{-\alpha}] = N_{\alpha+\beta,-\alpha} e_\beta.$$

Since $N_{\alpha,\beta} \neq 0$ it follows from Proposition 7.1 that $N_{\alpha+\beta,-\alpha} \neq 0$ and so $\beta \in \Phi(E_6)$, a contradiction. Hence $\alpha+\beta \in \Phi^+(E_7) - \Phi^+(E_6)$ and V is an E_6-module.

In order to determine the highest weight of V it is convenient to use the linear function

$$h: \sum_{i=1}^{8} \mathbb{R}\beta_i \to \mathbb{R}$$

determined by the property that $h(\alpha_i) = 1$ for each fundamental root α_i of E_8. Thus

$$h(\beta_i - \beta_{i+1}) = 1 \quad \text{for } i \in \{1, \ldots, 6\}$$
$$h(\beta_6 + \beta_7) = 1$$
$$h\left(-\tfrac{1}{2}\sum_{i=1}^{8}\beta_i\right) = 1.$$

Hence we have

$$h(\beta_1) = 6, \quad h(\beta_2) = 5, \quad h(\beta_3) = 4, \quad h(\beta_4) = 3, \quad h(\beta_5) = 2,$$
$$h(\beta_6) = 1, \quad h(\beta_7) = 0, \quad h(\beta_8) = -23.$$

Of our 27 roots the one with the highest h-value is $-\beta_1 - \beta_8$. This must therefore be a highest weight of V. Now the fundamental roots of E_6 are

```
β₃-β₄   β₄-β₅   β₅-β₆   β₆+β₇        -½ Σᵢ₌₁⁸ βᵢ
  o───────o───────o───────o──────────────o
                  │
                  o
                β₆-β₇
```

and $-\beta_1-\beta_8$ is orthogonal to all of them except $-\frac{1}{2}\sum_{i=1}^{8}\beta_i$. Moreover the scalar product $\{,\}$ satisfies

$$\left\{-\beta_1-\beta_8, -\frac{1}{2}\sum_{i=1}^{8}\beta_i\right\} = \frac{1}{2}\left\{-\frac{1}{2}\sum_{i=1}^{8}\beta_i, -\frac{1}{2}\sum_{i=1}^{8}\beta_i\right\}$$

thus $-\beta_1-\beta_8$ is the fundamental weight ω_8. Hence $L(\omega_8)$ is an irreducible direct summand of V. Since

$$\dim L(\omega_8) = \dim V = 27$$

we deduce that $V = L(\omega_8)$. □

In order to obtain the other 27-dimensional fundamental E_6-module we introduce the dual module. We recall that, given any L-module V, the dual space V^* of linear maps from V to $\mathbb{C}$ may be made into an L-module by the rule

$$(xf)v = -f(xv) \qquad x \in L, \quad f \in V^*, \quad v \in V.$$

The weights of V^* are the negatives of the weights of V. In the case of the 27-dimensional E_6-module V above, the highest weight of V^* is the negative of the lowest weight of V. The lowest weight of V is the one with the smallest value of h, i.e. $\beta_2-\beta_3$. Thus the highest weight of V^* is $\beta_3-\beta_2$. This is orthogonal to all fundamental roots of E_6 except for $\alpha_3 = \beta_3-\beta_4$. Since

$$\{\beta_3-\beta_2, \beta_3-\beta_4\} = \frac{1}{2}\{\beta_3-\beta_4, \beta_3-\beta_4\}$$

we deduce that $\beta_3-\beta_2 = \omega_3$. Hence $V^* = L(\omega_3)$.

Now the weight of V with second highest value of h is $\frac{1}{2}(-\beta_1+\beta_2+\beta_3+\beta_4+\beta_5+\beta_6+\beta_7-\beta_8)$ and the third highest is $\frac{1}{2}(-\beta_1+\beta_2+\beta_3+\beta_4+\beta_5-\beta_6-\beta_7-\beta_8)$. Thus the highest weight of $\Lambda^2 V$ is

$$(-\beta_1-\beta_8) + \frac{1}{2}(-\beta_1+\beta_2+\beta_3+\beta_4+\beta_5+\beta_6+\beta_7-\beta_8)$$
$$= \frac{1}{2}(-3\beta_1+\beta_2+\beta_3+\beta_4+\beta_5+\beta_6+\beta_7-3\beta_8).$$

13.8 Fundamental modules for exceptional algebras

By considering the scalar products $\{,\}$ of this weight with the fundamental roots of E_6 we see that this weight is ω_7. Since

$$\dim L(\omega_7) = \binom{27}{2} = \dim \Lambda^2 V$$

we deduce $\Lambda^2 V = L(\omega_7)$.

Similarly the highest weight of $\Lambda^3 V$ is

$$(-\beta_1 - \beta_8) + \tfrac{1}{2}(-\beta_1 + \beta_2 + \beta_3 + \beta_4 + \beta_5 + \beta_6 + \beta_7 - \beta_8)$$
$$+ \tfrac{1}{2}(-\beta_1 + \beta_2 + \beta_3 + \beta_4 + \beta_5 - \beta_6 - \beta_7 - \beta_8)$$
$$= -2\beta_1 + \beta_2 + \beta_3 + \beta_4 + \beta_5 - 2\beta_8.$$

We check by computing scalar products $\{,\}$ that this is the weight ω_5 of E_6. Since

$$\dim L(\omega_5) = \binom{27}{3} = \dim \Lambda^3 V$$

we deduce $\Lambda^3 V = L(\omega_5)$.

It may be shown similarly that

$$\Lambda^2 V^* = L(\omega_4) \quad \text{and} \quad \Lambda^3 V^* = \Lambda^3 V = L(\omega_5).$$

Finally $L(\omega_6)$ is the adjoint module. Thus the fundamental E_6-modules are

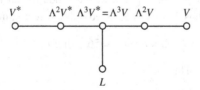

We now consider the simple Lie algebra E_7 and obtain a 56-dimensional fundamental module. The idea is similar to what we have seen for E_6.

Proposition 13.30 (a) *The number of positive roots of E_8 not in E_7 is 57.*
(b) *The subspace V of E_8 spanned by vectors e_α for such roots is a 57-dimensional E_7-module. V decomposes as the direct sum of a 56-dimensional fundamental module with a 1-dimensional module $L(0)$.*

Proof. We see from Section 8.7 that the positive roots of E_8 not in E_7 are

$$\beta_1 - \beta_j \quad j \in \{2, 3, 4, 5, 6, 7\}$$
$$\beta_1 + \beta_j \quad j \in \{2, 3, 4, 5, 6, 7\}$$
$$\beta_i - \beta_8 \quad i \in \{2, 3, 4, 5, 6, 7\}$$
$$-\beta_i - \beta_8 \quad i \in \{2, 3, 4, 5, 6, 7\}$$
$$\beta_1 - \beta_8$$
$$\frac{1}{2} \sum_{i=1}^{8} \varepsilon_i \beta_i \quad \prod \varepsilon_i = 1, \quad \varepsilon_8 = -1, \quad \varepsilon_1 = 1.$$

The number of such roots is 57.

Let V be the subspace of E_8 spanned by the e_α for this set of roots. The argument of Proposition 13.29 shows that V is an E_7-module. Now $\beta_1 - \beta_8$ is orthogonal to all fundamental roots of E_7 and it follows that

$$[e_\alpha e_{\beta_1 - \beta_8}] = 0 \quad \text{for all } \alpha \in \Phi(E_7).$$

Hence $\mathbb{C} e_{\beta_1 - \beta_8}$ is a 1-dimensional E_7-submodule of V. Let V' be the subspace spanned by the remaining e_α. The fact that $\beta_1 - \beta_8$ is orthogonal to all $\alpha \in \Phi(E_7)$ implies that $\beta_1 - \beta_8$ cannot be expressed in the form $\alpha + \beta$ where $\alpha \in \Phi(E_7), \beta \in \Phi^+(E_8)$. This shows that V' is an E_7-submodule of V. Its highest weight is obtained by picking the weight with the highest value of h, and this is $\beta_2 - \beta_8$. In fact the first few highest weights are

$$\beta_2 - \beta_8, \quad \beta_3 - \beta_8, \quad \beta_4 - \beta_8, \quad \beta_5 - \beta_8, \quad \ldots$$

By calculating scalar products $\{,\}$ with the fundamental roots of E_7 we see that $\beta_2 - \beta_8 = \omega_2$. Thus $L(\omega_2)$ is an irreducible direct summand of V'. Since

$$\dim L(\omega_2) = 56 = \dim V'$$

we have $V' = L(\omega_2)$. Thus

$$V = L(\omega_2) \oplus L(0). \qquad \square$$

We can obtain information about some of the other fundamental E_7-modules by considering exterior powers of V'. The highest weight of $\Lambda^2 V'$ is

$$(\beta_2 - \beta_8) + (\beta_3 - \beta_8) = \beta_2 + \beta_3 - 2\beta_8.$$

A calculation of scalar products $\{,\}$ shows that

$$\beta_2 + \beta_3 - 2\beta_8 = \omega_3.$$

13.8 Fundamental modules for exceptional algebras

Thus $\Lambda^2 V'$ contains $L(\omega_3)$ as an irreducible direct summand. But we know that
$$\dim L(\omega_3) = \binom{56}{2} - 1.$$

Thus
$$\Lambda^2 V' = L(\omega_3) \oplus L(0).$$

The highest weight of $\Lambda^3 V'$ is
$$(\beta_2 - \beta_8) + (\beta_3 - \beta_8) + (\beta_4 - \beta_8) = \beta_2 + \beta_3 + \beta_4 - 3\beta_8.$$

We have
$$\beta_2 + \beta_3 + \beta_4 - 3\beta_8 = \omega_4.$$

Thus $L(\omega_4)$ is an irreducible direct summand of $\Lambda^3 V'$. We know that
$$\dim L(\omega_4) = \binom{56}{3} - 56.$$

In fact we have
$$\Lambda^3 V' = L(\omega_4) \oplus L(\omega_2).$$

The highest weight of $\Lambda^4 V'$ is
$$(\beta_2 - \beta_8) + (\beta_3 - \beta_8) + (\beta_4 - \beta_8) + (\beta_5 - \beta_8) = \beta_2 + \beta_3 + \beta_4 + \beta_5 - 4\beta_8.$$

We have
$$\beta_2 + \beta_3 + \beta_4 + \beta_5 - 4\beta_8 = \omega_5.$$

We know that
$$\dim L(\omega_5) = \binom{56}{4} - \binom{56}{2}.$$

In fact it turns out that
$$\Lambda^4 V' = L(\omega_5) \oplus L(\omega_3) \oplus L(0).$$

Some of the remaining fundamental E_7-modules may be identified by means of the adjoint module. The highest root of E_7 is $-\beta_1 - \beta_8$ and we have $-\beta_1 - \beta_8 = \omega_8$. Thus we see that $L(\omega_8)$ is the adjoint E_7-module, since
$$\dim L(\omega_8) = 133 = \dim L.$$

The second highest root of E_7 is $\frac{1}{2}(-\beta_1+\beta_2+\beta_3+\beta_4+\beta_5+\beta_6+\beta_7-\beta_8)$. Thus the highest weight of $\Lambda^2 L$ is

$$(-\beta_1-\beta_8)+\tfrac{1}{2}(-\beta_1+\beta_2+\beta_3+\beta_4+\beta_5+\beta_6+\beta_7-\beta_8)$$
$$=\tfrac{1}{2}(-3\beta_1+\beta_2+\beta_3+\beta_4+\beta_5+\beta_6+\beta_7-3\beta_8).$$

We have

$$\tfrac{1}{2}(-3\beta_1+\beta_2+\beta_3+\beta_4+\beta_5+\beta_6+\beta_7-3\beta_8)=\omega_7.$$

Thus $L(\omega_7)$ is an irreducible direct summand of $\Lambda^2 L$. Since

$$\dim L(\omega_7) = \binom{133}{2} - 1$$

we have

$$\Lambda^2 L = L(\omega_7) \oplus L(0).$$

We next consider the simple Lie algebra E_8. The smallest dimension of a fundamental module for E_8 is

$$\dim L(\omega_1) = 248.$$

The highest root of E_8 is $\beta_1-\beta_8$, and we have $\beta_1-\beta_8=\omega_1$. Since $\dim L = 248$ we deduce that $L(\omega_1) = L$. Thus the fundamental module $L(\omega_1)$ is the adjoint module.

The description of the remaining fundamental modules of E_8 is considerably more complicated than in the other simple Lie algebras. We shall not discuss the details.

We now turn to the simple Lie algebra F_4 and show how to obtain the 26-dimensional fundamental module. This will be done by identifying F_4 with a subalgebra of E_6. We shall retain our previous numbering of the fundamental roots of E_6 given by

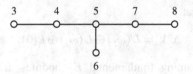

Let σ be the permutation of the vertices given by

$$\sigma = (3\ \ 8)(4\ \ 7)(5)(6).$$

13.8 Fundamental modules for exceptional algebras

Then σ gives a symmetry of the Dynkin diagram of E_6 with $\sigma^2 = 1$. We have

$$A_{\sigma(i)\sigma(j)} = A_{ij} \quad \text{for all } i, j.$$

Thus by Theorem 7.5 there is an automorphism of E_6, which we shall also call σ, satisfying

$$\sigma(e_i) = e_{\sigma(i)}$$
$$\sigma(f_i) = f_{\sigma(i)}$$
$$\sigma(h_i) = h_{\sigma(i)}.$$

Since e_i, f_i, h_i generate the Lie algebra, σ is determined by these conditions, and we have $\sigma^2 = 1$.

We may define a linear map on the real vector space spanned by the simple roots, also denoted by σ, to satisfy

$$\sigma(\alpha_i) = \alpha_{\sigma(i)}.$$

Then we have $\sigma(\Phi) = \Phi$. All the σ-orbits on Φ have size 1 or 2. Examination of the root system of E_6 shows there are 24 orbits of size 1 and 24 of size 2.

Proposition 13.31 *Let L be the simple Lie algebra E_6 and $\sigma : L \to L$ be the automorphism of order 2 given above. Then the subalgebra L^σ of σ-stable elements of L is isomorphic to F_4. The elements*

$$E_1 = e_6 \quad E_2 = e_5 \quad E_3 = e_4 + e_7 \quad E_4 = e_3 + e_8$$
$$F_1 = f_6 \quad F_2 = f_5 \quad F_3 = f_4 + f_7 \quad F_4 = f_3 + f_8$$
$$H_1 = h_6 \quad H_2 = h_5 \quad H_3 = h_4 + h_7 \quad H_4 = h_3 + h_8$$

are standard generators of F_4.

Proof. Let (A_{ij}) be the Cartan matrix of F_4 given by

$$A = \begin{pmatrix} 2 & -1 & 0 & 0 \\ -1 & 2 & -1 & 0 \\ 0 & -2 & 2 & -1 \\ 0 & 0 & -1 & 2 \end{pmatrix}.$$

It is straightforward to check that the elements E_i, F_i, H_i satisfy the relations

$$[H_i H_j] = 0$$
$$[H_i E_j] = A_{ij} E_j$$
$$[H_i F_j] = -A_{ij} F_j$$

$$[E_i F_i] = H_i$$
$$[E_i F_j] = 0 \quad \text{if } i \neq j$$
$$[E_i, \ldots [E_i E_j]] = 0 \quad \text{if } i \neq j$$
$$[F_i, \ldots [F_i F_j]] = 0 \quad \text{if } i \neq j$$

where the last two relations have $1 - A_{ij}$ factors E_i, F_i respectively. By Proposition 7.35 there is a homomorphism

$$\theta : F_4 \to L$$

whose image is the subalgebra generated by the elements E_i, F_i, H_i. Since $\theta \neq 0$ and F_4 is simple the image of θ is isomorphic to F_4.

We shall also show that $\operatorname{im} \theta = L^\sigma$. Since each E_i, F_i, H_i lies in L^σ we have $\operatorname{im} \theta \subset L^\sigma$. On the other hand consider the decomposition

$$L = H \oplus \sum_{\substack{\alpha \in \Phi \\ \sigma(\alpha) = \alpha}} \mathbb{C} e_\alpha \oplus \sum_{\substack{\alpha \in \Phi \\ \sigma(\alpha) \neq \alpha}} \left(\mathbb{C} e_\alpha + \mathbb{C} e_{\sigma(\alpha)} \right).$$

Each direct summand is σ-stable, thus L^σ is the direct sum of the σ-stable subspaces of the components. We have

$$\dim H^\sigma = 4$$
$$\dim (\mathbb{C} e_\alpha)^\sigma \leq 1 \quad \text{if } \sigma(\alpha) = \alpha$$
$$\dim (\mathbb{C} e_\alpha + \mathbb{C} e_{\sigma(\alpha)})^\sigma \leq 1 \quad \text{if } \sigma(\alpha) \neq \alpha.$$

Thus $\dim L^\sigma \leq 4 + 24 + 24 = 52$. But $\dim(\operatorname{im} \theta) = 52$, thus $\operatorname{im} \theta = L^\sigma$. Hence L^σ is isomorphic to F_4. $\square$

Now let V be the 27-dimensional fundamental module $L(\omega_8)$ for E_6 constructed in Proposition 13.29. Then V may be regarded as an F_4-module using our embedding of F_4 in E_6. We label the fundamental roots of F_4 by the diagram

```
  1      2      3      4
  o------o=>=o------o
```

Proposition 13.32 *The F_4-module V decomposes as*

$$V = L(\omega_4) \oplus L(0)$$

where $L(\omega_4)$ is the 26-dimensional fundamental module.

13.8 Fundamental modules for exceptional algebras

Proof. We determine the weights of the F_4-module V. We recall that the weights of V have form

$$\beta_2 - \beta_j \quad 3 \le j \le 7$$
$$\beta_2 + \beta_j \quad 3 \le j \le 7$$
$$-\beta_1 - \beta_8$$
$$\tfrac{1}{2} \sum_{i=1}^{8} \varepsilon_i \beta_i \quad \prod \varepsilon_i = 1, \quad \varepsilon_1 = -1, \quad \varepsilon_2 = 1, \quad \varepsilon_8 = -1.$$

Now the fundamental roots of E_8 are

$$\alpha_i = \beta_i - \beta_{i+1} \quad i = 1, \ldots, 6$$
$$\alpha_7 = \beta_6 + \beta_7$$
$$\alpha_8 = -\tfrac{1}{2} \sum_{i=1}^{8} \beta_i.$$

Also $\alpha_j(h_i) = A_{ij}$ $i, j \in \{1, \ldots, 8\}$ where (A_{ij}) is the Cartan matrix of E_8. It follows that the numbers $\beta_j(h_i)$ $i, j \in \{1, \ldots, 8\}$ are given by

$$\beta_i(h_i) = 1 \quad i = 1, \ldots, 7$$
$$\beta_{i+1}(h_i) = -1 \quad i = 1, \ldots, 6$$
$$\beta_i(h_8) = -\tfrac{1}{2} \quad i = 1, \ldots, 8$$
$$\beta_i(h_j) = 0 \quad \text{otherwise.}$$

Let H_1, H_2, H_3, H_4 be the fundamental coroots of F_4 defined above and $\omega_1, \omega_2, \omega_3, \omega_4$ the corresponding fundamental weights of F_4. Then $\omega_i(H_j) = \delta_{ij}$. By calculating the values $\beta_i(H_j)$ we deduce

$$\beta_1 = \beta_2 = \beta_8 = -\tfrac{1}{2}\omega_4$$
$$\beta_3 \qquad = \tfrac{1}{2}\omega_4$$
$$\beta_4 \qquad = \omega_3 - \tfrac{3}{2}\omega_4$$
$$\beta_5 \qquad = \omega_2 - \omega_3 - \tfrac{1}{2}\omega_4$$
$$\beta_6 \qquad = \omega_1 - \omega_2 + \omega_3 - \tfrac{1}{2}\omega_4$$
$$\beta_7 \qquad = -\omega_1 + \omega_3 - \tfrac{1}{2}\omega_4$$

when the β_i are regarded as weights for F_4. Hence the 27 weights of the F_4-module V are

$$\pm\{\omega_4, \omega_1-\omega_3, \omega_1-\omega_4, \omega_2-\omega_3, \omega_3-\omega_4, \omega_3-2\omega_4, \omega_1-\omega_2+\omega_3, \omega_1-\omega_2$$
$$+\omega_4, \omega_1-\omega_3+\omega_4, \omega_2-\omega_3-\omega_4, \omega_2-2\omega_3+\omega_4, \omega_1-\omega_2+\omega_3-\omega_4\}$$
$$\cup\{0,0,0\}.$$

The only dominant weight among these, excluding 0, is ω_4. Thus V has highest weight ω_4 and so $L(\omega_4)$ is an irreducible direct summand of V. Since

$$\dim V = 27, \quad \dim L(\omega_4) = 26$$

we have

$$V = L(\omega_4) \oplus L(0). \qquad \square$$

Using the relation $\alpha_i = \sum_j A_{ji}\omega_j$ in F_4 we see that

$$\omega_4 = \alpha_1 + 2\alpha_2 + 3\alpha_3 + 2\alpha_4.$$

This is the highest short root of F_4. All short roots of F_4 are transforms of this one under elements of the Weyl group W. Thus all 24 short roots of F_4 are weights of $L(\omega_4)$. So the weights of $L(\omega_4)$ are the 24 short roots together with 0 with multiplicity 2.

We now discuss the other fundamental modules for F_4. We first consider $L(\omega_1)$. The relations $\alpha_i = \sum_j A_{ji}\omega_j$ for F_4 show that

$$\omega_1 = 2\alpha_1 + 3\alpha_2 + 4\alpha_3 + 2\alpha_4.$$

We recall from Section 8.6 that

$$\alpha_1 = \beta_1 - \beta_2 \quad \alpha_2 = \beta_2 - \beta_3 \quad \alpha_3 = \beta_3 \quad \alpha_4 = \tfrac{1}{2}(-\beta_1 - \beta_2 - \beta_3 + \beta_4)$$

and so

$$\omega_1 = 2\alpha_1 + 3\alpha_2 + 4\alpha_3 + 2\alpha_4 = \beta_1 + \beta_4.$$

The long roots of F_4 have form $\pm\beta_i \pm \beta_j$ and, since $\beta_4 \succ \beta_1 \succ \beta_2 \succ \beta_3$, $\beta_1 + \beta_4$ is the highest root. Thus ω_1 is the highest root of F_4 and $L(\omega_1)$ is therefore the adjoint F_4-module.

The remaining fundamental modules $L(\omega_2), L(\omega_3)$ for F_4 satisfy

$$\dim L(\omega_2) = \binom{52}{2} - 52$$

$$\dim L(\omega_3) = \binom{26}{2} - 52.$$

13.8 Fundamental modules for exceptional algebras

It can be shown that $L(\omega_2), L(\omega_3)$ appear as irreducible direct summands of $\Lambda^2 L(\omega_1), \Lambda^2 L(\omega_4)$ respectively, and that

$$\Lambda^2 L(\omega_1) = L(\omega_1) \oplus L(\omega_2)$$
$$\Lambda^2 L(\omega_4) = L(\omega_1) \oplus L(\omega_3).$$

Finally we consider the simple Lie algebra G_2 and show how to obtain the 7-dimensional fundamental module. We do this by identifying G_2 with a subalgebra of D_4. The fundamental roots of D_4 will be numbered as in the diagram

Let σ be the permutation of the vertices given by

$$\sigma = (1 \ \ 3 \ \ 4)(2).$$

σ gives a symmetry of the Dynkin diagram with $\sigma^3 = 1$. Since

$$A_{\sigma(i)\sigma(j)} = A_{ij} \quad \text{for all } i, j$$

there exists by Theorem 7.5 an automorphism σ of D_4 satisfying

$$\sigma(e_i) = e_{\sigma(i)}$$
$$\sigma(f_i) = f_{\sigma(i)}$$
$$\sigma(h_i) = h_{\sigma(i)}.$$

We may also define a linear map σ on the vector space spanned by the simple roots, satisfying $\sigma(\alpha_i) = \alpha_{\sigma(i)}$. We have $\sigma(\Phi) = \Phi$. These are 6 σ-orbits of size 1 on Φ and 6 orbits of size 3.

Proposition 13.33 *Let L be the simple Lie algebra D_4 and $\sigma : L \to L$ be the automorphism of order 3 given above. Then the subalgebra L^σ of σ-stable elements of L is isomorphic to G_2.*

The elements

$$E_1 = e_2 \qquad E_2 = e_1 + e_3 + e_4$$
$$F_1 = f_2 \qquad F_2 = f_1 + f_3 + f_4$$
$$H_1 = h_2 \qquad H_2 = h_1 + h_3 + h_4$$

are standard generators of G_2.

Proof. The idea is the same as that for F_4 in E_6. The Cartan matrix of G_2 is

$$A = \begin{pmatrix} 2 & -1 \\ -3 & 2 \end{pmatrix}.$$

It is again straightforward to check that the elements $E_1, E_2, F_1, F_2, H_1, H_2$ satisfy the defining relations

$$[H_i H_j] = 0$$
$$[H_i E_j] = A_{ij} E_j$$
$$[H_i F_j] = -A_{ij} F_j$$
$$[E_i F_i] = H_i$$
$$[E_i F_j] = 0 \quad \text{if } i \neq j$$
$$[E_i, \ldots [E_i E_j]] = 0 \quad \text{if } i \neq j$$
$$[F_i, \ldots [F_i F_j]] = 0 \quad \text{if } i \neq j$$

where the last two relations have $1 - A_{ij}$ factors E_i, F_i respectively.

Thus by Proposition 7.35 there is a homomorphism $\theta : G_2 \to L$. The image im θ is isomorphic to G_2. We show im $\theta = L^\sigma$. Since E_i, F_i, H_i lie in L^σ we have im $\theta \subset L^\sigma$. Now consider the decomposition

$$L = H \oplus \sum_{\substack{\alpha \in \Phi \\ \sigma(\alpha) = \alpha}} \mathbb{C} e_\alpha \oplus \sum_{\substack{\alpha \in \Phi \\ \sigma(\alpha) \neq \alpha}} \left(\mathbb{C} e_\alpha + \mathbb{C} e_{\sigma(\alpha)} + \mathbb{C} e_{\sigma^2(\alpha)} \right).$$

Each direct summand is σ-stable, thus L^σ is the direct sum of the σ-stable subspaces of the components. We have

$$\dim H^\sigma = 2$$
$$\dim (\mathbb{C} e_\alpha)^\sigma \leq 1 \quad \text{if } \sigma(\alpha) = \alpha$$
$$\dim \left(\mathbb{C} e_\alpha + \mathbb{C} e_{\sigma(\alpha)} + \mathbb{C} e_{\sigma^2(\alpha)} \right)^\sigma \leq 1 \quad \text{if } \sigma(\alpha) \neq \alpha.$$

Thus $\dim L^\sigma \leq 2 + 6 + 6 = 14$. But $\dim(\text{im } \theta) = 14$, thus im $\theta = L^\sigma$. Hence L^σ is isomorphic to G_2. $\square$

Proposition 13.34 *Let V be the 8-dimensional natural D_4-module. Regard V as a G_2-module using the above embedding of G_2 in D_4. Then*

$$V = L(\omega_2) \oplus L(0)$$

where $L(\omega_2)$ is the 7-dimensional fundamental G_2-module.

13.8 Fundamental modules for exceptional algebras

Proof. We recall from Section 8.2 that in this 8-dimensional representation we have

$$e_1 = E_{12} - E_{-2-1}, \quad e_2 = E_{23} - E_{-3-2}, \quad e_3 = E_{34} - E_{-4-3}, \quad e_4 = E_{3-4} - E_{4-3}$$
$$f_1 = -E_{-1-2} + E_{21}, \quad f_2 = -E_{-2-3} + E_{32}, \quad f_3 = -E_{-3-4} + E_{43},$$
$$f_4 = -E_{-34} + E_{-43}.$$

Hence

$$h_1 = E_{11} - E_{22} - E_{-1-1} + E_{-2-2}$$
$$h_2 = E_{22} - E_{33} - E_{-2-2} + E_{-3-3}$$
$$h_3 = E_{33} - E_{44} - E_{-3-3} + E_{-4-4}$$
$$h_4 = E_{33} - E_{-4-4} - E_{-3-3} + E_{44}$$

and so

$$H_1 = E_{22} - E_{33} - E_{-2-2} + E_{-3-3}$$
$$H_2 = E_{11} - E_{22} + 2E_{33} - E_{-1-1} + E_{-2-2} - 2E_{-3-3}.$$

Let $v_1, v_2, v_3, v_4, v_{-1}, v_{-2}, v_{-3}, v_{-4}$ be the natural basis of V. Let ω_1, ω_2 be the fundamental weights for G_2. Since $\omega_i(H_j) = \delta_{ij}$ these basis vectors span weight spaces with weights

$$\omega_2, \quad \omega_1 - \omega_2, \quad -\omega_1 + 2\omega_2, \quad 0, \quad -\omega_2, \quad -\omega_1 + \omega_2, \quad \omega_1 - 2\omega_2, \quad 0$$

respectively. The highest weight is ω_2, thus $L(\omega_2)$ is an irreducible direct summand of V. We have

$$\dim V = 8, \quad \dim L(\omega_2) = 7$$

and so

$$V = L(\omega_2) \oplus L(0).$$

We note that $\omega_2 = \alpha_1 + 2\alpha_2$ is the highest short root of G_2. All short roots are transforms of this root by elements of the Weyl group, thus all six short roots are weights of $L(\omega_2)$. Thus the weights of $L(\omega_2)$ are the short roots together with 0.

Now we have

$$E_1 = E_{23} - E_{-3-2}, \quad E_2 = E_{12} + E_{34} + E_{3-4} - E_{-2-1} - E_{-4-3} - E_{4-3}$$
$$F_1 = -E_{-2-3} + E_{32}, \quad F_2 = -E_{-1-2} - E_{-3-4} - E_{-34} + E_{21} + E_{43} + E_{-43}.$$

It may be checked that the vector $v_4 - v_{-4}$ is annihilated by E_1, E_2, F_1, F_2 and so spans the 1-dimensional submodule $L(0)$. □

Finally we consider the other fundamental G_2-module $L(\omega_1)$. The relations $\alpha_i = \sum_j A_{ji} \omega_j$ show that

$$\omega_1 = 2\alpha_1 + 3\alpha_2.$$

This is the highest root of G_2. Therefore the fundamental module $L(\omega_1)$ is the 14-dimensional adjoint module.

14

Generalised Cartan matrices and Kac–Moody algebras

In 1967 V. G. Kac and R. V. Moody independently initiated the study of certain Lie algebras $L(A)$ associated with a generalised Cartan matrix A. An $n \times n$ matrix $A = (A_{ij})$ is called a **generalised Cartan matrix** if it satisfies the conditions

$$A_{ii} = 2 \quad \text{for } i = 1, \ldots, n$$

$$A_{ij} \in \mathbb{Z} \quad \text{and} \quad A_{ij} \leq 0 \quad \text{if } i \neq j$$

$$A_{ij} = 0 \quad \text{implies} \quad A_{ji} = 0.$$

The Cartan matrix of any finite dimensional simple Lie algebra is a generalised Cartan matrix, as shown in Section 6.4. We shall see that, in the special case when A is a Cartan matrix, the Lie algebra $L(A)$ constructed by Kac and Moody coincides with the finite dimensional simple Lie algebra with Cartan matrix A. However, the Lie algebra $L(A)$ can in general be infinite dimensional.

The term 'generalised Cartan matrix' will be abbreviated to GCM. The Lie algebra $L(A)$ associated to a GCM A will be called the Kac–Moody algebra associated to A. We shall explain the definition and some of the basic properties of $L(A)$ in the present chapter. In fact the introductory ideas do not use the fact that A is a GCM – we shall assume initially that A is any $n \times n$ matrix over $\mathbb{C}$.

14.1 Realisations of a square matrix

Let A be an $n \times n$ matrix over $\mathbb{C}$. A **realisation** of A is a triple $(H, \Pi, \Pi^\vee)$ where:

H is a finite dimensional vector space over $\mathbb{C}$

$\Pi^\vee = \{h_1, \ldots, h_n\}$ is a linearly independent subset of H

$\Pi = \{\alpha_1, \ldots, \alpha_n\}$ is a linearly independent subset of H^*

$\alpha_j(h_i) = A_{ij}$ for all i, j.

Proposition 14.1 *If (H, Π, Π^v) is a realisation of A then $\dim H \geq 2n - \operatorname{rank} A$.*

Proof. Let $\operatorname{rank} A = l$ and $\dim H = m$. We extend the set Π^v to give a basis $h_1, \ldots, h_m$ of H and extend Π to give a basis $\alpha_1, \ldots, \alpha_m$ of H^*. Consider the $m \times m$ matrix $(\alpha_j(h_i))$. This is non-singular so its rows are linearly independent. Thus the $n \times m$ matrix given by the first n rows has rank n. This matrix therefore has n linearly independent columns. Now the leading $n \times n$ submatrix is A, so has rank l. Thus the remaining $n \times (m - n)$ matrix has rank at least $n - l$. It follows that $m - n \geq n - l$, that is $m \geq 2n - l$. □

Definition *A **minimal realisation** of A is a realisation in which*

$$\dim H = 2n - \operatorname{rank} A.$$

Proposition 14.2 *Any $n \times n$ matrix over $\mathbb{C}$ has a minimal realisation.*

Proof. Since $\operatorname{rank} A = l$, A has a non-singular $l \times l$ submatrix. By reordering the rows and columns we obtain a matrix

$$\begin{matrix} l \\ n-l \end{matrix} \begin{pmatrix} A_{11} & A_{12} \\ A_{21} & A_{22} \end{pmatrix} \\ l n-l$$

in which A_{11} is non-singular. Let

$$C = \begin{pmatrix} A_{11} & A_{12} & O \\ A_{21} & A_{22} & I_{n-l} \\ O & I_{n-l} & O \end{pmatrix} \begin{matrix} l \\ n-l \\ n-l \end{matrix} \\ l n-l n-l.$$

Since $\det C = \pm \det A_{11} \neq 0$ we see that C is a non-singular $(2n - l) \times (2n - l)$ matrix. Let H be the vector space of all $(2n - l)$-tuples over $\mathbb{C}$. Define $\alpha_1, \ldots, \alpha_n \in H^*$ to be the first n coordinate functions

$$(\lambda_1, \ldots, \lambda_{2n-l}) \to \lambda_i \quad i = 1, \ldots, n.$$

Define $h_1, \ldots, h_n \in H$ to be the first n row vectors of C. Then $\alpha_1, \ldots, \alpha_n$ and $h_1, \ldots, h_n$ are linearly independent and we obtain a realisation of

$$\begin{pmatrix} A_{11} & A_{12} \\ A_{21} & A_{22} \end{pmatrix}$$

14.1 Realisations of a square matrix

with $\dim H = 2n - l$. By reordering $\alpha_1, \ldots, \alpha_n$ and $h_1, \ldots, h_n$ appropriately we obtain a minimal realisation of A. $\square$

Now let (H, Π, Π^v) and $(H', \Pi', (\Pi')^v)$ be two realisations of A. We say the realisations are isomorphic if there is an isomorphism of vector spaces

$$\phi : H \to H'$$

such that $\phi(h_i) = h'_i$ and $\phi^*(\alpha'_i) = \alpha_i$ where

$$\phi^* : (H')^* \to H^*$$

is the isomorphism induced by ϕ.

Proposition 14.3 *Any two minimal realisations of an $n \times n$ matrix A over $\mathbb{C}$ are isomorphic.*

Proof. Let (H, Π, Π^v) be the minimal realisation of A constructed in Proposition 14.2 and $(H', \Pi', (\Pi')^v)$ be another minimal realisation. We reorder the rows and columns of A as before to obtain

$$\begin{pmatrix} A_{11} & A_{12} \\ A_{21} & A_{22} \end{pmatrix}$$

where A_{11} is non-singular.

We complete $h'_1, \ldots, h'_n$ to a basis $h'_1, \ldots, h'_{2n-l}$ of H'. Then the matrix $(\alpha'_j(h'_i))$ for $i = 1, \ldots, 2n-l$; $j = 1, \ldots, n$ has form

$$\begin{pmatrix} A_{11} & A_{12} \\ A_{21} & A_{22} \\ B_1 & B_2 \end{pmatrix}.$$

Since $\alpha'_1, \ldots, \alpha'_n$ are linearly independent this matrix has rank n. Thus it has n linearly independent rows. Since rows $l+1, \ldots, n$ are linear combinations of rows $1, \ldots, l$ the matrix

$$\begin{pmatrix} A_{11} & A_{12} \\ B_1 & B_2 \end{pmatrix} \begin{matrix} l \\ n-l \end{matrix}$$
$$\quad\; l \quad\; n-l$$

must have linearly independent rows, so is non-singular.

We now extend $\alpha'_1, \ldots, \alpha'_n$ to $\alpha'_1, \ldots, \alpha'_{2n-l}$ so that the $(2n-l) \times (2n-l)$ matrix $(\alpha'_j(h'_i))$ is

$$\begin{pmatrix} A_{11} & A_{12} & O \\ A_{21} & A_{22} & I_{n-l} \\ B_1 & B_2 & O \end{pmatrix} \begin{matrix} l \\ n-l \\ n-l \end{matrix}$$
$$\begin{matrix} l & n-l & n-l. \end{matrix}$$

This matrix is non-singular, thus $\alpha'_1, \ldots, \alpha'_{2n-l}$ are a basis for $(H')^*$.

Since A_{11} is non-singular, by adding suitable linear combinations of the first l rows to the last $n-l$ rows we may achieve $B_1 = O$. Thus it is possible to choose $h'_{n+1}, \ldots, h'_{2n-l}$ so that $h'_1, \ldots, h'_{2n-l}$ are a basis of H' and

$$(\alpha'_j(h'_i)) = \begin{pmatrix} A_{11} & A_{12} & O \\ A_{21} & A_{22} & I_{n-l} \\ O & B'_2 & O \end{pmatrix}.$$

The matrix B'_2 must be non-singular since the whole matrix is non-singular.

We now make a further change to $h'_{n+1}, \ldots, h'_{2n-l}$ equivalent to left multiplying the above matrix by

$$\begin{pmatrix} I_l & O & O \\ O & I_{n-l} & O \\ O & O & (B'_2)^{-1} \end{pmatrix}.$$

Then we obtain

$$(\alpha'_j(h'_i)) = \begin{pmatrix} A_{11} & A_{12} & O \\ A_{21} & A_{22} & I_{n-l} \\ O & I_{n-l} & O \end{pmatrix}.$$

This is equal to the matrix C above. Thus the map $h_i \to h'_i$ gives an isomorphism $H \to H'$ which induces the isomorphism $(H')^* \to H^*$ given by $\alpha'_j \to \alpha_j$. This shows that the realisations $(H, \Pi, \Pi^\vee)$ and $(H', \Pi', (\Pi')^\vee)$ are isomorphic.

14.2 The Lie algebra $\tilde{L}(A)$ associated with a complex matrix

Let A be an $n \times n$ matrix over $\mathbb{C}$ with rank l. Let $(H, \Pi, \Pi^\vee)$ be a minimal realisation of A. Then we have

$$\dim H = 2n - l$$
$$\Pi^\vee = \{h_1, \ldots, h_n\} \subset H, \quad \Pi = \{\alpha_1, \ldots, \alpha_n\} \subset H^*$$
$$\alpha_j(h_i) = A_{ij}$$

14.2 The Lie algebra $\tilde{L}(A)$ associated with a complex matrix

We define a Lie algebra $\tilde{L}(A)$ by generators and relations.
Let $X = \{e_1, \ldots, e_n, f_1, \ldots, f_n, \tilde{x} \text{ for all } x \in H\}$ and let R be the following set of Lie words in X:

$$\tilde{x} - \lambda \tilde{y} - \mu \tilde{z} \quad \text{for all } x, y, z \in H, \quad \lambda, \mu \in \mathbb{C} \text{ with } x = \lambda y + \mu z$$

$$[\tilde{x}\tilde{y}] \quad \text{for all } x, y \in H$$

$$[e_i f_i] - \tilde{h}_i \quad \text{for } i = 1, \ldots, n$$

$$[e_i f_j] \quad \text{for all } i \neq j$$

$$[\tilde{x} e_i] - \alpha_i(x) e_i \quad \text{for all } x \in H \text{ and } i = 1, \ldots, n$$

$$[\tilde{x} f_i] + \alpha_i(x) f_i \quad \text{for all } x \in H \text{ and } i = 1, \ldots, n.$$

We define $\tilde{L}(A) = L(X ; R)$ to be the Lie algebra generated by the elements X subject to relations R.

Lemma 14.4 *If a different minimal realisation of A is chosen the Lie algebra $\tilde{L}(A)$ is the same up to isomorphism.*

Proof. This follows from Proposition 14.3. □

We note that if A is a Cartan matrix then $\tilde{L}(A)$ is the Lie algebra investigated earlier in Section 7.4 and Example 9.13. For in this case A is non-singular and H is the vector space with basis $h_i = [e_i f_i]$.

Proposition 14.5 *There is an automorphism $\tilde{\omega}$ of $\tilde{L}(A)$ uniquely determined by*

$$\tilde{\omega}(e_i) = -f_i, \quad \tilde{\omega}(f_i) = -e_i, \quad \tilde{\omega}(\tilde{x}) = -\tilde{x}$$

for all $x \in H$. Also $\tilde{\omega}^2 = 1$.

Proof. There is a map $\tilde{\omega} : X \to FL(X)$ given by the above formulae. By Proposition 9.9 there is a unique Lie algebra homomorphism $FL(X) \to FL(X)$ extending this map. We shall denote this map also by $\tilde{\omega}$. It satisfies $\tilde{\omega}^2 = 1$. Let $\langle R \rangle$ be the ideal of $FL(X)$ generated by the above set R of Lie words. By applying $\tilde{\omega}$ to the elements of R we see that $\tilde{\omega}(\langle R \rangle) \subset \langle R \rangle$. Thus we may define the induced map

$$\tilde{\omega} : FL(X)/\langle R \rangle \to FL(X)/\langle R \rangle.$$

Since $\tilde{\omega}^2 = 1$, $\tilde{\omega}$ is an automorphism of $\tilde{L}(A)$. □

Let $\tilde{H}$ be the subalgebra of $\tilde{L}(A)$ generated by the elements $\tilde{x}$ for all $x \in H$. Let $\tilde{N}$ be the subalgebra generated by $e_1, \ldots, e_n$ and $\tilde{N}^-$ the subalgebra generated by $f_1, \ldots, f_n$. Then we have

$$\tilde{\omega}(\tilde{H}) = \tilde{H}, \quad \tilde{\omega}(\tilde{N}) = \tilde{N}^-, \quad \tilde{\omega}(\tilde{N}^-) = \tilde{N}.$$

Now let V be an n-dimensional vector space over $\mathbb{C}$ with basis $v_1, \ldots, v_n$ and let

$$T(V) = \bigoplus_{s \geq 0} T^s(V)$$

be the tensor algebra of V. Thus $T^s(V)$ has basis

$$v_{i_1} \otimes \cdots \otimes v_{i_s} = v_{i_1} \ldots v_{i_s}$$

for all $i_1, \ldots, i_s \in \{1, \ldots, n\}$. For each linear map $\lambda \in H^*$ we define a map

$$\theta_\lambda : X \to \text{End } T(V).$$

It is sufficient to define the effect of these endomorphisms on the basis elements of $T(V)$. $T^0(V)$ has basis 1. We define

$$\theta_\lambda(\tilde{x}) \cdot 1 = \lambda(x) 1$$
$$\theta_\lambda(\tilde{x}) \cdot (v_{i_1} \ldots v_{i_s}) = (\lambda - \alpha_{i_1} - \cdots - \alpha_{i_s})(x) v_{i_1} \ldots v_{i_s}$$

for $x \in H$.

$$\theta_\lambda(f_j) \cdot 1 = v_j$$
$$\theta_\lambda(f_j) \cdot (v_{i_1} \ldots v_{i_s}) = v_j v_{i_1} \ldots v_{i_s}.$$

We define $\theta_\lambda(e_j)$ by induction on s as follows

$$\theta_\lambda(e_j) \cdot 1 = 0$$
$$\theta_\lambda(e_j) \cdot v_i = \delta_{ij} \lambda(h_j) 1$$
$$\theta_\lambda(e_j) \cdot (v_{i_1} \ldots v_{i_s}) = v_{i_1} (\theta_\lambda(e_j)(v_{i_2} \ldots v_{i_s}))$$
$$+ \delta_{ij}(\lambda - \alpha_{i_2} - \cdots - \alpha_{i_s})(h_j) v_{i_2} \ldots v_{i_s} \quad s > 1.$$

Proposition 14.6 *The above map* $\theta_\lambda : X \to \text{End } T(V)$ *can be extended to a Lie algebra homomorphism* $\tilde{L}(A) \to [\text{End } T(V)]$.

Proof. The idea of the proof is essentially the same as in Proposition 7.9. θ_λ can first be extended to a homomorphism

$$\theta_\lambda : FL(X) \to [\text{End } T(V)]$$

14.2 The Lie algebra $\tilde{L}(A)$ associated with a complex matrix

by Proposition 9.9. We have

$$\tilde{L}(A) \cong FL(X)/\langle R \rangle$$

and so in order to show that θ_λ induces a homomorphism $\tilde{L}(A) \to [\text{End } T(V)]$ we must verify that $\theta_\lambda(r) = 0$ for all $r \in R$.

The elements of R have form

$$\tilde{x} - \lambda \tilde{y} - \mu \tilde{z}$$
$$[\tilde{x}\tilde{y}]$$
$$[e_i f_i] - \tilde{h}_i$$
$$[e_i f_j] \quad i \neq j$$
$$[\tilde{x} e_i] - \alpha_i(x) e_i$$
$$[\tilde{x} f_i] + \alpha_i(x) f_i.$$

The relation $\theta_\lambda(r) = 0$ may be checked for each such $r \in R$ in a straightforward manner, just as in the proof of Proposition 7.9. □

Corollary 14.7 *The map $x \to \tilde{x}$ is an isomorphism of vector spaces $H \to \tilde{H}$.*

Proof. $\tilde{H}$ is the subalgebra of $\tilde{L}(A)$ generated by $\tilde{x}$ for all $x \in H$. However, these elements form a Lie algebra since

$$\tilde{x}_1 + \tilde{x}_2 = \widetilde{x_1 + x_2}$$
$$\lambda \tilde{x} = \widetilde{\lambda x}$$
$$[\tilde{x}_1 \tilde{x}_2] = 0.$$

Thus $\tilde{H} = \{\tilde{x} \; ; \; x \in H\}$.

Consider the map $H \to \tilde{H}$ given by $x \to \tilde{x}$. This is a homomorphism of Lie algebras. It is surjective. To show it is an isomorphism we must show it is also injective. Thus suppose $x \in H$ and $\tilde{x} = 0$. Then $\theta_\lambda(\tilde{x}) = 0$. Thus $\lambda(x) = 0$. Since this holds for all $\lambda \in H^*$ we may deduce that $x = 0$. □

We next consider the restriction of θ_λ to $\tilde{N}^-$. It is clear from the definition that this is independent of λ. We call it

$$\theta : \tilde{N}^- \to [\text{End } T(V)].$$

Now $\theta(f_i)$ is left multiplication by v_i. Thus, for any Lie word $w(f_1, \ldots, f_n)$ in $f_1, \ldots, f_n$, $\theta(w(f_1, \ldots, f_n))$ is left multiplication by $w(v_1, \ldots, v_n)$.

Proposition 14.8 $f_1, \ldots, f_n$ *generate* $\tilde{N}^-$ *freely, and so* $\tilde{N}^-$ *is isomorphic to* $FL(f_1, \ldots, f_n)$.

Proof. Define $\phi : \tilde{N}^- \to [T(V)]$ by $\phi(w) = \theta(w) \cdot 1$. Thus

$$\phi(w(f_1, \ldots, f_n)) = w(v_1, \ldots, v_n).$$

Then ϕ is a Lie algebra homomorphism, since

$$\phi[w(f_1, \ldots, f_n), w'(f_1, \ldots, f_n)] = [w(v_1, \ldots, v_n), w'(v_1, \ldots, v_n)]$$
$$= [\phi(w(f_1, \ldots, f_n)), \phi(w'(f_1, \ldots, f_n))].$$

Now $T(V) = F(v_1, \ldots, v_n)$, the free associative algebra on $v_1, \ldots, v_n$. Thus the free Lie algebra $FL(v_1, \ldots, v_n)$ lies in $[T(V)]$ and consists of all Lie words in $v_1, \ldots, v_n$. Thus $FL(v_1, \ldots, v_n)$ is the image of ϕ. Hence the homomorphism

$$\phi : \tilde{N}^- \to FL(v_1, \ldots, v_n)$$

is surjective. But there is a Lie algebra homomorphism

$$\phi' : FL(v_1, \ldots, v_n) \to \tilde{N}^-$$

with $\phi'(v_i) = f_i$. Moreover we have $\phi \circ \phi' = 1$ on $FL(v_1, \ldots, v_n)$ and $\phi' \circ \phi = 1$ on $\tilde{N}^-$. Thus ϕ, ϕ' are inverse isomorphisms and $\tilde{N}^-$ is isomorphic to $FL(f_1, \ldots, f_n)$. $\square$

Corollary 14.9 $e_1, \ldots, e_n$ *generate* $\tilde{N}$ *freely.*

Proof. Apply the automorphism $\tilde{w}$ of Proposition 14.5. We have $\tilde{w}(\tilde{N}^-) = \tilde{N}$ and $\tilde{w}(f_i) = -e_i$. Thus the result follows from Proposition 14.8. $\square$

Proposition 14.10 $\tilde{L}(A) = \tilde{N}^- \oplus \tilde{H} \oplus \tilde{N}$, *a direct sum of subspaces.*

Proof. The proof is similar to that of Proposition 7.12. We show that $I = \tilde{N}^- + \tilde{H} + \tilde{N}$ is an ideal of $\tilde{L}(A)$. It is sufficient to show that

$$\operatorname{ad} e_i \cdot I \subset I, \quad \operatorname{ad} f_i \cdot I \subset I, \quad \operatorname{ad} \tilde{x} \cdot I \subset I.$$

14.2 The Lie algebra $\tilde{L}(A)$ associated with a complex matrix

Since the defining relations show that

$$\operatorname{ad} e_i \cdot \tilde{H} \subset \tilde{N}, \quad \operatorname{ad} e_i \cdot \tilde{N} \subset \tilde{N}$$
$$\operatorname{ad} f_i \cdot \tilde{H} \subset \tilde{N}^-, \quad \operatorname{ad} f_i \cdot \tilde{N}^- \subset \tilde{N}^-$$
$$\operatorname{ad} \tilde{x} \cdot \tilde{H} = O, \quad \operatorname{ad} \tilde{x} \cdot \tilde{N} \subset \tilde{N}, \quad \operatorname{ad} \tilde{x} \cdot \tilde{N}^- \subset \tilde{N}^-$$

it is sufficient to check that

$$\operatorname{ad} f_i \cdot \tilde{N} \subset \tilde{H} + \tilde{N}$$
$$\operatorname{ad} e_i \cdot \tilde{N}^- \subset \tilde{H} + \tilde{N}^-.$$

We have

$$\operatorname{ad} f_i \cdot e_j = \delta_{ij} \tilde{h}_i \in \tilde{H} + \tilde{N}.$$

Suppose $w_1, w_2 \in \tilde{N}$ satisfy

$$\operatorname{ad} f_i \cdot w_1 \in \tilde{H} + \tilde{N}, \quad \operatorname{ad} f_i \cdot w_2 \in \tilde{H} + \tilde{N}.$$

Then

$$\operatorname{ad} f_i [w_1 w_2] = [\operatorname{ad} f_i \cdot w_1, w_2] + [w_1, \operatorname{ad} f_i \cdot w_2] \in \tilde{H} + \tilde{N}.$$

Thus $\operatorname{ad} f_i \cdot \tilde{N} \subset \tilde{H} + \tilde{N}$.

The relation $\operatorname{ad} e_i \cdot \tilde{N}^- \subset \tilde{H} + \tilde{N}^-$ follows similarly. Thus I is an ideal of $\tilde{L}(A)$ containing all the generators, and so $\tilde{L}(A) = \tilde{N}^- + \tilde{H} + \tilde{N}$.

In order to show the sum is direct we verify that if $w_- \in \tilde{N}^-, \tilde{x} \in \tilde{H}, w \in \tilde{N}$ satisfy

$$w_- + \tilde{x} + w = 0$$

then we have $w_- = 0, \tilde{x} = 0, w = 0$. Thus suppose $w_- + \tilde{x} + w = 0$. Then $\theta_\lambda(w_- + \tilde{x} + w)$ is the zero endomorphism of $T(V)$. In particular $\theta_\lambda(w_- + \tilde{x} + w) \cdot 1 = 0$. Now $\theta_\lambda(w_-) \cdot 1 = \phi(w_-)$, $\theta_\lambda(\tilde{x}) \cdot 1 = \lambda(x) 1$ and $\theta_\lambda(w) \cdot 1 = 0$. Hence

$$\phi(w_-) + \lambda(x) 1 = 0.$$

Now $\phi(w_-) \in \bigoplus_{s \geq 1} T^s(V)$ and $\lambda(x) 1 \in T^0(V)$. It follows that $\phi(w_-) = 0$ and $\lambda(x) 1 = 0$, that is $\lambda(x) = 0$. Since this holds for all $\lambda \in H^*$ we have $x = 0$. Hence $\tilde{x} = 0$.

Now $\phi : \tilde{N}^- \to FL(v_1, \ldots, v_n)$ is an isomorphism, and so $\phi(w_-) = 0$ implies $w_- = 0$. Finally $w_- + \tilde{x} + w = 0$ implies $w = 0$. Thus

$$\tilde{L}(A) = \tilde{N}^- \oplus \tilde{H} \oplus \tilde{N}.$$

$\square$

Let Q be the subgroup of H^* given by $Q = \{\alpha = k_1\alpha_1 + \cdots + k_n\alpha_n \ ; \ k_1, \ldots, k_n \in \mathbb{Z}\}$. Let $Q^+ = \{\alpha \neq 0 \in Q \ ; \ k_i \geq 0 \text{ for all } i\}$ and $Q^- = \{\alpha \neq 0 \in Q \ ; \ k_i \leq 0 \text{ for all } i\}$. For each $\alpha \in Q$ let

$$\tilde{L}_\alpha = \{y \in \tilde{L}(A) \ ; \ [\tilde{x}y] = \alpha(x)y \quad \text{for all } x \in H\}.$$

Proposition 14.11 (i) $\tilde{L}(A) = \oplus_{\alpha \in Q} \tilde{L}_\alpha$
(ii) $\dim \tilde{L}_\alpha$ is finite for all $\alpha \in Q$.
(iii) $\tilde{L}_0 = \tilde{H}$.
(iv) If $\alpha \neq 0$ then $\tilde{L}_\alpha = 0$ unless $\alpha \in Q^+$ or $\alpha \in Q^-$.
(v) $[\tilde{L}_\alpha \tilde{L}_\beta] \subset \tilde{L}_{\alpha+\beta}$ for all $\alpha, \beta \in Q$.

Proof. To show $\tilde{L}(A) = \sum_{\alpha \in Q} \tilde{L}_\alpha$ it is sufficient to show $\tilde{H} \subset \sum_{\alpha \in Q} \tilde{L}_\alpha$, $\tilde{N} \subset \sum_{\alpha \in Q} \tilde{L}_\alpha$, $\tilde{N}^- \subset \sum_{\alpha \in Q} \tilde{L}_\alpha$. It is clear that $\tilde{H} \subset \tilde{L}_0$. To show that $\tilde{N} \subset \sum_{\alpha \in Q^+} \tilde{L}_\alpha$ we observe that each Lie monomial w in $e_1, \ldots, e_n$ satisfies $[\tilde{x}w] = \alpha(x)w$ for all $x \in H$ and some $\alpha \in Q^+$. For

$$[\tilde{x}e_i] = \alpha_i(x)e_i$$

and if

$$[\tilde{x}w_1] = \beta(x)w_1, \quad [\tilde{x}w_2] = \gamma(x)w_2$$

we have

$$[\tilde{x}[w_1 w_2]] = (\beta + \gamma)(x)[w_1 w_2].$$

This shows $\tilde{N} \subset \sum_{\alpha \in Q^+} \tilde{L}_\alpha$ and similarly we have $\tilde{N}^- \subset \sum_{\alpha \in Q^-} \tilde{L}_\alpha$. Thus $\tilde{L}(A) = \sum_{\alpha \in Q} \tilde{L}_\alpha$.

In order to show that the sum is direct we show that

$$v_1 + \cdots + v_k = 0$$

for $v_i \in \tilde{L}_{\beta_i}$ with $\beta_1, \ldots, \beta_k$ distinct implies each $v_i = 0$. Suppose this is false. Choose the minimal value of k for which it is false. Suppose $v_1 + \cdots + v_k = 0$ for this value of k but that not each $v_i = 0$. Then

$$[\tilde{x}, v_1 + \cdots + v_k] = 0 \quad \text{for all } x \in H.$$

Thus

$$\beta_1(x)v_1 + \cdots + \beta_k(x)v_k = 0.$$

We also have

$$\beta_k(x)v_1 + \cdots + \beta_k(x)v_k = 0.$$

14.2 The Lie algebra $\tilde{L}(A)$ associated with a complex matrix

Hence

$$(\beta_1(x) - \beta_k(x))v_1 + \cdots + (\beta_{k-1}(x) - \beta_k(x))v_{k-1} = 0.$$

By the minimality of k we have

$$(\beta_i(x) - \beta_k(x))v_i = 0 \quad \text{for } i = 1, \ldots, k-1.$$

Since $\beta_i \neq \beta_k$ there exists $x \in H$ with $\beta_i(x) \neq \beta_k(x)$. Hence $v_i = 0$ for $i = 1, \ldots, k-1$. It follows that $v_k = 0$. This contradicts our assumption. Hence

$$\tilde{L}(A) = \bigoplus_{\alpha \in Q} \tilde{L}_\alpha.$$

Since $\tilde{L}(A) = \tilde{N}^- \oplus \tilde{H} \oplus \tilde{N}$ by Proposition 14.10 and

$$\tilde{N}^- \subset \sum_{\alpha \in Q^-} \tilde{L}_\alpha, \quad \tilde{H} \subset \tilde{L}_0, \quad \tilde{N} \subset \sum_{\alpha \in Q^+} \tilde{L}_\alpha$$

it follows that

$$\tilde{H} = \tilde{L}_0, \quad \tilde{N} = \sum_{\alpha \in Q^+} \tilde{L}_\alpha, \quad \tilde{N}^- = \sum_{\alpha \in Q^-} \tilde{L}_\alpha.$$

Also we have $\tilde{L}_\alpha = 0$ if $\alpha \neq 0$, $\alpha \notin Q^+$, $\alpha \notin Q^-$. The Jacobi identity shows that $[\tilde{L}_\alpha \tilde{L}_\beta] \subset \tilde{L}_{\alpha+\beta}$ for all $\alpha, \beta \in Q$.

Finally we show dim $\tilde{L}_\alpha$ is finite. We have dim $\tilde{L}_0 = 2n - l$. So let $\alpha \in Q^+$. Then $\tilde{L}_\alpha \subset \tilde{N}$. Now $\tilde{N}$ is spanned by Lie monomials in $e_1, \ldots, e_n$ and each Lie monomial lies in some $\tilde{L}_\alpha$. Let $\alpha = k_1 \alpha_1 + \cdots + k_n \alpha_n$ with $k_i \in \mathbb{Z}$ and $k_i \geq 0$. A Lie monomial lies in $\tilde{L}_\alpha$ if and only if e_i appears k_i times in it for each i. But there are only finitely many Lie monomials in which e_i appears k_i times for each i. Thus dim $\tilde{L}_\alpha$ is finite. A similar argument proves this when $\alpha \in Q^-$. We note in particular that

$$\dim \tilde{L}_{\alpha_i} = 1, \quad \dim \tilde{L}_{-\alpha_i} = 1$$
$$\dim \tilde{L}_{k\alpha_i} = 0, \quad \dim \tilde{L}_{-k\alpha_i} = 0 \quad \text{if } k > 1. \qquad \square$$

The following lemma will be needed in the proof of the next proposition.

Lemma 14.12 *Let H be a finite dimensional abelian Lie algebra and V be an H-module such that*

$$V = \bigoplus_{\lambda \in H^*} V_\lambda$$

where $V_\lambda = \{v \in V \ ; \ xv = \lambda(x)v \text{ for all } x \in H\}$. *Let U be a submodule of V. Then*

$$U = \bigoplus_{\lambda \in H^*} (U \cap V_\lambda).$$

Proof. Let $u \in U$. Then $u = u_1 + \cdots + u_m$ where $u_i \in V_{\lambda_i}$ and $\lambda_1, \ldots, \lambda_m$ are distinct elements of H^*. Let

$$H_{ij} = \{x \in H \ ; \ \lambda_i(x) = \lambda_j(x)\} \qquad \text{for } i \neq j.$$

H_{ij} is a subspace of H of codimension 1. Now $H \neq \bigcup_{i \neq j} H_{ij}$ since a finite dimensional vector space over $\mathbb{C}$ cannot be the union of finitely many proper subspaces. So we can find $x \in H$ with $\lambda_1(x), \ldots, \lambda_m(x)$ all distinct.

Let $\theta(x) : V \to V$ be the linear map given by $\theta(x)v = xv$. Then we have

$$u = u_1 + \cdots + u_m$$
$$\theta(x)u = \lambda_1(x)u_1 + \cdots + \lambda_m(x)u_m$$
$$\theta(x)^2 u = \lambda_1(x)^2 u_1 + \cdots + \lambda_m(x)^2 u_m$$
$$\vdots$$
$$\theta(x)^{m-1} u = \lambda_1(x)^{m-1} u_1 + \cdots + \lambda_m(x)^{m-1} u_m.$$

We have here m equations in $u_1, \ldots, u_m$ whose coefficients have non-zero determinant. Thus $u_1, \ldots, u_m$ may be expressed as linear combinations of $u, \theta(x)u, \theta(x)^2 u, \ldots, \theta(x)^{m-1} u$. These vectors all lie in U. Thus $u_i \in U \cap V_{\lambda_i}$. Thus we have shown that $U = \sum_{\lambda \in H^*} (U \cap V_\lambda)$ and the sum is direct because $\sum_{\lambda \in H^*} V_\lambda$ is a direct sum. $\square$

Proposition 14.13 *The algebra $\tilde{L}(A)$ contains a unique ideal I maximal with respect to $I \cap \tilde{H} = O$.*

Proof. Let J be any ideal of $\tilde{L}(A)$ with $J \cap \tilde{H} = O$. We have

$$\tilde{L}(A) = \bigoplus_{\alpha \in H^*} \tilde{L}_\alpha$$

by Proposition 14.11, and we consider $\tilde{L}(A)$ as an $\tilde{H}$-module. By Lemma 14.12 we have

$$J = \bigoplus_{\alpha \in H^*} (\tilde{L}_\alpha \cap J).$$

14.3 The Kac–Moody algebra $L(A)$

Now each $\tilde{L}_\alpha$ with $\alpha \neq 0$ lies in $\tilde{N}$ or in $\tilde{N}^-$. Thus
$$J = (\tilde{N}^- \cap J) \oplus (\tilde{N} \cap J).$$
In particular $J \subset \tilde{N}^- \oplus \tilde{N}$.

Now consider the ideal I of $\tilde{L}(A)$ generated by all ideals J with $J \cap \tilde{H} = O$. All such ideals J lie in $\tilde{N}^- \oplus \tilde{N}$, thus I lies in $\tilde{N}^- \oplus \tilde{N}$. Hence $I \cap \tilde{H} = O$. Thus I is the unique ideal of $\tilde{L}(A)$ maximal with respect to $I \cap \tilde{H} = O$. □

14.3 The Kac–Moody algebra $L(A)$

We now suppose that A is a GCM. Let $\tilde{L}(A)$ be the Lie algebra associated with A defined in Section 14.2 and I be the unique maximal ideal of $\tilde{L}(A)$ with $I \cap \tilde{H} = O$. Let $L(A)$ be defined by
$$L(A) = \tilde{L}(A)/I.$$
The Lie algebra $L(A)$ is called the **Kac–Moody algebra** with GCM A. We have a natural homomorphism $\theta : \tilde{L}(A) \to L(A)$. We define $N = \theta(\tilde{N})$ and $N^- = \theta(\tilde{N}^-)$.

Proposition 14.14 $L(A) = N^- \oplus \theta(\tilde{H}) \oplus N$. *Moreover the map* $\theta : \tilde{H} \to \theta(\tilde{H})$ *is an isomorphism.*

Proof. We know from the proof of Proposition 14.13 that
$$I = (\tilde{N}^- \cap I) \oplus (\tilde{N} \cap I).$$
Since $\tilde{L}(A) = \tilde{N}^- \oplus \tilde{H} \oplus \tilde{N}$ it follows that
$$L(A) = N^- \oplus \theta(\tilde{H}) \oplus N$$
and that $\theta : \tilde{H} \to \theta(\tilde{H})$ is an isomorphism. □

We recall from Corollary 14.7 that there is a natural isomorphism $H \to \tilde{H}$. Combining this with θ we obtain an isomorphism $H \to \theta(\tilde{H})$. We shall subsequently use this isomorphism to identify $\theta(\tilde{H})$ with H, and we shall write
$$L(A) = N^- \oplus H \oplus N.$$

In order to show that a given Lie algebra is isomorphic to $L(A)$ the following result is often useful.

Proposition 14.15 *Suppose we are given an $n \times n$ GCM $A = (A_{ij})$. Let L be a Lie algebra over $\mathbb{C}$ and H be a finite dimensional abelian subalgebra of L*

with $\dim H = 2n - \operatorname{rank} A$. Suppose $\Pi = \{\alpha_1, \ldots, \alpha_n\}$ is a linearly independent subset of H^* and $\Pi^v = \{h_1, \ldots, h_n\}$ a linearly independent subset of H satisfying $\alpha_j(h_i) = A_{ij}$.

Suppose also that $e_1, \ldots, e_n, f_1, \ldots, f_n$ are elements of L satisfying

$$[e_i f_i] = h_i$$
$$[e_i f_j] = 0 \quad \text{if } i \neq j$$
$$[x e_i] = \alpha_i(x) e_i \quad \text{for } x \in H$$
$$[x f_i] = -\alpha_i(x) f_i \quad \text{for } x \in H$$

Suppose that $e_1, \ldots, e_n, f_1, \ldots, f_n$ and H generate L and that L has no non-zero ideal J with $J \cap H = O$. Then L is isomorphic to the Kac–Moody algebra $L(A)$.

Proof. The elements $e_1, \ldots, e_n, f_1, \ldots, f_n$ and $x \in H$ generate L and satisfy all the defining relations of $\tilde{L}(A)$ given in Section 14.2. Thus there is a surjective Lie algebra homomorphism $\theta : \tilde{L}(A) \to L$ and L is isomorphic to $\tilde{L}(A)/\ker \theta$. The restriction map $\theta : \tilde{H} \to H$ is an isomorphism by Corollary 14.7, thus $\ker \theta \cap \tilde{H} = O$. It follows that $\ker \theta \subset I$, the largest ideal of $\tilde{L}(A)$ with $I \cap \tilde{H} = O$. In fact we have $\ker \theta = I$ since L has no non-zero ideal J with $J \cap H = O$. Hence

$$L \cong \tilde{L}(A)/I = L(A).$$

Corollary 14.16 *If A is a Cartan matrix then $L(A)$ is the finite dimensional semisimple Lie algebra with Cartan matrix A.*

Proof. In this case we have $\operatorname{rank} A = n$, so $\dim H = n$. The finite dimensional semisimple Lie algebra satisfies all the hypotheses of Proposition 14.15, so is isomorphic to the Kac–Moody algebra $L(A)$. □

This result shows that the theory of Kac–Moody algebras is an extension of the theory of finite dimensional semisimple Lie algebras, which we have already described.

We shall now describe some further basic properties of the Kac–Moody algebra $L(A)$. We shall denote the images of $e_i, h_i, f_i \in \tilde{L}(A)$ under the natural homomorphism $\tilde{L}(A) \to L(A)$ by $e_i, h_i, f_i \in L(A)$. This should not lead to confusion as we shall subsequently be concentrating on $L(A)$ rather than $\tilde{L}(A)$.

14.3 The Kac–Moody algebra L(A)

Proposition 14.17 *There is an automorphism ω of $L(A)$ satisfying $\omega^2 = 1$ determined by*

$$\omega(e_i) = -f_i, \quad \omega(f_i) = -e_i$$
$$\omega(x) = -x \quad \text{for all } x \in H.$$

Proof. By Proposition 14.5 $\tilde{L}(A)$ has an automorphism $\tilde{\omega}$ with $\tilde{\omega}^2 = 1$. Thus $\tilde{\omega}(I)$ is the unique maximal ideal with

$$\tilde{\omega}(I) \cap \tilde{\omega}(\tilde{H}) = 0.$$

But $\tilde{\omega}(\tilde{H}) = \tilde{H}$ so $\tilde{\omega}(I) = I$. Thus $\tilde{\omega}$ induces an automorphism ω of $\tilde{L}(A)/I = L(A)$ satisfying the stated conditions. $\square$

There is also an analogue of Proposition 14.11. For each $\alpha \in Q$ define L_α by

$$L_\alpha = \{ y \in L(A); \ [xy] = \alpha(x)y \quad \text{for all } x \in H \}.$$

Proposition 14.18 (i) $L(A) = \bigoplus_{\alpha \in Q} L_\alpha$
(ii) $\dim L_\alpha$ *is finite for all* $\alpha \in Q$.
(iii) $L_0 = H$
(iv) *If* $\alpha \neq 0$ *then* $L_\alpha = 0$ *unless* $\alpha \in Q^+$ *or* $\alpha \in Q^-$.
(v) $[L_\alpha L_\beta] \subset L_{\alpha+\beta}$ *for all* $\alpha, \beta \in Q$.

Proof. Let $\theta : \tilde{L}(A) \to L(A) = \tilde{L}(A)/I$ be the natural homomorphism. We have

$$\tilde{L}(A) = \bigoplus_{\alpha \in Q} \tilde{L}_\alpha \quad \text{by Proposition 14.11}.$$

Also

$$I = \bigoplus_{\alpha \in Q} (I \cap \tilde{L}_\alpha) \quad \text{by Lemma 14.12}.$$

It follows that

$$L(A) = \bigoplus_{\alpha \in Q} \theta(\tilde{L}_\alpha).$$

Now we clearly have $\theta(\tilde{L}_\alpha) \subset L_\alpha$, thus $L(A) = \sum_{\alpha \in Q} L_\alpha$. This sum is direct, just as in the proof of Proposition 14.11. It follows that $L(A) = \bigoplus_{\alpha \in Q} L_\alpha$ and that $L_\alpha = \theta(\tilde{L}_\alpha)$. Now

$$L(A) = N^- \oplus H \oplus N \quad \text{by Proposition 14.14}$$

and $N^- \subset \sum_{\alpha \in Q^-} L_\alpha$, $H \subset L_0$, $N \subset \sum_{\alpha \in Q^+} L_\alpha$, hence we have

$$N^- = \bigoplus_{\alpha \in Q^-} L_\alpha, \quad H = L_0, \quad N = \bigoplus_{\alpha \in Q^+} L_\alpha.$$

$\dim L_\alpha$ is finite because $L_\alpha = \theta(\tilde{L}_\alpha)$ and $\dim \tilde{L}_\alpha$ is finite. Finally $[L_\alpha L_\beta] \subset L_{\alpha+\beta}$ follows from the Jacobi identity. $\square$

Definitions H will be called a **Cartan subalgebra** of $L(A)$. This fits in with our previous terminology when A was a Cartan matrix. An element $\alpha \in H^*$ is called a **root** of $L(A)$ if $\alpha \neq 0$ and $L_\alpha \neq O$. Every root lies in Q^+ or Q^-. The roots in Q^+ are called **positive roots** and those in Q^- **negative roots**. If α is a root then L_α is called the **root space** of α. The dimension of L_α is called the **multiplicity** of α. When A is a Cartan matrix we recall that all roots have multiplicity 1. However, we shall see that this is not always the case when A is a GCM.

Proposition 14.19 (i) $\dim L_{\alpha_i} = 1$ and $\dim L_{-\alpha_i} = 1$.
(ii) If $k > 1$ then $\dim L_{k\alpha_i} = 0$, $\dim L_{-k\alpha_i} = 0$.

Proof. Since $L_{\alpha_i} = \theta(\tilde{L}_{\alpha_i})$ and $\dim \tilde{L}_{\alpha_i} = 1$ we have $\dim L_{\alpha_i} \leq 1$. If $\dim L_{\alpha_i} = 0$ we would have $e_i \in I = \ker \theta$. This would imply $[e_i f_i] = \tilde{h}_i \in I$, contrary to $I \cap \tilde{H} = O$. Thus $\dim L_{\alpha_i} = 1$. A similar argument gives $\dim L_{-\alpha_i} = 1$.

Since $\tilde{L}_{k\alpha_i} = O$ and $\tilde{L}_{-k\alpha_i} = O$ for $k > 1$ it follows that $L_{k\alpha_i} = O$ and $L_{-k\alpha_i} = O$. $\square$

$\alpha_1, \alpha_2, \ldots, \alpha_n$ are called the **fundamental roots** of $L(A)$, again in agreement with the earlier terminology when A is a Cartan matrix.

Remark 14.20 For a general $n \times n$ matrix A over $\mathbb{C}$ we constructed a minimal realisation $(H, \Pi, \Pi^\vee)$ where H is a vector space over $\mathbb{C}$ of dimension $2n - \text{rank } A$, $\Pi^\vee = \{h_1, \ldots, h_n\}$ is a linearly independent subset of H and $\Pi = \{\alpha_1, \ldots, \alpha_n\}$ is a linearly independent subset of H^* such that $\alpha_j(h_i) = A_{ij}$.

In the case when A is a GCM the matrix A is real and so we can find a real vector space $H_\mathbb{R}$, of dimension $2n - \text{rank } A$ over $\mathbb{R}$, contained in H such that $h_1, \ldots, h_n$ lie in $H_\mathbb{R}$ and are linearly independent and $\alpha_1, \ldots, \alpha_n$, when restricted to $H_\mathbb{R}^*$, remain linearly independent. In the construction of H, described in Proposition 14.2 as the vector space of all $(2n - l)$-tuples over $\mathbb{C}$, we define $H_\mathbb{R}$ as the subset of all $(2n - l)$-tuples over $\mathbb{R}$. The triple $(H_\mathbb{R}, \Pi, \Pi^\vee)$ with $\Pi^\vee \subset H_\mathbb{R}$ and $\Pi \subset H_\mathbb{R}^*$ is called a **real minimal realisation** of A.

14.3 The Kac–Moody algebra $L(A)$

We denote by $L(A)'$ the subalgebra of $L(A)$ generated by $e_1, \ldots, e_n$, $f_1, \ldots, f_n$.

Proposition 14.21 (i) L_α lies in $L(A)'$ for each root α of $L(A)$.
(ii) $L(A)' = (H \cap L(A)') \oplus \sum_{\alpha \neq 0} L_\alpha$.
(iii) $L(A)' = [L(A) L(A)]$.

Proof. We know from Proposition 14.18 that $L_\alpha \neq 0$ implies $\alpha \in Q^+$ or $\alpha \in Q^-$. If $\alpha \in Q^+$ then $L_\alpha \subset N$ and if $\alpha \in Q^-$ then $L_\alpha \subset N^-$. Since N is the subalgebra generated by $e_1, \ldots, e_n$ and N^- is the subalgebra generated by $f_1, \ldots, f_n$ we have $L_\alpha \subset L(A)'$ for each α.

Since $L(A) = H \oplus \sum_{\alpha \neq 0} L_\alpha$ and $L_\alpha \subset L(A)'$ we have

$$L(A)' = (H \cap L(A)') \oplus \sum_{\alpha \neq 0} L_\alpha.$$

It follows that $L(A) = L(A)' + H$. We also have $[H, L(A)'] \subset L(A)'$ and so $L(A)'$ is an ideal of $L(A)$. We have

$$L(A)/L(A)' \cong H / H \cap L(A)'$$

and so $L(A)/L(A)'$ is abelian. Hence $[L(A)L(A)] \subset L(A)'$. On the other hand we have $[e_i f_i] = h_i$, $[h_i e_i] = 2e_i$, $[h_i f_i] = -2f_i$ and so $e_i, f_i \in [L(A)L(A)]$. Thus $L(A)' \subset [L(A)L(A)]$ and we have equality.

15
The classification of generalised Cartan matrices

The structure of the Kac–Moody algebra $L(A)$ depends crucially on the GCM A. In the present chapter we shall discuss various possible types of GCM A which can occur.

15.1 A trichotomy for indecomposable GCMs

Two GCMs A, A' are called **equivalent** if they have the same degree n and there is a permutation σ of $1, \ldots, n$ such that

$$A'_{ij} = A_{\sigma(i)\sigma(j)} \quad \text{for all } i, j.$$

A GCM A is called **indecomposable** if it is not equivalent to a diagonal sum

$$\begin{pmatrix} A_1 & O \\ O & A_2 \end{pmatrix}$$

of smaller GCMs A_1, A_2. If A is a GCM so is its transpose A^t. Moreover A is indecomposable if and only if A^t is indecomposable.

We shall now define three particular types of GCM. Let $v = (v_1, \ldots, v_n)$ be a vector in $\mathbb{R}^n$. We write $v \geq 0$ if $v_i \geq 0$ for each i, and $v > 0$ if $v_i > 0$ for each i.

Definitions *A GCM A has finite type if*

(i) $\det A \neq 0$
(ii) *there exists $u > 0$ with $Au > 0$*
(iii) *$Au \geq 0$ implies $u > 0$ or $u = 0$.*

15.1 A trichotomy for indecomposable GCMs

The GCM A has **affine type** if

(i) corank $A = 1$ (i.e. rank $A = n-1$)
(ii) there exists $u > 0$ such that $Au = 0$
(iii) $Au \geq 0$ implies $Au = 0$.

The GCM A has **indefinite type** if

(i) there exists $u > 0$ such that $Au < 0$
(ii) $Au \geq 0$ and $u \geq 0$ imply $u = 0$.

All vectors u in these definitions are assumed to lie in $\mathbb{R}^n$, and are column vectors.

We aim to prove the following theorem.

Theorem 15.1 *Let A be an indecomposable GCM. Then exactly one of the following three possibilities holds*:

(a) *A has finite type*
(b) *A has affine type*
(c) *A has indefinite type.*

Moreover the type of A^t is the same as the type of A.

This section will be devoted to the proof of Theorem 15.1, which gives a trichotomy on the set of indecomposable GCMs.

We begin with a lemma on inequalities.

Lemma 15.2 *Let $v^i = (v_{i1}, \ldots, v_{in}) \in \mathbb{R}^n$ for $i = 1, \ldots, m$. Then there exist $x_1, \ldots, x_n \in \mathbb{R}$ with $\sum_{j=1}^n v_{ij} x_j > 0$ for $i = 1, \ldots, m$ if and only if $\lambda_1 v^1 + \cdots + \lambda_m v^m = 0$, $\lambda_i \geq 0$ implies $\lambda_i = 0$ for $i = 1, \ldots, m$.*

Proof. Suppose there exists a column vector $x = (x_1, \ldots, x_n)^t$ satisfying $v^i x > 0$ for all i. Suppose $\lambda_1 v^1 + \cdots + \lambda_m v^m = 0$ with all $\lambda_i \geq 0$. Then $\lambda_1 v^1 x + \cdots + \lambda_m v^m x = 0$. But $v^i x > 0$ and $\lambda_i \geq 0$, thus we have $\lambda_i = 0$ for all i.

Conversely suppose $\lambda_1 v^1 + \cdots + \lambda_m v^m = 0$, $\lambda_i \geq 0$ implies $\lambda_i = 0$ for all i. Let

$$S = \left\{ \sum_{i=1}^m \lambda_i v^i \; ; \; \lambda_i \geq 0, \; \sum_{i=1}^m \lambda_i = 1 \right\}.$$

Define $f : S \to \mathbb{R}$ by $f(y) = \|y\|$ where $\|y\| = \sqrt{y_1^2 + \cdots + y_n^2}$. Then S is a compact subset of $\mathbb{R}^n$ and f is a continuous function from S to $\mathbb{R}$. Thus $f(S)$ is a compact subset of $\mathbb{R}$. Hence there exists $x \in S$ with $\|x\| \leq \|x'\|$ for all $x' \in S$. Clearly $x \neq 0$ since the zero vector does not lie in S. We shall show

$v^i x > 0$ for all i as required. In fact we shall show that $(y, x) > 0$ for all $y \in S$, where $(y, x) = \sum y_i x_i$. This implies the required result since each v^i lies in S.

Now S is a convex subset of $\mathbb{R}^n$. We assume $y \neq x$, then $ty + (1-t)x \in S$ for all t with $0 \leq t \leq 1$. By the choice of x we have

$$(ty + (1-t)x, \quad ty + (1-t)x) \geq (x, x)$$

that is

$$t(y - x, y - x) + 2(y - x, x) \geq 0$$

for $0 < t \leq 1$. This implies $(y - x, x) \geq 0$, that is $(y, x) \geq (x, x) > 0$. □

We make use of this lemma in the following proposition.

Proposition 15.3 *Let M be an $m \times n$ matrix over $\mathbb{R}$. Suppose*

$$u \geq 0 \text{ and } M^t u \geq 0 \text{ imply } u = 0.$$

Then there exists $v > 0$ with $Mv < 0$.

Proof. Let $M = (m_{ij})$ and consider the following system of inequalities:

$$-\sum_{j=1}^{n} m_{ij} x_j > 0 \qquad i = 1, \ldots, m$$

$$x_j > 0 \qquad j = 1, \ldots, n.$$

We shall use Lemma 15.2 to show that these inequalities have a solution. Thus we consider an equation of form

$$\sum_{i=1}^{m} \lambda_i (-m_{i1}, \ldots, -m_{in}) + \sum_{j=1}^{n} \mu_j (0, \ldots, 1, \ldots, 0) = 0$$

with $\lambda_i \geq 0$, $\mu_j \geq 0$ for all i, j. Then

$$\sum_{i=1}^{m} \lambda_i m_{ij} = \mu_j.$$

Let $u = (\lambda_1, \ldots, \lambda_m)^t$. Then $M^t u = (\mu_1, \ldots, \mu_n)^t$. Thus we have $u \geq 0$ and $M^t u \geq 0$. This implies that $u = 0$. We also have $M^t u = 0$. Thus $\lambda_i = 0$ and $\mu_j = 0$ for all i, j. Hence Lemma 15.2 shows that the above inequalities have a solution. Thus there exists $v > 0$ with $Mv < 0$. □

15.1 A trichotomy for indecomposable GCMs

We now consider our three classes of GCM A. Let

$$S_F = \{A \, ; \, A \text{ has finite type}\}$$
$$S_A = \{A \, ; \, A \text{ has affine type}\}$$
$$S_I = \{A \, ; \, A \text{ has indefinite type}\}.$$

It is easy to see that no GCM can lie in more than one of these classes.

Lemma 15.4 $S_F \cap S_A = \phi$, $S_F \cap S_I = \phi$, $S_A \cap S_I = \phi$.

Proof. If $A \in S_F \cap S_A$ then $\det A \neq 0$ and corank $A = 1$, a contradiction.

If $A \in S_F \cap S_I$ there exists $u > 0$ with $Au > 0$. But $Au \geq 0$ and $u \geq 0$ imply $u = 0$, a contradiction.

If $A \in S_A \cap S_I$ there exists $u > 0$ with $Au = 0$. But $Au \geq 0$ and $u \geq 0$ imply $u = 0$, a contradiction. □

We must therefore show that each indecomposable GCM lies in one of the three classes.

Lemma 15.5 *Let A be an indecomposable GCM. Then $u \geq 0$ and $Au \geq 0$ imply that $u > 0$ or $u = 0$.*

Proof. Suppose $u \neq 0$ and $u \not> 0$. Then we can reorder $1, \ldots, n$ so that $u_i = 0$ for $i = 1, \ldots, s$ and $u_i > 0$ for $i = s+1, \ldots, n$. Let

$$A = \begin{pmatrix} P & Q \\ R & S \end{pmatrix} \begin{matrix} s \\ n-s \end{matrix}$$
$$\phantom{A = \begin{pmatrix}}\,s \;\; n-s$$

Now all entries of the block Q are ≤ 0 and if Q has an entry < 0 then Au has a negative coefficient, which is impossible. Thus $Q = 0$. This implies $R = 0$ by the definition of a GCM, thus A is decomposable, a contradiction. □

Now let A be an indecomposable GCM and define K_A by

$$K_A = \{u \, ; \, Au \geq 0\}.$$

K_A is a convex cone. We consider its intersection with the convex cone $\{u \, ; \, u \geq 0\}$. We shall distinguish between two cases:

$$\{u \, ; \, u \geq 0, Au \geq 0\} \neq \{0\}$$
$$\{u \, ; \, u \geq 0, Au \geq 0\} = \{0\}.$$

The first of these cases splits into two subcases, as is shown by the next lemma.

Lemma 15.6 *Suppose* $\{u\,;\, u \geq 0, Au \geq 0\} \neq \{0\}$. *Then just one of the following cases occurs*:

$$K_A \subset \{u\,;\, u > 0\} \cup \{0\}$$

$$K_A = \{u\,;\, Au = 0\} \quad \text{and} \quad K_A \text{ is a 1-dimensional subspace of } \mathbb{R}^n.$$

Proof. We know there exists $u \neq 0$ with $u \geq 0$ and $Au \geq 0$. By Lemma 15.5 this implies that $u > 0$. Suppose the first case does not hold. Then there exists $v \neq 0$ with $Av \geq 0$ such that some coordinate of v is ≤ 0. If $v \geq 0$ then $v > 0$ by Lemma 15.5, thus some coordinate of v is < 0.

We have $Au \geq 0$ and $Av \geq 0$, hence $A(tu + (1-t)v) \geq 0$ for $0 \leq t \leq 1$. Since all coordinates of u are positive and some coordinate of v is negative there exists t with $0 < t < 1$ such that $tu + (1-t)v \geq 0$ and some coordinate of $tu + (1-t)v$ is 0. But then $tu + (1-t)v = 0$ by a further use of Lemma 15.5. Thus v is a scalar multiple of u. We also have

$$0 = A(tu + (1-t)v) = tAu + (1-t)Av.$$

Since $Au \geq 0$, $Av \geq 0$ this implies that $Au = 0$, $Av = 0$.

Now let $w \in K_A$. Then $Aw \geq 0$. Either $w \geq 0$ or some coordinate of w is negative. If $w \geq 0$ then $w > 0$ or $w = 0$ by Lemma 15.5. Suppose $w > 0$. Then by the above argument with u replaced by w, v is a scalar multiple of w. Hence w is a scalar multiple of u. Now suppose some coordinate of w is negative. Then by the above argument with v replaced by w, w is a scalar multiple of u. Thus in all cases w is a scalar multiple of u. Hence K_A is the 1-dimensional subspace $\mathbb{R}u$. Thus we have shown that K_A is a 1-dimensional subspace of $\mathbb{R}^n$. We have also shown that if $w \in K_A$ then $Aw = 0$. Thus $K_A = \{w\,;\, Aw = 0\}$.

Thus if the first case does not hold the second case must hold. We note finally that the two cases cannot hold together since in the first case K_A cannot contain a 1-dimensional subspace of $\mathbb{R}^n$. □

We now identify the first case in Lemma 15.6 with the case of matrices of finite type.

Proposition 15.7 *Let A be an indecomposable GCM. Then the following conditions are equivalent*:

A has finite type

$$\{u\,;\, u \geq 0 \text{ and } Au \geq 0\} \neq \{0\} \quad \text{and} \quad K_A \subset \{u\,;\, u > 0\} \cup \{0\}.$$

15.1 A trichotomy for indecomposable GCMs

Proof. Suppose A has finite type. Then there exists $u > 0$ with $Au > 0$. Hence $\{u;\ u \geq 0$ and $Au \geq 0\} \neq \{0\}$. Also $\det A \neq 0$. Thus $\{u;\ Au = 0\}$ is not a 1-dimensional subspace of $\mathbb{R}^n$. Hence $K_A \subset \{u;\ u > 0\} \cup \{0\}$ by Lemma 15.6.

Conversely suppose $\{u;\ u \geq 0$ and $Au \geq 0\} \neq \{0\}$ and $K_A \subset \{u;\ u > 0\} \cup \{0\}$. Then there cannot exist $u \neq 0$ with $Au = 0$. For this would give a 1-dimensional subspace contained in K_A. Thus $\det A \neq 0$. Now there exists $u \neq 0$ with $u \geq 0$ and $Au \geq 0$. By Lemma 15.5 we have $u > 0$. If $Au > 0$, A has finite type. So suppose to the contrary that some coordinates of Au are zero and some are non-zero. We choose the numbering $1, \ldots, n$ so that the first s components of Au are 0 and the last $n - s$ are positive. Let

$$A = \begin{pmatrix} P & Q \\ R & S \end{pmatrix} \begin{matrix} s \\ n-s \end{matrix}$$
$$s n-s$$

Now the block Q satisfies $Q \neq O$ since A is indecomposable. We choose the numbering so that the first row of Q is not the zero vector. We have

$$Au = \begin{pmatrix} P & Q \\ R & S \end{pmatrix} \begin{pmatrix} u^1 \\ u^2 \end{pmatrix} = \begin{pmatrix} Pu^1 + Qu^2 \\ Ru^1 + Su^2 \end{pmatrix}.$$

Hence $Pu^1 + Qu^2 = 0$ and $Ru^1 + Su^2 > 0$. We also have $u^1 > 0$, $u^2 > 0$. Thus $Qu^2 \leq 0$ and the first coordinate of Qu^2 is < 0. Hence $Pu^1 \geq 0$ and the first coordinate of Pu^1 is > 0. Since $Ru^1 + Su^2 > 0$ we can choose $\varepsilon > 0$ such that $R(1+\varepsilon)u^1 + Su^2 > 0$.

We now consider, instead of our original vector $u = \begin{pmatrix} u^1 \\ u^2 \end{pmatrix}$, the vector $\begin{pmatrix} (1+\varepsilon)u^1 \\ u^2 \end{pmatrix}$. We have

$$\begin{pmatrix} (1+\varepsilon)u^1 \\ u^2 \end{pmatrix} > 0$$

$$A\begin{pmatrix} (1+\varepsilon)u^1 \\ u^2 \end{pmatrix} = \begin{pmatrix} Pu^1 + Qu^2 + \varepsilon Pu^1 \\ Ru^1 + Su^2 + \varepsilon Ru^1 \end{pmatrix} = \begin{pmatrix} \varepsilon Pu^1 \\ R(1+\varepsilon)u^1 + Su^2 \end{pmatrix}.$$

The first coordinate and the last $n - s$ coordinates of this vector are positive and the remaining coordinates are ≥ 0. Thus

$$A\begin{pmatrix} (1+\varepsilon)u^1 \\ u^2 \end{pmatrix} \geq 0$$

and the number of non-zero coordinates in this vector is greater than that in Au. We may now iterate this process, obtaining at each stage at least one more non-zero coordinate than we had before. We eventually obtain a vector $v > 0$ such that $Av > 0$. Thus A has finite type. $\square$

We next identify the second case in Lemma 15.6 with that of an affine GCM.

Proposition 15.8 *Let A be an indecomposable GCM. Then the following conditions are equivalent*:
(i) *A has affine type*
(ii) *$\{u;\ u \geq 0$ and $Au \geq 0\} \neq \{0\}$, $K_A = \{u;\ Au = 0\}$, and K_A is a 1-dimensional subspace of $\mathbb{R}^n$.*

Proof. Suppose A has affine type. Then there exists $u > 0$ with $Au = 0$. It follows that $\{u;\ u \geq 0$ and $Au \geq 0\} \neq \{0\}$. Also $\lambda u \in K_A$ for all $\lambda \in \mathbb{R}$. By Lemma 15.6 we see that $K_A = \{w;\ Aw = 0\}$ and that K_A is a 1-dimensional subspace of $\mathbb{R}^n$.

Conversely suppose the three conditions of (ii) are satisfied.

Then corank $A = 1$.

Also there exists $u \neq 0$ with $u \geq 0$ and $Au \geq 0$. By Lemma 15.5 we have $u > 0$. So there exists $u > 0$ with $Au \geq 0$. But $K_A = \{u;\ Au = 0\}$. Hence there exists $u > 0$ with $Au = 0$. Finally $Au \geq 0$ implies $Au = 0$. Thus A has affine type. □

Proposition 15.9 *Let A be an indecomposable GCM. Then*:

if A has finite type A^t has finite type
if A has affine type A^t has affine type.

Proof. To prove these results we shall make use of Proposition 15.3.

Suppose A has finite type. We show there does not exist $v > 0$ with $Av < 0$. For if $Av < 0$ then $A(-v) > 0$ and so $-v > 0$ or $-v = 0$. Hence $v < 0$ or $v = 0$. This contradicts $v > 0$. We may now apply Proposition 15.3 to show there exists $u \neq 0$ with $u \geq 0$ and $A^t u \geq 0$. So $\{u;\ u \geq 0$ and $A^t u \geq 0\} \neq \{0\}$. By Lemma 15.6 either

$$K_{A^t} \subset \{u;\ u > 0\} \cup \{0\}$$

or $K_{A^t} = \{u;\ A^t u = 0\}$ and this is a 1-dimensional subspace. Now $\det A \neq 0$ so $\det A^t \neq 0$. Thus the latter case cannot occur. The former case must therefore occur, so by Proposition 15.7 A^t has finite type.

Now suppose A has affine type. We again show there does not exist $v > 0$ with $Av < 0$. For $A(-v) > 0$ is impossible in the affine case.

15.1 A trichotomy for indecomposable GCMs

By Proposition 15.3 there exists $u \neq 0$ with $u \geq 0$ and $A^t u \geq 0$. So $\{u;\ u \geq 0$ and $A^t u \geq 0\} \neq \{0\}$. By Lemma 15.6 we may again conclude that either

$$K_{A^t} \subset \{u;\ u > 0\} \cup \{0\} \text{ or}$$
$$K_{A^t} = \{u;\ Au = 0\} \text{ and this is a 1-dimensional subspace.}$$

Now corank $A = 1$ so corank $A^t = 1$. This shows that we cannot have the first possibility. Thus the second possibility holds, and then by Proposition 15.8 we see that A^t has affine type. □

We may now identify the case not appearing in Lemma 15.6 with that of an indefinite GCM.

Proposition 15.10 *Let A be an indecomposable GCM. Then the following conditions are equivalent:*

$$A \text{ has indefinite type}$$
$$\{u;\ u \geq 0 \text{ and } Au \geq 0\} = \{0\}.$$

Proof. Suppose A has indefinite type. Then $u \geq 0$ and $Au \geq 0$ imply $u = 0$.

Conversely suppose $\{u;\ u \geq 0 \text{ and } Au \geq 0\} = \{0\}$. Then the same condition holds for A^t, i.e. $\{u;\ u \geq 0 \text{ and } A^t u \geq 0\} = \{0\}$. This follows from Lemma 15.6 and Propositions 15.7, 15.8 and 15.9. But then Proposition 15.3 shows that there exists $v > 0$ with $Av < 0$. Thus A has indefinite type. □

We are now able to achieve our aim of proving Theorem 15.1. For each indecomposable GCM A Lemma 15.6 shows that exactly one of the following conditions holds:

(a) $\{u;\ u \geq 0 \text{ and } Au \geq 0\} \neq \{0\}$ and $K_A \subset \{u;\ u > 0\} \cup \{0\}$.
(b) $\{u;\ u \geq 0 \text{ and } Au \geq 0\} \neq \{0\}$, $K_A = \{u;\ Au = 0\}$, and K_A is a 1-dimensional subspace.
(c) $\{u;\ u \geq 0 \text{ and } Au \geq 0\} = \{0\}$.

By Proposition 15.7 A satisfies (a) if and only if A has finite type. By Proposition 15.8 A satisfies (b) if and only if A has affine type. By Proposition 15.10 A satisfies (c) if and only if A has indefinite type. Thus we have the required trichotomy for GCMs. Moreover Proposition 15.9 shows that the type of A^t is the same as the type of A. This completes the proof of Theorem 15.1

Corollary 15.11 *Let A be an indecomposable GCM. Then:*

(a) *A has finite type if and only if there exists $u > 0$ with $Au > 0$.*
(b) *A has affine type if and only if there exists $u > 0$ with $Au = 0$.*
(c) *A has indefinite type if and only if there exists $u > 0$ with $Au < 0$.*

Proof. (a) Suppose $u > 0$ and $Au > 0$. A cannot have affine type as then $Au \geq 0$ would imply $Au = 0$. A cannot have indefinite type as then $u \geq 0$ and $Au \geq 0$ would imply $u = 0$. Thus A has finite type.
(b) Suppose $u > 0$ and $Au = 0$. A cannot have finite type since $\det A = 0$. A cannot have indefinite type since then $u \geq 0$ and $Au \geq 0$ would imply $u = 0$. Thus A has affine type.
(c) Suppose $u > 0$ and $Au < 0$. Then $A(-u) > 0$. A cannot have finite type as this would imply $-u > 0$ or $-u = 0$. A cannot have affine type since $A(-u) > 0$ would then imply $A(-u) = 0$. Thus A has indefinite type.

Remark 15.12 In proving the results of Section 15.1 we have assumed that A is a GCM. However, we have not used the full force of this assumption. Inspection of the proofs shows that we have nowhere assumed that $A_{ii} = 2$ or that $A_{ij} \in \mathbb{Z}$. This remark will be useful in some subsequent applications.

15.2 Symmetrisable generalised Cartan matrices

In this section we shall consider a special type of GCM which plays a key role in the theory of Kac–Moody algebras. These are the symmetrisable GCMs. Before giving the definition we obtain some preliminary results.

Let $A = (A_{ij})$ be a GCM with $i, j \in \{1, \ldots, n\}$ and let J be a subset of $\{1, \ldots, n\}$. Let $A_J = (A_{ij})$, $i, j \in J$. Then A_J is also a GCM, called a **principal minor** of A.

Lemma 15.13 (i) *Suppose A is an indecomposable GCM of finite type and A_J is an indecomposable principal minor of A. Then A_J also has finite type.*
(ii) *Suppose A is an indecomposable GCM of affine type and A_J is a proper indecomposable principal minor of A. Then A_J has finite type.*

15.2 Symmetrisable generalised Cartan matrices

Proof. (i) By passing to an equivalent GCM we may choose the numbering so that $J = \{1, \ldots, m\}$ for some $m \leq n$. Let $K = \{m+1, \ldots, n\}$. Let

$$A = \begin{pmatrix} P & Q \\ R & S \end{pmatrix} \begin{matrix} m \\ n-m \end{matrix}$$
$$\, m \;\; n-m$$

Now there exists $u > 0$ with $Au > 0$. Let $u = \binom{u_J}{u_K}$. Then

$$Au = \begin{pmatrix} P & Q \\ R & S \end{pmatrix} \begin{pmatrix} u_J \\ u_K \end{pmatrix} = \begin{pmatrix} Pu_J + Qu_K \\ Ru_J + Su_K \end{pmatrix}.$$

Since $Au > 0$ we have $Pu_J + Qu_K > 0$. However, $Qu_K \leq 0$ so $Pu_J > 0$. Thus there exists $u_J > 0$ with $A_J u_J > 0$. By Corollary 15.11 A_J has finite type.

(ii) As before we may assume $J = \{1, \ldots, m\}$. This time we have $m < n$. Let

$$A = \begin{pmatrix} P & Q \\ R & S \end{pmatrix} \begin{matrix} m \\ n-m \end{matrix} \quad \text{where } P = A_J$$
$$\, m \;\; n-m$$

Since A has affine type there exists $u > 0$ with $Au = 0$. We have

$$Au = \begin{pmatrix} P & Q \\ R & S \end{pmatrix} \begin{pmatrix} u_J \\ u_K \end{pmatrix} = \begin{pmatrix} Pu_J + Qu_K \\ Ru_J + Su_K \end{pmatrix}.$$

Hence $Pu_J + Qu_K = 0$. Now $Qu_K \leq 0$ so $Pu_J \geq 0$.

Suppose if possible that $Pu_J = 0$. Then $Qu_K = 0$, and since $u_K > 0$ this implies that $Q = O$. But then $R = O$ also and A is decomposable, a contradiction. Hence we have $u_J > 0$, $Pu_J \geq 0$, $Pu_J \neq 0$. This implies that $P = A_J$ cannot have affine type or indefinite type. Thus A_J has finite type. $\square$

We next describe our trichotomy in the special case in which the indecomposable GCM is symmetric.

Proposition 15.14 *Suppose A is a symmetric indecomposable GCM. Then:*

(a) *A has finite type if and only if A is positive definite.*
(b) *A has affine type if and only if A is positive semidefinite of corank 1.*
(c) *A has indefinite type if and only if A satisfies neither of these conditions.*

Proof. (a) Let A have finite type. Then there exists $u > 0$ with $Au > 0$. Hence for all $\lambda \geq 0$ we have $(A + \lambda I)u > 0$. Thus $A + \lambda I$ has finite type by Corollary 15.11. (Note that $A + \lambda I$ need not be a GCM, but the results of Section 15.1 can be applied to it by Remark 15.12.) Thus $\det(A + \lambda I) \neq 0$

when $\lambda \geq 0$, that is $\det(A - \lambda I) \neq 0$ when $\lambda \leq 0$. Now the eigenvalues of the real symmetric matrix A are all real. Thus all the eigenvalues of A must be positive. Hence A is positive definite.

Conversely suppose A is positive definite. Then $\det A \neq 0$ so A has finite or indefinite type. If A has indefinite type there exists $u > 0$ with $Au < 0$. But then $u^t A u < 0$, contradicting the fact that A is positive definite. Thus A must have finite type.

(b) Let A have affine type. Then there exists $u > 0$ with $Au = 0$. Hence for all $\lambda > 0$ we have $(A + \lambda I)u > 0$. Thus by Corollary 15.11 $A + \lambda I$ has finite type when $\lambda > 0$. (We are again using Remark 15.12 here.) Thus $\det(A + \lambda I) \neq 0$ when $\lambda > 0$, that is $\det(A - \lambda I) \neq 0$ when $\lambda < 0$. Thus all eigenvalues of A are non-negative. But A has corank 1 so 0 occurs as an eigenvalue with multiplicity 1, and the remaining eigenvalues are all positive. Hence A is positive semi-definite of corank 1.

Conversely suppose A is positive semi-definite of corank 1. Then $\det A = 0$ so A cannot have finite type. Suppose A has indefinite type. Then there exists $u > 0$ with $Au < 0$. Thus $u^t Au < 0$, which contradicts the fact that A is positive semi-definite. Thus A must have affine type.

(c) This follows from (a) and (b). $\square$

In general a GCM need not be symmetric, but it may nevertheless satisfy the weaker condition of being symmetrisable.

Definition *A GCM A is **symmetrisable** if there exists a non-singular diagonal matrix D and a symmetric matrix B such that $A = DB$.*

Lemma 15.15 *Let A be a GCM. Then A is symmetrisable if and only if*

$$A_{i_1 i_2} A_{i_2 i_3} \ldots A_{i_k i_1} = A_{i_2 i_1} A_{i_3 i_2} \ldots A_{i_1 i_k}$$

for all $i_1, i_2, \ldots, i_k \in \{1, \ldots, n\}$.

Proof. Suppose A is symmetrisable. Then $A = DB$ with $D = \mathrm{diag}(d_1, \ldots, d_n)$ and $B = (B_{ij})$. Thus $A_{ij} = d_i B_{ij}$. Hence

$$A_{i_1 i_2} \ldots A_{i_k i_1} = d_{i_1} \ldots d_{i_k} B_{i_1 i_2} \ldots B_{i_k i_1}$$

$$A_{i_2 i_1} \ldots A_{i_1 i_k} = d_{i_1} \ldots d_{i_k} B_{i_2 i_1} \ldots B_{i_1 i_k}$$

and these are equal since B is symmetric.

Conversely suppose

$$A_{i_1 i_2} \ldots A_{i_k i_1} = A_{i_2 i_1} \ldots A_{i_1 i_k}$$

for all $i_1, \ldots, i_k$. We may suppose A is indecomposable since the result in this case implies it for all A. Thus for each $i \in \{1, \ldots, n\}$ there exists a sequence

$$1 = j_1, j_2, \ldots, j_t = i$$

with

$$A_{j_1 j_2} \neq 0, \quad A_{j_2 j_3} \neq 0, \quad \ldots, \quad A_{j_{t-1} j_t} \neq 0.$$

We choose a number $d_1 \neq 0$ in $\mathbb{R}$. We wish to define d_i by

$$d_i = \frac{A_{j_t j_{t-1}} \cdots A_{j_2 j_1}}{A_{j_1 j_2} \cdots A_{j_{t-1} j_t}} d_1.$$

However, we must check that this definition of d_i depends only upon i and not on the sequence chosen from 1 to i. So let

$$1 = k_1, k_2, \ldots, k_u = i$$

be a second such sequence from 1 to i. We claim that

$$\frac{A_{j_t j_{t-1}} \cdots A_{j_2 j_1}}{A_{j_1 j_2} \cdots A_{j_{t-1} j_t}} = \frac{A_{k_u k_{u-1}} \cdots A_{k_2 k_1}}{A_{k_1 k_2} \cdots A_{k_{u-1} k_u}}$$

that is $A_{1 k_2} A_{k_2 k_3} \cdots A_{k_{u-1} i} A_{i j_{t-1}} \cdots A_{j_2 1} = A_{k_2 1} A_{k_3 k_2} \cdots A_{i k_{u-1}} A_{j_{t-1} i} \cdots A_{1 j_2}$. This is in fact one of the given conditions on the matrix A. Thus $d_i \in \mathbb{R}$ is well defined and $d_i \neq 0$. Let $D = \mathrm{diag}\,(d_1, \ldots, d_n)$. Define B_{ij} by $A_{ij} = d_i B_{ij}$. We show that $B_{ji} = B_{ij}$, that is $\frac{A_{ji}}{d_j} = \frac{A_{ij}}{d_i}$. If $A_{ij} = 0$ then $A_{ji} = 0$ also and the condition is satisfied. So suppose $A_{ij} \neq 0$. Let $1 = j_1, j_2, \ldots, j_t = i$ be a sequence from 1 to i of the type described above. Then $1 = j_1, j_2, \ldots, j_t, j$ is such a sequence from 1 to j. These sequences may be used to obtain d_i and d_j respectively, and we have

$$d_j = \frac{A_{ji}}{A_{ij}} d_i.$$

Thus $B_{ji} = B_{ij}$. Hence $A = DB$ where D is diagonal and non-singular, and B is symmetric. Thus A is symmetrisable.

Corollary 15.16 *Let A be a symmetrisable indecomposable GCM. Then A can be expressed in the form $A = DB$ where $D = \mathrm{diag}\,(d_1, \ldots, d_n)$, B is symmetric, with $d_1, \ldots, d_n > 0$ in $\mathbb{Z}$ and $B_{ij} \in \mathbb{Q}$. Also D is determined by these conditions up to a scalar multiple.*

Proof. We choose any $d_1 \in \mathbb{Q}$ with $d_1 > 0$. Then Lemma 15.15 shows that $d_i \in \mathbb{Q}$ and $d_i > 0$ for each i. Thus by multiplying by a positive scalar we may assume each $d_i \in \mathbb{Z}$ with $d_i > 0$. Also $B_{ij} = A_{ij}/d_i$ lies in $\mathbb{Q}$. The proof of Lemma 15.15 also shows that D is determined up to a scalar multiple. □

The following important result shows that indecomposable GCMs in the first two classes of our trichotomy are symmetrisable.

Theorem 15.17 *Let A be an indecomposable GCM of finite or affine type. Then A is symmetrisable.*

Proof. First suppose there is no set of integers $i_1, i_2, \ldots, i_k$ with $k \geq 3$ such that $i_1 \neq i_2, i_2 \neq i_3, \ldots, i_{k-1} \neq i_k, i_k \neq i_1$ and

$$A_{i_1 i_2} \neq 0, A_{i_2 i_3} \neq 0, \ldots, A_{i_{k-1} i_k} \neq 0, A_{i_k i_1} \neq 0.$$

Then Lemma 15.15 shows that A is symmetrisable.

Thus we suppose there is such a sequence $i_1, \ldots, i_k$ with $k \geq 3$ and we choose such a sequence with minimal possible value of k. We thus have

$$A_{i_r, i_s} \neq 0 \quad \text{if } (r, s) \in \{(1, 2), (2, 3), \ldots, (k, 1), (2, 1), (3, 2), \ldots, (1, k)\}.$$

The minimality of k shows that $A_{i_r, i_s} = 0$ if (r, s) does not lie in the above set. Otherwise there would be such a sequence with a smaller value of k.

Let $J = \{i_1, \ldots, i_k\}$. Then the principal minor A_J of A has form

$$A_J = \begin{pmatrix} 2 & -r_1 & 0 & \cdots & & 0 & -s_k \\ -s_1 & 2 & -r_2 & \cdot & & & 0 \\ 0 & -s_2 & 2 & \cdot & & & \cdot \\ \cdot & & \cdot & \cdot & \cdot & & \cdot \\ \cdot & & & \cdot & 2 & \cdot & 0 \\ 0 & & & & \cdot & 2 & -r_{k-1} \\ -r_k & 0 & \cdots & & 0 & -s_{k-1} & 2 \end{pmatrix}.$$

with $r_i, s_i \in \mathbb{Z}$ satisfying $r_i > 0, s_i > 0$. In particular we see that A_J is indecomposable. Now A_J must have finite or affine type by Lemma 15.13. Thus there exists $u > 0$ with $A_J u \geq 0$. Let $u = (u_1, \ldots, u_k)$. We define the $k \times k$ matrix M by

$$M = \text{diag}(u_1^{-1}, \ldots, u_k^{-1}) A_J \text{diag}(u_1, \ldots, u_k).$$

Then $M_{ij} = u_i^{-1}(A_J)_{ij} u_j$. Thus

$$\sum_j M_{ij} = u_i^{-1} \sum_j (A_J)_{ij} u_j \geq 0.$$

In particular we have $\sum_{i,j} M_{ij} \geq 0$. Now we have

$$M = \begin{pmatrix} 2 & -r_1' & 0 & \cdots & & 0 & -s_k' \\ -s_1' & 2 & -r_2' & \cdot & & & 0 \\ 0 & -s_2' & 2 & \cdot & \cdot & & \cdot \\ \cdot & & & & & & \cdot \\ \cdot & & & & & & \cdot \\ \cdot & & & & & & 0 \\ 0 & & & \cdot & \cdot & 2 & -r_{k-1}' \\ -r_k' & 0 & \cdot & \cdot & 0 & -s_{k-1}' & 2 \end{pmatrix}.$$

with $r_i' = u_i^{-1} r_i u_{i+1}$, $s_i' = u_{i+1}^{-1} s_i u_i$. (We define $u_{k+1} = u_1$.)
We note that $r_i' > 0$, $s_i' > 0$ and $r_i' s_i' = r_i s_i \in \mathbb{Z}$. We also have

$$\sum_{i,j} M_{ij} = 2k - (r_1' + s_1') - \cdots - (r_k' + s_k').$$

Now $\frac{r_i' + s_i'}{2} \geq \sqrt{r_i' s_i'} = \sqrt{r_i s_i} \geq 1$ hence $r_i' + s_i' \geq 2$. Since $\sum_{i,j} M_{ij} \geq 0$ we deduce that $r_i' + s_i' = 2$ and $r_i' s_i' = 1$. Hence $r_i s_i = 1$ and, since r_i, s_i are positive integers, we have $r_i = 1$, $s_i = 1$. It follows that

$$A_J = \begin{pmatrix} 2 & -1 & 0 & \cdots & & 0 & -1 \\ -1 & 2 & -1 & \cdot & & & 0 \\ 0 & -1 & 2 & \cdot & \cdot & & \cdot \\ \cdot & & & & & & \cdot \\ \cdot & & & & & & \cdot \\ \cdot & & & & & & 0 \\ 0 & & & \cdot & \cdot & 2 & -1 \\ -1 & 0 & \cdot & \cdot & 0 & -1 & 2 \end{pmatrix}.$$

Let $v = (1, \ldots, 1)$. Then $v > 0$ and $A_J v = 0$. Thus A_J has affine type by Corollary 15.11. Lemma 15.13 shows that this can only happen when $A_J = A$. Thus A is symmetric, in particular symmetrisable as required. □

We are now able to prove the following basic description of our trichotomy. It generalises the description previously obtained in Proposition 15.14 for symmetric indecomposable GCMs.

Theorem 15.18 *Let A be an indecomposable GCM. Then*:

(a) *A has finite type if and only if all its principal minors have positive determinant.*

(b) *A has affine type if and only if* $\det A = 0$ *and all proper principal minors have positive determinant.*

(c) *A has indefinite type if and only if A satisfies neither of these two conditions.*

Proof. (a) Suppose A has finite type. Then A is symmetrisable by Theorem 15.17, hence $A = DB$ where $D = \text{diag}(d_1, \ldots, d_n)$ with $d_i > 0$ and B is symmetric, by Corollary 15.16. The matrix B need not necessarily be a GCM, but Remark 15.12 shows that we can nevertheless define the type of B. Moreover Corollary 15.11 shows that A and B have the same type. Thus B is a symmetric indecomposable matrix of finite type, and so $\det B > 0$ by Proposition 15.14. It follows that $\det A > 0$ also. Now all principal minors of A also have finite type by Lemma 15.13. Thus these also have positive determinant.

Conversely suppose that all principal minors of A have positive determinant. Suppose there is a set of integers $i_1, \ldots, i_k$ with $k \geq 3$ such that $i_1 \neq i_2, i_2 \neq i_3, \ldots, i_k \neq i_1$ with

$$A_{i_1 i_2} A_{i_2 i_3} \ldots A_{i_k i_1} \neq 0.$$

Choose such a sequence with minimal possible k and let $J = \{i_1, \ldots, i_k\}$. Then the principal minor A_J of A has form

$$A_J = \begin{pmatrix} 2 & -1 & 0 & \cdot & \cdot & \cdot & 0 & -1 \\ -1 & 2 & -1 & \cdot & & & & 0 \\ 0 & -1 & 2 & \cdot & \cdot & & & \cdot \\ \cdot & \cdot & \cdot & \cdot & \cdot & & & \cdot \\ \cdot & & & & \cdot & \cdot & \cdot & \cdot \\ \cdot & & & & & \cdot & \cdot & 0 \\ 0 & & & & \cdot & \cdot & 2 & -1 \\ -1 & 0 & \cdot & \cdot & & 0 & -1 & 2 \end{pmatrix}.$$

by the proof of Theorem 15.17. But then $\det A_J = 0$, a contradiction. Thus there is no such sequence $i_1, \ldots, i_k$. By Lemma 15.15 A is symmetrisable. Hence $A = DB$ where $D = \text{diag}(d_1, \ldots, d_n)$ with $d_i > 0$ and B is symmetric. Again B need not be a GCM but we can nevertheless define its type using Remark 15.12 and, by Corollary 15.11, A and B have the same type. Now the principal minors of the symmetric matrix B all have positive determinant and so B is positive definite. Thus B has finite type by Proposition 15.14, and so A has finite type also.

(b) Now suppose A has affine type. Then $\det A = 0$. All proper principal minors of A have finite type by Lemma 15.13 and so have positive determinant by (a).

Suppose conversely that $\det A = 0$ and that all proper principal minors of A have positive determinant. Suppose there is a set of integers $i_1, \ldots, i_k$ with $k \geq 3$ such that $i_1 \neq i_2, \ldots, i_k \neq i_1$ with

$$A_{i_1 i_2} A_{i_2 i_3} \ldots A_{i_k i_1} \neq 0.$$

Choose such a sequence with minimal k, and let $J = \{i_1, \ldots, i_k\}$. Then the principal minor A_J has form

$$A_J = \begin{pmatrix} 2 & -1 & 0 & \cdot & \cdot & \cdot & 0 & -1 \\ -1 & 2 & -1 & \cdot & & & & 0 \\ 0 & -1 & 2 & \cdot & \cdot & & & \cdot \\ \cdot & \cdot & \cdot & & & & & \cdot \\ \cdot & & & \cdot & \cdot & \cdot & & \cdot \\ \cdot & & & & \cdot & \cdot & \cdot & 0 \\ 0 & & & & \cdot & \cdot & 2 & -1 \\ -1 & 0 & \cdot & \cdot & \cdot & 0 & -1 & 2 \end{pmatrix}.$$

as above. Since $\det A_J = 0$ we have $A_J = A$. But then A is affine since $Au = 0$ with $u = (1, \ldots, 1)$. Thus suppose there is no such sequence $i_1, \ldots, i_k$. Then A is symmetrisable by Lemma 15.15, and has form $A = DB$ where $D = \text{diag}(d_1, \ldots, d_n)$ with $d_i > 0$ and B is symmetric. Now $\det B = 0$ and all proper principal minors of B have positive determinant. This implies that the symmetric matrix B is positive semidefinite of corank 1. Hence B is of affine type by Proposition 15.14. Thus A has affine type also, by Corollary 15.11.

(c) This follows directly from (a) and (b). □

15.3 The classification of affine generalised Cartan matrices

In this section we shall determine explicitly which indecomposable GCMs lie in each class of our trichotomy. We begin with indecomposable GCMs of finite type.

Theorem 15.19 *Let A be an indecomposable GCM. Then A has finite type if and only if A is a Cartan matrix. Thus the indecomposable GCMs of finite type are those on the standard list* 6.12.

Proof. We recall from Sections 6.1 and 6.2 that a GCM is a Cartan matrix if and only if it satisfies the conditions:

(a) $A_{ij} \in \{0, -1, -2, -3\}$ for all $i \neq j$
(b) $A_{ij} = -2$ or -3 implies $A_{ji} = -1$
(c) the quadratic form

$$Q(x_1, \ldots, x_n) = 2\sum_{i=1}^{n} x_i^2 - \sum_{i \neq j} \sqrt{n_{ij}} x_i x_j$$

is positive definite, where $n_{ij} = A_{ij} A_{ji}$.

Suppose A is a Cartan matrix. Then $A_{ij} = 2\frac{\langle \alpha_i, \alpha_j \rangle}{\langle \alpha_i, \alpha_i \rangle}$. Let $D = \text{diag}(d_1, \ldots, d_n)$ where $d_i = \sqrt{\langle \alpha_i, \alpha_i \rangle}$. Then $(DAD^{-1})_{ij} = 2\frac{\langle \alpha_i, \alpha_j \rangle}{\sqrt{\langle \alpha_i, \alpha_i \rangle} \sqrt{\langle \alpha_j, \alpha_j \rangle}}$ and so DAD^{-1} is the matrix of the quadratic form $Q(x_1, \ldots, x_n)$. Since Q is positive definite $\det(DAD^{-1}) > 0$ and so $\det A > 0$.

Now any principal minor A_J of the Cartan matrix A is also a Cartan matrix. Hence $\det A_J > 0$ for all principal minors of A. Thus A has finite type by Theorem 15.18 (a).

Now suppose conversely that A has finite type. Suppose $i \neq j$ and consider the 2×2 principal minor

$$\begin{pmatrix} 2 & A_{ij} \\ A_{ji} & 2 \end{pmatrix}.$$

By Theorem 15.18 (a) the determinant of this minor is positive, hence $A_{ij} A_{ji} < 4$. Since A_{ij} and A_{ji} are both non-positive integers such that $A_{ij} = 0$ if and only if $A_{ji} = 0$ we deduce that $A_{ij} \in \{0, -1, -2, -3\}$ and that $A_{ij} \in \{-2, -3\}$ implies $A_{ji} = -1$.

Since A has finite type A is symmetrisable by Theorem 15.17. Thus $A = DB$ where $D = \text{diag}(d_1, \ldots, d_n)$, $d_i > 0$, and B is symmetric. Although B need not be a GCM we may define the type of B by using Remark 15.12. Thus B is indecomposable of finite type, and so B is positive definite by Proposition 15.14 (a). Let $y_i = \sqrt{d_i} x_i$. Then

$$Q(x_1, \ldots, x_n) = 2\sum_{i} x_i^2 - \sum_{i \neq j} \sqrt{n_{ij}} x_i x_j$$

15.3 The classification of affine generalised Cartan matrices

$$= \frac{2}{d_i} \sum_i y_i^2 - \sum_{i \ne j} \frac{1}{\sqrt{d_i}} \frac{1}{\sqrt{d_j}} \sqrt{(A_{ij}A_{ji})} y_i y_j$$

$$= \sum_i B_{ii} y_i^2 + \sum_{i \ne j} B_{ij} y_i y_j.$$

Since B is positive definite we see that $Q(x_1, \ldots, x_n)$ is positive definite. Thus A is a Cartan matrix. □

Having determined the indecomposable GCMs of finite type we next determine those of affine type. This will also determine those of indefinite type, as those remaining.

To each GCM A we define an associated diagram $\Delta(A)$ called the **Dynkin diagram** of A. This extends the definition of the Dynkin diagram of a Cartan matrix given in Section 6.2. The vertices of $\Delta(A)$ are labelled $1, \ldots, n$ where A is an $n \times n$ matrix. Suppose i, j are distinct vertices of $\Delta(A)$. We explain how i, j are joined in $\Delta(A)$. This depends on the pair (A_{ij}, A_{ji}). We recall that A_{ij} and A_{ji} lie in $\mathbb{Z}$, $A_{ij} \le 0$, $A_{ji} \le 0$ and $A_{ij} = 0$ if and only if $A_{ji} = 0$. The rules are as follows.

(a) If $A_{ij}A_{ji} = 0$ vertices i, j are not joined.
(b) If $A_{ij}A_{ji} = 1$ vertices i, j are joined by a single edge.
(c) If $A_{ij}A_{ji} = 2$, $A_{ij} = -1$, $A_{ji} = -2$ vertices i, j are joined by a double edge with an arrow pointing towards j.
(d) If $A_{ij}A_{ji} = 3$, $A_{ij} = -1$, $A_{ji} = -3$ vertices i, j are joined by a triple edge with an arrow pointing towards j.
(e) If $A_{ij}A_{ji} = 4$, $A_{ij} = -1$, $A_{ji} = -4$ vertices i, j are joined by a quadruple edge with an arrow pointing towards j.
(f) If $A_{ij}A_{ji} = 4$, $A_{ij} = -2$, $A_{ji} = -2$ vertices i, j are joined by a double edge with two arrows pointing away from i, j.

$$\underset{i \quad\quad j}{\circ\!\!=\!\!=\!\!\circ}$$

(g) If $A_{ij}A_{ji} \ge 5$ vertices i, j are joined by an edge with the numbers $|A_{ij}|, |A_{ji}|$ shown on it.

$$\underset{i \quad\quad j}{\circ \xrightarrow{|A_{ij}|, |A_{ji}|} \circ}$$

It is clear that the GCM A is determined by its Dynkin diagram $\Delta(A)$. Moreover A is indecomposable if and only if $\Delta(A)$ is connected.

We now consider a set of connected Dynkin diagrams called the affine list.

15.20 The affine list of Dynkin diagrams

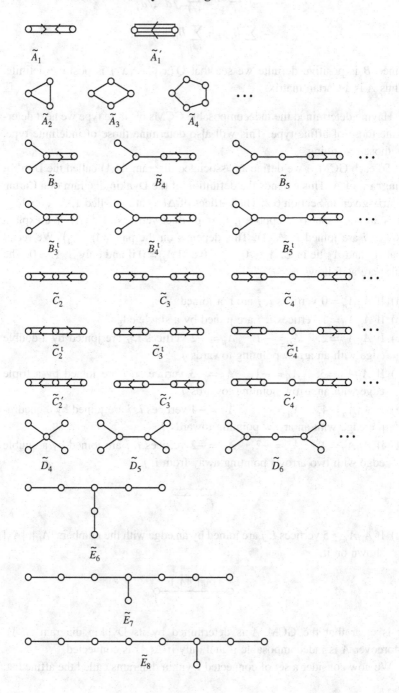

15.3 The classification of affine generalised Cartan matrices 355

We note that many, but not all, of the Dynkin diagrams on the affine list appeared in Lemma 6.8. We also note that every proper connected subdiagram of a Dynkin diagram on the affine list appears on list 6.11 of Dynkin diagrams of finite type. We shall call this the finite list.

Proposition 15.21 *Let A be a GCM whose Dynkin diagram lies on the affine list. Then* $\det A = 0$.

Proof. First suppose that $\Delta(A)$ has 2 vertices. Then either $\Delta(A) = \tilde{A}_1$ and $A = \begin{pmatrix} 2 & -2 \\ -2 & 2 \end{pmatrix}$ or $\Delta(A) = \tilde{A}'_1$ and $A = \begin{pmatrix} 2 & -4 \\ -1 & 2 \end{pmatrix}$. In either case $\det A = 0$.

Next suppose that $\Delta(A) = \tilde{A}_l$ for $l \geq 2$. Then the sum of all the rows of A is zero, and so $\det A = 0$.

In all other cases $\Delta(A)$ has a vertex, say 1, joined to just one other vertex, say 2. Moreover we can choose these vertices so that they are joined by a single or a double edge. In the case of a single edge we have

$$\det A = 2 \det B - \det C$$

where B is obtained from A by removing row and column 1, and C is obtained from B by removing row and column 2. This relation between determinants is obtained as in the proof of Theorem 6.7. The connected components of B and C are Cartan matrices of finite type, so their determinants are known from the proof of Theorem 6.7. In all cases this gives $\det A = 0$.

In the case when vertices 1, 2 are joined by a double edge we obtain

$$\det A = 2 \det B - 2 \det C$$

again as in the proof of Theorem 6.7. Again B, C have connected components of finite type so we know their determinants, and in each case we obtain $\det A = 0$.

Proposition 15.22 *Let A be a GCM whose Dynkin diagram lies on the affine list. Then A has affine type.*

Proof. By Proposition 15.21 we have $\det A = 0$. Also the Dynkin diagram of any proper principal minor has connected components on the finite list. Thus all proper principal minors have positive determinant. It follows that A has affine type by Theorem 15.18 (b). □

We shall now prove the converse.

Theorem 15.23 *Let A be an indecomposable GCM. Then A has affine type if and only if its Dynkin diagram $\Delta(A)$ lies on the affine list.*

Proof. Suppose A has affine type. Then every proper indecomposable principal minor of A has finite type, by Lemma 15.13 (ii). Thus all proper connected subdiagrams of $\Delta(A)$ lie on the finite list, by Theorem 15.19.

If $\Delta(A)$ has only one vertex A has finite type, so there is no possible affine A.

If $\Delta(A)$ has two vertices then

$$A = \begin{pmatrix} 2 & -a \\ -b & 2 \end{pmatrix}$$

where a, b are positive integers. Since $\det A = 0$ we have $ab = 4$. The possibilities are $(a, b) = (1, 4)(4, 1)(2, 2)$. Thus $\Delta(A) = \tilde{A}_1$ or $\tilde{A}_1'$.

Now suppose $\Delta(A)$ has at least three vertices. If $\Delta(A)$ contains a cycle then the proof of Theorem 15.17 shows that $\Delta(A) = \tilde{A}_l$ for some $l \geq 2$. Thus we suppose that $\Delta(A)$ contains no cycle. Since all the connected subdiagrams with two vertices lie on the finite list all edges of $\Delta(A)$ have one of the forms

○――○   ○⇒○   ○⇛○

Suppose $\Delta(A)$ has a triple edge ○⇛○ Then $\Delta(A)$ must have exactly three vertices, otherwise $\Delta(A)$ would have a proper connected subdiagram with three vertices containing a triple edge, whereas there is no such diagram on the finite list. Thus we have

$$A = \begin{pmatrix} 2 & -1 & 0 \\ -3 & 2 & -a \\ 0 & -b & 2 \end{pmatrix} \quad \text{or} \quad \begin{pmatrix} 2 & -3 & 0 \\ -1 & 2 & -a \\ 0 & -b & 2 \end{pmatrix}$$

where a, b are positive integers. Thus $\det A = 2(1 - ab)$. However, $\det A = 0$ and so $a = 1, b = 1$. Thus $\Delta(A) = \tilde{G}_2$ or $\tilde{G}_2^t$.

So we now suppose that $\Delta(A)$ has no triple edge. Now $\Delta(A)$ has at most two double edges, as every proper connected subdiagram appears on the finite list so has at most one. Suppose $\Delta(A)$ has two double edges.

15.3 The classification of affine generalised Cartan matrices

Then every edge which can be removed to give a connected subdiagram must be a double edge. This implies that $\Delta(A)$ must be one of $\tilde{C}_l, \tilde{C}_l^t, \tilde{C}_l'$.

Thus we suppose that $\Delta(A)$ has just one double edge. If $\Delta(A)$ has a branch point then no proper connected subdiagram can contain both a double edge and a branch point, since the subdiagram lies on the finite list. This implies that $\Delta(A)$ is $\tilde{B}_l$ or $\tilde{B}_l^t$.

Now suppose that $\Delta(A)$ has one double edge but no branch point. Then $\Delta(A)$ has form

with $a+b+2$ vertices. We have $a>0$ and $b>0$ since $\Delta(A)$ is not on the finite list. Also $b \leq 2$, otherwise there would be a proper subdiagram

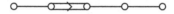

and $a \leq 2$, otherwise there would be a proper subdiagram

Thus the possibilities are

$$(a, b) = (1, 1), (2, 1), (1, 2), (2, 2).$$

The case $(a, b) = (1, 1)$ appears on the finite list so is not affine. The case $(a, b) = (2, 2)$ is impossible, since it would give proper subdiagrams as above. Thus $(a, b) = (2, 1)$ or $(1, 2)$ and $\Delta(A)$ is $\tilde{F}_4$ or $\tilde{F}_4^t$.

Thus we may now assume that $\Delta(A)$ has only single edges. Consider the branch points of $\Delta(A)$. Each branch point has at most four branches, otherwise there would be a proper subdiagram

which does not appear on the finite list. If there is a branch point with four branches then $\Delta(A) = \tilde{D}_4$, as otherwise there would again be a proper subdiagram $\tilde{D}_4$.

Thus we may assume that all branch points in $\Delta(A)$ have three branches. There cannot be more than two branch points, as otherwise there would be a proper connected subdiagram with two branch points which could not be on the finite list.

Suppose $\Delta(A)$ has 2 branch points. Then any proper connected subdiagram has only one branch point, and this implies that $\Delta(A) = \tilde{D}_l$ for some $l \geq 5$.

So suppose $\Delta(A)$ has just one branch point. Let the branch lengths be l_1, l_2, l_3 with $l_1 \leq l_2 \leq l_3$ so that there are $l_1 + l_2 + l_3 + 1$ vertices. We must have $l_1 \leq 2$, otherwise there would be a proper subdiagram

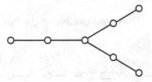

which is not on the finite list. Suppose $l_1 = 2$. Then we must have $l_2 = 2$ and $l_3 = 2$, otherwise there would again be a proper subdiagram as above. Thus $\Delta(A) = \tilde{E}_6$.

Thus we may assume $l_1 = 1$. Since $l_2 = 1$ would give a diagram of finite type we must have $l_2 \geq 2$. However, $l_2 \leq 3$ as otherwise there would be a proper subdiagram

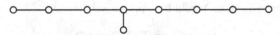

which is not on the finite list. Thus $l_2 = 2$ or 3.

Suppose $l_1 = 1$, $l_2 = 3$. Then we must have $l_3 = 3$, otherwise there would be a proper subdiagram

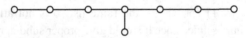

which is again not on the finite list. Thus $\Delta(A) = \tilde{E}_7$.

We may now suppose that $l_1 = 1$, $l_2 = 2$. Since the diagrams with $l_3 = 2, 3, 4$ are of finite type we must have $l_3 \geq 5$. But $l_3 \leq 5$ also, as otherwise there would be a proper subdiagram

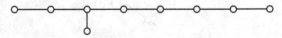

which is not on the finite list. Hence $l_3 = 5$ and $\Delta(A) = \tilde{E}_8$.

Finally if $\Delta(A)$ has only single edges and no branch points then it lies on the finite list so A cannot be affine.

Thus we have shown that whenever A is affine $\Delta(A)$ must appear on the affine list. This, together with proposition 15.22, completes the proof. □

A GCM A such that $\Delta(A)$ is on the affine list will be called an **affine Cartan matrix**.

15.3 The classification of affine generalised Cartan matrices 359

Corollary 15.24 *Let A be an indecomposable GCM. Then A has indefinite type if and only if its Dynkin diagram $\Delta(A)$ does not appear on the finite list or the affine list.*

Proof. This follows from Theorems 15.1, 15.19 and 15.23 □

16
The invariant form, Weyl group, and root system

We now turn to the study of the Kac–Moody algebra $L(A)$ associated with a GCM A.

16.1 The invariant bilinear form

We recall from Section 4.2 that when A is a Cartan matrix the corresponding finite dimensional Lie algebra $L(A)$ has a non-degenerate symmetric bilinear form

$$\langle,\rangle : L(A) \times L(A) \to \mathbb{C}$$

which is invariant in the sense that

$$\langle [xy], z \rangle = \langle x, [yz] \rangle$$

for $x, y, z \in L(A)$. The Killing form has these properties.

In the case of a GCM A we cannot define the Killing form on $L(A)$ as in the finite dimensional case. We can nevertheless ask whether there is a non-degenerate, symmetric, invariant bilinear form on $L(A)$. This is not always the case, but we shall show that such a form does exist when A is symmetrisable.

Thus suppose A is a symmetrisable GCM. Then $A = DB$ where D is diagonal and B is symmetric. Let $D = \mathrm{diag}\,(d_1, \ldots, d_n)$. Let $(H, \Pi, \Pi^\vee)$ be a minimal realisation of A, where $\Pi^\vee = \{h_1, \ldots, h_n\}$ is a linearly independent subset of H, $\Pi = \{\alpha_1, \ldots, \alpha_n\}$ is a linearly independent subset of H^*, $\alpha_j(h_i) = A_{ij}$ and $\dim H = 2n - l$ where $l = \mathrm{rank}\,A$.

Let H' be the subspace of H spanned by $h_1, \ldots, h_n$ and let H'' be a complementary subspace of H' in H. Then we have

$$H = H' \oplus H'' \qquad \dim H' = n, \quad \dim H'' = n - l.$$

16.1 The invariant bilinear form

We define a bilinear form $\langle , \rangle : H \times H \to \mathbb{C}$ by the rules:

$$\langle h_i, h_j \rangle = d_i d_j B_{ij} \qquad i, j = 1, \ldots, n$$
$$\langle h_i, x \rangle = \langle x, h_i \rangle = d_i \alpha_i(x) \qquad \text{for } x \in H''$$
$$\langle x, y \rangle = 0 \qquad \text{for } x, y \in H''.$$

This is evidently a symmetric bilinear form on H.

Proposition 16.1 *This form on H is non-degenerate.*

Proof. We have $A = DB$ where D is diagonal and non-singular and B is symmetric. We have rank $B = l$. We observe that the symmetric matrix B of rank l has a non-singular $l \times l$ principal minor. If $l = n$ we can take B itself as the principal minor, so suppose $l < n$. Then, for some i, the ith row of B is a linear combination of the remaining rows of B. Since B is symmetric the ith column of B is a linear combination of the remaining columns of B. Let B' be the $(n-1) \times n$ matrix obtained from B by removing the ith row. Then rank $B' = l$. Let B'' be the $(n-1) \times (n-1)$ matrix obtained from B' by removing the ith column. Then rank $B'' = l$. Now B'' is symmetric of degree $n-1$ and rank l. Thus by induction B'' has a non-singular $l \times l$ principal minor, and this is the required principal minor of B.

It follows that the symmetrisable matrix A has a non-singular $l \times l$ principal minor. For let B_J be non-singular where J is a subset of $\{1, \ldots, n\}$ with $|J| = l$. Then $A_J = D_J B_J$ where $D_J = \text{diag}\{d_j, j \in J\}$ with each $d_j \neq 0$. Since D_J is non-singular A_J is also non-singular.

We now consider the special case in which $J = \{1, \ldots, l\}$. Then A has form

$$A = \begin{pmatrix} A_{11} & A_{12} \\ A_{21} & A_{22} \end{pmatrix} \begin{matrix} l \\ n-l \end{matrix} \qquad A_{11} \text{ non-singular.}$$
$$\begin{matrix} l & n-l \end{matrix}$$

By Proposition 14.2 we may extend the linearly independent sets $h_1, \ldots, h_n \in H$, $\alpha_1, \ldots, \alpha_n \in H^*$, to bases $h_1, \ldots, h_{2n-l}$; $\alpha_1, \ldots, \alpha_{2n-l}$ such that $\alpha_j(h_i) = C_{ij}$ where

$$C = \begin{pmatrix} A_{11} & A_{12} & O \\ A_{21} & A_{22} & I \\ O & I & O \end{pmatrix} \begin{matrix} l \\ n-l \\ n-l \end{matrix} .$$
$$\begin{matrix} l & n-l & n-l \end{matrix}$$

Let

$$D = \begin{pmatrix} D_1 & O \\ O & D_2 \end{pmatrix} \begin{matrix} l \\ n-l \end{matrix}.$$
$$\quad\quad\; l \quad\; n-l$$

Then

$$C = \begin{pmatrix} D_1 B_{11} & D_1 B_{12} & O \\ D_2 B_{21} & D_2 B_{22} & I \\ O & I & O \end{pmatrix}.$$

The symmetric matrix M of the bilinear form $\langle h_i, h_j \rangle$ $\;\; i,j \in \{1,\ldots,2n-l\}$ is

$$M = \begin{pmatrix} D_1 B_{11} D_1 & D_1 B_{12} D_2 & O \\ D_2 B_{21} D_1 & D_2 B_{22} D_2 & D_2 \\ O & D_2 & O \end{pmatrix}.$$

This matrix is non-singular since

$$\det M = \pm (\det D_1)^2 (\det D_2)^2 \det B_{11} \neq 0.$$

Now suppose A is any $n \times n$ symmetrisable GCM of rank l. Then A has a non-singular $l \times l$ principal minor A_J for some $J \subset \{1,\ldots,n\}$. Let K be the complementary subset of J in $\{1,\ldots,n\}$ and $L = \{n+1,\ldots,2n-l\}$. Then there exists a realisation $h_1, \ldots, h_{2n-l}$; $\alpha_1, \ldots, \alpha_{2n-l}$ whose matrix $\alpha_j(h_i) = C_{ij}$ may be written symbolically in the form

$$C = \begin{pmatrix} A_J & A_{JK} & O \\ A_{KJ} & A_K & I \\ O & I & O \end{pmatrix} \begin{matrix} J \\ K \\ L \end{matrix}.$$
$$\quad\;\; J \quad\; K \quad\; L$$

Let

$$D = \begin{pmatrix} D_J & O \\ O & D_K \end{pmatrix} \begin{matrix} J \\ K \end{matrix}.$$
$$\quad\;\; J \quad\; K$$

Then

$$C = \begin{pmatrix} D_J B_J & D_J B_{JK} & O \\ D_K B_{KJ} & D_K B_K & I \\ O & I & O \end{pmatrix}.$$

This time the symmetric matrix M of the bilinear form $\langle h_i, h_j \rangle$ for $i, j \in \{1, \ldots, 2n-l\}$ is

$$M = \begin{pmatrix} D_J B_J D_J & D_J B_{JK} D_K & 0 \\ D_K B_{KJ} D_J & D_K B_K D_K & D_K \\ 0 & D_K & 0 \end{pmatrix}.$$

Since $\det M = \pm (\det D_J)^2 (\det D_K)^2 \det B_J \neq 0$ the bilinear form is non-degenerate on H. $\square$

Theorem 16.2 *Suppose A is a symmetrisable GCM. Then the Kac–Moody algebra $L(A)$ has a non-degenerate symmetric invariant bilinear form.*

Proof. We have $L(A) = \bigoplus_{\alpha \in Q} L_\alpha$. For $\alpha = m_1 \alpha_1 + \cdots + m_n \alpha_n \in Q$ we define the height of α by $\operatorname{ht} \alpha = m_1 + \cdots + m_n$. Then

$$L(A) = \bigoplus_{i \in \mathbb{Z}} L_i$$

where L_i is the direct sum of all L_α with $\operatorname{ht} \alpha = i$. Since $[L_\alpha L_\beta] \subset L_{\alpha+\beta}$ we have $[L_i L_j] \subset L_{i+j}$. Thus $L(A)$ may be considered in this way as a $\mathbb{Z}$-graded Lie algebra.

We define, for each integer $r \geq 0$,

$$L(r) = \bigoplus_{-r \leq i \leq r} L_i.$$

Then we have

$$H = L(0) \subset L(1) \subset L(2) \subset \cdots$$

and $\bigcup_{r \geq 0} L(r) = L(A)$.

We have already defined a symmetric bilinear form on $H = L(0)$. We shall extend this definition to give a symmetric bilinear form on $L(r)$ for $r = 1, 2, 3, \ldots$ thus eventually defining such a form on $L(A)$. We shall define the form on $L(r)$ by induction on r, assuming it is already defined on $L(r-1)$.

We begin with the case $r = 1$. We have

$$L(1) = \left(\bigoplus_{i=1}^{n} \mathbb{C} f_i \right) \oplus H \oplus \left(\bigoplus_{i=1}^{n} \mathbb{C} e_i \right).$$

We define a bilinear form $\langle , \rangle$ on $L(1)$ which is uniquely determined by the following rules:

$\langle , \rangle$ agrees with the form already defined on H

$\langle L_i, L_j \rangle = 0$ unless $i+j=0$

$\langle e_i, f_i \rangle = \langle f_i, e_i \rangle = d_i$

$\langle e_i, f_j \rangle = \langle f_j, e_i \rangle = 0$ if $i \neq j$.

This bilinear form on $L(1)$ is clearly symmetric. We show

$$\langle [xy], z \rangle = \langle x, [yz] \rangle$$

for all $x, y, z \in L(1)$. In showing this we may assume $x \in L_i$, $y \in L_j$, $z \in L_k$ for some $i, j, z \in \mathbb{Z}$ with $|i|, |j|, |k| \leq 1$. We may assume $i+j+k=0$ as otherwise both sides of our required equality are zero. The relation is known already when i, j, k are all 0. Thus we may assume i, j, k are $1, -1, 0$ in some order. There are six possible orders, but it is only necessary to check three of them as the other three follow from them. Thus we show

$$\langle [e_i h], f_j \rangle = \langle e_i, [hf_j] \rangle$$
$$\langle [he_i], f_j \rangle = \langle h, [e_i f_j] \rangle$$
$$\langle [hf_j], e_i \rangle = \langle h, [f_j e_i] \rangle$$

for $h \in H$. Both sides are zero in these relations if $i \neq j$. If $i=j$ the relations are valid because

$$\langle h_i, h \rangle = d_i \alpha_i(h) \quad \text{for all } h \in H.$$

This follows from the definition of the form $\langle , \rangle$ on H. Thus we have

$$\langle [xy], z \rangle = \langle x, [yz] \rangle \quad \text{for all } x, y, z \in L(1).$$

Now suppose inductively that a symmetric bilinear form has already been defined on $L(r-1)$ and satisfies:

$\langle L_i, L_j \rangle = 0$ unless $i+j=0$ for $|i|, |j| \leq r-1$

$\langle [xy], z \rangle = \langle x, [yz] \rangle$ for all $x \in L_i$, $y \in L_j$, $z \in L_k$ with $|i|, |j|, |k| \leq r-1$ and $i+j+k=0$.

16.1 The invariant bilinear form

We shall show this form can be extended to one on $L(r)$ with analogous properties. We extend the form to $L(r)$ by defining

$$\langle L_i, L_j \rangle = 0 \quad \text{unless } i+j = 0 \text{ for } |i|, |j| \le r.$$

We must also define $\langle x, y \rangle = \langle y, x \rangle$ for $x \in L_r$, $y \in L_{-r}$. We assume $r \ge 2$.

Now we have $L(A) = N^- \oplus H \oplus N$ with $H = L_0$, $N^- = \bigoplus_{i<0} L_i$, $N = \bigoplus_{i>0} L_i$. The algebra N^- is generated by $f_1, \ldots, f_n$, thus each element of N^- can be written as a Lie word in $f_1, \ldots, f_n$, so is a linear combination of Lie monomials in $f_1, \ldots, f_n$. An element of L_{-r} is a linear combination of Lie monomials in $f_1, \ldots, f_n$ such that the number of factors in each Lie monomial is r. If $r \ge 2$ each Lie monomial is the Lie product of Lie monomials of degree s, t say with $s + t = r$. It follows that each element $y \in L_{-r}$ can be written in the form

$$y = \sum_j [c_j d_j]$$

where $c_j \in L_{-u_j}$, $d_j \in L_{-v_j}$ with $u_j > 0$, $v_j > 0$ and $u_j + v_j = r$. The expression of y in this form need not be unique.

Given $x \in L_r$, $y \in L_{-r}$ we write $y = \sum_j [c_j d_j]$ as above and wish to define

$$\langle x, y \rangle = \sum_j \langle [x c_j], d_j \rangle.$$

The right-hand side is known since $[x c_j]$ and d_j lie in $L(r-1)$, so if there is a form of the required type on $L(r)$ it must satisfy the above relation in order to be invariant. However, the right-hand side appears to depend on the particular expression $y = \sum_j [c_j d_j]$ for y which need not be unique. We must therefore show that the right-hand side remains the same if a different such expression for y is chosen.

In a similar way we can write $x \in L_r$ in the form

$$x = \sum_i [a_i b_i]$$

where $a_i \in L_{s_i}$, $b_i \in L_{t_i}$ and $s_i > 0$, $t_i > 0$ with $s_i + t_i = r$. We shall show

$$\sum_i \langle a_i, [b_i y] \rangle = \sum_j \langle [x c_j], d_j \rangle.$$

This will imply that the right-hand side is independent of the given expression for y, and also that the left-hand side is independent of the given expression for x. In fact it is sufficient to show

$$\langle a_i, [b_i [c_j d_j]] \rangle = \langle [[a_i b_i] c_j], d_j \rangle.$$

Now

$$\langle[[a_ib_i]c_j],d_j\rangle = \langle[[a_ic_j]b_i],d_j\rangle - \langle[[b_ic_j]a_i],d_j\rangle$$
$$= \langle[a_ic_j],[b_id_j]\rangle - \langle[b_ic_j],[a_id_j]\rangle$$
$$= \langle[a_ic_j],[b_id_j]\rangle - \langle[a_id_j],[b_ic_j]\rangle$$
$$= \langle a_i,[c_j[b_id_j]]\rangle - \langle a_i,[d_j[b_ic_j]]\rangle$$
$$= \langle a_i,[b_i[c_jd_j]]\rangle$$

using the invariance of the form on $L(r-1)$. Hence our form $\langle x,y\rangle$ is now well defined on $L(r)$, where it is bilinear and symmetric.

We must now check that

$$\langle[xy],z\rangle = \langle x,[yz]\rangle$$

when $x \in L_i, y \in L_j, z \in L_k$ with $|i|,|j|,|k| \le r$ and $i+j+k=0$. This is known already by induction unless at least one of $|i|,|j|,|k|$ is equal to r.

It is impossible for all of $|i|,|j|,|k|$ to be equal to r since $i+j+k=0$. We suppose first that just one of $|i|,|j|,|k|$ is r. Then the other two are non-zero. If $|i|=r$ then

$$\langle x,[yz]\rangle = \langle[xy],z\rangle$$

by definition of the form on $L(r)$. Similarly if $|k|=r$ this relation also holds by definition. So suppose $|j|=r$. We may assume that y has the form $y=[ab]$ where $a \in L_s, b \in L_t, s+t=j$ and $0 < |s| < |j|, 0 < |t| < |j|$. Then

$$\langle[xy],z\rangle = \langle[x[ab]],z\rangle$$
$$= \langle[[bx]a],z\rangle + \langle[[xa]b],z\rangle$$
$$= \langle[bx],[az]\rangle + \langle[xa],[bz]\rangle$$
$$= \langle[xb],[za]\rangle + \langle[xa],[bz]\rangle$$
$$= \langle x,[b[za]]\rangle + \langle x,[a[bz]]\rangle$$
$$= \langle x,[[ab]z]\rangle$$
$$= \langle x,[yz]\rangle$$

using the invariance of the form on $L(r-1)$.

Now suppose that two of $|i|,|j|,|k|$ are equal to r. Then i,j,k are $r,-r,0$ in some order. Thus one of x,y,z lies in H.

Suppose $x \in H$. We may again assume $y = [ab]$ where $a \in L_s$, $b \in L_t$, $s+t = j$, $0 < |s| < |j|$, $0 < |t| < |j|$.
Then

$$\begin{aligned}
\langle [xy], z \rangle &= \langle [x[ab]], z \rangle = \langle [[xa]b], z \rangle - \langle [[xb]a], z \rangle \\
&= \langle [xa], [bz] \rangle - \langle [xb], [az] \rangle \quad \text{by definition of} \langle, \rangle \text{on } L(r) \\
&= \langle x, [a[bz]] \rangle - \langle x, [b[az]] \rangle \quad \text{by invariance on } L(r-1) \\
&= \langle x, [[ab]z] \rangle = \langle x, [yz] \rangle.
\end{aligned}$$

If $z \in H$ the result also holds by using the symmetry of the form.

Finally suppose $y \in H$. Then we may assume $z = [ab]$ where $a \in L_s$, $b \in L_t$, $s+t = k$, $0 < |s| < |k|$, $0 < |t| < |k|$. Then

$$\begin{aligned}
\langle x, [yz] \rangle &= \langle x, [y[ab]] \rangle = \langle x, [a[yb]] \rangle + \langle x, [[ya]b] \rangle \\
&= \langle [xa], [yb] \rangle + \langle [x[ya]], b \rangle \quad \text{by definition of} \langle, \rangle \text{ on } L(r) \\
&= \langle [[xa]y], b \rangle + \langle [x[ya]], b \rangle \quad \text{by invariance on } L(r-1) \\
&= \langle [[xy]a], b \rangle \\
&= \langle [xy], [ab] \rangle \quad \text{by definition of} \langle, \rangle \text{on } L(r) \\
&= \langle [xy], z \rangle.
\end{aligned}$$

We have therefore proved invariance when $x \in L_i$, $y \in L_j$, $z \in L_k$ with $|i|, |j|, |k| \le r$ and $i+j+k = 0$. It follows that invariance holds for all $x, y, z \in L(r)$. By induction the form is therefore invariant on $L(A)$.

Thus we have now defined a symmetric invariant bilinear form on $L(A)$. We show it is non-degenerate. Let I be the kernel of $\langle , \rangle$, i.e. the set of $x \in L(A)$ such that $\langle x, y \rangle = 0$ for all $y \in L(A)$. Since the form is invariant I is an ideal of $L(A)$. Since by Proposition 16.1 the form is non-degenerate on restriction to H we have $I \cap H = O$. But the Kac–Moody algebra $L(A)$ has no non-zero ideal I with $I \cap H = O$. Hence $I = O$ and the form is non-degenerate on $L(A)$. $\square$

Note The proof of this theorem shows that any symmetric invariant bilinear form on $L(A)$ is uniquely determined by its restriction to H.

Definition *The form constructed in Theorem 16.2 will be called the* **standard invariant** *form on $L(A)$.*

Corollary 16.3 *For each $i \in \mathbb{Z}$ the pairing $L_i \times L_{-i} \to \mathbb{C}$ given by $x, y \to \langle x, y \rangle$ is non-degenerate.*

Proof. Suppose $x \in L_i$ satisfies $\langle x, y \rangle = 0$ for all $y \in L_{-i}$. Since $\langle L_i, L_j \rangle = 0$ unless $i + j = 0$ we have $\langle x, y \rangle = 0$ for all $y \in L(A)$. Hence $x = 0$. □

Corollary 16.4 $\langle L_\alpha, L_\beta \rangle = 0$ *unless* $\alpha + \beta = 0$.

Proof. Suppose $\alpha + \beta \neq 0$ and let $x \in L_\alpha, y \in L_\beta$. Choose $h \in H$ with
$$(\alpha + \beta)(h) \neq 0.$$
Then
$$\langle [xh], y \rangle = \langle x, [hy] \rangle$$
implies
$$-\alpha(h)\langle x, y \rangle = \beta(h)\langle x, y \rangle$$
that is
$$(\alpha + \beta)(h)\langle x, y \rangle = 0.$$
Hence $\langle x, y \rangle = 0$. □

Since the form $\langle , \rangle$ is non-degenerate on H it determines a bijection $H^* \to H$ given by $\alpha \to h'_\alpha$ where
$$\langle h'_\alpha, h \rangle = \alpha(h) \qquad \text{for all } h \in H.$$

Corollary 16.5 (i) Suppose $x \in L_\alpha, y \in L_{-\alpha}$. Then $[xy] = \langle x, y \rangle h'_\alpha$.
(ii) *The pairing* $L_\alpha \times L_{-\alpha} \to \mathbb{C}$ *given by* $x, y \to \langle x, y \rangle$ *is non-degenerate.*
(iii) *For each* $x \in L_\alpha$ *with* $x \neq 0$ *there exists* $y \in L_{-\alpha}$ *with* $[xy] \neq 0$.

Proof. (i) Consider the element $[xy] - \langle x, y \rangle h'_\alpha \in H$. For all $h \in H$ we have
$$\langle [xy] - \langle x, y \rangle h'_\alpha, h \rangle = \langle [xy], h \rangle - \langle x, y \rangle \langle h'_\alpha, h \rangle$$
$$= \langle x, [yh] \rangle - \alpha(h)\langle x, y \rangle$$
$$= 0.$$
Since the form is non-degenerate on H we deduce that $[xy] - \langle x, y \rangle h'_\alpha = 0$.
(ii) Since the form is non-degenerate on $L(A)$ and $\langle L_\alpha, L_\beta \rangle = 0$ unless $\beta = -\alpha$ the pairing $L_\alpha \times L_{-\alpha} \to \mathbb{C}$ must be non-degenerate.
(iii) For each $x \in L_\alpha$ with $x \neq 0$ there exists $y \in L_{-\alpha}$ with $\langle x, y \rangle \neq 0$. Hence $[xy] \neq 0$ by (i). □

We now consider to what extent a non-degenerate symmetric invariant bilinear form on $L(A)$ is unique. The following proposition deals with this question.

16.1 The invariant bilinear form

Proposition 16.6 *Suppose A is an indecomposable symmetrisable GCM and $\{,\}$ is a non-degenerate symmetric invariant bilinear form on the Kac–Moody algebra $L(A)$. Then there exists a non-zero $\xi \in \mathbb{C}$ such that*

$$\{x, y\} = \xi \langle x, y \rangle \text{ for all } x, y \in L(A)'.$$

Thus such a form is determined on the subalgebra $L(A)'$ up to a non-zero constant.

Proof. The argument of Corollary 16.4 shows that $\{L_\alpha, L_\beta\} = 0$ whenever $\alpha + \beta \neq 0$. In particular we have $\{H, L_\alpha\} = 0$ whenever $\alpha \neq 0$. Since $L(A) = H \oplus \sum_{\alpha \neq 0} L_\alpha$ it follows that $\{,\}$ is non-degenerate on restriction to H. The form $\{,\}$ on $L(A)$ is determined by its restriction to H and by the map $L_\alpha \times L_{-\alpha} \to \mathbb{C}$ given by $x, y \to \{x, y\}$ for each $\alpha \in \Phi$. The argument of Corollary 16.5 shows that, for $x \in L_\alpha, y \in L_{-\alpha}$, we have

$$[xy] = \{x, y\} k'_\alpha$$

where k'_α is the unique element of H satisfying $\{k'_\alpha, h\} = \alpha(h)$ for all $h \in H$.

We therefore have

$$[L_\alpha L_{-\alpha}] = \mathbb{C} h'_\alpha = \mathbb{C} k'_\alpha$$

for each $\alpha \in \Phi$. Thus there exists a non-zero $\xi_\alpha \in \mathbb{C}$ with $h'_\alpha = \xi_\alpha k'_\alpha$. This implies that

$$\{h'_\alpha, h\} = \xi_\alpha \langle h'_\alpha, h \rangle \qquad \text{for all } h \in H$$

since both sides are equal to $\xi_\alpha \alpha(h)$. Let α_i, α_j be simple roots. Then we have

$$\left\{ h'_{\alpha_i}, h'_{\alpha_j} \right\} = \xi_{\alpha_i} \left\langle h'_{\alpha_i}, h'_{\alpha_j} \right\rangle$$

and so by the symmetry of the forms

$$\xi_{\alpha_i} \left\langle h'_{\alpha_i}, h'_{\alpha_j} \right\rangle = \xi_{\alpha_j} \left\langle h'_{\alpha_i}, h'_{\alpha_j} \right\rangle.$$

If $A_{ij} \neq 0$ then $\langle h'_{\alpha_i}, h'_{\alpha_j} \rangle \neq 0$ and we have $\xi_{\alpha_i} = \xi_{\alpha_j}$. If the GCM A is indecomposable this shows that there exists $\xi \neq 0$ in $\mathbb{C}$ such that $\xi_{\alpha_i} = \xi$ for all simple roots α_i. Thus

$$\{h'_{\alpha_i}, h\} = \xi \langle h'_{\alpha_i}, h \rangle \qquad \text{for all } h \in H.$$

Now for any $\alpha \in \Phi$ h'_α is a linear combination of the h'_{α_i}. Hence

$$\{h'_\alpha, h\} = \xi \langle h'_\alpha, h \rangle \qquad \text{for all } h \in H.$$

Thus $\xi_\alpha = \xi$ for all $\alpha \in \Phi$. Using the equations

$$[xy] = \{x, y\} k'_\alpha = \langle x, y \rangle h'_\alpha$$

for $x \in L_\alpha, y \in L_{-\alpha}$ we deduce that

$$\{x, y\} = \xi \langle x, y \rangle \qquad \text{for } x \in L_\alpha, y \in L_{-\alpha}.$$

Now $L(A)'$ was defined as the subalgebra of $L(A)$ generated by $e_1, \ldots, e_n$, $f_1, \ldots, f_n$. We recall from Proposition 14.21 that $L(A)' = [L(A)L(A)]$. It follows that $L(A)' \cap H$ is generated by $[L_\alpha L_{-\alpha}] = \mathbb{C} h'_\alpha$ for all $\alpha \in \Phi$. It follows that

$$\{h', h\} = \xi \langle h', h \rangle \qquad \text{for all } h, h' \in L(A)' \cap H$$

$$\{x, y\} = \xi \langle x, y \rangle \qquad \text{for all } x \in L_\alpha, \ y \in L_{-\alpha}$$

But $L(A)' = (L(A)' \cap H) \oplus \sum_{\alpha \neq 0} L_\alpha$ also by Proposition 14.21. Thus we see that

$$\{x, y\} = \xi \langle x, y \rangle \qquad \text{for all } x, y \in L(A)'. \qquad \square$$

Corollary 16.7 $L(A)' \cap H$ is the subspace of H spanned by $h_1, \ldots, h_n$.

Proof. We saw in the proof of Proposition 16.6 that $L(A)' \cap H$ is the subspace generated by the elements h'_α for all $\alpha \in \Phi$. Each h'_α is a linear combination of $h_1, \ldots, h_n$ and so the result follows. $\square$

Corollary 16.8 *Any non-degenerate symmetric invariant bilinear form on a finite dimensional simple Lie algebra is a constant multiple of the Killing form.*

Proof. Since $L(A)$ is simple we have $L(A)' = [L(A)L(A)] = L(A)$. Thus the given form is a constant multiple of the Killing form on the whole of $L(A)$. $\square$

Important comment on notation. In the case when $L(A)$ has finite type the standard invariant form is not the same as the Killing form. It is a constant multiple of the Killing form.

In our development of the theory of finite dimensional simple Lie algebras we have used the notation $\langle , \rangle$ to denote the Killing form. In the theory of Kac–Moody algebras the Killing form does not exist in general, but the standard invariant form exists whenever the Kac–Moody algebra is symmetrisable. In the subsequent development the notation $\langle , \rangle$ will denote the standard

16.2 The Weyl group of a Kac–Moody algebra

Lemma 16.9 *Let $x \in L(A)$ and J be the ideal of $L(A)$ generated by x. Then $J = \mathfrak{U}(L(A))x$.*

Proof. The adjoint representation of $L(A)$ gives a Lie algebra homomorphism $L(A) \to [\mathrm{End}\, L(A)]$. By Proposition 9.3 there is an associative algebra homomorphism
$$\mathfrak{U}(L(A)) \to \mathrm{End}\, L(A).$$
A subspace K of $L(A)$ satisfies $[L(A), K] \subset K$ if and only if $\mathfrak{U}(L(A))K \subset K$. Now we have $[L(A), J] \subset J$. Hence $\mathfrak{U}(L(A))J \subset J$. Since $x \in J$ we have $\mathfrak{U}(L(A))x \subset J$.

On the other hand $\mathfrak{U}(L(A))(\mathfrak{U}(L(A))x) = \mathfrak{U}(L(A))x$, thus
$$[L(A), \mathfrak{U}(L(A))x] \subset \mathfrak{U}(L(A))x.$$
Hence $\mathfrak{U}(L(A))x$ is an ideal of $L(A)$ containing x. Hence $\mathfrak{U}(L(A))x \supset J$. Thus we must have equality. $\square$

Proposition 16.10 *In $L(A)$ we have, for $i \ne j$, $(\mathrm{ad}\, e_i)^{1-A_{ij}} e_j = 0$ and $(\mathrm{ad}\, f_i)^{1-A_{ij}} f_j = 0$.*

Proof. We shall show $(\mathrm{ad}\, f_i)^{1-A_{ij}} f_j = 0$. The other relation holds similarly.

Let $x = (\mathrm{ad}\, f_i)^{1-A_{ij}} f_j \in N^-$. We shall show $[e_k, x] = 0$ for all $k = 1, \ldots, n$. Suppose this is so. Then the set of all $y \in L(A)$ with $[yx] = 0$ is a subalgebra containing $e_1, \ldots, e_n$, so contains N. Thus $[N, x] = 0$ and so $\mathfrak{U}(N)x = \mathbb{C}x$. Since $L(A) = N^- \oplus H \oplus N$ we have $\mathfrak{U}(L(A)) = \mathfrak{U}(N^-)\mathfrak{U}(H)\mathfrak{U}(N)$ by the PBW basis theorem. Hence
$$\mathfrak{U}(L(A))x = \mathfrak{U}(N^-)\mathfrak{U}(H)\mathfrak{U}(N)x = \mathfrak{U}(N^-)\mathfrak{U}(H)x.$$
Since $[H, N^-] \subset N^-$ we have $\mathfrak{U}(H)N^- \subset N^-$ and $\mathfrak{U}(H)x \subset N^-$. Thus
$$\mathfrak{U}(L(A))x \subset \mathfrak{U}(N^-)N^- \subset N^-.$$
Let $J = \mathfrak{U}(L(A))x$. This is the ideal of $L(A)$ generated by x, by Lemma 16.9. We have $J \subset N^-$ so $J \cap H = 0$. This implies $J = 0$ by definition of $L(A)$. Thus $x = 0$.

Thus in order to obtain the required result $x=0$ it is sufficient to show

$$\left[e_k, (\operatorname{ad} f_i)^{1-A_{ij}} f_j\right] = 0 \quad \text{when } i \neq j.$$

We first suppose $k \neq i$ and $k \neq j$. Then

$$\left[e_k, (\operatorname{ad} f_i)^t f_j\right] = \left[e_k, \left[f_i, (\operatorname{ad} f_i)^{t-1} f_j\right]\right]$$
$$= \left[[e_k, f_i], (\operatorname{ad} f_i)^{t-1} f_j\right] + \left[f_i, \left[e_k, (\operatorname{ad} f_i)^{t-1} f_j\right]\right]$$
$$= \left[f_i, \left[e_k, (\operatorname{ad} f_i)^{t-1} f_j\right]\right].$$

Repeating we obtain, for each t,

$$\left[e_k, (\operatorname{ad} f_i)^t f_j\right] = (\operatorname{ad} f_i)^t \left[e_k, f_j\right] = 0.$$

Next suppose $k = j$. Then

$$\left[e_j, (\operatorname{ad} f_i)^t f_j\right] = (\operatorname{ad} f_i)^t \left[e_j, f_j\right] = (\operatorname{ad} f_i)^t h_j$$

as above. If $1 - A_{ij} \geq 2$ then this shows that $\left[e_j, (\operatorname{ad} f_i)^{1-A_{ij}} f_j\right] = 0$. If $1 - A_{ij} = 1$ then $A_{ij} = 0$ so $[f_i, h_j] = A_{ji} f_i = 0$. Thus $\left[e_j, (\operatorname{ad} f_i)^{1-A_{ij}} f_j\right] = 0$ in this case also.

Finally we suppose $k = i$. Then

$$\left[e_i, (\operatorname{ad} f_i)^t f_j\right] = \left[e_i, \left[f_i, (\operatorname{ad} f_i)^{t-1} f_j\right]\right]$$
$$= \left[[e_i, f_i], (\operatorname{ad} f_i)^{t-1} f_j\right] + \left[f_i, \left[e_i, (\operatorname{ad} f_i)^{t-1} f_j\right]\right]$$
$$= \left[h_i, (\operatorname{ad} f_i)^{t-1} f_j\right] + \left[f_i, \left[e_i, (\operatorname{ad} f_i)^{t-1} f_j\right]\right]$$
$$= -((t-1)\alpha_i + \alpha_j)(h_i)(\operatorname{ad} f_i)^{t-1} f_j + \left[f_i, \left[e_i, (\operatorname{ad} f_i)^{t-1} f_j\right]\right]$$
$$= (-2(t-1) - A_{ij})(\operatorname{ad} f_i)^{t-1} f_j + \left[f_i, \left[e_i, (\operatorname{ad} f_i)^{t-1} f_j\right]\right].$$

Repeating, we obtain

$$(-2(t-1) - A_{ij})(\operatorname{ad} f_i)^{t-1} f_j + (-2(t-2) - A_{ij})(\operatorname{ad} f_i)^{t-1} f_j + \cdots$$
$$+ (-A_{ij})(\operatorname{ad} f_i)^{t-1} f_j = -t(t - 1 + A_{ij})(\operatorname{ad} f_i)^{t-1} f_j.$$

We now put $t = 1 - A_{ij}$. Then we have

$$\left[e_i, (\operatorname{ad} f_i)^{1-A_{ij}} f_j\right] = 0.$$

This completes the proof in all cases. $\square$

16.2 The Weyl group of a Kac–Moody algebra

Using this result of Proposition 16.10 we may deduce, as in Proposition 7.17, that the maps $\operatorname{ad} e_i$ and $\operatorname{ad} f_i$ are locally nilpotent. Then the proof of Proposition 3.4 shows that $\exp \operatorname{ad} e_i$ and $\exp \operatorname{ad}(-f_i)$ are automorphisms of $L(A)$. Let

$$n_i = \exp \operatorname{ad} e_i \cdot \exp \operatorname{ad}(-f_i) \cdot \exp \operatorname{ad} e_i \in \operatorname{Aut} L(A).$$

Proposition 16.11 $n_i(H) = H$. For $x \in H$ we have

$$n_i(x) = x - \alpha_i(x) h_i.$$

Proof. Let $x \in H$. Then

$$\exp \operatorname{ad} e_i \cdot x = (1 + \operatorname{ad} e_i) x = x + [e_i, x] = x - \alpha_i(x) e_i$$

$$\exp \operatorname{ad}(-f_i) \cdot (x - \alpha_i(x) e_i) = \left(1 - \operatorname{ad} f_i + \frac{(\operatorname{ad} f_i)^2}{2}\right)(x - \alpha_i(x) e_i)$$

$$= x - \alpha_i(x) e_i - [f_i, x] + \alpha_i(x)[f_i, e_i] + \tfrac{1}{2} \operatorname{ad} f_i([f_i, x] + \alpha_i(x) h_i)$$

$$= x - \alpha_i(x) e_i - \alpha_i(x) f_i - \alpha_i(x) h_i + \tfrac{1}{2} \alpha_i(x) \cdot 2 f_i$$

$$= x - \alpha_i(x) e_i - \alpha_i(x) h_i$$

$$\exp \operatorname{ad} e_i (x - \alpha_i(x) e_i - \alpha_i(x) h_i) = (1 + \operatorname{ad} e_i)(x - \alpha_i(x) e_i - \alpha_i(x) h_i)$$

$$= x - \alpha_i(x) e_i - \alpha_i(x) h_i + [e_i, x] - \alpha_i(x)[e_i, h_i]$$

$$= x - \alpha_i(x) e_i - \alpha_i(x) h_i - \alpha_i(x) e_i + 2\alpha_i(x) e_i$$

$$= x - \alpha_i(x) h_i.$$

This gives the required result. □

Proposition 16.12 *The map $s_i : H \to H$ induced by n_i satisfies $s_i^2 = 1$, $s_i(h_i) = -h_i$, $s_i(x) = x$ when $\langle h_i, x \rangle = 0$.*

Proof. This follows from $s_i(x) = x - \alpha_i(x) h_i$ together with $\alpha_i(h_i) = 2$ and $\langle h_i, x \rangle = d_i \alpha_i(x)$. □

The maps $s_i : H \to H$ are called **fundamental reflections**. The group W of non-singular linear transformations of H generated by $s_1, \ldots, s_n$ is called the **Weyl group** W of $L(A)$.

Proposition 16.13 *The bilinear form $\langle , \rangle$ on H is invariant under W.*

Proof. Let $x, y \in H$. Then

$$\langle s_i x, s_i y \rangle = \langle x - \alpha_i(x) h_i, y - \alpha_i(y) h_i \rangle$$
$$= \langle x, y \rangle - \alpha_i(x) \langle h_i, y \rangle - \alpha_i(y) \langle x, h_i \rangle + \alpha_i(x) \alpha_i(y) \langle h_i, h_i \rangle$$
$$= \langle x, y \rangle - \alpha_i(x) d_i \alpha_i(y) - \alpha_i(y) d_i \alpha_i(x) + \alpha_i(x) \alpha_i(y) \cdot 2 d_i$$
$$= \langle x, y \rangle. \qquad \square$$

We may also define an action of W on H^* by

$$(w\lambda) x = \lambda \left(w^{-1} x \right) \qquad \text{for } w \in W, \ \lambda \in H^*, \ x \in H.$$

This action is compatible with the isomorphism $H^* \to H$ given by $\lambda \to h'_\lambda$ where $\langle h'_\lambda, x \rangle = \lambda(x)$ for all $x \in H$. For suppose $w(\lambda) = \mu$ for $\lambda, \mu \in H^*$. Then

$$\langle w(h'_\lambda), x \rangle = \langle h'_\lambda, w^{-1}(x) \rangle = \lambda \left(w^{-1}(x) \right)$$
$$= (w\lambda) x = \mu(x) = \langle h'_\mu, x \rangle$$

for all $x \in H$. Thus $w(h'_\lambda) = h'_\mu$.

Proposition 16.14 *The action of s_i on H^* is given by*

$$s_i(\lambda) = \lambda - \lambda(h_i) \alpha_i.$$

Proof. Let $x \in H$. Then

$$(s_i \lambda) x = \lambda \left(s_i^{-1} x \right) = \lambda (s_i x) = \lambda (x - \alpha_i(x) h_i)$$
$$= \lambda(x) - \lambda(h_i) \alpha_i(x) = (\lambda - \lambda(h_i) \alpha_i) x. \qquad \square$$

In fact the Weyl group acts on the root system Φ of $L(A)$.

Proposition 16.15 *If $\alpha \in \Phi$, $w \in W$ then $w(\alpha) \in \Phi$. Moreover $\dim L_\alpha = \dim L_{w(\alpha)}$.*

Proof. The proof of Proposition 7.21 also applies in our present situation. $\square$

We shall now determine the order of the product $s_i s_j$ of two distinct fundamental reflections.

16.2 The Weyl group of a Kac–Moody algebra

Theorem 16.16 *Suppose $i \neq j$. Then the order of $s_i s_j \in W$ is:*

$$\begin{array}{ll} 2 & \text{if } A_{ij}A_{ji} = 0 \\ 3 & \text{if } A_{ij}A_{ji} = 1 \\ 4 & \text{if } A_{ij}A_{ji} = 2 \\ 6 & \text{if } A_{ij}A_{ji} = 3 \\ \infty & \text{if } A_{ij}A_{ji} \geq 4. \end{array}$$

Proof. The Weyl group W acts faithfully on H^*. Let K be the 2-dimensional subspace of H^* given by $K = \mathbb{C}\alpha_i + \mathbb{C}\alpha_j$. We have

$$s_i(\alpha_i) = -\alpha_i, \quad s_i(\alpha_j) = \alpha_j - A_{ij}\alpha_i$$
$$s_j(\alpha_i) = \alpha_i - A_{ji}\alpha_j, \quad s_j(\alpha_j) = -\alpha_j.$$

Thus the subgroup $\langle s_i, s_j \rangle$ of W acts on K. We obtain a 2-dimensional representation of $\langle s_i, s_j \rangle$ given by

$$s_i \to \begin{pmatrix} -1 & -A_{ij} \\ 0 & 1 \end{pmatrix}, \quad s_j \to \begin{pmatrix} 1 & 0 \\ -A_{ji} & -1 \end{pmatrix}, \quad s_i s_j \to \begin{pmatrix} -1 + A_{ij}A_{ji} & A_{ij} \\ -A_{ji} & -1 \end{pmatrix}.$$

Consider the order of this 2×2 matrix representing $s_i s_j$. Its characteristic polynomial is

$$\begin{vmatrix} \lambda + 1 - A_{ij}A_{ji} & -A_{ij} \\ A_{ji} & \lambda + 1 \end{vmatrix} = \lambda^2 + (2 - A_{ij}A_{ji})\lambda + 1.$$

The discriminant of this polynomial is

$$D = (2 - A_{ij}A_{ji})^2 - 4 = A_{ij}A_{ji}(A_{ij}A_{ji} - 4).$$

Thus there are two equal eigenvalues if $A_{ij}A_{ji} = 0$ or 4, two distinct complex eigenvalues if $A_{ij}A_{ji} = 1, 2$ or 3, and two distinct real eigenvalues if $A_{ij}A_{ji} > 4$.

Suppose $A_{ij}A_{ji} = 0$. Then $s_i s_j \to \begin{pmatrix} -1 & 0 \\ 0 & -1 \end{pmatrix}$ and the matrix has order 2.

Suppose $A_{ij}A_{ji} = 1$. Then the characteristic polynomial is $\lambda^2 + \lambda + 1$ so the eigenvalues are ω, ω^2 where $\omega = e^{2\pi i/3}$. Thus the matrix is similar to $\begin{pmatrix} \omega & 0 \\ 0 & \omega^2 \end{pmatrix}$ and so has order 3.

Suppose $A_{ij}A_{ji} = 2$. The characteristic polynomial is then $\lambda^2 + 1 = (\lambda - i)(\lambda + i)$. Thus the matrix is similar to $\begin{pmatrix} i & 0 \\ 0 & -i \end{pmatrix}$ and so has order 4.

Suppose $A_{ij}A_{ji} = 3$. The characteristic polynomial is $\lambda^2 - \lambda + 1 = (\lambda + \omega)(\lambda + \omega^2)$. Thus the matrix is similar to $\begin{pmatrix} -\omega & 0 \\ 0 & -\omega^2 \end{pmatrix}$ and so has order 6.

Suppose $A_{ij}A_{ji}=4$. The characteristic polynomial is $\lambda^2-2\lambda+1=(\lambda-1)^2$. The eigenvalues are 1, 1, so the matrix is similar to $\begin{pmatrix}1&\xi\\0&1\end{pmatrix}$ with $\xi\neq 0$. This matrix has infinite order.

Now suppose $A_{ij}A_{ji}>4$. Then the eigenvalues are real and their product is 1. They are also positive and unequal, so have form ξ, ξ^{-1} where $\xi>1$. Thus the matrix is similar to $\begin{pmatrix}\xi&0\\0&\xi^{-1}\end{pmatrix}$ so has infinite order.

We have so far considered the action of s_is_j on the 2-dimensional subspace K of H^*. We now consider the action of s_is_j on the whole of H^*. Let

$$K'=\{\lambda\in H^* \; ; \; \lambda(h_i)=0, \quad \lambda(h_j)=0\}.$$

Then $\dim K'=\dim H^*-2$. Let $\lambda\in K\cap K'$. Then $\lambda=\xi\alpha_i+\eta\alpha_j$ and

$$\lambda(h_i)=2\xi+\eta A_{ij}=0$$
$$\lambda(h_j)=\xi A_{ji}+2\eta=0.$$

Now $\begin{vmatrix}2&A_{ij}\\A_{ji}&2\end{vmatrix}=4-A_{ij}A_{ji}$. Thus if $A_{ij}A_{ji}\neq 4$ we have $\xi=0, \eta=0$ so $K\cap K'=0$. Then $H=K\oplus K'$. Now s_is_j acts trivially on K' since, for $\lambda\in K'$, we have

$$s_is_j(\lambda)=s_i(\lambda-\lambda(h_j)\alpha_j)=s_i(\lambda)=\lambda-\lambda(h_i)\alpha_i=\lambda.$$

Thus the order of s_is_j on H^* is equal to the order of s_is_j on K provided $A_{ij}A_{ji}\neq 4$. If $A_{ij}A_{ji}=4$ the order of s_is_j on K is infinite, so s_is_j has infinite order on H^*. □

We now define $l(w)$ and $n(w)$ for $w\in W$ in the same way as when $L(A)$ is finite dimensional. $l(w)$ is the minimal length of w as a product of generators $s_1,\ldots,s_n$, and $n(w)$ is the number of $\alpha\in\Phi^+$ with $w(\alpha)\in\Phi^-$. Then the proof of Theorem 5.15 also applies in our present situation and shows that W satisfies the deletion condition. Also the proof of Corollary 5.16 applies in our situation and shows that $l(w)=n(w)$. Finally the proof of Theorem 5.18 applies and shows that W is generated by $s_1,\ldots,s_n$ as a Coxeter group. Thus we have:

Theorem 16.17 *The Weyl group W of the Kac–Moody algebra $L(A)$ is a Coxeter group generated by $s_1,\ldots,s_n$ with relations*

$$s_i^2=1$$
$$(s_is_j)^2=1 \quad \text{if} \quad A_{ij}A_{ji}=0$$
$$(s_is_j)^3=1 \quad \text{if} \quad A_{ij}A_{ji}=1$$
$$(s_is_j)^4=1 \quad \text{if} \quad A_{ij}A_{ji}=2$$
$$(s_is_j)^6=1 \quad \text{if} \quad A_{ij}A_{ji}=3.$$

16.3 The roots of a Kac–Moody algebra

Let A be a GCM and $L(A)$ the corresponding Kac–Moody algebra. Then

$$L(A) = H \oplus \sum_{\alpha \in \Phi} L_\alpha$$

where $\Phi = \{\alpha \neq 0 \,;\, L_\alpha \neq O\}$. Φ is the set of roots of $L(A)$. We recall that $\Phi = \Phi^+ \cup \Phi^-$ where $\Phi^+ = \Phi \cap Q^+$ and $\Phi^- = \Phi \cap Q^-$. These are the positive and negative roots. $\Pi = \{\alpha_1, \ldots, \alpha_n\}$ is a subset of Φ^+ called the set of fundamental roots. The multiplicity of the root α is defined as $\dim L_\alpha$. We know from Proposition 14.19 that the fundamental roots $\alpha_1, \ldots, \alpha_n$ have multiplicity 1. We also know from Proposition 16.15 that the Weyl group W acts on Φ and preserves multiplicities.

Definition $\alpha \in \Phi$ *is called a **real root** if there exist $\alpha_i \in \Pi$ and $w \in W$ such that $\alpha = w(\alpha_i)$.*

$\alpha \in \Phi$ *is called an **imaginary root** if α is not real.*

We note that if α is a real root so is $-\alpha$. For let $\alpha = w(\alpha_i)$. Then $-\alpha = ws_i(\alpha_i)$. It follows that if α is an imaginary root so is $-\alpha$.

Proposition 16.18 *Let α be a real root. Then α has multiplicity 1. Also, for $k \in \mathbb{Z}$, $k\alpha$ is a root if and only if $k = \pm 1$.*

Proof. Since $\alpha = w(\alpha_i)$ and α_i has multiplicity 1, Proposition 16.15 implies that α has multiplicity 1. We also know from Proposition 14.19 that if $k > 1$ then $k\alpha_i$ is not a root. Since $k\alpha = w(k\alpha_i)$, $k\alpha$ is also not a root. □

We now consider the imaginary roots. Let Φ_{im}^+ be the set of positive imaginary roots.

Proposition 16.19 *If $\alpha \in \Phi_{\text{Im}}^+$ and $w \in W$ then $w(\alpha) \in \Phi_{\text{Im}}^+$.*

Proof. We know that W acts both on Φ and on the set Φ_{Re} of real roots. Hence W acts on the set Φ_{Im} of imaginary roots. We must show that an element $w \in W$ cannot change the sign of an imaginary root. Let

$$\alpha = \sum_{i=1}^n k_i \alpha_i \qquad k_i \geq 0.$$

Now at least two coefficients k_i must be positive. Otherwise α would be a multiple of some α_i and hence equal to α_i. But then α would be real, a contradiction. Now $s_i(\alpha) = \alpha - \alpha(h_i)\alpha_i$, thus $s_i(\alpha)$ contains at least one fundamental root with positive coefficient. Hence $s_i(\alpha) \in \Phi_{\text{Im}}^+$. Since $w \in W$ is a product of fundamental reflections s_i we have $w(\alpha) \in \Phi_{\text{Im}}^+$. □

We now introduce the fundamental chamber in the context of Kac–Moody algebras. We recall that in Section 12.3 the fundamental chamber was defined for finite dimensional semisimple Lie algebras. In the present context we begin with a GCM A and take a real minimal realisation $(H_\mathbb{R}, \Pi, \Pi^v)$ as in Remark 14.20. We then define the fundamental chamber as

$$C = \{\lambda \in H_\mathbb{R}^* \ ; \ \lambda(h_i) > 0 \ \text{ for } i = 1, \ldots, n\}.$$

Its closure is

$$\bar{C} = \{\lambda \in H_\mathbb{R}^* \ ; \ \lambda(h_i) \geq 0 \ \text{ for } i = 1, \ldots, n\}.$$

Proposition 16.20 *Suppose $\alpha \in \Phi_{\text{Im}}^+$. Then there exists $w \in W$ with $w(\alpha) \in -\bar{C}$.*

Proof. Consider the set of all elements $w(\alpha)$ for $w \in W$. These are all positive imaginary roots by Proposition 16.19. Let β be such a root for which $\text{ht}\,\beta$ is as small as possible. Let $\beta = \sum k_i \alpha_i$. Then $s_i(\beta) = \beta - \beta(h_i)\alpha_i$. Since $\text{ht}\,s_i(\beta) \geq \text{ht}\,\beta$ we have $\beta(h_i) \leq 0$. This holds for all i, thus $\beta \in -\bar{C}$. □

Proposition 16.21 *Let $\alpha \in \Phi$, $\alpha = \sum_{i=1}^n k_i \alpha_i$ and $\text{supp}\,\alpha = \{i \ ; \ k_i \neq 0\}$. Then $\text{supp}\,\alpha$ is connected.*

Proof. We may assume $\alpha \in \Phi^+$. Let $\text{supp}\,\alpha = J \subset \{1, \ldots, n\}$. Suppose if possible that J is disconnected, that is $J = J_1 \cup J_2$ with J_1, J_2 non-empty and $A_{ij} = 0$ for all $i \in J_1, j \in J_2$.

We shall show that $[e_i\, e_j] = 0$ for all $i \in J_1, j \in J_2$. We first show the weaker condition

$$[[e_i e_j] f_k] = 0 \quad \text{for } i \in J_1, \ j \in J_2, \ k = 1, \ldots, n.$$

We have

$$[[e_i e_j] f_k] = [e_i [e_j f_k]] + [[e_i f_k] e_j].$$

If $k \notin \{i, j\}$ then $[e_i f_k] = 0$ and $[e_j f_k] = 0$.

If $k = i$ then $[[e_i e_j] f_k] = [h_i e_j] = A_{ij} e_j = 0$.

If $k = j$ then $[[e_i e_j] f_k] = [e_i h_j] = -A_{ji} e_i = 0$.

Thus in all cases $[[e_i e_j] f_k] = 0$ for all k.

16.3 The roots of a Kac–Moody algebra

Write $x = [e_i e_j]$. The ideal of $L(A)$ generated by x is $\mathfrak{U}(L(A))x$ as in Lemma 16.9. Since $L(A) = N \oplus H \oplus N^-$ we have $\mathfrak{U}(L(A)) = \mathfrak{U}(N)\mathfrak{U}(H)\mathfrak{U}(N^-)$. Now N^- is generated by $f_1, \ldots, f_n$ and $[x f_i] = 0$ for each i, thus $\mathfrak{U}(N^-)x = \mathbb{C}x$. Since $[HN] \subset N$ we have $\mathfrak{U}(H)N \subset N$ and so $\mathfrak{U}(H)\mathfrak{U}(N^-)x \subset N$ since $x \in N$. Finally $\mathfrak{U}(N)\mathfrak{U}(H)\mathfrak{U}(N^-)x \subset \mathfrak{U}(N)N \subset N$. Thus $\mathfrak{U}(L(A))x$ is an ideal of $L(A)$ intersecting H in O. By definition of $L(A)$ this ideal must be O. In particular we have $x = 0$. Thus $[e_i e_j] = 0$ for all $i \in J_1, j \in J_2$.

We use this fact to obtain the required contradiction. Since $\alpha \in \Phi^+$ we have $L_\alpha \neq O$ and $L_\alpha \subset N$. The elements of L_α are Lie words in $e_1, \ldots, e_n$ of weight α, and so are linear combinations of Lie monomials in $e_1, \ldots, e_n$ of weight α. Thus there exists a non-zero Lie monomial m in $e_1, \ldots, e_n$ of weight α. We show that any such Lie monomial must be 0 since it contains factors e_i both with $i \in J_1$ and with $i \in J_2$. We can write $m = [m_1 \, m_2]$ where m_1, m_2 are shorter Lie monomials. If either m_1 or m_2 involves factors e_i both with $i \in J_1$ and with $i \in J_2$ we have $m_1 = 0$ or $m_2 = 0$ by induction. Otherwise all factors e_i of m_1 have $i \in J_1$ and all factors e_i of m_2 have $i \in J_2$, or vice versa. But then $[m_1 \, m_2] = 0$ since $[e_i e_j] = 0$ for all $i \in J_1, j \in J_2$. Thus $m = 0$ and we have the required contradiction. $\square$

In order to understand the imaginary roots it will by Proposition 16.20 be sufficient to understand the positive imaginary roots which lie in $-\bar{C}$, the negative of the closure of the fundamental chamber. Such roots satisfy the conditions:

$$\alpha \in Q^+, \alpha \neq 0, \operatorname{supp} \alpha \text{ is connected}, \alpha \in -\bar{C}.$$

It is a remarkable fact that, conversely, any element α satisfying these conditions is a positive imaginary root. Before being able to prove this we need a lemma.

Lemma 16.22 (i) *Suppose* $\alpha \in \Phi$, $\alpha \neq \pm \alpha_i$, *satisfies* $\alpha - \alpha_i \notin \Phi$ *and* $\alpha + \alpha_i \notin \Phi$. *Then* $\alpha(h_i) = 0$.
(ii) *Suppose* $\alpha \in \Phi$, $\alpha \neq -\alpha_i$, *satisfies* $\alpha + \alpha_i \notin \Phi$. *Then* $\alpha(h_i) \geq 0$.

Proof. (i) Since $\alpha \in \Phi$ we have $L_\alpha \neq O$. Let $x \in L_\alpha$ with $x \neq 0$. Let

$$n_i = \exp \operatorname{ad} e_i \cdot \exp \operatorname{ad}(-f_i) \cdot \exp \operatorname{ad} e_i \in \operatorname{Aut} L(A).$$

We show that $n_i x \in L_{s_i(\alpha)}$. For $[hx] = \alpha(h)x$ for all $h \in H$, hence $[n_i h, n_i x] = \alpha(h) n_i x$. Now $n_i(H) = H$ by Proposition 16.11 and so

$$[h', n_i x] = \alpha(n_i^{-1} h') n_i x \quad \text{for all } h' \in H.$$

We have $n_i^{-1} h' = s_i^{-1} h' = s_i h'$ also by Proposition 16.11. Thus $[h', n_i x] = \alpha(s_i h') n_i x = (s_i \alpha)(h') n_i x$ for all $h' \in H$. Hence $n_i x \in L_{s_i(\alpha)}$.

Now $\operatorname{ad} e_i \cdot x \in L_{\alpha + \alpha_i}$ and so $\operatorname{ad} e_i \cdot x = 0$ since $\alpha + \alpha_i \notin \Phi$ and $\alpha + \alpha_i \neq 0$. Thus $\exp \operatorname{ad} e_i \cdot x = x$. Also $\operatorname{ad} f_i \cdot x \in L_{\alpha - \alpha_i}$ so $\operatorname{ad} f_i \cdot x = 0$ since $\alpha - \alpha_i \notin \Phi$ and $\alpha - \alpha_i \neq 0$. Thus $\exp \operatorname{ad}(-f_i) \cdot x = x$. Hence $n_i x = x$. Since $x \in L_\alpha$ and $n_i x \in L_{s_i(\alpha)}$ we deduce $s_i(\alpha) = \alpha$. But $s_i(\alpha) = \alpha - \alpha(h_i) \alpha_i$ and so $\alpha(h_i) = 0$.

(ii) Again let x be a non-zero element of L_α. As before $\exp \operatorname{ad} e_i \cdot x = x$ since $\alpha + \alpha_i \notin \Phi$ and $\alpha + \alpha_i \neq 0$. We have

$$\exp \operatorname{ad}(-f_i) x = x - \operatorname{ad} f_i \cdot x + \frac{(\operatorname{ad} f_i)^2}{2!} x - \cdots \pm \frac{(\operatorname{ad} f_i)^p}{p!} x$$

where $(\operatorname{ad} f_i)^{p+1} x = 0$, since $\operatorname{ad}(-f_i)$ is locally nilpotent. Thus

$$n_i x = \exp \operatorname{ad} e_i \cdot \exp \operatorname{ad}(-f_i) \cdot x = \sum_{t \geq 0} \frac{(\operatorname{ad} e_i)^t}{t!} (x - \operatorname{ad} f_i x + \cdots).$$

Now $(\operatorname{ad} e_i)^{t+1} (\operatorname{ad} f_i)^t x = 0$ for each positive integer t, since $\alpha + \alpha_i \notin \Phi$ and $\alpha + \alpha_i \neq 0$. Hence $(\operatorname{ad} e_i)^k (\operatorname{ad} f_i)^t x = 0$ for all $k \geq t+1$. It follows by considering the above expression for $n_i x$ that

$$n_i x \in L_\alpha \oplus L_{\alpha - \alpha_i} \oplus \cdots \oplus L_{\alpha - p \alpha_i}.$$

However, $n_i x \in L_{s_i(\alpha)}$ as in (i). Thus $s_i(\alpha) = \alpha - \alpha(h_i) \alpha_i = \alpha - k \alpha_i$ for some k with $0 \leq k \leq p$. Hence $\alpha(h_i) \geq 0$. $\square$

We now define

$$K = \{\alpha \in Q^+,\ \alpha \neq 0,\ \operatorname{supp} \alpha \text{ is connected},\ \alpha \in -\bar{C}\}.$$

Proposition 16.23 $K \subset \Phi_{\operatorname{im}}^+$.

Proof. Let $\alpha \in K$. Then $\alpha = \sum_{i=1}^n k_i \alpha_i$ with each $k_i \geq 0$ and $k_i > 0$ for some i. Also $\operatorname{supp} \alpha = \{i\ ;\ k_i \neq 0\}$.

We define a set Ψ of roots by

$$\Psi = \left\{ \beta \in \Phi^+\ ;\ \beta = \sum_{i=1}^n m_i \alpha_i \quad \text{with } m_i \leq k_i \quad \text{for each } i \right\}.$$

Ψ is a finite non-empty set of positive roots, since it contains at least one fundamental root. We choose a root $\beta \in \Psi$ such that $\operatorname{ht} \beta$ is as large as possible. We aim to show that $\beta = \alpha$ and hence that $\alpha \in \Phi$. We shall show first that $\operatorname{supp} \beta = \operatorname{supp} \alpha$.

16.3 The roots of a Kac–Moody algebra

Suppose if possible that $\operatorname{supp}\beta \ne \operatorname{supp}\alpha$. Since $\operatorname{supp}\alpha$ is connected there exist $j \in \operatorname{supp}\alpha - \operatorname{supp}\beta$ and $j' \in \operatorname{supp}\beta$ such that $A_{jj'} \ne 0$. Let $\beta = \sum_{i=1}^{n} m_i \alpha_i$. Then $m_j = 0$. Now $\beta - \alpha_j \notin \Phi$ since $m_j = 0$ and $\beta + \alpha_j \notin \Phi$ by the maximality of $\operatorname{ht}\beta$. By Lemma 16.22 (i) we have $\beta(h_j) = 0$. But

$$\beta(h_j) = \sum_{i \in \operatorname{supp}\beta} m_i \alpha_i(h_j) = \sum_{i \in \operatorname{supp}\beta} m_i A_{ji} < 0$$

since $m_i > 0$, $A_{ji} \le 0$ and $A_{jj'} < 0$. This contradicts $\beta(h_j) = 0$. Thus $\operatorname{supp}\beta = \operatorname{supp}\alpha$.

Now we have $\alpha = \sum_{i=1}^{n} k_i \alpha_i$, $\beta = \sum_{i=1}^{n} m_i \alpha_i$ with $m_i \le k_i$. Let

$$J = \{i \in \operatorname{supp}\alpha \ ; \ k_i = m_i\}.$$

We aim to show that $J = \operatorname{supp}\alpha$ and so that $\beta = \alpha$.

Suppose if possible that $J \ne \operatorname{supp}\alpha$. Let $i \in \operatorname{supp}\alpha - J$. Then $m_i < k_i$. Hence $\beta + \alpha_i \notin \Phi$ by the maximality of $\operatorname{ht}\beta$. Thus $\beta(h_i) \ge 0$ by Lemma 16.22 (ii).

Let M be a connected component of $\operatorname{supp}\alpha - J$. Then $\beta(h_i) \ge 0$ for all $i \in M$. Let $\beta' = \sum_{i \in M} m_i \alpha_i$. Then

$$\beta'(h_i) = \beta(h_i) - \sum_{j \in \operatorname{supp}\alpha - M} m_j \alpha_j(h_i)$$

$$= \beta(h_i) - \sum_{j \in \operatorname{supp}\alpha - M} m_j A_{ij}.$$

If $i \in M$ then $\beta(h_i) \ge 0$, $m_j > 0$ (since $\operatorname{supp}\beta = \operatorname{supp}\alpha$) and $A_{ij} \le 0$. Hence $\beta'(h_i) \ge 0$. Since $\operatorname{supp}\alpha$ is connected there exists $i' \in M$ and $j' \in \operatorname{supp}\alpha - M$ with $A_{i'j'} \ne 0$. Then

$$\beta'(h_{i'}) = \beta(h_{i'}) - \sum_{j \in \operatorname{supp}\alpha - M} m_j A_{i'j}$$

We have $\beta'(h_{i'}) \ge 0$ as before; however, in fact we have $\beta'(h_{i'}) > 0$. The strict inequality holds since $m_{j'} > 0$ and $A_{i'j'} < 0$.

Let A_M be the principal minor (A_{ij}) with $i, j \in M$. Let u be the column vector with entries m_j for $j \in M$. Since

$$\beta'(h_i) = \sum_{j \in M} A_{ij} m_j \quad \text{for } i \in M$$

we have $u > 0$, $A_M u \ge 0$, $A_M u \ne 0$. Now we recall that if the indecomposable GCM A_M is of affine type then $A_M u \ge 0$ implies $A_M u = 0$. Also if A_M is of indefinite type then $A_M u \ge 0$ and $u \ge 0$ imply $u = 0$. Thus A_M cannot have affine type or indefinite type. Hence A_M has finite type.

Now let $\gamma = \sum_{i \in M} (k_i - m_i) \alpha_i$. We have $k_i - m_i > 0$ for all $i \in M$. We recall that

$$\alpha - \beta = \sum_{i \in \text{supp}\, \alpha - J} (k_i - m_i) \alpha_i.$$

Thus for $i \in M$ we have

$$(\alpha - \beta)(h_i) = \sum_{j \in \text{supp}\, \alpha - J} (k_j - m_j) A_{ij} = \sum_{j \in M} (k_j - m_j) A_{ij} = \gamma(h_i)$$

since M is a connected component of $\text{supp}\, \alpha - J$. Thus $\gamma(h_i) = \alpha(h_i) - \beta(h_i)$ for all $i \in M$. Now $\alpha(h_i) \leq 0$ since $\alpha \in K$ and $\beta(h_i) \geq 0$ since $i \in M$. Thus $\gamma(h_i) \leq 0$ for all $i \in M$.

Now let u be the column vector with entries $k_i - m_i$ for $i \in M$. Then we have $u > 0$ and $A_M u \leq 0$. Since A_M has finite type $A_M(-u) \geq 0$ implies $-u > 0$ or $-u = 0$, that is $u < 0$ or $u = 0$. This is a contradiction since $u > 0$. This contradiction shows that $J = \text{supp}\, \alpha$ and hence that $\beta = \alpha$. Thus $\alpha \in \Phi$. Since $\alpha \in Q^+$ we have $\alpha \in \Phi^+$. Thus $\alpha \in K$ implies $\alpha \in \Phi^+$. Now $\alpha \in K$ implies $2\alpha \in K$, so $2\alpha \in \Phi^+$. By Proposition 16.18 this implies that $\alpha \in \Phi^+_{\text{Im}}$. This completes the proof. $\square$

This remarkable proof, due to V. Kac, enables us to determine the set of all positive imaginary roots, and hence the set of all imaginary roots.

Theorem 16.24 *The set of positive imaginary roots of $L(A)$ is given by*

$$\Phi^+_{\text{Im}} = \cup_{w \in W} w(K)$$

where

$$K = \{\alpha \in Q^+\ ;\ \alpha \neq 0, \text{supp}\, \alpha \text{ is connected}, \alpha \in -\bar{C}\}.$$

The set of all imaginary roots is $\Phi^+_{\text{Im}} \cup (-\Phi^+_{\text{Im}})$.

Proof. This follows from Propositions 16.19, 16.20, 16.21 and 16.23. $\square$

Corollary 16.25 *Let $\alpha \in \Phi^+_{\text{Im}}$. Then $k\alpha \in \Phi^+_{\text{Im}}$ for all positive integers k.*

Proof. This follows from Theorem 16.24 and the fact that $\alpha \in K$ implies $k\alpha \in K$. $\square$

We next consider the real and imaginary roots of $L(A)$ when A is symmetrisable. Then $L(A)$ has an invariant bilinear form $\langle , \rangle$ described in Section 16.1. This form is non-degenerate on restriction to H, so determines

16.3 The roots of a Kac–Moody algebra

an isomorphism $H^* \to H$ under which $\lambda \to h'_\lambda$, where $\lambda(x) = \langle h'_\lambda, x \rangle$ for all $x \in H$. We can then transfer the bilinear form to H^* by defining

$$\langle \lambda, \mu \rangle = \langle h'_\lambda, h'_\mu \rangle.$$

In particular we can define $\langle \alpha, \alpha \rangle$ for $\alpha \in \Phi$.

Proposition 16.26 *Suppose A is a symmetrisable GCM. Then if α is a real root of $L(A)$ we have $\langle \alpha, \alpha \rangle > 0$. If α is an imaginary root then $\langle \alpha, \alpha \rangle \leq 0$.*

Proof. The form $\langle , \rangle$ on H is W-invariant by Proposition 16.13, so the induced form on H^* is also W-invariant. By definition of the form on H we have

$$\langle h_i, x \rangle = d_i \alpha_i(x) \qquad \text{for all } x \in H.$$

Hence $d_i \alpha_i \in H^*$ corresponds to $h_i \in H$ under our map $H^* \to H$. Thus

$$\langle \alpha_i, \alpha_i \rangle = \frac{1}{d_i^2} \langle h_i, h_i \rangle = \frac{2}{d_i}.$$

In particular $\langle \alpha_i, \alpha_i \rangle > 0$. Now each real root has form $w(\alpha_i)$ for some $w \in W$ and some i. Hence

$$\langle w(\alpha_i), w(\alpha_i) \rangle = \langle \alpha_i, \alpha_i \rangle > 0.$$

Now consider the imaginary roots. Let $\alpha \in K$. Then $\alpha = \sum k_i \alpha_i$ with each $k_i \geq 0$ and $\alpha(h_i) \leq 0$ for each i. Thus

$$\langle \alpha, \alpha \rangle = \sum k_i \langle \alpha, \alpha_i \rangle = \sum k_i \cdot \frac{1}{d_i} \alpha(h_i) \leq 0$$

since $k_i \geq 0$, $d_i > 0$, $\alpha(h_i) \leq 0$. Every positive imaginary root has form $w(\alpha)$ for some $w \in W$, $\alpha \in K$, thus

$$\langle w(\alpha), w(\alpha) \rangle = \langle \alpha, \alpha \rangle \leq 0.$$

For the negative imaginary roots we have

$$\langle -w(\alpha), -w(\alpha) \rangle = \langle w(\alpha), w(\alpha) \rangle \leq 0. \qquad \square$$

We next obtain information about the imaginary roots in the three cases of our trichotomy.

Theorem 16.27 *Let A be an indecomposable GCM.*

(i) *If A has finite type then $L(A)$ has no imaginary roots.*

(ii) *Suppose A has affine type. Then there exists $u > 0$ with $Au = 0$. The vector u is determined to within a scalar multiple. Thus there is a unique such u whose entries are positive integers with no common factor. Let $u = (a_1, \ldots, a_n)$. Let $\delta = a_1 \alpha_1 + \cdots + a_n \alpha_n$. Then the imaginary roots of $L(A)$ are the elements $k\delta$ for $k \in \mathbb{Z}, k \neq 0$.*

(iii) *Suppose A has indefinite type. Then there exists $\alpha \in \Phi_{\mathrm{Im}}^+$ such that $\alpha = \sum_{i=1}^n k_i \alpha_i$ with $k_i > 0$ and $\alpha(h_i) < 0$ for all $i = 1, \ldots, n$.*

Proof. (i) If A has finite type $L(A)$ is a finite dimensional simple Lie algebra by Theorem 15.19. Thus each root of $L(A)$ is real by Proposition 5.12.

(ii) Suppose A has affine type. We first consider the imaginary roots in K. Let $\alpha \in K$ satisfy $\alpha = \sum_{i=1}^n k_i \alpha_i$. Let v be the column vector $(k_1, \ldots, k_n)$. Then we have $v \geq 0$ and $Av \leq 0$, since $\alpha(h_i) \leq 0$ for each i. But in affine type $A(-v) \geq 0$ implies $A(-v) = 0$. Thus $v \neq 0$ and $Av = 0$.

We also have $u > 0$ and $Au = 0$. Since A has corank 1, v is a multiple of u. Since the coefficients of u have no common factor $v = ku$ for some $k \in \mathbb{Z}$ with $k > 0$. Thus $\alpha = k\delta$.

Now every positive imaginary root has form $w(\alpha)$ for some $\alpha \in K$, by Theorem 16.24. We have

$$s_i(\delta) = \delta - \delta(h_i) \alpha_i = \delta$$

since $\delta(h_i) = 0$ follows from $Au = 0$. It follows that $w(\delta) = \delta$ for each $w \in W$. Thus the only positive imaginary roots are the elements $k\delta$ with $k \in \mathbb{Z}, k > 0$. Hence the only imaginary roots are the $k\delta$ with $k \in \mathbb{Z}, k \neq 0$.

(iii) Suppose A has indefinite type. Then there exists $u > 0$ with $Au < 0$. Suppose $u = (k_1, \ldots, k_n)$. Let $\alpha = \sum_{i=1}^n k_i \alpha_i$. Then $\alpha \in K$ and $\alpha(h_i) < 0$ for all i. Thus α is a positive imaginary root of the required kind. $\square$

A significant consequence of the last result is as follows.

Corollary 16.28 *If A is an indecomposable GCM of affine or indefinite type then the dimension of $L(A)$ is infinite.*

Proof. In both cases $L(A)$ has an imaginary root α. Thus it has infinitely many imaginary roots $k\alpha$ for $k \in \mathbb{Z}, k \neq 0$, by Corollary 16.25. Since $L(A) = H \oplus \sum_{\alpha \in \Phi} L_\alpha$, $\dim L(A)$ must be infinite. $\square$

16.3 The roots of a Kac–Moody algebra

We next consider which of the imaginary roots of $L(A)$ when A is symmetrisable satisfy $\langle \alpha, \alpha \rangle = 0$.

Proposition 16.29 *Let A be symmetrisable and α be an imaginary root of $L(A)$. Then $\langle \alpha, \alpha \rangle = 0$ if and only if there exists $w \in W$ such that the support of $w(\alpha)$ has a diagram of affine type.*

Proof. First suppose $\langle \alpha, \alpha \rangle = 0$. We may assume without loss of generality that $\alpha \in \Phi^+$. Thus there exists $w \in W$ with $w(\alpha) \in K$, by Theorem 16.24. Let $\beta = w(\alpha)$. Then $\beta(h_i) \leq 0$ for all i. Let J be the support of β and $\beta = \sum_{i \in J} k_i \alpha_i$. Then J is connected. Now

$$\langle \beta, \beta \rangle = \sum_{i \in J} k_i \langle \beta, \alpha_i \rangle = \sum_{i \in J} \frac{k_i}{d_i} \beta(h_i).$$

Now $k_i > 0$, $d_i > 0$ and $\beta(h_i) \leq 0$ for all $i \in J$. Since we also have $\langle \beta, \beta \rangle = \langle w(\alpha), w(\alpha) \rangle = \langle \alpha, \alpha \rangle = 0$ we deduce that $\beta(h_i) = 0$ for all $i \in J$. Hence $\sum_{j \in J} k_j \alpha_j(h_i) = 0$, that is $\sum_{j \in J} A_{ij} k_j = 0$. Let u be the column vector with entries k_j for $j \in J$. Then $u > 0$ and $A_J u = 0$. Since A_J is an indecomposable GCM this implies that A_J has affine type, by Corollary 15.11.

Conversely suppose β is a positive imaginary root whose support J has a diagram of affine type. Then $L_\beta \neq O$ and so $L(A)$ contains a non-zero Lie monomial in $e_1, \ldots, e_n$ of weight β. The letters e_i in this Lie monomial all have $i \in J$. Thus the Lie monomial lies in $L(A_J)$ and so β is a root of $L(A_J)$. If β were a real root of $L(A_J)$ it would have form $w(\alpha_i)$ for some $w \in W(A_J)$ and $i \in J$, and so β would be a real root of $L(A)$. Thus β is an imaginary root of $L(A_J)$. Since $L(A_J)$ has affine type, $\beta = k\delta$ where δ is the element for $L(A_J)$ defined in Theorem 16.27 (ii). Let $\delta = \sum_{i \in J} a_i \alpha_i$. Then

$$\langle \delta, \delta \rangle = \sum_{i \in J} a_i \langle \delta, \alpha_i \rangle = \sum_{i \in J} \frac{a_i}{d_i} \delta(h_i) = 0$$

since $\delta(h_i) = 0$ for all $i \in J$. Thus $\langle \beta, \beta \rangle = 0$ also. Finally if α is any root of $L(A)$ satisfying $w(\alpha) = \beta$ for some $w \in W$, we have $\langle \alpha, \alpha \rangle = 0$ also. □

17
Kac–Moody algebras of affine type

17.1 Properties of the affine Cartan matrix

We now consider the Kac–Moody algebras $L(A)$ where A is a GCM of affine type. Let A be an $n \times n$ matrix of rank l. Then we know that $n = l+1$. We shall number the rows and columns of A by the integers $0, 1, \ldots, l$. There exists a unique vector $a = (a_0, a_1, \ldots, a_l)$ whose coordinates are positive integers with no common factor such that

$$A \begin{pmatrix} a_0 \\ a_1 \\ \vdots \\ a_l \end{pmatrix} = \begin{pmatrix} 0 \\ 0 \\ \vdots \\ 0 \end{pmatrix}.$$

The possible Dynkin diagrams of such matrices A were obtained on the affine list 15.20. We shall choose the numbering of the vertices in such a way that node 0 is the one in black in the diagram below. We also show in each diagram the integer a_i associated to each vertex.

17.1 The integers $a_0, a_1, \ldots, a_l$.

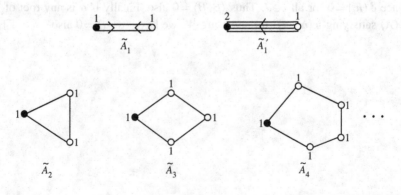

386

17.1 Properties of the affine Cartan matrix

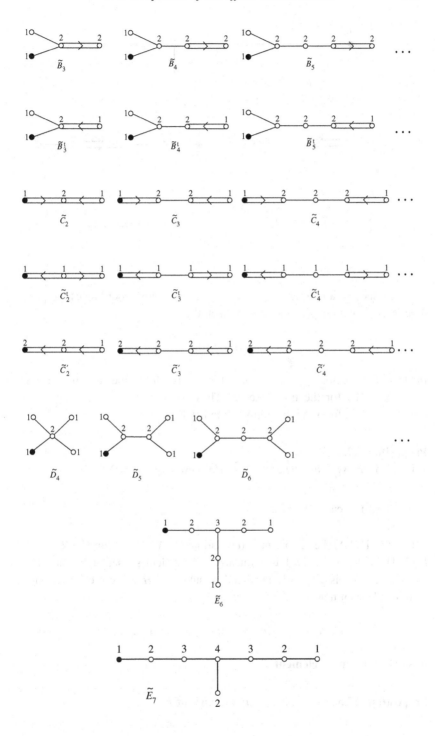

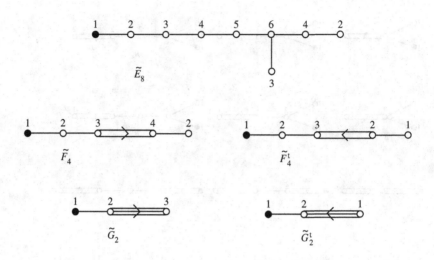

There exists also a unique vector $(c_0, c_1, \ldots, c_l)$ whose coordinates are positive integers with no common factor such that

$$(c_0, c_1, \ldots, c_l) A = (0, 0, \ldots, 0).$$

In fact the vector $(c_0, c_1, \ldots, c_l)$ for A is the same as the vector $(a_0, a_1, \ldots, a_l)$ for the transpose A^t. Thus the vector $(c_0, c_1, \ldots, c_l)$ may also be read off from the diagrams in the list 17.1.

Proposition 17.2 (i) $c_0 = 1$.
(ii) $a_0 = 1$ unless A has type $\tilde{C}_l^t$ or $\tilde{A}_1^t$. In these cases $a_0 = 2$.

Proof. This is clear from 17.2 □

Let (H, Π, Π^v) be a minimal realisation of A. Then $\dim H = 2n - l = l + 2$. $\Pi^v = \{h_0, h_1, \ldots, h_l\}$ is a linearly independent subset of H and $\Pi = \{\alpha_0, \alpha_1, \ldots, \alpha_l\}$ is a linearly independent subset of H^*. These exists an element $d \in H$ such that

$$\alpha_0(d) = 1 \quad \alpha_i(d) = 0 \quad \text{for } i = 1, \ldots, l.$$

d is called a **scaling element**.

Proposition 17.3 $h_0, h_1, \ldots, h_l, d$ is a basis of H.

17.1 Properties of the affine Cartan matrix

Proof. We must show that d is not a linear combination of $h_0, h_1, \ldots, h_l$. Suppose if possible that $d = \sum_{i=0}^{l} k_i h_i$. Then $\alpha_j(d) = \sum_{i=0}^{l} k_i \alpha_j(h_i) = \sum_{i=0}^{l} k_i A_{ij}$. Hence

$$\sum_{i=0}^{l} k_i (A_{i0}, \ldots, A_{il}) = (1, 0, \ldots, 0).$$

In particular, omitting the first column of A,

$$\sum_{i=0}^{l} k_i (A_{i1}, \ldots, A_{il}) = (0, \ldots, 0).$$

However, we also have

$$\sum_{i=0}^{l} c_i (A_{i1}, \ldots, A_{il}) = (0, \ldots, 0).$$

Since the $(l+1) \times l$ matrix (A_{ij}), $0 \le i \le l$, $1 \le j \le l$ has rank l, this implies that $(k_0, \ldots, k_l)$ is a scalar multiple of $(c_0, \ldots, c_l)$. But this would imply that

$$(k_0, k_1, \ldots, k_l) A = (0, 0, \ldots, 0)$$

a contradiction. $\square$

We now define an element $\gamma \in H^*$ determined uniquely by

$$\gamma(h_0) = 1 \qquad \gamma(h_i) = 0 \quad \text{for } i = 1, \ldots, l \qquad \gamma(d) = 0.$$

Proposition 17.4 $\alpha_0, \alpha_1, \ldots, \alpha_l, \gamma$ *is a basis of* H^*.

Proof. The $(l+2) \times (l+2)$ matrix obtained by applying these elements of H^* to the basis of H in Proposition 17.3 is

$$\begin{pmatrix} 2 & * & \cdots & * & 1 \\ * & & & 0 & 0 \\ \cdot & & A^0 & \cdot & 1 \\ \cdot & & & \cdot & \vdots \\ \cdot & & & \cdot & l \\ * & & & 0 & l+1 \\ 1 & 0 & \cdots & 0 & 0 \\ 0 & 1 & \cdots & l & l+1 \end{pmatrix}$$

where A^0 is a Cartan matrix of finite type. Thus $\det A^0 \ne 0$ and so the determinant of the above matrix is also non-zero. Hence $\alpha_0, \alpha_1, \ldots, \alpha_l, \gamma$ must be a basis of H^*. $\square$

We know that any indecomposable GCM of affine type is symmetrisable. We shall now express the affine Cartan matrix A in an explicit way as the product of a diagonal matrix D with positive diagonal entries and a symmetric matrix B. The diagonal entries of D are rational, but not necessarily integral.

Proposition 17.5 *We have $A = DB$ where $D = \text{diag}(d_0, d_1, \ldots, d_l)$ and B is symmetric, where $d_i = a_i/c_i$.*

Proof. By Theorem 15.17 there exists a diagonal matrix D with positive diagonal entries and a symmetric matrix B such that $A = DB$. Let $c = (c_0, c_1, \ldots, c_l)$ and $a^t = (a_0, a_1, \ldots, a_l)$. Then $Aa = 0$ so $DBa = 0$, and hence $Ba = 0$. Thus $a^t B = 0$. Also $cA = 0$ so $(cD)B = 0$. Since B has corank 1 cD must be a scalar multiple of a^t. In fact we can choose D so that $cD = a^t$, that is $d_i = a_i/c_i$. □

Now we have a non-degenerate bilinear form on H defined as in Proposition 16.1. This form satisfies

$$\langle h_i, h_j \rangle = d_i d_j B_{ij} = a_j c_j^{-1} A_{ij} \quad \text{for } i, j = 0, 1, \ldots, l$$

$$\langle h_0, d \rangle = d_0 \alpha_0(d) = a_0$$

$$\langle h_i, d \rangle = 0 \quad \text{for } i = 1, \ldots, l$$

$$\langle d, d \rangle = 0.$$

This standard invariant form on H defines a bijection $H^* \to H$ given by $\lambda \to h'_\lambda$ where $\lambda(x) = \langle h'_\lambda, x \rangle$ for all $x \in H$.

Proposition 17.6 *Under this bijection between H and H^*, $h_i \in H$ corresponds to $a_i c_i^{-1} \alpha_i \in H^*$ for $i = 0, 1, \ldots, l$ and $d \in H$ corresponds to $a_0 \gamma \in H^*$.*

Proof. For $j = 0, 1, \ldots, l$ we have

$$a_j c_j^{-1} \alpha_j (h_i) = d_j A_{ij} = \langle h_j, h_i \rangle \quad \text{for } i = 0, 1, \ldots, l$$

$$a_j c_j^{-1} \alpha_j (d) = d_j \alpha_j(d) = \langle h_j, d \rangle$$

thus $a_j c_j^{-1} \alpha_j \in H^*$ corresponds to $h_j \in H$. We also have

$$a_0 \gamma (h_i) = \langle d, h_i \rangle \quad \text{for } i = 0, 1, \ldots, l$$

$$a_0 \gamma(d) = \langle d, d \rangle$$

thus $a_0 \gamma \in H^*$ corresponds to $d \in H$. □

We may transfer the standard bilinear form from H to H^* using this bijection. The form on H^* is then given by

$$\langle \alpha_i, \alpha_j \rangle = a_i^{-1} c_i A_{ij} \qquad i, j = 0, 1, \ldots, l$$

$$\langle \alpha_0, \gamma \rangle = a_0^{-1}$$

$$\langle \alpha_i, \gamma \rangle = 0 \qquad i = 1, \ldots, l$$

$$\langle \gamma, \gamma \rangle = 0.$$

We note in particular that

$$A_{ij} = \frac{2 \langle \alpha_i, \alpha_j \rangle}{\langle \alpha_i, \alpha_i \rangle}.$$

Corollary 17.7 *Under the given bijection between H and H^*, $h_i \in H$ corresponds to $\frac{2\alpha_i}{\langle \alpha_i, \alpha_i \rangle} \in H^*$.*

Proof. This follows from Proposition 17.6. □

We now define an element $c \in H$ by $c = \sum_{i=0}^{l} c_i h_i$. Under the bijection $H \to H^*$ c corresponds to δ. For $\delta = \sum_{i=0}^{l} a_i \alpha_i$ and h_i corresponds to $a_i c_i^{-1} \alpha_i$ by Proposition 17.6.

Proposition 17.8 *The element c lies in the centre of $L(A)$. In fact the centre is 1-dimensional and consists of all scalar multiples of c.*

Proof. For each simple root α_j we have $\alpha_j(c) = \sum_{i=0}^{l} c_i \alpha_j(h_i) = \sum_{i=0}^{l} c_i A_{ij} = 0$. It follows that $\alpha(c) = 0$ for all $\alpha \in \Phi$. Now $L(A) = H \oplus \sum_{\alpha \in \Phi} L_\alpha$. Thus each element of $L(A)$ has form $h + \sum x_\alpha$ where $h \in H$, $x_\alpha \in L_\alpha$ and finitely many x_α are non-zero. Thus

$$\left[c, h + \sum_\alpha x_\alpha \right] = \sum_\alpha \alpha(c) x_\alpha = 0.$$

Hence c lies in the centre of $L(A)$.

Now let $h + \sum_\alpha x_\alpha$ be any element of the centre of $L(A)$. Then we have

$$\left[x, h + \sum_\alpha x_\alpha \right] = 0 \qquad \text{for all } x \in H.$$

Thus $\sum_\alpha \alpha(x) x_\alpha = 0$ for all $x \in H$. This implies $\alpha(x) x_\alpha = 0$ for all $x \in H$. Now for each $\alpha \in \Phi$ there exists $x \in H$ with $\alpha(x) \neq 0$. Hence $x_\alpha = 0$. This shows that the centre of $L(A)$ lies in H.

So let $h \in H$ lie in the centre of $L(A)$. By Proposition 17.3 we have

$$h = \sum_{i=0}^{l} \xi_i h_i + \xi d \quad \text{for } \xi_i, \xi \in \mathbb{C}.$$

Let $x \in L_{\alpha_j}$. Then $[hx] = \alpha_j(h)x$. Hence $\alpha_j(h) = 0$ for each $j = 0, 1, \ldots, l$. Thus

$$\sum_{i=0}^{l} \xi_i \alpha_j(h_i) + \xi \alpha_j(d) = 0$$

that is

$$\sum_{i=0}^{l} \xi_i A_{ij} = 0 \quad \text{for } j = 1, \ldots, l$$

and

$$\sum_{i=0}^{l} \xi_i A_{i0} = -\xi.$$

However, we have $\sum_{j=0}^{l} A_{ij} a_j = 0$, hence $A_{i0} = -a_0^{-1} \sum_{j=1}^{l} A_{ij} a_j$. Thus $\sum_{i=0}^{l} \xi_i A_{ij} = 0$ for $j = 1, \ldots, l$ implies $\sum_{i=0}^{l} \xi_i A_{i0} = 0$. Thus we deduce that $\xi = 0$, and so $h = \sum_{i=0}^{l} \xi_i h_i$. This in turn gives $\sum_{i=0}^{l} \xi_i A_{ij} = 0$ for $j = 0, 1, \ldots, l$. This implies that $(\xi_0, \xi_1, \ldots, \xi_l)$ is a scalar multiple of $(c_0, c_1, \ldots, c_l)$ since A is an $(l+1) \times (l+1)$ matrix of rank l. Thus h is a multiple of c. Thus the centre of $L(A)$ is the 1-dimensional subspace spanned by c. □

c is called the **canonical central element** of $L(A)$.

Summary

We will find it convenient to summarise in one place the properties of the various elements discussed in this section:

(a) $h_0, h_1, \ldots, h_l, d$ are a basis of H.
(b) $c = c_0 h_0 + \cdots + c_l h_l$ is the canonical central element.
(c) $\alpha_0, \alpha_1, \ldots, \alpha_l, \gamma$ are a basis of H^*.
(d) $\delta = a_0 \alpha_0 + \cdots + a_l \alpha_l$ is the basic imaginary root.
(e) The standard invariant form on H is given by

$$\langle h_i, h_j \rangle = a_j c_j^{-1} A_{ij} \quad i, j = 0, 1, \ldots, l$$

$$\langle h_0, d \rangle = a_0$$

$$\langle h_i, d \rangle = 0 \quad i = 1, \ldots, l$$

$$\langle d, d \rangle = 0.$$

17.1 Properties of the affine Cartan matrix

(f) The standard invariant form on H^* is given by

$$\langle \alpha_i, \alpha_j \rangle = a_i^{-1} c_i A_{ij} \qquad i, j = 0, 1, \ldots, l$$
$$\langle \alpha_0, \gamma \rangle = a_0^{-1}$$
$$\langle \alpha_i, \gamma \rangle = 0 \qquad i = 1, \ldots, l$$
$$\langle \gamma, \gamma \rangle = 0.$$

(g) The action of H^* on H is given by

$$\alpha_j(h_i) = A_{ij} \qquad i, j = 0, 1, \ldots, l$$
$$\alpha_0(d) = 1$$
$$\alpha_i(d) = 0 \qquad i = 1, \ldots, l$$
$$\gamma(h_0) = 1$$
$$\gamma(h_i) = 0 \qquad i = 1, \ldots, l$$
$$\gamma(d) = 0.$$

(h) The properties of the central element c.

$$\langle h_i, c \rangle = 0 \qquad i = 0, 1, \ldots, l$$
$$\langle d, c \rangle = a_0$$
$$\langle c, c \rangle = 0$$
$$\alpha_j(c) = 0 \qquad j = 0, 1, \ldots, l$$
$$\gamma(c) = 1.$$

(i) The properties of the imaginary root δ.

$$\langle \alpha_j, \delta \rangle = 0 \qquad j = 0, 1, \ldots, l$$
$$\langle \gamma, \delta \rangle = 1$$
$$\langle \delta, \delta \rangle = 0$$
$$\delta(h_i) = 0 \qquad i = 0, 1, \ldots, l$$
$$\delta(d) = a_0$$
$$\delta(c) = 0.$$

(j) Properties of the standard bijection $H \to H^*$.

$$h_i \to a_i c_i^{-1} \alpha_i \quad i = 0, 1, \ldots, l$$
$$d \to a_0 \gamma$$
$$c \to \delta.$$

17.2 The roots of an affine Kac–Moody algebra

Let A^0 be the matrix obtained from the affine Cartan matrix A by removing the row and the column 0. Then A^0 is an $l \times l$ Cartan matrix of finite type. By list 17.1 we see that A^0 is given in each case by the following list.

The underlying Cartan matrix A^0

A		A^0
$\tilde{A}_l$	$l \geq 1$	A_l
$\tilde{A}'_1$		A_1
$\tilde{B}_l$	$l \geq 3$	B_l
$\tilde{B}^t_l$	$l \geq 3$	C_l
$\tilde{C}_l$	$l \geq 2$	C_l
$\tilde{C}^t_l$	$l \geq 2$	B_l
$\tilde{C}'_l$	$l \geq 2$	C_l
$\tilde{D}_l$	$l \geq 4$	D_l
$\tilde{E}_l$	$l = 6, 7, 8$	E_l
$\tilde{F}_4$		F_4
$\tilde{F}^t_4$		F_4
$\tilde{G}_2$		G_2
$\tilde{G}^t_2$		G_2

Let Φ^0 be the set of roots of the finite dimensional Lie algebra $L(A^0)$. Φ^0 has a fundamental system $\Pi^0 = \{\alpha_1, \ldots, \alpha_l\}$. Let W^0 be the Weyl group of Φ^0. Then W^0 is generated by the fundamental reflections $s_1, \ldots, s_l$.

Now we know that the imaginary roots of $L(A)$ are the elements $k\delta$ with $k \in \mathbb{Z}$ and $k \neq 0$, by Theorem 16.27 (ii). (However, we do not yet know the multiplicities of these roots.) Thus we shall now consider the real roots of $L(A)$. These have the form $w(\alpha_i)$ for some $w \in W$ and $i = 0, 1, \ldots, l$. We consider the squared lengths $\langle \alpha, \alpha \rangle$ of the roots $\alpha \in \Phi_{\text{Re}}$. Since $\langle w(\alpha_i), w(\alpha_i) \rangle = \langle \alpha_i, \alpha_i \rangle$ the length of any real root is equal to the length of some fundamental

17.2 The roots of an affine Kac–Moody algebra

root. The relative lengths of the fundamental roots may be obtained from list 17.1 using the formulae

$$A_{ij} = 2\frac{\langle \alpha_i, \alpha_j \rangle}{\langle \alpha_i, \alpha_i \rangle}$$

$$\frac{\langle \alpha_j, \alpha_j \rangle}{\langle \alpha_i, \alpha_i \rangle} = \frac{A_{ij}}{A_{ji}}.$$

Proposition 17.9 (a) *If A is an affine Cartan matrix of types $\tilde{A}_l, \tilde{D}_l, \tilde{E}_6, \tilde{E}_7, \tilde{E}_8$ all the fundamental roots have the same length.*
(b) *If A has types $\tilde{B}_l, \tilde{B}_l^t, \tilde{C}_l, \tilde{C}_l^t, \tilde{F}_4, \tilde{F}_4^t$ there are fundamental roots of two different lengths. The ratio $\langle \beta, \beta \rangle / \langle \alpha, \alpha \rangle$ where α is short and β is long is 2.*
(c) *If A has type $\tilde{G}_2$ or $\tilde{G}_2^t$ there are fundamental roots of two different lengths with $\langle \beta, \beta \rangle / \langle \alpha, \alpha \rangle = 3$.*
(d) *If A has type $\tilde{A}_1'$ there are fundamental roots of two different lengths with $\langle \beta, \beta \rangle / \langle \alpha, \alpha \rangle = 4$.*
(e) *If A has type $\tilde{C}_l'$ there are fundamental roots of three different lengths, say α, β, γ, with $\langle \beta, \beta \rangle / \langle \alpha, \alpha \rangle = 2$ and $\langle \gamma, \gamma \rangle / \langle \beta, \beta \rangle = 2$.*

Proof. This is clear from list 17.1. □

We shall denote by $\Phi_{\text{Re},s}$ the set of short real roots, by $\Phi_{\text{Re},l}$ the set of long real roots and by $\Phi_{\text{Re},i}$ the set of real roots of intermediate length. The latter set is non-empty only when A has type $\tilde{C}_l'$ for some l. If all real roots have the same length we use the convention $\Phi_{\text{Re}} = \Phi_{\text{Re},s}$.

We now aim to characterise the set $\Phi_{\text{Re},s}$. We consider the possible values of $\langle \alpha, \alpha \rangle$ for $\alpha \in Q$. Let $\alpha = \sum_{i=0}^{l} k_i \alpha_i$. Then $\langle \alpha, \alpha \rangle = \sum_{i,j} k_i k_j \langle \alpha_i, \alpha_j \rangle$. Now $\langle \alpha_i, \alpha_j \rangle \in \mathbb{Q}$ for all i, j. Thus there exists $d \in \mathbb{Z}$ with $d > 0$ such that $\langle \alpha_i, \alpha_j \rangle \in \frac{1}{d}\mathbb{Z}$ for all i, j. Thus if $\langle \alpha, \alpha \rangle > 0$ then $\langle \alpha, \alpha \rangle \geq \frac{1}{d}$. Hence there exists $m > 0$ such that $m = \min \langle \alpha, \alpha \rangle$ for all $\alpha \in Q$ with $\langle \alpha, \alpha \rangle > 0$.

Proposition 17.10 *If $\alpha \in Q$ satisfies $\langle \alpha, \alpha \rangle = m$ then $\alpha \in Q^+$ or $\alpha \in Q^-$.*

Proof. Suppose if possible there exists $\alpha \in Q$ with $\langle \alpha, \alpha \rangle = m$ but $\alpha \notin Q^+$ and $\alpha \notin Q^-$. Then $\alpha = \beta - \gamma$ where $\beta, \gamma \in Q^+, \beta \neq 0, \gamma \neq 0$ and supp $\beta \cap$ supp $\gamma = \phi$. Hence

$$\langle \alpha, \alpha \rangle = \langle \beta, \beta \rangle + \langle \gamma, \gamma \rangle - 2 \langle \beta, \gamma \rangle$$

and $\langle \beta, \gamma \rangle \leq 0$ since supp $\beta \cap$ supp $\gamma = \phi$. Hence $\langle \alpha, \alpha \rangle \geq \langle \beta, \beta \rangle + \langle \gamma, \gamma \rangle$.

Now all proper connected principal minors of A have finite type. Thus, considering the connected components of $\operatorname{supp}\beta$, we have $\beta = \beta_1 + \cdots + \beta_r$ with $\operatorname{supp}\beta_i$ connected for each i, $\langle \beta_i, \beta_j \rangle = 0$ for $i \neq j$, and $\langle \beta_i, \beta_i \rangle > 0$. Thus

$$\langle \beta, \beta \rangle = \langle \beta_1, \beta_1 \rangle + \cdots + \langle \beta_r, \beta_r \rangle > 0.$$

Hence $\langle \beta, \beta \rangle \geq m$. Similarly $\langle \gamma, \gamma \rangle \geq m$. But then $\langle \alpha, \alpha \rangle \geq 2m$, a contradiction. Hence $\alpha \in Q^+$ or $\alpha \in Q^-$. □

Proposition 17.11 *Let A be an indecomposable GCM of finite or affine type. Then the set $\Phi_{\mathrm{Re},s}$ of short real roots of $L(A)$ is given by*

$$\Phi_{\mathrm{re},s} = \{ \alpha \in Q \ ; \ \langle \alpha, \alpha \rangle = m \}.$$

Proof. Suppose $\alpha \in Q$ satisfies $\langle \alpha, \alpha \rangle = m$. We show $\alpha \in \Phi_{\mathrm{Re}}$. By Proposition 17.10 $\alpha \in Q^+$ or $\alpha \in Q^-$. We may suppose $\alpha \in Q^+$. Consider the set

$$\{ w(\alpha) \ ; \ w \in W \} \cap Q^+.$$

We choose an element $\beta = \sum k_i \alpha_i$ in this set with $\operatorname{ht}\beta$ minimal. Then $\langle \beta, \beta \rangle = m$, so $\sum_i k_i \langle \alpha_i, \beta \rangle = m$. Since $k_i \geq 0$ and $m > 0$ there exists i with $\langle \alpha_i, \beta \rangle > 0$. Thus $\beta(h_i) = 2\frac{\langle \alpha_i, \beta \rangle}{\langle \alpha_i, \alpha_i \rangle} > 0$. Now $s_i(\beta) = \beta - \beta(h_i)\alpha_i$ so $\operatorname{ht} s_i(\beta) < \operatorname{ht}\beta$. By minimality of $\operatorname{ht}\beta$ we must have $s_i(\beta) \in Q^-$. But $\beta \in Q^+$, $s_i(\beta) \in Q^-$ imply $\beta = r\alpha_i$ for some $r \in \mathbb{Z}$ with $r > 0$. Since

$$\langle r\alpha_i, r\alpha_i \rangle = r^2 \langle \alpha_i, \alpha_i \rangle \geq r^2 m$$

we have $r = 1$. Thus $\beta = \alpha_i$ and $\langle \alpha_i, \alpha_i \rangle = m$. Hence $\beta \in \Phi_{\mathrm{Re},s}$ and so $\alpha \in \Phi_{\mathrm{Re},s}$ also.

Conversely if $\alpha \in \Phi_{\mathrm{Re},s}$ then $\alpha = w(\alpha_i)$ for some $w \in W$ and some i, and $\langle \alpha, \alpha \rangle = \langle \alpha_i, \alpha_i \rangle$. However, we have seen that the short fundamental roots have $\langle \alpha_i, \alpha_i \rangle = m$. Thus $\langle \alpha, \alpha \rangle = m$ also. □

We aim next to characterise the set $\Phi_{\mathrm{Re},l}$ of long real roots. In order to do this we compare the roots of $L(A)$ and $L(A^t)$. Here A can be any GCM.

Proposition 17.12 *If $(H, \Pi, \Pi^\vee)$ is a minimal realisation of the GCM A then $(H^*, \Pi^\vee, \Pi)$ is a minimal realisation of A^t.*

Proof. Let A be an $n \times n$ matrix of rank l. Then $\dim H = 2n - l$, $\Pi^\vee = \{h_1, \ldots, h_n\}$ is a linearly independent subset of H, $\Pi = \{\alpha_1, \ldots, \alpha_n\}$ is a linearly independent subset of H^*, and $\alpha_j(h_i) = A_{ij}$.

17.2 The roots of an affine Kac–Moody algebra

We now replace H by its dual space H^*. We still have $\dim H^* = 2n - l$, $(H^*)^*$ can be identified with H by means of the formula

$$h(\lambda) = \lambda(h) \quad \text{for } h \in H, \lambda \in H^*.$$

Since

$$h_j(\alpha_i) = \alpha_i(h_j) = A_{ji}$$

we see that $(H^*, \Pi^\vee, \Pi)$ is a minimal realisation of A^t. $\square$

Now suppose A is symmetrisable. Then we have an isomorphism between H and H^* induced by our standard invariant form. Under this isomorphism h_i corresponds to $\frac{2\alpha_i}{\langle \alpha_i, \alpha_i \rangle} = d_i \alpha_i$. For each real root $\alpha \in \Phi_{\text{Re}}$ we define the corresponding **coroot** $h_\alpha \in H$ to be the element of H corresponding to $\frac{2\alpha}{\langle \alpha, \alpha \rangle} \in H^*$. The element h_α can also be described by using the Weyl group. Since the W-actions on H and H^* are compatible with the above isomorphism, if $\alpha = w(\alpha_i)$ then $h_\alpha = w(h_i)$. For h_i corresponds to $\frac{2\alpha_i}{\langle \alpha_i, \alpha_i \rangle}$ and $\langle \alpha, \alpha \rangle = \langle \alpha_i, \alpha_i \rangle$. Thus the coroots h_α for $\alpha \in \Phi_{\text{Re}}$ for $L(A)$ may be interpreted as the real roots for $L(A^t)$. Moreover we have

$$\langle h_\alpha, h_\alpha \rangle = \left\langle \frac{2\alpha}{\langle \alpha, \alpha \rangle}, \frac{2\alpha}{\langle \alpha, \alpha \rangle} \right\rangle = \frac{4}{\langle \alpha, \alpha \rangle}.$$

Hence α is a short root for $L(A)$ if and only if h_α is a long root for $L(A^t)$. The fact that short roots give long coroots and long roots give short coroots is very useful. We shall apply this to characterise $\Phi_{\text{Re},l}$ in the case when A is of finite or affine type.

Proposition 17.13 *Let A be an indecomposable GCM of finite or affine type. Then the set $\Phi_{\text{Re},l}$ of long real roots of $L(A)$ is given by*

$$\Phi_{\text{Re},l} = \left\{ \alpha = \sum k_i \alpha_i \in Q \; ; \; \langle \alpha, \alpha \rangle = M, k_i \frac{\langle \alpha_i, \alpha_i \rangle}{\langle \alpha, \alpha \rangle} \in \mathbb{Z} \text{ for all } i \right\}$$

where $M = \max \{ \langle \alpha, \alpha \rangle \; ; \; \alpha \in \Phi_{\text{Re}} \}$.

Proof. We first show the long real roots satisfy the given conditions. Let $\alpha \in \Phi_{\text{Re},l}$. Then $\langle \alpha, \alpha \rangle = M$. Let $\alpha = \sum k_i \alpha_i$. Then

$$\frac{2\alpha}{\langle \alpha, \alpha \rangle} = \sum k_i \frac{\langle \alpha_i, \alpha_i \rangle}{\langle \alpha, \alpha \rangle} \frac{2\alpha_i}{\langle \alpha_i, \alpha_i \rangle}$$

and so
$$h_\alpha = \sum k_i \frac{\langle \alpha_i, \alpha_i \rangle}{\langle \alpha, \alpha \rangle} h_i.$$

This expresses a root for $L(A^t)$ as a linear combination of fundamental roots, thus the coefficients $k_i \frac{\langle \alpha_i, \alpha_i \rangle}{\langle \alpha, \alpha \rangle}$ lie in $\mathbb{Z}$.

Conversely suppose $\alpha \in Q$ satisfies the given conditions. Then $h_\alpha \in \sum \mathbb{Z} h_i$ and $\langle h_\alpha, h_\alpha \rangle = 4/M$. Now $4/M$ is the minimum possible value of $\langle \beta, \beta \rangle$ for all real roots β of $L(A^t)$. Thus by Proposition 17.11 h_α is a short root of $L(A^t)$. Hence α is a long root of $L(A)$. □

We next wish to characterise the set $\Phi_{\text{Re},i}$ of intermediate roots of $L(A)$ when A has type $\tilde{C}'_l$. We first need a lemma.

Lemma 17.14 (a) *Suppose A is an indecomposable GCM of finite or affine type. Then the set of all $\alpha = \sum k_i \alpha_i \in Q$ satisfying $k_i \frac{\langle \alpha_i, \alpha_i \rangle}{\langle \alpha, \alpha \rangle} \in \mathbb{Z}$ for all i is invariant under W.*
(b) *If $\alpha = \sum k_i \alpha_i \in Q$ satisfies $k_i \frac{\langle \alpha_i, \alpha_i \rangle}{\langle \alpha, \alpha \rangle} \in \mathbb{Z}$ for all i then $\alpha \in Q^+$ or $\alpha \in Q^-$.*

Proof. (a) Suppose α satisfies our condition. It is sufficient to show that $s_j(\alpha)$ satisfies it also. Now $s_j(\alpha) = \alpha - \alpha(h_j) \alpha_j$. Thus it is sufficient to show that
$$(k_j - \alpha(h_j)) \frac{\langle \alpha_j, \alpha_j \rangle}{\langle \alpha, \alpha \rangle} \in \mathbb{Z}$$
that is $\alpha(h_j) \frac{\langle \alpha_j, \alpha_j \rangle}{\langle \alpha, \alpha \rangle} \in \mathbb{Z}$. Now we have
$$\alpha(h_j) \frac{\langle \alpha_j, \alpha_j \rangle}{\langle \alpha, \alpha \rangle} = \sum_i k_i \alpha_i(h_j) \frac{\langle \alpha_j, \alpha_j \rangle}{\langle \alpha, \alpha \rangle}$$
$$= \sum_i k_i \alpha_j(h_i) \frac{\langle \alpha_i, \alpha_i \rangle}{\langle \alpha, \alpha \rangle} = \sum_i A_{ij} k_i \frac{\langle \alpha_i, \alpha_i \rangle}{\langle \alpha, \alpha \rangle} \in \mathbb{Z}$$

as required.
(b) Suppose the result is false. Then $\alpha = \beta - \gamma$ where $\beta, \gamma \in Q^+$, $\beta \neq 0$, $\gamma \neq 0$ and supp $\beta \cap$ supp $\gamma = \phi$. Then
$$\langle \alpha, \alpha \rangle = \langle \beta, \beta \rangle + \langle \gamma, \gamma \rangle - 2\langle \beta, \gamma \rangle \geq \langle \beta, \beta \rangle + \langle \gamma, \gamma \rangle.$$

17.2 The roots of an affine Kac–Moody algebra

Now $\beta = \sum_{i \in \text{supp} \beta} k_i \alpha_i$ so

$$\langle \beta, \beta \rangle = \sum_i k_i^2 \langle \alpha_i, \alpha_i \rangle + \sum_{i<j} 2 k_i k_j \langle \alpha_i, \alpha_j \rangle$$

and

$$\frac{\langle \beta, \beta \rangle}{\langle \alpha, \alpha \rangle} = \sum_i k_i \left(k_i \frac{\langle \alpha_i, \alpha_i \rangle}{\langle \alpha, \alpha \rangle} \right) + \sum_{i<j} A_{ij} k_j \left(k_i \frac{\langle \alpha_i, \alpha_i \rangle}{\langle \alpha, \alpha \rangle} \right).$$

Hence $\frac{\langle \beta, \beta \rangle}{\langle \alpha, \alpha \rangle} \in \mathbb{Z}$. Similarly we have $\frac{\langle \gamma, \gamma \rangle}{\langle \alpha, \alpha \rangle} \in \mathbb{Z}$.

Now all proper connected principal minors of A have finite type. Thus we have $\beta = \beta_1 + \cdots + \beta_r$ with $\text{supp}\,\beta_i$ connected for each i, $\langle \beta_i, \beta_j \rangle = 0$ for $i \neq j$, and $\langle \beta_i, \beta_i \rangle > 0$. Thus

$$\langle \beta, \beta \rangle = \sum_i \langle \beta_i, \beta_i \rangle > 0.$$

Similarly we can show $\langle \gamma, \gamma \rangle > 0$. Thus $\langle \alpha, \alpha \rangle > 0$ also. But now we have $\frac{\langle \beta, \beta \rangle}{\langle \alpha, \alpha \rangle} \in \mathbb{Z}$ so $\langle \beta, \beta \rangle \geq \langle \alpha, \alpha \rangle$, and $\frac{\langle \gamma, \gamma \rangle}{\langle \alpha, \alpha \rangle} \in \mathbb{Z}$ so $\langle \gamma, \gamma \rangle \geq \langle \alpha, \alpha \rangle$. Hence $\langle \alpha, \alpha \rangle \geq \langle \beta, \beta \rangle + \langle \gamma, \gamma \rangle \geq 2 \langle \alpha, \alpha \rangle$, a contradiction. □

We now suppose A is a GCM of affine type $\tilde{C}_l'$. The diagram of A is

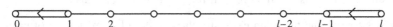

Let m' be defined by $\langle \alpha_i, \alpha_i \rangle = m'$ for $i = 1, \ldots, l-1$. Thus m' is the squared length of the intermediate roots.

Lemma 17.15 *Suppose A has type $\tilde{C}_l'$. Suppose $\alpha = \sum_{i=0}^l k_i \alpha_i \in Q$ satisfies $\langle \alpha, \alpha \rangle = m'$. Then $k_i \frac{\langle \alpha_i, \alpha_i \rangle}{\langle \alpha, \alpha \rangle} \in \mathbb{Z}$ for all i.*

Proof. The required condition is obvious for all $i \neq 0$ since $\langle \alpha_i, \alpha_i \rangle = m'$ for $i = 1, \ldots, l-1$ and $\langle \alpha_l, \alpha_l \rangle = 2m'$. We must therefore show $k_0 \frac{\langle \alpha_0, \alpha_0 \rangle}{\langle \alpha, \alpha \rangle} \in \mathbb{Z}$, that is that k_0 is even.

Now $\alpha = k_0 \alpha_0 + \sum_{i=1}^l k_i \alpha_i$, thus

$$\langle \alpha, \alpha \rangle = k_0^2 \langle \alpha_0, \alpha_0 \rangle + 2 k_0 k_1 \langle \alpha_0, \alpha_1 \rangle + \left\langle \sum_{i=1}^l k_i \alpha_i, \sum_{i=1}^l k_i \alpha_i \right\rangle$$

$$= k_0^2 \langle \alpha_0, \alpha_0 \rangle + k_0 k_1 A_{10} \langle \alpha_1, \alpha_1 \rangle + \sum_{i=1}^l k_i^2 \langle \alpha_i, \alpha_i \rangle + \sum_{\substack{i,j=1 \\ i<j}}^l k_i k_j A_{ij} \langle \alpha_i, \alpha_i \rangle.$$

Thus $\langle \alpha, \alpha \rangle \in k_0^2 \langle \alpha_0, \alpha_0 \rangle + \mathbb{Z}m'$. But $\langle \alpha, \alpha \rangle = m'$ so $k_0^2 \langle \alpha_0, \alpha_0 \rangle \in \mathbb{Z}m'$. Since $\langle \alpha_0, \alpha_0 \rangle = \frac{1}{2}m'$ we have $k_0^2/2 \in \mathbb{Z}$ and so k_0 is even, as required. □

We can now characterise $\Phi_{\text{Re},i}$.

Proposition 17.16 *Suppose A is a GCM of type $\tilde{C}_l'$. Then*
$$\Phi_{\text{Re},i} = \{\alpha \in Q \; ; \; \langle \alpha, \alpha \rangle = m'\}.$$

Proof. Let $\alpha \in Q$ satisfy $\langle \alpha, \alpha \rangle = m'$. By Lemma 17.15 $\alpha = \sum_{i=0}^{l} k_i \alpha_i$ with $k_i \frac{\langle \alpha_i, \alpha_i \rangle}{\langle \alpha, \alpha \rangle} \in \mathbb{Z}$ for each i. By Lemma 17.14 (b) $\alpha \in Q^+$ or $\alpha \in Q^-$. We may assume $\alpha \in Q^+$.

Consider the set
$$\{w(\alpha) \; ; \; w \in W\} \cap Q^+.$$

We choose an element $\beta = \sum_{i=0}^{l} k_i' \alpha_i$ in this set with $\operatorname{ht} \beta$ minimal. Then $\langle \beta, \beta \rangle = m'$ and so $\sum_{i=0}^{l} k_i' \langle \alpha_i, \beta \rangle = m'$. Since $m' > 0$ and $k_i' \geq 0$ there exists i with $\langle \alpha_i, \beta \rangle > 0$. Thus $\beta(h_i) = 2\frac{\langle \alpha_i, \beta \rangle}{\langle \alpha_i, \alpha_i \rangle} > 0$. Now $s_i(\beta) = \beta - \beta(h_i)\alpha_i$ so $\operatorname{ht} s_i(\beta) < \operatorname{ht} \beta$. By the minimality of $\operatorname{ht} \beta$, $s_i(\beta) \notin Q^+$. But $s_i(\beta) \in Q^+$ or Q^- by Lemma 17.14 (a) and (b). Thus $\beta \in Q^+$ and $s_i(\beta) \in Q^-$. Hence $\beta = r\alpha_i$ for some $r \in \mathbb{Z}$ with $r > 0$. Thus $\langle \beta, \beta \rangle = r^2 \langle \alpha_i, \alpha_i \rangle = m'$. However, $\langle \alpha_i, \alpha_i \rangle \geq \frac{1}{2}m'$ thus $r = 1$. Thus $\beta = \alpha_i \in \Phi_{\text{Re},i}$. It follows that $\alpha \in \Phi_{\text{Re},i}$ also. □

We are now able to obtain explicitly the set Φ_{Re} of all real roots of each affine Kac–Moody algebra individually. We recall that Φ^0 is the root system of the Lie algebra $L(A^0)$ of finite type obtained by removing vertex 0 from the diagram of A. We denote by Φ_s^0, Φ_l^0 the set of short and long roots in Φ^0. If all roots of Φ^0 have the same length we write $\Phi_s^0 = \Phi^0$.

Theorem 17.17 *The real roots of the affine Kac–Moody algebra $L(A)$ are as follows.*
(a) *If A is one of the types $\tilde{A}_l, \tilde{B}_l, \tilde{C}_l, \tilde{D}_l, \tilde{E}_6, \tilde{E}_7, \tilde{E}_8, \tilde{F}_4, \tilde{G}_2$ then $\Phi_{\text{Re}} = \{\alpha + r\delta \; ; \; \alpha \in \Phi^0, r \in \mathbb{Z}\}$.*
(b) *If A is one of the types $\tilde{B}_l^t, \tilde{C}_l^t, \tilde{F}_4^t$ then*
$$\Phi_{\text{Re},s} = \{\alpha + r\delta \; ; \; \alpha \in \Phi_s^0, r \in \mathbb{Z}\}$$
$$\Phi_{\text{Re},l} = \{\alpha + 2r\delta \; ; \; \alpha \in \Phi_l^0, r \in \mathbb{Z}\}.$$

(c) *If A is of type $\tilde{G}_2^t$ then*
$$\Phi_{\text{Re},s} = \{\alpha + r\delta \; ; \; \alpha \in \Phi_s^0, r \in \mathbb{Z}\}$$
$$\Phi_{\text{Re},l} = \{\alpha + 3r\delta \; ; \; \alpha \in \Phi_l^0, r \in \mathbb{Z}\}.$$

(d) If A is of type $\tilde{C}'_l$ then

$$\Phi_{\text{Re},s} = \{\tfrac{1}{2}(\alpha + (2r-1)\delta) \; ; \; \alpha \in \Phi^0_l, r \in \mathbb{Z}\}$$
$$\Phi_{\text{Re},i} = \{\alpha + r\delta \; ; \; \alpha \in \Phi^0_s, r \in \mathbb{Z}\}$$
$$\Phi_{\text{Re},l} = \{\alpha + 2r\delta \; ; \; \alpha \in \Phi^0_l, r \in \mathbb{Z}\}.$$

(e) If A is of type $\tilde{A}'_1$ then

$$\Phi_{\text{Re},s} = \{\tfrac{1}{2}(\alpha + (2r-1)\delta) \; ; \; \alpha \in \Phi^0, r \in \mathbb{Z}\}$$
$$\Phi_{\text{Re},l} = \{\alpha + 2r\delta \; ; \; \alpha \in \Phi^0, r \in \mathbb{Z}\}.$$

Proof. (i) Suppose first that A is not of type $\tilde{C}'_l$ or $\tilde{A}'_1$. Then $\Phi^0_s \subset \Phi_{\text{Re},s}$. Let $\alpha \in \Phi^0_s$. Then $\langle \alpha, \alpha \rangle = m$. Hence for $r \in \mathbb{Z}$ we have $\langle \alpha + r\delta, \alpha + r\delta \rangle = m$ since $\langle \alpha, \delta \rangle = 0$ and $\langle \delta, \delta \rangle = 0$. By Proposition 17.11 this implies $\alpha + r\delta \in \Phi_{\text{Re},s}$. Conversely suppose $\beta = \sum_{i=0}^l k_i \alpha_i \in \Phi_{\text{Re},s}$. We have $a_0 = 1$, thus $\delta = \alpha_0 + \sum_{i=1}^l a_i \alpha_i$. Hence $\beta - k_0 \delta = \sum_{i=1}^l (k_i - k_0 a_i) \alpha_i$. Thus $\langle \beta - k_0 \delta, \beta - k_0 \delta \rangle = \langle \beta, \beta \rangle = m$. Again by Proposition 17.11 we deduce $\beta - k_0 \delta \in \Phi^0_s$. Let $\alpha = \beta - k_0 \delta$. Then $\beta = \alpha + k_0 \delta$ for $\alpha \in \Phi^0_s, k_0 \in \mathbb{Z}$.

Thus the short roots in Φ have the required form. We now consider the long roots. We have $\Phi^0_l \subset \Phi_{\text{Re},l}$.

Let $\alpha \in \Phi^0_l$. Then $\langle \alpha, \alpha \rangle = M$ and so $\langle \alpha + s\delta, \alpha + s\delta \rangle = M$ for all $s \in \mathbb{Z}$. Let $\alpha = \sum_{i=1}^l k_i \alpha_i$. By Proposition 17.13 we have $k_i \frac{\langle \alpha_i, \alpha_i \rangle}{\langle \alpha, \alpha \rangle} \in \mathbb{Z}$ for $i = 1, \ldots, l$. The same proposition shows that $\alpha + s\delta \in \Phi_{\text{Re},l}$ if and only if $s a_i \frac{\langle \alpha_i, \alpha_i \rangle}{\langle \alpha, \alpha \rangle} \in \mathbb{Z}$ for $i = 0, 1, \ldots, l$. Now $\langle \alpha_i, \alpha_i \rangle = \frac{2c_i}{a_i}$, thus the condition is $\frac{2c_i}{\langle \alpha, \alpha \rangle} s \in \mathbb{Z}$ for $i = 0, 1, \ldots, l$. We note that $\langle \alpha_0, \alpha_0 \rangle = 2$ since $a_0 = 1$.

First suppose that α_0 is a long root, that is that we are in case (a). Then $\langle \alpha, \alpha \rangle = 2$ and so $\frac{2c_i s}{\langle \alpha, \alpha \rangle} = c_i s \in \mathbb{Z}$. Hence $\alpha + s\delta \in \Phi_{\text{Re},l}$ for all $s \in \mathbb{Z}$.

Conversely suppose $\beta = \sum_{i=0}^l k_i \alpha_i \in \Phi_{\text{Re},l}$. Then

$$\beta - k_0 \delta = \sum_{i=1}^l (k_i - k_0 a_i) \alpha_i$$

and we have $\langle \beta - k_0 \delta, \beta - k_0 \delta \rangle = \langle \beta, \beta \rangle = M$. Since $\beta \in \Phi_{\text{Re},l}$ we have $k_i \frac{\langle \alpha_i, \alpha_i \rangle}{\langle \beta, \beta \rangle} \in \mathbb{Z}$ for $i = 0, 1, \ldots, l$. We have $k_0 a_i \frac{\langle \alpha_i, \alpha_i \rangle}{\langle \beta, \beta \rangle} \in \mathbb{Z}$ also since $\langle \alpha_i, \alpha_i \rangle = \frac{2c_i}{a_i}$ and $\langle \beta, \beta \rangle = 2$. Hence $\beta - k_0 \delta \in \Phi^0_l$ by Proposition 17.13. Thus $\beta = \alpha + k_0 \delta$ for some $\alpha \in \Phi^0_l$ and $k_0 \in \mathbb{Z}$.

Next suppose that α_0 is a short root, i.e. that we are in case (b) or (c). Let $\frac{\langle \alpha, \alpha \rangle}{\langle \alpha_0, \alpha_0 \rangle} = p$. Then $p = 2$ in case (b) and $p = 3$ in case (c). Thus

$$\frac{2c_i}{\langle \alpha, \alpha \rangle} s = c_i \frac{s}{p} \quad \text{since } \langle \alpha, \alpha \rangle = 2p.$$

Since $c_0 = 1$ this lies in $\mathbb{Z}$ for all $i = 0, 1, \ldots, l$ if and only if s is divisible by p. Thus by Proposition 17.13 $\alpha + pr\delta \in \Phi_{\text{Re},1}$ for all $r \in \mathbb{Z}$.

Conversely suppose $\beta = \sum_{i=0}^{l} k_i \alpha_i \in \Phi_{\text{Re},1}$. Then

$$\beta - k_0 \delta = \sum_{i=1}^{l} (k_i - k_0 a_i) \alpha_i.$$

We have $\langle \beta - k_0 \delta, \beta - k_0 \delta \rangle = \langle \beta, \beta \rangle = M$. Since $\beta \in \Phi_{\text{Re},1}$ we have $k_i \frac{\langle \alpha_i, \alpha_i \rangle}{\langle \beta, \beta \rangle} \in \mathbb{Z}$ for $i = 0, 1, \ldots, l$. In particular $k_0 \frac{\langle \alpha_0, \alpha_0 \rangle}{\langle \beta, \beta \rangle} = \frac{k_0}{p} \in \mathbb{Z}$. We show $k_0 a_i \frac{\langle \alpha_i, \alpha_i \rangle}{\langle \beta, \beta \rangle} \in \mathbb{Z}$ for $i = 1, \ldots, l$. For $k_0 a_i \frac{\langle \alpha_i, \alpha_i \rangle}{\langle \beta, \beta \rangle} = \frac{k_0}{p} c_i \in \mathbb{Z}$ since $\langle \alpha_i, \alpha_i \rangle = \frac{2c_i}{a_i}$ and $\langle \beta, \beta \rangle = 2p$. Thus by Proposition 17.13 $\beta - k_0 \delta \in \Phi_l^0$. Let $\alpha = \beta - k_0 \delta$. Then $\beta = \alpha + pr\delta$ for some $\alpha \in \Phi_l^0, r \in \mathbb{Z}$.

We have thus proved the required result in cases (a), (b) and (c).

(ii) We now suppose that A has type $\tilde{C}_l'$. Then we have $\Phi_s^0 \subset \Phi_{\text{Re},1}$ and $\Phi_l^0 \subset \Phi_{\text{Re},1}$. First suppose $\alpha \in \Phi_s^0$. Then $\langle \alpha, \alpha \rangle = m'$ and so $\langle \alpha + r\delta, \alpha + r\delta \rangle = m'$. By Proposition 17.16 $\alpha + r\delta \in \Phi_{\text{Re},i}$ for all $\alpha \in \Phi_s^0, r \in \mathbb{Z}$.

Conversely suppose $\beta = \sum_{i=0}^{l} k_i \alpha_i \in \Phi_{\text{Re},i}$. Then $k_i \frac{\langle \alpha_i, \alpha_i \rangle}{\langle \beta, \beta \rangle} \in \mathbb{Z}$ for $i = 0, 1, \ldots, l$ by Lemma 17.15, in particular $k_0 \frac{\langle \alpha_0, \alpha_0 \rangle}{\langle \beta, \beta \rangle} = \frac{k_0}{2} \in \mathbb{Z}$. Now

$$\beta - \frac{k_0}{2} \delta = \sum_{i=1}^{l} \left(k_i - \frac{k_0}{2} a_i \right) \alpha_i.$$

We have $\langle \beta - \frac{k_0}{2} \delta, \beta - \frac{k_0}{2} \delta \rangle = m'$ and so by Proposition 17.11 $\beta - \frac{k_0}{2} \delta \in \Phi_s^0$. Let $\alpha = \beta - \frac{k_0}{2} \delta$. Then $\beta = \alpha + \frac{k_0}{2} \delta = \alpha + r\delta$ for some $\alpha \in \Phi_s^0, r \in \mathbb{Z}$.

We now turn from the intermediate roots to the long roots. Suppose $\alpha \in \Phi_l^0$. Then $\langle \alpha, \alpha \rangle = M$ and $\langle \alpha + s\delta, \alpha + s\delta \rangle = M$ for $s \in \mathbb{Z}$. Let $\alpha = \sum_{i=1}^{l} k_i \alpha_i$. Then $k_i \frac{\langle \alpha_i, \alpha_i \rangle}{\langle \alpha, \alpha \rangle} \in \mathbb{Z}$ for $i = 1, \ldots, l$. Now

$$\alpha + s\delta = \sum_{i=1}^{l} k_i \alpha_i + \sum_{i=0}^{l} s a_i \alpha_i.$$

We wish to know for which $s \in \mathbb{Z}$ we have

$$s a_i \frac{\langle \alpha_i, \alpha_i \rangle}{\langle \alpha, \alpha \rangle} \in \mathbb{Z} \quad \text{for all } i = 0, 1, \ldots, l.$$

Now $\langle \alpha_0, \alpha_0 \rangle = \frac{2c_0}{a_0} = 1$, thus $\langle \alpha, \alpha \rangle = 4$. Hence $s a_i \frac{\langle \alpha_i, \alpha_i \rangle}{\langle \alpha, \alpha \rangle} = \frac{c_i s}{2}$. Since $c_0 = 1$ this lies in $\mathbb{Z}$ for all $i = 0, 1, \ldots, l$ if and only if s is even. Thus by Proposition 17.13 $\alpha + 2r\delta \in \Phi_{\text{Re},1}$ for all $\alpha \in \Phi_l^0, r \in \mathbb{Z}$.

17.2 The roots of an affine Kac–Moody algebra

Conversely suppose $\beta = \sum_{i=0}^{l} k_i \alpha_i \in \Phi_{\text{Re},1}$. Then $k_i \frac{\langle \alpha_i, \alpha_i \rangle}{\langle \beta, \beta \rangle} \in \mathbb{Z}$ for $i = 0, 1, \ldots, l$, in particular $k_0 \frac{\langle \alpha_0, \alpha_0 \rangle}{\langle \beta, \beta \rangle} = \frac{k_0}{4} \in \mathbb{Z}$. Now

$$\beta - \frac{k_0}{2}\delta = \sum_{i=1}^{l} \left(k_i - \frac{k_0}{2} a_i \right) \alpha_i$$

satisfies $\langle \beta - \frac{k_0}{2}\delta, \beta - \frac{k_0}{2}\delta \rangle = M$. Also $\frac{k_0}{2} a_i \frac{\langle \alpha_i, \alpha_i \rangle}{\langle \beta, \beta \rangle} = \frac{k_0}{4} c_i \in \mathbb{Z}$. Thus by Proposition 17.13 we have $\beta - \frac{k_0}{2}\delta \in \Phi_l^0$. Let $\alpha = \beta - \frac{k_0}{2}\delta$. Then $\beta = \alpha + 2r\delta$ for some $\alpha \in \Phi_l^0, r \in \mathbb{Z}$.

We now consider the short roots of Φ. There is no root of Φ^0 of the same length as the short roots of Φ_{Re}. The squared length of the short roots of Φ_{Re} is one half that of the long roots of Φ^0. So suppose $\alpha \in \Phi_l^0$. We consider elements of form $\frac{1}{2}(\alpha + s\delta)$ where $s \in \mathbb{Z}$. We consider which of these elements lie in Q. Since the long roots of Φ^0 have form

$$\pm \{\alpha_l, 2\alpha_{l-1} + \alpha_l, \ldots, 2\alpha_1 + \cdots + 2\alpha_{l-1} + \alpha_l\}$$

and $\delta = 2\alpha_0 + 2\alpha_1 + \cdots + 2\alpha_{l-1} + \alpha_l$ we see that $\frac{1}{2}(\alpha + s\delta) \in Q$ if and only if s is odd. Thus we consider elements of Q of form $\frac{1}{2}(\alpha + (2r-1)\delta)$ with $r \in \mathbb{Z}$. We have

$$\langle \tfrac{1}{2}(\alpha + (2r-1)\delta), \tfrac{1}{2}(\alpha + (2r-1)\delta) \rangle = \tfrac{1}{4}\langle \alpha, \alpha \rangle = m.$$

By Proposition 17.11 this implies that $\frac{1}{2}(\alpha + (2r-1)\delta) \in \Phi_{\text{Re},s}$.

Conversely suppose $\beta = \sum_{i=0}^{l} k_i \alpha_i \in \Phi_{\text{Re},s}$. Then $k_i \frac{\langle \alpha_i, \alpha_i \rangle}{\langle \beta, \beta \rangle} \in \mathbb{Z}$ for $i = 0, 1, \ldots, l$. Then $2\beta - k_0\delta = \sum_{i=1}^{l}(2k_i - k_0 a_i)\alpha_i$. We have $\langle 2\beta - k_0\delta, 2\beta - k_0\delta \rangle = 4\langle \beta, \beta \rangle = 4$.

This is the squared length of the elements of Φ_l^0. We also have

$$\frac{k_0 a_i \langle \alpha_i, \alpha_i \rangle}{\langle 2\beta, 2\beta \rangle} = \frac{k_0}{2} c_i \in \mathbb{Z} \qquad \text{for } i = 1, \ldots, l$$

since $k_0 \in \mathbb{Z}$ and $c_i = 2$ for $i = 1, \ldots, l$. By Proposition 17.13 we have $2\beta - k_0\delta \in \Phi_l^0$. Let $\alpha = 2\beta - k_0\delta$. Then $\beta = \frac{1}{2}(\alpha + k_0\delta)$. Since $\beta \in Q$, k_0 is odd. Thus

$$\beta = \tfrac{1}{2}(\alpha + (2r-1)\delta) \qquad \text{for some } \alpha \in \Phi_l^0, r \in \mathbb{Z}.$$

(iii) Finally we suppose that A has type $\tilde{A}_1'$. The diagram of A is

with $a_0 = 2, a_1 = 1, c_0 = 1, c_1 = 2$. We also have

$$\langle \alpha_0, \alpha_0 \rangle = 1, \langle \alpha_1, \alpha_1 \rangle = 4.$$

Now $\Phi^0 \subset \Phi_{\text{Re},l}$. Let $\alpha \in \Phi^0$. Then $\langle \alpha + s\delta, \alpha + s\delta \rangle = M$. We can write $\alpha = k_1 \alpha_1$. Then $\alpha + s\delta = 2s\alpha_0 + (k_1 + s)\alpha_1$. We have $k_1 \frac{\langle \alpha_1, \alpha_1 \rangle}{\langle \alpha, \alpha \rangle} \in \mathbb{Z}$ and we consider which $s \in \mathbb{Z}$ have the property that $sa_i \frac{\langle \alpha_i, \alpha_i \rangle}{\langle \alpha, \alpha \rangle} \in \mathbb{Z}$ for $i = 0, 1$. Since $sa_i \frac{\langle \alpha_i, \alpha_i \rangle}{\langle \alpha, \alpha \rangle} = \frac{sc_i}{2}$ and $c_0 = 1$ this lies in $\mathbb{Z}$ for $i = 0, 1$ if and only if s is even. By Proposition 17.13 we deduce that $\alpha + 2r\delta \in \Phi_{\text{Re},l}$ for all $\alpha \in \Phi^0, r \in \mathbb{Z}$.

Conversely suppose $\beta = k_0 \alpha_0 + k_1 \alpha_1$ lies in $\Phi_{\text{Re},l}$. Then $\beta - \frac{k_0}{2}\delta = \left(k_1 - \frac{k_0 a_1}{2}\right) \alpha_1$. We have $\langle \beta - \frac{k_0}{2}\delta, \beta - \frac{k_0}{2}\delta \rangle = M$. Also $k_0 \frac{\langle \alpha_0, \alpha_0 \rangle}{\langle \beta, \beta \rangle} = \frac{k_0}{4} \in \mathbb{Z}$. By Proposition 17.11 we have $\beta - \frac{k_0}{2}\delta \in \Phi^0$. Let $\alpha = \beta - \frac{k_0}{2}\delta$. Then $\beta = \alpha + \frac{k_0}{2}\delta = \alpha + 2r\delta$ for some $\alpha \in \Phi^0, r \in \mathbb{Z}$.

We now consider the short roots. Suppose $\alpha \in \Phi^0$ and consider the element $\frac{1}{2}(\alpha + s\delta)$ for $s \in \mathbb{Z}$. Since $\alpha = \pm \alpha_1$ and $\delta = 2\alpha_0 + \alpha_1$ this element lies in Q if and only if s is odd. We have

$$\langle \tfrac{1}{2}(\alpha + (2r-1)\delta), \tfrac{1}{2}(\alpha + (2r-1)\delta) \rangle = \tfrac{1}{4}\langle \alpha, \alpha \rangle = 1.$$

This is the squared length of the short roots of $\Phi_{\text{Re},s}$. By Proposition 17.11 $\frac{1}{2}(\alpha + (2r-1)\delta) \in \Phi_{\text{Re},s}$ for all $r \in \mathbb{Z}$.

Conversely suppose $\beta = k_0 \alpha_0 + k_1 \alpha_1 \in \Phi_{\text{Re},s}$. Then

$$\beta - \frac{k_0}{2}\delta = \left(k_1 - \frac{k_0 a_1}{2}\right)\alpha_1.$$

We have $\langle 2\beta - k_0\delta, 2\beta - k_0\delta \rangle = 4\langle \beta, \beta \rangle = 4$. This is the squared length of the roots in Φ^0. So by Proposition 17.11 we have $2\beta - k_0\delta \in \Phi^0$. Let $\alpha = 2\beta - k_0\delta$. Then $\beta = \frac{1}{2}(\alpha + k_0\delta)$. Since $\beta \in Q$ k_0 must be odd. Hence $\beta = \frac{1}{2}(\alpha + (2r-1)\delta)$ for some $\alpha \in \Phi^0, r \in \mathbb{Z}$. This completes the proof. □

17.3 The Weyl group of an affine Kac–Moody algebra

Let A be an affine Cartan matrix and W the Weyl group of $L(A)$. Then $W = \langle s_0, s_1, \ldots, s_l \rangle$. The subgroup $W^0 = \langle s_1, \ldots, s_l \rangle$ is the Weyl group of the finite dimensional simple Lie algebra $L(A^0)$. In order to investigate the structure of W we introduce the element $\theta = \delta - a_0\alpha_0 = \sum_{i=1}^{l} a_i\alpha_i$. This element θ lies in $Q^0 = Q(A^0)$.

Proposition 17.18 (i) *If the affine Cartan matrix A is not of type $\tilde{B}_l^t, \tilde{C}_l^t, \tilde{F}_4^t, \tilde{G}_2^t$ then θ is the highest root of Φ^0.*
(ii) *If A is of type $\tilde{B}_l^t, \tilde{C}_l^t, \tilde{F}_4^t, \tilde{G}_2^t$ then θ is the highest short root of Φ^0.*

17.3 The Weyl group of an affine Kac–Moody algebra

Proof. We first show that $\theta \in \Phi^0$. We have $\langle \theta, \theta \rangle = \langle \delta - a_0\alpha_0, \delta - a_0\alpha_0 \rangle = a_0^2 \langle \alpha_0, \alpha_0 \rangle = 2a_0$. First suppose $a_0 = 1$ and α_0 is a long root. Then $\langle \theta, \theta \rangle = \langle \alpha_0, \alpha_0 \rangle = 2$. Also

$$a_i \frac{\langle \alpha_i, \alpha_i \rangle}{\langle \theta, \theta \rangle} = c_i \in \mathbb{Z}.$$

Thus $\theta \in \Phi^0_l$ by Proposition 17.13

Next suppose $a_0 = 2$. Then $\langle \theta, \theta \rangle = 4 \langle \alpha_0, \alpha_0 \rangle = 4$. Thus θ has the same squared length as a long root. Also $a_i \frac{\langle \alpha_i, \alpha_i \rangle}{\langle \theta, \theta \rangle} = \frac{c_i}{2}$. This lies in $\mathbb{Z}$ for $i = 1, \ldots, l$ since $c_i = 2$ for such values of i. Hence $\theta \in \Phi^0_l$ by Proposition 17.13.

Finally suppose $a_0 = 1$ and α_0 is a short root. This occurs for the cases in (ii). Then $\langle \theta, \theta \rangle = \langle \alpha_0, \alpha_0 \rangle$. Hence $\theta \in \Phi^0_s$ by Proposition 17.11.

Thus we have shown $\theta \in \Phi^0$ in all cases. We also have

$$\langle \theta, \alpha_i \rangle = \langle \delta - a_0\alpha_0, \alpha_i \rangle = -a_0 \langle \alpha_0, \alpha_i \rangle$$
$$= -\frac{a_0 A_{0i} \langle \alpha_0, \alpha_0 \rangle}{2} = -c_0 A_{0i} = -A_{0i}.$$

Thus $\langle \theta, \alpha_i \rangle \geq 0$ for $i = 1, \ldots, l$. Hence $\theta \in \bar{C}_0$, the closure of the fundamental chamber for Φ^0. This implies that θ is the highest root of Φ^0 in the cases in (i) and the highest short root of Φ^0 in the cases in (ii), by Proposition 12.9. □

Now let s_θ be the reflection corresponding to the root θ. Then $s_\theta : H^0 \to H^0$ is given by $s_\theta(h) = h - \theta(h)h_\theta$.

Lemma 17.19 *The coroot h_θ is given by $h_\theta = \frac{1}{a_0}(c - h_0)$.*

Proof. Since $\theta = \sum_{i=1}^l a_i\alpha_i$ we have $\frac{2\theta}{\langle \theta, \theta \rangle} = \sum_{i=1}^l a_i \frac{\langle \alpha_i, \alpha_i \rangle}{\langle \theta, \theta \rangle} \frac{2\alpha_i}{\langle \alpha_i, \alpha_i \rangle}$, hence

$$h_\theta = \sum_{i=1}^l a_i \frac{\langle \alpha_i, \alpha_i \rangle}{\langle \theta, \theta \rangle} h_i = \sum_{i=1}^l \frac{2c_i h_i}{2a_0}$$
$$= \frac{1}{a_0} \sum_{i=1}^l c_i h_i = \frac{1}{a_0}(c - h_0). \qquad \square$$

Now the affine Weyl group W is generated by W^0 and s_0, so is also generated by W^0 and $s_0 s_\theta$. We consider the action of $s_0 s_\theta$ on H.

Proposition 17.20 *Let $h \in H$. Then*

$$s_0 s_\theta(h) = h + \delta(h)h_\theta - \left(\langle h_\theta, h \rangle + \tfrac{1}{2} \langle h_\theta, h_\theta \rangle \delta(h) \right) c.$$

Proof.

$$s_0 s_\theta(h) = s_0\left(h - \theta(h)h_\theta\right) = h - \alpha_0(h)h_0 - \theta(h)\left(h_\theta - \alpha_0(h_\theta)h_0\right)$$
$$= h - \alpha_0(h)(c - a_0 h_\theta) - \theta(h)h_\theta + \theta(h)\alpha_0(h_\theta)(c - a_0 h_\theta)$$
$$= h + (a_0\alpha_0(h) - \theta(h) - a_0\theta(h)\alpha_0(h_\theta))h_\theta + (\theta(h)\alpha_0(h_\theta) - \alpha_0(h))c.$$

Now $\alpha_0(h_\theta) = \alpha_0\left(\frac{1}{a_0}(c - h_0)\right) = -\frac{2}{a_0}$. Thus

$$s_0 s_\theta(h) = h + (a_0\alpha_0(h) + \theta(h))h_\theta - \left(\frac{2}{a_0}\theta(h) + \alpha_0(h)\right)c$$
$$= h + \delta(h)h_\theta - \frac{1}{a_0}(\theta(h) + \delta(h))c$$
$$= h + \delta(h)h_\theta - \left(\langle h_\theta, h \rangle + \tfrac{1}{2}\langle h_\theta, h_\theta \rangle \delta(h)\right)c$$

since $\langle h_\theta, h \rangle = \frac{2\theta(h)}{\langle \theta, \theta \rangle} = \frac{1}{a_0}\theta(h)$ and $\langle h_\theta, h_\theta \rangle = \left\langle \frac{2\theta}{\langle \theta, \theta \rangle}, \frac{2\theta}{\langle \theta, \theta \rangle} \right\rangle = \frac{4}{\langle \theta, \theta \rangle} = \frac{2}{a_0}$. We define $t_{h_\theta} : H \to H$ by

$$t_{h_\theta}(h) = h + \delta(h)h_\theta - \left(\langle h_\theta, h \rangle + \tfrac{1}{2}\langle h_\theta, h_\theta \rangle \delta(h)\right)c.$$

Thus we have $s_0 s_\theta = t_{h_\theta}$. Hence W is generated by W^0 and t_{h_θ}.

More generally, for any $x \in H^0$ we define $t_x : H \to H$ by

$$t_x(h) = h + \delta(h)x - \left(\langle x, h \rangle + \tfrac{1}{2}\langle x, x \rangle \delta(h)\right)c. \qquad \square$$

Proposition 17.21 (i) $t_x t_y = t_{x+y}$ for all $x, y \in H^0$.
(ii) $w t_x w^{-1} = t_{w(x)}$ for all $w \in W^0$, $x \in H^0$.

Proof. The linear map $t_x : H \to H$ is uniquely determined by the properties

$$t_x(h) = h - \langle x, h \rangle c \quad \text{when } \delta(h) = 0$$
$$t_x(d) = d + a_0 x - \tfrac{1}{2} a_0 \langle x, x \rangle c$$

since $\delta(h_i) = 0$ and $\delta(d) = a_0$. If $\delta(h) = 0$ then

$$t_x t_y(h) = t_x(h - \langle y, h \rangle c) = h - \langle x, h \rangle c - \langle y, h \rangle (c - \langle x, c \rangle c)$$
$$= h - \langle x + y, h \rangle c \quad \text{since } \langle x, c \rangle = 0$$
$$= t_{x+y}(h).$$

Also

$$t_x t_y(d) = t_x \left(d + a_0 y - \tfrac{1}{2} a_0 \langle y, y \rangle c\right)$$
$$= d + a_0 x - \tfrac{1}{2} a_0 \langle x, x \rangle c + a_0 (y - \langle x, y \rangle c) - \tfrac{1}{2} a_0 \langle y, y \rangle (c - \langle x, c \rangle c)$$
$$= d + a_0(x+y) - a_0 \left(\tfrac{1}{2}\langle x, x \rangle + \langle x, y \rangle + \tfrac{1}{2}\langle y, y \rangle\right) c$$
$$= d + a_0(x+y) - a_0 \cdot \tfrac{1}{2}\langle x+y, x+y \rangle c$$
$$= t_{x+y}(d).$$

Thus $t_x t_y = t_{x+y}$ for all $x, y \in H^0$.

Now let $w \in W^0$, and $h \in H$ satisfy $\delta(h) = 0$. Then

$$w t_x w^{-1}(h) = w\left(w^{-1}(h) - \langle x, w^{-1}(h) \rangle c\right)$$

since $\delta\left(w^{-1}(h)\right) = (w\delta)(h) = \delta(h) = 0$. Thus

$$w t_x w^{-1}(h) = h - \langle w(x), h \rangle c = t_{w(x)}(h)$$

since $w(c) = c$. Also $w(d) = d$ for all $w \in W^0$ and so

$$w t_x w^{-1}(d) = w t_x(d) = w\left(d + a_0 x - \tfrac{1}{2}\langle x, x \rangle a_0 c\right)$$
$$= d + a_0 w(x) - \tfrac{1}{2}\langle w(x), w(x) \rangle a_0 c$$
$$= t_{w(x)}(d).$$

Hence $w t_x w^{-1} = t_{w(x)}$. $\square$

Let M be the additive subgroup (i.e. lattice) of H^0 generated by the elements $w(h_\theta)$ for all $w \in W^0$. Let $t(M) = \{t_m \; ; \; m \in M\}$.

Proposition 17.22 $W = t(M) W^0$ where $t(M)$ is normal in W and $t(M) \cap W^0 = 1$. Thus W is a semidirect product of $t(M)$ and W^0.

Proof. We know that $t_{h_\theta} \in W$, hence $w t_{h_\theta} w^{-1} = t_{w(h_\theta)} \in W$ for all $w \in W^0$. Thus $t(M)$ is a subgroup of W. Since W is generated by W^0 and t_{h_θ}, W is generated by $t(M)$ and W^0. But W^0 lies in the normaliser of $t(M)$ by Proposition 17.21 (ii). Thus $W = t(M) W^0$. Finally $t(M) \cap W^0 = 1$ since $t(M)$ is a free abelian group whereas W^0 is finite. $\square$

The lattice $M \subset H_{\mathbb{R}}^0$ will be important in understanding the affine Weyl group W. We shall now identify it in each case.

Proposition 17.23 (i) *If A is an affine Cartan matrix not of types $\tilde{B}_l^t, \tilde{C}_l^t, \tilde{F}_4^t, \tilde{G}_2^t$ then $M = \sum_{i=1}^l \mathbb{Z} h_i$.*
(ii) *If A has type $\tilde{B}_l^t, \tilde{C}_l^t, \tilde{F}_4^t, \tilde{G}_2^t$ then*

$$M = \sum_{\alpha_i \text{ short}} \mathbb{Z} h_i + \sum_{\alpha_i \text{ long}} p\mathbb{Z} h_i$$

where p is the ratio of the squared lengths of the long and short roots ($p=3$ for $\tilde{G}_2^t$ and $p=2$ in the other cases).

Proof. By Proposition 17.18 θ is a long root in the cases in (i) and a short root in the cases in (ii). Thus h_θ is a short coroot in (i) and a long coroot in (ii). Thus M is generated by all short coroots in (i) and by all long coroots in (ii). Now it follows from Proposition 8.18 that the set of all short coroots generates the coroot lattice $\sum_{i=1}^l \mathbb{Z} h_i$. But the set of all long coroots generates the sublattice with basis h_i for h_i long (i.e. α_i short) and ph_i for h_i short (i.e. α_i long). The result follows. □

We have been considering an action of the affine Weyl group W by linear transformations of the vector space H of dimension $l+2$. However, we now show that there is a simpler action of W by affine transformations on the real vector space $H_\mathbb{R}^0$ of dimension l. We recall that the group of affine transformations of a vector space is generated by the group of non-singular linear transformations and the group of translations.

We first define $H_{\mathbb{R},1} = \{h \in H_\mathbb{R} \; ; \; \delta(h) = 1\}$. The space $H_{\mathbb{R},1}$, although not a subspace of $H_\mathbb{R}$, is invariant under W. For

$$\delta(w(h)) = (w^{-1}\delta)(h) = \delta(h)$$

since $w(\delta) = \delta$. Now we have a decomposition

$$H_\mathbb{R} = H_\mathbb{R}^0 \oplus (\mathbb{R} c + \mathbb{R} d)$$

into subspaces of dimension l and 2 which are mutually orthogonal. For $\langle h_i, c \rangle = 0$ and $\langle h_i, d \rangle = 0$ for $i = 1, \ldots, l$. Since $\delta(h_i) = 0$ for $i = 1, \ldots, l$, $\delta(c) = 0$ and $\delta(d) = a_0$ the elements of $H_\mathbb{R}$ which lie in $H_{\mathbb{R},1}$ are those of form

$$\sum_{i=1}^l \lambda_i h_i + \lambda c + \frac{1}{a_0} d \qquad \lambda_i \in \mathbb{R}, \lambda \in \mathbb{R}.$$

Now $h \in H_{\mathbb{R},1}$ implies $h + \mu c \in H_{\mathbb{R},1}$ for $\mu \in \mathbb{R}$. Since $w(c) = c$ for all $w \in W$, W acts on the quotient space $H_{\mathbb{R},1}/\mathbb{R} c$. Also we have a bijective map

$$H_{\mathbb{R},1}/\mathbb{R} c \to H_\mathbb{R}^0$$

17.3 The Weyl group of an affine Kac–Moody algebra

given by

$$\mathbb{R}c + \sum_{i=1}^{l} \lambda_i h_i + \frac{1}{a_0} d \to \sum_{i=1}^{l} \lambda_i h_i$$

and this bijection may be used to define an action of W on $H_\mathbb{R}^0$.

Proposition 17.24 *The action of $W = t(M)W^0$ on $H_\mathbb{R}^0$ is as follows. The W^0-action on $H_\mathbb{R}^0$ is that previously considered. For $m \in M$, $h \in H_\mathbb{R}^0$ we have $t_m(h) = h + m$. Thus t_m acts on $H_\mathbb{R}^0$ as translation by m. Hence W acts on $H_\mathbb{R}^0$ as a group of affine transformations.*

Proof. If $w \in W^0$ then $w(c) = c$ and $w(d) = d$. This implies that the w-action on $H_\mathbb{R}^0$ defined above is the usual w-action. If $m \in M$, $h \in H_{\mathbb{R},1}$ then $t_m(h) = h + m + \mu c$ for some $\mu \in \mathbb{R}$. This induces an action of t_m on $H_\mathbb{R}^0$ given by $t_m(h) = h + m$. Thus t_m acts on $H_\mathbb{R}^0$ as translation by m. □

Corollary 17.25 *The action of W on $H_\mathbb{R}^0$ is faithful.*

Proof. Suppose $t_m w$, $w \in W^0$, acts trivially on $H_\mathbb{R}^0$. Then $t_m w(0) = 0$. This implies $m = 0$, that is $t_m = 1$. Hence $w \in W^0$ acts trivially on $H_\mathbb{R}^0$. Since W^0 acts faithfully on $H_\mathbb{R}^0$ this implies $w = 1$. □

Corollary 17.26 s_0 *acts on $H_\mathbb{R}^0$ as the reflection in the affine hyperplane*

$$L_{\theta,1} = \{h \in H_\mathbb{R}^0 \; ; \; \theta(h) = 1\}.$$

Proof. For $h \in H_\mathbb{R}^0$ we have

$$s_0(h) = t_{h_\theta} s_\theta(h) = h - \theta(h) h_\theta + h_\theta = h + (1 - \theta(h)) h_\theta.$$

This is the reflection in $L_{\theta,1}$. □

For each $\alpha \in \Phi^0$ and $k \in \mathbb{Z}$ let $L_{\alpha,k}$ be the affine hyperplane given by

$$L_{\alpha,k} = \{h \in H_\mathbb{R}^0 \; ; \; \alpha(h) = k\}.$$

Thus the generators $s_0, s_1, \ldots, s_l$ of the affine Weyl group W act on $H_\mathbb{R}^0$ as the reflections in the hyperplanes $L_{\theta,1}, L_{\alpha_1,0}, \ldots, L_{\alpha_l,0}$ respectively.

We now introduce a collection of affine hyperplanes whose corresponding affine reflections will lie in W.

Let $\mathfrak{L} = \{L_{\alpha,k} \; ; \; \alpha \in \Phi^0, k \in \mathbb{Z}, p$ divides k if α is a long root and A is one of $\tilde{B}_l^t, \tilde{C}_l^t, \tilde{F}_4^t, \tilde{G}_2^t\}$. Here as usual $p = 2$ in the first three cases and $p = 3$ for $\tilde{G}_2^t$.

Let $s_{\alpha,k}$ be the reflection in $L_{\alpha,k}$. Then $s_{\alpha,k}(h) = h + (k - \alpha(h))h_\alpha$. For
$$\frac{h + (h + (k - \alpha(h))h_\alpha)}{2} \in L_{\alpha,k}$$
and $h + (k - \alpha(h))h_\alpha$ differs from h by a multiple of h_α. Thus $s_{\alpha,k} = t_{kh_\alpha} s_\alpha$.

Proposition 17.27 *The reflection $s_{\alpha,k} \in W$ for all $L_{\alpha,k} \in \mathfrak{L}$. In fact $s_{\alpha,k} = s_{\alpha-k\delta}$.*

Proof. The reflection $s_{\alpha-k\delta} : H_\mathbb{R} \to H_\mathbb{R}$ is given by
$$s_{\alpha-k\delta}(h) = h - (\alpha - k\delta)(h)h_{\alpha-k\delta}.$$
Thus the restriction of $s_{\alpha-k\delta}$ to $H_{\mathbb{R},1}$ is given by
$$s_{\alpha-k\delta}(h) = h - (\alpha(h) - k)h_{\alpha-k\delta}.$$
Since $\delta \in H^*$ corresponds to $c \in H$ under our bijection between H and H^* we have $h_{\alpha-k\delta} = h_\alpha - \frac{2k}{\langle \alpha, \alpha \rangle} c$. Thus the action of $s_{\alpha-k\delta}$ on $H_{\mathbb{R},1}/\mathbb{R}c$ is $s_{\alpha-k\delta}(h) = h - (\alpha(h) - k)h_\alpha$ and the action on $H_\mathbb{R}^0$ is given by the same formula. Thus $s_{\alpha-k\delta} = s_{\alpha,k}$ on $H_\mathbb{R}^0$. Moreover we know from Theorem 17.17 that $s_{\alpha-k\delta} \in W$ whenever $L_{\alpha,k} \in \mathfrak{L}$. $\square$

We note that $L_{\theta,1}, L_{\alpha_1,0}, \ldots, L_{\alpha_l,0}$ all lie in $\mathfrak{L}$. For by Proposition 17.18 θ is a short root when A has one of the types $\tilde{B}_l^t, \tilde{C}_l^t, \tilde{F}_4^t, \tilde{G}_2^t$.

Definition *The connected components of the set $H_\mathbb{R}^0 - \bigcup_{L_{\alpha,k} \in \mathfrak{L}} L_{\alpha,k}$ are called alcoves.*

Proposition 17.28 *The set*
$$A = \{h \in H_\mathbb{R}^0 \; ; \; \alpha_i(h) > 0 \text{ for } i = 1, \ldots, l, \; \theta(h) < 1\}$$
is an alcove.

Proof. We show $A \cap L_{\alpha,k} = \phi$ for all $L_{\alpha,k} \in \mathfrak{L}$. Let $h \in A \cap L_{\alpha,k}$. We may assume $\alpha \in (\Phi^0)^+$. Suppose θ is a long root. Then $0 < \alpha(h) \leq \theta(h) < 1$ by Proposition 12.9 and so h cannot lie in $L_{\alpha,k}$ for $k \in \mathbb{Z}$. So suppose θ is a short root. If α is a short root we again have $0 < \alpha(h) \leq \theta(h) < 1$, so h cannot lie in $L_{\alpha,k}$ with $k \in \mathbb{Z}$. Thus suppose α is a long root. Then $\alpha(h) \leq \theta_l(h)$ where θ_l is the highest root of Φ^0. Let $\theta_l = \sum_{i=1}^l b_i \alpha_i$. Then we have

$$\theta_l = \sum_{i=1}^l b_i \alpha_i \text{ is the highest root of } \Phi^0$$

$$\theta = \sum_{i=1}^l a_i \alpha_i \text{ is the highest short root of } \Phi^0.$$

17.3 The Weyl group of an affine Kac–Moody algebra

By considering the coroot of the highest root, or by a case-by-case check, one may show

$$b_i = \begin{cases} a_i & \text{if } \alpha_i \text{ is long} \\ pa_i & \text{if } \alpha_i \text{ is short.} \end{cases}$$

In particular $b_i \le pa_i$ for all i. Hence

$$0 < \alpha(h) \le \theta_l(h) \le p\theta(h) < p.$$

Thus h cannot lie in $L_{\alpha,k}$ with $k \in \mathbb{Z}$ divisible by p.

Thus A lies in an alcove. But

$$\bar{A} = A \cup L_{\alpha_1,0} \cup \cdots \cup L_{\alpha_l,0} \cup L_{\theta,1}.$$

This shows that A cannot be properly contained in an alcove, since $L_{\alpha_1,0}, \ldots, L_{\alpha_l,0}, L_{\theta,1}$ lie in $\mathfrak{L}$. Thus A is an alcove. □

Let $\mathfrak{A}$ be the set of alcoves. We show that W acts on $\mathfrak{A}$. Since W is generated by $s_1, \ldots, s_l, s_\theta$ it is sufficient to prove the following lemma.

Lemma 17.29 (i) $s_i(L_{\alpha,k}) = L_{s_i(\alpha),k}$ for $i = 1, \ldots, l$.
(ii) $s_\theta(L_{\alpha,k}) = L_{s_0(\alpha),k+\alpha(h_\theta)}$. Also if $L_{\alpha,k} \in \mathfrak{L}$ then $L_{s_0(\alpha),k+\alpha(h_\theta)} \in \mathfrak{L}$.

Proof. (i) Let $h \in H_\mathbb{R}^0$. Then $h \in L_{\alpha,k}$ if and only if $\alpha(h) = k$, and this is equivalent to $(s_i(\alpha))(s_i(h)) = k$, that is $s_i(h) \in L_{s_i(\alpha),k}$. Thus $s_i(L_{\alpha,k}) = L_{s_i(\alpha),k}$.
(ii) $s_\theta(h) = s_0 t_{h_\theta}(h) = s_0(h + h_\theta) = s_0(h) + s_0(h_\theta)$. Thus $\alpha(h) = k$ if and only if $(s_0(\alpha))(s_0(h)) = k$, that is $(s_0(\alpha))(s_0(h) + s_0(h_\theta)) = k + \alpha(h_\theta)$. It follows that $s_\theta(L_{\alpha,k}) = L_{s_0(\alpha),k+\alpha(h_\theta)}$.

Now suppose $L_{\alpha,k} \in \mathfrak{L}$. Then k is divisible by p if α is a long root and $A \in \{\tilde{B}_l^t, \tilde{C}_l^t, \tilde{F}_4^t, \tilde{G}_2^t\}$. If we are not in this special case then $L_{s_0(\alpha),k+\alpha(h_\theta)} \in \mathfrak{L}$ since $\alpha(h_\theta) \in \mathbb{Z}$. So suppose A is one of the above four possibilities and α is a long root. We know p divides k and must show p divides $\alpha(h_\theta)$. Now $h_\theta = \frac{1}{a_0}(c - h_0)$. We have $a_0 = 1$ in the given cases and $\alpha(c) = 0$, thus $\alpha(h_\theta) = -\alpha(h_0)$. Let $\alpha = \sum_{i=1}^l k_i \alpha_i$. Then $\alpha(h_0) = \sum_{i=1}^l k_i \alpha_i(h_0) = \sum_{i=1}^l A_{0i} k_i$. There is precisely one $i \in \{1, \ldots, l\}$ with $A_{0i} \ne 0$. For this i, $A_{0i} = -2$ in type $\tilde{C}_l^t$ and $A_{0i} = -1$ in the other cases. In the latter cases α_i is a short root. Thus

$$k_i \frac{\langle \alpha_i, \alpha_i \rangle}{\langle \alpha, \alpha \rangle} = \frac{k_i}{p} \in \mathbb{Z}.$$

This shows that p divides $\sum_{i=1}^{l} A_{0i} k_i$, and so p divides $\alpha(h_\theta)$ in all cases. Thus $L_{s_0(\alpha), k+\alpha(h_\theta)} \in \mathfrak{L}$. □

Corollary 17.30 *If $w \in W$, $A' \in \mathfrak{A}$ then $w(A') \in \mathfrak{A}$.*

Proof. This follows from the definition of alcoves, together with the fact that the elements of W permute the affine hyperplanes in $\mathfrak{L}$. □

We define $L_i = L_{\alpha_i, 0}$ for $i = 1, \ldots, l$ and $L_0 = L_{\theta, 1}$. Thus $L_0, L_1, \ldots, L_l$ are the walls bounding the alcove A and $s_0, s_1, \ldots, s_l$ are the reflections in $L_0, L_1, \ldots, L_l$ respectively.

Given $w \in W$ we say that L_i separates the alcoves A and $w(A)$ if these alcoves lie on opposite sides of L_i.

Lemma 17.31 L_i *separates A and $w(A)$ if and only if $l(w) = l(s_i w) + 1$.*

Proof. First suppose $w' \in W$ has the property that $w'(A)$ lies on the same side of L_i as A but $w' s_j(A)$ lies on the opposite side of L_i to A. Then $w'(A)$, $w' s_j(A)$ lie on opposite sides of L_i so A, $s_j(A)$ lie on opposite sides of $w'^{-1}(L_i)$. This implies $w'^{-1}(L_i) = L_j$ so $L_i = w'(L_j)$. Hence $s_i = w' s_j w'^{-1}$ and $w' s_j = s_i w'$.

Now suppose $w \in W$ is such that $w(A)$ is on the opposite side of L_i to A. Let $w = s_{i_1} \ldots s_{i_r}$ be a reduced expression for w. Then there exists $q \geq 1$ such that $s_{i_1} \ldots s_{i_{q-1}}(A)$ lies on the same side of L_i as A but $s_{i_1} \ldots s_{i_q}(A)$ lies on the opposite side of L_i. Then we have

$$s_{i_1} \ldots s_{i_{q-1}} s_{i_q} = s_i s_{i_1} \ldots s_{i_{q-1}}$$

as above. Hence

$$s_i w = s_i s_{i_1} \ldots s_{i_r} = s_{i_1} \ldots s_{i_{q-1}} s_{i_{q+1}} \ldots s_{i_r}$$

and so $l(s_i w) < l(w)$.

If $w(A)$ is on the same side of L_i as A then $s_i w(A)$ is on the opposite side. Hence $l(s_i \cdot s_i w) < l(s_i w)$, that is $l(s_i w) > l(w)$. □

Theorem 17.32 *The map $w \to w(A)$ is a bijection between the elements of the affine Weyl group W and the set $\mathfrak{A}$ of alcoves.*

Proof. Given any alcove $A' \in \mathfrak{A}$ we can find a sequence of alcoves

$$A = A_1, A_2, \ldots, A_r = A'$$

17.3 The Weyl group of an affine Kac–Moody algebra

such that A_i is obtained from A_{i-1} by reflection in a common wall. Such reflections lie in W by Proposition 17.27. Hence $A' = w(A)$ for some $w \in W$. Thus the map $w \to w(A)$ is surjective.

Next suppose $w(A) = w'(A)$. Then $w'^{-1}w(A) = A$. We show $w'^{-1}w = 1$. If this is not so then

$$w'^{-1}w = s_i w'' \quad \text{with} \quad l\left(w'^{-1}w\right) = l\left(s_i w'^{-1}w\right) + 1$$

for some i. By Lemma 17.31 L_i separates A and $w'^{-1}w(A)$. This is a contradiction so $w'^{-1}w = 1$ and $w = w'$. □

Theorem 17.33 *The closure $\bar{A}$ of A is a fundamental region for the action of the affine Weyl group W on $H_\mathbb{R}^0$, i.e. each W-orbit on $H_\mathbb{R}^0$ intersects $\bar{A}$ in exactly one point.*

Proof. Each point in $H_\mathbb{R}^0$ lies in the closure $\overline{A'}$ of some alcove A'. By Theorem 17.32 $A' = w(A)$ for some $w \in W$. Thus the W-orbit of the given point intersects $\bar{A}$.

Now suppose $x, y \in \bar{A}$ satisfy $y = w(x)$ for $w \in W$. We shall show $x = y$ by induction on $l(w)$. If $l(w) = 0$ then $w = 1$ so $x = y$. So suppose $l(w) > 0$. Then $w = s_i w'$ with $l(s_i w) < l(w)$. By Lemma 17.31 L_i separates A and $w(A)$. Thus $\bar{A} \cap w(\bar{A}) \subset L_i$. Now $y \in \bar{A} \cap w(\bar{A})$ hence $y \in L_i$. Thus $s_i(y) = y$. But then $s_i(y) = w'(x)$ so $y = w'(x)$. Since $l(w') < l(w)$ we deduce $x = y$ by induction. □

Remark 17.34 We may also define an action of the affine Weyl group W on H^* in a way which is compatible with the bijection $H \to H^*$ determined by the standard invariant form $\langle , \rangle$ on H. Under this bijection the element $h_\theta \in H$ corresponds to $\frac{1}{a_0}\theta \in H^*$.

For each $\alpha \in (H^0)^*$ we may define $t_\alpha : H^* \to H^*$ by

$$t_\alpha(\lambda) = \lambda + \lambda(c)\alpha - (\langle \lambda, \alpha \rangle + \tfrac{1}{2}\langle \alpha, \alpha \rangle \lambda(c))\delta.$$

Then we have $s_0 s_\theta = t_{(1/a_0)\theta}$ on H^*. Moreover we have $t_\alpha t_\beta = t_{\alpha+\beta}$ and $w t_\alpha w^{-1} = t_{w(\alpha)}$ for $w \in W^0$. It follows that we have a semidirect decomposition $W = t(M^*) W^0$ where $t(M^*)$ is the set of t_α for $\alpha \in M^*$ and M^* is the sublattice of $(H_\mathbb{R}^0)^*$ spanned by $w\left(\frac{1}{a_0}\theta\right)$ for all $w \in W^0$.

The lattice M^* is given explicitly as follows.

$$M^* = \sum_{i=1}^{l} \mathbb{Z}\alpha_i \quad \text{for types } \tilde{A}_l, \tilde{D}_l, \tilde{E}_6, \tilde{E}_7, \tilde{E}_8$$

$$M^* = \sum_{\alpha_i \text{ long}} \mathbb{Z}\alpha_i + \sum_{\alpha_i \text{ short}} p\mathbb{Z}\alpha_i \quad \text{for types } \tilde{B}_l, \tilde{C}_l, \tilde{F}_4, \tilde{G}_2$$

$$M^* = \sum_{i=1}^{l} \mathbb{Z}\alpha_i \quad \text{for types } \tilde{B}_l^t, \tilde{C}_l^t, \tilde{F}_4^t, \tilde{G}_2^t$$

$$M^* = \sum_{\alpha_i \text{ long}} \tfrac{1}{2}\mathbb{Z}\alpha_i + \sum_{\alpha_i \text{ short}} \mathbb{Z}\alpha_i \quad \text{for type } \tilde{C}_l'$$

$$M^* = \tfrac{1}{2}\mathbb{Z}\alpha_1 \quad \text{for type } \tilde{A}_1'.$$

Now the affine Weyl group W acts on the subset

$$H_{\mathbb{R},1}^* = \{\lambda \in H_\mathbb{R}^* \ ; \ \lambda(c) = 1\}$$

and this induces an action on the orbit space $H_{\mathbb{R},1}^*/\mathbb{R}\delta$. However, there is a natural bijection between this orbit space and $\left(H_\mathbb{R}^0\right)^*$. This defines a W-action on $\left(H_\mathbb{R}^0\right)^*$. The W_0-action on $\left(H_\mathbb{R}^0\right)^*$ is just as before, and the remaining generator s_0 of W acts as the reflection in the affine hyperplane

$$L_{h_\theta, 1/a_0}^* = \left\{\lambda \in \left(H_\mathbb{R}^0\right)^* \ ; \ \lambda(h_\theta) = 1/a_0\right\}.$$

The element $t_\alpha, \alpha \in M^*$, acts on $\left(H_\mathbb{R}^0\right)^*$ as translation by α, thus W acts on $\left(H_\mathbb{R}^0\right)^*$ as a group of affine transformations.

We may also introduce alcove geometry in $\left(H_\mathbb{R}^0\right)^*$. We define a set $\mathfrak{L}^*$ of affine hyperplanes in $\left(H_\mathbb{R}^0\right)^*$ as follows.

$$\mathfrak{L}^* = \left\{L_{h_\alpha, k}^* \ ; \ \alpha \in \Phi^0, \ k \text{ as below}\right\}$$

where $L_{h_\alpha, k}^* = \left\{\lambda \in \left(H_\mathbb{R}^0\right)^* \ ; \ \lambda(h_\alpha) = k\right\}$. The number k runs through the set given as follows.

For types $\tilde{A}_l, \tilde{D}_l, \tilde{E}_6, \tilde{E}_7, \tilde{E}_8 \quad k \in \mathbb{Z}$.

For types $\tilde{B}_l, \tilde{C}_l, \tilde{F}_4, \tilde{G}_2 \quad k \in \mathbb{Z}$ if α is long

$\qquad\qquad\qquad\qquad\qquad\quad k \in p\mathbb{Z}$ if α is short.

For types $\tilde{B}_l^t, \tilde{C}_l^t, \tilde{F}_4^t, \tilde{G}_2^t \quad k \in \mathbb{Z}$.

For type $\tilde{C}_l' \quad k \in \tfrac{1}{2}\mathbb{Z}$ if α is long

$\qquad\qquad\quad k \in \mathbb{Z}$ if α is short.

For type $\tilde{A}_1' \quad k \in \tfrac{1}{2}\mathbb{Z}$.

17.3 The Weyl group of an affine Kac–Moody algebra

Then the elements of the affine Weyl group W permute the set $\mathfrak{L}^*$ of affine hyperplanes. The connected components of

$$H_{\mathbb{R}}^0 - \bigcup_{L^* \in \mathfrak{L}^*} L^*$$

are called the alcoves of $H_{\mathbb{R}}^0$. The set A^* given by

$$A^* = \left\{ \lambda \in \left(H_{\mathbb{R}}^0\right)^* \; ; \; \lambda(h_i) > 0 \text{ for } i=1,\ldots,l, \quad \lambda(h_\theta) < 1/a_0 \right\}$$

is an alcove called the **fundamental alcove**. The group W acts on the alcoves and the map $w \to w(A^*)$ is a bijective correspondence between elements of W and alcoves. Moreover the closure $\overline{A^*}$ is a fundamental region for the W-action on $\left(H_{\mathbb{R}}^0\right)^*$.

We omit the proofs of these facts, which are entirely analogous to the corresponding results for the W-action on H, or may be deduced from these.

18
Realisations of affine Kac–Moody algebras

18.1 Loop algebras and central extensions

Let A^0 be an indecomposable Cartan matrix of finite type. We have $A^0 = \left(A^0_{ij}\right)$ for $i, j = 1, \ldots, l$. Let $L^0 = L\left(A^0\right)$ be the finite dimensional simple Lie algebra with Cartan matrix A^0. We may construct an $(l+1) \times (l+1)$ affine Cartan matrix A from A^0 by adding an additional row and column, labelled by 0, as follows. Let $\theta = \sum_{i=1}^{l} a_i \alpha_i$ be the highest root of L^0 and $h_\theta = \sum_{i=1}^{l} c_i h_i$ be the coroot of θ. We then define A by:

$$A_{ij} = A^0_{ij} \quad \text{if } i, j \in \{1, \ldots, l\}$$

$$A_{i0} = -\sum_{j=1}^{l} a_j A^0_{ij} \quad \text{if } i \in \{1, \ldots, l\}$$

$$A_{0j} = -\sum_{i=1}^{l} c_i A^0_{ij} \quad \text{if } j \in \{1, \ldots, l\}$$

$$A_{00} = 2.$$

Proposition 18.1 *A is an affine Cartan matrix. The type of A is as follows.*

Type of $A^0 : A_l, B_l, C_l, D_l, E_6, E_7, E_8, F_4, G_2$

Type of $A : \tilde{A}_l, \tilde{B}_l, \tilde{C}_l, \tilde{D}_l, \tilde{E}_6, \tilde{E}_7, \tilde{E}_8, \tilde{F}_4, \tilde{G}_2$

Proof. We have

$$A \begin{pmatrix} a_0 \\ a_1 \\ \vdots \\ a_l \end{pmatrix} = \begin{pmatrix} 0 \\ 0 \\ \vdots \\ 0 \end{pmatrix} \quad \text{where } a_0 = 1.$$

For $\sum_{j=0}^{l} A_{ij}a_j = A_{i0} + \sum_{j=1}^{l} A_{ij}a_j$. If $i \neq 0$ this is 0 by definition. If $i=0$ we have

$$\sum_{j=0}^{l} A_{0j}a_j = 2 + \sum_{j=1}^{l} A_{0j}a_j = 2 - \sum_{i=1}^{l}\sum_{j=1}^{l} c_i A_{ij}^0 a_j.$$

However, $\sum_{i=1}^{l} \sum_{j=1}^{l} c_i A_{ij}^0 a_j = \left(\sum_{j=1}^{l} a_j \alpha_j\right)\left(\sum_{i=1}^{l} c_i h_i\right) = \theta(h_\theta) = 2$, thus $\sum_{j=0}^{l} A_{0j}a_j = 0$.

A similar argument shows that

$$(c_0 c_1 \ldots c_l) A = (00 \ldots 0) \qquad \text{where } c_0 = 1.$$

Now A is determined by A^0 and the relations

$$A \begin{pmatrix} a_0 \\ a_1 \\ \vdots \\ a_l \end{pmatrix} = \begin{pmatrix} 0 \\ 0 \\ \vdots \\ 0 \end{pmatrix}, \qquad (c_0 c_1 \ldots c_l) A = (0 \ldots 0).$$

But the affine Cartan matrix A of type $\tilde{L}_0$, where L_0 is A_l, B_l, C_l, D_l, E_6, E_7, E_8, F_4, G_2 gives A^0 when row and column 0 are removed, and satisfies the above two relations, by Proposition 17.18 and Lemma 17.19. Thus our given matrix A is the affine Cartan matrix of type $\tilde{L}_0$. □

Definition *An affine Cartan matrix A is of **untwisted type** if it is one of*

$$\tilde{A}_l, \tilde{B}_l, \tilde{C}_l, \tilde{D}_l, \tilde{E}_6, \tilde{E}_7, \tilde{E}_8, \tilde{F}_4, \tilde{G}_2.$$

Since any affine Cartan matrix A of untwisted type can be constructed as above from a Cartan matrix A^0 of finite type by the addition of an extra row and column, it seems natural to ask whether the affine Kac–Moody algebra $L(A)$ can be constructed in some way from the finite dimensional simple Lie algebra $L^0 = L(A^0)$. We shall now describe a method of doing this.

Let $\mathbb{C}[t, t^{-1}]$ be the ring of Laurent polynomials $\sum_{i \in \mathbb{Z}} \zeta_i t^i$ for $\zeta_i \in \mathbb{C}$ with finitely many $\zeta_i \neq 0$. Let

$$\mathfrak{L}(L^0) = \mathbb{C}[t, t^{-1}] \otimes_\mathbb{C} L^0.$$

Then $\mathfrak{L}(L^0)$ may be made into a Lie algebra in a unique way satisfying

$$[p \otimes x, q \otimes y] = pq \otimes [xy]$$

for $p, q \in \mathbb{C}[t, t^{-1}]$, $x, y \in L^0$. This Lie algebra $\mathfrak{L}(L^0)$ is called the **loop algebra** of L^0.

We now wish to construct a 1-dimensional central extension of $\mathfrak{L}(L^0)$.

Lemma 18.2 *Let L be a Lie algebra over $\mathbb{C}$ and $\tilde{L}$ be the set of elements $x + \lambda c$ with $x \in L$ and $\lambda \in \mathbb{C}$. Let $\kappa : L \times L \to \mathbb{C}$ be a bilinear map satisfying*

$$\kappa(y, x) = -\kappa(x, y) \quad \text{for } x, y \in L$$

$$\kappa([xy], z) + \kappa([yz], x) + \kappa([zx], y) = 0 \quad \text{for } x, y, z \in L.$$

(κ is called a 2-cocycle on L.) Then the Lie multiplication

$$[x + \lambda c, y + \mu c] = [xy] + \kappa(x, y)c$$

makes $\tilde{L}$ into a Lie algebra.

Proof. This is elementary. The two relations satisfied by κ give anticommutativity and the Jacobi identity on $\tilde{L}$. □

We note that $\tilde{L}$ is a 1-dimensional central extension of L, i.e. there is a surjective homomorphism

$$\theta : \tilde{L} \to L$$

given by $\theta(x + \lambda c) = x$, such that $\dim(\ker \theta) = 1$ and $\ker \theta$ lies in the centre of $\tilde{L}$.

We apply this idea to construct a 1-dimensional central extension of $\mathfrak{L}(L^0)$ by taking a 2-cocycle on $\mathfrak{L}(L^0)$. Let $\langle , \rangle$ be the invariant bilinear form on L^0 satisfying $\langle h_\theta, h_\theta \rangle = 2$. Since an invariant bilinear form is determined up to a scalar multiple on L^0, this condition determines it uniquely. In fact this form on L^0 is the restriction to L^0 of the standard invariant form on $L = L(A)$, since for the standard form we have $\langle \theta, \theta \rangle = 2$ as in Proposition 17.18, hence

$$\langle h_\theta, h_\theta \rangle = \left\langle \frac{2\theta}{\langle \theta, \theta \rangle}, \frac{2\theta}{\langle \theta, \theta \rangle} \right\rangle = \frac{4}{\langle \theta, \theta \rangle} = 2.$$

We next define a bilinear form

$$\langle , \rangle_t : \mathfrak{L}(L^0) \times \mathfrak{L}(L^0) \to \mathbb{C}[t, t^{-1}]$$

by $\langle p \otimes x, q \otimes y \rangle_t = pq \langle x, y \rangle$. We define the residue function

$$\text{Res} : \mathbb{C}[t, t^{-1}] \to \mathbb{C}$$

by $\text{Res}\left(\sum \zeta_i t^i\right) = \zeta_{-1}$.

18.1 Loop algebras and central extensions

Lemma 18.3 *The function* $\kappa : \mathfrak{L}(L^0) \times \mathfrak{L}(L^0) \to \mathbb{C}$ *defined by*

$$\kappa(a,b) = \text{Res}\left\langle \frac{da}{dt}, b \right\rangle_t$$

is a 2-cocycle on $\mathfrak{L}(L^0)$.

Proof. To show that κ is anticommutative it is sufficient to verify that

$$\kappa(t^i \otimes x, t^j \otimes y) = -\kappa(t^j \otimes y, t^i \otimes x).$$

Now

$$\kappa(t^i \otimes x, t^j \otimes y) = \text{Res}\langle it^{i-1} \otimes x, t^j \otimes y \rangle_t$$
$$= \text{Res}\left(it^{i+j-1}\langle x, y \rangle \right)$$
$$= \begin{cases} i\langle x, y \rangle & \text{if } i+j=0 \\ 0 & \text{if } i+j \neq 0. \end{cases}$$

The anticommutativity follows.

We also need

$$\kappa\left([t^i \otimes x, t^j \otimes y], t^k \otimes z\right) + \kappa\left([t^j \otimes y, t^k \otimes z], t^i \otimes x\right)$$
$$+ \kappa\left([t^k \otimes z, t^i \otimes x], t^j \otimes y\right) = 0.$$

Now we have

$$\kappa\left([t^i \otimes x, t^j \otimes y], t^k \otimes z\right) = \kappa\left(t^{i+j} \otimes [xy], t^k \otimes z\right)$$
$$= \text{Res}\langle (i+j)t^{i+j-1} \otimes [xy], t^k \otimes z \rangle_t$$
$$= \text{Res}\left((i+j)t^{i+j+k-1}\langle [xy], z \rangle \right)$$
$$= \begin{cases} (i+j)\langle [xy], z \rangle & \text{if } i+j+k=0 \\ 0 & \text{if } i+j+k \neq 0. \end{cases}$$

If $i+j+k \neq 0$ the required property is clear. If $i+j+k=0$ the required sum is

$$-k\langle [xy], z \rangle - i\langle [yz], x \rangle - j\langle [zx], y \rangle$$
$$= -k\langle [xy], z \rangle - i\langle [xy], z \rangle - j\langle [xy], z \rangle$$
$$= 0$$

since the form is symmetric and invariant. $\square$

We may therefore construct the 1-dimensional central extension $\tilde{\mathfrak{L}}(L^0)$ of $\mathfrak{L}(L^0)$ given by

$$\tilde{\mathfrak{L}}(L^0) = \mathfrak{L}(L^0) \oplus \mathbb{C}c$$

whose Lie multiplication is given by

$$[a+\lambda c, b+\mu c] = [a,b]_0 + \kappa(a,b)c$$

where $a, b \in \mathfrak{L}(L^0)$ and $[a,b]_0$ is the Lie product of a, b in $\mathfrak{L}(L^0)$.

We next wish to adjoin to $\tilde{\mathfrak{L}}(L^0)$ an element d which acts on $\tilde{\mathfrak{L}}(L^0)$ as a derivation.

Lemma 18.4 *The map* $\Delta : \tilde{\mathfrak{L}}(L^0) \to \tilde{\mathfrak{L}}(L^0)$ *given by* $\Delta(a+\lambda c) = t\frac{da}{dt}$ *for* $a \in \mathfrak{L}(L^0), \lambda \in \mathbb{C}$, *is a derivation.*

Proof. Since $[a+\lambda c, b+\mu c] = [a,b]_0 + \kappa(a,b)c$ we must show that

$$t\frac{d}{dt}[a,b]_0 = \left[t\frac{da}{dt}, b+\mu c\right] + \left[a+\lambda c, t\frac{db}{dt}\right]$$

that is

$$t\left[\frac{da}{dt}, b\right]_0 + t\left[a, \frac{db}{dt}\right]_0 = t\left[\frac{da}{dt}, b\right]_0 + \kappa\left(t\frac{da}{dt}, b\right)c + \left[a, t\frac{db}{dt}\right]_0$$
$$+ \kappa\left(a, t\frac{db}{dt}\right)c$$

that is $\kappa\left(t\frac{da}{dt}, b\right) + \kappa\left(a, t\frac{db}{dt}\right) = 0$. It is sufficient to prove this when $a = p \otimes x$, $b = q \otimes y$ with $p, q \in \mathbb{C}[t, t^{-1}]$, $x, y \in L^0$. Then

$$\kappa\left(t\frac{da}{dt}, b\right) + \kappa\left(a, t\frac{db}{dt}\right) = \kappa\left(t\frac{dp}{dt} \otimes x, q \otimes y\right) + \kappa\left(p \otimes x, t\frac{dq}{dt} \otimes y\right)$$

$$= \kappa\left(p \otimes x, t\frac{dq}{dt} \otimes y\right) - \kappa\left(q \otimes y, t\frac{dp}{dt} \otimes x\right)$$

$$= \text{Res}\left\langle \frac{dp}{dt} \otimes x, t\frac{dq}{dt} \otimes y \right\rangle_t - \text{Res}\left\langle \frac{dq}{dt} \otimes y, t\frac{dp}{dt} \otimes x \right\rangle_t$$

$$= \text{Res}\left(t\frac{dp}{dt}\frac{dq}{dt}\langle x, y \rangle\right) - \text{Res}\left(t\frac{dp}{dt}\frac{dq}{dt}\langle x, y \rangle\right)$$

$$= 0. \qquad \square$$

We now define $\hat{\mathfrak{L}}(L^0)$ by

$$\hat{\mathfrak{L}}(L^0) = \tilde{\mathfrak{L}}(L^0) \oplus \mathbb{C}d$$

and make $\hat{\mathfrak{L}}(L^0)$ into a Lie algebra by defining the Lie product as

$$[a+\lambda d, b+\mu d] = [a,b] + \lambda\Delta(b) - \mu\Delta(a).$$

This is clearly skew-symmetric, and the Jacobi identity follows from the fact that Δ is a derivation. In particular we have

$$[(t^i \otimes x) + \lambda c + \mu d, (t^j \otimes y) + \lambda' c + \mu' d]$$
$$= (t^{i+j} \otimes [xy]) + \mu j (t^i \otimes y) - \mu' i (t^i \otimes x) + \delta_{i,-j} i \langle x, y \rangle c$$

for $x, y \in L^0$, $\lambda, \mu, \lambda', \mu' \in \mathbb{C}$.

18.2 Realisations of untwisted affine Kac–Moody algebras

We aim to show that $\hat{\mathfrak{L}}(L^0)$ is isomorphic to the affine Kac–Moody algebra $L(A)$. Thus $L(A)$ can be constructed from $L^0 = L(A^0)$ by the following procedure. First form the loop algebra $\mathfrak{L}(L^0)$. Then form the 1-dimensional central extension $\tilde{\mathfrak{L}}(L^0)$. Finally extend this Lie algebra by a derivation to give $\hat{\mathfrak{L}}(L^0)$.

Theorem 18.5 *Let $L^0 = L(A^0)$ be a finite dimensional simple Lie algebra and let A be the untwisted affine Cartan matrix obtained from A^0 as in Section 18.1. Then $L(A)$ is isomorphic to $\hat{\mathfrak{L}}(L^0)$.*

Proof. We shall define elements $e_0, e_1, \ldots, e_l$; $f_0, f_1, \ldots, f_l$; $h_0, h_1, \ldots, h_l$ in $\hat{\mathfrak{L}}(L^0)$ with the aim of using Proposition 14.15 to show that our Lie algebra is isomorphic to $L(A)$.

Let $E_1, \ldots, E_l$; $F_1, \ldots, F_l$; $H_1, \ldots, H_l$ be corresponding generators of L^0. We define

$$e_i = 1 \otimes E_i, \quad f_i = 1 \otimes F_i, \quad h_i = 1 \otimes H_i$$

for $i = 1, \ldots, l$. Then $[e_i f_i] = h_i$ for each i. We must also define $e_0, f_0, h_0 \in \hat{\mathfrak{L}}(L^0)$.

We consider the root spaces $L^0_\theta, L^0_{-\theta}$ where θ is the highest root of L^0. We have $\dim L^0_\theta = \dim L^0_{-\theta} = 1$, and the map $L^0_\theta \times L^0_{-\theta} \to \mathbb{C}$ given by the invariant bilinear form $\langle , \rangle$ on L^0 defined in Section 18.1 is non-degenerate. Let ω^0 be the automorphism of L^0 satisfying $\omega^0(E_i) = -F_i$, $\omega^0(F_i) = -E_i$. Then $\omega^0(L^0_\theta) = L^0_{-\theta}$. We claim it is possible to choose elements $F_0 \in L^0_\theta$, $E_0 \in L^0_{-\theta}$ such that $\omega^0(F_0) = -E_0$ and $\langle F_0, E_0 \rangle = 1$. First choose any non-zero element $F'_0 \in L^0_\theta$ and let $E'_0 = -\omega^0(F'_0)$. Let $\langle F'_0, E'_0 \rangle = \xi$. Then we have $\xi \neq 0$.

Now let $F_0 = \lambda F_0'$ and $E_0 = \lambda E_0'$ for $\lambda \in \mathbb{C}$ with $\lambda \neq 0$. Then we have $E_0 = -\omega^0(F_0)$ and $\langle F_0, E_0 \rangle = \lambda^2 \xi$. By a suitable choice of $\lambda \in \mathbb{C}$ we can ensure that $\lambda^2 \xi = 1$.

We now define $e_0 = t \otimes E_0$ and $f_0 = t^{-1} \otimes F_0$. Let H^0 be the subspace of L^0 spanned by $h_1, \ldots, h_l$ and
$$H = (1 \otimes H^0) \otimes \mathbb{C}c \oplus \mathbb{C}d.$$

We define $h_0 \in H$ by
$$h_0 = (1 \otimes (-H_\theta)) + c.$$

Then we have
$$[e_0 f_0] = [t \otimes E_0, t^{-1} \otimes F_0] = (1 \otimes [E_0 F_0]) + \langle E_0, F_0 \rangle c.$$

But
$$[E_0 F_0] = \langle E_0, F_0 \rangle H'_{-\theta} = H'_{-\theta} = H_{-\theta} = -H_\theta$$

by Corollary 16.5, since $\langle \theta, \theta \rangle = 2$. Thus
$$[e_0 f_0] = (1 \otimes (-H_\theta)) + c = h_0.$$

We also define elements $\alpha_0, \alpha_1, \ldots, \alpha_l \in H^*$. We have elements $\alpha_1, \ldots, \alpha_l \in (H^0)^*$ and we extend these to H^* by saying that $\alpha_i(c) = \alpha_i(d) = 0$ for $i = 1, \ldots, l$. We also define $\theta \in H^*$ similarly, saying that $\theta(c) = \theta(d) = 0$. Let $\delta \in H^*$ be the element defined by
$$\delta(x) = 0 \quad \text{for } x \in H^0, \quad \delta(c) = 0, \quad \delta(d) = 1.$$

We then define $\alpha_0 \in H^*$ by $\alpha_0 = -\theta + \delta$.

We now show that $(H, \Pi, \Pi^\vee)$ is a realisation of A where $\Pi = \{\alpha_0, \alpha_1, \ldots, \alpha_l\}$ and $\Pi^\vee = \{h_0, h_1, \ldots, h_l\}$. Π is linearly independent since $\alpha_1, \ldots, \alpha_l$ are linearly independent and $\alpha_0(d) \neq 0$, $\alpha_i(d) = 0$ for $i = 1, \ldots, l$. $\Pi^\vee$ is linearly independent since $h_1, \ldots, h_l$ are linearly independent and h_0 involves c whereas h_i does not for $i = 1, \ldots, l$.

We show that $\alpha_j(h_i) = A_{ij}$ for $i, j \in \{0, 1, \ldots, l\}$. This is clear if $i \neq 0$, $j \neq 0$. Also for $i \neq 0$ we have

$$\alpha_0(h_i) = -\theta(h_i) + \delta(h_i) = -\theta(h_i) = -\sum_{j=1}^{l} a_j \alpha_j(h_i) = -\sum_{j=1}^{l} A_{ij} a_j = A_{i0}.$$

Similarly for $j \neq 0$ we have

$$\alpha_j(h_0) = \alpha_j(-h_\theta + c) = -\alpha_j(h_\theta) = -\sum_{i=1}^{l} c_i \alpha_j(h_i) = -\sum_{i=1}^{l} c_i A_{ij} = A_{0j}.$$

18.2 Realisations of untwisted affine Kac–Moody algebras

Finally $\alpha_0(h_0) = (-\theta + \delta)(-h_\theta + c) = \theta(h_\theta) = 2$. Thus $(H, \Pi, \Pi^\vee)$ is a realisation of A.

We next verify the relations

$$[e_i f_i] = h_i$$
$$[e_i f_j] = 0 \quad \text{if } i \neq j$$
$$[x e_i] = \alpha_i(x) e_i \quad \text{for } x \in H$$
$$[x f_i] = -\alpha_i(x) f_i \quad \text{for } x \in H$$

where $i = 0, 1, \ldots, l$. These relations certainly hold when $i \neq 0$ and $j \neq 0$. We have shown above that $[e_0 f_0] = h_0$. For $i \neq 0$ we have

$$[e_i f_0] = [1 \otimes E_i, t^{-1} \otimes F_0] = t^{-1} \otimes [E_i F_0] = 0$$

since $F_0 \in L_\theta^0$ and θ is the highest root of L^0. Similarly for $j \neq 0$ we have

$$[e_0 f_j] = [t \otimes E_0, 1 \otimes F_j] = t \otimes [E_0 F_j] = 0.$$

Now let $x = x_0 + \lambda c + \mu d \in H$ where $x_0 \in H^0$ and $\lambda, \mu \in \mathbb{C}$. Then

$$\alpha_0(x) = -\theta(x) + \delta(x) = -\theta(x_0) + \mu$$

since $\theta(c) = \theta(d) = 0$, $\delta(x_0) = \delta(c) = 0$, $\delta(d) = 1$. Also

$$[x e_0] = [x_0 + \lambda c + \mu d, t \otimes E_0] = (t \otimes [x_0 E_0]) + \mu(t \otimes E_0)$$
$$= -\theta(x_0)(t \otimes E_0) + \mu(t \otimes E_0)$$
$$= \alpha_0(x) e_0.$$

Similarly one shows $[x f_0] = -\alpha_0(x) f_0$. Thus the required relations are all satisfied.

We show next that $e_0, e_1, \ldots, e_l, f_0, f_1, \ldots, f_l$ and H generate $\hat{\mathfrak{L}}(L^0)$. Let M be the subalgebra of $\hat{\mathfrak{L}}(L^0)$ generated by this subset. Since $E_1, \ldots, E_l, F_1, \ldots, F_l$ generate L^0, $e_1, \ldots, e_l, f_1, \ldots, f_l$ generate $1 \otimes L^0$. Thus $1 \otimes L^0 \subset M$.

Let $I^0 = \{x \in L^0 ; t \otimes x \in M\}$. Since $e_0 = t \otimes E_0$ we have $E_0 \in I^0$ so $I^0 \neq 0$. Also if $x \in I^0$, $y \in L^0$ then $[xy] \in I^0$ since

$$t \otimes [xy] = [t \otimes x, 1 \otimes y] \in M.$$

Thus I^0 is a non-zero ideal of L^0. Since L^0 is simple we have $I^0 = L^0$. Thus $t \otimes x \in M$ for all $x \in L^0$. We may now use the relation

$$[t \otimes x, t^{k-1} \otimes y] = t^k \otimes [xy]$$

to deduce by induction on k that $t^k \otimes x \in M$ for all $x \in L^0$ and all $k > 0$. In an analogous way, starting with $f_0 = t^{-1} \otimes F_0$ we can show $t^{-k} \otimes x \in M$ for all $x \in L^0$ and all $k > 0$. Now

$$\hat{\mathfrak{L}}(L^0) = H + (1 \otimes L^0) + \sum_{k>0}(t^k \otimes L^0) + \sum_{k<0}(t^k \otimes L^0)$$

hence $M = \hat{\mathfrak{L}}(L^0)$.

It remains to show that $\hat{\mathfrak{L}}(L^0)$ has no non-zero ideal J with $J \cap H = 0$. Let

$$L = \hat{\mathfrak{L}}(L^0) = H \oplus \sum_{(i,\alpha)\neq(0,0)} (t^i \otimes (L^0)_\alpha)$$

summed over $i \in \mathbb{Z}$, $\alpha \in (H^0)^*$ with $(i, \alpha) \neq (0, 0)$. We claim that this is the weight space decomposition of L with respect to H. For let $h \in H$, $x \in (L^0)_\alpha$. Then $h = h_0 + \lambda c + \mu d$ with $h_0 \in H^0$, $\lambda, \mu \in \mathbb{C}$. Thus

$$[h, t^i \otimes x] = [h_0 + \lambda c + \mu d, t^i \otimes x] = (t^i \otimes [h_0 x]) + \mu i (t^i \otimes x)$$
$$= (\alpha(h_0) + \mu i)(t^i \otimes x)$$
$$= (\alpha(h) + i\delta(h))(t^i \otimes x)$$
$$= (\alpha + i\delta)(h)(t^i \otimes x)$$

since $\alpha(h) = \alpha(h_0)$, $\delta(h) = \mu$. Thus $t^i \otimes x$ is a weight vector with weight $\alpha + i\delta$. Thus we have

$$L = L_0 \oplus \sum_{(\alpha,i)\neq(0,0)} L_{\alpha+i\delta}$$

where $L_0 = H$ and $L_{\alpha+i\delta} = t^i \otimes (L^0)_\alpha$.

Let J be a non-zero ideal of L with $J \cap H = O$. By Lemma 14.12 we have

$$J = (L_0 \cap J) \oplus \sum_{(\alpha,i)\neq(0,0)} (L_{\alpha+i\delta} \cap J).$$

Since $L_0 \cap J = O$ we have $L_{\alpha+i\delta} \cap J \neq O$ for some (α, i). Let $t^i \otimes x \in J$ for $x \in (L^0)_\alpha$ with $x \neq 0$. Then there exists $y \in (L^0)_{-\alpha}$ with $\langle x, y \rangle \neq 0$. Thus

$$[t^i \otimes x, t^{-i} \otimes y] = [xy] + i\langle x, y \rangle c$$

lies in $J \cap H$, and hence

$$[xy] + i\langle x, y \rangle c = 0.$$

Since $[xy] \in H^0$ and $\langle x, y \rangle \neq 0$ we must have $i = 0$. But this implies $[xy] = 0$, whereas we have

$$[xy] = \langle x, y \rangle h'_\alpha \neq 0$$

by Corollary 16.5. This gives us the required contradiction. Thus $J \cap H = O$ implies $J = O$.

We have now verified all the conditions of Proposition 14.15. We may therefore deduce that $\hat{\mathfrak{L}}(L^0)$ is isomorphic to $L(A)$. □

We can deduce from this theorem the multiplicities of the imaginary roots of $L(A)$. These multiplicities were not obtained in Chapter 17. We recall from Theorem 16.27 (ii) that the imaginary roots of $L(A)$ have form $k\delta$ where $k \in \mathbb{Z}$ and $k \neq 0$.

Corollary 18.6 *Let A be an indecomposable affine GCM of untwisted type. Then the multiplicity of each imaginary root $k\delta, k \neq 0$, is $l = $ rank A.*

Proof. We use the realisation $L(A) = \hat{\mathfrak{L}}(L^0)$. The weight space decomposition of $\hat{\mathfrak{L}}(L^0)$ shows that the root space for the root $k\delta$ is $t^k \otimes H^0$. The multiplicity of $k\delta$ is the dimension of this root space, which is dim $H^0 = l$. □

We now make some comments on the isomorphism between $L(A)$ and $\hat{\mathfrak{L}}(L^0)$ which we have obtained. Firstly the standard invariant form on $L(A)$ maps under this isomorphism to the form on $\hat{\mathfrak{L}}(L^0)$ given as follows:

$$\langle t^i \otimes x, t^j \otimes y \rangle = 0 \quad \text{if } j \neq -i, \quad \text{for } x, y \in L^0$$
$$\langle t^i \otimes x, t^{-i} \otimes y \rangle = \langle x, y \rangle$$
$$\langle t^i \otimes x, c \rangle = 0$$
$$\langle t^i \otimes x, d \rangle = 0$$
$$\langle c, c \rangle = 0$$
$$\langle d, d \rangle = 0$$
$$\langle c, d \rangle = 1.$$

For it is readily checked that the form defined in this way on $\hat{\mathfrak{L}}(L^0)$ is invariant. Moreover we also see that the above form on the subspace $(1 \otimes H^0) \oplus \mathbb{C}c \oplus \mathbb{C}d$ of $\hat{\mathfrak{L}}(L^0)$ agrees with the standard invariant form on the subspace H of $L(A)$ under our isomorphism between these subspaces. However, the proof of Theorem 16.2 shows that a symmetric invariant bilinear form on $L(A)$ is uniquely determined by its restriction to H. Thus the above form on $\hat{\mathfrak{L}}(L^0)$ corresponds to the standard invariant form on $L(A)$.

We also observe that the element $c \in \hat{\mathfrak{L}}(L^0)$ corresponds to the canonical central element in $L(A)$ under the isomorphism. For we have

$$h_0 = (1 \otimes -H_\theta) + c \quad \text{in } \hat{\mathfrak{L}}(L^0)$$

and hence $\sum_{i=0}^{l} c_i h_i = c$ in $\hat{\mathfrak{L}}(L^0)$. It follows that the image of c under the isomorphism is the canonical central element of $L(A)$.

Also, since $\alpha_0(d) = 1$, $\alpha_i(d) = 0$ for $i = 1, \ldots, l$ the element $d \in \hat{\mathfrak{L}}(L^0)$ corresponds under the isomorphism to an analogous scaling element d for $L(A)$.

18.3 Some graph automorphisms of affine algebras

We now wish to find realisations of the remaining affine Kac–Moody algebras $L(A)$ where A has type $\tilde{B}_l^t$, $\tilde{C}_l^t$, $\tilde{F}_4^t$, $\tilde{G}_2^t$, $\tilde{A}_1'$ or $\tilde{C}_l'$. These are called the twisted affine Kac–Moody algebras. We shall obtain realisations for them as fixed point subalgebras of certain automorphisms of untwisted Kac–Moody algebras. Before doing so, however, we consider the graph automorphisms of the untwisted algebras which fix the vertex 0 and therefore arise from graph automorphisms of the corresponding finite dimensional simple Lie algebras. The graph automorphisms of the finite dimensional simple Lie algebras were considered in Section 9.5. We recall from Theorem 9.19 that if σ is a graph automorphism of the finite dimensional simple Lie algebra $L(A)$ then $L(A)^\sigma$ is isomorphic to the simple Lie algebra $L(A^1)$ where A^1 is obtained from A as follows.

A	:	A_{2k}	A_{2k-1}	D_{k+1}	D_4	E_6
Order of σ	:	2	2	2	3	2
A^1	:	B_k	C_k	B_k	G_2	F_4

We shall now prove an analogous result to Theorem 9.19 for affine algebras.

Theorem 18.7 *Let A be an affine Cartan matrix of type $\tilde{A}_{2k-1}$, $\tilde{D}_{k+1}$, $\tilde{D}_4$ or $\tilde{E}_6$ and let σ be a graph automorphism of the Kac–Moody algebra $L(A)$ which fixes vertex 0 and has order 2, 2, 3, 2 respectively. Let A^0 be the corresponding finite Cartan matrix and A^1 be the finite Cartan matrix associated with A^0 as above. Let $\tilde{A}^1$ be the untwisted affine Cartan matrix obtained from A^1. Then $L(A)^\sigma$ is isomorphic to $L(\tilde{A}^1)$.*

Specifically we have

$$L(\tilde{A}_{2k-1})^\sigma \cong L(\tilde{C}_k)$$
$$L(\tilde{D}_{k+1})^\sigma \cong L(\tilde{B}_k)$$
$$L(\tilde{D}_4)^\sigma \cong L(\tilde{G}_2)$$
$$L(\tilde{E}_6)^\sigma \cong L(\tilde{F}_4)$$

18.3 Some graph automorphisms of affine algebras

Proof. The algebra $L(A^0)$ has a Cartan decomposition

$$L(A^0) = H^0 \oplus \sum_{\alpha \in \Phi^0} L_\alpha^0.$$

Thus $L(A)$ has a corresponding decomposition

$$L(A) = H^0 \oplus \mathbb{C}c \oplus \mathbb{C}d \oplus \sum_{k \neq 0}(t^k \otimes H^0) + \sum_{k,\alpha}(t^k \otimes L_\alpha^0).$$

Similarly we have decompositions

$$L(A^1) = H^1 \oplus \sum_{\alpha \in \Phi^1} L_\alpha^1$$

$$L(\tilde{A}^1) = H^1 \oplus \mathbb{C}c \oplus \mathbb{C}d \oplus \sum_{k \neq 0}(t^k \otimes H^1) \oplus \sum_{k,\alpha}(t^k \otimes L_\alpha^1).$$

Consider the graph automorphism $\sigma : L(A) \to L(A)$. We have

$$\sigma(H^0) = H^0, \quad \sigma(t^k \otimes H^0) = t^k \otimes H^0, \quad \sigma(t^k \otimes L_\alpha^0) = t^k \otimes L_{\sigma(\alpha)}^0,$$

$$\sigma(c) = c, \quad \sigma(d) = d.$$

Hence

$$L(A)^\sigma = (H^0)^\sigma \oplus \mathbb{C}c \oplus \mathbb{C}d \oplus \sum_{k \neq 0}(t^k \otimes (H^0)^\sigma) + \sum_{k,S}(t^k \otimes (L_S^0)^\sigma)$$

where S is an equivalence class of roots in Φ^0 and $L_S^0 = \sum_{\alpha \in S} L_\alpha^0$ (cf. Proposition 9.18). Now the isomorphism $L(A^0)^\sigma \to L(A^1)$ of Theorem 9.19 gives rise to bijective maps

$$(H^0)^\sigma \to H^1$$
$$(L_S^0)^\sigma \to L_\alpha^1 \quad \text{where } \alpha \in \Phi^1 \text{ corresponds to } S$$
$$t^k \otimes (H^0)^\sigma \to t^k \otimes H^1$$
$$t^k \otimes (L_S^0)^\sigma \to t^k \otimes L_\alpha^1.$$

These maps, together with $c \to c, d \to d$, determine a bijective map $\phi : L(A)^\sigma \to L(\tilde{A}^1)$. We wish to show this map is an isomorphism.

Under this bijection ϕ the subalgebra $(H^0)^\sigma \oplus \mathbb{C}c \oplus \mathbb{C}d$ maps to $H^1 \oplus \mathbb{C}c \oplus \mathbb{C}d$ and both are abelian. The action of $(H^0)^\sigma$ on $t^k \otimes (H^0)^\sigma$ and $t^k \otimes (L_S^0)^\sigma$ agrees with the action of H^1 on $t^k \otimes H^1$ and $t^k \otimes L_\alpha^1$ respectively. The element c lies in the centre on both sides. The action of d on $t^k \otimes (H^0)^\sigma$ and $t^k \otimes (L_S^0)^\sigma$ (i.e. multiplication by k) agrees with the action of d on $t^k \otimes H^1$ and $t^k \otimes L_\alpha^1$ respectively. Thus it is sufficient to compare the multiplication of the root spaces on both sides. These multiplications are trivially preserved by ϕ unless we take two roots whose sum is 0. So suppose $x, y \in (H^0)^\sigma$ and x_1, y_1 are the corresponding elements of H^1. We have

$$[t^k \otimes x, t^{-k} \otimes y] = k\langle x, y\rangle_0 c$$
$$[t^k \otimes x_1, t^{-k} \otimes y_1] = k\langle x_1, y_1\rangle_1 c$$

where $\langle , \rangle_0$ is the standard invariant form on $L(A^0)$ and $\langle , \rangle_1$ the standard invariant form on $L(A^1)$.

Also if $x \in (L_S^0)^\sigma$, $y \in (L_{-S}^0)^\sigma$ and x_1, y_1 are the corresponding elements of $L_\alpha^1, L_{-\alpha}^1$ then we have

$$[t^k \otimes x, t^{-k} \otimes y] = [xy] + k\langle x, y\rangle_0 c$$
$$[t^k \otimes x_1, t^{-k} \otimes y_1] = [x_1 y_1] + k\langle x_1, y_1\rangle_1 c.$$

Thus to show that ϕ is an isomorphism it is sufficient to show that the isomorphism $L(A^0)^\sigma \to L(A^1)$ preserves the standard invariant form, that is if $x \to x_1, y \to y_1$ then $\langle x, y\rangle_0 = \langle x_1, y_1\rangle_1$. Since any two symmetric invariant bilinear forms on a finite dimensional simple Lie algebra are proportional it is sufficient to check this for just one non-zero value. To do this we choose a 1-element orbit (i) of σ on $\{1, \ldots, l\}$. Such a 1-element orbit exists in all the cases being considered. Then we have an element $h_i \in L(A^0)^\sigma$ mapping to an element $h_i \in L(A^1)$. We have

$$\langle h_i, h_i\rangle_0 = 2 \quad \text{and} \quad \langle h_i, h_i\rangle_1 = 2d_i = 2a_i/c_i.$$

A glance at the values of a_i, c_i for $L(A^1)$ for i coming from 1-element orbits of σ shows that $a_i = c_i$ in these cases, so $d_i = 1$. Hence $\langle h_i, h_i\rangle_0 = \langle h_i, h_i\rangle_1$ and it follows that the isomorphism $L(A^0)^\sigma \to L(A^1)$ preserves the standard invariant forms. This completes the proof. □

Note The reader will have noticed that the case $L(\tilde{A}_{2k})^\sigma$ has not been included in this theorem. The above proof breaks down in this case because σ has no 1-element orbit on $\{1, \ldots, l\}$. The diagrams of A^0, A^1 are as shown.

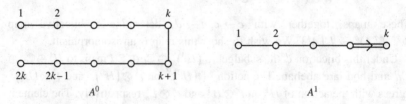

In fact, if we take the σ-orbit $(k, k+1)$ on $\{1, \ldots, 2k\}$, then the isomorphism $L(A^0)^\sigma \to L(A^1)$ of Theorem 9.19 maps

$$2(h_k + h_{k+1}) \in L(A^0)^\sigma \quad \text{to} \quad h_k \in L(A^1).$$

We have

$$\langle h_k, h_k \rangle_0 = 2, \quad \langle h_{k+1}, h_{k+1} \rangle_0 = 2, \quad \langle h_k, h_{k+1} \rangle_0 = -1.$$

Thus

$$\langle 2(h_k + h_{k+1}), 2(h_k + h_{k+1}) \rangle_0 = 8.$$

On the other hand

$$\langle h_k, h_k \rangle_1 = 2\frac{a_k}{c_k} = 2d_k = 4.$$

Thus

$$\langle 2(h_k + h_{k+1}), 2(h_k + h_{k+1}) \rangle_0 \neq \langle h_k, h_k \rangle_1$$

and so the isomorphism between $L(A_{2k})^\sigma$ and $L(B_k)$ does not preserve the standard invariant form. It does not therefore lead to an isomorphism between $L(\tilde{A}_{2k})^\sigma$ and $L(\tilde{B}_k)$ in the manner described in Theorem 18.7.

18.4 Realisations of twisted affine algebras

In order to obtain realisations of the twisted Kac–Moody algebras $L(A)$ where A has types $\tilde{B}_l^t, \tilde{C}_l^t, \tilde{F}_4^t, \tilde{G}_2^t, \tilde{A}_1', \tilde{C}_l'$ we must consider the fixed point subalgebras of so-called twisted graph automorphisms.

Let $L^0 = L(A^0)$ be a finite dimensional simple Lie algebra and $\sigma : L^0 \to L^0$ be a graph automorphism of L^0. Then σ extends to a graph automorphism of $\hat{\mathfrak{L}}(L^0) = \mathfrak{L}(L^0) \oplus \mathbb{C}c \oplus \mathbb{C}d$ given by

$$\sigma(t^i \otimes x) = t^i \otimes \sigma(x) \quad \text{for } x \in L^0$$

$$\sigma(c) = c, \quad \sigma(d) = d.$$

Suppose σ has order m and let $\delta = e^{2\pi i/m}$. Then we may define an automorphism τ of $\hat{\mathfrak{L}}(L^0)$ by

$$\tau(t^i \otimes x) = \delta^{-i} t^i \otimes \sigma(x) \quad \text{for } x \in L^0$$

$$\tau(c) = c, \quad \tau(d) = d.$$

τ is called a **twisted graph automorphism** of $\hat{\mathfrak{L}}(L^0)$, and also has order m. In fact $m = 2$ or 3 in the cases which can arise. We shall consider the fixed point subalgebras $\hat{\mathfrak{L}}(L^0)^\tau$. In order to do so we first obtain more information about the action of σ on L^0.

Proposition 18.8 (i) Let L^0 be a simple Lie algebra of type A_{2l-1}, D_{l+1} or E_6 and σ be a graph automorphism of L^0 of order 2. Let $(L^0)_{-1}$ be the eigenspace of σ on L^0 with eigenvalue -1. Then

$$L^0 = (L^0)^\sigma \oplus (L^0)_{-1}$$

and $(L^0)_{-1}$ is an irreducible $(L^0)^\sigma$-module.

(ii) Let L^0 have type D_4 and σ be a graph automorphism of L^0 of order 3. Let $(L^0)_\omega, (L^0)_{\omega^2}$ be the eigenspaces of σ with eigenvalues ω, ω^2 where $\omega = e^{2\pi i/3}$. Then

$$L^0 = (L^0)^\sigma \oplus (L^0)_\omega \oplus (L^0)_{\omega^2}$$

and $(L^0)_\omega, (L^0)_{\omega^2}$ are both irreducible $(L^0)^\sigma$-modules.

Proof. Let $x \in (L^0)^\sigma, y \in (L^0)_\varepsilon$ where ε is an eigenvalue of σ. Then

$$\sigma[xy] = [\sigma(x), \sigma(y)] = \varepsilon[xy].$$

Thus $[xy] \in (L^0)_\varepsilon$ and so $(L^0)_\varepsilon$ is an $(L^0)^\sigma$-module.

Suppose first that σ has order 2. Let (α, β) be a 2-element orbit of σ on Φ^0 and $E_\alpha, E_\beta \in L^0$ be root vectors such that $\sigma(E_\alpha) = E_\beta$. Then $E_\alpha - E_\beta \in (L^0)_{-1}$ and the weight spaces of $(L^0)_{-1}$ are spanned by such elements for all 2-element orbits (α, β). The roots $\alpha, \beta \in (H^0)^*$ have the same restriction to $((H^0)^\sigma)^*$ and $\alpha|_{((H^0)^\sigma)^*}$ is the weight of $E_\alpha - E_\beta$. The highest weight of the $(L^0)^\sigma$-module $(L^0)_{-1}$ is obtained from the highest 2-element orbit (α, β). Let us choose the labellings

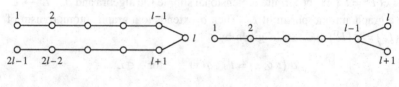

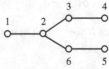

for the Dynkin diagrams of A_{2l-1}, D_{l+1}, E_6. Then the highest 2-element orbits are

$(\alpha_1 + \alpha_2 + \cdots + \alpha_{2l-2}, \quad \alpha_2 + \cdots + \alpha_{2l-2} + \alpha_{2l-1})$ for A_{2l-1}

$(\alpha_1 + \alpha_2 + \cdots + \alpha_{l-1} + \alpha_l, \quad \alpha_1 + \alpha_2 + \cdots + \alpha_{l-1} + \alpha_{l+1})$ for D_{l+1}

$(\alpha_1 + 2\alpha_2 + 2\alpha_3 + \alpha_4 + \alpha_5 + \alpha_6, \quad \alpha_1 + 2\alpha_2 + \alpha_3 + \alpha_4 + \alpha_5 + 2\alpha_6)$ for E_6.

18.4 Realisations of twisted affine algebras

In these three cases the subalgebra $(L^0)^\sigma$ has type C_l, B_l or F_4 respectively by Theorem 9.19. We choose the labellings

for these Dynkin diagrams. Thus the highest weights for the $(L^0)^\sigma$-modules $(L^0)_{-1}$ are

$$\alpha_1 + 2\alpha_2 + \cdots + 2\alpha_{l-1} + \alpha_l \quad \text{for } C_l$$

$$\alpha_1 + \alpha_2 + \cdots + \alpha_{l-1} + \alpha_l \quad \text{for } B_l$$

$$\alpha_1 + 2\alpha_2 + 3\alpha_3 + 2\alpha_4 \quad \text{for } F_4.$$

Using the equation $\alpha_i = \sum_j A_{ji} \omega_j$ we see that these highest weights are ω_2 for C_l, ω_1 for B_l and ω_4 for F_4.

Now $\dim (L^0)_{-1} = \dim L^0 - \dim (L^0)^\sigma$ and this is $2l^2 - l - 1 = \binom{2l}{2} - 1$ for C_l, $2l+1$ for B_l, and 26 for F_4. However, we also have

$$\dim L(\omega_2) = \binom{2l}{2} - 1 \quad \text{in } C_l$$

$$\dim L(\omega_1) = 2l+1 \quad \text{in } B_l$$

$$\dim L(\omega_4) = 26 \quad \text{in } F_4$$

by Weyl's dimension formula. Thus in each case $\dim (L^0)_{-1}$ is the dimension of the irreducible module with the appropriate highest weight. Thus $(L^0)_{-1}$ is isomorphic to this irreducible $(L^0)^\sigma$-module.

Now suppose σ has order 3. Then L^0 has type D_4. Let (α, β, γ) be a 3-element orbit of σ on Φ^0 and $E_\alpha, E_\beta, E_\gamma$ be root vectors such that $\sigma(E_\alpha) = E_\beta$, $\sigma(E_\beta) = E_\gamma$. Then we have

$$E_\alpha + \omega^2 E_\beta + \omega E_\gamma \in (L^0)_\omega$$

$$E_\alpha + \omega E_\beta + \omega^2 E_\gamma \in (L^0)_{\omega^2}$$

where $\omega = e^{2\pi i/3}$, and the weight spaces of the $(L^0)^\sigma$-modules $(L^0)_\omega$ and $(L^0)_{\omega^2}$ are spanned by such vectors for all 3-element orbits. We choose the labelling

for the Dynkin diagram of D_4. The highest 3-element orbit of σ on Φ^0 is then

$$(\alpha_1 + \alpha_2 + \alpha_3, \quad \alpha_1 + \alpha_2 + \alpha_4, \quad \alpha_1 + \alpha_3 + \alpha_4).$$

The subalgebra $(L^0)^\sigma$ has type G_2, for which we take the labelling

$$\underset{1}{\circ}\!\!\Rrightarrow\!\!\underset{2}{\circ}$$

Thus the highest weights of the G_2-modules $(L^0)_\omega$ and $(L^0)_{\omega^2}$ are both $\alpha_1+2\alpha_2$. Now in G_2 we have $\alpha_1+2\alpha_2=\omega_2$ and $\dim L(\omega_2)=7$. However, we also have

$$\dim (L^0)_\omega = \dim (L^0)_{\omega^2} = \frac{1}{2}\left(\dim L^0 - \dim (L^0)^\sigma\right) = 7.$$

Thus the G_2-modules $(L^0)_\omega$ and $(L^0)_{\omega^2}$ are both irreducible and isomorphic to $L(\omega_2)$. □

Theorem 18.9 *Let L^0 be a simple Lie algebra of type A_{2l-1}, D_{l+1}, E_6 or D_4 and let σ be a graph automorphism of L^0 of order 2, 2, 2, 3 respectively. Let τ be the corresponding twisted graph automorphism of $\hat{\mathfrak{L}}(L^0)$. Then the fixed point subalgebra $\hat{\mathfrak{L}}(L^0)^\tau$ is isomorphic to a twisted affine Kac–Moody algebra.*

Explicitly we have

$$\hat{\mathfrak{L}}(\tilde{A}_{2l-1})^\tau \cong L(\tilde{B}_l^t)$$

$$\hat{\mathfrak{L}}(\tilde{D}_{l+1})^\tau \cong L(\tilde{C}_l^t)$$

$$\hat{\mathfrak{L}}(\tilde{E}_6)^\tau \cong L(\tilde{F}_4^t)$$

$$\hat{\mathfrak{L}}(\tilde{D}_4)^\tau \cong L(\tilde{G}_2^t).$$

Proof. The method of proof is broadly similar to that of Theorem 18.5 giving the realisations of the untwisted affine Kac–Moody algebras. The basic idea is to show that the given subalgebra of τ-invariant elements satisfies the conditions of Proposition 14.15, and is therefore isomorphic to the appropriate twisted affine Kac–Moody algebra.

We have

$$\hat{\mathfrak{L}}(L^0) = \sum_{k\in\mathbb{Z}}(t^k\otimes L^0)\oplus\mathbb{C}c\oplus\mathbb{C}d.$$

If σ has order 2 we have

$$\hat{\mathfrak{L}}(L^0)^\tau = \sum_{k\in\mathbb{Z}}(t^{2k}\otimes (L^0)^\sigma)\oplus\sum_{k\in\mathbb{Z}}(t^{2k+1}\otimes (L^0)_{-1})\oplus\mathbb{C}c\oplus\mathbb{C}d$$

whereas if σ has order 3

$$\hat{\mathfrak{L}}(L^0)^\tau = \sum_{k\in\mathbb{Z}}(t^{3k}\otimes (L^0)^\sigma)\oplus\sum_{k\in\mathbb{Z}}(t^{3k+1}\otimes (L^0)_\omega)\oplus\sum_{k\in\mathbb{Z}}(t^{3k+2}\otimes (L^0)_{\omega^2})$$
$$\oplus\mathbb{C}c\oplus\mathbb{C}d.$$

18.4 Realisations of twisted affine algebras

Let $E_1, \ldots, E_k, F_1, \ldots, F_k, H_1, \ldots, H_k$ be standard generators of L^0. We wish to define analogous elements

$$e_0, e_1, \ldots, e_l, \quad f_0, f_1, \ldots, f_l, \quad h_0, h_1, \ldots, h_l \quad \text{in } \hat{\mathfrak{L}}(L^0)^\tau.$$

We pick a representative $\theta^0 \in \Phi^0$ of the highest 2- or 3-element σ-orbit on Φ^0. Specifically we have

$$\theta^0 = \alpha_1 + \alpha_2 + \cdots + \alpha_{2l-2} \quad \text{in } A_{2l-1}$$
$$\theta^0 = \alpha_1 + \alpha_2 + \cdots + \alpha_l \quad \text{in } D_{l+1}$$
$$\theta^0 = \alpha_1 + 2\alpha_2 + 2\alpha_3 + \alpha_4 + \alpha_5 + \alpha_6 \quad \text{in } E_6.$$

The elements e_i, f_i, h_i are then chosen as follows.

Type A_{2l-1}

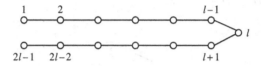

$$e_1 = 1 \otimes (E_1 + E_{2l-1}), \ldots, e_{l-1} = 1 \otimes (E_{l-1} + E_{l+1}), \quad e_l = 1 \otimes E_l$$
$$f_1 = 1 \otimes (F_1 + F_{2l-1}), \ldots, f_{l-1} = 1 \otimes (F_{l-1} + F_{l+1}), \quad f_l = 1 \otimes F_l$$
$$h_1 = 1 \otimes (H_1 + H_{2l-1}), \ldots, h_{l-1} = 1 \otimes (H_{l-1} + H_{l+1}), \quad h_l = 1 \otimes H_l$$
$$e_0 = t \otimes \left(F_{\theta^0} - F_{\sigma(\theta^0)}\right), \quad f_0 = t^{-1} \otimes \left(E_{\theta^0} - E_{\sigma(\theta^0)}\right)$$
$$h_0 = 1 \otimes \left(-H_{\theta^0} - H_{\sigma(\theta^0)}\right) + 2c.$$

Type D_{l+1}

$$e_1 = 1 \otimes E_1, \ldots, e_{l-1} = 1 \otimes E_{l-1}, \quad e_l = 1 \otimes (E_l + E_{l+1})$$
$$f_1 = 1 \otimes F_1, \ldots, f_{l-1} = 1 \otimes F_{l-1}, \quad f_l = 1 \otimes (F_l + F_{l+1})$$
$$h_1 = 1 \otimes H_1, \ldots, h_{l-1} = 1 \otimes H_{l-1}, \quad h_l = 1 \otimes (H_l + H_{l+1})$$
$$e_0 = t \otimes \left(F_{\theta^0} - F_{\sigma(\theta^0)}\right), \quad f_0 = t^{-1} \otimes \left(E_{\theta^0} - E_{\sigma(\theta^0)}\right)$$
$$h_0 = 1 \otimes \left(-H_{\theta^0} - H_{\sigma(\theta^0)}\right) + 2c.$$

Type E_6

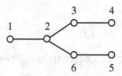

$e_1 = 1 \otimes E_1, \quad e_2 = 1 \otimes E_2, \quad e_3 = 1 \otimes (E_3 + E_6), \quad e_4 = 1 \otimes (E_4 + E_5)$

$f_1 = 1 \otimes F_1, \quad f_2 = 1 \otimes F_2, \quad f_3 = 1 \otimes (F_3 + F_6), \quad f_4 = 1 \otimes (F_4 + F_5)$

$h_1 = 1 \otimes H_1, \quad h_2 = 1 \otimes H_2, \quad h_3 = 1 \otimes (H_3 + H_6), \quad h_4 = 1 \otimes (H_4 + H_5)$

$e_0 = t \otimes \left(F_{\theta^0} - F_{\sigma(\theta^0)} \right), \quad f_0 = t^{-1} \otimes \left(E_{\theta^0} - E_{\sigma(\theta^0)} \right)$

$h_0 = 1 \otimes \left(-H_{\theta^0} - H_{\sigma(\theta^0)} \right) + 2c.$

Type D_4

$e_1 = 1 \otimes E_1, \quad e_2 = 1 \otimes (E_2 + E_3 + E_4)$

$f_1 = 1 \otimes F_1, \quad f_2 = 1 \otimes (F_2 + F_3 + F_4)$

$h_1 = 1 \otimes H_1, \quad h_2 = 1 \otimes (H_2 + H_3 + H_4)$

$e_0 = t \otimes \left(F_{\theta^0} + \omega^2 F_{\sigma(\theta^0)} + \omega F_{\sigma^2(\theta^0)} \right)$

$f_0 = t^{-1} \otimes \left(E_{\theta^0} + \omega E_{\sigma(\theta^0)} + \omega^2 E_{\sigma^2(\theta^0)} \right)$

$h_0 = 1 \otimes \left(-H_{\theta^0} - H_{\sigma(\theta^0)} - H_{\sigma^2(\theta^0)} \right) + 3c.$

Let $H \subset \hat{\mathfrak{L}}(L^0)$ be given by $H = (1 \otimes H^0) \oplus \mathbb{C}c \oplus \mathbb{C}d$. We define maps $\alpha_0, \alpha_1, \ldots, \alpha_l : H^\sigma \to \mathbb{C}$. We have roots $\alpha_i \in H^*$ and such roots in the same σ-orbit have the same restriction to H^σ. We define $\alpha_1, \ldots, \alpha_l \in (H^\sigma)^*$ to be the restrictions of the corresponding roots in H^*. We also define $\alpha_0 \in (H^\sigma)^*$ by $\alpha_0 = -\theta^0 + \delta$.

Let $\hat{\mathfrak{L}}(L^0) \cong L(A)$ as in Theorem 18.5. We wish to show that $\hat{\mathfrak{L}}(L^0)^\tau \cong L(A')$ where A' is the affine Cartan matrix of type given below:

A :	$\tilde{A}_{2l-1}$	$\tilde{D}_{l+1}$	$\tilde{E}_6$	$\tilde{D}_4$
A' :	$\tilde{B}_l^t$	$\tilde{C}_l^t$	$\tilde{F}_4^t$	$\tilde{G}_2^t$

18.4 Realisations of twisted affine algebras

We shall first show that (H^σ, Π, Π^v) is a realisation of A' where

$$\Pi^v = \{h_0, h_1, \ldots, h_l\} \subset H^\sigma, \quad \Pi = \{\alpha_0, \alpha_1, \ldots, \alpha_l\} \subset (H^\sigma)^*.$$

We know from Theorem 9.19 that $\alpha_j(h_i) = A'_{ij}$ for $i, j \in \{1, \ldots, l\}$. This $l \times l$ matrix is non-singular and so $h_1, \ldots, h_l$ and $\alpha_1, \ldots, \alpha_l$ are linearly independent. The element h_0 involves c whereas $h_1, \ldots, h_l$ do not, thus $h_0, h_1, \ldots, h_l$ are linearly independent. We have $\alpha_i(d) = 0$ for $i = 1, \ldots, l$ but $\alpha_0(d) \neq 0$, thus $\alpha_0, \alpha_1, \ldots, \alpha_l$ are linearly independent. We must show that

$$\alpha_j(h_0) = A'_{0j} \quad j = 1, \ldots, l$$

$$\alpha_0(h_i) = A'_{i0} \quad i = 1, \ldots, l$$

$$\alpha_0(h_0) = 2.$$

We recall the integers a_i, c_i associated with the affine Cartan matrix A', which are as follows.

$a_0, a_1, \ldots, a_l$ 	$c_0, c_1, \ldots, c_l$

We then note that the following significant equations hold in each of the cases being considered:

$$\sum_{i=0}^{l} a_i \alpha_i = \delta$$

$$\sum_{i=0}^{l} c_i h_i = mc \quad \text{where } m \text{ is the order of } \sigma.$$

We then have

$$\alpha_0(h_i) = -\sum_{j=1}^{l} a_j \alpha_j(h_i) = -\sum_{j=1}^{l} A'_{ij} a_j = A'_{i0} a_0 = A'_{i0}$$

for $i=1,\ldots,l$ since $a_0=1$ in the cases being considered. Similarly

$$\alpha_j(h_0) = -\sum_{i=1}^{l} c_i \alpha_j(h_i) = -\sum_{i=1}^{l} c_i A'_{ij} = c_0 A'_{0j} = A'_{0j}$$

for $j=1,\ldots,l$. Also

$$\alpha_0(h_0) = \left(-\theta^0 + \delta\right)\left(1 \otimes \left(H_{\theta^0} - H_{\sigma(\theta^0)} - \cdots\right) + mc\right) = \theta^0(H_{\theta^0}) = 2.$$

We also note that A' is an $(l+1) \times (l+1)$ matrix of rank l and that $\dim H^\sigma = l+2$. Thus we have shown that (H^σ, Π, Π^v) is a realisation of A'.

We next verify the relations necessary for applying Proposition 14.15. We first show that

$$[h_i e_j] = A'_{ij} e_j \qquad [h_i f_j] = -A'_{ij} f_j$$

for $i, j \in \{0, 1, \ldots, l\}$. These are known for $i, j \in \{1, \ldots, l\}$ by Theorem 9.19. So we must check

$$[h_0 e_j] = A'_{0j} e_j, \quad [h_0 f_j] = -A'_{0j} f_j \qquad j=1,\ldots,l$$
$$[h_i e_0] = A'_{i0} e_0, \quad [h_i f_0] = -A'_{i0} f_0 \qquad i=1,\ldots,l$$
$$[h_0 e_0] = 2e_0, \quad [h_0 f_0] = -2f_0.$$

For $j=1,\ldots,l$ we have

$$[h_0 e_j] = -\sum_{i=1}^{l} c_i [h_i e_j] = \left(-\sum_{i=1}^{l} c_i A'_{ij}\right) e_j = A'_{0j} e_j$$

and similarly for $i=1,\ldots,l$ we have $[h_0 f_j] = -A'_{0j} f_j$. Also

$$[h_i e_0] = \left[h_i, t \otimes \left(F_{\theta^0} + \varepsilon^{-1} F_{\sigma(\theta^0)} + \cdots\right)\right]$$
$$= t \otimes -\theta^0(h_i)\left(F_{\theta^0} + \varepsilon^{-1} F_{\sigma(\theta^0)} + \cdots\right)$$
$$= -\theta^0(h_i) e_0 = \left(-\sum_{j=1}^{l} a_j \alpha_j(h_i)\right) e_0 = \left(-\sum_{j=1}^{l} A'_{ij} a_j\right) e_0$$
$$= A'_{i0} e_0.$$

Similarly we have $[h_i f_0] = -A'_{i0} f_0$. We also have

$$[h_0 e_0] = \left[1 \otimes \left(-H_{\theta^0} - H_{\sigma(\theta^0)} - \cdots\right), t \otimes \left(F_{\theta^0} + \varepsilon^{-1} F_{\sigma(\theta^0)} + \cdots\right)\right]$$
$$= 2t \otimes \left(F_{\theta^0} + \varepsilon^{-1} F_{\sigma(\theta^0)} + \cdots\right) = 2e_0.$$

Similarly we have $[h_0 f_0] = -2f_0$.

18.4 Realisations of twisted affine algebras

Finally we have relations

$$[ce_i] = 0 = \alpha_i(c)e_i \qquad i = 1, \ldots, l$$
$$[cf_i] = 0 = -\alpha_i(c)f_i \qquad i = 1, \ldots, l$$
$$[de_i] = 0 = \alpha_i(d)e_i \qquad i = 1, \ldots, l$$
$$[df_i] = 0 = -\alpha_i(d)f_i \qquad i = 1, \ldots, l$$
$$[de_0] = e_0 = \alpha_0(d)e_0$$
$$[df_0] = -f_0 = -\alpha_0(d)f_0$$

Since $H^\tau = \left(1 \otimes (H^0)^\sigma\right) \oplus \mathbb{C}c \oplus \mathbb{C}d$ we have now verified all relations necessary for applying Proposition 14.15.

We next show that the elements $e_0, e_1, \ldots, e_l, f_0, f_1, \ldots, f_l$ together with H^τ generate $\hat{\mathfrak{L}}(L^0)^\tau$. We know that $e_1, \ldots, e_l, f_1, \ldots, f_l$ generate $(L^0)^\sigma$ by Theorem 9.19. Since

$$\hat{\mathfrak{L}}(L^0)^\tau = \sum_{k \in \mathbb{Z}} \left(t^{2k} \otimes (L^0)^\sigma\right) \oplus \sum_{k \in \mathbb{Z}} \left(t^{2k+1} \otimes (L^0)_{-1}\right) \oplus \mathbb{C}c \oplus \mathbb{C}d$$

when σ has order 2 and

$$\hat{\mathfrak{L}}(L^0)^\tau = \sum_{k \in \mathbb{Z}} \left(t^{3k} \otimes (L^0)^\sigma\right) \oplus \sum_{k \in \mathbb{Z}} \left(t^{3k+1} \otimes (L^0)_\omega\right) \oplus \sum_{k \in \mathbb{Z}} \left(t^{3k+2} \otimes (L^0)_{\omega^2}\right) \oplus \mathbb{C}c \oplus \mathbb{C}d$$

when σ has order 3, it is sufficient to show that the subspaces $\left(t^{2k} \otimes (L^0)^\sigma\right)$ for $k \neq 0$ and $t^{2k+1} \otimes (L^0)_{-1}$ lie in the subalgebra generated by the above elements when σ has order 2, and the subspaces $\left(t^{3k} \otimes (L^0)^\sigma\right)$ for $k \neq 0$, $\left(t^{3k+1} \otimes (L^0)_\omega\right)$, $\left(t^{3k+2} \otimes (L^0)_{\omega^2}\right)$ lie in this subalgebra when σ has order 3.

Let M be the subalgebra of $\hat{\mathfrak{L}}(L^0)^\tau$ generated by $e_0, e_1, \ldots, e_l$, $f_0, f_1, \ldots, f_l, H^\tau$. Suppose first that σ has order 2. We have

$$e_0 = t \otimes \left(F_{\theta^0} - F_{\sigma(\theta^0)}\right) \in M \text{ and } F_{\theta^0} - F_{\sigma(\theta^0)} \in (L^0)_{-1}.$$

Now if $x \in (L^0)^\sigma, y \in (L^0)_{-1}$ then

$$[1 \otimes x, t \otimes y] = t \otimes [xy] \in t \otimes (L^0)_{-1}.$$

Thus the elements $y \in (L^0)_{-1}$ for which $t \otimes y \in M$ form an $(L^0)^\sigma$-submodule of $(L^0)_{-1}$. However, $(L^0)_{-1}$ is an irreducible $(L^0)^\sigma$-module by Proposition 18.8. Thus $t \otimes (L^0)_{-1}$ lies in M. Now we can find elements $x, y \in (L^0)_{-1}$ such that $[xy] \neq 0$. Then $[t \otimes x, t \otimes y] = t^2 \otimes [xy]$ is a non-zero element of M. However, the set of $z \in (L^0)^\sigma$ such that $t^2 \otimes z \in M$ is an ideal of $(L^0)^\sigma$ and $(L^0)^\sigma$ is a simple Lie algebra. Thus $t^2 \otimes (L^0)^\sigma$ lies in M. The relations

$$[t^2 \otimes x, t^{2k} \otimes y] = t^{2k+2} \otimes [xy] \quad x, y \in (L^0)^\sigma$$
$$[t^2 \otimes x, t^{2k+1} \otimes y] = t^{2k+3} \otimes [xy] \quad x \in (L^0)^\sigma, y \in (L^0)_{-1}$$

can then be used to show by induction on k that $t^{2k} \otimes (L^0)^\sigma \subset M$ and $t^{2k+1} \otimes (L^0)_{-1} \subset M$ when $k > 0$. Starting with f_0 instead of e_0 will similarly show this when $k < 0$. Thus $M = \hat{\mathfrak{L}}(L^0)^\tau$.

Now suppose that σ has order 3. We have

$$e_0 = t \otimes \left(F_{\theta^0} + \omega^2 F_{\sigma(\theta^0)} + \omega F_{\sigma^2(\theta^0)} \right)$$
$$F_{\theta^0} + \omega^2 F_{\sigma(\theta^0)} + \omega F_{\sigma^2(\theta^0)} \in (L^0)_\omega.$$

An argument similar to the above shows that $t \otimes (L^0)_\omega \subset M$. Now there exist elements $x, y \in (L^0)_\omega$ with $[xy] \ne 0$. Then

$$[t \otimes x, t \otimes y] = t^2 \otimes [xy] \in t^2 \otimes (L^0)_{\omega^2}.$$

We can then show as above that $t^2 \otimes (L^0)_{\omega^2} \subset M$. There exist elements $x \in (L^0)_\omega, y \in (L^0)_{\omega^2}$ with $[xy] \ne 0$. Then

$$[t \otimes x, t^2 \otimes y] = t^3 \otimes [xy] \in t^3 \otimes (L^0)^\sigma.$$

We then see as above that $t^3 \otimes (L^0)^\sigma \subset M$. Induction on k can then be used to see that the subspaces

$$t^{3k} \otimes (L^0)^\sigma, \quad t^{3k+1} \otimes (L^0)_\omega, \quad t^{3k+2} \otimes (L^0)_{\omega^2}$$

all lie in M when $k > 0$. A similar result is obtained when $k < 0$ starting with f_0 instead of e_0. Thus $M = \hat{\mathfrak{L}}(L^0)^\tau$. Hence in all cases the elements $e_0, e_1, \ldots, e_l, f_0, f_1, \ldots, f_l$ and H^τ generate $\hat{\mathfrak{L}}(L^0)^\tau$.

Finally we must show that $\hat{\mathfrak{L}}(L^0)^\tau$ has no non-zero ideal J with $J \cap H^\tau = O$. To see this we decompose $\hat{\mathfrak{L}}(L^0)^\tau$ into root spaces with respect to H^τ. We first suppose σ has order 2. For each 1-element orbit (α) of σ on Φ^0 we choose $E_\alpha \in L_\alpha^0$. We showed in the proof of Theorem 9.19 that $\sigma(E_\alpha) = E_\alpha$. For each 2-element orbit (α, β) of σ on Φ^0 we choose $E_\alpha \in L_\alpha^0, E_\beta \in L_\beta^0$ such that $\sigma(E_\alpha) = E_\beta$. Then $\hat{\mathfrak{L}}(L^0)^\tau$ is the direct sum of H^τ and the following weight spaces.

18.4 Realisations of twisted affine algebras

$t^{2k} \otimes (H^0)^\sigma$ with weight $2k\delta$

$t^{2k+1} \otimes (H^0)_{-1}$ with weight $(2k+1)\delta$

$t^{2k} \otimes \mathbb{C} E_\alpha$ with weight $\alpha + 2k\delta$ where (α) is a 1-element orbit

$t^{2k} \otimes \mathbb{C}(E_\alpha + E_\beta)$ with weight $\alpha + 2k\delta$ where (α, β) is a 2-element orbit

$t^{2k+1} \otimes \mathbb{C}(E_\alpha - E_\beta)$ with weight $\alpha + (2k+1)\delta$ where (α, β) is a 2-element orbit.

By Lemma 14.12 the ideal J is the direct sum of its intersections with these weight spaces. Thus J has non-zero intersection with one of these weight spaces. Taking a non-zero element x in this weight space and in J we can find an element y in the negative weight space such that $[xy]$ is a non-zero element of H^τ, and this contradicts $J \cap H^\tau = 0$.

When σ has order 3 a similar argument can be applied. This time the weight spaces are

$t^{3k} \otimes (H^0)^\sigma$ with weight $3k\delta$

$t^{3k+1} \otimes (H^0)_\omega$ with weight $(3k+1)\delta$

$t^{3k+2} \otimes (H^0)_{\omega^2}$ with weight $(3k+2)\delta$

$t^{3k} \otimes \mathbb{C} E_\alpha$ with weight $\alpha + 3k\delta$ where (α) is a 1-element orbit

$t^{3k} \otimes \mathbb{C}(E_\alpha + E_\beta + E_\gamma)$ with weight $\alpha + 3k\delta$ where (α, β, γ) is a 3-element orbit

$t^{3k+1} \otimes \mathbb{C}(E_\alpha + \omega^2 E_\beta + \omega E_\gamma)$ with weight $\alpha + (3k+1)\delta$ where (α, β, γ) is a 3-element orbit

$t^{3k+2} \otimes \mathbb{C}(E_\alpha + \omega E_\beta + \omega^2 E_\gamma)$ with weight $\alpha + (3k+2)\delta$ where (α, β, γ) is a 3-element orbit.

Any non-zero ideal J with $J \cap H^\tau = 0$ must have non-zero intersection with one of these weight spaces. We can then multiply it by an element of the negative weight space to give a non-zero element of H^τ, and this contradicts $J \cap H^\tau = 0$. Thus we deduce that $J = 0$.

We have now verified all the conditions of Proposition 14.15 and can conclude that $\hat{\mathfrak{L}}(L^0)^\tau$ is isomorphic to $L(A')$. $\square$

As a corollary we obtain the multiplicities of the imaginary roots of the affine Kac–Moody algebras of types $\tilde{B}_l^t, \tilde{C}_l^t, \tilde{F}_4^t, \tilde{G}_2^t$.

Corollary 18.10 *The multiplicities of the imaginary roots are as follows.*

Type $\tilde{B}_l^t$. *The roots $2k\delta$ ($k \ne 0$) have multiplicity l and the roots $(2k+1)\delta$ have multiplicity $l-1$.*

Type $\tilde{C}_l^t$ *The roots $2k\delta$ ($k \ne 0$) have multiplicity l and the roots $(2k+1)\delta$ have multiplicity 1.*

Type $\tilde{F}_4^t$ *The roots $2k\delta$ ($k \ne 0$) have multiplicity 4 and the roots $(2k+1)\delta$ have multiplicity 2.*

Type $\tilde{G}_2^t$ *The roots $3k\delta$ ($k \ne 0$) have multiplicity 2 and the roots $(3k+1)\delta$ and $(3k+2)\delta$ have multiplicity 1.*

Proof. For types $\tilde{B}_l^t, \tilde{C}_l^t, \tilde{F}_4^t$ Theorem 18.9 shows that the multiplicity of $2k\delta$ ($k \ne 0$) is $\dim(H^0)^\sigma$, which is equal to l. The multiplicity of $(2k+1)\delta$ is $\dim(H^0)_{-1}$, which is $l-1, 1, 2$ in the three cases respectively.

For type $\tilde{G}_2^t$ the multiplicity of $3k\delta$ ($k \ne 0$) is $\dim(H^0)^\sigma = 2$, and the multiplicities of $(3k+1)\delta$ and $(3k+2)\delta$ are $\dim(H^0)_\omega = \dim(H^0)_{\omega^2} = 1$. □

We now make some comments on the isomorphism between $L(A')$ and $\hat{\mathfrak{L}}(L^0)^\tau$ which we have obtained. The standard invariant form on $L(A')$ does not map under this isomorphism to the restriction of the standard invariant form on $\hat{\mathfrak{L}}(L^0)$. For let (i) be a 1-element σ-orbit on $\{1, \ldots, l\}$. Such an orbit exists in each of the cases. Then $h_i \in L(A')$ corresponds to $1 \otimes H_i \in \hat{\mathfrak{L}}(L^0)^\tau$. We have

$$\langle 1 \otimes H_i, 1 \otimes H_i \rangle = 2$$

$$\langle h_i, h_i \rangle' = 2a_i/c_i.$$

However, we may check that $c_i = ma_i$ for all 1-element σ-orbits, hence

$$\langle h_i, h_i \rangle' = 2/m$$

where m is the order of σ. Thus

$$\langle 1 \otimes H_i, 1 \otimes H_i \rangle = m \langle h_i, h_i \rangle'.$$

Also the isomorphism does not map the element $c \in \hat{\mathfrak{L}}(L^0)^\tau$ to the canonical central element $c' \in L(A')$. We noted in the proof of Theorem 18.9 that

$$\sum_{i=0}^{l} c_i h_i = mc \quad \text{in } \hat{\mathfrak{L}}(L^0)^\tau$$

and hence our isomorphism maps mc to c'.

However, the scaling element $d \in \hat{\mathfrak{L}}(L^0)^\tau$ maps to a scaling element $d' \in L(A')$. For we have $\alpha_0 = -\theta^0 + \delta$ and so $\alpha_0(d) = \delta(d) = 1$. Also $\alpha_i(d) = 0$ for $i = 1, \ldots, l$.

18.4 Realisations of twisted affine algebras

We also see that, if $x, y \in \hat{\mathscr{L}}(L^0)^\tau$ map to $x', y' \in L(A')$, then

$$\langle x, y \rangle = m \langle x', y' \rangle'$$

where $\langle , \rangle$ and $\langle , \rangle'$ are the standard invariant forms. This is true for $x', y' \in L(A')'$ since any two symmetric invariant forms are proportional on $L(A')'$. However, it is also true when $y = d, y' = d'$ since

$$\langle h_i, d \rangle = 0 \quad \text{for } i = 1, \ldots, l$$

$$\langle c, d \rangle = 1$$

$$\langle d, d \rangle = 0$$

Thus it is true for all $x', y' \in L(A')$. □

We now wish to obtain realisations of the remaining twisted Kac–Moody algebras of type $\tilde{C}'_l$ for $l \geq 2$ and $\tilde{A}'_1$. These will be obtained as fixed point subalgebras of the untwisted Kac–Moody algebra of type $\tilde{A}_{2l}$ under the twisted graph automorphism τ. We begin by recalling from Theorem 9.19 that the fixed point subalgebra of the finite dimensional Lie algebra $L(A_{2l})$ under its graph automorphism σ is given by

$$L(A_{2l})^\sigma \cong L(B_l).$$

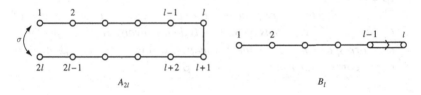

In order to show that $L(\tilde{A}_{2l})^\tau \cong L(\tilde{C}'_l)$ if $l \geq 2$ and $L(\tilde{A}_2)^\tau \cong L(\tilde{A}'_1)$ we shall compare the diagrams

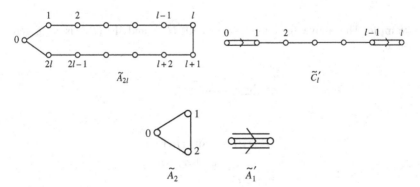

We note that the numbering of the vertices of the graph for $\tilde{C}'_l$ is not the same as the numbering previously used when $\tilde{C}'_l$ was constructed from C_l by adding an extra vertex labelled 0. Here we are starting from the finite dimensional simple Lie algebra B_l rather than C_l. In Theorem 17.17 (d) and (e) we obtained the roots of $\tilde{C}'_l$ and $\tilde{A}'_1$ in terms of those of C_l. For our present purpose we require these roots in terms of those of B_l.

We consider the diagram of $\tilde{C}'_l$ labelled as follows.

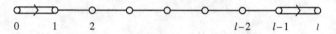

and let B_l be the subdiagram obtained by omitting vertex 0 and C_l be the subdiagram obtained by omitting vertex l. We recall that

$$\delta = \alpha_0 + 2\alpha_1 + \cdots + 2\alpha_{l-1} + 2\alpha_l.$$

The following lemma will be useful by relating the roots of B_l and C_l.

Lemma 18.11 (i) *Each long positive root of C_l involves α_0. Each short positive root of B_l involves α_l. There is a bijective correspondence $\alpha \leftrightarrow \beta$ between long positive roots of C_l and short positive roots of B_l satisfying $\alpha + 2\beta = \delta$.*

(ii) *There is a bijective correspondence $\alpha \leftrightarrow \beta$ between short positive roots of C_l and long positive roots of B_l. If α, β do not involve α_0, α_l respectively this correspondence is the identity map. If α involves α_0 and β involves α_l the correspondence is given by $\alpha + \beta = \delta$.*

Proof. This follows immediately from expressing the roots of B_l and C_l in terms of the fundamental roots $\alpha_1, \ldots, \alpha_l$ and $\alpha_0, \alpha_1, \ldots, \alpha_{l-1}$ respectively. □

Example The above bijection between $\Phi^+(C_3)$ and $\Phi^+(B_3)$ is as given below.

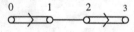

$\Phi^+(C_3)$	$\Phi^+(B_3)$
α_0	$\alpha_1 + \alpha_2 + \alpha_3$
$\alpha_0 + 2\alpha_1$	$\alpha_2 + \alpha_3$

18.4 Realisations of twisted affine algebras

$$\begin{array}{cc} \alpha_0+2\alpha_1+2\alpha_2 & \alpha_3 \\ \alpha_1 & \alpha_1 \\ \alpha_2 & \alpha_2 \\ \alpha_1+\alpha_2 & \alpha_1+\alpha_2 \\ \alpha_0+\alpha_1 & \alpha_1+2\alpha_2+2\alpha_3 \\ \alpha_0+\alpha_1+\alpha_2 & \alpha_1+\alpha_2+2\alpha_3 \\ \alpha_0+2\alpha_1+\alpha_2 & \alpha_2+2\alpha_3 \end{array}$$

By using Lemma 18.11 together with Theorem 17.17 we may express the real roots of $\tilde{C}'_l$ in terms of the roots of B_l.

Proposition 18.12 (i) *The real roots of* $L(\tilde{C}'_l)$, $l \geq 2$, *are*

$$\Phi_{\text{Re},s} = \{\alpha + r\delta \; ; \; \alpha \in \Phi^0_s, r \in \mathbb{Z}\}$$
$$\Phi_{\text{Re},i} = \{\alpha + r\delta \; ; \; \alpha \in \Phi^0_l, r \in \mathbb{Z}\}$$
$$\Phi_{\text{Re},l} = \{2\alpha + (2r+1)\delta \; ; \; \alpha \in \Phi^0_s, r \in \mathbb{Z}\}$$

where $\Phi_{\text{Re},s}, \Phi_{\text{Re},i}, \Phi_{\text{Re},l}$ *are the short, intermediate and long roots respectively, and* Φ^0_s, Φ^0_l *are the short and long roots of* B_l.
(ii) *The real roots of* $L(\tilde{A}'_1)$ *are*

$$\Phi_{\text{Re},s} = \{\alpha + r\delta \; ; \; \alpha \in \Phi^0, r \in \mathbb{Z}\}$$
$$\Phi_{\text{Re},l} = \{2\alpha + (2r+1)\delta \; ; \; \alpha \in \Phi^0, r \in \mathbb{Z}\}$$

where Φ^0 *is the root system of type* A_1 *obtained from the short fundamental root of* $\tilde{A}'_1$.

Proof. (i) We know from Theorem 17.17 that

$$\Phi_{\text{Re},s} = \left\{\frac{1}{2}(\alpha + (2r-1)\delta) \; ; \; \alpha \in \Phi^0_l(C_l), r \in \mathbb{Z}\right\}$$
$$\Phi_{\text{Re},i} = \{\alpha + r\delta \; ; \; \alpha \in \Phi^0_s(C_l), r \in \mathbb{Z}\}$$
$$\Phi_{\text{Re},l} = \{\alpha + 2r\delta \; ; \; \alpha \in \Phi^0_l(C_l), r \in \mathbb{Z}\}.$$

We make use of the bijection $\alpha \leftrightarrow \beta, -\alpha \leftrightarrow -\beta$ of Lemma 18.11 where $\alpha \in \Phi^+(C_l), \beta \in \Phi^+(B_l)$. For each $\alpha \in \Phi^0(C_l)$ we choose the corresponding

$\beta \in \Phi^0(B_l)$. First suppose $\alpha \in \Phi_l^0(C_l)$. The corresponding $\beta \in \Phi_s^0(B_l)$ is given by $\alpha + 2\beta = \delta$ if α is positive and $\alpha + 2\beta = -\delta$ if α is negative. Thus

$$\frac{1}{2}(\alpha + (2r-1)\delta) = \begin{cases} -\beta + r\delta & \text{if } \alpha \text{ is positive} \\ -\beta + (r-1)\delta & \text{if } \alpha \text{ is negative} \end{cases}$$

$$\alpha + 2r\delta = \begin{cases} -2\beta + (2r+1)\delta & \text{if } \alpha \text{ is positive} \\ -2\beta + (2r-1)\delta & \text{if } \alpha \text{ is negative} \end{cases}$$

This gives the required formulae for $\Phi_{\text{Re},s}$ and $\Phi_{\text{Re},l}$ as r runs through $\mathbb{Z}$. Next consider $\Phi_{\text{Re},i}$. If $\alpha \in \Phi_s^0(C_l)$ does not involve α_0 we have $\beta = \alpha \in \Phi_1^0(B_l)$. If $\alpha \in \Phi_s^0(C_l)$ does involve α_0 the corresponding $\beta \in \Phi_1^0(B_l)$ satisfies $\alpha + \beta = \delta$ if α is positive and $\alpha + \beta = -\delta$ if α is negative. Then

$$\alpha + r\delta = \beta + r\delta \quad \text{in the first case}$$

$$\alpha + r\delta = \begin{cases} -\beta + (r+1)\delta & \text{if } \alpha \text{ is positive} \\ -\beta + (r-1)\delta & \text{if } \alpha \text{ is negative} \end{cases}$$

in the second case. This gives the required formula for $\Phi_{\text{Re},i}$ as r runs through $\mathbb{Z}$.

(ii) In type $\tilde{A}_1'$ the argument is exactly the same except that $\Phi^0 = \Phi_s^0$ has type A_1 and Φ_1^0 is empty. Thus $\Phi_{\text{Re},i}$ is empty in this case. $\square$

We shall next prove the analogue of Proposition 18.8 in our present case.

Proposition 18.13 *Let L^0 be the simple Lie algebra of type A_{2l} and σ be its graph automorphism of order 2. Let $(L^0)_{-1}$ be the eigenspace of σ on L^0 with eigenvalue -1. Then $L^0 = (L^0)^\sigma \oplus (L^0)_{-1}$. The eigenspace $(L^0)_{-1}$ is an irreducible $(L^0)^\sigma$-module. The algebra $(L^0)^\sigma$ is isomorphic to $L(B_l)$ and $(L^0)_{-1}$ is isomorphic to its irreducible module $L(2\omega_1)$.*

Proof. It is clear that $(L^0)_{-1}$ is an $(L^0)^\sigma$-module and that $L^0 = (L^0)^\sigma \oplus (L^0)_{-1}$. We have

$$\dim L^0 = \dim L(A_{2l}) = 2l(2l+2)$$

$$\dim (L^0)^\sigma = \dim L(B_l) = l(2l+1).$$

Thus $\dim (L^0)_{-1} = \dim L^0 - \dim (L^0)^\sigma = l(2l+3)$. Let $\theta = \alpha_1 + \cdots + \alpha_{2l}$ be the highest root of A_{2l}. Then a corresponding root vector E_θ lies in $(L^0)_{-1}$ and gives the highest weight of $(L^0)_{-1}$. It gives rise to the weight $2\alpha_1 + \cdots + 2\alpha_l$ of $L(B_l)$. By considering the Cartan matrix of B_l we see that $\alpha_1 + \cdots + \alpha_l = \omega_1$ and so the highest weight of the $L(B_l)$-module $(L^0)_{-1}$ is $2\omega_1$.

Weyl's dimension formula shows that $\dim L(2\omega_1) = l(2l+3)$. Since this is also the dimension of $(L^0)_{-1}$ we see that $(L^0)_{-1}$ is an irreducible $L(B_l)$-module isomorphic to $L(2\omega_1)$. □

By using this result we can prove the analogue of Theorem 18.9 in the A_{2l} case.

Theorem 18.14 *Let L^0 be a simple Lie algebra of type A_{2l} with $l \geq 2$ and σ be its graph automorphism of order 2. Let τ be the corresponding twisted graph automorphism of $\hat{\mathfrak{L}}(L^0)$. Then the fixed point subalgebra $\hat{\mathfrak{L}}(L^0)^\tau$ is isomorphic to $L(\tilde{C}'_l)$.*

When $l = 1$ the fixed point subalgebra is isomorphic to $L(\tilde{A}'_1)$.

Proof. The general idea of the proof is like that of Theorems 18.5 and 18.9. We aim to obtain the result by applying Proposition 14.15.

We first suppose $l \geq 2$. We have

$$\hat{\mathfrak{L}}(L^0) = \sum_{k \in \mathbb{Z}} (t^k \otimes L^0) \oplus \mathbb{C}c \oplus \mathbb{C}d$$

$$\hat{\mathfrak{L}}(L^0)^\tau = \sum_{k \in \mathbb{Z}} (t^{2k} \otimes (L^0)^\sigma) + \sum_{k \in \mathbb{Z}} (t^{2k+1} \otimes (L^0)_{-1}) \oplus \mathbb{C}c \oplus \mathbb{C}d.$$

Let $E_1, \ldots, E_{2l}, F_1, \ldots, F_{2l}, H_1, \ldots, H_{2l}$ be standard generators of L^0. We wish to define analogous elements $e_0, e_1, \ldots, e_l, f_0, f_1, \ldots, f_l, h_0, h_1, \ldots, h_l$ in $\hat{\mathfrak{L}}(L^0)^\tau$. These are chosen as follows.

$$e_1 = 1 \otimes (E_1 + E_{2l}), \quad \ldots, \quad e_{l-1} = 1 \otimes (E_{l-1} + E_{l+2}), \quad e_l = 1 \otimes \sqrt{2}(E_l + E_{l+1})$$

$$f_1 = 1 \otimes (F_1 + F_{2l}), \quad \ldots, \quad f_{l-1} = 1 \otimes (F_{l-1} + F_{l+2}), \quad f_l = 1 \otimes \sqrt{2}(F_l + F_{l+1})$$

$$h_1 = 1 \otimes (H_1 + H_{2l}), \quad \ldots, \quad h_{l-1} = 1 \otimes (H_{l-1} + H_{l+2}), \quad h_l = 1 \otimes 2(H_l + H_{l+1})$$

$$e_0 = t \otimes F_\theta, \quad f_0 = t^{-1} \otimes E_\theta, \quad h_0 = -(1 \otimes H_\theta) + c$$

where θ is the highest root of L^0.

Let $H \subset \hat{\mathfrak{L}}(L^0)$ be given by $H = (1 \otimes H^0) \oplus \mathbb{C}c \oplus \mathbb{C}d$. We define maps $\alpha_0, \alpha_1, \ldots, \alpha_l : H^\sigma \to \mathbb{C}$. For $i = 1, \ldots, l$ these are the restrictions of the roots $\alpha_i : H \to \mathbb{C}$. For $i = 0$ we define $\alpha_0 = -\theta + \delta$.

Let $\Pi^\vee = \{h_0, h_1, \ldots, h_l\} \subset H^\sigma$ and $\Pi = \{\alpha_0, \alpha_1, \ldots, \alpha_l\} \subset (H^\sigma)^*$. We show that $(H^\sigma, \Pi, \Pi^\vee)$ is a realisation of the Cartan matrix A' of type $\tilde{C}'_l$.

We know from Theorem 9.19 that $\alpha_j(h_i) = A'_{ij}$ for $i, j \in \{1, \ldots, l\}$. This $l \times l$ matrix is the Cartan matrix of type B_l. In particular it is non-singular. Thus $h_1, \ldots, h_l$ and $\alpha_1, \ldots, \alpha_l$ are linearly independent. Now h_0 involves c whereas $h_1, \ldots, h_l$ do not, thus $h_0, h_1, \ldots, h_l$ are linearly independent.

Also we have $\alpha_0(d) \neq 0$ and $\alpha_i(d) = 0$ for $i = 1, \ldots, l$ thus $\alpha_0, \alpha_1, \ldots, \alpha_l$ are linearly independent. We must show in addition

$$\alpha_j(h_0) = A'_{0j} \quad j = 1, \ldots, l$$
$$\alpha_0(h_i) = A'_{i0} \quad i = 1, \ldots, l$$
$$\alpha_0(h_0) = 2.$$

We recall that the integers a_i, c_i associated with the affine Cartan matrix A' are

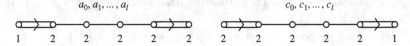

In particular we have $a_0 = 1$, $c_0 = 2$. (The change of labelling explains the fact that c_0 is not 1, as it usually is.) We note that

$$\sum_{i=0}^{l} a_i \alpha_i = \delta$$

$$\sum_{i=0}^{l} c_i h_i = 2c.$$

We then have

$$\alpha_0(h_i) = -\sum_{j=1}^{l} a_j \alpha_j(h_i) = -\sum_{j=1}^{l} A'_{ij} a_j = A'_{i0} a_0 = A'_{i0}$$

$$\alpha_j(h_0) = -\frac{1}{2}\sum_{i=1}^{l} c_i \alpha_j(h_i) = -\frac{1}{2}\sum_{i=1}^{l} c_i A'_{ij} = \frac{1}{2} c_0 A'_{0j} = A'_{0j}$$

$$\alpha_0(h_0) = (-\theta + \delta)((1 \otimes -H_\theta) + c) = \theta(H_\theta) = 2.$$

We observe that A' is an $(l+1) \times (l+1)$ matrix of rank l and that $\dim H^\sigma = l + 2$. Thus we have shown that $(H^\sigma, \Pi, \Pi^\vee)$ is a realisation of A'.

We next verify the relations

$$[h_i e_j] = A'_{ij} e_j \quad [h_i f_j] = -A'_{ij} f_j$$

for $i, j \in \{0, 1, \ldots, l\}$. We know this already for $i, j \in \{1, \ldots, l\}$ by Theorem 9.19. Thus we must verify

$$[h_0 e_j] = A'_{0j} e_j \quad [h_0 f_j] = -A'_{0j} f_j$$
$$[h_i e_0] = A'_{i0} e_0 \quad [h_i f_0] = -A'_{i0} f_0$$
$$[h_0 e_0] = 2 e_0 \quad [h_0 f_0] = -2 f_0.$$

18.4 Realisations of twisted affine algebras

Now

$$[h_0 e_j] = -\frac{1}{2}\sum_{i=1}^{l} c_i [h_i e_j] = \left(-\frac{1}{2}\sum_{i=1}^{l} c_i A'_{ij}\right) e_j = A'_{0j} e_j$$

and similarly $[h_0 f_j] = -A'_{0j} f_j$. Also

$$[h_i e_0] = [h_i, t \otimes F_\theta] = t \otimes (-\theta(h_i) F_\theta) = -\theta(h_i) e_0$$

$$= \left(-\sum_{j=1}^{l} a_j \alpha_j(h_i)\right) e_0 = \left(-\sum_{j=1}^{l} A'_{ij} a_j\right) e_0 = A'_{i0} e_0.$$

Similarly we have $[h_i f_0] = -A'_{i0} f_0$. We also have

$$[h_0 e_0] = [-(1 \otimes H_\theta) + c, t \otimes F_\theta] = -t \otimes [H_\theta F_\theta]$$
$$= 2t \otimes F_\theta = 2e_0$$

and similarly $[h_0 f_0] = -2f_0$. Finally we have

$$[c e_i] = \alpha_i(c) e_i = 0 \qquad i = 0, 1, \ldots, l$$
$$[c f_i] = -\alpha_i(c) f_i = 0 \qquad i = 0, 1, \ldots, l$$
$$[d e_i] = \alpha_i(d) e_i = 0 \qquad i = 1, \ldots, l$$
$$[d f_i] = -\alpha_i(d) f_i = 0 \qquad i = 1, \ldots, l$$
$$[d e_0] = \alpha_0(d) e_0 = e_0$$
$$[d f_0] = -\alpha_0(d) f_0 = -f_0.$$

Since $H^\tau = (1 \otimes (H^0)^\sigma) \oplus \mathbb{C}c \oplus \mathbb{C}d$ we have verified all relations necessary for the application of Proposition 14.15.

We next show that the elements $e_0, e_1, \ldots, e_l, f_0, f_1, \ldots, f_l$ together with H^τ generate $\hat{\mathfrak{L}}(L^0)^\tau$. By Theorem 9.19 $e_1, \ldots, e_l, f_1, \ldots, f_l$ generate $(L^0)^\sigma$. Since $\hat{\mathfrak{L}}(L^0)^\tau = \sum_{k \in \mathbb{Z}} (t^{2k} \otimes (L^0)^\sigma) \oplus \sum_{k \in \mathbb{Z}} (t^{2k+1} \otimes (L^0)_{-1}) \oplus \mathbb{C}c \oplus \mathbb{C}d$ it is sufficient to show that the subspaces

$$(t^{2k} \otimes (L^0)^\sigma) \quad \text{for } k \neq 0 \quad \text{and} \quad (t^{2k+1} \otimes (L^0)_{-1})$$

lie in the subalgebra M generated by $e_0, e_1, \ldots, e_l, f_0, f_1, \ldots, f_l, H^\tau$.

Now $e_0 = t \otimes F_\theta$ lies in M and $F_\theta \in (L^0)_{-1}$. If $x \in (L^0)^\sigma, y \in (L^0)_{-1}$ then

$$[1 \otimes x, t \otimes y] = t \otimes [xy] \in t \otimes (L^0)_{-1}.$$

Thus the elements $y \in (L^0)_{-1}$ for which $t \otimes y \in M$ form an $(L^0)^\sigma$-submodule of $(L^0)_{-1}$. This submodule contains F_θ so is non-zero. Since $(L^0)_{-1}$ is an irreducible $(L^0)^\sigma$-module by Proposition 18.13 this submodule is the whole of $(L^0)_{-1}$. Thus $t \otimes (L^0)_{-1}$ lies in M.

Now we can find elements $x, y \in (L^0)_{-1}$ with $[xy] \neq 0$. (For example, $x = F_{\alpha_l} - F_{\alpha_{l+1}}, y = E_{\alpha_l + \alpha_{l+1}}$.) Thus $[t \otimes x, t \otimes y] = t^2 \otimes [xy]$ is a non-zero element of M. However, the set of $z \in (L^0)^\sigma$ for which $t^2 \otimes z \in M$ is an ideal of $(L^0)^\sigma$ since

$$[t^2 \otimes z, 1 \otimes w] = t^2 \otimes [zw] \quad \text{for } w \in (L^0)^\sigma.$$

Since $[xy] \in (L^0)^\sigma$ this is a non-zero ideal of $(L^0)^\sigma$ and since $(L^0)^\sigma$ is a simple Lie algebra it is the whole of $(L^0)^\sigma$. Thus $t^2 \otimes (L^0)^\sigma$ lies in M.

Now the relations

$$[t^2 \otimes x, t^{2k} \otimes y] = t^{2k+2} \otimes [xy], \quad x, y \in (L^0)^\sigma$$
$$[t^2 \otimes x, t^{2k+1} \otimes y] = t^{2k+3} \otimes [xy], \quad x \in (L^0)^\sigma, y \in (L^0)_{-1}$$

can be used to show by induction on k that $t^{2k} \otimes (L^0)^\sigma \subset M$ and

$$t^{2k+1} \otimes (L^0)_{-1} \subset M \quad \text{when } k > 0.$$

Starting with f_0 instead of e_0 will similarly show this for $k < 0$. Thus $M = \hat{\mathfrak{L}}(L^0)^\tau$ as required.

Finally we must show that $\hat{\mathfrak{L}}(L^0)^\tau$ has no non-zero ideal J with $J \cap H^\tau = O$. To see this we decompose $\hat{\mathfrak{L}}(L^0)^\tau$ into weight spaces with respect to H^τ. Any non-zero ideal J with $J \cap H^\tau = O$ must have non-zero intersection with one of these weight spaces, by Lemma 14.12. Let x be a non-zero element in such an intersection. Then there exists y in the weight space corresponding to the negative of this weight such that $[xy] \neq 0$, by Corollary 16.5. But then $[xy] \in J \cap H^\tau$ and so $J \cap H^\tau \neq O$, a contradiction. Hence $J = O$.

We have now verified all the hypotheses of the recognition theorem Proposition 14.15 and so can conclude that $\hat{\mathfrak{L}}(L^0)^\tau$ is isomorphic to $L(\tilde{C}'_l)$.

We now consider the case $l = 1$. This time the graphs are

and the Cartan matrix A' of $\tilde{A}'_1$ is

$$A' = \begin{pmatrix} 2 & -1 \\ -4 & 2 \\ 0 & 1 \end{pmatrix} \begin{matrix} 0 \\ 1. \end{matrix}$$

18.4 Realisations of twisted affine algebras

The elements $e_0, e_1, f_0, f_1, h_0, h_1$ of $\hat{\mathfrak{L}}(L^0)^\tau$ are

$$e_1 = 1 \otimes \sqrt{2}(E_1 + E_2), \quad f_1 = 1 \otimes \sqrt{2}(F_1 + F_2), \quad h_1 = 1 \otimes 2(H_1 + H_2)$$

$$e_0 = t \otimes F_\theta, \quad f_0 = t^{-1} \otimes E_\theta, \quad h_0 = -(1 \otimes H_\theta) + c$$

where $\theta = \alpha_1 + \alpha_2$ is the highest root of A_2. We have

$$\check{\Pi} = \{h_0, h_1\} \qquad \Pi = \{\alpha_0, \alpha_1\}$$

where $\alpha_0 = -\theta + \delta$ and $\alpha_1 \in (H^\sigma)^*$ is the restriction of $\alpha_1 \in H^*$. The integers a_0, a_1, c_0, c_1 for $\tilde{A}'_1$ are

$$a_0 = 1, \quad a_1 = 2, \quad c_0 = 2, \quad c_1 = 1.$$

We have

$$a_0 \alpha_0 + a_1 \alpha_1 = \delta$$
$$c_0 h_0 + c_1 h_1 = 2c.$$

We can then check that $(H^\sigma, \Pi, \Pi^\vee)$ is a realisation of A'. We also check the relations

$$[h_0 e_0] = 2e_0, \quad [h_0 e_1] = -e_1, \quad [h_1 e_0] = -4e_0, \quad [h_1 e_1] = 2e_1$$

$$[h_0 f_0] = -2f_0, \quad [h_0 f_1] = f_1, \quad [h_1 f_0] = 4f_0, \quad [h_1 f_1] = -2f_1.$$

The facts that $H^\tau, e_0, e_1, f_0, f_1$ generate $\hat{\mathfrak{L}}(L^0)^\tau$ and that $\hat{\mathfrak{L}}(L^0)^\tau$ has no non-zero ideal J with $J \cap H^\tau = 0$ are proved just as before. Thus applying Proposition 14.15 shows that $\hat{\mathfrak{L}}(L^0)^\tau$ is isomorphic to $L(\tilde{A}'_1)$. □

We shall describe explicitly the weight space decomposition of $\hat{\mathfrak{L}}(L^0)^\tau$ with respect to H^τ. We recall from Proposition 9.18 that there is a bijective correspondence between roots of $(L^0)^\sigma = L(B_l)$ and equivalence classes of roots of $L^0 = L(A_{2l})$. Each equivalence class has 2 or 3 elements. Equivalence classes with 2 elements have form (α, β) where $\sigma(\alpha) = \beta, \sigma(\beta) = \alpha$ and $\alpha + \beta$ is not a root. Equivalence classes with 3 elements have form $(\alpha, \beta, \alpha + \beta)$ where $\sigma(\alpha) = \beta, \sigma(\beta) = \alpha$ and $\sigma(\alpha + \beta) = \alpha + \beta$. Equivalence classes with 2 elements correspond to long roots of B_l and equivalence classes with 3 elements correspond to short roots of B_l. For each 2-element equivalence class we can choose root vectors E_α, E_β with $\sigma(E_\alpha) = E_\beta$. For each 3-element equivalence class we choose root vectors $E_\alpha, E_\beta, E_{\alpha+\beta}$ with $\sigma(E_\alpha) = E_\beta$ and $[E_\alpha E_\beta] = E_{\alpha+\beta}$. Then

$$\sigma(E_{\alpha+\beta}) = \sigma[E_\alpha E_\beta] = [E_\beta E_\alpha] = -E_{\alpha+\beta}$$

thus $E_{\alpha+\beta} \in (L^0)_{-1}$.

The Lie algebra $\hat{\mathfrak{L}}(L^0)^\tau$ is the direct sum of H^τ and the following weight spaces.

$t^{2k} \otimes (H^0)^\sigma$ with weight $2k\delta$

$t^{2k+1} \otimes (H^0)_{-1}$ with weight $(2k+1)\delta$

$t^{2k} \otimes \mathbb{C}(E_\alpha + E_\beta)$ with weight $\alpha + 2k\delta$ for each 2 element equivalence class (α, β)

$t^{2k+1} \otimes \mathbb{C}(E_\alpha - E_\beta)$ with weight $\alpha + (2k+1)\delta$ for each 2 element equivalence class (α, β)

$t^{2k} \otimes \mathbb{C}(E_\alpha + E_\beta)$ with weight $\alpha + 2k\delta$ for each 3 element equivalence class $(\alpha, \beta, \alpha+\beta)$

$t^{2k+1} \otimes \mathbb{C}(E_\alpha - E_\beta)$ with weight $\alpha + (2k+1)\delta$ for each 3 element equivalence class $(\alpha, \beta, \alpha+\beta)$

$t^{2k+1} \otimes \mathbb{C}E_{\alpha+\beta}$ with weight $2\alpha + (2k+1)\delta$ for each 3 element equivalence class $(\alpha, \beta, \alpha+\beta)$.

The weights listed above correspond to the roots of $L(\tilde{C}'_l)$ as described in Proposition 18.12.

In the case $l = 1$ the weight spaces of $L(\tilde{A}_2)^\tau$ are

$t^{2k} \otimes \mathbb{C}(H_1 + H_2)$ with weight $2k\delta$

$t^{2k+1} \otimes \mathbb{C}(H_1 - H_2)$ with weight $(2k+1)\delta$

$t^{2k} \otimes \mathbb{C}(E_1 + E_2)$ with weight $\alpha_1 + 2k\delta$

$t^{2k+1} \otimes \mathbb{C}(E_1 - E_2)$ with weight $\alpha_1 + (2k+1)\delta$

$t^{2k+1} \otimes \mathbb{C}E_{\alpha_1+\alpha_2}$ with weight $2\alpha_1 + (2k+1)\delta$

$t^{2k} \otimes \mathbb{C}(F_1 + F_2)$ with weight $-\alpha_1 + 2k\delta$

$t^{2k+1} \otimes \mathbb{C}(F_1 - F_2)$ with weight $-\alpha_1 + (2k+1)\delta$

$t^{2k+1} \otimes \mathbb{C}F_{\alpha_1+\alpha_2}$ with weight $-2\alpha_1 + (2k+1)\delta$.

Corollary 18.15 (i) *The multiplicities of the imaginary roots $k\delta$ of $\tilde{C}'_l$ are equal to l.*
(ii) *The multiplicities of the imaginary roots $k\delta$ of $\tilde{A}'_1$ are equal to 1.*

Proof. (i) The multiplicity of $2k\delta$ is $\dim (H^0)^\sigma$, which is equal to l. The multiplicity of $(2k+1)\delta$ is $\dim (H^0)_{-1}$, which is also equal to l.
(ii) The same applies to $\tilde{A}'_1$ when $l = 1$. □

18.4 Realisations of twisted affine algebras

We note that the isomorphism between $L(A')$ and $\hat{\mathcal{L}}(A_{2l})^\tau$ does not map the standard invariant form $\langle,\rangle'$ on $L(A')$ to the restriction of the standard invariant form $\langle,\rangle$ on $\hat{\mathcal{L}}(A_{2l})$. For $h_l \in L(A')$ corresponds to $1 \otimes 2(H_l + H_{l+1})$ in $\hat{\mathcal{L}}(A_{2l})$. We have

$$\langle h_l, h_l \rangle' = 2 \frac{a_l}{c_l} = 4$$

$$\langle 1 \otimes 2(H_l + H_{l+1}), 1 \otimes 2(H_l + H_{l+1}) \rangle = 4 \langle H_l + H_{l+1}, H_l + H_{l+1} \rangle = 8.$$

Thus the form is not preserved by the isomorphism.

Also the canonical central element $c \in \hat{\mathcal{L}}(A_{2l})^\tau$ does not map to the canonical central element $c' \in L(A')$. Since we showed that $\sum_{i=0}^{l} c_i h_i = 2c$ it follows that $2c$ corresponds to c' under our isomorphism.

Comments on notation

An alternative notation is sometimes given to the affine Kac–Moody algebras of twisted type, based on the results of this chapter. The twisted affine algebra can be specified by the type of the untwisted affine algebra from which it is obtained, together with the order of the automorphism of which it is the fixed point subalgebra. This is the notation used by Kac in his book *Infinite Dimensional Lie Algebras*. The alternative notation in each case is shown below.

$\tilde{B}_l^t$	$^2\tilde{A}_{2l-1}$	$l \geq 3$
$\tilde{C}_l^t$	$^2\tilde{D}_{l+1}$	$l \geq 2$
$\tilde{F}_4^t$	$^2\tilde{E}_6$	
$\tilde{G}_2^t$	$^3\tilde{D}_4$	
$\tilde{C}_l'$	$^2\tilde{A}_{2l}$	$l \geq 2$
$\tilde{A}_1'$	$^2\tilde{A}_2$	

19
Some representations of symmetrisable Kac–Moody algebras

19.1 The category $\mathcal{O}$ of $L(A)$-modules

We now turn to the representation theory of Kac–Moody algebras. We shall not consider arbitrary representations, but restrict attention to those in the category $\mathcal{O}$ introduced by Bernstein, Gelfand and Gelfand. Let

$$L(A) = N^- \oplus H \oplus N$$

be a Kac–Moody algebra and V be an $L(A)$-module. We say that V is an object in the category $\mathcal{O}$ if the following conditions are satisfied:

(i) $V = \bigoplus_{\lambda \in H^*} V_\lambda$ where $V_\lambda = \{v \in V \,;\, xv = \lambda(x)v \text{ for all } x \in H\}$
(ii) $\dim V_\lambda$ is finite for each $\lambda \in H^*$
(iii) there exists a finite set $\lambda_1, \ldots, \lambda_s \in H^*$ such that each λ with $V_\lambda \neq O$ satisfies $\lambda \prec \lambda_i$ for some $i \in \{1, \ldots, s\}$.

The morphisms in category $\mathcal{O}$ are the homomorphisms of $L(A)$-modules.

Thus each module in $\mathcal{O}$ is a direct sum of its weight spaces and these weight spaces are finite dimensional. Moreover all the weights are bounded above by finitely many elements of H^*.

We now give some examples of modules in category $\mathcal{O}$. For each $\lambda \in H^*$ we may define the Verma module $M(\lambda)$ with highest weight λ. This is defined in a manner analogous to that in which we defined Verma modules for finite dimensional Lie algebras in Section 10.1. Let $\mathfrak{U}(L(A))$ be the universal enveloping algebra of $L(A)$ and K_λ be the left ideal of $\mathfrak{U}(L(A))$ generated by N and $x - \lambda(x)$ for all $x \in H$. Thus

$$K_\lambda = \mathfrak{U}(L(A))N + \sum_{x \in H} \mathfrak{U}(L(A))(x - \lambda(x)).$$

Then $M(\lambda) = \mathfrak{U}(L(A))/K_\lambda$ is an $L(A)$-module called the Verma module with highest weight λ. Let $m_\lambda \in M(\lambda)$ be defined by $m_\lambda = 1 + K_\lambda$. Then, just as in Theorem 10.6, we see that each element of $M(\lambda)$ is uniquely expressible in the form um_λ for some $u \in \mathfrak{U}(N^-)$. Also, as in Theorem 10.7, we have

$$M(\lambda) = \bigoplus_{\mu \in H^*} M(\lambda)_\mu$$

$$M(\lambda)_\mu \neq 0 \quad \text{if and only if } \mu \prec \lambda$$

$$\dim M(\lambda)_\mu = \mathfrak{P}(\lambda - \mu).$$

This shows that $M(\lambda) \in \mathcal{O}$. The finite set of weights giving an upper bound for all weights can be taken in this case to have just one element λ.

Lemma 19.1 (i) *If $V \in \mathcal{O}$ and U is a submodule of V then $U \in \mathcal{O}$ and $V/U \in \mathcal{O}$.*
(ii) *If $V_1, V_2 \in \mathcal{O}$ then $V_1 \oplus V_2 \in \mathcal{O}$ and $V_1 \otimes V_2 \in \mathcal{O}$.*

Proof. (i) We have $V = \bigoplus_{\lambda \in H^*} V_\lambda$. The argument of Theorem 10.9 shows that $U_\lambda = U \cap V_\lambda$ and $U = \bigoplus_{\lambda \in H^*} U_\lambda$. It follows that $U \in \mathcal{O}$. Moreover we have $(V/U)_\lambda = V_\lambda/U_\lambda$ and $V/U = \bigoplus_{\lambda \in H^*} (V/U)_\lambda$. It follows that $V/U \in \mathcal{O}$.
(ii) We have $(V_1 \oplus V_2)_\lambda = (V_1)_\lambda \oplus (V_2)_\lambda$ and $V_1 \oplus V_2 = \bigoplus_{\lambda \in H^*} (V_1 \oplus V_2)_\lambda$. It follows that $V_1 \oplus V_2 \in \mathcal{O}$.

Now consider $V_1 \otimes V_2$. We have

$$V_1 = \bigoplus_{\lambda_1 \in H^*} (V_1)_{\lambda_1}, \quad V_2 = \bigoplus_{\lambda_2 \in H^*} (V_2)_{\lambda_2}$$

thus

$$V_1 \otimes V_2 = \bigoplus_{\lambda_1, \lambda_2} \left((V_1)_{\lambda_1} \otimes (V_2)_{\lambda_2}\right).$$

Now $(V_1)_{\lambda_1} \otimes (V_2)_{\lambda_2} \subset (V_1 \otimes V_2)_{\lambda_1 + \lambda_2}$. Hence $V_1 \otimes V_2 = \bigoplus_{\lambda \in H^*} (V_1 \otimes V_2)_\lambda$ where

$$(V_1 \otimes V_2)_\lambda = \sum_{\substack{\lambda_1, \lambda_2 \\ \lambda_1 + \lambda_2 = \lambda}} \left((V_1)_{\lambda_1} \otimes (V_2)_{\lambda_2}\right).$$

Now there exist $\xi_i \in H^*, i = 1, \ldots, s_1$, such that $\lambda_1 \prec \xi_i$ for some i. Also there exist $\eta_j \in H^*, j = 1, \ldots, s_2$ such that $\lambda_2 \prec \eta_j$ for some j. Thus $\lambda = \lambda_1 + \lambda_2 \prec \xi_i + \eta_j$ for some pair (i, j). We have

$$(\xi_i + \eta_j) - \lambda = (\xi_i - \lambda_1) + (\eta_j - \lambda_2).$$

The expressions $(\xi_i + \eta_j) - \lambda$, $\xi_i - \lambda_1$, $\eta_j - \lambda_2$ are all non-negative integral combinations of the fundamental roots. Thus for given i, j the $(\xi_i + \eta_j) - \lambda$ has only finitely many such decompositions. It follows that for each λ with $(V_1 \otimes V_2)_\lambda \neq 0$ there exist only finitely many pairs λ_1, λ_2 with $\lambda_1 + \lambda_2 = \lambda$, $(V_1)_{\lambda_1} \neq 0$, $(V_2)_{\lambda_2} \neq 0$. It follows from this that $V_1 \otimes V_2 \in \mathcal{O}$. □

Now each $L(A)$-module $V \in \mathcal{O}$ admits a character ch V. We recall from Section 12.1 that ch V is the function from H^* to $\mathbb{Z}$ defined by

$$(\operatorname{ch} V)(\lambda) = \dim V_\lambda.$$

We also recall from Section 12.1 the definition of the ring $\mathfrak{R}$ of functions from H^* to $\mathbb{Z}$. A function $f : H^* \to \mathbb{Z}$ lies in $\mathfrak{R}$ if there exists a finite set $\lambda_1, \ldots, \lambda_s \in H^*$ such that

$$\operatorname{Supp} f \subset S(\lambda_1) \cup \cdots \cup S(\lambda_s)$$

where $S(\lambda) = \operatorname{Supp}(\operatorname{ch} M(\lambda))$. It follows from the definition of category $\mathcal{O}$ that ch $V \in \mathfrak{R}$ for all $V \in \mathcal{O}$.

In Proposition 12.4 we obtained a formula for the character of a Verma module for a finite dimensional semisimple Lie algebra. We now generalise this result to Verma modules for Kac–Moody algebras. We recall that the function $e_\lambda : H^* \to \mathbb{Z}$ was defined by $e_\lambda(\lambda) = 1$ and $e_\lambda(\mu) = 0$ if $\lambda \neq \mu$. The characteristic functions e_λ lie in $\mathfrak{R}$ and any function $f \in \mathfrak{R}$ can be written in the form

$$f = \sum_{\lambda \in H^*} f(\lambda) e_\lambda$$

where the sum may be infinite.

Proposition 19.2 *Let $M(\lambda)$ be a Verma module for the Kac–Moody algebra $L(A)$. Then*

$$\operatorname{ch} M(\lambda) = \frac{e_\lambda}{\prod_{\alpha \in \Phi^+} (1 - e_{-\alpha})^{m_\alpha}}$$

where m_α is the multiplicity of α.

Proof. We use the fact that the map $u \to u m_\lambda$ is a bijection between $\mathfrak{U}(N^-)$ and $M(\lambda)$. This bijection maps the weight space $\mathfrak{U}(N^-)_{-\mu}$ to the weight space $M(\lambda)_{\lambda - \mu}$.

For each $\alpha \in \Phi^+$ we have $\dim (N^-)_{-\alpha} = m_\alpha$. Let $e_{-\alpha,i}$, $1 \le i \le m_\alpha$, be a basis of $(N^-)_{-\alpha}$. We choose an order on these basis elements for all α, i. We then obtain a PBW-basis of $\mathfrak{U}(N^-)$ consisting of all products

$$\prod_\alpha \prod_{i=1}^{m_\alpha} e_{-\alpha,i}^{n_{\alpha,i}}$$

19.1 The category $\mathcal{O}$ of $L(A)$-modules

with $n_{\alpha,i} \in \mathbb{Z}$ and $n_{\alpha,i} \geq 0$. Thus the weight space $\mathfrak{U}(N^-)_{-\mu}$ has a basis consisting of the above elements which satisfy

$$\sum_{\alpha \in \Phi^+} \left(\sum_{i=1}^{m_\alpha} n_{\alpha,i} \right) \alpha = \mu.$$

This shows that the character of $\mathfrak{U}(N^-)$ is

$$\operatorname{ch} \mathfrak{U}(N^-) = \prod_{\alpha \in \Phi^+} \left(1 + e_{-\alpha} + e_{-\alpha}^2 + \cdots\right)^{m_\alpha}$$

since the number of times $e_{-\mu}$ appears on the right-hand side is the number of sets $(n_{\alpha,i})$ of non-negative integers such that

$$\sum_{\alpha \in \Phi^+} \left(\sum_{i=1}^{m_\alpha} n_{\alpha,i} \right) \alpha = \mu.$$

Hence the character of $M(\lambda)$ is

$$\operatorname{ch} M(\lambda) = e_\lambda \prod_{\alpha \in \Phi^+} \left(1 + e_{-\alpha} + e_{-\alpha}^2 + \cdots\right)^{m_\alpha}.$$

Now the element $1 + e_{-\alpha} + e_{-\alpha}^2 + \cdots \in \mathfrak{R}$ has inverse $1 - e_{-\alpha} \in \mathfrak{R}$. Thus we have

$$\operatorname{ch} M(\lambda) = \frac{e_\lambda}{\prod_{\alpha \in \Phi^+} (1 - e_{-\alpha})^{m_\alpha}} \qquad \square$$

Of course in the special case when $L(A)$ is finite dimensional this formula reduces to that obtained in Proposition 12.4. In the general case there are two differences – the roots need not have multiplicity 1 and the product over the positive roots can be an infinite product.

Now the Verma module $M(\lambda)$ for $L(A)$ has a unique maximal submodule $J(\lambda)$, just as in the proof of Theorem 10.9. We define

$$L(\lambda) = M(\lambda)/J(\lambda).$$

Then $L(\lambda)$ is an irreducible $L(A)$-module in the category $\mathcal{O}$.

Proposition 19.3 *The modules $L(\lambda)$ for $\lambda \in H^*$ are the only irreducible modules in category $\mathcal{O}$.*

Proof. Let V be an irreducible $L(A)$-module with $V \in \mathcal{O}$. The definition of $\mathcal{O}$ shows that V has a maximal weight λ under the partial ordering $\prec$. Let $v_\lambda \in V$ be a weight vector with weight λ. Then $xv_\lambda = 0$ for $x \in N$ and $xv_\lambda = \lambda(x)v_\lambda$ for $x \in H$.

We may define a map from the Verma module $M(\lambda)$ into V as follows. Each element of $M(\lambda)$ has a unique expression of form um_λ for $u \in \mathfrak{U}(N^-)$. Let $\theta : M(\lambda) \to V$ be defined by $\theta(um_\lambda) = uv_\lambda$ for $u \in \mathfrak{U}(N^-)$. It may then be shown, just as in the proof of Proposition 10.13, that θ is a homomorphism of $\mathfrak{U}(L(A))$-modules. The image of θ is a submodule of V containing v_λ, so is the whole of V since V is irreducible. Thus the kernel of θ is a maximal submodule of $M(\lambda)$, so must be $J(\lambda)$. Thus V is isomorphic to $M(\lambda)/J(\lambda) = L(\lambda)$. □

Now in Theorem 12.16 we showed that each Verma module $M(\lambda)$ for a finite dimensional semisimple Lie algebra L has a finite composition series. The proof of this result made extensive use of the fact that L is finite dimensional, and the result does not carry over to Verma modules for Kac–Moody algebras $L(A)$. For example it can be shown that the Verma module $M(0)$ has no irreducible submodule when $L(A)$ is infinite dimensional.

We would nevertheless like to define the multiplicity $[V : L(\lambda)]$ of the irreducible module $L(\lambda)$ in the module $V \in \mathcal{O}$. If V had a finite composition series $[V : L(\lambda)]$ would be the number of composition factors isomorphic to $L(\lambda)$ in a given composition series, and this would be independent of the choice of composition series by the Jordan–Hölder theorem. However, V does not in general have a finite composition series. Even so, Kac found a way of defining the multiplicity $[V : L(\lambda)]$. This makes use of the following lemma.

Lemma 19.4 *Let $V \in \mathcal{O}$ and $\lambda \in H^*$. Then V has a filtration*

$$V = V_0 \supset V_1 \supset \cdots \supset V_t = O$$

of finite length by means of a sequence of submodules such that each factor V_{i-1}/V_i either is isomorphic to $L(\mu)$ for some $\mu \succ \lambda$ or has the property that $(V_{i-1}/V_i)_\mu = O$ for all $\mu \succ \lambda$.

Proof. The definition of $\mathcal{O}$ shows that V has only finitely many weights μ with $\mu \succ \lambda$. Thus

$$a(V, \lambda) = \sum_{\mu \succ \lambda} \dim V_\mu$$

is finite. We shall prove the lemma by induction on $a(V, \lambda)$. If $a(V, \lambda) = 0$ then $V = V_0 \supset V_1 = O$ is the required filtration. So suppose $a(V, \lambda) > 0$. Then V has a weight μ with $\mu \succ \lambda$. We may choose a maximal weight μ with $\mu \succ \lambda$. Let $v_\mu \in V$ be a weight vector with weight μ. Then $xv_\mu = 0$ for $x \in N$ and $xv_\mu = \mu(x)v_\mu$ for $x \in H$. Let $U = \mathfrak{U}(L(A))v_\mu$ be the submodule of V

generated by v_μ. We then have a map $\theta : M(\mu) \to U$ defined by $\theta(um_\mu) = uv_\mu$ for $u \in \mathfrak{U}(N^-)$, and θ is a homomorphism of $\mathfrak{U}(L(A))$-modules as before, as shown in the proof of Proposition 10.13. Moreover θ is surjective. Thus U is isomorphic to a factor module of the Verma module $M(\mu)$, and so has a unique maximal submodule $\bar{U}$. We also have

$$U/\bar{U} \cong M(\mu)/J(\mu) \cong L(\mu).$$

Now consider the filtration

$$V \supset U \supset \bar{U} \supset O.$$

We have $a(\bar{U}, \lambda) < a(V, \lambda)$ and $a(V/U, \lambda) < a(V, \lambda)$, since the weight $\mu \succ \lambda$ appears in $U/\bar{U}$. Thus by induction we obtain filtrations for the modules $\bar{U} \in \mathcal{O}$ and $V/U \in \mathcal{O}$ of the required kind, and these may be combined to give the required filtration of V. □

Lemma 19.5 *Let $V \in \mathcal{O}$ and $\lambda \in H^*$. Consider filtrations of the type given in Lemma 19.4 with respect to λ. Let $\mu \in H^*$ satisfy $\mu \succ \lambda$. Then the number of factors $L(\mu)$ in such a filtration is independent of the choice of filtration and also of the choice of λ.*

Proof. We first observe that a filtration with respect to λ is also a filtration with respect to μ when $\mu \succ \lambda$. Also the multiplicity of $L(\mu)$ in such a filtration is the same whether it is regarded as a filtration with respect to λ or μ. Thus to prove the lemma it will be sufficient to take two filtrations with respect to μ and show that $L(\mu)$ has the same multiplicity in each.

The following variant of the proof of the Jordan–Hölder theorem achieves this. Let

$$V = V_0 \supset V_1 \supset \cdots \supset V_{l_1} = O \qquad (19.1)$$

$$V = V_0' \supset V_1' \supset \cdots \supset V_{l_2}' = O \qquad (19.2)$$

be two such filtrations of lengths l_1, l_2. We shall use induction on $\min(l_1, l_2)$.

Suppose first that $\min(l_1, l_2) = 1$. Then either V is irreducible and the two filtrations are identical, or μ is not a weight of V and $L(\mu)$ does not appear in either filtration.

Thus suppose $\min(l_1, l_2) > 1$. We suppose first that $V_1 = V_1'$. We then consider the two filtrations

$$V_1 \supset \cdots \supset V_{l_1} = O$$

$$V_1' \supset \cdots \supset V_{l_2}' = O$$

of V_1. By induction they give the same multiplicity for $L(\mu)$, and the filtrations for V are obtained by adding the additional factor V/V_1 which is the same for both.

We may therefore suppose that $V_1 \ne V_1'$. Suppose first that one contains the other, say $V_1 \subset V_1'$. Then V/V_1 is not irreducible and so μ is not a weight of V/V_1. Thus neither V/V_1 nor V/V_1' is isomorphic to $L(\mu)$. Let

$$V_1 \supset U_1 \supset \cdots \supset U_m = O$$

be a filtration of V_1 of the required type with respect to μ. We then consider the filtrations

$$V \supset V_1 \supset U_1 \supset \cdots \supset U_m = O \tag{19.3}$$

$$V \supset V_1' \supset V_1 \supset U_1 \supset \cdots \supset U_m = O. \tag{19.4}$$

These are filtrations of V of the required type with respect to μ. $L(\mu)$ has the same multiplicity in filtrations (19.1), (19.3) since they have the same leading term V_1. Similarly $L(\mu)$ has the same multiplicity in filtrations (19.2), (19.4). So $L(\mu)$ has the same multiplicity in filtrations (19.3), (19.4) since none of V/V_1, V/V_1', V_1'/V_1 is isomorphic to $L(\mu)$. Thus $L(\mu)$ has the same multiplicity in filtrations (19.1), (19.2) as required.

We may therefore assume that neither of V_1, V_1' is contained in the other. Let $U = V_1 \cap V_1'$ and choose a filtration of U of the required kind with respect to μ. This has form

$$U \supset U_1 \supset \cdots \supset U_m = O.$$

We then consider the filtrations

$$V \supset V_1 \supset U \supset U_1 \supset \cdots \supset U_m = O \tag{19.5}$$

$$V \supset V_1' \supset U \supset U_1 \supset \cdots \supset U_m = O. \tag{19.6}$$

These are filtrations of V of the required type with respect to μ. This is clear since

$$V_1/U \cong (V_1 + V_1')/V_1', \quad V_1'/U \cong (V_1 + V_1')/V_1.$$

Now $L(\mu)$ has the same multiplicity in filtrations (19.1), (19.5) and the same multiplicity in filtrations (19.2), (19.6) since the leading terms are the same. It is therefore sufficient to show that $L(\mu)$ has the same multiplicity in filtrations (19.5), (19.6). These filtrations differ only in the first two factors. If $V_1 + V_1' = V$ then we have

$$V/V_1 \cong V_1'/U, \quad V/V_1' \cong V_1/U$$

as required. If $V_1 + V_1' \neq V$ then V/V_1 and V/V_1' are not irreducible. In this case μ is not a weight of V/V_1 or V/V_1', so is not a weight of V_1/U. Thus none of V/V_1, V_1/U, V/V_1', V_1'/U is isomorphic to $L(\mu)$. This completes the proof. □

Definition The **multiplicity** of $L(\mu)$ in a filtration of $V \in \mathcal{O}$ of the type considered in Lemmas 19.4 and 19.5 will be denoted by $[V : L(\mu)]$.

Of course this agrees with the previous definition of $[V : L(\mu)]$ in the case when V has a composition series of finite length.

Proposition 19.6 Let $V \in \mathcal{O}$. Then

$$\operatorname{ch} V = \sum_{\lambda \in H^*} [V : L(\lambda)] \operatorname{ch} L(\lambda).$$

Proof. Both sides are functions $H^* \to \mathbb{Z}$. We have $(\operatorname{ch} V)(\mu) = \dim V_\mu$ and the right-hand side evaluated at μ is

$$\sum_{\lambda \in H^*} [V : L(\lambda)] \dim L(\lambda)_\mu.$$

We choose a filtration of V with respect to μ of the type given in Lemma 19.4. Each factor either is isomorphic to $L(\lambda)$ for some $\lambda \succ \mu$ or does not contain μ as a weight. The multiplicity of $L(\lambda)$ as a factor is $[V : L(\lambda)]$. Hence we have

$$\dim V_\mu = \sum_\lambda [V : L(\lambda)] \dim L(\lambda)_\mu$$

summed over all $\lambda \succ \mu$. We may in fact take the sum over all $\lambda \in H^*$ since $\dim L(\lambda)_\mu = 0$ unless $\lambda \succ \mu$. □

19.2 The generalised Casimir operator

We recall from Section 11.6 that, if L is a finite dimensional semisimple Lie algebra, the Casimir element of the centre of the enveloping algebra $\mathfrak{U}(L)$ plays an important role in the representation theory of L. If $x_1, \ldots, x_m$ are any basis of L and $y_1, \ldots, y_m$ are the dual basis with respect to the Killing form the Casimir element is given by

$$\sum x_i y_i \in \mathfrak{U}(L).$$

We showed in Proposition 11.36 that the Casimir element acts on a Verma module $M(\lambda)$ for L as scalar multiplication by $\langle \lambda + \rho, \lambda + \rho \rangle - \langle \rho, \rho \rangle$ where $\langle , \rangle$ is the Killing form and ρ is, as usual, the element of H^* given by $\rho(h_i) = 1$ for $i = 1, \ldots, l$.

Now let $L(A)$ be a Kac–Moody algebra where A is symmetrisable. We cannot define an analogous Casimir element $\sum x_i y_i$ in $\mathfrak{U}(L(A))$ since the sum will in general be infinite and make no sense. It was shown by Kac, however, that it is possible to define an operator $c : V \to V$ on any $L(A)$-module V in category $\mathcal{O}$ which has properties analogous to the action of the Casimir element for finite dimensional algebras. In order to define Kac' operator on V we recall the formula for the Casimir element of a finite dimensional algebra given in Proposition 11.35. Let $h'_1, \ldots, h'_l$ be a basis of H and $h''_1, \ldots, h''_l$ be the dual basis of H with respect to the Killing form of L. Choose elements $e_\alpha \in L_\alpha, f_\alpha \in L_{-\alpha}$ such that $[e_\alpha f_\alpha] = h'_\alpha$ for each $\alpha \in \Phi^+$. Then the Casimir element of $\mathfrak{U}(L)$ is given by

$$\sum_{i=1}^{l} h'_i h''_i + \sum_{\alpha \in \Phi^+} h'_\alpha + 2 \sum_{\alpha \in \Phi^+} f_\alpha e_\alpha.$$

Since $\sum_{\alpha \in \Phi^+} \alpha = 2\rho$ this element can also be written

$$\sum_{i=1}^{l} h'_i h''_i + 2h'_\rho + 2 \sum_{\alpha \in \Phi^+} f_\alpha e_\alpha$$

where $h'_\rho \in H$ satisfies $\rho(x) = \langle h'_\rho, x \rangle$ for all $x \in H$.

We wish to define an analogous element for the symmetrisable Kac–Moody algebra $L(A)$. The root space L_α of $L(A)$ need not be 1-dimensional, so we choose a basis $e_\alpha^{(1)}, e_\alpha^{(2)}, \ldots$ for L_α. Instead of using the Killing form we use the standard invariant bilinear form on $L(A)$. (In the case when $L(A)$ is finite dimensional this is a scalar multiple of the Killing form.) We recall from Corollary 16.5 that the pairing $L_\alpha \times L_{-\alpha} \to \mathbb{C}$ given by $x, y \to \langle x, y \rangle$ is non-degenerate. Thus we may choose a corresponding dual basis $f_\alpha^{(1)}, f_\alpha^{(2)}, \ldots$ for $L_{-\alpha}$ such that

$$\langle e_\alpha^{(i)}, f_\alpha^{(j)} \rangle = \delta_{ij}.$$

We choose a basis $h'_1, h'_2, \ldots$ of H and let $h''_1, h''_2, \ldots$ be the dual basis of H satisfying $\langle h'_i, h''_j \rangle = \delta_{ij}$. Since the fundamental coroots $h_1, \ldots, h_n \in H$ are linearly independent there exists $\rho \in H^*$ such that $\rho(h_i) = 1$ for $i = 1, \ldots, n$. However, ρ is not in general uniquely determined by this condition. So we choose any element $\rho \in H^*$ satisfying $\rho(h_i) = 1$ for $i = 1, \ldots, n$. We then have a corresponding element $h'_\rho \in H$ such that $\rho(x) = \langle h'_\rho, x \rangle$ for all $x \in H$. We then consider the expression

$$\sum_i h'_i h''_i + 2h'_\rho + 2 \sum_{\alpha \in \Phi^+} \sum_i f_\alpha^{(i)} e_\alpha^{(i)}.$$

This element does not make sense as an element of $\mathfrak{U}(L(A))$ in general since the sum over $\alpha \in \Phi^+$ may be infinite. However, if V is an $L(A)$-module in

19.2 The generalised Casimir operator

$\mathcal{O}$ we know that $\operatorname{ch} V \in \mathfrak{R}$ and so there exist only finitely many $\alpha \in \Phi^+$ such that $L_\alpha V \neq 0$. Thus the operator $\Omega : V \to V$ given by

$$\Omega = \sum_i h'_i h''_i + 2h'_\rho + 2 \sum_{\alpha \in \Phi^+} \sum_i f_\alpha^{(i)} e_\alpha^{(i)}$$

is well defined. It is straightforward to check that this operator $\Omega : V \to V$ does not depend on the choice of dual bases $h'_1, h'_2, \ldots \quad h''_1, h''_2, \ldots$ of H or on the choice of dual bases $e_\alpha^{(i)}, f_\alpha^{(i)}$ for L_α and $L_{-\alpha}$. It may, however, depend upon the choice of ρ.

Definition *The operator $\Omega : V \to V$ for $V \in \mathcal{O}$ is called the **generalised Casimir operator** on V with respect to ρ.*

In the case of a finite dimensional semisimple Lie algebra the Casimir element lies in the centre of the universal enveloping algebra. We shall prove an analogous result in the present situation, i.e. that the generalised Casimir operator commutes with the action on $V \in \mathcal{O}$ of any element of $\mathfrak{U}(L(A))$. We first need some preliminary results.

Lemma 19.7 *Let $\alpha, \beta, \beta - \alpha \in \Phi$. Suppose $e_\alpha^{(i)}, f_\alpha^{(i)}$ are dual bases of $L_\alpha, L_{-\alpha}$ and $e_\beta^{(i)}, f_\beta^{(i)}$ are dual bases of $L_\beta, L_{-\beta}$. Let $x \in L_{\beta - \alpha}$. Then in the vector space $L(A) \otimes L(A)$ we have*

$$\sum_i f_\alpha^{(i)} \otimes [x, e_\alpha^{(i)}] = \sum_i \left[f_\beta^{(i)}, x \right] \otimes e_\beta^{(i)}.$$

Proof. We note that both sides lie in the subspace $L_{-\alpha} \otimes L_\beta$. We define a bilinear form on $L(A) \otimes L(A)$, uniquely determined by

$$\langle x_1 \otimes y_1, x_2 \otimes y_2 \rangle = \langle x_1, x_2 \rangle \langle y_1, y_2 \rangle.$$

Since the standard invariant form is non-degenerate on $L(A)$ this bilinear form will be non-degenerate on $L(A) \otimes L(A)$.

Let $a \otimes b \in L_\gamma \otimes L_\delta$. Then the scalar products of both sides of the required equation with $a \otimes b$ are zero unless $\gamma = \alpha$ and $\delta = -\beta$. We therefore suppose $\gamma = \alpha$ and $\delta = -\beta$ and consider the scalar products

$$\left\langle \sum_i f_\alpha^{(i)} \otimes [x, e_\alpha^{(i)}], a \otimes b \right\rangle$$

$$\left\langle \sum_i \left[f_\beta^{(i)}, x \right] \otimes e_\beta^{(i)}, a \otimes b \right\rangle.$$

We have

$$\left\langle \sum_i f_\alpha^{(i)} \otimes [x, e_\alpha^{(i)}], a \otimes b \right\rangle = \sum_i \langle f_\alpha^{(i)}, a \rangle \langle [x, e_\alpha^{(i)}], b \rangle$$

$$= -\sum_i \langle f_\alpha^{(i)}, a \rangle \langle e_\alpha^{(i)}, [xb] \rangle$$

$$= -\langle a, [xb] \rangle$$

since $e_\alpha^{(i)}, f_\alpha^{(i)}$ are dual bases of $L_\alpha, L_{-\alpha}$. Similarly

$$\left\langle \sum_i [f_\beta^{(i)}, x] \otimes e_\beta^{(i)}, a \otimes b \right\rangle = \sum_i \langle [f_\beta^{(i)}, x], a \rangle \langle e_\beta^{(i)}, b \rangle$$

$$= \sum_i \langle f_\beta^{(i)}, [x, a] \rangle \langle e_\beta^{(i)}, b \rangle$$

$$= \langle [xa], b \rangle$$

$$= -\langle a, [xb] \rangle.$$

Thus the two sides of our equation have the same scalar product with each $a \otimes b \in L_\alpha \otimes L_{-\beta}$. Since the form is non-degenerate on $L(A) \otimes L(A)$ this shows the two sides are equal. □

Corollary 19.8 *In the enveloping algebra* $\mathfrak{U}(L(A))$ *we have*

$$\sum_i f_\alpha^{(i)} [x, e_\alpha^{(i)}] = \sum_i [f_\beta^{(i)}, x] e_\beta^{(i)}.$$

Proof. We apply the natural homomorphism from the tensor algebra $T(L(A))$ to $\mathfrak{U}(L(A))$. The result then follows from Lemma 19.7. □

Theorem 19.9 *Let* $u \in \mathfrak{U}(L(A))$ *and* $V \in \mathcal{O}$. *Then the maps* $\Omega : V \to V$ *and* $u : V \to V$ *commute.*

Proof. The algebra $\mathfrak{U}(L(A))$ is generated by e_i, f_i for $i = 1, \ldots, n$ and the elements of H. If $x \in H$ then x commutes with each term $f_\alpha^{(i)} e_\alpha^{(i)}$ in $\mathfrak{U}(L(A))$ since this term has weight 0. Thus $x : V \to V$ commutes with $f_\alpha^{(i)} e_\alpha^{(i)} : V \to V$ and hence with $\Omega : V \to V$. It is therefore sufficient to show that $\Omega : V \to V$ commutes with $e_i : V \to V$ and $f_i : V \to V$.

19.2 The generalised Casimir operator

We consider the element $\left[\sum_j f_\alpha^{(j)} e_\alpha^{(j)}, e_i\right]$ of $\mathfrak{U}(L(A))$. We have

$$\left[\sum_j f_\alpha^{(j)} e_\alpha^{(j)}, e_i\right] = \sum_j [f_\alpha^{(j)}, e_i] e_\alpha^{(j)} + \sum_j f_\alpha^{(j)} [e_\alpha^{(j)}, e_i]$$

$$= \sum_j [f_\alpha^{(j)}, e_i] e_\alpha^{(j)} - \sum_j f_\alpha^{(j)} [e_i, e_\alpha^{(j)}]$$

$$= \sum_j [f_\alpha^{(j)}, e_i] e_\alpha^{(j)} - \sum_j [f_{\alpha+\alpha_i}^{(j)}, e_i] e_{\alpha+\alpha_i}^{(j)}$$

by Corollary 19.8. If $\alpha + \alpha_i \notin \Phi$ the second term is interpreted as 0. We show that

$$\sum_{\substack{\alpha \in \Phi^+ \\ \alpha \neq \alpha_i}} \left(\sum_j f_\alpha^{(j)} e_\alpha^{(j)}\right) : V \to V$$

commutes with $e_i : V \to V$. We have

$$\left[\sum_{\substack{\alpha \in \Phi^+ \\ \alpha \neq \alpha_i}} \left(\sum_j f_\alpha^{(j)} e_\alpha^{(j)}\right), e_i\right] = \sum_{\substack{\alpha \in \Phi^+ \\ \alpha \neq \alpha_i}} \left[\sum_j f_\alpha^{(j)}, e_i\right] e_\alpha^{(j)} - \sum_{\substack{\alpha \in \Phi^+ \\ \alpha \neq \alpha_i}} \sum_j [f_{\alpha+\alpha_i}^{(j)}, e_i] e_{\alpha+\alpha_i}^{(j)}$$

on V. If $\alpha - \alpha_i \notin \Phi$ then $[\sum_j f_\alpha^{(j)}, e_i] = 0$. Thus we may assume $\alpha = \beta + \alpha_i$ in the first term with $\beta \in \Phi^+$ and get

$$\sum_{\substack{\beta \in \Phi^+ \\ \beta \neq \alpha_i}} \left[\sum_j f_{\beta+\alpha_i}^{(j)}, e_i\right] e_{\beta+\alpha_i}^{(j)} - \sum_{\substack{\alpha \in \Phi^+ \\ \alpha \neq \alpha_i}} \sum_j [f_{\alpha+\alpha_i}^{(j)}, e_i] e_{\alpha+\alpha_i}^{(j)} = 0.$$

Since $\Omega = \sum_j h'_j h''_j + 2h'_\rho + 2 \sum_{\alpha \in \Phi^+} \sum_j f_\alpha^{(j)} e_\alpha^{(j)}$ on V it is now sufficient to show that $\sum_j h'_j h''_j + 2h'_\rho + 2f_i e_i$ commutes with e_i on V. In fact these elements commute in $\mathfrak{U}(L(A))$. For we have

$$\left[\sum_j h'_j h''_j, e_i\right] = \sum_j [h'_j, e_i] h''_j + \sum_j h'_j [h''_j, e_i]$$

$$= \sum_j \alpha_i(h'_j) e_i h''_j + \sum_j \alpha_i(h''_j) h'_j e_i$$

$$= e_i \left(\sum_j \alpha_i(h'_j) h''_j + \sum_j \alpha_i(h''_j) h'_j\right) + \left(\sum_j \alpha_i(h''_j) \alpha_i(h'_j)\right) e_i$$

$$= e_i \left(\sum_j \langle h'_{\alpha_i}, h'_j\rangle h''_j + \sum_j \langle h'_{\alpha_i}, h''_j\rangle h'_j\right) + \left(\sum_j \langle h'_{\alpha_i}, h''_j\rangle \langle h'_{\alpha_i}, h'_j\rangle\right) e_i.$$

Since $h'_1, h'_2, \ldots$ and $h''_1, h''_2, \ldots$ are dual bases of H we have

$$\sum_j \langle h'_{\alpha_i}, h'_j \rangle h''_j = \sum_j \langle h'_{\alpha_i}, h''_j \rangle h'_j = h'_{\alpha_i}$$

and

$$\sum_j \langle h'_{\alpha_i}, h''_j \rangle \langle h'_{\alpha_i}, h'_j \rangle = \langle h'_{\alpha_i}, h'_{\alpha_i} \rangle = \langle \alpha_i, \alpha_i \rangle.$$

Hence

$$\left[\sum_j h'_j h''_j, e_i \right] = 2 e_i h'_{\alpha_i} + \langle \alpha_i, \alpha_i \rangle e_i.$$

Secondly we have

$$[2h'_\rho, e_i] = 2\alpha_i(h'_\rho) e_i = 2 \langle h'_{\alpha_i}, h'_\rho \rangle e_i = 2\rho(h'_{\alpha_i}) e_i = \langle \alpha_i, \alpha_i \rangle e_i$$

since $h_i = \dfrac{2 h'_{\alpha_i}}{\langle h'_{\alpha_i}, h'_{\alpha_i} \rangle}$ and $\rho(h_i) = 1$, hence

$$\rho(h'_{\alpha_i}) = \frac{\langle h'_{\alpha_i}, h'_{\alpha_i} \rangle}{2} = \frac{\langle \alpha_i, \alpha_i \rangle}{2}.$$

Thirdly we have

$$[2 f_i e_i, e_i] = 2 [f_i, e_i] e_i = -2 [e_i, f_i] e_i.$$

We recall that e_i, f_i were chosen so that $\langle e_i, f_i \rangle = 1$. By Corollary 16.5 this implies $[e_i f_i] = h'_{\alpha_i}$. Hence

$$[2 f_i e_i, e_i] = -2 h'_{\alpha_i} e_i = -2 e_i h'_{\alpha_i} - 2 \alpha_i(h'_{\alpha_i}) e_i$$
$$= -2 e_i h'_{\alpha_i} - 2 \langle \alpha_i, \alpha_i \rangle e_i.$$

Thus we have shown:

$$\left[\sum_j h'_j h''_j, e_i \right] = 2 e_i h'_{\alpha_i} + \langle \alpha_i, \alpha_i \rangle e_i$$

$$[2 h'_\rho, e_i] = \langle \alpha_i, \alpha_i \rangle e_i$$

$$[2 f_i e_i, e_i] = -2 e_i h'_{\alpha_i} - 2 \langle \alpha_i, \alpha_i \rangle e_i.$$

Hence

$$\left[\sum_j h'_j h''_j + 2 h'_\rho + 2 f_i e_i, e_i \right] = 0.$$

19.2 The generalised Casimir operator

Thus we have shown that $\Omega : V \to V$ commutes with $e_i : V \to V$. The proof that $\Omega : V \to V$ commutes with $f_i : V \to V$ is similar. Using the fact that

$$\left[\sum_j f_\alpha^{(j)} e_\alpha^{(j)}, f_i\right] = \sum_j [f_\alpha^{(j)}, f_i] e_\alpha^{(j)} + \sum_j f_\alpha^{(j)} [e_\alpha^{(j)}, f_i]$$

$$= \sum_j f_{\alpha+\alpha_i}^{(j)} \left[f_i, e_{\alpha+\alpha_i}^{(j)}\right] - \sum_j f_\alpha^{(j)} [f_i, e_\alpha^{(j)}]$$

we deduce as before that

$$\sum_{\substack{\alpha \in \Phi^+ \\ \alpha \neq \alpha_i}} \left(\sum_j f_\alpha^{(j)} e_\alpha^{(j)}\right) : V \to V$$

commutes with $f_i : V \to V$. We also obtain

$$\left[\sum_j h'_j h''_j, f_i\right] = -2 f_i h'_{\alpha_i} + \langle \alpha_i, \alpha_i \rangle f_i$$

$$[2h'_\rho, f_i] = -\langle \alpha_i, \alpha_i \rangle f_i$$

$$[2f_i e_i, f_i] = 2 f_i h'_{\alpha_i}.$$

Hence

$$\left[\sum_j h'_j h''_j + 2h'_\rho + 2f_i e_i, f_i\right] = 0.$$

Thus $\Omega : V \to V$ commutes with $f_i : V \to V$ and the proof is complete. □

We next describe the action of the generalised Casimir operator Ω on a Verma module.

Proposition 19.10 Ω *acts on the Verma module* $M(\lambda)$ *as scalar multiplication by* $\langle \lambda + \rho, \lambda + \rho \rangle - \langle \rho, \rho \rangle$.

Proof. Let m_λ be a highest weight vector of $M(\lambda)$. Then

$$\Omega m_\lambda = \left(\sum_j h'_j h''_j + 2h'_\rho + 2 \sum_{\alpha \in \Phi^+} \sum_j f_\alpha^{(j)} e_\alpha^{(j)}\right) m_\lambda$$

$$= \left(\sum_j \lambda(h'_j) \lambda(h''_j) + 2\lambda(h'_\rho)\right) m_\lambda.$$

Now $\sum_j \lambda(h'_j) \lambda(h''_j) = \langle \lambda, \lambda \rangle$ and $\lambda(h'_\rho) = \langle \lambda, \rho \rangle$. Hence

$$\Omega m_\lambda = (\langle \lambda + \rho, \lambda + \rho \rangle - \langle \rho, \rho \rangle) m_\lambda.$$

Now each element of $M(\lambda)$ has form um_λ for some $u \in \mathfrak{U}(N^-)$. Thus

$$\Omega(um_\lambda) = u(\Omega m_\lambda) = (\langle \lambda + \rho, \lambda + \rho \rangle - \langle \rho, \rho \rangle) um_\lambda,$$

by Theorem 19.9. Hence Ω acts on $M(\lambda)$ as scalar multiplication by

$$\langle \lambda + \rho, \lambda + \rho \rangle - \langle \rho, \rho \rangle. \qquad \square$$

Corollary 19.11 Ω *acts on the irreducible $L(A)$-module $L(\lambda)$ as scalar multiplication by* $\langle \lambda + \rho, \lambda + \rho \rangle - \langle \rho, \rho \rangle$ $\qquad \square$

Note Proposition 19.10 is the analogue of Proposition 11.36 for finite dimensional semisimple Lie algebras. In Proposition 11.36 the invariant form which appeared was the Killing form whereas in Proposition 19.10 and Corollary 19.11 it is the standard invariant form. The difference is explained by the fact that the Casimir element in the enveloping algebra of a finite dimensional semisimple Lie algebra was defined in terms of the Killing form, whereas the generalised Casimir operator was defined in terms of the standard invariant form.

19.3 Kac' character formula

Let X be the set of integral weights $\lambda \in H^*$, that is the set of all λ such that $\lambda(h_i) \in \mathbb{Z}$ for $i = 1, \ldots, n$. Let X^+ be the subset of dominant integral weights, that is the set of weights $\lambda \in X$ such that $\lambda(h_i) \geq 0$ for all i. In this section we shall prove a formula due to Kac for the character of the irreducible $L(A)$-module $L(\lambda)$ when $\lambda \in X^+$. The reason for the restriction to weights in X^+ lies in the fact that the modules $L(\lambda)$ for $\lambda \in X^+$ are integrable.

Definition *An $L(A)$-module V is called* **integrable** *if*

$$V = \bigoplus_{\lambda \in H^*} V_\lambda$$

and if $e_i : V \to V$ and $f_i : V \to V$ are locally nilpotent for all $i = 1, \ldots, n$.

Proposition 19.12 *The adjoint module $L(A)$ is integrable.*

Proof. The proof of Proposition 7.17 carries over to the present situation.
$\qquad \square$

19.3 Kac' character formula

Proposition 19.13 *Let V be an integrable $L(A)$-module. Then $\dim V_\lambda = \dim V_{w(\lambda)}$ for each $\lambda \in H^*$ and each $w \in W$.*

Proof. Since the Weyl group W of $L(A)$ is generated by the elements s_i it is sufficient to show that $\dim V_\lambda = \dim V_{s_i(\lambda)}$.

We may regard V as a module for the 3-dimensional simple subalgebra $\langle e_i, h_i, f_i \rangle$ of $L(A)$. Let $v \in V_\lambda$ and consider the $\langle e_i, h_i, f_i \rangle$-submodule generated by v. The vectors

$$v, \quad e_i v, \quad e_i^2 v, \quad \ldots, \quad e_i^{r-1} v$$

lie in this submodule, where r is the smallest positive integer with $e_i^r v = 0$. The vectors $f_i^b e_i^a v$ also lie in this submodule, and there are only finitely many (a, b) for which such a vector is non-zero. Each such vector is a weight vector in V. However, the relation

$$e_i f_i^n = f_i^n e_i + n f_i^{n-1} (h_i - (n-1))$$

obtained in the proof of Theorem 10.20 shows that the subspace spanned by all vectors $f_i^b e_i^a v$ is an $\langle e_i, h_i, f_i \rangle$-submodule. Hence every weight vector $v \in V$ lies in a finite dimensional $\langle e_i, h_i, f_i \rangle$-submodule which is also an H-module, i.e. it is an $\langle e_i, H, f_i \rangle$-submodule.

Now let U be the subspace of V given by

$$U = \sum_{k \in \mathbb{Z}} V_{\lambda + k\alpha_i}.$$

U is clearly an $\langle e_i, H, f_i \rangle$-submodule of V. The $\langle e_i, H, f_i \rangle$-submodule generated by each weight vector is finite dimensional, thus U is a sum of finite dimensional $\langle e_i, H, f_i \rangle$-submodules. Now $\langle e_i, h_i, f_i \rangle$ is a 3-dimensional simple Lie algebra of type A_1. Thus every finite dimensional $\langle e_i, h_i, f_i \rangle$-module is a direct sum of finite dimensional irreducible $\langle e_i, h_i, f_i \rangle$-modules, by the complete reducibility theorem, Theorem 12.20. The weight spaces involved in such a decomposition of an H-invariant $\langle e_i, h_i, f_i \rangle$-module can be chosen as weight spaces for H, as in the proof of Theorem 10.20, thus every finite dimensional H-invariant $\langle e_i, h_i, f_i \rangle$-module is a direct sum of finite dimensional H-invariant irreducible $\langle e_i, h_i, f_i \rangle$-modules. Thus U is a sum of finite dimensional H-invariant irreducible $\langle e_i, h_i, f_i \rangle$-submodules, so is a direct sum of certain of these submodules. However, for each of these irreducible submodules M we have

$$\dim M_\lambda = \dim M_{s_i(\lambda)}$$

by Proposition 10.22. It follows that

$$\dim V_\lambda = \dim V_{s_i(\lambda)}$$

and the required result follows. □

Proposition 19.14 *Let $L(A)$ be a symmetrisable Kac–Moody algebra and $L(\lambda)$ be an irreducible $L(A)$-module in the category $\mathcal{O}$. Then $L(\lambda)$ is integrable if and only if λ is dominant and integral.*

Proof. Suppose first that $L(\lambda)$ is integrable. Let v_λ be a highest weight vector in $L(\lambda)$. Then $f_i^r v_\lambda = 0$ for some r. Consider the vectors

$$v_\lambda, \quad f_i v_\lambda, \quad \ldots, \quad f_i^{r-1} v_\lambda.$$

Since $e_i f_i^n = f_i^n e_i + n f_i^{n-1}(h_i - (n-1))$ for each n we see that these vectors span an $\langle e_i, h_i, f_i \rangle$-submodule of $L(\lambda)$. The highest weight of this finite dimensional $\langle e_i, h_i, f_i \rangle$-module is λ. But the highest weight of any finite dimensional module for a finite dimensional simple Lie algebra is dominant and integral. Thus $\lambda(h_i) \in \mathbb{Z}$ and $\lambda(h_i) \geq 0$. Since this holds for all i, λ is dominant and integral.

Now suppose conversely that $\lambda(h_i) \in \mathbb{Z}$ and $\lambda(h_i) \geq 0$ for each i. Then we have

$$f_i^{\lambda(h_i)+1} v_\lambda = 0$$

as in the proof of Theorem 10.20. Now each element of $L(\lambda)$ has form uv_λ for some $u \in L(A)$. We have

$$f_i^N(uv_\lambda) = \sum_{k=0}^{N} \binom{N}{k} \left((\operatorname{ad} f_i)^k u\right) \left(f_i^{N-k} v_\lambda\right).$$

Now $(\operatorname{ad} f_i)^k u = 0$ for k sufficiently large since $L(A)$ is integrable, by Proposition 19.12. Also $f_i^{N-k} v_\lambda = 0$ for $N-k$ sufficiently large, as shown above. Thus $f_i^N(uv_\lambda) = 0$ for N sufficiently large, and so $f_i : L(\lambda) \to L(\lambda)$ is locally nilpotent. The fact that $e_i : L(\lambda) \to L(\lambda)$ is locally nilpotent follows from the fact that $L(\lambda)$ lies in category $\mathcal{O}$. Thus $L(\lambda)$ is integrable. □

As before we write

$$X^+ = \{\lambda \in H^* \; ; \; \lambda(h_i) \in \mathbb{Z}, \lambda(h_i) \geq 0 \quad \text{for each } i\}.$$

19.3 Kac' character formula

We now turn to Kac' character formula for $\operatorname{ch} L(\lambda)$ when $\lambda \in X^+$. We recall from Proposition 19.2 that the character of the corresponding Verma module $M(\lambda)$ is given by

$$\operatorname{ch} M(\lambda) = \frac{e_\lambda}{\Delta}$$

where $\Delta = \prod_{\alpha \in \Phi^+} (1 - e_{-\alpha})^{m_\alpha}$ and m_α is the multiplicity of α.

We begin with a lemma.

Lemma 19.15 *Let $X^{++} = \{\lambda \in X \; ; \; \lambda(h_i) > 0 \text{ for all } i\}$. Suppose $\xi \in X^{++}$, $\eta \in X^+$ satisfy $\eta \prec \xi$ and $\langle \eta, \eta \rangle = \langle \xi, \xi \rangle$. Then $\eta = \xi$.*

Proof. Since $\eta \prec \xi$ we have $\xi - \eta = \sum_{i=1}^n k_i \alpha_i$ with $k_i \in \mathbb{Z}$ and $k_i \geq 0$. Thus

$$\langle \xi, \xi \rangle - \langle \eta, \eta \rangle = \langle \xi + \eta, \xi - \eta \rangle$$
$$= \sum_i k_i \langle \xi + \eta, \alpha_i \rangle = \sum_i k_i \frac{\langle \alpha_i, \alpha_i \rangle}{2} (\xi + \eta)(h_i).$$

Now $\langle \alpha_i, \alpha_i \rangle > 0$ and $(\xi + \eta)(h_i) > 0$. Hence $\langle \xi, \xi \rangle - \langle \eta, \eta \rangle = 0$ implies that $k_i = 0$ for each i. Thus $\xi = \eta$. □

Theorem 19.16 *(Kac' character formula). Let $L(A)$ be a symmetrisable Kac–Moody algebra and $L(\lambda)$ be an irreducible $L(A)$-module with $\lambda \in X^+$. Then*

$$\operatorname{ch} L(\lambda) = \frac{\sum_{w \in W} \varepsilon(w) e_{w(\lambda + \rho) - \rho}}{\prod_{\alpha \in \Phi^+} (1 - e_{-\alpha})^{m_\alpha}}.$$

(This is an equality in the ring $\mathfrak{R}$.)

Proof. By Proposition 19.6 we have

$$\operatorname{ch} M(\lambda) = \sum_{\mu \in H^*} [M(\lambda) : L(\mu)] \operatorname{ch} L(\mu).$$

Now all μ for which $[M(\lambda) : L(\mu)] \neq 0$ satisfy $\mu \prec \lambda$. For $L(\mu)$ appears as a factor in some filtration of $M(\lambda)$, so μ is a weight of $M(\lambda)$.

We consider the action of the generalised Casimir operator Ω on $M(\lambda)$. By Proposition 19.10 Ω acts on $M(\lambda)$ as scalar multiplication by $\langle \lambda + \rho, \lambda + \rho \rangle - \langle \rho, \rho \rangle$. Similarly by Corollary 19.11 Ω acts on $L(\mu)$ as scalar multiplication by $\langle \mu + \rho, \mu + \rho \rangle - \langle \rho, \rho \rangle$. Thus if $[M(\lambda) : L(\mu)] \neq 0$ we must have

$$\langle \lambda + \rho, \lambda + \rho \rangle - \langle \rho, \rho \rangle = \langle \mu + \rho, \mu + \rho \rangle - \langle \rho, \rho \rangle$$

that is $\langle \lambda+\rho, \lambda+\rho \rangle = \langle \mu+\rho, \mu+\rho \rangle$. Thus

$$\operatorname{ch} M(\lambda) = \sum_\mu [M(\lambda) : L(\mu)] \operatorname{ch} L(\mu)$$

summed over all $\mu \prec \lambda$ with $\langle \mu+\rho, \mu+\rho \rangle = \langle \lambda+\rho, \lambda+\rho \rangle$. If we take a total ordering on the weights μ satisfying $\mu \prec \lambda$ and $\langle \mu+\rho, \mu+\rho \rangle = \langle \lambda+\rho, \lambda+\rho \rangle$ which is compatible with the partial ordering $\prec$ these equations can be written

$$\operatorname{ch} M(\lambda) = \sum_\mu a_{\lambda\mu} \operatorname{ch} L(\mu)$$

where $(a_{\lambda\mu})$ is an infinite matrix with non-negative integer entries such that $a_{\lambda\lambda} = 1$ and $a_{\lambda\mu} = 0$ for all entries below the diagonal. Such a matrix $(a_{\lambda\mu})$ can be inverted to give a matrix $(b_{\lambda\mu})$ with $b_{\lambda\mu} \in \mathbb{Z}$, $b_{\lambda\lambda} = 1$ and $b_{\lambda\mu} = 0$ for entries below the diagonal. Thus we have

$$\operatorname{ch} L(\lambda) = \sum_\mu b_{\lambda\mu} \operatorname{ch} M(\mu)$$

$$= \sum_\mu b_{\lambda\mu} \frac{e_\mu}{\Delta}.$$

Thus $\Delta \operatorname{ch} L(\lambda) = \sum_\mu b_{\lambda\mu} e_\mu$ and $e_\rho \Delta \operatorname{ch} L(\lambda) = \sum_\mu b_{\lambda\mu} e_{\mu+\rho}$.

We consider the action of the Weyl group on the functions which appear here. Since s_i transforms α_i to $-\alpha_i$ and $\Phi^+ - \{\alpha_i\}$ into itself we have

$$s_i(e_\rho \Delta) = s_i \left(e_\rho (1 - e_{-\alpha_i}) \prod_{\alpha \in \Phi^+ - \{\alpha_i\}} (1 - e_{-\alpha})^{m_\alpha} \right)$$

$$= e_{\rho-\alpha_i} (1 - e_{\alpha_i}) \prod_{\alpha \in \Phi^+ - \{\alpha_i\}} (1 - e_{-\alpha})^{m_\alpha}$$

$$= -e_\rho \Delta$$

since $s_i(\rho) = \rho - \alpha_i$. Hence

$$w(e_\rho \Delta) = \varepsilon(w) \, e_\rho \Delta \qquad \text{for all } w \in W.$$

Also by Proposition 19.13 we have

$$w(\operatorname{ch} L(\lambda)) = \operatorname{ch} L(\lambda) \qquad \text{for all } w \in W,$$

since $L(\lambda)$ is integrable. It follows that

$$w\left(\sum_\mu b_{\lambda\mu} e_{\mu+\rho} \right) = \varepsilon(w) \sum_\mu b_{\lambda\mu} e_{\mu+\rho}.$$

This implies that
$$b_{\lambda\mu} = \varepsilon(w)b_{\lambda\nu}$$
where $w(\mu+\rho) = \nu+\rho$.

Suppose μ is a weight for which $b_{\lambda\mu} \ne 0$. Consider the set of all weights ν for which $w(\mu+\rho) = \nu+\rho$ for some $w \in W$. All such weights satisfy $b_{\lambda\nu} \ne 0$ and we have $\nu \prec \lambda$. Among all such weights ν we can choose one for which the height of $\lambda - \nu$ is minimal. Then $\nu + \rho$ must lie in X^+. For if there existed an i for which $(\nu+\rho)(h_i) < 0$ we would have
$$s_i w(\mu+\rho) = s_i(\nu+\rho) = \nu+\rho - (\nu+\rho)(h_i)\alpha_i$$
contradicting the minimality of ht $(\lambda - \nu)$. Hence $\nu+\rho \in X^+$. We also have
$$\langle \nu+\rho, \nu+\rho \rangle = \langle \mu+\rho, \mu+\rho \rangle = \langle \lambda+\rho, \lambda+\rho \rangle.$$

Thus we have
$$\lambda+\rho \in X^{++}, \quad \nu+\rho \in X^+, \quad \nu+\rho \prec \lambda+\rho$$
and $\langle \nu+\rho, \nu+\rho \rangle = \langle \lambda+\rho, \lambda+\rho \rangle$. By Lemma 19.15 this implies $\nu = \lambda$. Hence every weight μ for which $b_{\lambda\mu} \ne 0$ satisfies $\mu+\rho = w(\lambda+\rho)$ for some $w \in W$. But then
$$b_{\lambda\mu} = \varepsilon(w)b_{\lambda\lambda} = \varepsilon(w).$$

Hence
$$e_\rho \Delta \operatorname{ch} L(\lambda) = \sum_{w \in W} \varepsilon(w) e_{w(\lambda+\rho)}.$$

If follows that
$$\operatorname{ch} L(\lambda) = \frac{\sum_{w \in W} \varepsilon(w) e_{w(\lambda+\rho)-\rho}}{\Delta}.$$

(We note that $\frac{1}{\Delta} = e_{-\lambda} \operatorname{ch} M(\lambda)$ lies in $\mathfrak{R}$.)

Corollary 19.17 (*Kac' denominator formula*). *For a symmetrisable Kac-Moody algebra we have*
$$e_\rho \prod_{\alpha \in \Phi^+} (1-e_{-\alpha})^{m_\alpha} = \sum_{w \in W} \varepsilon(w) e_{w(\rho)}.$$

Proof. $L(0)$ is the 1-dimensional trivial module with $\operatorname{ch} L(0) = e_0$. Hence
$$\Delta = \Delta e_0 = \sum_{w \in W} \varepsilon(w) e_{w(\rho) - \rho}$$
and so $e_\rho \Delta = \sum_{w \in W} \varepsilon(w) e_{w(\rho)}$.

Corollary 19.18 *(Alternative form of Kac' character formula).* Let $L(\lambda), \lambda \in X^+$, be an irreducible module for a symmetrisable Kac–Moody algebra. Then
$$\operatorname{ch} L(\lambda) = \frac{\sum_{w \in W} \varepsilon(w) e_{w(\lambda + \rho)}}{\sum_{w \in W} \varepsilon(w) e_{w(\rho)}}.$$

Proof. This follows from Theorem 19.16 and Corollary 19.17. □

Note Kac' character formula and denominator formula appear very similar to Weyl's character and denominator formulae for finite dimensional semisimple Lie algebras. However, the nature of Kac' formulae is in fact rather different, since they involve in general infinite sums over the elements of W and infinite products over the positive roots.

Theorem 19.19 *Let $L(A)$ be a symmetrisable Kac–Moody algebra and $\lambda \in X^+$. Then $L(\lambda) = M(\lambda)/J(\lambda)$ where $J(\lambda)$ is the submodule of $M(\lambda)$ generated by elements $f_i^{\lambda(h_i)+1} m_\lambda$ for $i = 1, \ldots, n$.*

Proof. Let $K(\lambda)$ be the submodule of $M(\lambda)$ generated by the elements $f_i^{\lambda(h_i)+1} m_\lambda$. We know that $L(\lambda) = M(\lambda)/J(\lambda)$ where $J(\lambda)$ is the unique maximal submodule of $M(\lambda)$, and wish to show that $K(\lambda) = J(\lambda)$. Now we have
$$f_i^{\lambda(h_i)+1} v_\lambda = 0 \quad \text{where } v_\lambda = J(\lambda) + m_\lambda$$
as in the proof of Proposition 19.14 (the detailed argument is given in Theorem 10.20). Thus $f_i^{\lambda(h_i)+1} m_\lambda \in J(\lambda)$ and so $K(\lambda) \subset J(\lambda)$.

Let $V(\lambda) = M(\lambda)/K(\lambda)$. Then $V(\lambda)$ is an $L(A)$-module in the category $\mathcal{O}$, so
$$\operatorname{ch} V(\lambda) = \sum_{\mu \prec \lambda} [V(\lambda) : L(\mu)] \operatorname{ch} L(\mu)$$
by Proposition 19.6. We also have
$$\operatorname{ch} L(\mu) = \sum_{\nu \prec \mu} b_{\mu\nu} \operatorname{ch} M(\nu).$$

19.3 Kac' character formula

Hence

$$\text{ch } V(\lambda) = \sum_{\mu \prec \lambda} c_{\lambda\mu} \text{ ch } M(\mu)$$

for certain $c_{\lambda\mu} \in \mathbb{Z}$. By considering the action of the generalised Casimir operator Ω on $M(\mu)$ and on $V(\lambda)$ and using Proposition 19.10 we have

$$\text{ch } V(\lambda) = \sum_{\substack{\mu \prec \lambda \\ \langle \mu+\rho, \mu+\rho \rangle = \langle \lambda+\rho, \lambda+\rho \rangle}} c_{\lambda\mu} \text{ ch } M(\mu).$$

Now let $v'_\lambda = K(\lambda) + m_\lambda$ be the highest weight vector of $V(\lambda)$. Then we have

$$f_i^{\lambda(h_i)+1} v_\lambda = 0.$$

It follows, as in the proof of Proposition 19.14, that $f_i : V(\lambda) \to V(\lambda)$ is locally nilpotent. Since $V(\lambda) \in \mathcal{O}$, $e_i : V(\lambda) \to V(\lambda)$ is locally nilpotent. Hence $V(\lambda)$ is an integrable $L(A)$-module. Thus

$$w(\text{ch } V(\lambda)) = \text{ch } V(\lambda) \qquad \text{for all } w \in W$$

by Proposition 19.13. We then have

$$e_\rho \Delta \text{ ch } V(\lambda) = \sum_{\substack{\mu \prec \lambda \\ \langle \mu+\rho, \mu+\rho \rangle = \langle \lambda+\rho, \lambda+\rho \rangle}} c_{\lambda\mu} e_{\mu+\rho}$$

by Proposition 19.2. It then follows exactly as in the proof of Theorem 19.16 that every weight μ for which $c_{\lambda\mu} \neq 0$ satisfies $\mu + \rho = w(\lambda + \rho)$ for some $w \in W$, and that then $c_{\lambda\mu} = \varepsilon(w)$. Hence

$$e_\rho \Delta \text{ ch } V(\lambda) = \sum_{w \in W} \varepsilon(w) e_{w(\lambda+\rho)}$$

and so ch $V(\lambda) = $ ch $L(\lambda)$ by Theorem 19.16. Since $L(\lambda)$ is a factor module of $V(\lambda)$ this can only happen if $V(\lambda) = L(\lambda)$. Thus $K(\lambda) = J(\lambda)$ as required. $\square$

We now recall that for finite dimensional semisimple Lie algebras the partition function $\mathfrak{P}$ was defined as follows. If $\xi \in H^*$ $\mathfrak{P}(\xi)$ is the number of ways of writing ξ as a sum of positive roots, i.e. as the number of sets of non-negative integers r_α, $\alpha \in \Phi^+$, such that $\xi = \sum_{\alpha \in \Phi^+} r_\alpha \alpha$.

For Kac–Moody algebras we define the generalised partition function $\mathfrak{K}$ as follows. If $\xi \in H^*$ $\mathfrak{K}(\xi)$ is the number of ways of writing ξ as a sum of positive roots, each such root α being taken m_α times, i.e. as the number of sets of non-negative integers $r_{\alpha,i}$ for $\alpha \in \Phi^+$ and $1 \le i \le m_\alpha$ such that

$$\xi = \sum_{\alpha \in \Phi^+} \sum_{i=1}^{m_\alpha} r_{\alpha,i} \alpha.$$

We then have an analogue for symmetrisable Kac–Moody algebras of Kostant's multiplicity formula Theorem 12.18.

Proposition 19.20 *Let $L(A)$ be a symmetrisable Kac–Moody algebra and let $\lambda \in \chi^+$. Then for each weight μ of $L(\lambda)$ we have*

$$\dim L(\lambda)_\mu = \sum_{w \in W} \varepsilon(w) \mathfrak{K}(w(\lambda+\rho) - (\mu+\rho)).$$

Proof. By Proposition 19.2 we have

$$\operatorname{ch} M(\lambda) = e_\lambda \prod_{\alpha \in \Phi^+} (1 + e_{-\alpha} + e_{-2\alpha} + \cdots)^{m_\alpha}.$$

By definition of $\mathfrak{K}$ we have

$$\prod_{\alpha \in \Phi^+} (1 + e_\alpha + e_{2\alpha} + \cdots)^{m_\alpha} = \sum_{\beta \in Q^+} \mathfrak{K}(\beta) e_\beta.$$

Thus $\operatorname{ch} M(\lambda) = e_\lambda \sum_{\beta \in Q^+} \mathfrak{K}(\beta) e_{-\beta}$. It follows that

$$\operatorname{ch} L(\lambda) = \sum_{w \in W} \varepsilon(w) \operatorname{ch} M(w(\lambda+\rho) - \rho)$$

$$= \sum_{w \in W} \sum_{\beta \in Q^+} \varepsilon(w) e_{w(\lambda+\rho)-\rho} \mathfrak{K}(\beta) e_{-\beta}$$

$$= \sum_{w \in W} \sum_{\beta \in Q^+} \varepsilon(w) e_{w(\lambda+\rho)-\rho-\beta} \mathfrak{K}(\beta)$$

$$= \sum_\mu \sum_{w \in W} \varepsilon(w) \mathfrak{K}(w(\lambda+\rho) - (\mu+\rho)) e_\mu.$$

Hence the multiplicity of μ as a weight of $L(\lambda)$ is

$$\sum_{w \in W} \varepsilon(w) \mathfrak{K}(w(\lambda+\rho) - (\mu+\rho)). \qquad \square$$

19.4 Generators and relations for symmetrisable algebras

We recall that the Kac–Moody algebra $L(A)$ was not defined in terms of generators and relations. The larger algebra $\tilde{L}(A)$ was defined by generators and relations and its quotient $L(A)$ is given as $\tilde{L}(A)/I$ where I is the largest ideal of $\tilde{L}(A)$ satisfying $I \cap \tilde{H} = O$. It is natural to ask what additional relations are required to pass from $\tilde{L}(A)$ to $L(A)$. We shall answer this in the case when the GCM A is symmetrisable.

We first require some preliminary results on enveloping algebras and modules in category $\mathcal{O}$.

19.4 Generators and relations for symmetrisable algebras

Proposition 19.21 *Let $\theta : L \to L'$ be a surjective homomorphism of Lie algebras with kernel K. Let $\phi : \mathfrak{U}(L) \to \mathfrak{U}(L')$ be the corresponding homomorphism between enveloping algebras. Then the kernel of ϕ is $K\mathfrak{U}(L)$.*

Proof. Since K is an ideal of L, $K\mathfrak{U}(L)$ is a 2-sided ideal of $\mathfrak{U}(L)$. For $[kx] \in K$ for $k \in K, x \in L$ and so $kx = xk + [kx]$ in $\mathfrak{U}(L)$. Thus $K\mathfrak{U}(L) = \mathfrak{U}(L)K$ and $K\mathfrak{U}(L)$ is a 2-sided ideal of $\mathfrak{U}(L)$. Thus $K\mathfrak{U}(L) \subset \ker \phi$.

Conversely we have a homomorphism

$$\alpha : \mathfrak{U}(L)/K\mathfrak{U}(L) \to \mathfrak{U}(L')$$

induced by ϕ. We consider the Lie algebra $[\mathfrak{U}(L)/K\mathfrak{U}(L)]$. We shall define a map $L' \to [\mathfrak{U}(L)/K\mathfrak{U}(L)]$ as follows. Given $x' \in L'$ we choose $x_1 \in L$ with $\theta(x_1) = x'$. Then $x_1 \in \mathfrak{U}(L)$ gives rise to $\bar{x}_1 \in [\mathfrak{U}(L)/K\mathfrak{U}(L)]$. We show that the map $x' \to \bar{x}_1$ is well defined. Suppose $x_2 \in L$ also satisfies $\theta(x_2) = x'$. Then $\bar{x}_2 \in [\mathfrak{U}(L)/K\mathfrak{U}(L)]$. Now $\theta(x_1) = \theta(x_2)$ so $x_1 - x_2 \in K$. Hence $\bar{x}_1 = \overline{x_1 - x_2} + \bar{x}_2 = \bar{x}_2$. Thus our map is well defined and is clearly a Lie algebra homomorphism. By the universal property of enveloping algebras there is a homomorphism

$$\beta : \mathfrak{U}(L') \to \mathfrak{U}(L)/K\mathfrak{U}(L)$$

compatible with our homomorphism of Lie algebras

$$L' \to [\mathfrak{U}(L)/K\mathfrak{U}(L)].$$

It is readily checked that α, β are inverse homomorphisms, and thus isomorphisms. Hence the homomorphism $\phi : \mathfrak{U}(L) \to \mathfrak{U}(L')$ has kernel $K\mathfrak{U}(L)$. $\square$

The 2-sided ideal $L\mathfrak{U}(L)$ of $\mathfrak{U}(L)$ will be denoted by $\mathfrak{U}(L)^+$. We have

$$\mathfrak{U}(L) = \mathbb{C}1 \oplus \mathfrak{U}(L)^+.$$

Proposition 19.22 $L \cap (\mathfrak{U}(L)^+)^2 = [LL]$.

Proof. Since $L \subset \mathfrak{U}(L)^+$ and, for $x, y \in L$, $[xy] = xy - yx$ we see that

$$[LL] \subset L \cap (\mathfrak{U}(L)^+)^2.$$

Conversely let $\bar{L} = L/[LL]$. We have a natural homomorphism $\mathfrak{U}(L) \to \mathfrak{U}(\bar{L})$ under which $L \cap (\mathfrak{U}(L)^+)^2$ maps to $\bar{L} \cap (\mathfrak{U}(\bar{L})^+)^2$. Now $\bar{L}$ is an abelian Lie algebra so $\mathfrak{U}(\bar{L})$ is a polynomial algebra. In such a polynomial algebra it is evident that

$$\bar{L} \cap \left(\mathfrak{U}(\bar{L})^+\right)^2 = 0.$$

It follows that $L \cap (\mathfrak{U}(L)^+)^2$ lies in the kernel of $L \to \bar{L}$ and so

$$L \cap \left(\mathfrak{U}(L)^+\right)^2 \subset [LL]. \qquad \square$$

Proposition 19.23 *Let K be a subalgebra of the Lie algebra L. Then $K \cap K\mathfrak{U}(L)^+ = [KK]$.*

Proof. Since $K \subset \mathfrak{U}(K)^+$ we have $[KK] \subset K\mathfrak{U}(K)^+$, using $[xy] = xy - yx$. Hence $[KK] \subset K \cap K\mathfrak{U}(L)^+$. To prove the converse we use the PBW basis theorem. Let $\{k_i\}$ be a basis of K, and extend it to a basis $\{k_i, u_j\}$ of L. Then all finite products of the form $\prod k_i^{m_i} u_j^{n_j}$ with $m_i \geq 0$, $n_j \geq 0$ form a basis of $\mathfrak{U}(L)$ and the subset $\prod k_i^{m_i}$ with $m_i \geq 0$ is a basis for $\mathfrak{U}(K)$. The monomials $\prod k_i^{m_i} u_j^{n_j}$ with $\sum m_i + \sum n_j \geq 1$ form a basis of $\mathfrak{U}(L)^+$ and those with $\sum m_i + \sum n_j \geq 2$ and $\sum m_i \geq 1$ form a basis of $K\mathfrak{U}(L)^+$. Now

$$K \cap K\mathfrak{U}(L)^+ \subset \mathfrak{U}(K) \cap K\mathfrak{U}(L)^+.$$

A linear combination of monomials $\prod k_i^{m_i} u_j^{n_j}$ lies in $\mathfrak{U}(K)$ if and only if all such monomials have $n_j = 0$. Thus each element of $K \cap K\mathfrak{U}(L)^+$ is a linear combination of such monomials with all $n_j = 0$ and $\sum m_i \geq 2$. Hence

$$K \cap K\mathfrak{U}(L)^+ \subset \left(\mathfrak{U}(K)^+\right)^2 \cap K$$

and so $K \cap K\mathfrak{U}(L)^+ \subset [KK]$ by Proposition 19.22. $\qquad \square$

We next need some further properties of Verma modules. We recall that for $\lambda \in H^*$ the Verma module $M(\lambda)$ for $L(A)$ is given by

$$M(\lambda) = \mathfrak{U}(L(A))/K_\lambda$$

where $K_\lambda = \mathfrak{U}(L(A))N + \sum_{x \in H} \mathfrak{U}(L(A))(x - \lambda(x))$. The Verma module $M(\lambda)$ can also be described as a tensor product. Let B be the subalgebra of $L(A)$ given by $B = N + H$.

Lemma 19.24 *$M(\lambda)$ is isomorphic to the $\mathfrak{U}(L)$-module $\mathfrak{U}(L) \otimes_{\mathfrak{U}(B)} \mathbb{C}v_\lambda$, where $\mathbb{C}v_\lambda$ is the 1-dimensional B-module with N in the kernel and H acting by the weight λ.*

19.4 Generators and relations for symmetrisable algebras

Proof. There is a bijection
$$\mathfrak{U}(L) \otimes_{\mathfrak{U}(B)} \mathbb{C}v_\lambda \to \mathfrak{U}(N^-) \otimes_{\mathbb{C}} \mathbb{C}v_\lambda$$
given as follows. Since $L = N^- \oplus B$ we have a bijection
$$\mathfrak{U}(L) \to \mathfrak{U}(N^-) \otimes_{\mathbb{C}} \mathfrak{U}(B).$$
Thus we have bijections
$$\mathfrak{U}(L) \otimes_{\mathfrak{U}(B)} \mathbb{C}v_\lambda \to (\mathfrak{U}(N^-) \otimes_{\mathbb{C}} \mathfrak{U}(B)) \otimes_{\mathfrak{U}(B)} \mathbb{C}v_\lambda$$
$$\to \mathfrak{U}(N^-) \otimes_{\mathbb{C}} \left(\mathfrak{U}(B) \otimes_{\mathfrak{U}(B)} \mathbb{C}v_\lambda \right) \to \mathfrak{U}(N^-) \otimes_{\mathbb{C}} \mathbb{C}v_\lambda.$$
The $\mathfrak{U}(L)$-action on $\mathfrak{U}(N^-) \otimes_{\mathbb{C}} \mathbb{C}v_\lambda$ is given as follows. Let $u' \in \mathfrak{U}(L)$ and $u \in \mathfrak{U}(N^-)$. Then
$$u'u = \sum_i a_i b_i \quad \text{where } a_i \in \mathfrak{U}(N^-), b_i \in \mathfrak{U}(B).$$
We have $u'(u \otimes v_\lambda) = (\sum_i \lambda(b_i) a_i) \otimes v_\lambda$. On the other hand we know that each element of $M(\lambda)$ is expressible uniquely as um_λ for $u \in \mathfrak{U}(N^-)$. Moreover for $u' \in \mathfrak{U}(L)$ we have
$$u'(um_\lambda) = \left(\sum_i \lambda(b_i) a_i \right) m_\lambda.$$
Thus there is a $\mathfrak{U}(L)$-module isomorphism between $\mathfrak{U}(L) \otimes_{\mathfrak{U}(B)} \mathbb{C}v_\lambda$ and $M(\lambda)$. $\square$

We may also define a module $\tilde{M}(\lambda)$ for the larger Lie algebra $\tilde{L}(A)$ by
$$\tilde{M}(\lambda) = \mathfrak{U}(\tilde{L}) \otimes_{\mathfrak{U}(\tilde{B})} \mathbb{C}v_\lambda$$
where $\lambda \in H^*$ and $\tilde{B} = \tilde{N} + H$.

Lemma 19.25 *For $\lambda \in H^*$ there is an isomorphism of $\mathfrak{U}(L)$-modules*
$$\mathfrak{U}(L) \otimes_{\mathfrak{U}(\tilde{L})} \tilde{M}(\lambda) \cong M(\lambda).$$

Proof. We have a sequence of bijections
$$\mathfrak{U}(L) \otimes_{\mathfrak{U}(\tilde{L})} \tilde{M}(\lambda) = \mathfrak{U}(L) \otimes_{\mathfrak{U}(\tilde{L})} \left(\mathfrak{U}(\tilde{L}) \otimes_{\mathfrak{U}(\tilde{B})} \mathbb{C}v_\lambda \right)$$
$$\to \left(\mathfrak{U}(L) \otimes_{\mathfrak{U}(\tilde{L})} \mathfrak{U}(\tilde{L}) \right) \otimes_{\mathfrak{U}(\tilde{B})} \mathbb{C}v_\lambda \to \mathfrak{U}(L) \otimes_{\mathfrak{U}(\tilde{B})} \mathbb{C}v_\lambda.$$
Now we have a natural homomorphism $\mathfrak{U}(\tilde{B}) \to \mathfrak{U}(B)$ with kernel K which acts trivially on $\mathfrak{U}(L)$ and on $\mathbb{C}v_\lambda$. Thus we have a bijection
$$\mathfrak{U}(L) \otimes_{\mathfrak{U}(\tilde{B})} \mathbb{C}v_\lambda \to \mathfrak{U}(L) \otimes_{\mathfrak{U}(B)} \mathbb{C}v_\lambda = M(\lambda).$$
The above bijections are isomorphisms of $\mathfrak{U}(L)$-modules. $\square$

We next require further information about modules in the category $\mathcal{O}$. The following definition turns out to be very useful.

Definition Let V be an $L(A)$-module with $V \in \mathcal{O}$. A vector $v \in V$ is called **primitive** if

(i) v is a weight vector
(ii) there exists a submodule $U \subset V$ such that $v \notin U$ but $Nv \subset U$.

Lemma 19.26 *A module $V \in \mathcal{O}$ is generated as an $L(A)$-module by its primitive vectors.*

Proof. Let V' be the submodule generated by the primitive vectors in V. Suppose $V' \neq V$. Consider the factor module V/V'. This factor module lies in $\mathcal{O}$ so contains a weight vector $\bar{v} \neq 0$ of maximal weight with respect to $\prec$. Thus $N\bar{v} = 0$. Let v be a weight vector in V such that $v \to \bar{v}$. Then $v \notin V'$ and $Nv \subset V'$. Thus v is a primitive vector not in V', a contradiction. $\square$

In fact the following stronger result is true.

Proposition 19.27 *A module $V \in \mathcal{O}$ is generated as a $\mathfrak{U}(N^-)$-module by its primitive vectors.*

Proof. We first show that if $v \in V$ is a weight vector which is not primitive then $v \in \mathfrak{U}(N^-)\mathfrak{U}(N)^+ v$. For consider the $\mathfrak{U}(L)$-submodule of V generated by Nv. We have

$$\mathfrak{U}(L)Nv = \mathfrak{U}(N^-)\mathfrak{U}(H)\mathfrak{U}(N)Nv$$
$$= \mathfrak{U}(N^-)\mathfrak{U}(N)Nv \quad \text{since } v \text{ is a weight vector}$$
$$= \mathfrak{U}(N^-)\mathfrak{U}(N)^+ v.$$

Now let U be the $\mathfrak{U}(N^-)$-submodule generated by the primitive vectors in V. We wish to show $U = V$. We shall assume $U \neq V$ and obtain a contradiction. For each weight vector $v \in V$ we have

$$\mathfrak{U}(L)v = \mathfrak{U}(N^-)\mathfrak{U}(H)\mathfrak{U}(N)v = \mathfrak{U}(N^-)\mathfrak{U}(N)v$$
$$= \mathfrak{U}(N^-)(\mathbb{C}1 + \mathfrak{U}(N)^+)v$$
$$= \mathfrak{U}(N^-)v + \mathfrak{U}(N^-)\mathfrak{U}(N)^+ v.$$

We can deduce from this that V is generated as $\mathfrak{U}(L)$-module by U and the $\mathfrak{U}(N^-)$-submodule generated by $\mathfrak{U}(N)^+ v$ for all primitive $v \in V$. Since

19.4 Generators and relations for symmetrisable algebras

$U \ne V$ there exists a primitive v such that $\mathfrak{U}(N)^+ v \not\subset U$. Let v have weight λ. Then there exists a weight vector $u_1 \in \mathfrak{U}(N)^+$ with $u_1 v \notin U$. So $u_1 v$ is not primitive in V. Thus we have

$$u_1 v \in \mathfrak{U}(N^-)\mathfrak{U}(N)^+ u_1 v.$$

Hence $\mathfrak{U}(N)^+ u_1 v \not\subset U$. So there exists a weight vector $u_2 \in \mathfrak{U}(N)^+$ with $u_2 u_1 v \notin U$.

Continuing in this way we obtain a sequence of weight vectors $u_1, u_2, u_3, \ldots$ in $\mathfrak{U}(N)^+$ such that $u_k \ldots u_1 v \notin U$ for each k. Let the weight of u_i be μ_i. Then the weight of $u_k \ldots u_1 v$ is $\lambda + \mu_1 + \cdots + \mu_k$. We have

$$\lambda \prec \lambda + \mu_1 \prec \lambda + \mu_1 + \mu_2 \prec \cdots .$$

But $V \in \mathcal{O}$ and so such a sequence of weights must terminate after finitely many steps. This gives the required contradiction. $\square$

We now consider the module $\bigoplus_{i=1}^n M(-\alpha_i)$ in $\mathcal{O}$.

Proposition 19.28 *Every primitive vector in the module $\bigoplus_{i=1}^n M(-\alpha_i)$ has weight $-\alpha$ where $\langle \alpha, \alpha \rangle = 2\langle \rho, \alpha \rangle$.*

Proof. Let v be a primitive vector in $\bigoplus_{i=1}^n M(-\alpha_i)$ of weight $-\alpha$. Then there is a submodule U of $\bigoplus_{i=1}^n M(-\alpha_i)$ such that $v \notin U$ and $Nv \subset U$. Write $v = v_1 + \cdots + v_n$ where $v_i \in M(-\alpha_i)$ and let $v \to \bar{v}$ where $\bar{v} \in (\bigoplus M(-\alpha_i))/U$. We consider the action of the generalised Casimir operator Ω on the module $\bigoplus M(-\alpha_i)$. By Proposition 19.10

$$\Omega v_i = (\langle -\alpha_i + \rho, -\alpha_i + \rho \rangle - \langle \rho, \rho \rangle) v_i$$
$$= (\langle \alpha_i, \alpha_i \rangle - 2 \langle \rho, \alpha_i \rangle) v_i.$$

Now $\rho(h_i) = 1$ so $\left\langle \rho, \dfrac{2\alpha_i}{\langle \alpha_i, \alpha_i \rangle} \right\rangle = 1$. Hence $\langle \alpha_i, \alpha_i \rangle = 2 \langle \rho, \alpha_i \rangle$ and so $\Omega v_i = 0$. Thus Ω acts as 0 on $\bigoplus M(-\alpha_i)$. Hence Ω acts as 0 on $(\bigoplus M(-\alpha_i))/U$ and $\Omega \bar{v} = 0$. But $\bar{v}$ has weight $-\alpha$ and so

$$\Omega \bar{v} = (\langle -\alpha + \rho, -\alpha + \rho \rangle - \langle \rho, \rho \rangle) \bar{v}$$
$$= (\langle \alpha, \alpha \rangle - 2\langle \rho, \alpha \rangle) \bar{v}.$$

Thus $\langle \alpha, \alpha \rangle = 2\langle \rho, \alpha \rangle$ as required. $\square$

We shall now start to see the relevance of the preliminary results which we have obtained. We concentrate on the kernel I of the natural homomorphism $\tilde{L}(A) \to L(A)$. We recall that $I = I^- \oplus I^+$ where $I^- \subset \tilde{N}^-$ and

$I^+ \subset \tilde{N}$. We have $I^- = \bigoplus_{\alpha \in Q^-} I_\alpha^-$ by Lemma 14.12. Since $\dim \tilde{L}(A)_{-\alpha_i} = \dim L(A)_{-\alpha_i} = 1$ we have $I_{-\alpha_i}^- = 0$. Thus $I^- = \bigoplus_{\alpha \in Q^-, \alpha \neq -\alpha_i} I_\alpha^-$. Hence each element of I^- has form $\sum_{i=1}^n u_i f_i$ where $u_i \in \mathfrak{U}(\tilde{N}^-)^+$. It is in fact uniquely expressible in this form since $\tilde{N}^-$ is the free Lie algebra on $f_1, \ldots, f_n$ by Proposition 14.8 and so $\mathfrak{U}(\tilde{N}^-)$ is the free associative algebra on $f_1, \ldots, f_n$ by Proposition 9.10.

Proposition 19.29 Let $\theta : \tilde{L}(A) \to L(A)$ be the natural homomorphism with kernel $I = I^- \oplus I^+$. Then there is a homomorphism of $\tilde{L}$-modules

$$I^- \to \bigoplus_{i=1}^n M(-\alpha_i)$$

given by $\sum_{i=1}^n u_i f_i \to \sum_{i=1}^n \theta(u_i) m_{-\alpha_i}$. The kernel of this homomorphism is $[I^- I^-]$.

Proof. We begin with the $\tilde{L}$-module

$$\tilde{M}(\lambda) = \mathfrak{U}(\tilde{L}) \otimes_{\mathfrak{U}(\tilde{B})} \mathbb{C} v_\lambda \quad \text{for } \lambda \in H^*.$$

The module $\tilde{M}(\lambda)$ has highest weight vector $\tilde{m}_\lambda = 1 \otimes v_\lambda$ and, just as for the Verma module $M(\lambda)$, each element of $\tilde{M}(\lambda)$ is uniquely expressible in the form $u\tilde{m}_\lambda$ for $u \in \mathfrak{U}(\tilde{N}^-)$. We take the special case $\lambda = 0$. Then $u \to u\tilde{m}_0$ is a bijection between $\mathfrak{U}(\tilde{N}^-)$ and $\tilde{M}(0)$.

Now $\mathfrak{U}(\tilde{N}^-)$ is freely generated by $f_1, \ldots, f_n$ so

$$\mathfrak{U}(\tilde{N}^-) = \mathbb{C} 1 \oplus \mathfrak{U}(\tilde{N}^-) f_1 \oplus \cdots \oplus \mathfrak{U}(\tilde{N}^-) f_n.$$

Thus $\bigoplus_{i=1}^n \mathfrak{U}(\tilde{N}^-) f_i$ is a $\mathfrak{U}(\tilde{N}^-)$-submodule of codimension 1 in $\mathfrak{U}(\tilde{N}^-)$. It corresponds to the subspace $\bigoplus_{i=1}^n \mathfrak{U}(\tilde{N}^-) f_i \tilde{m}_0$ of codimension 1 in $\tilde{M}(0)$. Let

$$\tilde{J}(0) = \bigoplus_{i=1}^n \mathfrak{U}(\tilde{N}^-) f_i \tilde{m}_0.$$

Then $\tilde{J}(0)$ is a $\mathfrak{U}(\tilde{L})$-submodule of $\tilde{M}(0)$. For it is clearly invariant under $\mathfrak{U}(\tilde{N}^-)$ and $\mathfrak{U}(H)$, but also

$$e_i f_i \tilde{m}_0 = f_i e_i \tilde{m}_0 + h_i \tilde{m}_0 = 0$$

$$e_j f_i \tilde{m}_0 = f_i e_j \tilde{m}_0 = 0 \quad \text{if } j \neq i.$$

Thus $\mathfrak{U}(\tilde{N})^+ f_i \tilde{m}_0 = 0$ and so $\mathfrak{U}(\tilde{N}) f_i \tilde{m}_0 = \mathbb{C} f_i \tilde{m}_0$. Hence

$$\mathfrak{U}(\tilde{L}) f_i \tilde{m}_0 = \mathfrak{U}(\tilde{N}^-) \mathfrak{U}(H) \mathfrak{U}(\tilde{N}) f_i \tilde{m}_0$$

$$= \mathfrak{U}(\tilde{N}^-) \mathfrak{U}(H) f_i \tilde{m}_0 = \mathfrak{U}(\tilde{N}^-) f_i \tilde{m}_0.$$

19.4 Generators and relations for symmetrisable algebras

Thus $\tilde{J}(0) = \bigoplus_{i=1}^{n} \mathfrak{U}(\tilde{L}) f_i \tilde{m}_0$, which is a $\mathfrak{U}(\tilde{L})$-submodule of $\tilde{M}(0)$. Now $f_i \tilde{m}_0$ has weight $-\alpha_i$ and the map

$$\mathfrak{U}(\tilde{N}^-) f_i \tilde{m}_0 \to \tilde{M}(-\alpha_i)$$

$$u_i f_i \tilde{m}_0 \to u_i \tilde{m}_{-\alpha_i}$$

is an isomorphism of $\mathfrak{U}(\tilde{L})$-modules. Thus $\tilde{J}(0)$ is isomorphic to $\bigoplus_{i=1}^{n} \tilde{M}(-\alpha_i)$ as $\mathfrak{U}(\tilde{L})$-modules. It then follows from Lemma 19.25 that

$$\mathfrak{U}(L) \otimes_{\mathfrak{U}(\tilde{L})} \tilde{J}(0) \cong \bigoplus_{i=1}^{n} M(-\alpha_i)$$

as $\mathfrak{U}(\tilde{L})$-modules, or as $\mathfrak{U}(L)$-modules.

We now consider the map

$$\phi : I^- \to \mathfrak{U}(L) \underset{\mathfrak{U}(\tilde{L})}{\otimes} \tilde{J}(0)$$

given by $x \to 1 \otimes x \tilde{m}_0$. We note that $x \tilde{m}_0$ lies in $\tilde{J}(0)$ since $I^- \subset \tilde{N}^-$. We show that ϕ is a homomorphism of $\tilde{L}$-modules. To see this let $y \in \tilde{L}$. Then

$$[y, x] \to 1 \otimes [y, x] \tilde{m}_0$$
$$= 1 \otimes (yx - xy) \tilde{m}_0$$
$$= 1 \otimes y(x \tilde{m}_0) - 1 \otimes x(y \tilde{m}_0).$$

Now we have $x \tilde{m}_0 \in \tilde{J}(0)$ and $y \tilde{m}_0 \in \tilde{J}(0)$. Thus

$$[y, x] \to \theta(y) \otimes (x \tilde{m}_0) - \theta(x) \otimes (y \tilde{m}_0)$$
$$= \theta(y) \otimes (x \tilde{m}_0) \quad \text{since } \theta(x) = 0,$$
$$= y(1 \otimes x \tilde{m}_0).$$

This shows that ϕ is a homomorphism of $\tilde{L}$-modules. Moreover $[I^- I^-]$ lies in the kernel of ϕ. For if $x, y \in I^-$ we have $[y, x] \to 0$ as above, since $\theta(x) = \theta(y) = 0$. Thus we have a homomorphism of $\tilde{L}$-modules

$$\phi : I^- \to \bigoplus_{i=1}^{n} M(-\alpha_i)$$

with $\sum_{i=1}^{n} u_i f_i \to \sum_{i=1}^{n} \theta(u_i) m_{-\alpha_i}$ where $u_i \in \mathfrak{U}(\tilde{N}^-)^+$.

We determine the kernel K of ϕ. We know that $[I^- I^-] \subset K$ and prove the reverse inclusion. Let $\sum u_i f_i \in K$ where $u_i \in \mathfrak{U}(\tilde{N}^-)^+$. Then $\sum \theta(u_i) m_{-\alpha_i} = 0$. This implies $\theta(u_i) m_{-\alpha_i} = 0$ for each i and then that $\theta(u_i) = 0$ for each i. Now the homomorphism of Lie algebras $\theta : \tilde{N}^- \to N^-$ gives rise to a homomorphism of enveloping algebras $\mathfrak{U}(\tilde{N}^-) \to \mathfrak{U}(N^-)$ with kernel $I^- \mathfrak{U}(\tilde{N}^-)$

by Proposition 19.21. Thus $u_i \in I^-\mathfrak{U}(\tilde{N}^-)$ and so $\sum u_i f_i \in I^-\mathfrak{U}(\tilde{N}^-)^+$. Hence $K \subset I^- \cap I^-\mathfrak{U}(\tilde{N}^-)^+$. However, $I^- \cap I^-\mathfrak{U}(\tilde{N}^-)^+ = [I^-I^-]$ by Proposition 19.23. Hence $K \subset [I^-I^-]$. Thus the kernel of our homomorphism is $[I^-I^-]$. □

We now come to our description of $L(A)$ by generators and relations.

Theorem 19.30 *Let $L(A)$ be a symmetrisable Kac–Moody algebra. Then $L(A) = \tilde{L}(A)/J$ where J is the ideal of $\tilde{L}(A)$ generated by the elements $(\operatorname{ad} e_i)^{1-A_{ij}} e_j$ and $(\operatorname{ad} f_i)^{1-A_{ij}} f_j$ for all $i \neq j$. Thus we obtain a system of generators and relations for $L(A)$ by taking generators and relations for $\tilde{L}(A)$ and adding the further relations*

$$(\operatorname{ad} e_i)^{1-A_{ij}} e_j = 0, \quad (\operatorname{ad} f_i)^{1-A_{ij}} f_j = 0$$

for all $i \neq j$.

Proof. Let J be the ideal of $\tilde{L}(A)$ generated by the elements $(\operatorname{ad} e_i)^{1-A_{ij}} e_j$ and $(\operatorname{ad} f_i)^{1-A_{ij}} f_j$. We have $L(A) = \tilde{L}(A)/I$ and $J \subset I$ by Proposition 16.10. We wish to show that $I = J$.

We shall suppose if possible that $I \neq J$ and obtain a contradiction. Let $\bar{I} = I/J$. Then $\bar{I} \neq 0$ and $\bar{I} = \bar{I}^+ \oplus \bar{I}^-$ where

$$\bar{I}^+ = \bigoplus_{\alpha \in Q^+} \bar{I}_\alpha, \quad \bar{I}^- = \bigoplus_{\alpha \in Q^-} \bar{I}_\alpha$$

since the analogous property holds for I. The automorphism $\tilde{\omega}$ of $\tilde{L}(A)$ given in Proposition 14.5 satisfies $\tilde{\omega}(I) = I$ and $\tilde{\omega}(J) = J$ so induces an automorphism on $\bar{I} = I/J$. This automorphism satisfies $\tilde{\omega}(\bar{I}^+) = \bar{I}^-$. Hence $\bar{I}^+ = 0$ if and only if $\bar{I}^- = 0$. Since $\bar{I} = \bar{I}^+ \oplus \bar{I}^-$ and $\bar{I} \neq 0$ we must have $\bar{I}^- \neq 0$.

We know from Section 16.2 that the Weyl group W acts on the weights of $\tilde{L}(A)/I$ and that weights in the same W-orbit have the same multiplicity. The same argument can be applied to $\tilde{L}(A)/J$ to give a similar result. Since

$$\dim \left(\tilde{L}(A)/J\right)_\alpha = \dim \left(\tilde{L}(A)/I\right)_\alpha + \dim (I/J)_\alpha$$

we see that W acts on the weights of $\bar{I}$ and that weights in the same W-orbit have the same multiplicity. In fact W acts on the weights of $\bar{I}^-$ since if $\alpha \in Q^-$ is a weight of $\bar{I}^-$ then $s_i(\alpha) \in Q^-$ also, since $-\alpha_i$ is not a weight of I.

We choose a weight $\alpha = \sum_{i=1}^n k_i \alpha_i \in Q^+$ such that $\bar{I}_{-\alpha} \neq 0$ and α has minimal possible height $\sum k_i$. Since $\bar{I}^-_{s_i(\alpha)} \neq 0$ we have $\operatorname{ht} s_i(\alpha) \geq \operatorname{ht} \alpha$. Since

$$s_i(\alpha) = \alpha - 2\frac{\langle \alpha_i, \alpha \rangle}{\langle \alpha_i, \alpha_i \rangle} \alpha_i$$

19.4 Generators and relations for symmetrisable algebras 483

we have $\langle \alpha_i, \alpha \rangle \leq 0$. Since $\alpha = \sum k_i \alpha_i$ with each $k_i \geq 0$ we deduce that $\langle \alpha, \alpha \rangle \leq 0$. On the other hand we have

$$2\langle \rho, \alpha \rangle = \sum_i k_i \langle \rho, 2\alpha_i \rangle = \sum_i k_i \langle \alpha_i, \alpha_i \rangle > 0.$$

Thus $\langle \alpha, \alpha \rangle \leq 0$ and $2\langle \rho, \alpha \rangle > 0$. In particular $\langle \alpha, \alpha \rangle \neq 2\langle \rho, \alpha \rangle$. Thus the weights $-\alpha$ for $\bar{I}^-$ for which α has minimal height satisfy $\langle \alpha, \alpha \rangle \neq 2\langle \rho, \alpha \rangle$.

We now recall from Proposition 19.29 that $I^-/[I^-, I^-]$ is isomorphic as $\tilde{L}$-module to a submodule of $\bigoplus_{i=1}^{n} M(-\alpha_i)$. By Proposition 19.28 all primitive vectors in $\bigoplus_{i=1}^{n} M(-\alpha_i)$ have a weight $-\alpha$ satisfying $\langle \alpha, \alpha \rangle = 2\langle \rho, \alpha \rangle$. Thus all primitive vectors of $I^-/[I^-I^-]$ have weight $-\alpha$ satisfying $\langle \alpha, \alpha \rangle = 2\langle \rho, \alpha \rangle$. Now $I^-/[I^-I^-]$ is generated as an N^--module by its primitive vectors, by Proposition 19.27. Thus $I^-/[I^-I^-]$ is generated as an $\tilde{N}^-$-module by its weight vectors with weight $-\alpha$ satisfying $\langle \alpha, \alpha \rangle = 2\langle \rho, \alpha \rangle$. (Recall that $\tilde{N}^-/I^- \cong N^-$.)

We claim the same is true of I^-. Let K be the $\tilde{N}^-$-submodule of I^- generated by all weight vectors with weight $-\alpha$ satisfying $\langle \alpha, \alpha \rangle = 2\langle \rho, \alpha \rangle$. Then $([I^-I^-]+K)/[I^-I^-]$ has the same property in $I^-/[I^-I^-]$, thus $[I^-I^-]+K=I^-$. Suppose if possible that $K \neq I^-$. Then I^-/K is an $\tilde{N}^-$-module whose weights are non-zero elements of Q^-. Consider the submodule $[I^-/K, I^-/K]$ of I^-/K. This is an $\tilde{N}^-$-module whose weights have form $\beta + \gamma$ where β, γ are weights of I^-/K. Thus if α is a weight of I^-/K for which $|\text{ht } \alpha|$ is minimal then α cannot be a weight of $[I^-/K, I^-/K]$. Thus

$$[I^-/K, I^-/K] \neq I^-/K$$

and this gives $K + [I^-, I^-] \neq I^-$, a contradiction. Thus I^- is generated as $\tilde{N}^-$-module by its weight vectors with weight $-\alpha$ satisfying $\langle \alpha, \alpha \rangle = 2\langle \rho, \alpha \rangle$. The same must therefore be true of $\bar{I}^-$. However, we have seen above that the weights $-\alpha$ of $\bar{I}^-$ for which $\text{ht } \alpha$ is minimal do not satisfy $\langle \alpha, \alpha \rangle = 2\langle \rho, \alpha \rangle$. This implies that the set of weight vectors with weight $-\alpha$ satisfying $\langle \alpha, \alpha \rangle = 2\langle \rho, \alpha \rangle$ cannot generate $\bar{I}^-$ as $\tilde{N}^-$-module. This gives the required contradiction. □

20
Representations of affine Kac–Moody algebras

20.1 Macdonald's identities

We now consider Kac' denominator formula

$$\prod_{\alpha \in \Phi^+} (1 - e_{-\alpha})^{m_\alpha} = \sum_{w \in W} \varepsilon(w) e_{w(\rho) - \rho}$$

in the special case when L is an affine Kac–Moody algebra.

We assume first that L is an untwisted affine algebra. Then $L = \hat{\mathfrak{L}}(L^0)$ where L^0 is a finite dimensional simple Lie algebra with root system Φ^0 and Weyl group W^0. We recall from Theorems 17.18 and 16.27, and Corollary 18.6 that

$$\Phi = \{\alpha + n\delta \; ; \; \alpha \in \Phi^0, n \in \mathbb{Z}\} \cup \{n\delta \; ; \; n \in \mathbb{Z}, n \neq 0\}$$

and that $\alpha + n\delta$ has multiplicity 1 and $n\delta$ has multiplicity l. Also

$$\Phi^+ = \{\alpha + n\delta \; ; \; \alpha \in \Phi^0, n > 0\} \cup (\Phi^0)^+ \cup \{n\delta \; ; \; n > 0\}.$$

Thus the left-hand side of the denominator formula can be expressed as

$$\prod_{\alpha \in (\Phi^0)^+} (1 - e_{-\alpha}) \prod_{n > 0} \left\{ (1 - e_{-n\delta})^l \prod_{\alpha \in \Phi^0} (1 - e_{-\alpha - n\delta}) \right\}$$

We also recall from Remark 17.34 that $W = t(M^*) W^0$ where M^* is the lattice given by

$$M^* = \begin{cases} \sum_{i=1}^{l} \mathbb{Z}\alpha_i & \text{for types } \tilde{A}_l, \tilde{D}_l, \tilde{E}_6, \tilde{E}_7, \tilde{E}_8 \\ \sum_{\alpha_i \text{ long}} \mathbb{Z}\alpha_i + \sum_{\alpha_i \text{ short}} p\mathbb{Z}\alpha_i & \text{for types } \tilde{B}_l, \tilde{C}_l, \tilde{F}_4, \tilde{G}_2 \end{cases}$$

20.1 Macdonald's identities

and $t(M^*)$ is the set of $t_\alpha : H^* \to H^*$ for $\alpha \in M^*$ given by

$$t_\alpha(\lambda) = \lambda + \lambda(c)\alpha - (\langle \lambda, \alpha \rangle + \tfrac{1}{2}\langle \alpha, \alpha \rangle \lambda(c))\delta.$$

In calculating the right-hand side of the denominator formula we recall that

$$H = H^0 \oplus (\mathbb{C}c + \mathbb{C}d)$$
$$H^* = (H^0)^* \oplus (\mathbb{C}\gamma + \mathbb{C}\delta)$$

where $(H^0)^*$ is embedded in H^* by assuming $\lambda(c) = 0$, $\lambda(d) = 0$ for $\lambda \in (H^0)^*$.

Lemma 20.1 *Let $\lambda \in H^*$. Then*

$$\lambda = \lambda^0 + \lambda(c)\gamma + a_0^{-1}\lambda(d)\delta$$

where $\lambda^0 \in (H^0)^*$.

Proof. Let $\lambda = \lambda^0 + r\gamma + s\delta$ where $\lambda^0 \in (H^0)^*$. Then $\lambda(c) = \lambda^0(c) + r\gamma(c) + s\delta(c)$. But we have $\alpha_i(c) = 0$ for $i = 1, \ldots, l$ hence $\lambda^0(c) = 0$. Also $\delta(c) = 0$ and $\gamma(c) = 1$. Hence $r = \lambda(c)$. Also $\lambda(d) = \lambda^0(d) + r\gamma(d) + s\delta(d)$. We know $\alpha_i(d) = 0$ for $i = 1, \ldots, l$ thus $\lambda^0(d) = 0$. Also $\gamma(d) = 0$ and $\delta(d) = a_0$. Hence $\lambda(d) = a_0 s$ and $s = a_0^{-1}\lambda(d)$. $\square$

Of course in the untwisted case we have $a_0 = 1$.

We recall that $\rho \in H^*$ satisfies $\rho(h_i) = 1$ for $i = 0, 1, \ldots, l$ and $\rho(d) = 0$. In particular we have $\rho(c) = c_0 + c_1 + \cdots + c_l$.

Definition *The number $h = a_0 + a_1 + \cdots + a_l$ is called the **Coxeter number** of L. The number $h^\vee = c_0 + c_1 + \cdots + c_l$ is called the **dual Coxeter number** of L.*

We note that if $L = \hat{\mathcal{L}}(L^0)$ is of untwisted type then the Coxeter number of L is equal to the Coxeter number of L^0.

The values of h and h^v are given in the following table.

Type of L	h	h^v
$\tilde{A}_l$	$l+1$	$l+1$
$\tilde{B}_l$	$2l$	$2l-1$
$\tilde{C}_l$	$2l$	$l+1$
$\tilde{D}_l$	$2l-2$	$2l-2$
$\tilde{E}_6$	12	12
$\tilde{E}_7$	18	18
$\tilde{E}_8$	30	30
$\tilde{F}_4$	12	9
$\tilde{G}_2$	6	4
$\tilde{B}_l^t$	$2l-1$	$2l$
$\tilde{C}_l^t$	$l+1$	$2l$
$\tilde{F}_4^t$	9	12
$\tilde{G}_2^t$	4	6
$\tilde{A}_1'$	3	3
$\tilde{C}_l'$	$2l+1$	$2l+1$

Lemma 20.2 $\rho = \rho^0 + h^v \gamma$ where $\rho^0 \in (H^0)^*$ satisfies $\rho^0(h_i) = 1$ for $i = 1, \ldots, l$.

Proof. This follows from Lemma 20.1 since $\rho(c) = h^v$ and $\rho(d) = 0$. □

We now consider the right-hand side of the denominator formula. Let $w \in W$ have form $w = w^0 t_\alpha$ where $w^0 \in W^0$ and $\alpha \in M^*$. Then

$$w(\rho) - \rho = w^0 t_\alpha(\rho) - \rho$$

$$= w^0 \left(\rho + h^v \alpha - \left(\langle \rho, \alpha \rangle + \frac{1}{2} \langle \alpha, \alpha \rangle h^v \right) \delta \right) - \rho$$

$$= w^0(\rho) - \rho + h^v w^0(\alpha) - \left(\langle \rho, \alpha \rangle + \frac{1}{2} \langle \alpha, \alpha \rangle h^v \right) \delta$$

$$= w^0(\rho^0) - \rho^0 + h^v w^0(\alpha) - \left(\langle \rho^0, \alpha \rangle + \frac{1}{2} \langle \alpha, \alpha \rangle h^v \right) \delta$$

since $w^0(\gamma) = \gamma$ and $\langle \gamma, \alpha \rangle = 0$ for all $\alpha \in M^*$

$$= w^0 (h^v \alpha + \rho^0) - \rho^0 - \frac{(\langle \rho^0 + h^v \alpha, \rho^0 + h^v \alpha \rangle - \langle \rho^0, \rho^0 \rangle)}{2h^v} \delta.$$

20.1 Macdonald's identities

For convenience we shall write, for $\lambda \in (H^0)^*$,

$$c(\lambda) = \langle \lambda + \rho^0, \lambda + \rho^0 \rangle - \langle \rho^0, \rho^0 \rangle.$$

We recall from Corollary 19.11 that when λ is dominant and integral the generalised Casimir operator acts on the irreducible module with highest weight λ as scalar multiplication by $c(\lambda)$.

We also write, for $\lambda \in (H^0)^*$,

$$\chi^0(\lambda) = \frac{\sum_{w \in W^0} \varepsilon(w) e_{w(\lambda+\rho^0)-\rho^0}}{\sum_{w \in W^0} \varepsilon(w) e_{w(\rho^0)-\rho^0}}.$$

We recall from Theorem 12.17 that when λ is dominant and integral $\chi^0(\lambda)$ is the character of the irreducible L^0-module $L(\lambda)$. However, $c(\lambda)$ and $\chi^0(\lambda)$ are now defined for all $\lambda \in (H^0)^*$. Then we have, writing $e(\lambda)$ instead of e_λ for convenience:

$$\sum_{w \in W} \varepsilon(w) e(w(\rho) - \rho)$$

$$= \sum_{\alpha \in M^*} \sum_{w^0 \in W^0} \varepsilon(w^0) e(w^0(h^\vee \alpha + \rho^0) - \rho^0) e\left(\frac{-c(h^\vee \alpha)}{2h^\vee}\delta\right)$$

$$= \sum_{w^0 \in W^0} \varepsilon(w^0) e(w^0(\rho^0) - \rho^0) \sum_{\alpha \in M^*} \chi^0(h^\vee \alpha) e\left(\frac{-c(h^\vee \alpha)}{2h^\vee}\delta\right)$$

$$= \prod_{\alpha \in (\Phi^0)^+} (1 - e_{-\alpha}) \sum_{\alpha \in M^*} \chi^0(h^\vee \alpha) e\left(\frac{-c(h^\vee \alpha)}{2h^\vee}\delta\right)$$

by Weyl's denominator formula.

We now put $q = e_{-\delta}$ and equate the left- and right-hand sides of Kac' denominator formula. We obtain the following result.

Theorem 20.3 *(Macdonald's identity for untwisted affine Kac–Moody algebras).*

$$\prod_{n>0} \left\{ (1-q^n)^l \prod_{\alpha \in \Phi^0} (1-q^n e_{-\alpha}) \right\} = \sum_{\alpha \in M^*} \chi^0(h^\vee \alpha) q^{c(h^\vee \alpha)/2h^\vee}$$

where

$$M^* = \begin{cases} \sum_{i=1}^l \mathbb{Z}\alpha_i & \text{for types } \tilde{A}_l, \tilde{D}_l, \tilde{E}_6, \tilde{E}_7, \tilde{E}_8 \\ \sum_{\alpha_i \text{ long}} \mathbb{Z}\alpha_i + \sum_{\alpha_i \text{ short}} p\mathbb{Z}\alpha_i & \text{for } \tilde{B}_l, \tilde{C}_l, \tilde{F}_4, \tilde{G}_2. \end{cases} \qquad \square$$

We next wish to state Macdonald's identities for the twisted affine Kac–Moody algebras. The left-hand side of the identity is obtained from a knowledge of the real and imaginary roots together with the multiplicities of the imaginary roots. The real roots are given in Theorem 17.8 and the multiplicities of the imaginary roots in Corollaries 18.10 and 18.15. The right-hand side of the identity looks the same as before – the only change being that the appropriate lattice M^* must be taken in each case. The appropriate lattice was described in Remark 17.34.

Theorem 20.4 (*Macdonald's identity for twisted affine Kac–Moody algebras*).
(a) *The left-hand side of the identity is given as follows.*

$$\tilde{B}_l^t \quad \prod_{n>0} \left\{ (1-q^{2n})^l (1-q^{2n-1})^{l-1} \prod_{\alpha \in \Phi_s^0} (1-q^n e_{-\alpha}) \prod_{\alpha \in \Phi_l^0} (1-q^{2n} e_{-\alpha}) \right\}$$

$$\tilde{C}_l^t \quad \prod_{n>0} \left\{ (1-q^{2n})^l (1-q^{2n-1}) \prod_{\alpha \in \Phi_s^0} (1-q^n e_{-\alpha}) \prod_{\alpha \in \Phi_l^0} (1-q^{2n} e_{-\alpha}) \right\}$$

$$\tilde{F}_4^t \quad \prod_{n>0} \left\{ (1-q^{2n})^4 (1-q^{2n-1})^2 \prod_{\alpha \in \Phi_s^0} (1-q^n e_{-\alpha}) \prod_{\alpha \in \Phi_l^0} (1-q^{2n} e_{-\alpha}) \right\}$$

$$\tilde{G}_2^t \quad \prod_{n>0} \left\{ (1-q^{3n})^2 (1-q^{3n-1})(1-q^{3n-2}) \right.$$
$$\left. \prod_{\alpha \in \Phi_s^0} (1-q^n e_{-\alpha}) \prod_{\alpha \in \Phi_l^0} (1-q^{3n} e_{-\alpha}) \right\}$$

$$\tilde{C}_l' \quad \prod_{n>0} \left\{ (1-q^n)^l \prod_{\alpha \in \Phi_s^0} (1-q^n e_{-\alpha}) \right.$$
$$\left. \prod_{\alpha \in \Phi_l^0} \left(1-q^{\frac{2n-1}{2}} e_{-\frac{1}{2}\alpha}\right) (1-q^{2n} e_{-\alpha}) \right\}$$

$$\tilde{A}_1' \quad \prod_{n>0} \left\{ (1-q^n) \prod_{\alpha \in \Phi^0} \left(1-q^{\frac{2n-1}{2}} e_{-1/2\alpha}\right)(1-q^{2n} e_{-\alpha}) \right\}.$$

(b) *The right-hand side of the identity is*

$$\sum_{\alpha \in M^*} \chi^0(h_\alpha^\vee) q^{c(h^\vee \alpha)/2h^\vee}$$

where

$$M^* = \begin{cases} \sum_{i=1}^{l} \mathbb{Z}\alpha_i & \text{for types } \tilde{B}_l^t, \tilde{C}_l^t, \tilde{F}_4^t, \tilde{G}_2^t \\ \sum_{\alpha_i \text{ long}} \tfrac{1}{2}\mathbb{Z}\alpha_i + \sum_{\alpha_i \text{ short}} \mathbb{Z}\alpha_i & \text{for type } \tilde{C}_l' \\ \tfrac{1}{2}\mathbb{Z}\alpha_1 & \text{for type } \tilde{A}_1'. \end{cases}$$

We now give some examples to illustrate Macdonald's identity. Suppose first that L has type $\tilde{A}_1$. Then L^0 has type A_1, $\Phi^0 = \{\alpha_1, -\alpha_1\}$, $h^\vee = 2$ and $M^* = \mathbb{Z}\alpha_1$. Moreover $\langle \alpha_1, \alpha_1 \rangle = 2\frac{c_1}{a_1} = 2$ and $\rho^0 = \tfrac{1}{2}\alpha_1$. Let $z = e_{-\alpha_1}$. The left-hand side of Macdonald's identity is

$$\prod_{n>0} (1-q^n)(1-q^n z)(1-q^n z^{-1}).$$

Now $\chi^0(n\alpha_1) = \dfrac{e_{n\alpha_1} - e_{-(n+1)\alpha_1}}{1 - e_{-\alpha_1}} = \dfrac{z^{-n} - z^{n+1}}{1-z}$. We also have

$$c(2n\alpha_1) = \langle (2n+\tfrac{1}{2})\alpha_1, (2n+\tfrac{1}{2})\alpha_1 \rangle - \langle \tfrac{1}{2}\alpha_1, \tfrac{1}{2}\alpha_1 \rangle$$
$$= 4n(2n+1).$$

Thus the right-hand side of Macdonald's identity is

$$\sum_{n \in \mathbb{Z}} \frac{z^{-2n} - z^{2n+1}}{1-z} q^{n(2n+1)}.$$

This can be written in the convenient form

$$\frac{1}{1-z} \sum_{n \in \mathbb{Z}} \left(z^{-2n} q^{n(2n+1)} - z^{2n+1} q^{n(2n+1)} \right) = \frac{1}{1-z} \sum_{m \in \mathbb{Z}} (-1)^m z^m q^{m(m-1)/2}.$$

Multiplying both sides of the identity by $1-z$ we obtain:

Proposition 20.5 (*Macdonald's identity for type* $\tilde{A}_1$).

$$\prod_{n>0} (1-q^n)(1-q^{n-1}z)(1-q^n z^{-1}) = \sum_{m \in \mathbb{Z}} (-1)^m z^m q^{m(m-1)/2}.$$

This is a classical identity known as **Jacobi's triple product identity**.

As a second example we suppose L has type $\tilde{A}'_1$. Then L^0 has type A_1 and $\Phi^0 = \{\alpha_1, -\alpha_1\}$ as before, but we now have $h^\vee = 3$ and $M^* = \frac{1}{2}\mathbb{Z}\alpha_1$. We have $a_0 = 2, a_1 = 1, c_0 = 1, c_1 = 2$, thus

$$\langle \alpha_1, \alpha_1 \rangle = \frac{2c_1}{a_1} = 4 \quad \text{and} \quad \delta = 2\alpha_0 + \alpha_1.$$

We write $e_{-\alpha_0} = z$. Then $e_{-\alpha_1} = z^{-2}q$. The left-hand side of Macdonald's identity is

$$\prod_{n>0} (1-q^n)(1-q^n z^{-1})(1-q^{n-1} z)(1-q^{2n+1} z^{-2})(1-q^{2n-1} z^2).$$

We have

$$c\left(\tfrac{3}{2}n\alpha_1\right) = \left(\tfrac{9}{4}n^2 + \tfrac{3}{2}n\right)\langle \alpha_1, \alpha_1 \rangle = 9n^2 + 6n.$$

Also

$$\chi^0\left(\tfrac{3}{2}n\alpha_1\right) = \frac{(z^{-2}q)^{-\frac{3}{2}n} - (z^{-2}q)^{\frac{3}{2}n+1}}{1 - z^{-2}q}.$$

Thus the right-hand side of the identity is

$$\sum_{n \in \mathbb{Z}} \frac{\left((z^{-2}q)^{-\frac{3}{2}n} - (z^{-2}q)^{\frac{3}{2}n+1}\right)}{1 - z^{-2}q} q^{\frac{n(3n+2)}{2}}$$

$$= \frac{1}{1-z^{-2}q}\left(\sum_{n \in \mathbb{Z}} z^{3n} q^{\frac{n(3n-1)}{2}} - \sum_{n \in \mathbb{Z}} z^{-3n-2} q^{\frac{(n+1)(3n+2)}{2}}\right)$$

$$= \frac{1}{1-z^{-2}q}\left(\sum_{n \in \mathbb{Z}} z^{3n} q^{\frac{n(3n-1)}{2}} - \sum_{n \in \mathbb{Z}} z^{-3n+1} q^{\frac{n(3n-1)}{2}}\right)$$

$$= \frac{1}{1-z^{-2}q} \sum_{n \in \mathbb{Z}} \left(z^{3n} - z^{-3n+1}\right) q^{\frac{n(3n-1)}{2}}.$$

We multiply both sides of the identity by $1 - z^{-2}q$ and obtain

Proposition 20.6 (*Macdonald's identity for type $\tilde{A}'_1$*).

$$\prod_{n>0} (1-q^n)(1-q^n z^{-1})(1-q^{n-1} z)(1-q^{2n-1} z^{-2})(1-q^{2n-1} z^2)$$

$$= \sum_{n \in \mathbb{Z}} \left(z^{3n} - z^{-3n+1}\right) q^{\frac{n(3n-1)}{2}}.$$

This is also a classical identity known as the **quintuple product identity**.

20.2 Specialisations of Macdonald's identities

We can obtain some striking identities, simpler than the original Macdonald identities, by specialising the latter identities in various ways. One way of specialising is simply to replace e_α by 1 for all $\alpha \in \Phi^0$. When this is done the expression $\chi^0(\lambda)$ is replaced by $d^0(\lambda)$ where

$$d^0(\lambda) = \frac{\prod_{\alpha \in (\Phi^0)^+} \langle \lambda + \rho^0, \alpha \rangle}{\prod_{\alpha \in (\Phi^0)^+} \langle \rho^0, \alpha \rangle}.$$

This is shown in Theorem 12.19. The identities obtained by specialisation in this way involve Euler's ϕ-function

$$\phi(q) = (1-q)(1-q^2)(1-q^3)\ldots$$

If we specialise the identity of Theorem 20.3 we obtain the following.

Theorem 20.7 (*Macdonald's ϕ-function identity*).

$$\phi(q)^{\dim L^0} = \sum_{\alpha \in M^*} d^0(h^\vee \alpha) q^{c(h_\alpha^\vee)/2h^\vee}.$$

Proof. The left-hand side of the specialised identity is

$$\phi(q)^{l+|\Phi^0|} = \phi(q)^{\dim L^0}.$$

On the right-hand side $\chi^0(h^\vee \alpha)$ specialises to $d^0(h^\vee \alpha)$. $\square$

We give some examples of this ϕ-function identity.

Type $\tilde{A}_1$

$$\phi(q)^3 = \sum_{n_1 \in \mathbb{Z}} (4n_1+1) q^{n_1(2n_1+1)}.$$

Type $\tilde{A}_2$

$$\phi(q)^8 = \sum_{(n_1,n_2) \in \mathbb{Z}^2} \tfrac{1}{2}(6n_1-3n_2+1)(-3n_1+6n_2+1)(3n_1+3n_2+2)$$

$$\times q^{3n_1^2-3n_1 n_2+3n_2^2+n_1+n_2}.$$

Type $\tilde{C}_2$

$$\phi(q)^{10} = \sum_{(n_1,n_2) \in \mathbb{Z}^2} (12n_1-6n_2+1)(-6n_1+6n_2+1)(2n_2+1)(3n_1+1)$$

$$\times q^{6n_1^2-6n_1 n_2+3n_2^2+n_1+n_2}.$$

Type $\tilde{G}_2$

$$\phi(q)^{14} = \sum_{(n_1,n_2)\in\mathbb{Z}^2} \tfrac{1}{15}(8n_1-12n_2+1)(-12n_1+24n_2+1)(3n_1-3n_2+1)$$
$$\times (12n_2+5)(-2n_1+6n_2+1)(4n_1+3)$$
$$\times q^{4n_1^2-12n_1n_2+12n_2^2+n_1+n_2}.$$

We next specialise the identities of Theorem 20.4 for twisted affine Kac–Moody algebras.

Theorem 20.8 *(Macdonald's twisted ϕ-function identities).*

(a) *The left-hand side of the identity is given as follows.*

$$\begin{array}{ll}\tilde{B}_l^t & \phi(q)^{2l^2-l-1}\phi(q^2)^{2l+1} \\ \tilde{C}_l^t & \phi(q)^{2l+1}\phi(q^2)^{2l^2-l-1} \\ \tilde{F}_4^t & \phi(q)^{26}\phi(q^2)^{26} \\ \tilde{G}_2^t & \phi(q)^7\phi(q^3)^7 \\ \tilde{C}_l' & \phi(q^{\frac{1}{2}})^{2l}\phi(q)^{2l^2-3l}\phi(q^2)^{2l} \\ \tilde{A}_1' & \phi(q^{\frac{1}{2}})^2\phi(q)^{-1}\phi(q^2)^2 \end{array}$$

(b) *The right-hand side of the identity is*

$$\sum_{\alpha\in M^*} d^0(h^\vee\alpha)\, q^{c(h^\vee\alpha)/2h^\vee}$$

where M^* is as in Theorem 20.4 (b). □

We give some examples of twisted ϕ-function identities.

Type $\tilde{A}_1'$

$$\phi(q^{\frac{1}{2}})^2 \phi(q)^{-1} \phi(q^2)^2 = \sum_{n_1\in\mathbb{Z}} (3n_1+1)\, q^{\frac{1}{2}n_1(3n_1+2)}.$$

Type $\tilde{C}_2^t$

$$\phi(q)^5\phi(q^2)^5 = \sum_{(n_1,n_2)\in\mathbb{Z}^2} \tfrac{1}{3}(8n_1-4n_2+1)(-8n_1+8n_2+1)$$
$$\times (8n_1+3)(2n_2+1)$$
$$\times q^{8n_1^2-8n_1n_2+4n_2^2+2n_1+n_2}.$$

20.2 Specialisations of Macdonald's identities

Type $\tilde{C}'_2$

$$\phi\left(q^{\frac{1}{2}}\right)^4 \phi(q)^2 \phi\left(q^2\right)^4 = \sum_{(n_1,n_2)\in\mathbb{Z}^2} \tfrac{1}{6}(10n_1 - 5n_2 + 1)(-5n_1 + 5n_2 + 1)$$
$$\times (5n_2 + 3)(5n_1 + 2)$$
$$\times q^{5n_1^2 - 5n_1 n_2 + \frac{5}{2}n_2^2 + n_1 + n_2}.$$

Type $\tilde{G}^t_2$

$$\phi(q)^7 \phi\left(q^3\right)^7 = \sum_{(n_1,n_2)\in\mathbb{Z}^2} \tfrac{1}{10}(12n_1 - 18n_2 + 1)(-6n_1 + 12n_2 + 1)$$
$$\times (-3n_1 + 9n_2 + 2)(6n_1 + 5)(3n_1 - 3n_2 + 1)$$
$$\times (2n_2 + 1) q^{6n_1^2 - 18n_1 n_2 + 18n_2^2 + n_1 + 3n_2}.$$

Another possibility to obtain specialised identities in one variable from Macdonald's identity is to apply a homomorphism

$$\theta : \mathbb{C}\left[\left[e_{-\alpha_0}, e_{-\alpha_1}, \ldots, e_{-\alpha_l}\right]\right] \to \mathbb{C}[[q]]$$

between rings of formal power series, given by

$$\theta\left(e_{-\alpha_0}\right) = q^{s_0}, \theta\left(e_{-\alpha_1}\right) = q^{s_1}, \ldots, \theta\left(e_{-\alpha_l}\right) = q^{s_l}$$

where $s_0, s_1, \ldots, s_l$ are non-negative integers. Of course under such a specialisation $e_{-\delta}$ would be mapped to $q^{a_0 s_0 + \cdots + a_l s_l}$, so that q would have to be replaced by this power of q in our earlier description of Macdonald's identity.

For example in type $\tilde{A}_1$ we obtain the following.

Proposition 20.9 (*Macdonald's 1-variable identity for $\tilde{A}_1$*).

$$\prod_{n>0}\left(1 - q^{(s_0+s_1)n}\right)\left(1 - q^{s_0(n-1)+s_1 n}\right)\left(1 - q^{s_0 n + s_1(n-1)}\right)$$
$$= \sum_{m\in\mathbb{Z}} (-1)^m q^{s_0 \frac{m(m-1)}{2} + s_1 \frac{m(m+1)}{2}}. \qquad \square$$

We mention some explicit examples of this identity. If $(s_0, s_1) = (1, 1)$ we obtain

$$\frac{\phi(q)^2}{\phi(q^2)} = \sum_{m\in\mathbb{Z}} (-1)^m q^{m^2}$$

that is

$$(1-q)^2 (1-q^2)(1-q^3)^2 (1-q^4)(1-q^5)^2 (1-q^6) \cdots$$
$$= 1 - 2q + 2q^4 - 2q^9 + 2q^{16} - \cdots.$$

This is a classical formula of Gauss.

Next consider the example given by $(s_0, s_1) = (2, 1)$. Then we obtain

$$\phi(q) = \sum_{m \in \mathbb{Z}} (-1)^m q^{\frac{m(3m-1)}{2}}$$

that is

$$(1-q)(1-q^2)(1-q^3)(1-q^4)(1-q^5)(1-q^6)\cdots$$
$$= 1 - q - q^2 + q^5 + q^7 - q^{12} - q^{15} + q^{22} + q^{26} - \cdots.$$

This is a well known formula of Euler.

Many additional formulae can be obtained by taking different values of (s_0, s_1) or different affine Kac–Moody algebras.

20.3 Irreducible modules for affine algebras

We next consider the weights of the irreducible modules $L(\lambda), \lambda \in X^+$, for the affine Kac–Moody algebra $L(A)$. We recall that X is the set of $\lambda \in H^*$ with $\lambda(h_i) \in \mathbb{Z}$ for $i = 0, 1, \ldots, l$ and X^+ is the set of $\lambda \in X$ with $\lambda(h_i) \geq 0$ for $i = 0, 1, \ldots, l$.

It is convenient to introduce the **fundamental weights** $\omega_0, \omega_1, \ldots, \omega_l$. ω_i is the element of X^+ defined by

$$\omega_i(h_j) = \delta_{ij} \qquad \omega_i(d) = 0.$$

Since the imaginary root δ satisfies

$$\delta(h_j) = 0 \qquad \delta(d) = 1$$

we see that $\omega_0, \omega_1, \ldots, \omega_l, \delta$ form a basis of H^*.

If $\lambda \in H^*$ satisfies

$$\lambda = \xi_0 \omega_0 + \xi_1 \omega_1 + \cdots + \xi_l \omega_l + \xi \delta$$

then λ lies in X if and only if $\xi_i \in \mathbb{Z}$ for $i = 0, 1, \ldots, l$. ξ can be any element of $\mathbb{C}$. Also $\lambda \in X^+$ if and only if $\xi_i \in \mathbb{Z}$ and $\xi_i \geq 0$ for $i = 0, 1, \ldots, l$.

Now every weight μ of $L(\lambda)$ has form $\mu = \lambda - m_0 \alpha_0 - m_1 \alpha_1 - \cdots - m_l \alpha_l$ for certain $m_i \in \mathbb{Z}$ with $m_i \geq 0$. Since $\alpha_i(c) = 0$ for $i = 0, 1, \ldots, l$ we have $\mu(c) = \lambda(c)$. Thus all the weights μ of $L(\lambda)$ have the same value of $\mu(c)$. Since $c = c_0 h_0 + c_1 h_1 + \cdots + c_l h_l$ we have $\lambda(c) = \sum_{i=0}^{l} c_i \lambda(h_i)$ and so, for $\lambda \in X^+$, $\lambda(c)$ is a non-negative integer. The integer $\lambda(c)$ is called the **level** of the module $L(\lambda)$.

20.3 Irreducible modules for affine algebras

Proposition 20.10 *If $L(\lambda)$ has level 0 then $\lambda = \xi \delta$ for some $\xi \in \mathbb{C}$ and $\dim L(\lambda) = 1$.*

Proof. If $\lambda(c) = 0$ then $\lambda(h_i) = 0$ for $i = 0, 1, \ldots, l$. Writing $\lambda = \xi_0 \omega_0 + \cdots + \xi_l \omega_l + \xi \delta$ we see that $\xi_i = 0$ for $i = 0, 1, \ldots, l$, hence $\lambda = \xi \delta$. Kac' character formula then shows that $\operatorname{ch} L(\xi \delta) = e_{\xi \delta}$. Thus $\dim L(\xi \delta) = 1$. □

Since the modules $L(\lambda)$ of level 0 are trivial 1-dimensional modules we shall subsequently concentrate on modules $L(\lambda), \lambda \in X^+$, of level greater than 0.

If μ is a weight of $L(\lambda)$ then so is $w(\mu)$ for any $w \in W$, by Proposition 19.13. We take a $w \in W$ for which the height of $\lambda - w(\mu)$ is minimal and put $\nu = w(\mu)$. Since $s_i(\nu) = \nu - \nu(h_i) \alpha_i$ the minimality of the height shows that $\nu(h_i) \geq 0$. Thus for any weight μ of $L(\lambda)$ there exists $w \in W$ with $w(\mu) \in X^+$. We now show that the converse is true also.

Theorem 20.11 *Let $\lambda \in X^+$ have $\lambda(c) > 0$. Then $\mu \in X$ is a weight of $L(\lambda)$ if and only if there exists $w \in W$ such that $w(\mu) \in X^+$ and $w(\mu) \prec \lambda$.*

Proof. It will be sufficient to show that if $\mu \in X^+$ with $\mu \prec \lambda$ then μ is a weight of $L(\lambda)$. The proof of this is non-trivial and reminiscent of that of Proposition 16.23.

Let $\mu = \lambda - \alpha$ where $\alpha = \sum k_i \alpha_i$ and each $k_i \geq 0$. We may assume $k_i > 0$ for some i. $\operatorname{supp} \alpha$ is the set of i for which $k_i > 0$. We first show that every connected component of $\operatorname{supp} \alpha$ contains an i with $\lambda(h_i) > 0$. Suppose if possible there exists a connected component S of $\operatorname{supp} \alpha$ with $\lambda(h_i) = 0$ for all $i \in S$. We have

$$L(\lambda)_\mu \subset \mathfrak{U}(N^-)_{-\alpha} v_\lambda$$

where v_λ is a highest weight vector of $L(\lambda)$ and, by the PBW basis theorem, $\mathfrak{U}(N^-)_{-\alpha}$ is spanned by elements of the form

$$\prod_{\beta \in \Phi^+} e_{-\beta}^{k_\beta}$$

where $k_\beta \geq 0, \sum k_\beta \beta = \alpha$, and each β involves fundamental roots which all lie in the same connected component of $\operatorname{supp} \alpha$. (We recall from Proposition 16.21 that $\operatorname{supp} \beta$ is connected.) Now the $e_{-\beta}$ with fundamental roots in different connected components of $\operatorname{supp} \alpha$ commute with one another, so we may bring the $e_{-\beta}$ with fundamental roots in S to the right of the above product. But for such β we have $e_{-\beta} v_\lambda = 0$.

For $f_i v_\lambda = 0$ for each $i \in S$ by Theorem 19.19, since $\lambda(h_i) = 0$. It follows that

$$\mathfrak{U}(N^-)_{-\alpha} v_\lambda = 0$$

and so $L(\lambda)_\mu = 0$, a contradiction. Hence there exists $i \in S$ with $\lambda(h_i) > 0$.

Now let Ψ be defined by

$$\Psi = \{\gamma \in Q^+ \,;\, \gamma \prec \alpha, \lambda - \gamma \text{ is a weight of } L(\lambda)\}.$$

The set Ψ is finite. Let $\beta \in \Psi$ be an element of maximal height. Then $\beta \prec \alpha$. We aim to show that $\beta = \alpha$ and hence that $\lambda - \alpha$ is a weight of $L(\lambda)$. Let $\beta = \sum m_i \alpha_i$ with each $m_i \geq 0$. We have $\alpha = \sum k_i \alpha_i$ with $m_i \leq k_i$ for each i.

Let $I = \{0, 1, \ldots, l\}$ and J be the subset of I given by $J = \{i \in I \,;\, k_i = m_i\}$. We aim to show that $J = I$ and so that $\beta = \alpha$. Suppose if possible that $J \neq I$. Consider the non-empty subset of I given by $\operatorname{supp}\alpha - (\operatorname{supp}\alpha \cap J)$. This set splits into connected components. Let M be a connected component of $\operatorname{supp}\alpha - (\operatorname{supp}\alpha \cap J)$. Let $i \in M$. Then $\lambda - \beta$ is a weight of $L(\lambda)$ but $\lambda - \beta - \alpha_i$ is not. Thus $(\lambda - \beta)(h_i) \leq 0$. Also $\mu(h_i) \geq 0$ since $\mu \in X^+$ and so $(\lambda - \alpha)(h_i) \geq 0$. Thus we have

$$\alpha(h_i) \leq \lambda(h_i) \leq \beta(h_i).$$

Let $\gamma = \sum_{j \in M} (k_j - m_j) \alpha_j$. We have $k_j - m_j > 0$ for all $j \in M$. We also have

$$\gamma(h_i) = \sum_{j \in M} (k_j - m_j) A_{ij}.$$

However, $\gamma(h_i) = (\alpha - \beta)(h_i)$ since $\operatorname{supp}(\alpha - \beta) = \operatorname{supp}\alpha - J$ and M is a connected component of $\operatorname{supp}\alpha - J$. Thus $\gamma(h_i) \leq 0$ for each $i \in M$.

Let A_M be the principal minor (A_{ij}) for $i, j \in M$. Let u be the column vector with entries $k_i - m_i$ for $i \in M$. Then we have $u > 0$ and $A_M u \leq 0$. If M has finite type $A_M(-u) \geq 0$ would imply $-u > 0$ or $-u = 0$. Thus M does not have finite type. Since M is a subset of I which has affine type we must have $M = I$ by Lemma 15.13. Thus $\operatorname{supp}\alpha = I$ and $J = \phi$. But then, for all $i \in I$, $\lambda - \beta$ is a weight of $L(\lambda)$ but $\lambda - \beta - \alpha_i$ is not. Thus $(\lambda - \beta)(h_i) \leq 0$ for all $i \in I$. Hence $\alpha(h_i) \leq \lambda(h_i) \leq \beta(h_i)$ for all $i \in I$. We now have $u > 0$ and $Au \leq 0$. Since A is affine we can deduce $Au = 0$. This shows that $\alpha(h_i) = \beta(h_i)$ for all $i \in I$. Hence $\alpha(h_i) = \lambda(h_i)$ for all $i \in I$, that is $\mu(h_i) = 0$ for each i. But then we have $\mu(c) = 0$, and so $\lambda(c) = 0$, a contradiction. □

Corollary 20.12 *If μ is a weight of $L(\lambda)$ then $\mu - \delta$ is also a weight.*

20.3 Irreducible modules for affine algebras

Proof. Since μ is a weight there exists $w \in W$ such that $w(\mu) \in X^+$. Then $w(\mu - \delta) = w(\mu) - \delta \in X^+$. Since $w(\mu) - \delta \prec \lambda$ it follows from Theorem 20.11 that $w(\mu) - \delta$ is a weight of $L(\lambda)$. Hence $\mu - \delta$ is also a weight. $\square$

It follows from this corollary that $\mu - i\delta$ is a weight for all positive integers i. On the other hand there exist only finitely many positive integers i such that $\mu + i\delta \prec \lambda$.

Definition *A weight μ of $L(\lambda)$ is called a **maximal weight** if $\mu + \delta$ is not a weight.*

Corollary 20.13 *For each weight μ of $L(\lambda)$ there are a unique maximal weight ν and a unique non-negative integer i such that $\mu = \nu - i\delta$.*

Proof. Consider the sequence $\mu, \mu + \delta, \mu + 2\delta, \ldots$ There exists i such that $\mu + i\delta$ is a weight of $L(\lambda)$ but $\mu + (i+1)\delta$ is not a weight. Let $\nu = \mu + i\delta$. Then ν is a maximal weight of $L(\lambda)$ and $\mu = \nu - i\delta$.

If $\mu = \nu' - i'\delta$ where ν' is a maximal weight and i' a non-negative integer we show $\nu = \nu'$ and $i = i'$. Otherwise we may assume $i < i'$. Then $\nu' = \nu + (i' - i)\delta$ is a weight. By Corollary 20.12 $\nu + \delta$ is also a weight. Thus ν is not a maximal weight and we have a contradiction. $\square$

A **string of weights** of $L(\lambda)$ is a set

$$\nu, \nu - \delta, \nu - 2\delta, \ldots$$

where ν is a maximal weight. Each weight lies in a unique string of weights. Thus it is natural to consider the set of maximal weights of $L(\lambda)$.

Proposition 20.14 *The set of maximal weights of $L(\lambda)$, $\lambda \in X^+$, is invariant under the Weyl group.*

Proof. Let $w \in W$. Then μ is a weight if and only if $w(\mu)$ is a weight. Thus if μ is a maximal weight $w(\mu)$ is a weight but $w(\mu) + \delta = w(\mu + \delta)$ is not a weight. Thus $w(\mu)$ is a maximal weight. $\square$

Corollary 20.15 *Each maximal weight of $L(\lambda)$, $\lambda \in X^+$, has form $w(\mu)$ where $w \in W$ and μ is a dominant maximal weight.*

We shall therefore consider the set of dominant maximal weights of $L(\lambda)$. We shall show that $L(\lambda)$ has only finitely many dominant maximal weights.

We recall from Section 17.3 that the fundamental alcove $A^* \subset (H_\mathbb{R}^0)^*$ was defined by

$$A^* = \left\{ \lambda \in (H_\mathbb{R}^0)^* \; ; \; \lambda(h_i) > 0 \quad \text{for } i = 1, \ldots, l \; ; \quad \lambda(h_\theta) < \frac{1}{a_0} \right\}$$
$$= \left\{ \lambda \in (H_\mathbb{R}^0)^* \; ; \; \langle \lambda, \alpha_i \rangle > 0 \quad \text{for } i = 1, \ldots, l \; ; \quad \langle \lambda, \theta \rangle < 1 \right\}.$$

Its closure $\overline{A^*}$ is a fundamental region for the action of W on $(H_\mathbb{R}^0)^*$.

We also recall that

$$H^* = (H^0)^* \oplus (\mathbb{C}\gamma + \mathbb{C}\delta)$$

where $(H^0)^*$ is embedded in H^* by assuming $\lambda(c) = 0$, $\lambda(d) = 0$ for $\lambda \in (H^0)^*$. By Lemma 20.1 we have, for $\lambda \in H^*$,

$$\lambda = \lambda^0 + \lambda(c)\gamma + a_0^{-1}\lambda(d)\delta$$

where $\lambda^0 \in (H^0)^*$. Let $Q^0 \subset (H^0)^*$ be the set of λ^0 given by λ in the root lattice $Q \subset H^*$.

Proposition 20.16 *Let $\lambda \in X^+$ have level $\lambda(c) = k > 0$. Then the map $\mu \to \mu^0$ gives a bijection between the set of dominant maximal weights of $L(\lambda)$ and $(\lambda^0 + Q^0) \cap \overline{kA^*}$.*

Proof. Let μ be a dominant maximal weight of $L(\lambda)$. Then $\mu = \lambda - m_0 \alpha_0 - \cdots - m_l \alpha_l$ for certain $m_i \in \mathbb{Z}$ with $m_i \geq 0$. Hence $\mu^0 = \lambda^0 - (m_0 \alpha_0 + \cdots + m_l \alpha_l)^0$ and so $\mu^0 \in \lambda^0 + Q^0$.

Now $\mu = \mu^0 + k\gamma + a_0^{-1}\mu(d)\delta$. Since $\mu \in X^+$ we have $\mu(h_i) \geq 0$ for $i = 0, 1, \ldots, l$. Now $\gamma(h_i) = \delta(h_i) = 0$ for $i = 1, \ldots, l$ and so $\mu^0(h_i) \geq 0$ for $i = 1, \ldots, l$. We also have

$$\langle \mu^0, \theta \rangle = \langle \mu, \theta \rangle = \langle \mu, \delta - a_0 \alpha_0 \rangle = \mu(c) - \mu(h_0) = k - \mu(h_0).$$

Since $\mu(h_0) \geq 0$ we have $\langle \mu^0, \theta \rangle \leq k$. Thus $\mu^0 \in \overline{kA^*}$.

Hence $\mu \to \mu^0$ maps dominant maximal weights of $L(\lambda)$ into $(\lambda^0 + Q^0) \cap \overline{kA^*}$. We wish to show this map is bijective. We first show it is surjective. Let $\nu \in (\lambda^0 + Q^0) \cap \overline{kA^*}$. Then, since $\alpha_i^0 = \alpha_i$ for $i = 1, \ldots, l$ and

$$\alpha_0^0 = (-a_0^{-1}\theta + a_0^{-1}\delta)^0 = -a_0^{-1}\theta$$

we have

$$\nu = \lambda^0 + k_1 \alpha_1 + \cdots + k_l \alpha_l - k_0 a_0^{-1}\theta$$

20.3 Irreducible modules for affine algebras

for certain $k_0, k_1, \ldots, k_l \in \mathbb{Z}$. Since $\theta = a_1\alpha_1 + \cdots + a_l\alpha_l$ we have

$$v = \lambda^0 + \left(m - k_0 a_0^{-1}\right)\theta - (ma_1 - k_1)\alpha_1 - \cdots - (ma_l - k_l)\alpha_l.$$

We choose $m \in \mathbb{Z}$ with $m \geq k_i/a_i$ for $i = 0, 1, \ldots, l$. Then

$$v = \lambda^0 + m_0\left(a_0^{-1}\theta\right) - m_1\alpha_1 - \cdots - m_l\alpha_l$$

where $m_i = ma_i - k_i$ for $i = 0, 1, \ldots, l$. Thus the m_i are non-negative integers for $i = 0, 1, \ldots, l$. Let $\mu = \lambda - m_0\alpha_0 - \cdots - m_l\alpha_l$. Then

$$\mu^0 = \lambda^0 + m_0\left(a_0^{-1}\theta\right) - m_1\alpha_1 - \cdots - m_l\alpha_l = v.$$

We show that $\mu \in X^+$. We have $\mu(h_i) = \mu^0(h_i) = v(h_i) \geq 0$ for $i = 1, \ldots, l$. Also $\mu(h_0) = k - \langle \mu^0, \theta \rangle = k - \langle v, \theta \rangle \geq 0$. Hence $\mu \in X^+$ and $\mu \prec \lambda$. Thus μ is a dominant weight of $L(\lambda)$ by Theorem 20.11. Hence we have shown that $v = \mu^0$ for some dominant weight μ of $L(\lambda)$. By replacing μ by the maximal weight in the chain of weights containing μ we may assume that μ is a dominant maximal weight. Thus our map is surjective.

To show the map is injective let μ, μ' be dominant maximal weights of $L(\lambda)$ with $\mu^0 = (\mu')^0$. We have

$$\mu = \mu^0 + k\gamma + a_0^{-1}\mu(d)\delta$$
$$\mu' = (\mu')^0 + k\gamma + a_0^{-1}\mu'(d)\delta$$

hence $\mu - \mu' = a_0^{-1}(\mu(d) - \mu'(d))\delta$. Now $\lambda - \mu \in Q$ and $\lambda - \mu' \in Q$ hence $\mu - \mu' \in Q$ and $a_0^{-1}(\mu(d) - \mu'(d))\delta \in Q$. This shows that $a_0^{-1}(\mu(d) - \mu'(d)) \in \mathbb{Z}$. Thus $\mu = \mu' + r\delta$ for some $r \in \mathbb{Z}$. Since μ, μ' are both maximal weights we must have $r = 0$. Thus $\mu' = \mu$. $\square$

Corollary 20.17 *The set of dominant maximal weights of $L(\lambda)$, $\lambda \in X^+$, is finite.*

Proof. Q^0 is a lattice in $(H^0)^*$, that is a free abelian subgroup whose rank is the dimension of $(H^0)^*$. $\lambda^0 + Q^0$ is a coset of this lattice. On the other hand the set $\overline{kA^*}$ is bounded. Hence the intersection $(\lambda^0 + Q^0) \cap \overline{kA^*}$ must be finite. Thus the set of dominant maximal weights is also finite, by Proposition 20.16. $\square$

We now have a procedure for describing all weights of $L(\lambda)$, $\lambda \in X^+$. First determine the finite set $(\lambda^0 + Q^0) \cap \overline{kA^*}$ where $k = \lambda(c)$. For each element v in this finite set there is a unique dominant maximal weight μ of $L(\lambda)$ with $\mu^0 = v$. This gives the set of all dominant maximal weights. By applying elements of the Weyl group to these we obtain all maximal weights. Finally by subtracting positive integral multiples of δ from the maximal weights we obtain all weights of $L(\lambda)$.

We next consider the weights in a string

$$\mu, \mu - \delta, \mu - 2\delta, \ldots$$

We wish to show that the multiplicities of these weights form an increasing function as we move down the string, i.e. that $m_{\mu-(i+1)\delta} \geq m_{\mu-i\delta}$ for all $i \geq 0$. In order to do this we consider $L(\lambda)$ as a T-module where T is the subalgebra of $L(A)$ given by

$$T = \cdots \oplus L_{-2\delta} \oplus L_{-\delta} \oplus H \oplus L_\delta \oplus L_{2\delta} \oplus \cdots.$$

Thus T is spanned by H and the root spaces for the imaginary roots. The algebra T has a triangular decomposition

$$T = T^- \oplus H \oplus T^+$$

where $T^- = \sum_{i>0} L_{-i\delta}$, $T^+ = \sum_{i>0} L_{i\delta}$. One can define the category $\mathcal{O}$ of T-modules in a manner analogous to that in Section 19.1. One can also define Verma modules for T. If $\lambda \in H^*$ we define

$$M(\lambda) = \mathfrak{U}(T)/\mathfrak{U}(T)T^+ + \sum_{x \in H} \mathfrak{U}(T)(x - \lambda(x)).$$

This is the Verma module for T with highest weight λ. There is a bijection $\mathfrak{U}(T^-) \to M(\lambda)$ given by $u \to u m_\lambda$ where $m_\lambda \in M(\lambda)$ is the image of $1 \in \mathfrak{U}(T)$.

We shall investigate properties of Verma modules for T by considering the expression

$$\Omega_0 = 2 \sum_{i>0} \sum_j e_{-i\delta}^{(j)} e_{i\delta}^{(j)}$$

where $e_{i\delta}^{(j)}$ is a basis for $L_{i\delta}$ and $e_{-i\delta}^{(j)}$ is the dual basis for $L_{-i\delta}$. Thus

$$\left\langle e_{i\delta}^{(j)}, e_{-i\delta}^{(k)} \right\rangle = \delta_{jk}$$

and $\left[e_{i\delta}^{(j)}, e_{-i\delta}^{(k)} \right] = \delta_{jk} ic$ by Corollary 16.5.

Although the expression for Ω_0 is an infinite sum the action of Ω_0 on any T-module in category $\mathcal{O}$ is well defined, since all but a finite number of the terms will act as zero.

Lemma 20.18 *Let $\lambda \in H^*$ and $M(\lambda)$ be the associated Verma module for T. Let $u \in \mathfrak{U}(T)_{m\delta}$ where $m \in \mathbb{Z}$ and $m \neq 0$. Then $\Omega_0 u - u \Omega_0$ acts on $M(\lambda)$ in the same way as $-2\lambda(c) m u$.*

20.3 Irreducible modules for affine algebras

Proof. u is a linear combination of products of elements, each in $T_{r\delta}$ for some r with $r \neq 0$.

First suppose $u \in T_{r\delta}$. We assume that u is one of the basis elements $u = e_{r\delta}^{(j)}$. Then u commutes with all $e_{i\delta}^{(k)}, e_{-i\delta}^{(k)}$ except for $e_{-r\delta}^{(j)}$. Thus

$$\Omega_0 u - u\Omega_0 = 2\left(e_{-r\delta}^{(j)} e_{r\delta}^{(j)} e_{r\delta}^{(j)} - e_{r\delta}^{(j)} e_{-r\delta}^{(j)} e_{r\delta}^{(j)}\right)$$
$$= -2rce_{r\delta}^{(j)} = -2ruc = -2r\lambda(c)u$$

on $M(\lambda)$. The same will then apply to any $u \in T_{r\delta}$.

Next suppose $u = u_1 u_2$ where

$$\Omega_0 u_1 - u_1 \Omega_0 = -2\lambda(c)r_1 u_1$$
$$\Omega_0 u_2 - u_2 \Omega_0 = -2\lambda(c)r_2 u_2 \quad \text{on } M(\lambda)$$

Then

$$\Omega_0 u - u\Omega_0 = \Omega_0 u_1 u_2 - u_1 u_2 \Omega_0$$
$$= u_1 \Omega_0 u_2 - 2\lambda(c)r_1 u - u_1 \Omega_0 u_2 - 2\lambda(c)r_2 u$$
$$= -2\lambda(c)(r_1 + r_2)u \quad \text{on } M(\lambda).$$

The required result then follows for arbitrary $u \in \mathfrak{U}(T)_{m\delta}$ by taking linear combinations of such repeated products. $\square$

Proposition 20.19 *Let $\lambda \in H^*$ satisfy $\lambda(c) > 0$. Then the Verma module $M(\lambda)$ for T is irreducible.*

Proof. Suppose if possible that $M(\lambda)$ has a proper submodule K. Let v be a highest weight vector of K. Then $v \in M(\lambda)_{\lambda - m\delta}$ for some $m \in \mathbb{Z}$ with $m > 0$. Thus $v = um_\lambda$ for some $u \in \mathfrak{U}(T^-)_{-m\delta}$. We consider the actions

$$\Omega_0 : M(\lambda) \to M(\lambda) \qquad u : M(\lambda) \to M(\lambda).$$

By Lemma 20.18 we have

$$(\Omega_0 u - u\Omega_0) m_\lambda = 2\lambda(c) mum_\lambda.$$

Thus $\Omega_0 v - u(\Omega_0 m_\lambda) = 2\lambda(c)mv$. Now $\Omega_0 m_\lambda = 0$ and $\Omega_0 v = 0$ since m_λ and v are highest weight vectors in $M(\lambda)$ and K respectively. Thus $2\lambda(c)mv = 0$. But $v \neq 0, m > 0, \lambda(c) > 0$ and so we have a contradiction. Thus $M(\lambda)$ is irreducible. $\square$

We now consider the structure of $L(\lambda)$ as a T-module.

Proposition 20.20 *Suppose $\lambda \in X^+$ with $\lambda(c) > 0$. Then the T-module $L(\lambda)$ is completely reducible. Its irreducible components are Verma modules for T.*

Proof. Let U be the subspace of $L(\lambda)$ given by

$$U = \{v \in L(\lambda) \; ; \; T^+ v = 0\}.$$

Let B be a basis of U. We may choose B to be a basis of weight vectors of U, i.e. so that each element of B lies in a weight space $L(\lambda)_\mu$. Suppose $v \in B$ has weight μ. Then $T^+ v = 0$ and $xv = \mu(x)v$ for $x \in H$, hence $Tv = T^- v$. Let $M(\mu)$ be the Verma module for T with highest weight μ. Then we have a homomorphism of T-modules $M(\mu) \to Tv$ given by $um_\mu \to uv$ for $u \in T^-$. Now $\mu = \lambda - i\delta$ for some $i \geq 0$ hence $\mu(c) = \lambda(c) > 0$. Thus the Verma module $M(\mu)$ for T is irreducible by Proposition 20.19. Hence the homomorphism $M(\mu) \to Tv$ is an isomorphism and so Tv is a Verma module for T. Let $V = \sum_{v \in B} Tv$. We claim that this sum of T-modules is a direct sum. For consider

$$Tv \cap \sum_{\substack{v' \in B \\ v' \neq v}} Tv'.$$

Since the Verma module Tv is irreducible we have $Tv \cap U = \mathbb{C} v$. We also have

$$\left(\sum_{\substack{v' \in B \\ v' \neq v}} Tv' \right) \cap U = \sum_{v' \neq v} \mathbb{C} v'.$$

Since $v \notin \sum_{v' \neq v} \mathbb{C} v'$ we see that Tv is not contained in $\sum_{v' \neq v} Tv'$. Again, since Tv is irreducible we have $Tv \cap \sum_{v' \neq v} Tv' = O$. Hence $V = \bigoplus_{v \in B} Tv$. Thus V is a direct sum of Verma modules for T.

We wish to show that $V = L(\lambda)$. We suppose if possible that $V \neq L(\lambda)$. We consider the T-module $L(\lambda)/V$. Since $L(\lambda) = \bigoplus_\mu L(\lambda)_\mu$ and $V = \bigoplus_\mu V_\mu$ we have $L(\lambda)/V = \bigoplus_\mu (L(\lambda)/V)_\mu$. As $L(\lambda)/V$ is assumed to be non-zero we can find a weight μ of $L(\lambda)/V$ such that $\mu + i\delta$ is not a weight for any $i > 0$. Then $T^+(L(\lambda)/V)_\mu = O$, that is $T^+ L(\lambda)_\mu \subset V$.

We now consider the map $\Omega_0 : L(\lambda) \to L(\lambda)$. Since the action of Ω_0 preserves weight spaces we have $\Omega_0 : L(\lambda)_\mu \to L(\lambda)_\mu$. The weight space

$L(\lambda)_\mu$ is finite dimensional, so decomposes into a direct sum of generalised eigenspaces of Ω_0, given by

$$L(\lambda)_\mu = \bigoplus_{\zeta \in \mathbb{C}} \left(L(\lambda)_\mu\right)_\zeta$$

where $(\Omega_0 - \zeta 1)^k = 0$ on $\left(L(\lambda)_\mu\right)_\zeta$ for some k. Since $L(\lambda)_\mu$ does not lie in V there exists $\zeta \in \mathbb{C}$ such that $\left(L(\lambda)_\mu\right)_\zeta$ does not lie in V. We choose $v \in \left(L(\lambda)_\mu\right)_\zeta$ with $v \notin V$. Then

$$(\Omega_0 - \zeta 1)^k v = 0$$

and $\Omega_0 v \in V$, since $T^+ L(\lambda)_\mu \subset V$. If $\zeta \neq 0$ the polynomials $(t - \zeta)^k$ and t are coprime so we could deduce $v \in V$, a contradiction. Hence $\zeta = 0$ and $\Omega_0^k v = 0$.

Now $T^+ v \neq 0$ since $v \notin V$. So there exist $m > 0$ and $u \in \mathfrak{U}(T^+)_{m\delta}$ with $uv \neq 0$ and $T^+(uv) = 0$. Let $v' = uv$. Then $v' \neq 0$ and $\Omega_0 v' = 0$.

Now all the weights ν of $L(\lambda)$ satisfy $\nu(c) = \lambda(c)$. Thus we may apply the argument of Lemma 20.18 to $L(\lambda)$ and obtain

$$\Omega_0 u - u\Omega_0 = -2\lambda(c)mu \quad \text{on } L(\lambda).$$

Then

$$\Omega_0 uv - u\Omega_0 v = -2\lambda(c)muv$$

that is

$$(\Omega_0 + 2\lambda(c)m) v' = u\Omega_0 v.$$

It follows that

$$(\Omega_0 + 2\lambda(c)m)^2 v' = (\Omega_0 + 2\lambda(c)m)(u\Omega_0 v) = u\left(\Omega_0^2 v\right)$$

and continuing thus we obtain

$$(\Omega_0 + 2\lambda(c)m)^k v' = u\left(\Omega_0^k v\right) = 0.$$

But $\lambda(c) > 0$ and $m > 0$, thus the polynomials $(t + 2\lambda(c)m)^k$ and t are coprime. Thus $(\Omega_0 + 2\lambda(c)m)^k v' = 0$ and $\Omega_0 v' = 0$ imply $v' = 0$, a contradiction.

Thus we have obtained our required contradiction and can deduce that $V = L(\lambda)$ and $L(\lambda)$ is the direct sum of the irreducible T-modules Tv for $v \in B$, each of which is isomorphic to a Verma module for T. $\square$

Proposition 20.21 *Let μ be a weight of $L(\lambda)$ where $\lambda \in X^+$ and $\lambda(c) > 0$. Then the multiplicities of the weights $\mu, \mu - \delta$ satisfy $m_{\mu-\delta} \geq m_\mu$.*

Proof. This follows from Proposition 20.20. We choose a non-zero element $x \in L(A)_{-\delta}$. Consider the action of x on the T-module $L(\lambda)$. This T-module is a direct sum of Verma modules for T. Since $x \in T^-$, x acts on each Verma module for T injectively. Thus x acts on $L(\lambda)$ injectively. We have a map

$$L(\lambda)_\mu \to L(\lambda)_{\mu-\delta}$$
$$v \mapsto xv$$

which is injective, and so

$$\dim L(\lambda)_{\mu-\delta} \geq \dim L(\lambda)_\mu$$

that is $m_{\mu-\delta} \geq m_\mu$ as required. $\square$

Thus the multiplicities form an increasing sequence as we move down a string of weights for $L(\lambda)$.

20.4 The fundamental modules for $L(\tilde{A}_1)$

We now give an example of the situation described in Section 20.3. We consider the affine Kac–Moody algebra of type $\tilde{A}_1$. This has diagram

0 ⇔ 1

and Cartan matrix

$$A = \begin{pmatrix} 2 & -2 \\ -2 & 2 \end{pmatrix}.$$

We consider the irreducible modules $L(\omega_0)$, $L(\omega_1)$ where ω_0, ω_1 are the fundamental weights. By symmetry we need only determine the character of one of these. We shall consider the module $L(\omega_0)$.

In type $\tilde{A}_1$ we have $\delta = \alpha_0 + \alpha_1$ and $c = h_0 + h_1$, that is

$$a_0 = 1, \quad a_1 = 1, \quad c_0 = 1, \quad c_1 = 1.$$

We recall that

$$H^* = (H^0)^* \oplus (\mathbb{C}\gamma + \mathbb{C}\delta)$$

and that

$$\lambda = \lambda^0 + \lambda(c)\gamma + a_0^{-1}\lambda(d)\delta \quad \text{by Lemma 20.1.}$$

The root lattice Q is given by

$$Q = \mathbb{Z}\alpha_0 + \mathbb{Z}\alpha_1$$

20.4 The fundamental modules for $L(\tilde{A}_1)$

and we have $\alpha_0^0 = -\alpha_1$, $\alpha_1^0 = \alpha_1$. We have $(H^0)^* = \mathbb{C}\alpha_1$ and the lattice $Q^0 \subset (H^0)^*$ is given by $Q^0 = \mathbb{Z}\alpha_1$. We have $\theta = \alpha_1$ and $h_\theta = \frac{1}{a_0}(c - h_0) = h_1$. The closure of the fundamental alcove is given by

$$\bar{A}^* = \{\lambda \in (H_\mathbb{R}^0)^* \; ; \; \lambda(h_1) \geq 0, \lambda(h_\theta) \leq 1\}$$
$$= \{\lambda \in (H_\mathbb{R}^0)^* \; ; \; 0 \leq \lambda(h_1) \leq 1\}.$$

We have $\omega_0 = \gamma$ and $\gamma^0 = 0$. Thus

$$(\gamma^0 + Q^0) \cap \bar{A}^* = \{m\alpha_1 \; ; \; m \in \mathbb{Z}, 0 \leq 2m \leq 1\}$$
$$= \{0\}.$$

Thus by Proposition 20.16 the module $L(\gamma)$ has only one dominant maximal weight, which must be the highest weight γ. The other maximal weights are the transforms of γ under the affine Weyl group $W = \langle s_0, s_1 \rangle$. We have

$$s_0(\alpha_0) = -\alpha_0 \qquad s_0(\alpha_1) = 2\alpha_0 + \alpha_1$$
$$s_1(\alpha_0) = \alpha_0 + 2\alpha_1 \qquad s_1(\alpha_1) = -\alpha_1.$$

The action of s_0, s_1 on the basis γ, α_1, δ of H^* is given by

$$s_0(\gamma) = \gamma + \alpha_1 - \delta \qquad s_0(\alpha_1) = -\alpha_1 + 2\delta \qquad s_0(\delta) = \delta$$
$$s_1(\gamma) = \gamma \qquad s_1(\alpha_1) = -\alpha_1 \qquad s_1(\delta) = \delta.$$

The affine Weyl group W is an infinite dihedral group and has a semidirect product decomposition

$$W = t(Q^0) W^0 = W^0 t(Q^0)$$

where $Q^0 = M^* = \mathbb{Z}\alpha_1$ and $W^0 = \{1, s_1\}$. The translation t_μ for $\mu \in Q^0$ is given by

$$t_\mu(\lambda) = \lambda + \lambda(c)\mu - \left(\langle \lambda, \mu \rangle + \frac{1}{2}\langle \mu, \mu \rangle \lambda(c)\right)\delta$$

which in the present case gives

$$t_{m\alpha_1}(\gamma) = \gamma + m\alpha_1 - m^2\delta$$
$$t_{m\alpha_1}(\alpha_1) = \alpha_1 - 2m\delta$$
$$t_{m\alpha_1}(\delta) = \delta$$

for $m \in \mathbb{Z}$. The stabiliser of γ in W is W^0 and the maximal weights in $L(\gamma)$ have the form $\gamma + m\alpha_1 - m^2\delta$ for $m \in \mathbb{Z}$. The set of all weights of $L(\gamma)$ is $\gamma + m\alpha_1 - m^2\delta - k\delta$ for $m \in \mathbb{Z}, k \in \mathbb{Z}$ and $k \geq 0$. The weights $\gamma + m\alpha_1 - m^2\delta$

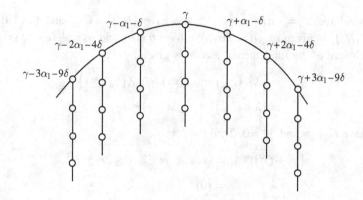

Figure 20.1 Maximal weights in $L(\gamma)$

have multiplicity 1, and $\gamma + m\alpha_1 - m^2\delta - k\delta$ has multiplicity depending only on k (i.e. independent of m). The weights are shown in Figure 20.1.

We shall determine the multiplicities of these weights. We use Kac' character formula

$$\operatorname{ch} L(\gamma) = \frac{\sum_{w \in W} \varepsilon(w) e_{w(\gamma+\rho)-\rho}}{\prod_{\alpha \in \Phi^+} (1 - e_{-\alpha})^{m_\alpha}}.$$

Now

$$\sum_{w \in W} \varepsilon(w) e_{w(\gamma+\rho)-\rho} = \sum_{w^0 \in W^0} \sum_{\mu \in \mathbb{Z}\alpha_1} \varepsilon(w^0) e_{w^0 t_\mu(\gamma+\rho)-\rho}$$

$$= \sum_{w^0 \in W^0} \varepsilon(w^0) \sum_{n \in \mathbb{Z}} e_{w^0 t_{n\alpha_1}(\gamma+\rho)-\rho}.$$

Now $\rho = \rho^0 + 2\gamma$ by Lemma 20.2 where $\rho^0 = \frac{1}{2}\alpha_1$. Thus $\rho = 2\gamma + \frac{1}{2}\alpha_1$ and $\gamma + \rho = 3\gamma + \frac{1}{2}\alpha_1$. Hence

$$t_{n\alpha_1}(\gamma+\rho) = 3\gamma + \left(3n + \frac{1}{2}\right)\alpha_1 - (3n^2 + n)\delta$$

so $t_{n\alpha_1}(\gamma+\rho) - \rho = \gamma + 3n\alpha_1 - (3n^2 + n)\delta$. Also

$$s_1 t_{n\alpha_1}(\gamma+\rho) = 3\gamma - (3n + \frac{1}{2})\alpha_1 - (3n^2 + n)\delta$$

so $s_1 t_{n\alpha_1}(\gamma+\rho) - \rho = \gamma - (3n+1)\alpha_1 - (3n^2 + n)\delta$. Thus

$$\sum_{w \in W} \varepsilon(w) e_{w(\gamma+\rho)-\rho} = e_\gamma \sum_{n \in \mathbb{Z}} \left(e_{3n\alpha_1} - e_{-(3n+1)\alpha_1}\right) e_{-(3n^2+n)\delta}.$$

We write $e_{-\alpha_1} = z$ and $e_{-\delta} = q^{1/2}$. Then our expression is

$$e_\gamma \sum_{n \in \mathbb{Z}} \left(z^{-3n} - z^{3n+1}\right) q^{n(3n+1)/2}.$$

20.4 The fundamental modules for $L(\tilde{A}_1)$

Now we may factorise this expression by using Macdonald's identity for type $\tilde{A}'_1$. By Proposition 20.6 it is equal to

$$e_\gamma \prod_{n>0} (1-q^n)(1-q^n z^{-1})(1-q^{n-1}z)(1-q^{2n-1}z^{-2})(1-q^{2n-1}z^2)$$

$$= e_\gamma(1-z) \prod_{n>0} (1-q^n)(1-q^n z^{-1})(1-q^n z)(1-q^{2n-1}z^{-2})(1-q^{2n-1}z^2)$$

$$= e_\gamma(1-z) \prod_{n>0} (1-q^n)(1-q^n z^{-1}) \left(1-q^{\frac{2n-1}{2}}z^{-1}\right)(1-q^n z)\left(1-q^{\frac{2n-1}{2}}z\right)$$

$$\times \left(1+q^{\frac{2n-1}{2}}z^{-1}\right)\left(1+q^{\frac{2n-1}{2}}z\right)$$

$$= e_\gamma(1-z) \prod_{k>0} (1-q^{k/2}z^{-1})(1-q^{k/2}z) \prod_{n>0}(1-q^n)$$

$$\times \left(1+q^{\frac{2n-1}{2}}z^{-1}\right)\left(1+q^{\frac{2n-1}{2}}z\right).$$

We now make use of Macdonald's identity for type $\tilde{A}_1$. By Proposition 20.5 this asserts that

$$\prod_{n>0}(1-q^n)(1-q^{n-1}z')(1-q^n z'^{-1}) = \sum_{n>0}(-1)^n \left(z'^n - z'^{-(n-1)}\right) q^{\frac{n(n-1)}{2}}.$$

Putting $z' = -z^{-1}q^{\frac{1}{2}}$ we obtain

$$\prod_{n>0}(1-q^n)\left(1+q^{\frac{2n-1}{2}}z^{-1}\right)\left(1+q^{\frac{2n-1}{2}}z\right) = \sum_{n>0}\left(z^{-n}q^{n^2/2} + z^{n-1}q^{(n-1)^2/2}\right)$$

$$= \sum_{n\in\mathbb{Z}} z^{-n}q^{n^2/2}.$$

Hence

$$\operatorname{ch} L(\gamma) = \frac{\sum_{w\in W} \varepsilon(w) e_{w(\gamma+\rho)-\rho}}{(1-z)\prod_{k>0}(1-q^{k/2}z^{-1})(1-q^{k/2}z)(1-q^{k/2})}$$

$$= \frac{e_\gamma \sum_{n\in\mathbb{Z}} z^{-n}q^{n^2/2}}{\prod_{k>0}(1-q^{k/2})} = \frac{\sum_{n\in\mathbb{Z}} e_{\gamma+n\alpha_1-n^2\delta}}{\prod_{k>0}(1-e_{-k\delta})}.$$

Now

$$\frac{1}{\prod_{k>0}(1-e_{-k\delta})} = \prod_{k>0}(1+e_{-k\delta}+e_{-2k\delta}+\cdots)$$

$$= \sum_{k\geq 0} p(k) e_{-k\delta}$$

where $p(k)$ is the number of partitions of k. Thus

$$\operatorname{ch} L(\gamma) = \sum_{n\in\mathbb{Z}} \sum_{k\geq 0} p(k) e_{\gamma+n\alpha_1-n^2\delta-k\delta}.$$

Hence we have proved

Proposition 20.22 *The weights of the fundamental module $L(\gamma)$ for $L(\tilde{A}_1)$ are $\gamma + n\alpha_1 - n^2\delta - k\delta$ for $n \in \mathbb{Z}$ and $k \geq 0$. This weight has multiplicity $p(k)$.* □

We note in particular that all the maximal weights $\gamma + n\alpha_1 - n^2\delta$ have multiplicity 1 and that the multiplicity of the weight $\mu - k\delta$ in the string with maximal weight μ depends only upon k and not on μ.

20.5 The basic representation

The module $L(\omega_0)$ for an affine Kac–Moody algebra $L(A)$ gives the so-called basic representation of $L(A)$. Since $\omega_0 = \gamma$ we have described the character of the basic representation of $L(\tilde{A}_1)$. We shall state without proof some generalisations of this character formula to other types of affine Kac–Moody algebras. For simplicity we shall concentrate on those of types $\tilde{A}_l$, $\tilde{D}_l$ and $\tilde{E}_l$.

Theorem 20.23 *The basic representation $L(\gamma)$ for the Kac–Moody algebra $L(A)$ of types $\tilde{A}_l, \tilde{D}_l, \tilde{E}_6, \tilde{E}_7, \tilde{E}_8$ has the following properties:*

(a) *γ is the unique dominant maximal weight of $L(\gamma)$.*

(b) *The set of all maximal weights is*
$$\{\gamma + \mu - \tfrac{1}{2}\langle \mu, \mu \rangle \delta \text{ for } \mu \in Q^0\}.$$

(c) *The set of all weights is*
$$\{\gamma + \mu - \tfrac{1}{2}\langle \mu, \mu \rangle \delta - k\delta \text{ for } \mu \in Q^0, \ k \in \mathbb{Z}, \ k \geq 0\}.$$

(d) *The character of the basic representation is*
$$\operatorname{ch} L(\gamma) = \frac{\sum_{\mu \in Q^0} e_{\gamma + \mu - \tfrac{1}{2}\langle \mu, \mu \rangle \delta}}{\left(\prod_{k>0}(1 - q^k)\right)^l}$$

where $q = e^{-\delta}$.

(e) *The multiplicity of the weight $\gamma + \mu - \tfrac{1}{2}\langle \mu, \mu \rangle \delta - k\delta$ is $p_l(k)$, the number of partitions of k into l colours. We have*
$$\frac{1}{(\prod_{k>0}(1-q^k))^l} = \sum_{k \geq 0} p_l(k) q^k.$$
□

20.5 The basic representation

The proof of this theorem can be found in the book of Kac, *Infinite-Dimensional Lie Algebras*, third edition, Chapter 12.

We shall also describe without proof how to obtain a realisation of the basic representation $L(\gamma)$ of $L(A)$ in types $\tilde{A}_l, \tilde{D}_l, \tilde{E}_l$. We first make some comments on differential operators. Let $R = \mathbb{C}[x_1, x_2, x_3, \ldots]$ be the polynomial ring over $\mathbb{C}$ in countably many variables and $\hat{R} = \mathbb{C}[[x_1, x_2, x_3, \ldots]]$ be the ring of formal power series in these variables. We shall consider differential operators on R with values in $\hat{R}$. An example is the partial derivative $\partial/\partial x_i$ or, more generally, the divided power $\frac{1}{m_i!}(\partial/\partial x_i)^{m_i}$. We also have finite products $\prod_i \frac{1}{m_i!}(\partial/\partial x_i)^{m_i}$ where $m = (m_1, m_2, m_3, \ldots)$ satisfies the conditions that $m_i \in \mathbb{Z}$, $m_i \geq 0$, and $m_i > 0$ for only finitely many i. We define $D_m : R \to \hat{R}$ by

$$D_m = \prod_i \frac{1}{m_i!}(\partial/\partial x_i)^{m_i}.$$

We also allow such operators combined with multiplication by elements of $\hat{R}$. Thus

$$\sum_m P_m D_m : R \to \hat{R}$$

is a differential operator, where $P_m \in \hat{R}$ and the sum over m will in general be infinite. $\sum_m P_m D_m$ is a linear map from R to $\hat{R}$. In fact each linear map from R to $\hat{R}$ has this form, as we now show.

Proposition 20.24 *Each linear map from R to $\hat{R}$ can be written as $\sum_m P_m D_m$ for a unique set of elements $P_m \in \hat{R}$.*

Proof. Let $M_m \in R$ be the monomial $M_m = \prod_i x_i^{m_i}$. The monomials M_m form a basis for R. We have $D_m(M_k) = 0$ unless $k_i \geq m_i$ for each i. We write this condition as $k \geq m$. We write $k > m$ if $k \geq m$ and $k \neq m$. We also have

$$D_m(M_k) = \binom{k}{m} M_{k-m} \quad \text{if } k \geq m$$

where $\binom{k}{m} = \prod_i \binom{k_i}{m_i}$ and $\binom{0}{0} = 1$.

Let $\Delta : R \to \hat{R}$ be the linear map given by $\Delta(M_m) = Q_m \in \hat{R}$. We show Δ is uniquely expressible in the form $\sum P_m D_m$. We have

$$\sum_m P_m D_m(M_k) = \sum_{m \leq k} P_m \binom{k}{m} M_{k-m}$$

$$= P_k + \sum_{m < k} P_m \binom{k}{m} M_{k-m}.$$

The condition we require on the P_m is that

$$P_k + \sum_{m<k} P_m \binom{k}{m} M_{k-m} = Q_k \quad \text{for all } k.$$

In particular $P_0 = Q_0$. Assuming inductively that P_m is uniquely determined for all $m < k$ we conclude that

$$P_k = Q_k - \sum_{m<k} P_m \binom{k}{m} M_{k-m}$$

is uniquely determined. □

Thus the set of differential operators from R to $\hat{R}$ is the set of all linear maps from R to $\hat{R}$.

We shall now consider certain special kinds of differential operators. Let $\lambda = (\lambda_1, \lambda_2, \lambda_3, \ldots)$ where $\lambda_i \in \mathbb{C}$. Here there may be infinitely many non-zero λ_i. Define $T_\lambda : R \to \hat{R}$ by

$$T_\lambda f(x_1, x_2, x_3, \ldots) = f(x_1 + \lambda_1, x_2 + \lambda_2, x_3 + \lambda_3, \ldots).$$

T_λ is clearly a linear map from R to $\hat{R}$. It may be written as a differential operator by using the Taylor expansion. We have

$$f(x_1 + \lambda_1, x_2 + \lambda_2, x_3 + \lambda_3, \ldots)$$

$$= \sum_m \frac{\lambda_1^{m_1} \lambda_2^{m_2} \lambda_3^{m_3}}{m_1! \, m_2! \, m_3!} \cdots (\partial/\partial x_1)^{m_1} (\partial/\partial x_2)^{m_2} (\partial/\partial x_3)^{m_3} \cdots f(x_1, x_2, x_3, \ldots)$$

$$= \sum_m \left(\prod_i \lambda_i^{m_i} \right) D_m f(x_1, x_2, x_3, \ldots).$$

Thus $T_\lambda = \sum_m \left(\prod_i \lambda_i^{m_i} \right) D_m$. The operator T_λ may also be written in the following convenient form. We have

$$f(x_1 + \lambda_1, x_2 + \lambda_2, x_3 + \lambda_3, \ldots)$$

$$= \left(\sum_{m_1} \frac{\lambda_1^{m_1}}{m_1!} (\partial/\partial x_1)^{m_1} \right) \left(\sum_{m_2} \frac{\lambda_2^{m_2}}{m_2!} (\partial/\partial x_2)^{m_2} \right) \cdots f(x_1, x_2, x_3, \ldots)$$

$$= \exp(\lambda_1 \partial/\partial x_1) \exp(\lambda_2 \partial/\partial x_2) \cdots f(x_1, x_2, x_3, \ldots)$$

$$= \exp(\lambda_1 \partial/\partial x_1 + \lambda_2 \partial/\partial x_2 + \ldots) f(x_1, x_2, x_3, \ldots).$$

Thus $T_\lambda = \exp\left(\sum_i \lambda_i (\partial/\partial x_i) \right)$.

20.5 The basic representation

Lemma 20.25 *Suppose* $D : R \to \hat{R}$ *is a linear map which satisfies* $[x_i, D] = \lambda_i D$ *for all i, that is*

$$x_i Df - D(x_i f) = \lambda_i Df \quad \text{for all } f \in R.$$

Then

$$D = D(1) \exp\left(-\sum_i \lambda_i \frac{\partial}{\partial x_i}\right).$$

Proof. We shall show $Df = D(1) \exp(-\sum_i \lambda_i (\partial/\partial x_i)) f$ for all monomials $f \in R$, using induction on the degree of f. If f has degree 0 then $f = c \in \mathbb{C}$ and we have

$$D(1) \exp\left(-\sum_i \lambda_i \frac{\partial}{\partial x_i}\right) c = D(1) c = D(c).$$

Assuming the result for a monomial f we prove it for $x_i f$. We have

$$D(x_i f) = (x_i - \lambda_i) Df = (x_i - \lambda_i) D(1) \exp\left(-\sum_i \lambda_i \frac{\partial}{\partial x_i}\right) f.$$

On the other hand

$$D(1) \exp\left(-\sum_i \lambda_i \frac{\partial}{\partial x_i}\right)(x_i f) = D(1) T_{-\lambda}(x_i f) = D(1)(x_i - \lambda_i) T_{-\lambda} f$$

$$= D(1)(x_i - \lambda_i) \exp\left(-\sum_i \lambda_i \frac{\partial}{\partial x_i}\right) f.$$

Thus the lemma is proved. □

Lemma 20.26 *Suppose* $D: R \to \hat{R}$ *is a linear map which satisfies* $[\partial/\partial x_i, D] = \mu_i D$ *for all i, that is*

$$(\partial/\partial x_i)(Df) - D(\partial f/\partial x_i) = \mu_i Df \quad \text{for all } f \in R.$$

Then

$$D(1) = c \exp\left(\sum_i \mu_i x_i\right) \quad \text{for some } c \in \mathbb{C}.$$

Proof. Consider the element $\exp\left(\sum_i -\mu_i x_i\right) Df \in \hat{R}$. We have

$$\frac{\partial}{\partial x_i}\left(\exp\left(\sum_i -\mu_i x_i\right) Df\right)$$

$$= -\mu_i \exp\left(\sum_i -\mu_i x_i\right) Df + \exp\left(\sum_i -\mu_i x_i\right) \frac{\partial}{\partial x_i}(Df)$$

$$= \exp\left(\sum_i -\mu_i x_i\right) (\partial/\partial x_i - \mu_i) Df.$$

Thus by the assumption of the lemma we have

$$\left(\exp\left(\sum_i -\mu_i x_i\right) D\right) \partial f/\partial x_i = \exp\left(\sum_i -\mu_i x_i\right) (\partial/\partial x_i - \mu_i) Df$$

$$= \frac{\partial}{\partial x_i}\left(\exp\left(\sum_i -\mu_i x_i\right) Df\right).$$

Write $\Delta = \exp\left(\sum_i -\mu_i x_i\right) D$. Then we have $\Delta \partial f/\partial x_i = (\partial/\partial x_i)(\Delta f)$ for each i and f. In particular we may put $f = 1$ and obtain $(\partial/\partial x_i)(\Delta(1)) = 0$. Thus $\Delta(1) = c$ for some $c \in \mathbb{C}$. Hence

$$D(1) = \exp\left(\sum_i \mu_i x_i\right) \Delta(1) = c \exp\left(\sum_i \mu_i x_i\right)$$

as required. □

Proposition 20.27 *The set of all differential operators $D : R \to \hat{R}$ satisfying the conditions $[x_i, D] = \lambda_i D$ and $[\partial/\partial x_i, D] = \mu_i D$ for $\lambda_i, \mu_i \in \mathbb{C}$ forms a 1-dimensional vector space with basis*

$$\exp\left(\sum \mu_i x_i\right) \exp\left(-\sum \lambda_i \frac{\partial}{\partial x_i}\right).$$

Proof. This follows from Lemmas 20.25 and 20.26. □

Definition *Differential operators $D : R \to \hat{R}$ of the form*

$$\exp\left(\sum \mu_i x_i\right) \exp\left(-\sum \lambda_i \frac{\partial}{\partial x_i}\right)$$

*for $\lambda_i, \mu_i \in \mathbb{C}$ are called **vertex operators**.*

Now let $L = \hat{\mathfrak{L}}(L^0)$ be an affine Kac–Moody algebra of type $\tilde{A}_l, \tilde{D}_l$ or $\tilde{E}_l$. Let

$$T^- = \bigoplus_{j<0} L_{j\delta}.$$

20.5 The basic representation

Then T^- has a basis $t^j \otimes h_i$ for $i = 1, \ldots, l$ and $j < 0$. Consider the symmetric algebra $S(T^-)$. This is isomorphic to the polynomial ring over $\mathbb{C}$ in the variables $t^j \otimes h_i$.

Let Q^0 be the subgroup of H^0 generated by $h_1, \ldots, h_l$. We shall write Q^0 multiplicatively, so that its elements have form $h_1^{m_1} \ldots h_l^{m_l}$ with $m_1, \ldots, m_l \in \mathbb{Z}$. Let $\mathbb{C}[Q^0]$ be the group algebra of Q^0 over $\mathbb{C}$. Elements of $\mathbb{C}[Q^0]$ have form

$$\sum_{m_1, \ldots, m_l \in \mathbb{Z}} \lambda_{m_1, \ldots, m_l} h_1^{m_1} \ldots h_l^{m_l}.$$

$\mathbb{C}[Q^0]$ is isomorphic to the algebra of Laurent polynomials over $\mathbb{C}$ in the variables $h_1, \ldots, h_l$.

We now form the tensor product

$$V = S(T^-) \otimes \mathbb{C}[Q^0].$$

V is isomorphic to the algebra

$$\mathbb{C}[h_1, \ldots, h_l, h_1^{-1}, \ldots, h_l^{-1}, \quad t^j \otimes h_i]$$

for $i = 1, \ldots, l$ and $j < 0$.

We define certain maps $h_\alpha(n) : V \to V$ out of which vertex operators will be constructed. For $n \in \mathbb{Z}$ with $n > 0$, $h_\alpha(n)$ is the derivation of V uniquely determined by the conditions

$$t^{-n} \otimes h_i \to n \langle h_i, h_\alpha \rangle$$

$$t^j \otimes h_i \to 0 \quad \text{for } j \neq -n$$

$$h_i \to 0.$$

For $n \in \mathbb{Z}$ with $n < 0$, $h_\alpha(n) : V \to V$ is multiplication by $(t^n \otimes h_\alpha) \otimes 1$.

We now consider the expression

$$\exp\left(\sum_{n<0} -\frac{h_\alpha(n)}{n} z^{-n}\right) \exp\left(\sum_{n>0} -\frac{h_\alpha(n)}{n} z^{-n}\right)$$

where z is an indeterminate. We first observe that $\exp\left(\sum_{n>0} -\frac{h_\alpha(n)}{n} z^{-n}\right)$ maps V into $\mathbb{C}[z^{-1}] \otimes V$. To see this we observe that each element $v \in V$ is a finite linear combination of monomials

$$M_m = \prod_{\substack{i,j \\ j<0}} (t^j \otimes h_i)^{m_{ij}} \prod_i h_i^{m_i}$$

where $\boldsymbol{m} = (m_{ij}, m_i)$ with $m_i \in \mathbb{Z}$, $m_{ij} \in \mathbb{Z}$, $m_{ij} \geq 0$.

Let $d_1(m) = \sum_{i,j} m_{ij}$. Then the derivation $h_\alpha(n)$, $n > 0$, transforms M_m into a linear combination of monomials in which $d_1(m)$ is decreased by 1 and $\prod_i h_i^{m_i}$ remains unchanged. Thus a succession of $d_1(m) + 1$ derivations $h_\alpha(n)$ for various $n > 0$ annihilates M_m. Also, for a given monomial M_m, $h_\alpha(n)$ annihilates M_m for all but finitely many $n > 0$. Thus in the expression

$$\exp\left(\sum_{n>0} -\frac{h_\alpha(n)}{n} z^{-n}\right) v \qquad v \in V$$

only finitely many terms $-\frac{h_\alpha(n)}{n} z^{-n}$ act on v and only a finite set of products of such terms can act on v to give a non-zero element. Thus we have

$$\exp\left(\sum_{n>0} -\frac{h_\alpha(n)}{n} z^{-n}\right) : V \to \mathbb{C}[z^{-1}] \otimes V.$$

We shall modify this operator in the following way. Let

$$\varepsilon : Q^0 \times Q^0 \to \{\pm 1\}$$

be the function defined by

$\varepsilon(h_i, h_i) = -1$

$\varepsilon(h_i, h_j) = 1 \quad \text{if } A_{ij} = 0$

$\varepsilon(h_i, h_j)$ is given by $\left[E_{\alpha_i} E_{\alpha_j}\right] = \varepsilon(h_i, h_j) E_{\alpha_i + \alpha_j}$ if $A_{ij} = -1$

$\varepsilon(h + h', h'') = \varepsilon(h, h'') \varepsilon(h', h'')$

$\varepsilon(h, h' + h'') = \varepsilon(h, h') \varepsilon(h, h'').$

Given $\alpha \in \Phi^0$ we define a map $\varepsilon^\alpha \in \operatorname{End} V$ by

$$\varepsilon^\alpha : P \otimes h \to P \otimes \varepsilon(h_\alpha, h) h$$

where $P \in S(T^-)$, $h \in Q^0$. We also define $e^\alpha \in \operatorname{End} V$ by

$$e^\alpha : P \otimes h \to P \otimes h_\alpha h$$

and $z^\alpha \in \operatorname{End}\left(\mathbb{C}[z, z^{-1}] \otimes V\right)$ by

$$z^\alpha : (P \otimes h) z^i \to (P \otimes h) z^{i + \langle h_\alpha, h \rangle}.$$

We now define, for $\alpha \in \Phi^0$, the operator $Y_\alpha(z)$ on V by

$$Y_\alpha(z) = \exp\left(\sum_{n<0} -\frac{h_\alpha(n)}{n} z^{-n}\right) \exp\left(\sum_{n>0} -\frac{h_\alpha(n)}{n} z^{-n}\right) e^\alpha z^\alpha \varepsilon^\alpha.$$

20.5 The basic representation

We claim that $Y_\alpha(z)$ can be written in the form

$$Y_\alpha(z) = \sum_{j \in \mathbb{Z}} \Gamma_\alpha(j) z^{-j-1}$$

where $\Gamma_\alpha(j) \in \text{End } V$. In order to see this we consider the effect of $Y_\alpha(z)$ on a monomial M_m in V. We have

$$e^\alpha z^\alpha \varepsilon^\alpha M_m \in z^{n_\alpha} \otimes V$$

where $n_\alpha = \sum_i m_i \langle h_\alpha, h_i \rangle$. Thus

$$\exp\left(\sum_{n>0} -\frac{h_\alpha(n)}{n} z^{-n}\right) e^\alpha z^\alpha \varepsilon^\alpha M_m \in \sum_{k=0}^{K} z^{n_\alpha - k} \otimes V$$

for some $K > 0$, and

$$\exp\left(\sum_{n<0} -\frac{h_\alpha(n)}{n} z^{-n}\right) \exp\left(\sum_{n>0} -\frac{h_\alpha(n)}{n} z^{-n}\right) e^\alpha z^\alpha \varepsilon^\alpha M_m$$

$$\in \sum_{k' \geq 0} \sum_{k=0}^{K} z^{n_\alpha - k + k'} \otimes V.$$

Thus to obtain z^{-j-1} on the right-hand side of $Y_\alpha(z) M_m$ we must have $n_\alpha - k + k' = -j - 1$. For each value of k there is at most one $k' \geq 0$ satisfying this. Since only finitely many k arise, only finitely many k' can arise for given j. This shows that only finitely many terms $-\frac{h_\alpha(n)}{n} z^{-n}$ in $\exp\left(\sum_{n<0} -\frac{h_\alpha(n)}{n} z^{-n}\right)$ are involved in $\Gamma_\alpha(j)$ and only finitely many products of such terms are involved. Thus we have

$$\Gamma_\alpha(j) : V \to V$$

and

$$Y_\alpha(z) = \sum_{j \in \mathbb{Z}} \Gamma_\alpha(j) z^{-j-1}$$

with $\Gamma_\alpha(j) \in \text{End } V$.

Now the vector space V can be regarded as an L-module giving the basic representation of L. In order to describe the L-action on V we introduce some further notation. We have defined $h_\alpha(n) \in \text{End } V$ for $n > 0$ and $n < 0$. We now define $h(0) \in \text{End } V$ for any $h \in H^0$. In contrast to the $h_\alpha(n)$ for $n \neq 0$, which act non-trivially on $S(T^-)$ and trivially on $\mathbb{C}[Q^0]$, $h(0)$ acts trivially on $S(T^-)$ and non-trivially on $\mathbb{C}[Q^0]$. We define, for $h \in H^0$, $h(0) : V \to V$ by

$$h(0) : P \otimes h_\alpha \to P \otimes \langle h_\alpha, h \rangle h_\alpha$$

for $P \in S(T^-)$, $\alpha \in Q^0$.

Let $h'_1, \ldots, h'_l$ be a basis for H^0 and $h''_1, \ldots, h''_l$ be the dual basis satisfying $\langle h'_i, h''_j \rangle = \delta_{ij}$. We define $D_0 \in \text{End } V$ by

$$D_0 = \sum_{i=1}^{l} \frac{1}{2} h'_i(0) h''_i(0) + \sum_{n \geq 1} h'_i(-n) h''_i(n).$$

since, for $v \in V$, $h''_i(n) v = 0$ for all but finitely many $n > 0$. D_0 lies in End V and is readily seen to be independent of the choice of basis of H^0.

We now have the definitions necessary to describe the action of L on V which gives the basic representation.

Theorem 20.28 *The vector space $V = S(T^-) \otimes \mathbb{C}[Q^0]$ is a module for the Kac–Moody algebra $L(A)$ of type $\tilde{A}_l$, $\tilde{D}_l$ or $\tilde{E}_l$ giving the basic representation under the following action $L(A) \to \text{End } V$:*

$$t^n \otimes H_\alpha \to H_\alpha(n) \quad \text{for } \alpha \in \Phi^0, n \in \mathbb{Z}$$

$$t^n \otimes E_\alpha \to \Gamma_\alpha(n) \quad \text{for } \alpha \in \Phi^0, n \in \mathbb{Z}$$

$$c \to 1_V$$

$$d \to -D_0.$$

The proof of this result can be found in the book of Kac, *Infinite-Dimensional Lie Algebras*, third edition, Chapter 14.

The highest weight vector of V is the element $1 \otimes 1$. The first 1 is the unit element of the symmetric algebra $S(T^-)$ and the second 1 is the unit element of the lattice Q^0 written multiplicatively, which is the unit element of the group algebra $\mathbb{C}[Q^0]$. This vector $1 \otimes 1$ is annihilated by the generators e_i, $i = 0, 1, \ldots, l$ of $L(A)$. To see this we recall that

$$e_i = 1 \otimes E_i \quad i = 1, \ldots, l \qquad e_0 = t \otimes E_0$$

with $E_0 \in L^0_{-\theta}$. We have

$$e_i(1 \otimes 1) = \Gamma_{\alpha_i}(0)(1 \otimes 1) \quad i = 1, \ldots, l$$

which is the coefficient of z^{-1} in $Y_{\alpha_i}(z)(1 \otimes 1)$. Also

$$e_0(1 \otimes 1) = \Gamma_{-\theta}(1)(1 \otimes 1)$$

which is the coefficient of z^{-2} in $Y_{-\theta}(z)(1 \otimes 1)$. Recalling that

$$Y_\alpha(z) = \exp\left(\sum_{n<0} \frac{-h_\alpha(n)}{n} z^{-n}\right) \exp\left(\sum_{n>0} \frac{-h_\alpha(n)}{n} z^{-n}\right) e^\alpha z^\alpha \varepsilon^\alpha$$

20.5 The basic representation

we first note that $e^\alpha z^\alpha \varepsilon^\alpha (1 \otimes 1) = 1 \otimes h_\alpha$. Now negative powers of z in $Y_\alpha(z)(1 \otimes 1)$ can only arise from derivations $h_\alpha(n)$ with $n > 0$. However, $h_\alpha(n)(1 \otimes h_\alpha) = 0$ for all $n > 0$ since any derivation annihilates the unit element $1 \in S(T^-)$. Thus we have

$$\Gamma_{\alpha_i}(0)(1 \otimes 1) = 0 \qquad \text{for } i = 1, \ldots, l$$

$$\Gamma_{-\theta}(1)(1 \otimes 1) = 0.$$

Hence $e_i(1 \otimes 1) = 0$ for all $i = 0, 1, \ldots, l$.

We now check how the elements $h_0, h_1, \ldots, h_l \in H$ act on $1 \otimes 1$. We have $h_i = 1 \otimes H_i$ for $i = 1, \ldots, l$. Thus

$$h_i(1 \otimes 1) = H_{\alpha_i}(0)(1 \otimes 1) = 1 \otimes \langle 0, h_i \rangle 1 = 0.$$

(We note that in the scalar product $\langle , \rangle$ the elements of Q^0 are written additively so that the unit element will be 0.) We also have

$$c = h_0 + c_1 h_1 + \cdots + c_l h_l.$$

Thus

$$h_0(1 \otimes 1) = c(1 \otimes 1) = 1 \otimes 1.$$

Hence we have

$$h_i(1 \otimes 1) = \gamma(h_i)(1 \otimes 1) \qquad \text{for } i = 0, 1, \ldots, l$$

since $\gamma(h_i) = 0$ for $i = 1, \ldots, l$ and $\gamma(h_0) = 1$. The highest weight vector $v_\gamma = 1 \otimes 1$ is often called the **vacuum vector** of the basic representation.

The realisation of the basic representation given by the module $S(T^-) \otimes \mathbb{C}[Q^0]$ is called the homogeneous realisation. It is one of a number of descriptions of the basic representation.

The basic representation is of great importance in a number of applications of the theory of affine Kac–Moody algebras in mathematics and physics. For example applications to the theory of differential equations are described in Kac' book, Chapter 14. There are also particularly interesting applications in the area of mathematical physics. Vertex operators arose in the context of dual resonance models, which subsequently developed into string theory, and the representation theory of affine Kac–Moody algebras plays a key role in string theory. This involves the calculus of vertex operators. The theory of modular forms also plays a key role.

Readers wishing to learn more about the relations between Kac–Moody algebras and string theory may wish to study the 30-page introduction to the book of Frenkel, Lepowsky and Meurman, *Vertex Operator Algebras and the*

Monster which also explains the connections with modular forms and sporadic simple groups such as the Monster. The book by Kac on *Vertex Algebras for Beginners* is a useful introduction to the calculus of vertex operators. This whole area relating mathematics and physics is of great current interest and seems certain to continue its rapid development.

21
Borcherds Lie algebras

21.1 Definition and examples of Borcherds algebras

A theory of generalised Kac–Moody algebras was introduced by R. Borcherds in 1988. The purpose for which these algebras were introduced was as part of Borcherds' proof of the Conway–Norton conjectures on the representation theory of the Monster simple group, for which Borcherds was awarded a Fields Medal in 1998. These generalised Kac–Moody algebras are now frequently called Borcherds algebras. A detailed discussion of Borcherds algebras, including proofs of all the assertions, is beyond the scope of this volume. However, we shall include the definition of Borcherds algebras and the statements of the main results about their structure and representation theory, but without detailed proofs. In fact many of the results are quite similar to those we have already obtained about Kac–Moody algebras. However, the theory of Borcherds algebras includes examples which are quite different from Kac–Moody algebras. The best known such example is the Monster Lie algebra, which we shall describe in Section 21.3.

We begin with the definition of a Borcherds algebra. A Lie algebra L over $\mathbb{R}$ is called a **Borcherds algebra** if it satisfies the following four axioms:

(i) L has a $\mathbb{Z}$-grading
$$L = \bigoplus_{i \in \mathbb{Z}} L_i$$
such that $\dim L_i$ is finite for all $i \neq 0$, and L is diagonalisable with respect to L_0. (Note that $\dim L_0$ need not be finite.)

(ii) There exists an automorphism $\omega : L \to L$ such that
$$\omega^2 = 1$$
$$\omega(L_i) = L_{-i} \quad \text{for all } i \in \mathbb{Z}$$
$$\omega = -1 \quad \text{on } L_0/L_0 \cap Z(L).$$

(iii) There is an invariant bilinear form

$$\langle , \rangle : L \times L \to \mathbb{R}$$

such that

$\langle x, y \rangle = 0$ if $x \in L_i$, $y \in L_j$ and $i+j \neq 0$

$\langle wx, wy \rangle = \langle x, y \rangle$ for all $x, y \in L$

$-\langle x, wx \rangle > 0$ for $x \in L_i$ with $i \neq 0$, $x \neq 0$.

(iv) $L_0 \subset [LL]$.

We observe some consequences of these axioms. In the first place it can be shown that $[L_0 L_0] = 0$, that is L_0 is abelian.

To describe a further consequence we define, for $x, y \in L$,

$$\langle x, y \rangle_0 = -\langle x, \omega y \rangle.$$

The scalar product $\langle , \rangle_0 : L \times L \to \mathbb{R}$ is called the contravariant bilinear form. We now restrict the contravariant form to one of the graded components L_i with $i \neq 0$ and have $\langle , \rangle_0 : L_i \times L_i \to \mathbb{R}$. Let $x \in L_i$. Then

$$\langle x, x \rangle_0 = -\langle x, wx \rangle > 0 \quad \text{if } x \neq 0.$$

Thus the contravariant form is positive definite on each graded component L_i for $i \neq 0$ (though not necessarily on L_0).

We now give some examples of Borcherds algebras. We begin with a symmetric matrix $\mathfrak{A}$ over $\mathbb{R}$ which is either finite or countable. Thus

$$\mathfrak{A} = (a_{ij}) \quad i, j \in I$$

with $a_{ij} \in \mathbb{R}$ and I either finite or countable. We shall assume that the matrix $\mathfrak{A}$ satisfies the conditions

$a_{ij} \leq 0$ if $i \neq j$

if $a_{ii} > 0$ then $2a_{ij}/a_{ii} \in \mathbb{Z}$ for all j.

Proposition 21.1 *There is a Borcherds algebra L associated to a symmetric matrix $\mathfrak{A}$ satisfying the above conditions which is defined as follows by generators and relations.*

L is generated by elements e_i, f_i, h_{ij} $i, j \in I$

21.1 Definition and examples of Borcherds algebras

subject to relations

$$[e_i f_j] = h_{ij}$$
$$[h_{ij} h_{kl}] = 0$$
$$[h_{ij} e_k] = \delta_{ij} a_{ik} e_k$$
$$[h_{ij} f_k] = -\delta_{ij} a_{ik} f_k$$

$(\operatorname{ad} e_i)^n e_j = 0$, $(\operatorname{ad} f_i)^n f_j = 0$ if $a_{ii} > 0$, $i \neq j$ and $n = 1 - 2a_{ij}/a_{ii}$

$[e_i e_j] = 0$, $[f_i f_j] = 0$ if $a_{ii} \leq 0$, $a_{jj} \leq 0$ and $a_{ij} = 0$.

The Borcherds algebra L defined by generators and relations in this way is called the **universal Borcherds algebra** associated with the symmetric matrix $\mathfrak{A}$. Its structure as a Borcherds algebra can be described as follows. Its involutary automorphism ω is given by

$$\omega(e_i) = -f_i, \quad \omega(f_i) = -e_i, \quad \omega(h_{ij}) = -h_{ji}.$$

Its invariant bilinear form is uniquely determined by the condition

$$\langle e_i, f_i \rangle = 1 \quad \text{for all } i \in I.$$

In particular, if we write $h_i = h_{ii}$, then $[e_i f_i] = h_i$ and

$$\langle h_i, h_j \rangle = \langle [e_i f_i], h_j \rangle = \langle e_i, [f_i h_j] \rangle = \langle e_i, a_{ij} f_i \rangle = a_{ij}.$$

Thus

$$\langle h_i, h_j \rangle = a_{ij} \text{ for all } i, j \in I.$$

There are many ways of defining an appropriate grading on this Borcherds algebra. For each $i \in \mathbb{Z}$ let $n_i \in \mathbb{Z}$ satisfy $n_i > 0$. Then there is a $\mathbb{Z}$-grading on L uniquely determined by the conditions

$$e_i \in L_{n_i}, \quad f_i \in L_{-n_i}.$$

Further examples of Borcherds algebras can be obtained from a universal Borcherds algebra as follows. The axiom $[h_{ij} h_{kl}] = 0$ shows that the subalgebra generated by all elements h_{ij}, $i, j \in I$, is abelian. If $i \neq j$ then h_{ij} lies in the centre of L since $[h_{ij}, e_k] = 0$ and $[h_{ij}, f_k] = 0$ for all $k \in I$. Thus the subalgebra generated by the h_{ij} for $i \neq j$ lies in the centre. It can be shown that the centre Z of L satisfies

$$\langle h_{ij} ; \ i, j \in I, i \neq j \rangle \subset Z \subset \langle h_{ij} ; \ i, j \in I \rangle.$$

In fact the Jacobi identity

$$[[e_i f_j] h_k] + [[f_j h_k] e_i] + [[h_k e_i] f_j] = 0$$

shows that

$$[h_k, [e_i f_j]] = (a_{ki} - a_{kj})[e_i f_j].$$

It can be shown that, as a consequence of this, $h_{ij} = 0$ unless $a_{ki} = a_{kj}$ for all $k \in I$, i.e. unless the ith and jth columns of $\mathfrak{A}$ are identical.

Proposition 21.2 *Let L be a universal Borcherds algebra and I be an ideal of L contained in the centre Z of L. Then L/I retains the structure of a Borcherds algebra.* □

The $\mathbb{Z}$-grading, involutary automorphism and invariant bilinear form on L/I are readily obtained from those on L. □

We now obtain still further Borcherds algebras. Starting from a universal Borcherds algebra L we factor out an ideal I of L contained in the centre Z of L. Then L/I is still a Borcherds algebra. We write $\bar{L} = L/I$.

An inner derivation of $\bar{L}$ is one of form $x \to [xy]$ for some $y \in \bar{L}$, and an outer derivation is a derivation which either is zero or is not an inner derivation Let

$$\bar{L}^* = \mathrm{Hom}(\bar{L}, \mathbb{R})$$

and $A \subset \bar{L}^*$ be an abelian Lie algebra of outer derivations of $\bar{L}$. We suppose also that

$$[\bar{e}_i x] \in \mathbb{R} \bar{e}_i, \quad [\bar{f}_i x] \in \mathbb{R} \bar{f}_i$$

for all $x \in A$ where $\bar{e}_i, \bar{f}_i$ are images of e_i, f_i under the natural homomorphism $L \to \bar{L}$.

Proposition 21.3 *Let L be a universal Borcherds algebra and I be an ideal of L contained in the centre Z of L. Let $\bar{L} = L/I$. Let A be an abelian Lie algebra of outer derivations of $\bar{L}$ and let $\bar{L} + A$ be the semidirect product of $\bar{L}$ by A whose elements have form $x + a$ with $x \in \bar{L}, a \in A$ where*

$$[x+a, y+b] = [xy] + a(y) - b(x).$$

Suppose that

$$[\bar{e}_i x] \in \mathbb{R} \bar{e}_i, \quad [\bar{f}_i x] \in \mathbb{R} \bar{f}_i$$

for all $x \in A$. Then $\bar{L} + A$ retains the structure of a Borcherds algebra in which $A \subset (\bar{L} + A)_0$. □

The $\mathbb{Z}$-grading, involutary automorphism and invariant bilinear form on $\bar{L} + A$ are easily obtained from those of $\bar{L}$. □

We have now constructed a family of Borcherds algebras which includes all universal Borcherds algebras, all quotients of such by ideals contained in the centre, and all semidirect products of such quotients by an abelian Lie algebra of outer derivations with suitable properties.

This turns out to give all possible Borcherds algebras, as is shown by the next theorem.

Theorem 21.4 *Let L be a Borcherds algebra. Then there is a unique universal Borcherds algebra L_u and a homomorphism*

$$f : L_u \to L$$

(not necessarily unique) such that $\ker f$ is an ideal in the centre of L_u, $\operatorname{im} f$ is an ideal of L, and L is the semidirect product of $\operatorname{im} f$ with an abelian Lie algebra of outer derivations lying in the 0-graded component of L and preserving all subspaces $\mathbb{R}\bar{e}_i$ and $\mathbb{R}\bar{f}_i$.

The homomorphism f preserves the grading, involutary automorphism, and bilinear form. □

Now that we have obtained the complete set of Borcherds algebras in this way, we explore their relationship with symmetrisable Kac–Moody algebras. It turns out that every symmetrisable Kac–Moody algebra over $\mathbb{R}$ gives rise to a universal Borcherds algebra, which is the subalgebra of the Kac–Moody algebra obtained by generators and relations prior to the extension of the Cartan subalgebra by an abelian Lie algebra of outer derivations.

Theorem 21.5 *Let L be a symmetrisable Kac–Moody algebra with GCM $A = (A_{ij})$. Thus there exists a diagonal matrix $D = \operatorname{diag}(d_1, \ldots, d_n)$ with each $d_i \in \mathbb{Z}$, $d_i > 0$ such that DA is symmetric. Let $\mathfrak{A} = (a_{ij})$ be given by*

$$a_{ij} = \frac{d_i A_{ij}}{2}.$$

Then we have $a_{ij} = a_{ji}$ and $a_{ii} = d_i$. Thus $a_{ij} \leq 0$ if $i \neq j$ and a_{ii} is a positive integer. Moreover $2a_{ij}/a_{ii} = A_{ij} \in \mathbb{Z}$.

Then the symmetric matrix (a_{ij}) satisfies the conditions needed to construct a Borcherds algebra, and the universal Borcherds algebra with symmetric matrix $\mathfrak{A}$ coincides with the subalgebra of the Kac–Moody algebra L obtained by generators and relations prior to the adjunction of the abelian Lie algebra of outer derivations. □

In this way every symmetrisable Kac–Moody algebra determines a certain subalgebra which is a universal Borcherds algebra. In fact the main points of difference between symmetrisable Kac–Moody algebras and universal Borcherds algebras are that, in a Borcherds algebra:

I may be countably infinite rather than finite

a_{ii} may not be positive and need not lie in $\mathbb{Z}$

$2a_{ij}/a_{ii}$ is only assumed to lie in $\mathbb{Z}$ when $a_{ii} > 0$.

21.2 Representations of Borcherds algebras

We now introduce the root system and Weyl group of a Borcherds algebra.

We suppose first that L is a universal Borcherds algebra. The root lattice Q of L is the free abelian group with basis α_i for $i \in I$. We have a symmetric bilinear form

$$Q \times Q \to \mathbb{R}$$

given by $(\alpha_i, \alpha_j) \to \langle \alpha_i, \alpha_j \rangle = a_{ij}$.

The basis elements α_i of Q are called the **fundamental roots**. The set of fundamental roots is denoted by Π. We have a grading

$$L = \bigoplus_{\alpha \in Q} L_\alpha$$

determined by the conditions

$$e_i \in L_{\alpha_i}, \quad f_i \in L_{-\alpha_i}.$$

An element $\alpha \in Q$ is called a **root** of L if $\alpha \neq 0$ and $L_\alpha \neq O$. α is called a **positive root** if α is a sum of fundamental roots. For any root α either α or $-\alpha$ is positive. We have

$$\Phi = \Phi^+ \cup \Phi^-$$

where Φ is the set of roots and Φ^+, Φ^- are the subsets of positive and negative roots respectively. We say that $\alpha \in \Phi$ is a **real root** if $\langle \alpha, \alpha \rangle > 0$ and $\alpha \in \Phi$ is an **imaginary root** if $\langle \alpha, \alpha \rangle \leq 0$.

We next introduce the Weyl group W of the universal Borcherds algebra L. W is the group of isometries of the root lattice Q generated by the reflections s_i corresponding to the real fundamental roots. We have

$$s_i(\alpha_j) = \alpha_j - 2 \frac{\langle \alpha_i, \alpha_j \rangle}{\langle \alpha_i, \alpha_i \rangle} \alpha_i = \alpha_j - 2 \frac{a_{ij}}{a_{ii}} \alpha_i.$$

We recall that $2a_{ij}/a_{ii} \in \mathbb{Z}$ since $a_{ii} > 0$.

21.2 Representations of Borcherds algebras

Let H be the abelian subalgebra of L generated by the elements h_{ij} for all $i, j \in I$. We have a map

$$Q \to H$$

under which α_i maps to h_i, which is a homomorphism of abelian groups and preserves the scalar product. However, this map need not necessarily be injective.

So for we have assumed that L is a universal Borcherds algebra. However, if L is an arbitrary Borcherds algebra there is an associated universal Borcherds algebra L_u given by Theorem 21.4. Then the root system of L is defined to be the root system of L_u and the Weyl group of L is defined to be the Weyl group of L_u.

This theory of Borcherds algebras is thus very similar to the theory of Kac–Moody algebras. The main difference is that for Borcherds algebras there can exist imaginary fundamental roots, and that the Weyl group is generated by the reflections with respect to the real fundamental roots only.

We now turn to the representation theory of Borcherds algebras. We define the set X of integral weights by

$$X = \left\{ \lambda \in Q \otimes \mathbb{R}; \, 2\frac{\langle \lambda, \alpha_i \rangle}{\langle \alpha_i, \alpha_i \rangle} \in \mathbb{Z} \quad \text{for all } \alpha_i \in \Pi_{\text{Re}} \right\}.$$

Here Π_{Re} is the set of real fundamental roots. We recall that $\langle \alpha_i, \alpha_i \rangle > 0$ when $\alpha_i \in \Pi_{\text{Re}}$. We define the subset $X^+ \subset X$ of dominant integral weights by

$$X^+ = \{ \lambda \in X; \langle \lambda, \alpha_i \rangle \geq 0 \quad \text{for all } \alpha_i \in \Pi \}.$$

In a manner very similar to that we have described for Kac–Moody algebras in Chapter 19 it is possible to define an irreducible module $L(\lambda)$ for the Borcherds algebra L associated to any dominant integral weight λ. $L(\lambda)$ is called the irreducible L-module with highest weight λ.

Now Borcherds proved a character formula for $L(\lambda)$ analogous to Kac' character formula Theorem 19.16 for Kac–Moody algebras.

Theorem 21.6 (*Borcherds' character formula*). *Let L be a Borcherds algebra, λ a dominant integral weight and $L(\lambda)$ the corresponding irreducible L-module with highest weight λ. Then the character of $L(\lambda)$ is given by*

$$\operatorname{ch} L(\lambda) = \frac{\sum\limits_{w \in W} \varepsilon(w) w \left(\sum\limits_{\Psi} (-1)^{|\Psi|} e(\lambda + \rho - \sum \Psi) \right)}{e(\rho) \prod\limits_{\alpha \in \Phi^+} (1 - e(-\alpha))^{m_\alpha}}$$

where $m_\alpha = \dim L_\alpha$, Ψ runs over all finite subsets of mutually orthogonal imaginary fundamental roots, and ρ is any element of $Q \otimes \mathbb{R}$ such that

$$\langle \rho, \alpha_i \rangle = \tfrac{1}{2} \langle \alpha_i, \alpha_i \rangle$$

for all real fundamental roots α_i.

As usual this character is interpreted as

$$\mathrm{ch}\, L(\lambda) = \sum_\mu \left(\dim L(\lambda)_\mu\right) e(\mu)$$

where $\mu \to e(\mu)$ is an isomorphism between the additive group of weights and the corresponding multiplicative group.

(In fact there may not exist a vector $\rho \in Q \otimes \mathbb{R}$ such that $\langle \rho, \alpha_i \rangle = \tfrac{1}{2} \langle \alpha_i, \alpha_i \rangle$ for all real fundamental roots α_i of a general Borcherds algebra. But if there is no such $\rho \in Q \otimes \mathbb{R}$, ρ may still be defined as the homomorphism from Q to $\mathbb{R}$ taking α_i to $\tfrac{1}{2} \langle \alpha_i, \alpha_i \rangle$ for all $i \in I$, and the character formula can be interpreted accordingly.) □

In the special case $\lambda = 0$, $L(\lambda)$ is the trivial 1-dimensional L-module. Then Borcherds' character formula reduces to the following identity.

Theorem 21.7 (*Borcherds' denominator formula*).

$$e(\rho) \prod_{\alpha \in \Phi^+} (1 - e(-\alpha))^{m_\alpha} = \sum_{w \in W} \varepsilon(w) w \left(e(\rho) \sum_\Psi (-1)^{|\Psi|} e\!\left(-\sum \Psi\right) \right). \quad \square$$

By substituting Borcherds' denominator formula into Theorem 21.6 we obtain a second form of Borcherds' character formula.

Theorem 21.8 (*Borcherds' character formula, second form*). With the notation of Theorem 21.6 we have

$$\mathrm{ch}\, L(\lambda) = \frac{\sum_{w \in W} \varepsilon(w) w \left(\sum_\Psi (-1)^{|\Psi|} e(\lambda + \rho - \sum \Psi) \right)}{\sum_{w \in W} \varepsilon(w) w \left(e(\rho) \sum_\Psi (-1)^{|\Psi|} e(-\sum \Psi) \right)}.$$

Comments on the proof of Borcherds' character formula

We shall not give the proof of Borcherds' character formula in detail, since the ideas are quite similar to those which arise in the proof of Kac' character formula for Kac–Moody algebras. However we shall say enough to explain where the additional term

$$\sum_\Psi (-1)^{|\Psi|} e(-\sum \Psi)$$

comes from, where Ψ runs over all finite sets of mutually orthogonal imaginary fundamental roots. Of course in Kac–Moody algebras there are no imaginary fundamental roots so the only possible subset Ψ is the empty set. The additional term then becomes $e(0)$, the identity element of $e(Q)$, and disappears from the formula.

For a Borcherds algebra L we have

$$\Pi = \Pi_{\text{Re}} \cup \Pi_{\text{Im}}$$

where Π_{Re} is the set of real fundamental roots and Π_{Im} is the set of imaginary fundamental roots. We also define

$$\Phi_{\text{Re}} = W(\Pi_{\text{Re}}), \quad \Phi_{\text{Im}} = \Phi - \Phi_{\text{Re}}$$

to be the sets of real and imaginary roots respectively. We recall from Theorem 16.24 that, in a Kac–Moody algebra,

$$\Phi_{\text{Im}}^+ = \bigcup_{w \in W} w(K)$$

where $K = \{\alpha \in Q^+ \; ; \; \alpha \neq 0, \; \operatorname{supp} \alpha \text{ is connected}, \; -\alpha \in \bar{C}\}$ and

$$\bar{C} = \{\lambda \in Q \otimes \mathbb{R} \; ; \; \langle \lambda, \alpha_i \rangle \geq 0 \text{ for all } \alpha_i \in \Pi_{\text{Re}}\}.$$

There is an analogous result for Borcherds algebras given as follows.

Theorem 21.9 *The set of positive imaginary roots of a Borcherds algebra is given by*

$$\Phi_{\text{Im}}^+ = \bigcup_{w \in W} w(K)$$

where K is given by

$$K = \{\alpha \in Q^+; \alpha \neq 0, \; -\alpha \in \bar{C}, \; \operatorname{supp} \alpha \text{ is connected}\}$$
$$- \{j\alpha_i \; ; \; j \in \mathbb{Z}, \; j \geq 2, \; \alpha_i \in \Pi_{\text{Im}}\}.$$

Proof. Omitted. The idea is generally similar to that of Theorem 16.24. It is clearly necessary to exclude positive multiples $j\alpha_i$ of imaginary fundamental roots with $j \geq 2$ since these vectors satisfy the conditions required for belonging to K, but cannot be roots since there is no possible root vector giving rise to such a root. □

Following closely the proof of Kac' character formula we obtain

$$e(\rho) \prod_{\alpha \in \Phi^+} (1 - e(-\alpha))^{m_\alpha} \operatorname{ch} L(\lambda) = \sum_\mu c_\mu e(\mu + \rho)$$

summed over all weights μ such that $\mu \prec \lambda$ and $\langle \mu+\rho, \mu+\rho \rangle = \langle \lambda+\rho, \lambda+\rho \rangle$, where $c_\mu \in \mathbb{Z}$ and both sides are skew-symmetric under the action of the Weyl group W. (See the proof of Theorem 19.16.)

We now define a certain partial sum S of terms on the right-hand side.

Let $S = \sum_\mu c_\mu e(\mu+\rho)$, summed over all weights μ satisfying $\mu \prec \lambda$, $\langle \mu+\rho, \mu+\rho \rangle = \langle \lambda+\rho, \lambda+\rho \rangle$ and $\langle \mu+\rho, \alpha_i \rangle \geq 0$ for all $\alpha_i \in \Pi_{\text{Re}}$.

Since $\mu \prec \lambda$ we have

$$\mu = \lambda - \sum k_i \alpha_i \quad \text{for some } k_i \in \mathbb{Z}, k_i > 0, \alpha_i \in \Pi.$$

Since $\langle \mu+\rho, \mu+\rho \rangle = \langle \lambda+\rho, \lambda+\rho \rangle$ we have

$$\langle (\lambda+\rho) - (\mu+\rho), (\lambda+\rho) + (\mu+\rho) \rangle = 0$$

that is $\langle \sum k_i \alpha_i, \lambda + \mu + 2\rho \rangle = 0$. This implies

$$\sum k_i \langle \alpha_i, \lambda \rangle + \sum k_i \langle \alpha_i, \mu + 2\rho \rangle = 0.$$

We can deduce several consequences from this equation. We note first that $\langle \alpha_i, \lambda \rangle \geq 0$ since λ is dominant. Also for $\alpha_i \in \Pi_{\text{Re}}$ we have

$$\langle \alpha_i, \mu + 2\rho \rangle = \langle \alpha_i, \mu+\rho \rangle + \langle \alpha_i, \rho \rangle = \langle \alpha_i, \mu+\rho \rangle + \tfrac{1}{2}\langle \alpha_i, \alpha_i \rangle > \langle \alpha_i, \mu+\rho \rangle.$$

Now $\langle \alpha_i, \mu+\rho \rangle \geq 0$ by definition of S, so $\langle \alpha_i, \mu+2\rho \rangle > 0$. On the other hand, for $\alpha_i \in \Pi_{\text{Im}}$ we have

$$\langle \alpha_i, \mu+2\rho \rangle = \langle \alpha_i, \mu+\alpha_i \rangle = \langle \alpha_i, \lambda - \sum k'_j \alpha_j \rangle \quad \text{for some } k'_j > 0.$$

Thus $\langle \alpha_i, \mu+2\rho \rangle \geq 0$ since $\langle \alpha_i, \lambda \rangle \geq 0, \langle \alpha_i, \alpha_j \rangle \leq 0$ if $i \neq j$, and $\langle \alpha_i, \alpha_i \rangle \leq 0$. We now collect these results together and put them into the equation

$$\sum k_i \langle \alpha_i, \lambda \rangle + \sum k_i \langle \alpha_i, \mu + 2\rho \rangle = 0.$$

The conclusion is that $\langle \alpha_i, \lambda \rangle = 0$ and $\langle \alpha_i, \mu+2\rho \rangle = 0$ for all α_i in the sum $\sum k_i \alpha_i$. This in turn implies that each such $\alpha_i \in \Pi_{\text{Im}}$. But then

$$\langle \alpha_i, \mu+2\rho \rangle = \langle \alpha_i, \lambda \rangle - \sum k'_j \langle \alpha_i, \alpha_j \rangle = -\sum k'_j \langle \alpha_i, \alpha_j \rangle.$$

21.2 Representations of Borcherds algebras

Since $k'_j > 0$ and $\langle \alpha_i, \alpha_j \rangle \leq 0$ this implies that $\langle \alpha_i, \alpha_j \rangle = 0$. In fact $k'_j = k_j$ if $j \neq i$ and $k'_i = k_i - 1$. So if $i \neq j$ we have $\langle \alpha_i, \alpha_j \rangle = 0$ for all α_i, α_j in the sum $\sum k_i \alpha_i$ with $i \neq j$. In other words,

$$\lambda - \mu = \sum k_i \alpha_i$$

is a linear combination of mutually orthogonal imaginary fundamental roots all of which are orthogonal to λ.

Now we have

$$\langle \mu, \alpha_j \rangle = \langle \lambda, \alpha_j \rangle - \sum_i k_i \langle \alpha_i, \alpha_j \rangle \geq 0$$

for all $\alpha_j \in \Pi_{\mathrm{Re}}$ since $\langle \lambda, \alpha_j \rangle \geq 0$, $k_i > 0$ and $\langle \alpha_i, \alpha_j \rangle \leq 0$ since $\alpha_i \in \Pi_{\mathrm{Im}}$, $\alpha_j \in \Pi_{\mathrm{Re}}$ so $i \neq j$. Thus $\mu \in \bar{C}$. It follows that $\mu + \rho \in C$ since

$$\langle \mu + \rho, \alpha_j \rangle = \langle \mu, \alpha_j \rangle + \tfrac{1}{2} \langle \alpha_j, \alpha_j \rangle > \langle \mu, \alpha_j \rangle$$

so $\langle \mu + \rho, \alpha_j \rangle > 0$. Thus $\mu + \rho$ lies in the fundamental chamber C. Since our sum

$$\sum_{\substack{\mu \\ \mu \prec \lambda \\ \langle \mu+\rho, \mu+\rho \rangle = \langle \lambda+\rho, \lambda+\rho \rangle}} c_\mu e(\mu + \rho)$$

is skew-symmetric under the action of W, this sum must be equal to

$$\sum_{w \in W} \varepsilon(w) w(S)$$

since S is the partial sum including all summands $c_\mu e(\mu + \rho)$ for which $\mu + \rho$ lies in $\bar{C}$.

We shall now determine S. Let the module $L(\lambda)$ have highest weight vector v_λ. If $\lambda(h_i) = 0$ then $f_i v_\lambda = 0$ by the analogue in Borcherds algebras of the proof of Theorem 10.20. If all α_i in the sum $\sum k_i \alpha_i$ satisfied $\lambda(h_i) = 0$ then we would have $f_i v_\lambda = 0$ for all such i and this would imply that $\lambda - \sum k_i \alpha_i$ could not be a weight of $L(\lambda)$. So if $\lambda - \sum k_i \alpha_i$ is a weight not equal to λ we must have $\lambda(h_i) \neq 0$ for some i in this sum. Thus $\langle \lambda, \alpha_i \rangle \neq 0$ for some i in this sum. On the other hand we know that $\mu = \lambda - \sum k_i \alpha_i$ where all α_i in the sum satisfy $\langle \lambda, \alpha_i \rangle = 0$. This implies that the weight

$$\mu + \rho = (\lambda + \rho) - \sum k_i \alpha_i$$

can only arise from the term $e(\lambda)$ in $\operatorname{ch} L(\lambda)$ in the formula

$$e(\rho) \prod_{\alpha \in \Phi^+} (1 - e(-\alpha))^{m_\alpha} \operatorname{ch} L(\lambda) = \sum_\mu e_\mu e(\mu + \rho).$$

So all terms on the right of this formula which lie in S arise from
$$e(\lambda+\rho) \prod_{\alpha\in\Phi^+} (1-e(-\alpha))^{m_\alpha}$$
on the left.

We consider which roots $\alpha \in \Phi^+$ in the formula
$$e(\lambda+\rho) \prod_{\alpha\in\Phi^+} (1-e(-\alpha))^{m_\alpha}$$
can contribute to give $\mu+\rho = (\lambda+\rho) - \sum k_i \alpha_i$. All such roots $\alpha \in \Phi^+$ must be linear combinations of the fundamental roots α_i arising in the sum $\sum k_i \alpha_i$. But such α_i are mutually orthogonal imaginary fundamental roots. So sums of two or more such α_i do not have connected support, so cannot be roots by Theorem 21.9. Also each $\alpha_i \in \Pi$ has $m_{\alpha_i} = 1$ since the corresponding root space is spanned by e_i. Thus a weight $\mu+\rho$ giving a term on the right which lies in S must arise from
$$e(\lambda+\rho) \prod_{\alpha_i \in \Pi_{\text{Im}}} (1-e(-\alpha_i)) = e(\lambda+\rho) \sum_\Psi (-1)^{|\Psi|} e(-\sum \Psi)$$
summed over all finite sets Ψ of mutually orthogonal imaginary fundamental roots. Thus we have
$$S = \sum_\Psi (-1)^{|\Psi|} e(\lambda+\rho-\sum \Psi)$$
and so
$$e(\rho) \prod_{\alpha\in\Phi^+} (1-e(-\alpha))^{m_\alpha} \operatorname{ch} L(\lambda) = \sum_{w\in W} \varepsilon(w) w \left(\sum_\Psi (-1)^{|\Psi|} e(\lambda+\rho-\sum \Psi) \right)$$
as required.

This argument therefore explains the difference between Kac' character formula and Borcherds' character formula, and where the extra term in Borcherds' character formula comes from.

21.3 The Monster Lie algebra

In this final section we shall show that, although Borcherds algebras have many properties which seem quite similar to those of Kac–Moody algebras, they include examples which behave in a very different way from Kac–Moody algebras. The example we have in mind is the Monster Lie algebra. The definition and properties of the Monster Lie algebra are closely related to the properties of a certain modular function j, so we shall begin by describing the definition and significance of this function.

21.3 The Monster Lie algebra

We first recall the action of the group $SL_2(\mathbb{R})$ on the upper half plane H. Let

$$SL_2(\mathbb{R}) = \left\{ \begin{pmatrix} a & b \\ c & d \end{pmatrix} ; ad - bc = 1, \quad a, b, c, d \in \mathbb{R} \right\}$$

$$H = \{\tau \in \mathbb{C} ; \operatorname{Im} \tau > 0\}.$$

The group $SL_2(\mathbb{R})$ acts on H by

$$\begin{pmatrix} a & b \\ c & d \end{pmatrix} \tau = \frac{a\tau + b}{c\tau + d}$$

since if $\operatorname{Im} \tau > 0$ we have $\operatorname{Im} \left(\frac{a\tau+b}{c\tau+d} \right) > 0$. In particular the subgroup $SL_2(\mathbb{Z})$ acts on H. Since

$$\begin{pmatrix} -1 & 0 \\ 0 & -1 \end{pmatrix} \tau = \tau$$

we see that $PSL_2(\mathbb{Z}) = SL_2(\mathbb{Z})/(\pm I_2)$ acts on H. $PSL_2(\mathbb{Z})$ is called the **modular group**.

We denote by $H/SL_2(\mathbb{Z})$ the set of orbits. Since $\begin{pmatrix} 1 & 1 \\ 0 & 1 \end{pmatrix} \in SL_2(\mathbb{Z})$ the elements τ and $\tau + 1$ of H lie in the same orbit. Thus each orbit intersects

$$\{\tau \in H ; -\tfrac{1}{2} \leq \operatorname{Re} \tau \leq \tfrac{1}{2}\}.$$

Again we have $\begin{pmatrix} 0 & 1 \\ -1 & 0 \end{pmatrix} \in SL_2(\mathbb{Z})$ and so the elements $\tau, -1/\tau \in H$ lie in the same orbit. Thus each orbit intersects

$$\{\tau \in H ; |\tau| \geq 1\}.$$

In fact we can obtain a fundamental region for the action of $SL_2(\mathbb{Z})$ on H by taking the region

$$\{\tau \in H ; -\tfrac{1}{2} \leq \operatorname{Re} \tau \leq \tfrac{1}{2}, \quad |\tau| \geq 1\}$$

and identifying the points $\tau, \tau + 1$ for $\operatorname{Re} \tau = -\tfrac{1}{2}$ and the points $\sigma, -1/\sigma$ for $|\sigma| = 1$. The fundamental region is illustrated in Figure 21.1.

Having made the above identifications we obtain a set intersecting each orbit in just one point. The set $H/SL_2(\mathbb{Z})$ of orbits has the structure of a compact Riemann surface with one point removed. This is a Riemann surface of genus 0, i.e. a Riemann sphere. When we remove one point from it we

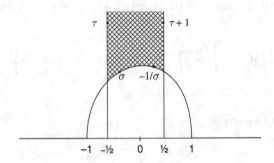

Figure 21.1 Fundamental region

obtain a subset which can be identified with $\mathbb{C}$. Thus we have an isomorphism of Riemann surfaces

$$H/SL_2(\mathbb{Z}) \to \mathbb{C}.$$

This can be extended to an isomorphism of compact Riemann surfaces by adding the point $i\infty$ on the left and ∞ on the right. Thus we have an isomorphism

$$(H/SL_2(\mathbb{Z})) \cup \{i\infty\} \to S^2 = \mathbb{C} \cup \{\infty\}$$

under which $i\infty$ maps to ∞. Such an isomorphism of Riemann surfaces is not uniquely determined. However, if j is any such isomorphism any other must have the form $a(j+b)$ where a, b are constants and $a \neq 0$. Such a map determines a map from H to $\mathbb{C}$ constant on orbits. This map will also be denoted by j. j is a modular function, i.e. a function invariant under the action of the modular group.

Since $\tau, \tau+1$ lie in the same orbit we have $j(\tau) = j(\tau+1)$, thus j is periodic. This implies that j has a Fourier expansion of form

$$j(\tau) = \sum_{n \in \mathbb{Z}} c_n e^{2\pi i n \tau}.$$

We write $q = e^{2\pi i \tau}$. Then we have

$$j(\tau) = \sum_{n \in \mathbb{Z}} c_n q^n.$$

We shall now describe such a function j. In order to do so we first introduce some modular forms. A function $f : H \to \mathbb{C}$ is called a modular form of weight k if

$$f\left(\frac{a\tau+b}{c\tau+d}\right) = (c\tau+d)^k f(\tau)$$

21.3 The Monster Lie algebra

for all $\begin{pmatrix} a & b \\ c & d \end{pmatrix} \in SL_2(\mathbb{Z})$. We give two examples of modular forms. The first is an example of a so-called Eisenstein series. For each positive integer n let

$$\sigma_3(n) = \sum_{d|n} d^3$$

summed over all divisors of n, and let

$$E_4(\tau) = 1 + 240 \sum_{n \geq 1} \sigma_3(n) q^n.$$

Thus

$$E_4(\tau) = 1 + 240q + 2160q^2 + \cdots.$$

This function is known to be a modular form of weight 4.

Secondly define Δ by

$$\Delta(\tau) = q \prod_{n \geq 1} (1 - q^n)^{24}.$$

Then

$$\Delta(\tau) = q - 24q^2 + 252q^3 - \cdots.$$

This is called Dedekind's Δ-function and is known to be a modular form of weight 12.

We now define $j : H \to \mathbb{C}$ by

$$j(\tau) = \frac{E_4(\tau)^3}{\Delta(\tau)}.$$

This is a modular form of weight 0, i.e. a modular function, and so is constant on orbits of $SL_2(\mathbb{Z})$ on H. We have

$$j(\tau) = q^{-1} + 744 + 196884q + 21493760q^2 + \cdots$$

and j has a simple pole at $\tau = i\infty$, i.e. $q = 0$.

j gives an isomorphism of Riemann surfaces

$$j : H/SL_2(\mathbb{Z}) \to \mathbb{C}$$

which extends to an isomorphism of compact Riemann surfaces

$$j : (H/SL_2(\mathbb{Z})) \cup \{i\infty\} \to S^2 = \mathbb{C} \cup \{\infty\}.$$

Any other such isomorphism has the form $a(j+b)$ where a, b are constants and $a \neq 0$. In particular there is just one such isomorphism with leading coefficient 1 and constant term 0. We shall call this the canonical isomorphism. This is the function

$$j(\tau) - 744 = q^{-1} + \sum_{n \geq 1} c_n q^n$$

where $c_1 = 196\,884$, $c_2 = 21\,493\,760$, etc. All the c_n are positive integers.

We are now ready to introduce the Monster Lie algebra. We first define a countable symmetric matrix $\mathfrak{A}$. $\mathfrak{A}$ is defined as a block matrix, with blocks of rows and columns parametrised by the natural numbers $\mathbb{N} = \{0, 1, 2, 3, \ldots\}$. Let B_{ij} be the (i, j)-block of $\mathfrak{A}$. The number of rows in B_{ij} is 1 if $i = 0$ and c_i if $i \neq 0$ where c_i is the coefficient of q^i in the modular function j. Similarly the number of columns in B_{ij} is 1 if $j = 0$ and c_j if $j \neq 0$. All the matrix entries in a given block B_{ij} are equal to one another. These entries are given as follows.

The single entry in block B_{00} is 2.

All entries in block B_{0n} for $n \neq 0$ are $-(n-1)$.

All entries in block B_{mn} for $m \neq 0, n \neq 0$ are $-(m+n)$.

These conditions determine the matrix $\mathfrak{A}$. We have

$$\mathfrak{A} = \begin{array}{c} \uparrow \\ 196\,884 \\ \downarrow \\ \uparrow \\ 21\,493\,760 \\ \downarrow \end{array} \begin{array}{|c|ccc|ccc|c|} \hline 2 & 0 & \cdots & 0 & -1 & \cdots & -1 & -2 & \cdots \\ \hline 0 & -2 & \cdots & -2 & -3 & \cdots & -3 & -4 & \cdots \\ \vdots & \vdots & & \vdots & \vdots & & \vdots & \vdots & \\ 0 & -2 & \cdots & -2 & -3 & \cdots & -3 & -4 & \cdots \\ \hline -1 & -3 & \cdots & -3 & -4 & \cdots & -4 & -5 & \cdots \\ \vdots & \vdots & & \vdots & \vdots & & \vdots & \vdots & \\ -1 & -3 & \cdots & -3 & -4 & \cdots & -4 & -5 & \cdots \\ \hline -2 & -4 & \cdots & -4 & -5 & \cdots & -5 & -6 & \cdots \\ \vdots & \vdots & & \vdots & \vdots & & \vdots & \vdots & \\ \end{array}$$

Let $L(\mathfrak{A})$ be the universal Borcherds algebra determined by the countable matrix $\mathfrak{A}$ as in Proposition 21.1. Let Z be the centre of $L(\mathfrak{A})$ and H be the subalgebra of $L(\mathfrak{A})$ generated by all elements $h_{ij} = [e_i f_j]$. We know from Section 21.1 that

$$\langle h_{ij} ; i \neq j \rangle \subset Z \subset H.$$

It is also clear that $h_i - h_j \in Z$ where $h_i = h_{ii}$, $h_j = h_{jj}$ and i, j are in the same block, since columns i, j of $\mathfrak{A}$ are then identical. Z is in fact generated by the

21.3 The Monster Lie algebra

elements h_{ij} for $i \ne j$ and $h_i - h_j$ where i, j are in the same block. Moreover Z is an ideal of $L(\mathfrak{A})$.

Let $\mathfrak{M} = L(\mathfrak{A})/Z$. $\mathfrak{M}$ is called the **Monster Lie algebra**. We shall now determine some properties of $\mathfrak{M}$.

Let the blocks of rows of $\mathfrak{A}$ be $B_0, B_1, B_2, \ldots$ with $|B_0| = 1$, $|B_n| = c_n$ for $n \ge 1$. We choose one $i \in I$ out of each block, such that i lies in the block B_i. We then consider the elements h_i for such elements $i \in I$. Thus we have elements $h_0, h_1, h_2, \ldots \in H$. Then the elements

$$2h_2 + h_0 - 3h_1$$

$$2h_3 + 2h_0 - 4h_1$$

$$2h_4 + 3h_0 - 5h_1$$

$$\vdots$$

all lie in Z, since the corresponding linear combinations of the rows of $\mathfrak{A}$ are all zero vectors.

Let $h_i \to \bar{h}_i$ under the natural homomorphism $L(\mathfrak{A}) \to \mathfrak{M}$. Then we have

$$2\bar{h}_2 = 3\bar{h}_1 - \bar{h}_0$$

$$2\bar{h}_3 = 4\bar{h}_1 - 2\bar{h}_0$$

$$2\bar{h}_4 = 5\bar{h}_1 - 3\bar{h}_0$$

$$\vdots$$

Let $\mathfrak{M}_0 = H/Z$. $\mathfrak{M}_0$ is the Cartan subalgebra of $\mathfrak{M}$, being the image of H under the natural homomorphism. We see from the above relations that $\mathfrak{M}_0$ is spanned by $\bar{h}_0$ and $\bar{h}_1$. Moreover $\bar{h}_0$ and $\bar{h}_1$ are linearly independent since this is true of the first two rows of $\mathfrak{A}$. Thus $\bar{h}_0, \bar{h}_1$ form a basis of $\mathfrak{M}_0$ and we have

$$\dim \mathfrak{M}_0 = 2.$$

In fact we find it more convenient to choose the basis b_0, b_1 of $\mathfrak{M}_0$ given by

$$b_0 = \frac{\bar{h}_0 + \bar{h}_1}{2}, \quad b_1 = \frac{-\bar{h}_0 + \bar{h}_1}{2}.$$

Thus $\mathfrak{M}_0 = \mathbb{R} b_0 + \mathbb{R} b_1$. The scalar product on $\mathfrak{M}_0$ is given by

$$\langle \bar{h}_0, \bar{h}_0 \rangle = a_{00} = 2$$

$$\langle \bar{h}_1, \bar{h}_1 \rangle = a_{11} = -2$$

$$\langle \bar{h}_0, \bar{h}_1 \rangle = \langle \bar{h}_1, \bar{h}_0 \rangle = a_{01} = 0.$$

It follows that
$$\langle b_0, b_0 \rangle = 0, \quad \langle b_1, b_1 \rangle = 0, \quad \langle b_0, b_1 \rangle = -1.$$

Hence
$$\langle mb_0 + nb_1, m'b_0 + n'b_1 \rangle = -(mn' + nm').$$

We now regard the Monster Lie algebra $\mathfrak{M}$ as a module over its Cartan subalgebra $\mathfrak{M}_0$. Let $m, n \in \mathbb{Z}$ and define $\mathfrak{M}_{(m,n)}$ by
$$\mathfrak{M}_{(m,n)} = \{x \in \mathfrak{M} \; ; \; [b_0 x] = mb_0, \; [b_1 x] = nb_1\}.$$

Then one can show that $\mathfrak{M}_{(0,0)} = \mathfrak{M}_0$ and
$$\mathfrak{M} = \bigoplus \mathfrak{M}_{(m,n)} \quad \text{for } (m,n) \in \mathbb{Z} \times \mathbb{Z}.$$

Moreover we have

$\dim \mathfrak{M}_{(m,n)} = c_{mn} \quad$ if $m \neq 0, n \neq 0$

$\dim \mathfrak{M}_{(0,0)} = 2$

$\dim \mathfrak{M}_{(m,0)} = \dim \mathfrak{M}_{(0,n)} = 0 \quad$ if $m \neq 0, n \neq 0$.

(These results follow from the 'no-ghost' theorem of Goddard and Thorn in string theory! A statement and proof of this theorem in an algebraic context can be found in E. Jurisich, *Journal of Pure and Applied Algebra* **126** (1998), 233–266).

Thus the graded components $\mathfrak{M}_{(m,n)}$ of the Monster Lie algebra $\mathfrak{M}$ are as shown in Table 21.1. In this table V_n is a vector space of dimension c_n if $n \geq 1$ and V_{-1} is a vector space of dimension 1.

Table 21.1 *Graded components $\mathfrak{M}_{(m,n)}$ of the Monster Lie Algebra.*

⋯	0	0	0	0	0	V_4	V_8	V_{12}	V_{16}	⋯
	0	0	0	0	0	V_3	V_6	V_9	V_{12}	
	0	0	0	0	0	V_2	V_4	V_6	V_8	
	0	0	0	V_{-1}	0	V_1	V_2	V_3	V_4	
	0	0	0	0	$\mathbb{R}^2$	0	0	0	0	
	V_4	V_3	V_2	V_1	0	V_{-1}	0	0	0	
	V_8	V_6	V_4	V_2	0	0	0	0	0	
	V_{12}	V_9	V_6	V_3	0	0	0	0	0	
⋯	V_{16}	V_{12}	V_8	V_4	0	0	0	0	0	⋯

21.3 The Monster Lie algebra

We now consider the roots of the Monster Lie algebra $\mathfrak{M}$. Since $\mathfrak{M} = L(\mathfrak{A})/Z$ we recall from Section 21.2 that the root lattice of $\mathfrak{M}$ is defined to be the root lattice of $L(\mathfrak{A})$. The fundamental roots of $\mathfrak{M}$ are the α_i for $i \in I$. We have a homomorphism $Q \to H$ under which α_i maps to h_i. We pointed out in Section 21.2 that this homomorphism is not in general injective. In the Monster Lie algebra it is far from injective, as α_i, α_j have the same image if and only if i, j lie in the same block of I.

We have

$$\langle \alpha_0, \alpha_0 \rangle = 2$$

$$\langle \alpha_i, \alpha_i \rangle = -2m \qquad \text{if } i \neq 0 \text{ and } i \in B_m.$$

Thus $\Pi_{\text{re}} = \{\alpha_0\}$ and $\Pi_{\text{im}} = \{\alpha_i \; ; \; i \neq 0\}$. Hence the Monster Lie algebra $\mathfrak{M}$ has just one real fundamental root and countably many imaginary fundamental roots.

The Weyl group W of $\mathfrak{M}$ is generated by the fundamental reflections corresponding to the real fundamental roots. Thus $W = \langle s_0 \rangle$, and so W has order 2. Thus $\mathfrak{M}$ has an infinite number of fundamental roots while at the same time having a very small Weyl group isomorphic to the cyclic group of order 2.

Finally we shall consider Borcherds' denominator formula for the Monster Lie algebra $\mathfrak{M}$. This formula plays an important role in Borcherds' proof of the Conway–Norton conjectures. We recall from Theorem 21.7 that Borcherds' denominator formula is given by

$$e(\rho) \prod_{\alpha \in \Phi^+} (1 - e(-\alpha))^{m_\alpha} = \sum_{w \in W} \varepsilon(w) w \left(e(\rho) \sum_{\Psi} (-1)^{|\Psi|} e(-\sum \Psi) \right)$$

where $\rho \in Q \otimes \mathbb{R}$ is any vector satisfying

$$\langle \rho, \alpha_i \rangle = \tfrac{1}{2} \langle \alpha_i, \alpha_i \rangle \qquad \text{for all } i \in I.$$

Now we have

$$\bar{h}_n = \frac{(n+1)\bar{h}_1 - (n-1)\bar{h}_0}{2} = b_0 + nb_1.$$

Thus we can identify a fundamental root α_i in the block B_n with its image $b_0 + nb_1$ in the Cartan subalgebra $\mathfrak{M}_0$ of $\mathfrak{M}$ provided we remember that there will be c_n different such fundamental roots α_i with a given image $b_0 + nb_1$.

We may take $\rho = \dfrac{(-\alpha_0 - \alpha_1)}{2}$, since if $\alpha_i \in B_n$ we have

$$\langle \rho, \alpha_i \rangle = \left\langle \dfrac{-\bar{h}_0 - \bar{h}_1}{2}, b_0 + nb_1 \right\rangle = \langle -b_0, b_0 + nb_1 \rangle = n$$

whereas

$$\langle \alpha_i, \alpha_i \rangle = \langle b_0 + nb_1, b_0 + nb_1 \rangle = -2n.$$

Thus $\langle \rho, \alpha_i \rangle = \tfrac{1}{2} \langle \alpha_i, \alpha_i \rangle$.

Hence we shall use this vector ρ in Borcherds' denominator formula. Using the natural homomorphisms

$$Q \to H \to \mathfrak{M}_0 = H/Z$$
$$\alpha_i \to h_i \to \bar{h}_i$$

we may interpret Borcherds' denominator formula in the integral group ring of $e(\mathfrak{M}_0)$ rather than the integral group ring of $e(Q)$. Bearing this in mind we define

$$p = e^{b_0}, \quad q = e^{b_1}.$$

Then $e^\rho = p^{-1}$ and so the left-hand side of the denominator identity is

$$p^{-1} \prod_{\substack{m > 0 \\ n \in \mathbb{Z}}} (1 - p^m q^n)^{c_{mn}}$$

since the positive roots $\alpha \in \Phi^+$ are the elements of Q^+ which map to elements of $\mathfrak{M}_0$ of the form $mb_0 + nb_1$ with $m > 0$ and $n \in \mathbb{Z}$, and the number of $\alpha \in \Phi^+$ mapping to $mb_0 + nb_1$ is $\dim \mathfrak{M}_{(m,n)} = c_{mn}$.

We now consider the right-hand side of the denominator identity. We recall that, for $\alpha \in Q$, $\varepsilon(\alpha) = (-1)^k$ where α is the sum of k orthogonal imaginary fundamental roots and $\varepsilon(\alpha) = 0$ otherwise. In the case of the Monster Lie algebra $\mathfrak{M}$ no two imaginary fundamental roots are orthogonal since

$$\langle b_0 + mb_1, b_0 + nb_1 \rangle = -(m+n).$$

Thus the elements $\alpha \in Q$ contributing to $\sum \varepsilon(\alpha) e^\alpha$ are $\alpha = 0$ with $\varepsilon(\alpha) = 1$ and the imaginary simple roots in Q. These map to elements of form $b_0 + nb_1 \in \mathfrak{M}_0$. There are c_n such roots $\alpha \in Q$ mapping to $b_0 + nb_1$ and they all give $\varepsilon(\alpha) = -1$. Thus

$$\sum_{\alpha \in Q} \varepsilon(\alpha) e^\alpha = 1 - \sum_{n > 0} c_n p q^n.$$

21.3 The Monster Lie algebra

Now the Weyl group W has order 2 and consists of the elements 1 and s_0. We have $s_0(p) = q$ and $s_0(q) = p$. Thus the right-hand side of the denominator identity is

$$p^{-1}\left(1 - \sum_{n>0} c_n pq^n\right) - q^{-1}\left(1 - \sum_{n>0} c_n qp^n\right)$$

$$= \left(p^{-1} + \sum_{n>0} c_n p^n\right) - \left(q^{-1} + \sum_{n>0} c_n q^n\right)$$

$$= j(p) - j(q).$$

Thus we have obtained the following result.

Theorem 21.10 *Borcherds' denominator identity for the Monster Lie algebra $\mathfrak{M}$ asserts that:*

$$p^{-1} \prod_{\substack{m>0 \\ n \in \mathbb{Z}}} (1 - p^m q^n)^{c_{mn}} = j(p) - j(q)$$

where c_n is the coefficient of q^n in the modular function j. □

In fact this identity was proved by Borcherds from first principles and used subsequently to prove that the fundamental roots of $\mathfrak{M}$ map to the elements

$$b_0 - b_1, b_0 + b_1, b_0 + 2b_1, b_0 + 3b_1, \ldots$$

of $\mathfrak{M}_0$.

Further information about results stated without proof in this chapter can be found in the papers of R. Borcherds 'Generalised Kac–Moody algebras', *Journal of Algebra* **115** (1988), 501–512, and 'Monstrous moonshine and monstrous Lie superalgebras', *Inventiones Mathematiae* **109** (1992), 405–444.

Appendix

Summary pages – explanation

There follow a number of summary pages, one for each Lie algebra of finite or affine type, giving basic properties of the Lie algebra in question. The information given differs to some extent between the Lie algebras of finite type and those of affine type.

In the case of the algebras of finite type we give the name of the algebra, the Dynkin diagram with the labelling we have chosen for its vertices, the Cartan matrix, the dimension of the Lie algebra, its Coxeter number, the order of its Weyl group W and the degrees of the basic polynomial invariants of W. We also give information about its root system. The roots are most conveniently described in terms of a basis $\beta_1, \ldots, \beta_m$ of mutually orthogonal basis vectors all of the same length. In several cases it is convenient to choose m greater than the rank l of the Lie algebra, so that the root system lies in a proper subspace of the vector space spanned by $\beta_1, \ldots, \beta_m$. In the cases when there are roots of two different lengths the long roots and short roots are both described. The extended Dynkin diagram is given and the root lattice described in terms of the above orthogonal basis. The fundamental weights are given, as is the index of the root lattice in the weight lattice. Finally the standard invariant forms on $H_\mathbb{R}$ and $H_\mathbb{R}^*$ are described, and the constant is given which converts the standard invariant form on $H_\mathbb{R}$ into the Killing form.

The labelling given here for the vertices of the Dynkin diagrams of types E_6 and E_7 differs from that used in Chapter 8, where it was convenient to describe the root systems of type E_6, E_7 or E_8 together in Section 8.7.

In the case of the Lie algebras of affine type we have given two names for each algebra which we have called the Dynkin name and the Kac name. The Dynkin name describes the Dynkin diagram of the algebra whereas the Kac name, introduced at the end of Chapter 18, indicates whether the Lie algebra is of untwisted or twisted type, and in the case of those of twisted type indicates the type of the untwisted affine algebra from which it is obtained, together with the order of the automorphism of which it is the fixed point subalgebra. This Kac notation is entirely consistent with the notation normally used to describe the twisted Chevalley groups.

The Dynkin diagram with chosen labelling is given, together with the generalised Cartan matrix and the integers $a_0, a_1, \ldots, a_l$ and $c_0, c_1, \ldots, c_l$. The central element c, the basic imaginary root δ, and the elements $\theta \in \left(H_\mathbb{R}^0\right)^*$ and $h_\theta \in H_\mathbb{R}^0$ which play an important role in the theory of affine algebras are written down explicitly. The Coxeter number and dual Coxeter number are given.

The type of the finite dimensional Lie algebra L^0 obtained by removing vertex 0 from the Dynkin diagram is given. The root system Φ is described in terms of the root system Φ^0 of L^0. The real and imaginary roots are given separately and the multiplicities of the imaginary roots are given. (The real roots all have multiplicity 1.) In order to clarify the action of the affine Weyl group we describe the lattices $M \subset H_\mathbb{R}^0$ and $M^* \subset \left(H_\mathbb{R}^0\right)^*$ which give rise to the translations in the affine Weyl group. We also describe the fundamental alcoves $A \subset H_\mathbb{R}^0$ and $A^* \subset \left(H_\mathbb{R}^0\right)^*$ whose closures give fundamental regions for the action of the affine Weyl group. Then we describe the fundamental weights in terms of the fundamental weights of L^0, and the standard invariant forms on H and on H^*.

For the affine algebras of types $\tilde{C}'_l, l \geq 2$, and $\tilde{A}'_1$ we have given two different descriptions, corresponding to two choices of the vertex of the Dynkin diagram labelled by 0. (The algebra $\tilde{A}'_1$ behaves just like $\tilde{C}'_l$ when $l=1$.) Both descriptions are useful, as is shown in Section 18.4. The first description is the conventional description in which the associated finite dimensional algebra has type C_l, and which is discussed in Chapter 17. The second description is the one used to obtain the realisation of $\tilde{C}'_l$ as $^2\tilde{A}_{2l}$ in Section 18.4. Here the associated finite dimensional algebra has type B_l. A word of caution is necessary in deriving the results appearing in the second description. In these cases we have $c_0 = 2$. Thus we cannot apply results from Chapter 17 uncritically to these cases, since $c_0 = 1$ is assumed in Chapter 17. Instead we have the following situation.

$$\theta = \delta - a_0 \alpha_0 = \sum_{i=1}^{l} a_i \alpha_i$$

satisfies $\langle \theta, \theta \rangle = 2a_0 c_0$. We also have

$$h_\theta = \frac{1}{a_0 c_0}(c - c_0 h_0) = \frac{1}{a_0 c_0} \sum_{i=1}^{l} c_i h_i.$$

Under the natural bijection $H \leftrightarrow H^*$ we have $h_i \leftrightarrow a_i c_i^{-1} \alpha_i$, $h_\theta \leftrightarrow a_0^{-1} c_0^{-1} \theta$, $d \leftrightarrow a_0 c_0^{-1} \gamma$. In addition we have

$$\langle h_0, d \rangle = a_0 c_0^{-1}, \ \langle \alpha_0, \gamma \rangle = a_0^{-1} c_0.$$

The lattices M, M^* are given as follows in these cases. M is the lattice generated by $w(h_\theta)$ for all $w \in W^0$, and M^* is the lattice generated by $w(a_0^{-1} c_0^{-1} \theta)$ for all $w \in W^0$. The alcove A is bounded by the affine hyperplane $\theta(h) = 1$ and the alcove A^* is bounded by the affine hyperplane $\lambda(h_\theta) = \frac{1}{a_0 c_0}$.

In fact in this second description θ turns out to be $2\theta_s$ where θ_s is the highest short root, and h_θ is $\frac{1}{2} h_{\theta_s}$.

NAME A_l

Dynkin diagram with labelling.

$$\underset{1}{\circ}\!\!-\!\!\underset{2}{\circ}\!\!-\!\!\underset{3}{\circ}\!\!-\!\!\circ\!\!-\!\!\circ\!\!-\!\!\circ\!\!-\!\!\underset{l-1}{\circ}\!\!-\!\!\underset{l}{\circ}$$

Cartan matrix

$$\begin{array}{c c} & \begin{matrix} 1 & 2 & 3 & \cdots & \cdots & l-1 & l \end{matrix} \\ \begin{matrix} 1 \\ 2 \\ 3 \\ \cdot \\ \cdot \\ \cdot \\ l-1 \\ l \end{matrix} & \begin{pmatrix} 2 & -1 & & & & & \\ -1 & 2 & -1 & & & & \\ & -1 & 2 & \cdot & & & \\ & & \cdot & \cdot & \cdot & & \\ & & & \cdot & \cdot & \cdot & \\ & & & & \cdot & 2 & -1 \\ & & & & & -1 & 2 \end{pmatrix} \end{array}$$

Dimension. $\dim L = l(l+2)$.

Coxeter number. $h = l+1$.

Order of the Weyl group. $|W| = (l+1)!$

Degrees of the basic polynomial invariants of W.

$$\{d_1, d_2, \ldots, d_l\} = \{2, 3, \ldots, l+1\}.$$

Number of roots. $|\Phi| = l(l+1)$.

The fundamental roots in terms of an orthogonal basis.

$$\alpha_1 = \beta_1 - \beta_2, \quad \alpha_2 = \beta_2 - \beta_3, \quad \ldots, \quad \alpha_l = \beta_l - \beta_{l+1}.$$

The root system.

$$\Phi = \{\beta_i - \beta_j \,;\; i,j = 1, \ldots, l+1, \; i \neq j\}.$$

The highest root. $\theta = \beta_1 - \beta_{l+1}$.

The extended Dynkin diagram, for $l \geq 2$.

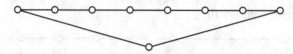

The root lattice $\quad Q = \sum \mathbb{Z}\alpha_i$.

$$Q = \left\{ \sum_{i=1}^{l+1} \xi_i \beta_i \; ; \; \xi_i \in \mathbb{Z}, \; \sum \xi_i = 0 \right\}.$$

The fundamental weights.

$$\omega_i = \frac{1}{l+1}((l+1-i)(\beta_1 + \cdots + \beta_i) - i(\beta_{i+1} + \cdots + \beta_{l+1})) \qquad i = 1, \ldots, l.$$

The index of the root lattice in the weight lattice. $\quad |X:Q| = l+1. \qquad X/Q$ is cyclic.

The standard invariant form on $H_\mathbb{R}$.

$$\langle h_i, h_j \rangle = A_{ij}.$$

The standard invariant form on $H_\mathbb{R}^*$.

$$\langle \alpha_i, \alpha_j \rangle = A_{ij}.$$

The Killing form on $H_\mathbb{R}$.

$$\langle x, y \rangle_K = \frac{1}{b} \langle x, y \rangle$$

where $b = 2(l+1)$.

NAME B_l

Dynkin diagram with labelling.

$$\underset{1}{\circ}\!\!-\!\!\underset{2}{\circ}\!\!-\!\!\underset{3}{\circ}\!\!-\!\!\circ\!\!-\!\!\circ\!\!-\!\!\circ\!\!-\!\!\underset{l-1}{\circ}\!\!\Rightarrow\!\!\underset{l}{\circ}$$

Cartan matrix

$$\begin{array}{c} \\ 1 \\ 2 \\ 3 \\ \cdot \\ \cdot \\ l-2 \\ l-1 \\ l \end{array}\begin{pmatrix} 1 & 2 & 3 & \cdot & \cdot & l-2 & l-1 & l \\ 2 & -1 & & & & & & \\ -1 & 2 & -1 & & & & & \\ & -1 & 2 & \cdot & & & & \\ & & \cdot & \cdot & \cdot & & & \\ & & & \cdot & \cdot & \cdot & & \\ & & & & \cdot & 2 & -1 & \\ & & & & & -1 & 2 & -1 \\ & & & & & & -2 & 2 \end{pmatrix}$$

Dimension. $\dim L = l(2l+1)$.

Coxeter number. $h = 2l$.

Order of the Weyl group. $|W| = 2^l \cdot l!$

Degrees of the basic polynomial invariants of W.

$$\{d_1, d_2, \ldots, d_l\} = \{2, 4, \ldots, 2l\}.$$

Number of roots. $|\Phi| = 2l^2$.

The fundamental roots in terms of an orthogonal basis.

$$\alpha_1 = \beta_1 - \beta_2, \quad \alpha_2 = \beta_2 - \beta_3, \quad \ldots, \quad \alpha_{l-1} = \beta_{l-1} - \beta_l, \quad \alpha_l = \beta_l.$$

The root system. $\Phi = \Phi_l \cup \Phi_s$ where

$$\Phi_l = \{\pm\beta_i \pm \beta_j \,;\; i, j = 1, \ldots, l, \; i \neq j\}$$
$$\Phi_s = \{\pm\beta_i \,;\; i = 1, \ldots, l\}.$$

The highest root. $\theta_l = \beta_1 + \beta_2$.

The highest short root. $\theta_s = \beta_1$.

The extended Dynkin diagram.

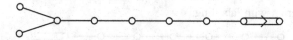

The root lattice $\quad Q = \sum \mathbb{Z} \alpha_i$.

$$Q = \left\{ \sum_{i=1}^{l} \xi_i \beta_i \ ; \ \xi_i \in \mathbb{Z} \right\}.$$

The fundamental weights.

$$\omega_i = \beta_1 + \cdots + \beta_i \qquad i = 1, \ldots, l-1$$
$$\omega_l = \tfrac{1}{2}(\beta_1 + \cdots + \beta_l).$$

The index of the root lattice in the weight lattice. $\quad |X:Q| = 2.$

The symmetrising matrix $D = \mathrm{diag}\,(d_i)$.

$$d_i = 1 \quad i = 1, \ldots, l-1 \qquad d_l = 2.$$

The standard invariant form on $H_\mathbb{R}$.

$$\langle h_i, h_j \rangle = A_{ij} d_j.$$

The standard invariant form on $H_\mathbb{R}^*$.

$$\langle \alpha_i, \alpha_j \rangle = d_i^{-1} A_{ij}.$$

The Killing form on $H_\mathbb{R}$.

$$\langle x, y \rangle_K = \frac{1}{b} \langle x, y \rangle$$

where $b = 4l - 2$.

NAME C_l

Dynkin diagram with labelling.

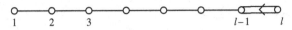

Cartan matrix

$$
\begin{array}{c}
\\
1\\
2\\
3\\
\cdot\\
\cdot\\
l-2\\
l-1\\
l
\end{array}
\begin{pmatrix}
2 & -1 & & & & & & \\
-1 & 2 & -1 & & & & & \\
& -1 & 2 & \cdot & & & & \\
& & \cdot & \cdot & \cdot & & & \\
& & & \cdot & \cdot & \cdot & & \\
& & & & 2 & -1 & & \\
& & & & -1 & 2 & -2 & \\
& & & & & -1 & 2 &
\end{pmatrix}
$$

with column labels $1, 2, 3, \ldots, l-2, l-1, l$.

Dimension. $\dim L = l(2l+1)$.

Coxeter number. $h = 2l$.

Order of the Weyl group. $|W| = 2^l \cdot l!$

Degrees of the basic polynomial invariants of W.

$$\{d_1, d_2, \ldots, d_l\} = \{2, 4, \ldots, 2l\}.$$

Number of roots. $|\Phi| = 2l^2$.

The fundamental roots in terms of an orthogonal basis.

$$\alpha_1 = \beta_1 - \beta_2, \quad \alpha_2 = \beta_2 - \beta_3, \quad \ldots, \quad \alpha_{l-1} = \beta_{l-1} - \beta_l, \quad \alpha_l = 2\beta_l.$$

The root system. $\Phi = \Phi_l \cup \Phi_s$ where

$$\Phi_l = \{\pm 2\beta_i \ ; \ i = 1, \ldots, l\}$$
$$\Phi_s = \{\pm \beta_i \pm \beta_j \ ; \ i, j = 1, \ldots, l \ \ i \neq j\}.$$

The highest root. $\theta_l = 2\beta_1$.

The highest short root. $\theta_s = \beta_1 + \beta_2$.

The extended Dynkin diagram.

The root lattice $\quad Q = \sum \mathbb{Z}\alpha_i$.

$$Q = \left\{ \sum_{i=1}^{l} \xi_i \beta_i \; ; \; \xi_i \in \mathbb{Z}, \; \sum \xi_i \text{ even} \right\}.$$

The fundamental weights

$$\omega_i = \beta_1 + \cdots + \beta_i \qquad i = 1, \ldots, l.$$

The index of the root lattice in the weight lattice. $\quad |X:Q| = 2.$

The symmetrising matrix $D = \operatorname{diag}(d_i)$.

$$d_i = 2 \quad i = 1, \ldots, l-1 \qquad d_l = 1.$$

The standard invariant form on $H_\mathbb{R}$.

$$\langle h_i, h_j \rangle = A_{ij} d_j.$$

The standard invariant form on $H_\mathbb{R}^*$.

$$\langle \alpha_i, \alpha_j \rangle = d_i^{-1} A_{ij}.$$

The Killing form on $H_\mathbb{R}$.

$$\langle x, y \rangle_K = \frac{1}{b} \langle x, y \rangle$$

where $b = 2(l+1)$.

NAME D_l

Dynkin diagram with labelling.

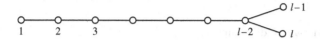

Cartan matrix

$$\begin{array}{c c} & \begin{array}{cccccccc} 1 & 2 & 3 & \cdot & \cdot & \cdot & l-2 & l-1 & l \end{array} \\ \begin{array}{c} 1 \\ 2 \\ 3 \\ \cdot \\ \cdot \\ \cdot \\ l-2 \\ l-1 \\ l \end{array} & \left(\begin{array}{ccccccccc} 2 & -1 & & & & & & & \\ -1 & 2 & -1 & & & & & & \\ & -1 & 2 & \cdot & & & & & \\ & & \cdot & \cdot & \cdot & & & & \\ & & & \cdot & \cdot & \cdot & & & \\ & & & & \cdot & \cdot & \cdot & & \\ & & & & & \cdot & 2 & -1 & -1 \\ & & & & & & -1 & 2 & \\ & & & & & & -1 & & 2 \end{array} \right) \end{array}$$

Dimension. $\dim L = l(2l-1)$.

Coxeter number. $h = 2l - 2$.

Order of the Weyl group. $|W| = 2^{l-1} l!$

Degrees of the basic polynomial invariants of W.

$$\{d_1, d_2, \ldots, d_l\} = \{2, 4, \ldots, 2l-2, l\}.$$

Number of roots. $|\Phi| = 2l(l-1)$.

The fundamental roots in terms of an orthogonal basis.

$$\alpha_1 = \beta_1 - \beta_2, \quad \alpha_2 = \beta_2 - \beta_3, \quad \ldots, \quad \alpha_{l-2} = \beta_{l-2} - \beta_{l-1},$$
$$\alpha_{l-1} = \beta_{l-1} - \beta_l, \quad \alpha_l = \beta_{l-1} + \beta_l.$$

The root system.
$$\Phi = \{\pm\beta_i \pm \beta_j \, ; \quad i,j=1,\ldots,l \quad i \neq j\}.$$

The highest root. $\theta = \beta_1 + \beta_2.$

The extended Dynkin diagram.

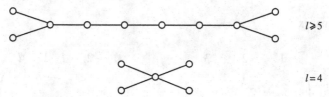

The root lattice $Q = \sum \mathbb{Z}\alpha_i.$
$$Q = \left\{ \sum_{i=1}^{l} \xi_i \beta_i \, ; \; \xi_i \in \mathbb{Z}, \; \sum \xi_i \text{ even} \right\}.$$

The fundamental weights.
$$\omega_i = \beta_1 + \cdots + \beta_i \qquad i=1,\ldots,l-2$$
$$\omega_{l-1} = \tfrac{1}{2}(\beta_1 + \cdots + \beta_{l-2} + \beta_{l-1} - \beta_l)$$
$$\omega_l = \tfrac{1}{2}(\beta_1 + \cdots + \beta_{l-2} + \beta_{l-1} + \beta_l).$$

The index of the root lattice in the weight lattice. $|X:Q| = 4.$

X/Q is cyclic if l is odd and non-cyclic if l is even.

The standard invariant form on $H_\mathbb{R}$.
$$\langle h_i, h_j \rangle = A_{ij}.$$

The standard invariant form on $H_\mathbb{R}^*$.
$$\langle \alpha_i, \alpha_j \rangle = A_{ij}.$$

The Killing form on $H_\mathbb{R}$.
$$\langle x, y \rangle_K = \frac{1}{b} \langle x, y \rangle$$

where $b = 4(l-1).$

NAME E_6

Dynkin diagram with labelling.

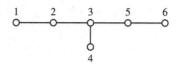

Cartan matrix

$$\begin{pmatrix} & 1 & 2 & 3 & 4 & 5 & 6 \\ 1 & 2 & -1 & 0 & 0 & 0 & 0 \\ 2 & -1 & 2 & -1 & 0 & 0 & 0 \\ 3 & 0 & -1 & 2 & -1 & -1 & 0 \\ 4 & 0 & 0 & -1 & 2 & 0 & 0 \\ 5 & 0 & 0 & -1 & 0 & 2 & -1 \\ 6 & 0 & 0 & 0 & 0 & -1 & 2 \end{pmatrix}$$

Dimension. $\dim L = 78$.

Coxeter number. $h = 12$.

Order of the Weyl group. $|W| = 2^7 \cdot 3^4 \cdot 5$

Degrees of the basic polynomial invariants of W.

$$\{d_1, d_2, \ldots, d_6\} = \{2, 5, 6, 8, 9, 12\}.$$

Number of roots. $|\Phi| = 72$.

The fundamental roots in terms of an orthogonal basis.

$\beta_1, \beta_2, \beta_3, \beta_4, \beta_5, \beta_6, \beta_7, \beta_8$ orthogonal basis.

$\alpha_1 = \beta_1 - \beta_2$, $\alpha_2 = \beta_2 - \beta_3$, $\alpha_3 = \beta_3 - \beta_4$, $\alpha_4 = \beta_4 - \beta_5$,

$\alpha_5 = \beta_4 + \beta_5$, $\alpha_6 = -\frac{1}{2}(\beta_1 + \beta_2 + \beta_3 + \beta_4 + \beta_5 + \beta_6 + \beta_7 + \beta_8)$.

The root system.
$$\Phi = \{\pm\beta_i \pm \beta_j ; \; i,j = 1,2,3,4,5 \; i \neq j\}$$
$$\cup \left\{ \tfrac{1}{2} \sum_{i=1}^{8} \varepsilon_i \beta_i ; \; \varepsilon_i = \pm 1, \prod_{i=1}^{8} \varepsilon_i = 1, \; \varepsilon_6 = \varepsilon_7 = \varepsilon_8 \right\}.$$

The highest root.
$$\theta = \tfrac{1}{2}(\beta_1 + \beta_2 + \beta_3 + \beta_4 - \beta_5 - \beta_6 - \beta_7 - \beta_8).$$

The extended Dynkin diagram.

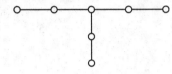

The root lattice $Q = \sum \mathbb{Z}\alpha_i$.
$$Q = \left\{ \sum_{i=1}^{8} \xi_i \beta_i ; \; 2\xi_i \in \mathbb{Z}, \xi_i - \xi_j \in \mathbb{Z}, \sum_{i=1}^{8} \xi_i \in 2\mathbb{Z}, i,j = 1,\ldots,8 \; \xi_6 = \xi_7 = \xi_8 \right\}.$$

The fundamental weights.
$$\omega_1 = \beta_1 - \tfrac{1}{3}(\beta_6 + \beta_7 + \beta_8)$$
$$\omega_2 = \beta_1 + \beta_2 - \tfrac{2}{3}(\beta_6 + \beta_7 + \beta_8)$$
$$\omega_3 = \beta_1 + \beta_2 + \beta_3 - (\beta_6 + \beta_7 + \beta_8)$$
$$\omega_4 = \tfrac{1}{2}(\beta_1 + \beta_2 + \beta_3 + \beta_4 - \beta_5 - \beta_6 - \beta_7 - \beta_8)$$
$$\omega_5 = \tfrac{1}{2}(\beta_1 + \beta_2 + \beta_3 + \beta_4 + \beta_5) - \tfrac{5}{6}(\beta_6 + \beta_7 + \beta_8)$$
$$\omega_6 = -\tfrac{2}{3}(\beta_6 + \beta_7 + \beta_8).$$

The index of the root lattice in the weight lattice. $|X:Q| = 3$.

The standard invariant form on $H_\mathbb{R}$.
$$\langle h_i, h_j \rangle = A_{ij}.$$

The standard invariant form on $H_\mathbb{R}^*$.
$$\langle \alpha_i, \alpha_j \rangle = A_{ij}.$$

The Killing form on $H_\mathbb{R}$.
$$\langle x, y \rangle_K = \frac{1}{b} \langle x, y \rangle$$

where $b = 24$.

NAME E_7

Dynkin diagram with labelling.

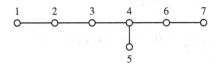

Cartan matrix

$$\begin{pmatrix} & 1 & 2 & 3 & 4 & 5 & 6 & 7 \\ 1 & 2 & -1 & 0 & 0 & 0 & 0 & 0 \\ 2 & -1 & 2 & -1 & 0 & 0 & 0 & 0 \\ 3 & 0 & -1 & 2 & -1 & 0 & 0 & 0 \\ 4 & 0 & 0 & -1 & 2 & -1 & -1 & 0 \\ 5 & 0 & 0 & 0 & -1 & 2 & 0 & 0 \\ 6 & 0 & 0 & 0 & -1 & 0 & 2 & -1 \\ 7 & 0 & 0 & 0 & 0 & 0 & -1 & 2 \end{pmatrix}$$

Dimension. $\dim L = 133$.

Coxeter number. $h = 18$.

Order of the Weyl group. $|W| = 2^{10} \cdot 3^4 \cdot 5 \cdot 7$.

Degrees of the basic polynomial invariants of W.

$$\{d_1, d_2, \ldots, d_7\} = \{2, 6, 8, 10, 12, 14, 18\}.$$

Number of roots. $|\Phi| = 126$.

The fundamental roots in terms of an orthogonal basis.

$\beta_1, \beta_2, \beta_3, \beta_4, \beta_5, \beta_6, \beta_7, \beta_8$ orthogonal basis.

$\alpha_1 = \beta_1 - \beta_2$, $\alpha_2 = \beta_2 - \beta_3$, $\alpha_3 = \beta_3 - \beta_4$, $\alpha_4 = \beta_4 - \beta_5$, $\alpha_5 = \beta_5 - \beta_6$,
$\alpha_6 = \beta_5 + \beta_6$, $\alpha_7 = -\frac{1}{2}(\beta_1 + \beta_2 + \beta_3 + \beta_4 + \beta_5 + \beta_6 + \beta_7 + \beta_8)$.

The root system.
$$\Phi = \{\pm\beta_i \pm \beta_j \; ; \;\; i, j = 1, 2, 3, 4, 5, 6 \;\; i \neq j\}$$
$$\cup \{\pm(\beta_7 + \beta_8)\}$$
$$\cup \left\{ \frac{1}{2} \sum_{i=1}^{8} \varepsilon_i \beta_i \; ; \;\; \varepsilon_i = \pm 1, \prod_{i=1}^{8} \varepsilon_i = 1, \varepsilon_7 = \varepsilon_8 \right\}.$$

The highest root. $\theta = -\beta_7 - \beta_8$.

The extended Dynkin diagram

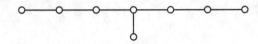

The root lattice $Q = \sum \mathbb{Z}\alpha_i$.
$$Q = \left\{ \sum_{i=1}^{8} \xi_i \beta_i \; ; \;\; 2\xi_i \in \mathbb{Z}, \; \xi_i - \xi_j \in \mathbb{Z}, \; \sum_{i=1}^{8} \xi_i \in 2\mathbb{Z}, \; i, j = 1, \ldots, 8 \;\; \xi_7 = \xi_8 \right\}.$$

The fundamental weights.
$$\omega_1 = \beta_1 - \tfrac{1}{2}(\beta_7 + \beta_8)$$
$$\omega_2 = \beta_1 + \beta_2 - (\beta_7 + \beta_8)$$
$$\omega_3 = \beta_1 + \beta_2 + \beta_3 - \tfrac{3}{2}(\beta_7 + \beta_8)$$
$$\omega_4 = \beta_1 + \beta_2 + \beta_3 + \beta_4 - 2(\beta_7 + \beta_8)$$
$$\omega_5 = \tfrac{1}{2}(\beta_1 + \beta_2 + \beta_3 + \beta_4 + \beta_5 - \beta_6) - (\beta_7 + \beta_8)$$
$$\omega_6 = \tfrac{1}{2}(\beta_1 + \beta_2 + \beta_3 + \beta_4 + \beta_5 + \beta_6) - \tfrac{3}{2}(\beta_7 + \beta_8)$$
$$\omega_7 = -(\beta_7 + \beta_8).$$

The index of the root lattice in the weight lattice. $|X:Q| = 2$.

The standard invariant form on $H_\mathbb{R}$.
$$\langle h_i, h_j \rangle = A_{ij}.$$

The standard invariant form on $H_\mathbb{R}^*$.
$$\langle \alpha_i, \alpha_j \rangle = A_{ij}.$$

The Killing form on $H_\mathbb{R}$.
$$\langle x, y \rangle_K = \frac{1}{b} \langle x, y \rangle.$$

where $b = 36$.

NAME E_8

Dynkin diagram with labelling.

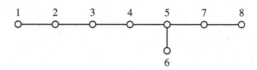

Cartan matrix

$$\begin{pmatrix} & 1 & 2 & 3 & 4 & 5 & 6 & 7 & 8 \\ 1 & 2 & -1 & 0 & 0 & 0 & 0 & 0 & 0 \\ 2 & -1 & 2 & -1 & 0 & 0 & 0 & 0 & 0 \\ 3 & 0 & -1 & 2 & -1 & 0 & 0 & 0 & 0 \\ 4 & 0 & 0 & -1 & 2 & -1 & 0 & 0 & 0 \\ 5 & 0 & 0 & 0 & -1 & 2 & -1 & -1 & 0 \\ 6 & 0 & 0 & 0 & 0 & -1 & 2 & 0 & 0 \\ 7 & 0 & 0 & 0 & 0 & -1 & 0 & 2 & -1 \\ 8 & 0 & 0 & 0 & 0 & 0 & 0 & -1 & 2 \end{pmatrix}$$

Dimension. $\dim L = 248$.

Coxeter number. $h = 30$.

Order of the Weyl group. $|W| = 2^{14} \cdot 3^5 \cdot 5^2 \cdot 7$.

Degrees of the basic polynomial invariants of W.

$$\{d_1, d_2, \ldots, d_8\} = \{2, 8, 12, 14, 18, 20, 24, 30\}.$$

Number of roots. $|\Phi| = 240$.

The fundamental roots in terms of an orthogonal basis.

$$\alpha_1 = \beta_1 - \beta_2, \quad \alpha_2 = \beta_2 - \beta_3, \quad \alpha_3 = \beta_3 - \beta_4, \quad \alpha_4 = \beta_4 - \beta_5,$$
$$\alpha_5 = \beta_5 - \beta_6, \quad \alpha_6 = \beta_6 - \beta_7, \quad \alpha_7 = \beta_6 + \beta_7,$$
$$\alpha_8 = -\tfrac{1}{2}(\beta_1 + \beta_2 + \beta_3 + \beta_4 + \beta_5 + \beta_6 + \beta_7 + \beta_8).$$

Appendix

The root system.
$$\Phi = \{\pm\beta_i \pm \beta_j \, ; \quad i,j = 1,2,3,4,5,6,7,8 \quad i \neq j\}$$
$$\cup \left\{\tfrac{1}{2}\sum_{i=1}^{8} \varepsilon_i \beta_i \, ; \quad \varepsilon_i = \pm 1, \; \prod_{i=1}^{8} \varepsilon_i = 1\right\}.$$

The highest root. $\theta = \beta_1 - \beta_8$.

The extended Dynkin diagram

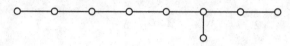

The root lattice $Q = \sum \mathbb{Z}\alpha_i$.
$$Q = \left\{\sum_{i=1}^{8} \xi_i \beta_i \, ; \; 2\xi_i \in \mathbb{Z}, \; \xi_i - \xi_j \in \mathbb{Z}, \; \sum_{i=1}^{8} \xi_i \in 2\mathbb{Z} \quad i,j = 1,\ldots,8\right\}.$$

The fundamental weights.
$$\omega_1 = \beta_1 - \beta_8$$
$$\omega_2 = \beta_1 + \beta_2 - 2\beta_8$$
$$\omega_3 = \beta_1 + \beta_2 + \beta_3 - 3\beta_8$$
$$\omega_4 = \beta_1 + \beta_2 + \beta_3 + \beta_4 - 4\beta_8$$
$$\omega_5 = \beta_1 + \beta_2 + \beta_3 + \beta_4 + \beta_5 - 5\beta_8$$
$$\omega_6 = \tfrac{1}{2}(\beta_1 + \beta_2 + \beta_3 + \beta_4 + \beta_5 + \beta_6 - \beta_7) - \tfrac{5}{2}\beta_8$$
$$\omega_7 = \tfrac{1}{2}(\beta_1 + \beta_2 + \beta_3 + \beta_4 + \beta_5 + \beta_6 + \beta_7) - \tfrac{7}{2}\beta_8$$
$$\omega_8 = -2\beta_8.$$

The index of the root lattice in the weight lattice. $|X:Q| = 1$.

The standard invariant form on $H_{\mathbb{R}}$.
$$\langle h_i, h_j \rangle = A_{ij}.$$

The standard invariant form on $H_{\mathbb{R}}^*$.
$$\langle \alpha_i, \alpha_j \rangle = A_{ij}.$$

The Killing form on $H_{\mathbb{R}}$.
$$\langle x, y \rangle_K = \frac{1}{b} \langle x, y \rangle$$
where $b = 60$.

NAME F_4

Dynkin diagram with labelling.

Cartan matrix

$$\begin{pmatrix} & 1 & 2 & 3 & 4 \\ 1 & 2 & -1 & 0 & 0 \\ 2 & -1 & 2 & -1 & 0 \\ 3 & 0 & -2 & 2 & -1 \\ 4 & 0 & 0 & -1 & 2 \end{pmatrix}$$

Dimension. $\dim L = 52$.

Coxeter number. $h = 12$.

Order of the Weyl group. $|W| = 2^7 \cdot 3^2$.

Degrees of the basic polynomial invariants of W.

$$\{d_1, d_2, d_3, d_4\} = \{2, 6, 8, 12\}.$$

Number of roots. $|\Phi| = 48$.

The fundamental roots in terms of an orthogonal basis.

$$\alpha_1 = \beta_1 - \beta_2, \quad \alpha_2 = \beta_2 - \beta_3, \quad \alpha_3 = \beta_3, \quad \alpha_4 = \tfrac{1}{2}(-\beta_1 - \beta_2 - \beta_3 + \beta_4).$$

The root system. $\Phi = \Phi_l \cup \Phi_s$ where

$$\Phi_l = \{\pm \beta_i \pm \beta_j \ ; \ i, j = 1, 2, 3, 4 \ \ i \neq j\}$$

$$\Phi_s = \{\pm \beta_i \ \ i = 1, 2, 3, 4\} \cup \left\{ \tfrac{1}{2} \sum_{i=1}^{4} \varepsilon_i \beta_i \ ; \ \varepsilon_i = \pm 1 \right\}.$$

The highest root. $\theta_l = \beta_1 + \beta_4$.

The highest short root. $\theta_s = \beta_4$.

The extended Dynkin diagram.

The root lattice $Q = \sum \mathbb{Z}\alpha_i$.

$$Q = \left\{ \sum_{i=1}^{4} \xi_i \beta_i \; ; \; 2\xi_i \in \mathbb{Z}, \; \xi_i - \xi_j \in \mathbb{Z} \; i,j = 1,2,3,4 \right\}.$$

The fundamental weights.

$$\omega_1 = \beta_1 + \beta_4$$
$$\omega_2 = \beta_1 + \beta_2 + 2\beta_4$$
$$\omega_3 = \tfrac{1}{2}(\beta_1 + \beta_2 + \beta_3 + 3\beta_4)$$
$$\omega_4 = \beta_4.$$

The index of the root lattice in the weight lattice. $\quad |X:Q| = 1.$

The symmetrising matrix $D = \operatorname{diag}(d_i)$.

$$d_1 = 1, \quad d_2 = 1, \quad d_3 = 2, \quad d_4 = 2.$$

The standard invariant form on $H_\mathbb{R}$.

$$\langle h_i, h_j \rangle = A_{ij} d_j.$$

The standard invariant form on $H_\mathbb{R}^*$.

$$\langle \alpha_i, \alpha_j \rangle = d_i^{-1} A_{ij}.$$

The Killing form on $H_\mathbb{R}$.

$$\langle x, y \rangle_K = \frac{1}{b} \langle x, y \rangle$$

where $b = 18$.

NAME G_2

Dynkin diagram with labelling.

$$\underset{1}{\circ}\!\!\Rrightarrow\!\!\underset{2}{\circ}$$

Cartan matrix

$$\begin{array}{c} \;\;1\;\;\;\;\;\;2 \\ \begin{array}{c}1\\2\end{array}\!\!\begin{pmatrix} 2 & -1 \\ -3 & 2 \end{pmatrix} \end{array}$$

Dimension. $\dim L = 14$.

Coxeter number. $h = 6$.

Order of the Weyl group. $|W| = 12$.

Degrees of the basic polynomial invariants of W.

$$\{d_1, d_2\} = \{2, 6\}.$$

Number of roots. $|\Phi| = 12$.

The fundamental roots in terms of an orthogonal basis.

$\beta_1, \beta_2, \beta_3$ orthogonal basis.

$$\alpha_1 = -2\beta_1 + \beta_2 + \beta_3, \quad \alpha_2 = \beta_1 - \beta_2.$$

The root system. $\Phi = \Phi_l \cup \Phi_s$ where

$$\Phi_l = \{\pm(-2\beta_1 + \beta_2 + \beta_3), \;\pm(\beta_1 - 2\beta_2 + \beta_3), \;\pm(\beta_1 + \beta_2 - 2\beta_3)\}$$
$$\Phi_s = \{\pm(\beta_1 - \beta_2), \;\pm(\beta_2 - \beta_3), \;\pm(\beta_1 - \beta_3)\}.$$

The highest root. $\theta_l = -\beta_1 - \beta_2 + 2\beta_3$.

The highest short root. $\theta_s = -\beta_2 + \beta_3$.

The extended Dynkin diagram.

$$\circ\!\!\!-\!\!\!-\!\!\!\circ\!\!\Rrightarrow\!\!\circ$$

The root lattice $Q = \sum \mathbb{Z}\alpha_i$.

$$Q = \left\{ \sum_{i=1}^{3} \xi_i \beta_i \;;\; \xi_i \in \mathbb{Z}, \; \xi_1 + \xi_2 + \xi_3 = 0 \right\}.$$

The fundamental weights.
$$\omega_1 = -\beta_1 - \beta_2 + 2\beta_3$$
$$\omega_2 = -\beta_2 + \beta_3.$$

The index of the root lattice in the weight lattice. $\quad |X:Q|=1.$

The symmetrising matrix $D = \operatorname{diag}(d_i)$.
$$d_1 = 1, \quad d_2 = 3.$$

The standard invariant form on $H_\mathbb{R}$.
$$\langle h_i, h_j \rangle = A_{ij} d_j.$$

The standard invariant form on $H_\mathbb{R}^*$.
$$\langle \alpha_i, \alpha_j \rangle = d_i^{-1} A_{ij}.$$

The Killing form on $H_\mathbb{R}$.
$$\langle x, y \rangle_K = \frac{1}{b} \langle x, y \rangle$$

where $b = 8$.

DYNKIN NAME $\tilde{A}_1$ KAC NAME $\tilde{A}_1$

Dynkin diagram with labelling.

$$\underset{0}{\circ}\!\!\Rrightarrow\!\!\underset{1}{\circ}$$

Generalised Cartan matrix.

$$\begin{array}{c} 01 \\ \begin{array}{c}0\\1\end{array}\!\!\left(\begin{array}{cc}2 & -2 \\ -2 & 2\end{array}\right)\end{array}$$

The integers $a_0, a_1, \ldots, a_l$.

$$\underset{}{\overset{1}{\circ}}\!\!\Rrightarrow\!\!\underset{}{\overset{1}{\circ}}$$

The integers $c_0, c_1, \ldots, c_l$.

$$\underset{}{\overset{1}{\circ}}\!\!\Rrightarrow\!\!\underset{}{\overset{1}{\circ}}$$

The central element c.

$$c = h_0 + h_1.$$

The basic imaginary root δ.

$$\delta = \alpha_0 + \alpha_1.$$

The element $h_\theta \in H_{\mathbb{R}}^0$.

$$h_\theta = h_1.$$

The element $\theta \in \left(H_{\mathbb{R}}^0\right)^*$.

$$\theta = \alpha_1.$$

The Coxeter number. $h = 2$.

The dual Coxeter number. $h^v = 2$.

The Lie algebra L^0. $L^0 = A_1$.

The lattice $M \subset H_{\mathbb{R}}^0$.

$$M = \mathbb{Z} h_1.$$

The lattice $M^* \subset \left(H_\mathbb{R}^0\right)^*$.

$$M^* = \mathbb{Z}\alpha_1.$$

The fundamental alcove $A \subset H_\mathbb{R}^0$.

$$A = \{h \in H_\mathbb{R}^0 \; ; \; \alpha_1(h) > 0, \; \alpha_1(h) < 1\}.$$

The fundamental alcove $A^* \subset \left(H_\mathbb{R}^0\right)^*$.

$$A^* = \{\lambda \in \left(H_\mathbb{R}^0\right)^* \; ; \; \lambda(h_1) > 0, \; \lambda(h_1) < 1\}.$$

The root system Φ in terms of the root system Φ^0 of L^0.

$$\Phi_{\text{Re}} = \{\alpha + r\delta \; ; \; \alpha \in \Phi^0, \; r \in \mathbb{Z}\}$$

$$\Phi_{\text{Im}} = \{k\delta \; ; \; k \in \mathbb{Z}, \; k \neq 0\} \qquad \text{Multiplicity 1.}$$

The fundamental weights $\omega_i \in H_\mathbb{R}^*$ $i = 0, 1$ in terms of the fundamental weights $\bar{\omega}_i$, $i = 1$, of L^0.

$$\omega_0 = \gamma, \quad \omega_1 = \bar{\omega}_1 + \gamma.$$

The standard invariant form on H.

$$\langle h_i, h_j \rangle = A_{ij} \qquad i, j = 0, 1$$

$$\langle h_0, d \rangle = 1, \quad \langle h_1, d \rangle = 0$$

$$\langle d, d \rangle = 0.$$

The standard invariant form on H^*.

$$\langle \alpha_i, \alpha_j \rangle = A_{ij} \qquad i, j = 0, 1$$

$$\langle \alpha_0, \gamma \rangle = 1, \quad \langle \alpha_1, \gamma \rangle = 0$$

$$\langle \gamma, \gamma \rangle = 0.$$

DYNKIN NAME $\tilde{A}'_1$ KAC NAME $^2\tilde{A}_2$
(1st description)

Dynkin diagram with labelling.

$$\underset{0 \quad\quad 1}{\Longrightarrow}$$

Generalised Cartan matrix.

$$\begin{array}{c} \quad 0 \quad\quad 1 \\ \begin{array}{c}0\\1\end{array}\!\!\left(\begin{array}{cc} 2 & -1 \\ -4 & 2 \end{array}\right)\end{array}$$

The integers $a_0, a_1, \ldots, a_l$.

$$\underset{1 \quad\quad 2}{\Longrightarrow}$$

The integers $c_0, c_1, \ldots, c_l$.

$$\underset{2 \quad\quad 1}{\Longrightarrow}$$

The central element c.
$$c = 2h_0 + h_1.$$

The basic imaginary root δ.
$$\delta = \alpha_0 + 2\alpha_1.$$

The element $h_\theta \in H^0_{\mathbb{R}}$.
$$h_\theta = \tfrac{1}{2} h_1.$$

The element $\theta \in \left(H^0_{\mathbb{R}}\right)^*$
$$\theta = 2\alpha_1.$$

The Coxeter number. $h = 3.$

The dual Coxeter number. $h^v = 3.$

The Lie algebra L^0. $L^0 = A_1.$

The lattice $M \subset H^0_{\mathbb{R}}$.
$$M = \tfrac{1}{2}\mathbb{Z} h_1.$$

The lattice $M^* \subset (H_\mathbb{R}^0)^*$

$$M^* = \mathbb{Z}\alpha_1.$$

The fundamental alcove $A \subset H_\mathbb{R}^0$.

$$A = \{h \in H_\mathbb{R}^0 \; ; \; \alpha_1(h) > 0, \; 2\alpha_1(h) < 1\}.$$

The fundamental alcove $A^* \subset (H_\mathbb{R}^0)^*$

$$A^* = \{\lambda \in (H_\mathbb{R}^0)^* \; ; \; \lambda(h_1) > 0, \; \lambda(h_1) < 1\}.$$

The root system Φ in terms of the root system Φ^0 of L^0.

$$\Phi_{\text{Re},s} = \{\alpha + r\delta \; ; \; \alpha \in \Phi^0, \; r \in \mathbb{Z}\}$$
$$\Phi_{\text{Re},l} = \{2\alpha + (2r+1)\delta \; ; \; \alpha \in \Phi^0, \; r \in \mathbb{Z}\}$$
$$\Phi_{\text{Im}} = \{k\delta \; ; \; k \in \mathbb{Z}, \; k \neq 0\} \quad \text{Multiplicity 1.}$$

The fundamental weights $\omega_i \in H_\mathbb{R}^* \; i = 0, 1, \ldots, l$ in terms of the fundamental weights $\bar{\omega}_i \; i = 1, \ldots, l$ of L^0.

$$\omega_0 = 2\gamma, \quad \omega_1 = \bar{\omega}_1 + \gamma.$$

The standard invariant form on H.

$$\langle h_i, h_j \rangle = a_j c_j^{-1} A_{ij} \quad i,j = 0, 1$$
$$\langle h_0, d \rangle = \tfrac{1}{2}, \quad \langle h_1, d \rangle = 0$$
$$\langle d, d \rangle = 0.$$

The standard invariant form on H^*

$$\langle \alpha_i, \alpha_j \rangle = a_i^{-1} c_i A_{ij} \quad i,j = 0, 1$$
$$\langle \alpha_0, \gamma \rangle = 2, \quad \langle \alpha_1, \gamma \rangle = 0$$
$$\langle \gamma, \gamma \rangle = 0.$$

DYNKIN NAME $\tilde{A}'_1$ KAC NAME $^2\tilde{A}_2$
(2nd description)

Dynkin diagram with labelling.

$$\underset{0\quad\ 1}{\Longleftarrow}$$

Generalised Cartan matrix.

$$\begin{array}{c c}& \begin{array}{c c}0 & 1\end{array}\\ \begin{array}{c}0\\ 1\end{array} & \begin{pmatrix} 2 & -4\\ -1 & 2\end{pmatrix}\end{array}$$

The integers $a_0, a_1, \ldots, a_l$.

$$\underset{}{\overset{2\quad\ 1}{\Longleftarrow}}$$

The integers $c_0, c_1, \ldots, c_l$.

$$\underset{}{\overset{1\quad\ 2}{\Longleftarrow}}$$

The central element c.

$$c = h_0 + 2h_1.$$

The basic imaginary root δ.

$$\delta = 2\alpha_0 + \alpha_1.$$

The element $h_\theta \in H^0_\mathbb{R}$.

$$h_\theta = h_1.$$

The element $\theta \in (H^0_\mathbb{R})^*$.

$$\theta = \alpha_1.$$

The Coxeter number. $h = 3$.

The dual Coxeter number. $h^\vee = 3$.

The Lie algebra L^0. $L^0 = A_1$.

The lattice $M \subset H^0_\mathbb{R}$.

$$M = \mathbb{Z}h_1.$$

The lattice $M^* \subset (H_\mathbb{R}^0)^*$.

$$M^* = \tfrac{1}{2}\mathbb{Z}\alpha_1.$$

The fundamental alcove $A \subset H_\mathbb{R}^0$.

$$A = \{h \in H_\mathbb{R}^0 \ ; \ \alpha_1(h) > 0, \ \alpha_1(h) < 1\}.$$

The fundamental alcove $A^* \subset (H_\mathbb{R}^0)^*$.

$$A^* = \{\lambda \in (H_\mathbb{R}^0)^* \ ; \ \lambda(h_1) > 0, \ 2\lambda(h_1) < 1\}.$$

The root system Φ in terms of the root system Φ^0 of L^0.

$$\Phi_{\text{Re},s} = \{\tfrac{1}{2}(\alpha + (2r-1)\delta) \ ; \ \alpha \in \Phi^0, \ r \in \mathbb{Z}\}$$

$$\Phi_{\text{Re},l} = \{\alpha + 2r\delta \ ; \ \alpha \in \Phi^0, \ r \in \mathbb{Z}\}$$

$$\Phi_{\text{Im}} = \{k\delta \ ; \ k \in \mathbb{Z}, \ k \neq 0\} \text{ Multiplicity 1}.$$

The fundamental weights $\omega_i \in H_\mathbb{R}^*$ $i = 0, 1, \ldots, l$ in terms of the fundamental weights $\bar{\omega}_i$ $i = 1, \ldots, l$ of L^0.

$$\omega_0 = \gamma, \quad \omega_1 = \bar{\omega}_1 + 2\gamma.$$

The standard invariant form on H.

$$\langle h_i, h_j \rangle = a_j c_j^{-1} A_{ij} \qquad i, j = 0, 1$$

$$\langle h_0, d \rangle = 2, \quad \langle h_1, d \rangle = 0$$

$$\langle d, d \rangle = 0.$$

The standard invariant form on H^*.

$$\langle \alpha_i, \alpha_j \rangle = a_i^{-1} c_i A_{ij} \qquad i, j = 0, 1$$

$$\langle \alpha_0, \gamma \rangle = \tfrac{1}{2}, \quad \langle \alpha_1, \gamma \rangle = 0$$

$$\langle \gamma, \gamma \rangle = 0.$$

DYNKIN NAME $\tilde{A}_l$ KAC NAME $\tilde{A}_l$ $l \geq 2$

Dynkin diagram with labelling.

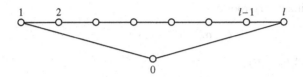

Generalised Cartan matrix.

$$\begin{array}{c} \\ 0 \\ 1 \\ 2 \\ \cdot \\ \cdot \\ \cdot \\ l-1 \\ l \end{array} \begin{array}{c} \begin{matrix} 0 & 1 & 2 & \cdot & \cdot & \cdot & l-1 & l \end{matrix} \\ \left(\begin{matrix} 2 & -1 & & & & & & -1 \\ -1 & 2 & -1 & & & & & \\ & -1 & 2 & \cdot & & & & \\ & & \cdot & \cdot & \cdot & & & \\ & & & \cdot & \cdot & \cdot & & \\ & & & & \cdot & 2 & -1 & \\ & & & & & -1 & 2 & -1 \\ -1 & & & & & & -1 & 2 \end{matrix} \right) \end{array}$$

The integers $a_0, a_1, \ldots, a_l$.

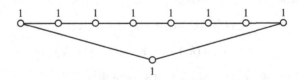

The integers $c_0, c_1, \ldots, c_l$.

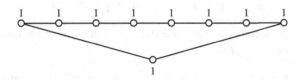

The central element c.

$$c = h_0 + h_1 + \cdots + h_{l-1} + h_l.$$

The basic imaginary root δ.

$$\delta = \alpha_0 + \alpha_1 + \cdots + \alpha_{l-1} + \alpha_l.$$

The element $h_\theta \in H_{\mathbb{R}}^0$.

$$h_\theta = h_1 + h_2 + \cdots + h_{l-1} + h_l.$$

The element $\theta \in (H_{\mathbb{R}}^0)^*$.

$$\theta = \alpha_1 + \alpha_2 + \cdots + \alpha_{l-1} + \alpha_l.$$

The Coxeter number. $\quad h = l+1$.

The dual Coxeter number. $\quad h^\vee = l+1$.

The Lie algebra L^0. $\quad L^0 = A_l$.

The lattice $M \subset H_{\mathbb{R}}^0$.

$$M = \mathbb{Z}h_1 + \mathbb{Z}h_2 + \cdots + \mathbb{Z}h_{l-1} + \mathbb{Z}h_l.$$

The lattice $M^* \subset (H_{\mathbb{R}}^0)^*$.

$$M^* = \mathbb{Z}\alpha_1 + \mathbb{Z}\alpha_2 + \cdots + \mathbb{Z}\alpha_{l-1} + \mathbb{Z}\alpha_l.$$

The fundamental alcove $A \subset H_{\mathbb{R}}^0$.

$$A = \{h \in H_{\mathbb{R}}^0 \;;\; \alpha_i(h) > 0 \text{ for } i = 1, \ldots, l$$
$$\alpha_1(h) + \alpha_2(h) + \cdots + \alpha_{l-1}(h) + \alpha_l(h) < 1\}$$

The fundamental alcove $A^* \subset (H_{\mathbb{R}}^0)^*$.

$$A^* = \{\lambda \in (H_{\mathbb{R}}^0)^* \;;\; \lambda(h_i) > 0 \text{ for } i = 1, \ldots, l$$
$$\lambda(h_1) + \lambda(h_2) + \cdots + \lambda(h_{l-1}) + \lambda(h_l) < 1\}.$$

The root system Φ in terms of the root system Φ^0 of L^0.

$$\Phi_{\text{Re}} = \{\alpha + r\delta \;;\; \alpha \in \Phi^0, r \in \mathbb{Z}\}$$
$$\Phi_{\text{Im}} = \{k\delta \;;\; k \in \mathbb{Z}, k \neq 0\} \text{ Multiplicity } l.$$

The fundamental weights $\omega_i \in H_{\mathbb{R}}^*$ $i = 0, 1, \ldots, l$ in terms of the fundamental weights $\bar{\omega}_i$, $i = 1, \ldots, l$ of L^0.

$$\omega_0 = \gamma, \quad \omega_1 = \bar{\omega}_1 + \gamma, \quad \omega_2 = \bar{\omega}_2 + \gamma, \quad \ldots, \quad \omega_{l-1} = \bar{\omega}_{l-1} + \gamma, \quad \omega_l = \bar{\omega}_l + \gamma.$$

The standard invariant form on H.

$$\langle h_i, h_j \rangle = A_{ij} \qquad i, j = 0, 1, \ldots, l$$
$$\langle h_0, d \rangle = 1, \quad \langle h_i, d \rangle = 0 \qquad i = 1, \ldots, l$$
$$\langle d, d \rangle = 0.$$

The standard invariant form on H^*.

$$\langle \alpha_i, \alpha_j \rangle = A_{ij} \qquad i, j = 0, 1, \ldots, l$$
$$\langle \alpha_0, \gamma \rangle = 1, \quad \langle \alpha_i, \gamma \rangle = 0 \qquad i = 1, \ldots, l$$
$$\langle \gamma, \gamma \rangle = 0.$$

DYNKIN NAME $\tilde{B}_l$ KAC NAME $\tilde{B}_l$ $l \geq 3$

Dynkin diagram with labelling.

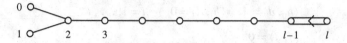

Generalised Cartan matrix.

$$
\begin{pmatrix}
 & 0 & 1 & 2 & 3 & \cdots & & & l-1 & l \\
0 & 2 & & -1 & & & & & & \\
1 & & 2 & -1 & & & & & & \\
2 & -1 & -1 & 2 & -1 & & & & & \\
3 & & & -1 & 2 & \cdot & & & & \\
\cdot & & & & \cdot & \cdot & \cdot & & & \\
\cdot & & & & & \cdot & \cdot & \cdot & & \\
\cdot & & & & & & \cdot & 2 & -1 & \\
l-1 & & & & & & & -1 & 2 & -1 \\
l & & & & & & & & -2 & 2
\end{pmatrix}
$$

The integers $a_0, a_1, \ldots, a_l$.

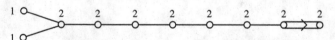

The integers $c_0, c_1, \ldots, c_l$.

The central element c.

$$c = h_0 + h_1 + 2h_2 + 2h_3 + \cdots + 2h_{l-1} + h_l.$$

DYNKIN NAME $\tilde{B}_l$ KAC NAME $\tilde{B}_l$ $l \geq 3$

The basic imaginary root δ.

$$\delta = \alpha_0 + \alpha_1 + 2\alpha_2 + 2\alpha_3 + \cdots + 2\alpha_{l-1} + 2\alpha_l.$$

The element $h_\theta \in H_\mathbb{R}^0$.

$$h_\theta = h_1 + 2h_2 + 2h_3 + \cdots + 2h_{l-1} + h_l.$$

The element $\theta \in (H_\mathbb{R}^0)^*$.

$$\theta = \alpha_1 + 2\alpha_2 + 2\alpha_3 + \cdots + 2\alpha_{l-1} + 2\alpha_l.$$

The Coxeter number. $h = 2l$.

The dual Coxeter number. $h^\vee = 2l - 1$.

The Lie algebra L^0. $L^0 = B_l$.

The lattice $M \subset H_\mathbb{R}^0$.

$$M = \mathbb{Z}h_1 + \mathbb{Z}h_2 + \cdots + \mathbb{Z}h_{l-1} + \mathbb{Z}h_l.$$

The lattice $M^* \subset (H_\mathbb{R}^0)^*$.

$$M^* = \mathbb{Z}\alpha_1 + \mathbb{Z}\alpha_2 + \cdots + \mathbb{Z}\alpha_{l-1} + 2\mathbb{Z}\alpha_l.$$

The fundamental alcove $A \subset H_\mathbb{R}^0$.

$$A = \{h \in H_\mathbb{R}^0 \;;\; \alpha_i(h) > 0 \text{ for } i = 1, \ldots, l$$
$$\alpha_1(h) + 2\alpha_2(h) + 2\alpha_3(h) + \cdots + 2\alpha_{l-1}(h) + 2\alpha_l(h) < 1\}.$$

The fundamental alcove $A^* \subset (H_\mathbb{R}^0)^*$.

$$A^* = \{\lambda \in (H_\mathbb{R}^0)^* \;;\; \lambda(h_i) > 0 \text{ for } i = 1, \ldots, l$$
$$\lambda(h_1) + 2\lambda(h_2) + 2\lambda(h_3) + \cdots + 2\lambda(h_{l-1}) + \lambda(h_l) < 1\}.$$

The root system Φ in terms of the root system Φ^0 of L^0.

$$\Phi_{\text{Re}} = \{\alpha + r\delta \;;\; \alpha \in \Phi^0, \; r \in \mathbb{Z}\}$$
$$\Phi_{\text{Im}} = \{k\delta \;;\; k \in \mathbb{Z}, \; k \neq 0\} \qquad \text{Multiplicity } l.$$

The fundamental weights $\omega_i \in H_\mathbb{R}^*$ $i = 0, \ldots, l$ in terms of the fundamental weights $\bar{\omega}_i$, $i = 1, \ldots, l$ of L^0.

$$\omega_0 = \gamma, \quad \omega_1 = \bar{\omega}_1 + \gamma, \quad \omega_2 = \bar{\omega}_2 + 2\gamma, \ldots, \omega_{l-1} = \bar{\omega}_{l-1} + 2\gamma, \quad \omega_l = \bar{\omega}_l + \gamma.$$

The standard invariant form on H.

$$\langle h_i, h_j \rangle = a_j c_j^{-1} A_{ij} \quad i,j = 0, 1, \ldots, l$$
$$\langle h_0, d \rangle = 1, \quad \langle h_i, d \rangle = 0 \quad i = 1, \ldots, l$$
$$\langle d, d \rangle = 0.$$

The standard invariant form on H^*.

$$\langle \alpha_i, \alpha_j \rangle = a_i^{-1} c_i A_{ij} \quad i,j = 0, 1, \ldots, l$$
$$\langle \alpha_0, \gamma \rangle = 1, \quad \langle \alpha_i, \gamma \rangle = 0 \quad i = 1, \ldots, l$$
$$\langle \gamma, \gamma \rangle = 0.$$

DYNKIN NAME $\tilde{B}_l^t$ KAC NAME $^2\tilde{A}_{2l-1}$ $l \geq 3$

Dynkin diagram with labelling.

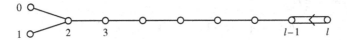

Generalised Cartan matrix.

$$\begin{pmatrix} & 0 & 1 & 2 & 3 & \cdots & & l-1 & l \\ 0 & 2 & & -1 & & & & & \\ 1 & & 2 & -1 & & & & & \\ 2 & -1 & -1 & 2 & -1 & & & & \\ 3 & & & -1 & 2 & \cdot & & & \\ \cdot & & & & \cdot & \cdot & \cdot & & \\ \cdot & & & & & \cdot & \cdot & \cdot & \\ \cdot & & & & & & \cdot & 2 & -1 \\ l-1 & & & & & & & -1 & 2 & -2 \\ l & & & & & & & & -1 & 2 \end{pmatrix}$$

The integers $a_0, a_1, \ldots, a_l$.

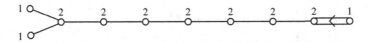

The integers $c_0, c_1, \ldots, c_l$.

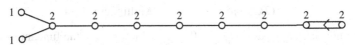

The central element c.

$$c = h_0 + h_1 + 2h_2 + 2h_3 + \cdots + 2h_{l-1} + 2h_l.$$

The basic imaginary root δ.
$$\delta = \alpha_0 + \alpha_1 + 2\alpha_2 + 2\alpha_3 + \cdots + 2\alpha_{l-1} + \alpha_l.$$

The element $h_\theta \in H_\mathbb{R}^0$.
$$h_\theta = h_1 + 2h_2 + 2h_3 + \cdots + 2h_{l-1} + 2h_l.$$

The element $\theta \in (H_\mathbb{R}^0)^*$.
$$\theta = \alpha_1 + 2\alpha_2 + 2\alpha_3 + \cdots + 2\alpha_{l-1} + \alpha_l.$$

The Coxeter number. $h = 2l - 1.$

The dual Coxeter number. $h^\vee = 2l.$

The Lie algebra L^0. $L^0 = C_l.$

The lattice $M \subset H_\mathbb{R}^0$.
$$M = \mathbb{Z}h_1 + \mathbb{Z}h_2 + \cdots + \mathbb{Z}h_{l-1} + 2\mathbb{Z}h_l.$$

The lattice $M^* \subset (H_\mathbb{R}^0)^*$.
$$M^* = \mathbb{Z}\alpha_1 + \mathbb{Z}\alpha_2 + \cdots + \mathbb{Z}\alpha_{l-1} + \mathbb{Z}\alpha_l.$$

The fundamental alcove $A \subset H_\mathbb{R}^0$.
$$A = \{h \in H_\mathbb{R}^0 \; ; \; \alpha_i(h) > 0 \text{ for } i = 1, \ldots, l$$
$$\alpha_1(h) + 2\alpha_2(h) + 2\alpha_3(h) + \cdots + 2\alpha_{l-1}(h) + \alpha_l(h) < 1\}.$$

The fundamental alcove $A^* \subset (H_\mathbb{R}^0)^*$.
$$A^* = \{\lambda \in (H_\mathbb{R}^0)^* \; ; \; \lambda(h_i) > 0 \text{ for } i = 1, \ldots, l$$
$$\lambda(h_1) + 2\lambda(h_2) + 2\lambda(h_3) + \cdots + 2\lambda(h_{l-1}) + 2\lambda(h_l) < 1\}.$$

The root system Φ in terms of the root system Φ^0 of L^0.
$$\Phi_{\text{Re},s} = \{\alpha + r\delta \; ; \; \alpha \in \Phi_s^0, \; r \in \mathbb{Z}\}$$
$$\Phi_{\text{Re},l} = \{\alpha + 2r\delta \; ; \; \alpha \in \Phi_l^0, \; r \in \mathbb{Z}\}.$$
$$\Phi_{\text{Im}} = \{2k\delta \; ; \; k \in \mathbb{Z}, \; k \neq 0\} \qquad \text{Multiplicity } l$$
$$\cup \{(2k+1)\delta \; ; \; k \in \mathbb{Z}\} \qquad \text{Multiplicity } l - 1.$$

The fundamental weights $\omega_i \in H_\mathbb{R}^*$ $i = 0, 1, \ldots, l$ in terms of the fundamental weights $\overline{\omega}_i$, $i = 1, \ldots, l$ of L^0.
$$\omega_0 = \gamma, \quad \omega_1 = \overline{\omega}_1 + \gamma, \quad \omega_2 = \overline{\omega}_2 + 2\gamma, \quad \ldots,$$
$$\omega_{l-1} = \overline{\omega}_{l-1} + 2\gamma, \quad \omega_l = \overline{\omega}_l + 2\gamma.$$

DYNKIN NAME $\tilde{B}_l^t$ KAC NAME $^2\tilde{A}_{2l-1}$ $l \geq 3$

The standard invariant form on H.

$$\langle h_i, h_j \rangle = a_j c_j^{-1} A_{ij} \quad i, j = 0, 1, \ldots, l$$
$$\langle h_0, d \rangle = 1, \quad \langle h_i, d \rangle = 0 \quad i = 1, \ldots, l$$
$$\langle d, d \rangle = 0.$$

The standard invariant form on H^*.

$$\langle \alpha_i, \alpha_j \rangle = a_i^{-1} c_i A_{ij} \quad i, j = 0, 1, \ldots, l$$
$$\langle \alpha_0, \gamma \rangle = 1, \quad \langle \alpha_i, \gamma \rangle = 0 \quad i = 1, \ldots, l$$
$$\langle \gamma, \gamma \rangle = 0.$$

DYNKIN NAME $\tilde{C}_l$ KAC NAME $\tilde{C}_l$ $l \geq 2$

Dynkin diagram with labelling.

Generalised Cartan matrix

$$\begin{array}{c} \\ 0 \\ 1 \\ 2 \\ \\ \\ \\ l-2 \\ l-1 \\ l \end{array} \begin{pmatrix} 0 & 1 & 2 & \cdots & l-2 & l-1 & l \\ 2 & -1 & & & & & \\ -2 & 2 & -1 & & & & \\ & -1 & 2 & \cdot & & & \\ & & \cdot & \cdot & \cdot & & \\ & & & \cdot & \cdot & \cdot & \\ & & & & \cdot & \cdot & \cdot \\ & & & \cdot & 2 & -1 & \\ & & & & -1 & 2 & -2 \\ & & & & & -1 & 2 \end{pmatrix}$$

The integers $a_0, a_1, \ldots, a_l$

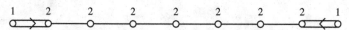

The integers $c_0, c_1, \ldots, c_l$

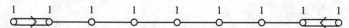

The central element c.

$$c = h_0 + h_1 + h_2 + \cdots + h_{l-1} + h_l.$$

The basic imaginary root δ.

$$\delta = \alpha_0 + 2\alpha_1 + 2\alpha_2 + \cdots + 2\alpha_{l-1} + \alpha_l.$$

The element $h_\theta \in H_\mathbb{R}^0$.

$$h_\theta = h_1 + h_2 + \cdots + h_{l-1} + h_l.$$

The element $\theta \in (H_\mathbb{R}^0)^*$.

$$\theta = 2\alpha_1 + 2\alpha_2 + \cdots + 2\alpha_{l-1} + \alpha_l.$$

The Coxeter number. $h = 2l$.

The dual Coxeter number. $h^v = l+1$.

The Lie algebra L^0. $L^0 = C_l$.

The lattice $M \subset H_\mathbb{R}^0$.

$$M = \mathbb{Z}h_1 + \mathbb{Z}h_2 + \cdots + \mathbb{Z}h_{l-1} + \mathbb{Z}h_l.$$

The lattice $M^* \subset (H_\mathbb{R}^0)^*$.

$$M^* = 2\mathbb{Z}\alpha_1 + 2\mathbb{Z}\alpha_2 + \cdots + 2\mathbb{Z}\alpha_{l-1} + \mathbb{Z}\alpha_l.$$

The fundamental alcove $A \subset H_\mathbb{R}^0$.

$$A = \{h \in H_\mathbb{R}^0 \ ; \ \alpha_i(h) > 0 \ \text{ for } i=1,\ldots,l$$
$$2\alpha_1(h) + 2\alpha_2(h) + \cdots + 2\alpha_{l-1}(h) + \alpha_l(h) < 1\}.$$

The fundamental alcove $A^* \subset (H_\mathbb{R}^0)^*$.

$$A^* = \{\lambda \in (H_\mathbb{R}^0)^* \ ; \ \lambda(h_i) > 0 \ \text{ for } i=1,\ldots,l$$
$$\lambda(h_1) + \lambda(h_2) + \cdots + \lambda(h_{l-1}) + \lambda(h_l) < 1\}.$$

The root system Φ in terms of the root system Φ^0 of L^0.

$$\Phi_{\text{Re}} = \{\alpha + r\delta \ ; \ \alpha \in \Phi^0, \ r \in \mathbb{Z}\}$$
$$\Phi_{\text{Im}} = \{k\delta \ ; \ k \in \mathbb{Z}, \ k \neq 0\} \qquad \text{Multiplicity } l.$$

The fundamental weights $\omega_i \in H_\mathbb{R}^*$ $i = 0, 1, \ldots, l$ in terms of the fundamental weights $\bar{\omega}_i$ $i = 1, \ldots, l$ of L^0.

$$\omega_0 = \gamma, \quad \omega_1 = \bar{\omega}_1 + \gamma, \quad \omega_2 = \bar{\omega}_2 + \gamma, \quad \ldots, \quad \omega_{l-1} = \bar{\omega}_{l-1} + \gamma, \quad \omega_l = \bar{\omega}_l + \gamma.$$

The standard invariant form on H.

$$\langle h_i, h_j \rangle = a_j c_j^{-1} A_{ij} \qquad i, j = 0, 1, \ldots, l$$
$$\langle h_0, d \rangle = 1, \quad \langle h_i, d \rangle = 0 \qquad i = 1, \ldots, l$$
$$\langle d, d \rangle = 0.$$

The standard invariant form on H^*.

$$\langle \alpha_i, \alpha_j \rangle = a_i^{-1} c_i A_{ij} \quad i,j = 0, 1, \ldots, l$$
$$\langle \alpha_0, \gamma \rangle = 1, \quad \langle \alpha_i, \gamma \rangle = 0 \quad i = 1, \ldots, l$$
$$\langle \gamma, \gamma \rangle = 0.$$

DYNKIN NAME $\tilde{C}_l^t$ KAC NAME $^2\tilde{D}_{l+1}$ $l \geq 2$

Dynkin diagram with labelling.

Generalised Cartan matrix.

$$\begin{pmatrix} & 0 & 1 & 2 & & & & l-2 & l-1 & l \\ 0 & 2 & -2 & & & & & & & \\ 1 & -1 & 2 & -1 & & & & & & \\ 2 & & -1 & 2 & \cdot & & & & & \\ & & & \cdot & \cdot & \cdot & & & & \\ & & & & \cdot & \cdot & \cdot & & & \\ & & & & & \cdot & \cdot & \cdot & & \\ l-2 & & & & & & \cdot & 2 & -1 & \\ l-1 & & & & & & & -1 & 2 & -1 \\ l & & & & & & & & -2 & 2 \end{pmatrix}$$

The integers $a_0, a_1, \ldots, a_l$.

[diagram with all labels 1]

The integers $c_0, c_1, \ldots, c_l$.

[diagram with labels 1, 2, 2, 2, 2, 2, 2, 2, 2, 1]

The central element c.

$$c = h_0 + 2h_1 + 2h_2 + \cdots + 2h_{l-1} + h_l.$$

The basic imaginary root δ.

$$\delta = \alpha_0 + \alpha_1 + \alpha_2 + \cdots + \alpha_{l-1} + \alpha_l.$$

The element $h_\theta \in H_{\mathbb{R}}^0$.
$$h_\theta = 2h_1 + 2h_2 + \cdots + 2h_{l-1} + h_l.$$

The element $\theta \in (H_{\mathbb{R}}^0)^*$.
$$\theta = \alpha_1 + \alpha_2 + \cdots + \alpha_{l-1} + \alpha_l.$$

The Coxeter number. $\quad h = l + 1.$

The dual Coxeter number. $\quad h^\vee = 2l.$

The Lie algebra L^0. $\quad L^0 = B_l.$

The lattice $M \subset H_{\mathbb{R}}^0$.
$$M = 2\mathbb{Z}h_1 + 2\mathbb{Z}h_2 + \cdots + 2\mathbb{Z}h_{l-1} + \mathbb{Z}h_l.$$

The lattice $M^* \subset (H_{\mathbb{R}}^0)^*$.
$$M^* = \mathbb{Z}\alpha_1 + \mathbb{Z}\alpha_2 + \cdots + \mathbb{Z}\alpha_{l-1} + \mathbb{Z}\alpha_l.$$

The fundamental alcove $A \subset H_{\mathbb{R}}^0$.
$$A = \{h \in H_{\mathbb{R}}^0 \; ; \; \alpha_i(h) > 0 \text{ for } i = 1, \ldots, l$$
$$\alpha_1(h) + \alpha_2(h) + \cdots + \alpha_{l-1}(h) + \alpha_l(h) < 1\}.$$

The fundamental alcove $A^* \subset (H_{\mathbb{R}}^0)^*$.
$$A^* = \{\lambda \in (H_{\mathbb{R}}^0)^* \; ; \; \lambda(h_i) > 0 \text{ for } i = 1, \ldots, l$$
$$2\lambda(h_1) + 2\lambda(h_2) + \cdots + 2\lambda(h_{l-1}) + \lambda(h_l) < 1\}.$$

The root system Φ in terms of the root system Φ^0 of L^0.
$$\Phi_{\text{Re},s} = \{\alpha + r\delta \; ; \; \alpha \in \Phi_s^0, \; r \in \mathbb{Z}\}$$
$$\Phi_{\text{Re},l} = \{\alpha + 2r\delta \; ; \; \alpha \in \Phi_l^0, \; r \in \mathbb{Z}\}.$$
$$\Phi_{\text{Im}} = \{2k\delta \; ; \; k \in \mathbb{Z}, \; k \neq 0\} \qquad \text{Multiplicity } l$$
$$\cup \{(2k+1)\delta \; ; \; k \in \mathbb{Z}\} \qquad \text{Multiplicity } 1.$$

The fundamental weights $\omega_i \in H_{\mathbb{R}}^* \;\; i = 0, 1, \ldots, l$ in terms of the fundamental weights $\bar{\omega}_i, \;\; i = 1, \ldots, l$ of L^0.

$$\omega_0 = \gamma, \quad \omega_1 = \bar{\omega}_1 + 2\gamma, \quad \omega_2 = \bar{\omega}_2 + 2\gamma, \quad \ldots, \omega_{l-1} = \bar{\omega}_{l-1} + 2\gamma, \quad \omega_l = \bar{\omega}_l + \gamma.$$

DYNKIN NAME $\tilde{C}_l^t$ KAC NAME $^2\tilde{D}_{l+1}$ $l \geq 2$

The standard invariant form on H.

$$\langle h_i, h_j \rangle = a_j c_j^{-1} A_{ij} \quad i, j = 0, 1, \ldots, l$$
$$\langle h_0, d \rangle = 1, \quad \langle h_i, d \rangle = 0 \quad i = 1, \ldots, l$$
$$\langle d, d \rangle = 0.$$

The standard invariant form on H^*.

$$\langle \alpha_i, \alpha_j \rangle = a_i^{-1} c_i A_{ij} \quad i, j = 0, 1, \ldots, l$$
$$\langle \alpha_0, \gamma \rangle = 1, \quad \langle \alpha_i, \gamma \rangle = 0 \quad i = 1, \ldots, l$$
$$\langle \gamma, \gamma \rangle = 0.$$

582 Appendix

DYNKIN NAME $\tilde{C}'_l$ **KAC NAME** $^2\tilde{A}_{2l}$ $l \geq 2$
(1st description)

Dynkin diagram with labelling.

Generalised Cartan matrix.

$$\begin{pmatrix}
 & 0 & 1 & 2 & \cdot & \cdot & \cdot & l-2 & l-1 & l \\
0 & 2 & -2 & & & & & & & \\
1 & -1 & 2 & -1 & & & & & & \\
2 & & -1 & 2 & \cdot & & & & & \\
\cdot & & & \cdot & \cdot & \cdot & & & & \\
\cdot & & & & \cdot & \cdot & \cdot & & & \\
\cdot & & & & & \cdot & \cdot & \cdot & & \\
l-2 & & & & & & \cdot & 2 & -1 & \\
l-1 & & & & & & & -1 & 2 & -2 \\
l & & & & & & & & -1 & 2
\end{pmatrix}$$

The integers $a_0, a_1, \ldots, a_l$.

The integers $c_0, c_1, \ldots, c_l$.

The central element c.

$$c = h_0 + 2h_1 + 2h_2 + \cdots + 2h_{l-1} + 2h_l.$$

The basic imaginary root δ.

$$\delta = 2\alpha_0 + 2\alpha_1 + 2\alpha_2 + \cdots + 2\alpha_{l-1} + \alpha_l.$$

DYNKIN NAME $\tilde{C}'_l$ KAC NAME $^2\tilde{A}_{2l}$ $l \geq 2$

The element $h_\theta \in H^0_\mathbb{R}$.

$$h_\theta = h_1 + h_2 + \cdots + h_{l-1} + h_l.$$

The element $\theta \in \left(H^0_\mathbb{R}\right)^*$.

$$\theta = 2\alpha_1 + 2\alpha_2 + \cdots + 2\alpha_{l-1} + \alpha_l.$$

The Coxeter number. $h = 2l + 1.$

The dual Coxeter number. $h^v = 2l + 1.$

The Lie algebra L^0. $L^0 = C_l.$

The lattice $M \subset H^0_\mathbb{R}$.

$$M = \mathbb{Z}h_1 + \mathbb{Z}h_2 + \cdots + \mathbb{Z}h_{l-1} + \mathbb{Z}h_l.$$

The lattice $M^* \subset \left(H^0_\mathbb{R}\right)^*$.

$$M^* = \mathbb{Z}\alpha_1 + \mathbb{Z}\alpha_2 + \cdots + \mathbb{Z}\alpha_{l-1} + \tfrac{1}{2}\mathbb{Z}\alpha_l.$$

The fundamental alcove $A \subset H^0_\mathbb{R}$.

$$A = \{h \in H^0_\mathbb{R} \; ; \; \alpha_i(h) > 0 \text{ for } i = 1, \ldots, l$$
$$2\alpha_1(h) + 2\alpha_2(h) + \cdots + 2\alpha_{l-1}(h) + \alpha_l(h) < 1\}.$$

The fundamental alcove $A^* \subset \left(H^0_\mathbb{R}\right)^*$.

$$A^* = \{\lambda \in \left(H^0_\mathbb{R}\right)^* \; ; \; \lambda(h_i) > 0 \text{ for } i = 1, \ldots, l$$
$$2\lambda(h_1) + 2\lambda(h_2) + \cdots + 2\lambda(h_{l-1}) + 2\lambda(h_l) < 1\}.$$

The root system Φ in terms of the root system Φ^0 of L^0.

$$\Phi_{\text{Re},s} = \{\tfrac{1}{2}(\alpha + (2r-1)\delta) \; ; \; \alpha \in \Phi^0_l, \; r \in \mathbb{Z}\}$$
$$\Phi_{\text{Re},i} = \{\alpha + r\delta \; ; \; \alpha \in \Phi^0_s, \; r \in \mathbb{Z}\}$$
$$\Phi_{\text{Re},l} = \{\alpha + 2r\delta \; ; \; \alpha \in \Phi^0_l, \; r \in \mathbb{Z}\}$$
$$\Phi_{\text{Im}} = \{k\delta \; ; \; k \in \mathbb{Z}, \; k \neq 0\} \quad \text{Multiplicity } l.$$

The fundamental weights $\omega_i \in H^*_\mathbb{R}$ $i = 0, 1, \ldots, l$ in terms of the fundamental weights $\bar{\omega}_i$ $i = 1, \ldots, l$ of L^0.

$$\omega_0 = \gamma, \quad \omega_1 = \bar{\omega}_1 + 2\gamma, \quad \omega_2 = \bar{\omega}_2 + 2\gamma, \quad \ldots,$$
$$\omega_{l-1} = \bar{\omega}_{l-1} + 2\gamma, \quad \omega_l = \bar{\omega}_l + 2\gamma.$$

The standard invariant form on H.

$$\langle h_i, h_j \rangle = a_j c_j^{-1} A_{ij} \quad i, j = 0, 1, \ldots, l$$
$$\langle h_0, d \rangle = 2, \quad \langle h_i, d \rangle = 0 \quad i = 1, \ldots, l$$
$$\langle d, d \rangle = 0.$$

The standard invariant form on H^*.

$$\langle \alpha_i, \alpha_j \rangle = a_i^{-1} c_i A_{ij} \quad i, j = 0, 1, \ldots, l$$
$$\langle \alpha_0, \gamma \rangle = \tfrac{1}{2}, \quad \langle \alpha_i, \gamma \rangle = 0 \quad i = 1, \ldots, l$$
$$\langle \gamma, \gamma \rangle = 0.$$

DYNKIN NAME $\tilde{C}'_l$ KAC NAME $^2\tilde{A}_{2l}$ $l \geq 2$
(2nd description)

Dynkin diagram with labelling.

Generalised Cartan matrix.

$$\begin{pmatrix} & 0 & 1 & 2 & \cdots & l-2 & l-1 & l \\ 0 & 2 & -1 & & & & & \\ 1 & -2 & 2 & -1 & & & & \\ 2 & & -1 & 2 & \cdot & & & \\ \cdot & & & \cdot & \cdot & \cdot & & \\ \cdot & & & & \cdot & \cdot & \cdot & \\ \cdot & & & & & \cdot & & \\ l-2 & & & & \cdot & 2 & -1 & \\ l-1 & & & & & -1 & 2 & -1 \\ l & & & & & & -2 & 2 \end{pmatrix}$$

The integers $a_0, a_1, \ldots, a_l$.

The integers $c_0, c_1, \ldots, c_l$.

The central element c.

$$c = 2h_0 + 2h_1 + 2h_2 + \cdots + 2h_{l-1} + h_l.$$

The basic imaginary root δ.

$$\delta = \alpha_0 + 2\alpha_1 + 2\alpha_2 + \cdots + 2\alpha_{l-1} + 2\alpha_l.$$

The element $h_\theta \in H_\mathbb{R}^0$.

$$h_\theta = h_1 + h_2 + \cdots + h_{l-1} + \tfrac{1}{2} h_l.$$

The element $\theta \in (H_\mathbb{R}^0)^*$.

$$\theta = 2\alpha_1 + 2\alpha_2 + \cdots + 2\alpha_{l-1} + 2\alpha_l.$$

The Coxeter number. $h = 2l+1$.

The dual Coxeter number. $h^\vee = 2l+1$.

The Lie algebra L^0. $L^0 = B_l$.

The lattice $M \subset H_\mathbb{R}^0$.

$$M = \mathbb{Z} h_1 + \mathbb{Z} h_2 + \cdots + \mathbb{Z} h_{l-1} + \tfrac{1}{2}\mathbb{Z} h_l.$$

The lattice $M^* \subset (H_\mathbb{R}^0)^*$.

$$M^* = \mathbb{Z}\alpha_1 + \mathbb{Z}\alpha_2 + \cdots + \mathbb{Z}\alpha_{l-1} + \mathbb{Z}\alpha_l.$$

The fundamental alcove $A \subset H_\mathbb{R}^0$.

$$A = \{h \in H_\mathbb{R}^0 \ ; \ \alpha_i(h) > 0 \ \text{ for } i=1,\ldots,l$$
$$2\alpha_1(h) + 2\alpha_2(h) + \cdots + 2\alpha_{l-1}(h) + 2\alpha_l(h) < 1\}.$$

The fundamental alcove $A^* \subset (H_\mathbb{R}^0)^*$.

$$A^* = \{\lambda \in (H_\mathbb{R}^0)^* \ ; \ \lambda(h_i) > 0 \ \text{ for } i=1,\ldots,l$$
$$2\lambda(h_1) + 2\lambda(h_2) + \cdots + 2\lambda(h_{l-1}) + \lambda(h_l) < 1\}.$$

The root system Φ in terms of the root system Φ^0 of L^0.

$$\Phi_{\text{Re},s} = \{\alpha + r\delta \ ; \ \alpha \in \Phi_s^0, \ r \in \mathbb{Z}\}$$
$$\Phi_{\text{Re},i} = \{\alpha + r\delta \ ; \ \alpha \in \Phi_l^0, \ r \in \mathbb{Z}\}$$
$$\Phi_{\text{Re},l} = \{2\alpha + (2r+1)\delta \ ; \ \alpha \in \Phi_s^0, \ r \in \mathbb{Z}\}$$
$$\Phi_{\text{Im}} = \{k\delta \ ; \ k \in \mathbb{Z}, \ k \neq 0\} \qquad \text{Multiplicity } l.$$

The fundamental weights $\omega_i \in H_\mathbb{R}^*$ $i=0,1,\ldots,l$ in terms of the fundamental weights $\bar{\omega}_i$ $i=1,\ldots,l$ of L^0.

$$\omega_0 = 2\gamma, \quad \omega_1 = \bar{\omega}_1 + 2\gamma, \quad \omega_2 = \bar{\omega}_2 + 2\gamma, \quad \ldots,$$
$$\omega_{l-1} = \bar{\omega}_{l-1} + 2\gamma, \quad \omega_l = \bar{\omega}_l + \gamma.$$

DYNKIN NAME $\tilde{C}'_l$ KAC NAME $^2\tilde{A}_{2l}$ $l \geq 2$

The standard invariant form on H.

$$\langle h_i, h_j \rangle = a_j c_j^{-1} A_{ij} \quad i, j = 0, 1, \ldots, l$$
$$\langle h_0, d \rangle = \tfrac{1}{2}, \quad \langle h_i, d \rangle = 0 \quad i = 1, \ldots, l$$
$$\langle d, d \rangle = 0.$$

The standard invariant form on H^*.

$$\langle \alpha_i, \alpha_j \rangle = a_i^{-1} c_i A_{ij} \quad i, j = 0, 1, \ldots, l$$
$$\langle \alpha_0, \gamma \rangle = 2, \quad \langle \alpha_i, \gamma \rangle = 0 \quad i = 1, \ldots, l$$
$$\langle \gamma, \gamma \rangle = 0.$$

DYNKIN NAME $\tilde{D}_4$ KAC NAME $\tilde{D}_4$

Dynkin diagram with labelling.

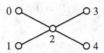

Generalised Cartan matrix.

$$\begin{array}{c} \\ 0 \\ 1 \\ 2 \\ 3 \\ 4 \end{array} \begin{pmatrix} 0 & 1 & 2 & 3 & 4 \\ 2 & 0 & -1 & 0 & 0 \\ 0 & 2 & -1 & 0 & 0 \\ -1 & -1 & 2 & -1 & -1 \\ 0 & 0 & -1 & 2 & 0 \\ 0 & 0 & -1 & 0 & 2 \end{pmatrix}$$

The integers $a_0, a_1, \ldots, a_l$.

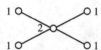

The integers $c_0, c_1, \ldots, c_l$.

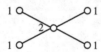

The central element c.

$$c = h_0 + h_1 + 2h_2 + h_3 + h_4.$$

The basic imaginary root δ.

$$\delta = \alpha_0 + \alpha_1 + 2\alpha_2 + \alpha_3 + \alpha_4.$$

The element $h_\theta \in H_{\mathbb{R}}^0$.

$$h_\theta = h_1 + 2h_2 + h_3 + h_4.$$

The element $\theta \in (H_{\mathbb{R}}^0)^*$.

$$\theta = \alpha_1 + 2\alpha_2 + \alpha_3 + \alpha_4.$$

The Coxeter number. $h = 6.$

The dual Coxeter number. $h^v = 6.$

DYNKIN NAME $\tilde{D}_4$ KAC NAME $\tilde{D}_4$

The Lie algebra L^0. $L^0 = D_4$.

The lattice $M \subset H_{\mathbb{R}}^0$.
$$M = \mathbb{Z}h_1 + \mathbb{Z}h_2 + \mathbb{Z}h_3 + \mathbb{Z}h_4.$$

The lattice $M^* \subset (H_{\mathbb{R}}^0)^*$.
$$M^* = \mathbb{Z}\alpha_1 + \mathbb{Z}\alpha_2 + \mathbb{Z}\alpha_3 + \mathbb{Z}\alpha_4.$$

The fundamental alcove $A \subset H_{\mathbb{R}}^0$.
$$A = \{h \in H_{\mathbb{R}}^0 \; ; \; \alpha_i(h) > 0 \text{ for } i = 1, \ldots, 4$$
$$\alpha_1(h) + 2\alpha_2(h) + \alpha_3(h) + \alpha_4(h) < 1\}.$$

The fundamental alcove $A^* \subset (H_{\mathbb{R}}^0)^*$.
$$A^* = \{\lambda \in (H_{\mathbb{R}}^0)^* \; ; \; \lambda(h_i) > 0 \text{ for } i = 1, \ldots, l$$
$$\lambda(h_1) + 2\lambda(h_2) + \lambda(h_3) + \lambda(h_4) < 1\}.$$

The root system Φ in terms of the root system Φ^0 of L^0.
$$\Phi_{\text{Re}} = \{\alpha + r\delta \; ; \; \alpha \in \Phi^0, \; r \in \mathbb{Z}\}$$
$$\Phi_{\text{Im}} = \{k\delta \; ; \; k \in \mathbb{Z}, \; k \neq 0\} \qquad \text{Multiplicity 4.}$$

The fundamental weights $\omega_i \in H_{\mathbb{R}}^*$ $i = 0, 1, \ldots l$ in terms of the fundamental weights $\bar{\omega}_i$ $i = 1, \ldots, l$ of L^0.

$\omega_0 = \gamma, \quad \omega_1 = \bar{\omega}_1 + \gamma, \quad \omega_2 = \bar{\omega}_2 + 2\gamma, \quad \omega_3 = \bar{\omega}_3 + \gamma, \quad \omega_4 = \bar{\omega}_4 + \gamma.$

The standard invariant form on H.
$$\langle h_i, h_j \rangle = A_{ij} \qquad i, j = 0, 1, 2, 3, 4$$
$$\langle h_0, d \rangle = 1, \quad \langle h_i, d \rangle = 0 \qquad i = 1, 2, 3, 4$$
$$\langle d, d \rangle = 0.$$

The standard invariant form on H^*.
$$\langle \alpha_i, \alpha_j \rangle = A_{ij} \qquad i, j = 0, 1, 2, 3, 4$$
$$\langle \alpha_0, \gamma \rangle = 1, \quad \langle \alpha_i, \gamma \rangle = 0 \qquad i = 1, 2, 3, 4$$
$$\langle \gamma, \gamma \rangle = 0.$$

DYNKIN NAME $\tilde{D}_l$ KAC NAME $\tilde{D}_l$ $l \geq 5$

Dynkin diagram with labelling.

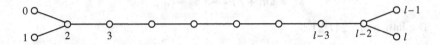

Generalised Cartan matrix.

$$\begin{pmatrix}
 & 0 & 1 & 2 & 3 & \cdots & & l-3 & l-2 & l-1 & l \\
0 & 2 & & -1 & & & & & & & \\
1 & & 2 & -1 & & & & & & & \\
2 & -1 & -1 & 2 & -1 & & & & & & \\
3 & & & -1 & 2 & \cdot & & & & & \\
 & & & & & \ddots & & & & & \\
 & & & & & & \ddots & & & & \\
l-3 & & & & & & \cdot & 2 & -1 & & \\
l-2 & & & & & & & -1 & 2 & -1 & -1 \\
l-1 & & & & & & & & -1 & 2 & \\
l & & & & & & & & -1 & & 2
\end{pmatrix}$$

The integers $a_0, a_1, \ldots, a_l$.

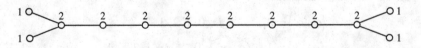

The integers $c_0, c_1, \ldots, c_l$.

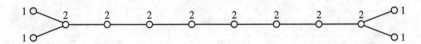

DYNKIN NAME $\tilde{D}_l$ KAC NAME $\tilde{D}_l$ $l \geq 5$

The central element c.
$$c = h_0 + h_1 + 2h_2 + 2h_3 + \cdots + 2h_{l-2} + h_{l-1} + h_l.$$

The basic imaginary root δ.
$$\delta = \alpha_0 + \alpha_1 + 2\alpha_2 + 2\alpha_3 + \cdots + 2\alpha_{l-2} + \alpha_{l-1} + \alpha_l.$$

The element $h_\theta \in H_\mathbb{R}^0$.
$$h_\theta = h_1 + 2h_2 + 2h_3 + \cdots + 2h_{l-2} + h_{l-1} + h_l.$$

The element $\theta \in (H_\mathbb{R}^0)^*$.
$$\theta = \alpha_1 + 2\alpha_2 + 2\alpha_3 + \cdots + 2\alpha_{l-2} + \alpha_{l-1} + \alpha_l.$$

The Coxeter number. $h = 2l - 2.$

The dual Coxeter number. $h^\vee = 2l - 2.$

The Lie algebra L^0. $L^0 = D_l.$

The lattice $M \subset H_\mathbb{R}^0$.
$$M = \mathbb{Z}h_1 + \mathbb{Z}h_2 + \cdots + \mathbb{Z}h_{l-1} + \mathbb{Z}h_l.$$

The lattice $M^* \subset (H_\mathbb{R}^0)^*$.
$$M^* = \mathbb{Z}\alpha_1 + \mathbb{Z}\alpha_2 + \cdots + \mathbb{Z}\alpha_{l-1} + \mathbb{Z}\alpha_l.$$

The fundamental alcove $A \subset H_\mathbb{R}^0$.
$$A = \{h \in H_\mathbb{R}^0 \ ; \ \alpha_i(h) > 0 \ \text{for} \ i = 1, \ldots, l$$
$$\alpha_1(h) + 2\alpha_2(h) + 2\alpha_3(h) + \cdots + 2\alpha_{l-2}(h) + \alpha_{l-1}(h) + \alpha_l(h) < 1\}.$$

The fundamental alcove $A^* \subset (H_\mathbb{R}^0)^*$.
$$A^* = \{\lambda \in (H_\mathbb{R}^0)^* \ ; \ \lambda(h_i) > 0 \ \text{for} \ i = 1, \ldots, l$$
$$\lambda(h_1) + 2\lambda(h_2) + 2\lambda(h_3) + \cdots + 2\lambda(h_{l-2}) + \lambda(h_{l-1}) + \lambda(h_l) < 1\}.$$

The root system Φ in terms of the root system Φ^0 of L^0.
$$\Phi_{\text{Re}} = \{\alpha + r\delta \ ; \ \alpha \in \Phi^0, \ r \in \mathbb{Z}\}$$
$$\Phi_{\text{Im}} = \{k\delta \ ; \ k \in \mathbb{Z}, \ k \neq 0\} \qquad \text{Multiplicity } l.$$

The fundamental weights $\omega_i \in H_\mathbb{R}^*$, $i = 0, 1, \ldots, l$ in terms of the fundamental weights $\bar{\omega}_i$, $i = 1, \ldots, l$ of L^0.
$$\omega_0 = \gamma, \quad \omega_1 = \bar{\omega}_1 + \gamma, \quad \omega_2 = \bar{\omega}_2 + 2\gamma, \quad \ldots,$$
$$\omega_{l-2} = \bar{\omega}_{l-2} + 2\gamma, \quad \omega_{l-1} = \bar{\omega}_{l-1} + \gamma, \quad \omega_l = \bar{\omega}_l + \gamma.$$

The standard invariant form on H.

$$\langle h_i, h_j \rangle = A_{ij} \qquad i, j = 0, 1, \ldots, l$$
$$\langle h_0, d \rangle = 1, \quad \langle h_i, d \rangle = 0 \qquad i = 1, \ldots, l$$
$$\langle d, d \rangle = 0.$$

The standard invariant form on H^*.

$$\langle \alpha_i, \alpha_j \rangle = A_{ij} \qquad i, j = 0, 1, \ldots, l$$
$$\langle \alpha_0, \gamma \rangle = 1, \quad \langle \alpha_i, \gamma \rangle = 0 \qquad i = 1, \ldots, l$$
$$\langle \gamma, \gamma \rangle = 0.$$

DYNKIN NAME $\tilde{E}_6$ KAC NAME $\tilde{E}_6$

Dynkin diagram with labelling.

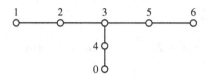

Generalised Cartan matrix.

$$\begin{pmatrix} & 0 & 1 & 2 & 3 & 4 & 5 & 6 \\ 0 & 2 & 0 & 0 & 0 & -1 & 0 & 0 \\ 1 & 0 & 2 & -1 & 0 & 0 & 0 & 0 \\ 2 & 0 & -1 & 2 & -1 & 0 & 0 & 0 \\ 3 & 0 & 0 & -1 & 2 & -1 & -1 & 0 \\ 4 & -1 & 0 & 0 & -1 & 2 & 0 & 0 \\ 5 & 0 & 0 & 0 & -1 & 0 & 2 & -1 \\ 6 & 0 & 0 & 0 & 0 & 0 & -1 & 2 \end{pmatrix}$$

The integers $a_0, a_1, \ldots, a_l$.

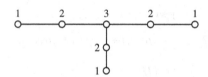

The integers $c_0, c_1, \ldots, c_l$.

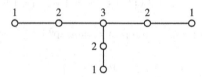

The central element c.

$$c = h_0 + h_1 + 2h_2 + 3h_3 + 2h_4 + 2h_5 + h_6.$$

The basic imaginary root δ.

$$\delta = \alpha_0 + \alpha_1 + 2\alpha_2 + 3\alpha_3 + 2\alpha_4 + 2\alpha_5 + \alpha_6.$$

The element $h_\theta \in H^0_\mathbb{R}$.

$$h_\theta = h_1 + 2h_2 + 3h_3 + 2h_4 + 2h_5 + h_6.$$

The element $\theta \in \left(H^0_\mathbb{R}\right)^*$.

$$\theta = \alpha_1 + 2\alpha_2 + 3\alpha_3 + 2\alpha_4 + 2\alpha_5 + \alpha_6.$$

The Coxeter number. $h = 12$.

The dual Coxeter number. $h^\vee = 12$.

The Lie algebra L^0. $L^0 = E_6$.

The lattice $M \subset H^0_\mathbb{R}$.

$$M = \mathbb{Z}h_1 + \mathbb{Z}h_2 + \mathbb{Z}h_3 + \mathbb{Z}h_4 + \mathbb{Z}h_5 + \mathbb{Z}h_6.$$

The lattice $M^* \subset \left(H^0_\mathbb{R}\right)^*$.

$$M^* = \mathbb{Z}\alpha_1 + \mathbb{Z}\alpha_2 + \mathbb{Z}\alpha_3 + \mathbb{Z}\alpha_4 + \mathbb{Z}\alpha_5 + \mathbb{Z}\alpha_6.$$

The fundamental alcove $A \subset H^0_\mathbb{R}$.

$$A = \{h \in H^0_\mathbb{R} \ ; \ \alpha_i(h) > 0 \ \text{ for } i = 1, \ldots, 6$$
$$\alpha_1(h) + 2\alpha_2(h) + 3\alpha_3(h) + 2\alpha_4(h) + 2\alpha_5(h) + \alpha_6(h) < 1\}.$$

The fundamental alcove $A^* \subset \left(H^0_\mathbb{R}\right)^*$.

$$A^* = \{\lambda \in \left(H^0_\mathbb{R}\right)^* \ ; \ \lambda(h_i) > 0 \ \text{ for } i = 1, \ldots, 6$$
$$\lambda(h_1) + 2\lambda(h_2) + 3\lambda(h_3) + 2\lambda(h_4) + 2\lambda(h_5) + \lambda(h_6) < 1\}.$$

The root system Φ in terms of the root system Φ^0 of L^0.

$$\Phi_{\text{Re}} = \{\alpha + r\delta \ ; \ \alpha \in \Phi^0, \ r \in \mathbb{Z}\}$$
$$\Phi_{\text{Im}} = \{k\delta \ ; \ k \in \mathbb{Z}, \ k \neq 0\} \qquad \text{Multiplicity 6.}$$

The fundamental weights $\omega_i \in H^*_\mathbb{R}$, $i = 0, \ldots, 6$ in terms of the fundamental weights $\bar\omega_i$, $i = 1, \ldots, 6$ of L^0.

$$\omega_0 = \gamma, \quad \omega_1 = \bar\omega_1 + \gamma, \quad \omega_2 = \bar\omega_2 + 2\gamma, \quad \omega_3 = \bar\omega_3 + 3\gamma,$$
$$\omega_4 = \bar\omega_4 + 2\gamma, \quad \omega_5 = \bar\omega_5 + 2\gamma, \quad \omega_6 = \bar\omega_6 + \gamma.$$

The standard invariant form on H.

$$\langle h_i, h_j \rangle = A_{ij} \qquad i, j = 0, \ldots, 6$$
$$\langle h_0, d \rangle = 1, \quad \langle h_i, d \rangle = 0 \qquad i = 1, \ldots, 6$$
$$\langle d, d \rangle = 0.$$

The standard invariant form on H^*.

$$\langle \alpha_i, \alpha_j \rangle = A_{ij} \qquad i, j = 0, \ldots, 6$$
$$\langle \alpha_0, \gamma \rangle = 1, \quad \langle \alpha_i, \gamma \rangle = 0 \qquad i = 1, \ldots, 6$$
$$\langle \gamma, \gamma \rangle = 0.$$

DYNKIN NAME $\tilde{E}_7$ KAC NAME $\tilde{E}_7$

Dynkin diagram with labelling.

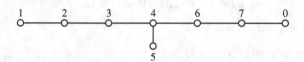

Generalised Cartan matrix.

$$\begin{array}{c c} & \begin{array}{cccccccc} 0 & 1 & 2 & 3 & 4 & 5 & 6 & 7 \end{array} \\ \begin{array}{c} 0 \\ 1 \\ 2 \\ 3 \\ 4 \\ 5 \\ 6 \\ 7 \end{array} & \left(\begin{array}{cccccccc} 2 & 0 & 0 & 0 & 0 & 0 & 0 & -1 \\ 0 & 2 & -1 & 0 & 0 & 0 & 0 & 0 \\ 0 & -1 & 2 & -1 & 0 & 0 & 0 & 0 \\ 0 & 0 & -1 & 2 & -1 & 0 & 0 & 0 \\ 0 & 0 & 0 & -1 & 2 & -1 & -1 & 0 \\ 0 & 0 & 0 & 0 & -1 & 2 & 0 & 0 \\ 0 & 0 & 0 & 0 & -1 & 0 & 2 & -1 \\ -1 & 0 & 0 & 0 & 0 & 0 & -1 & 2 \end{array} \right) \end{array}$$

The integers $a_0, a_1, \ldots, a_l$.

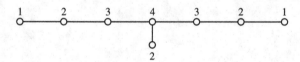

The integers $c_0, c_1, \ldots, c_l$.

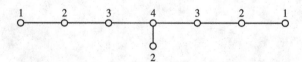

The central element c.

$$c = h_0 + h_1 + 2h_2 + 3h_3 + 4h_4 + 2h_5 + 3h_6 + 2h_7.$$

The basic imaginary root δ.

$$\delta = \alpha_0 + \alpha_1 + 2\alpha_2 + 3\alpha_3 + 4\alpha_4 + 2\alpha_5 + 3\alpha_6 + 2\alpha_7.$$

The element $h_\theta \in H_\mathbb{R}^0$.

$$h_\theta = h_1 + 2h_2 + 3h_3 + 4h_4 + 2h_5 + 3h_6 + 2h_7.$$

The element $\theta \in (H_\mathbb{R}^0)^*$.

$$\theta = \alpha_1 + 2\alpha_2 + 3\alpha_3 + 4\alpha_4 + 2\alpha_5 + 3\alpha_6 + 2\alpha_7.$$

The Coxeter number. $h = 18$.

The dual Coxeter number. $h^\vee = 18$.

The Lie algebra L^0. $L^0 = E_7$.

The lattice $M \subset H_\mathbb{R}^0$.

$$M = \mathbb{Z}h_1 + \mathbb{Z}h_2 + \mathbb{Z}h_3 + \mathbb{Z}h_4 + \mathbb{Z}h_5 + \mathbb{Z}h_6 + \mathbb{Z}h_7.$$

The lattice $M^* \subset (H_\mathbb{R}^0)^*$.

$$M^* = \mathbb{Z}\alpha_1 + \mathbb{Z}\alpha_2 + \mathbb{Z}\alpha_3 + \mathbb{Z}\alpha_4 + \mathbb{Z}\alpha_5 + \mathbb{Z}\alpha_6 + \mathbb{Z}\alpha_7.$$

The fundamental alcove $A \subset H_\mathbb{R}^0$.

$$A = \{h \in H_\mathbb{R}^0 \ ; \ \alpha_i(h) > 0 \ \text{ for } i = 1, \ldots, 7$$
$$\alpha_1(h) + 2\alpha_2(h) + 3\alpha_3(h) + 4\alpha_4(h) + 2\alpha_5(h) + 3\alpha_6(h) + 2\alpha_7(h) < 1\}.$$

The fundamental alcove $A^* \subset (H_\mathbb{R}^0)^*$.

$$A^* = \{\lambda \in (H_\mathbb{R}^0)^* \ ; \ \lambda(h_i) > 0 \ \text{ for } i = 1, \ldots, 7$$
$$\lambda(h_1) + 2\lambda(h_2) + 3\lambda(h_3) + 4\lambda(h_4) + 2\lambda(h_5) + 3\lambda(h_6) + 2\lambda(h_7) < 1\}.$$

The root system Φ in terms of the root system Φ^0 of L^0.

$$\Phi_{\text{Re}} = \{\alpha + r\delta \ ; \ \alpha \in \Phi^0, \ r \in \mathbb{Z}\}$$
$$\Phi_{\text{Im}} = \{k\delta \ ; \ k \in \mathbb{Z}, \ k \neq 0\} \qquad \text{Multiplicity 7.}$$

The fundamental weights $\omega_i \in H_\mathbb{R}^*$, $i = 0, \ldots, 7$ in terms of the fundamental weights $\bar{\omega}_i$, $i = 1, \ldots, 7$ of L^0.

$$\omega_0 = \gamma, \quad \omega_1 = \bar{\omega}_1 + \gamma, \quad \omega_2 = \bar{\omega}_2 + 2\gamma, \quad \omega_3 = \bar{\omega}_3 + 3\gamma, \quad \omega_4 = \bar{\omega}_4 + 4\gamma,$$
$$\omega_5 = \bar{\omega}_5 + 2\gamma, \quad \omega_6 = \bar{\omega}_6 + 3\gamma, \quad \omega_7 = \bar{\omega}_7 + 2\gamma.$$

The standard invariant form on H.

$$\langle h_i, h_j \rangle = A_{ij} \qquad i, j = 0, \ldots, 7$$
$$\langle h_0, d \rangle = 1, \quad \langle h_i, d \rangle = 0 \qquad i = 1, \ldots, 7$$
$$\langle d, d \rangle = 0.$$

The standard invariant form on H^*.

$$\langle \alpha_i, \alpha_j \rangle = A_{ij} \qquad i, j = 0, \ldots, 7$$
$$\langle \alpha_0, \gamma \rangle = 1, \quad \langle \alpha_i, \gamma \rangle = 0 \qquad i = 1, \ldots, 7$$
$$\langle \gamma, \gamma \rangle = 0.$$

DYNKIN NAME $\tilde{E}_8$ KAC NAME $\tilde{E}_8$

Dynkin diagram with labelling.

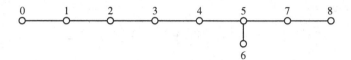

Generalised Cartan matrix.

$$\begin{array}{c|ccccccccc} & 0 & 1 & 2 & 3 & 4 & 5 & 6 & 7 & 8 \\ \hline 0 & 2 & -1 & 0 & 0 & 0 & 0 & 0 & 0 & 0 \\ 1 & -1 & 2 & -1 & 0 & 0 & 0 & 0 & 0 & 0 \\ 2 & 0 & -1 & 2 & -1 & 0 & 0 & 0 & 0 & 0 \\ 3 & 0 & 0 & -1 & 2 & -1 & 0 & 0 & 0 & 0 \\ 4 & 0 & 0 & 0 & -1 & 2 & -1 & 0 & 0 & 0 \\ 5 & 0 & 0 & 0 & 0 & -1 & 2 & -1 & -1 & 0 \\ 6 & 0 & 0 & 0 & 0 & 0 & -1 & 2 & 0 & 0 \\ 7 & 0 & 0 & 0 & 0 & 0 & -1 & 0 & 2 & -1 \\ 8 & 0 & 0 & 0 & 0 & 0 & 0 & 0 & -1 & 2 \end{array}$$

The integers $a_0, a_1, \ldots, a_l$.

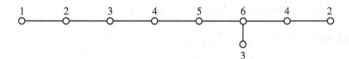

The integers $c_0, c_1, \ldots, c_l$.

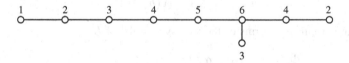

The central element c.

$$c = h_0 + 2h_1 + 3h_2 + 4h_3 + 5h_4 + 6h_5 + 3h_6 + 4h_7 + 2h_8.$$

The basic imaginary root δ.

$$\delta = \alpha_0 + 2\alpha_1 + 3\alpha_2 + 4\alpha_3 + 5\alpha_4 + 6\alpha_5 + 3\alpha_6 + 4\alpha_7 + 2\alpha_8.$$

The element $h_\theta \in H_\mathbb{R}^0$.

$$h_\theta = 2h_1 + 3h_2 + 4h_3 + 5h_4 + 6h_5 + 3h_6 + 4h_7 + 2h_8.$$

The element $\theta \in (H_\mathbb{R}^0)^*$.

$$\theta = 2\alpha_1 + 3\alpha_2 + 4\alpha_3 + 5\alpha_4 + 6\alpha_5 + 3\alpha_6 + 4\alpha_7 + 2\alpha_8.$$

The Coxeter number. $h = 30$.

The dual Coxeter number. $h^\vee = 30$.

The Lie algebra L^0. $L^0 = E_8$.

The lattice $M \subset H_\mathbb{R}^0$.

$$M = \mathbb{Z}h_1 + \mathbb{Z}h_2 + \mathbb{Z}h_3 + \mathbb{Z}h_4 + \mathbb{Z}h_5 + \mathbb{Z}h_6 + \mathbb{Z}h_7 + \mathbb{Z}h_8.$$

The lattice $M^* \subset (H_\mathbb{R}^0)^*$.

$$M^* = \mathbb{Z}\alpha_1 + \mathbb{Z}\alpha_2 + \mathbb{Z}\alpha_3 + \mathbb{Z}\alpha_4 + \mathbb{Z}\alpha_5 + \mathbb{Z}\alpha_6 + \mathbb{Z}\alpha_7 + \mathbb{Z}\alpha_8.$$

The fundamental alcove $A \subset H_\mathbb{R}^0$.

$$A = \{h \in H_\mathbb{R}^0 \ ; \ \alpha_i(h) > 0 \ \text{ for } i = 1, \ldots, 8$$
$$2\alpha_1(h) + 3\alpha_2(h) + 4\alpha_3(h) + 5\alpha_4(h) + 6\alpha_5(h)$$
$$+ 3\alpha_6(h) + 4\alpha_7(h) + 2\alpha_8(h) < 1\}.$$

The fundamental alcove $A^* \subset (H_\mathbb{R}^0)^*$.

$$A^* = \{\lambda \in (H_\mathbb{R}^0)^* \ ; \ \lambda(h_i) > 0 \ \text{ for } i = 1, \ldots, 8$$
$$2\lambda(h_1) + 3\lambda(h_2) + 4\lambda(h_3) + 5\lambda(h_4) + 6\lambda(h_5)$$
$$+ 3\lambda(h_6) + 4\lambda(h_7) + 2\lambda(h_8) < 1\}.$$

The root system Φ in terms of the root system Φ^0 of L^0.

$$\Phi_{\text{Re}} = \{\alpha + r\delta \ ; \ \alpha \in \Phi^0, \ r \in \mathbb{Z}\}$$
$$\Phi_{\text{Im}} = \{k\delta \ ; \ k \in \mathbb{Z}, \ k \neq 0\} \qquad \text{Multiplicity 8.}$$

The fundamental weights $\omega_i \in H_\mathbb{R}^*$, $i = 0, \ldots, 8$ in terms of the fundamental weights $\bar{\omega}_i$, $i = 1, \ldots, 8$ of L^0.

$$\omega_0 = \gamma, \quad \omega_1 = \bar{\omega}_1 + 2\gamma, \quad \omega_2 = \bar{\omega}_2 + 3\gamma, \quad \omega_3 = \bar{\omega}_3 + 4\gamma, \quad \omega_4 = \bar{\omega}_4 + 5\gamma,$$
$$\omega_5 = \bar{\omega}_5 + 6\gamma, \quad \omega_6 = \bar{\omega}_6 + 3\gamma, \quad \omega_7 = \bar{\omega}_7 + 4\gamma, \quad \omega_8 = \bar{\omega}_8 + 2\gamma.$$

The standard invariant form on H.

$$\langle h_i, h_j \rangle = A_{ij} \quad i,j = 0, \ldots, 8$$
$$\langle h_0, d \rangle = 1, \quad \langle h_i, d \rangle = 0 \quad i = 1, \ldots, 8$$
$$\langle d, d \rangle = 0.$$

The standard invariant form on H^*.

$$\langle \alpha_i, \alpha_j \rangle = A_{ij} \quad i,j = 0, \ldots, 8$$
$$\langle \alpha_0, \gamma \rangle = 1, \quad \langle \alpha_i, \gamma \rangle = 0 \quad i = 1, \ldots, 8$$
$$\langle \gamma, \gamma \rangle = 0.$$

DYNKIN NAME $\tilde{F}_4$ KAC NAME $\tilde{F}_4$

Dynkin diagram with labelling.

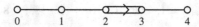

Generalised Cartan matrix.

$$\begin{array}{c} 0 1 2 3 4 \\ \begin{array}{c}0\\1\\2\\3\\4\end{array}\left(\begin{array}{rrrrr} 2 & -1 & 0 & 0 & 0 \\ -1 & 2 & -1 & 0 & 0 \\ 0 & -1 & 2 & -1 & 0 \\ 0 & 0 & -2 & 2 & -1 \\ 0 & 0 & 0 & -1 & 2 \end{array}\right) \end{array}$$

The integers $a_0, a_1, \ldots, a_l$.

$$\overset{1}{\circ}\!\!-\!\!\overset{2}{\circ}\!\!-\!\!\overset{3}{\circ}\!\!\Rightarrow\!\!\overset{4}{\circ}\!\!-\!\!\overset{2}{\circ}$$

The integers $c_0, c_1, \ldots, c_l$.

$$\overset{1}{\circ}\!\!-\!\!\overset{2}{\circ}\!\!-\!\!\overset{3}{\circ}\!\!\Rightarrow\!\!\overset{2}{\circ}\!\!-\!\!\overset{1}{\circ}$$

The central element c.

$$c = h_0 + 2h_1 + 3h_2 + 2h_3 + h_4.$$

The basic imaginary root δ.

$$\delta = \alpha_0 + 2\alpha_1 + 3\alpha_2 + 4\alpha_3 + 2\alpha_4.$$

The element $h_\theta \in H_\mathbb{R}^0$.

$$h_\theta = 2h_1 + 3h_2 + 2h_3 + h_4.$$

The element $\theta \in (h_\mathbb{R}^0)^*$.

$$\theta = 2\alpha_1 + 3\alpha_2 + 4\alpha_3 + 2\alpha_4.$$

The Coxeter number. $h = 12$.

The dual Coxeter number. $h^v = 9$.

The Lie algebra L^0. $L^0 = F_4$.

The lattice $M \subset H_\mathbb{R}^0$.

$$M = \mathbb{Z}h_1 + \mathbb{Z}h_2 + \mathbb{Z}h_3 + \mathbb{Z}h_4.$$

DYNKIN NAME $\tilde{F}_4$ KAC NAME $\tilde{F}_4$

The lattice $M^* \subset (H_\mathbb{R}^0)^*$.

$$M^* = \mathbb{Z}\alpha_1 + \mathbb{Z}\alpha_2 + 2\mathbb{Z}\alpha_3 + 2\mathbb{Z}\alpha_4.$$

The fundamental alcove $A \subset H_\mathbb{R}^0$.

$$A = \{h \in H_\mathbb{R}^0 \; ; \; \alpha_i(h) > 0 \text{ for } i = 1, \ldots, 4$$
$$2\alpha_1(h) + 3\alpha_2(h) + 4\alpha_3(h) + 2\alpha_4(h) < 1\}.$$

The fundamental alcove $A^* \subset (H_\mathbb{R}^0)^*$

$$A^* = \{\lambda \in (H_\mathbb{R}^0)^* \; ; \; \lambda(h_i) > 0 \text{ for } i = 1, \ldots, 4$$
$$2\lambda(h_1) + 3\lambda(h_2) + 2\lambda(h_3) + \lambda(h_4) < 1\}.$$

The root system Φ in terms of the root system Φ^0 of L^0.

$$\Phi_{\text{Re}} = \{\alpha + r\delta \; ; \; \alpha \in \Phi^0, \; r \in \mathbb{Z}\}$$
$$\Phi_{\text{Im}} = \{k\delta \; ; \; k \in \mathbb{Z}, \; k \neq 0\} \qquad \text{Multiplicity 4.}$$

The fundamental weights $\omega_i \in H_\mathbb{R}^*$ $i = 0, \ldots, l$ in terms of the fundamental weights $\bar{\omega}_i$, $i = 1, \ldots, l$ of L^0.

$$\omega_0 = \gamma, \quad \omega_1 = \bar{\omega}_1 + 2\gamma, \quad \omega_2 = \bar{\omega}_2 + 3\gamma, \quad \omega_3 = \bar{\omega}_3 + 2\gamma, \quad \omega_4 = \bar{\omega}_4 + \gamma.$$

The standard invariant form on H.

$$\langle h_i, h_j \rangle = a_j c_j^{-1} A_{ij} \qquad i, j = 0, \ldots, l$$
$$\langle h_0, d \rangle = 1, \quad \langle h_i, d \rangle = 0 \qquad i = 1, \ldots, l$$
$$\langle d, d \rangle = 0.$$

The standard invariant form on H^*.

$$\langle \alpha_i, \alpha_j \rangle = a_i^{-1} c_i A_{ij} \qquad i, j = 0, \ldots, l$$
$$\langle \alpha_0, \gamma \rangle = 1, \quad \langle \alpha_i, \gamma \rangle = 0 \qquad i = 1, \ldots, l$$
$$\langle \gamma, \gamma \rangle = 0.$$

DYNKIN NAME $\tilde{F}_4^t$ KAC NAME $^2\tilde{E}_6$

Dynkin diagram with labelling.

$$\underset{0}{\circ}\!\!-\!\!\underset{1}{\circ}\!\!-\!\!\underset{2}{\circ}\!\!\Leftarrow\!\!\underset{3}{\circ}\!\!-\!\!\underset{4}{\circ}$$

Generalised Cartan matrix.

$$\begin{array}{c}01234\\\begin{array}{c}0\\1\\2\\3\\4\end{array}\!\!\left(\begin{array}{rrrrr}2&-1&0&0&0\\-1&2&-1&0&0\\0&-1&2&-2&0\\0&0&-1&2&-1\\0&0&0&-1&2\end{array}\right)\end{array}$$

The integers $a_0, a_1, \ldots, a_l$.

$$\underset{1}{\circ}\!\!-\!\!\underset{2}{\circ}\!\!-\!\!\underset{3}{\circ}\!\!\Leftarrow\!\!\underset{2}{\circ}\!\!-\!\!\underset{1}{\circ}$$

The integers $c_0, c_1, \ldots, c_l$.

$$\underset{1}{\circ}\!\!-\!\!\underset{2}{\circ}\!\!-\!\!\underset{3}{\circ}\!\!\Leftarrow\!\!\underset{4}{\circ}\!\!-\!\!\underset{2}{\circ}$$

The central element c.

$$c = h_0 + 2h_1 + 3h_2 + 4h_3 + 2h_4.$$

The basic imaginary root δ.

$$\delta = \alpha_0 + 2\alpha_1 + 3\alpha_2 + 2\alpha_3 + \alpha_4.$$

The element $h_\theta \in H_\mathbb{R}^0$.

$$h_\theta = 2h_1 + 3h_2 + 4h_3 + 2h_4.$$

The element $\theta \in (H_\mathbb{R}^0)^*$.

$$\theta = 2\alpha_1 + 3\alpha_2 + 2\alpha_3 + \alpha_4.$$

The Coxeter number. $h = 9$.

The dual Coxeter number. $h^v = 12$.

The Lie algebra L^0. $L^0 = F_4$.

The lattice $M \subset H_\mathbb{R}^0$.

$$M = \mathbb{Z}h_1 + \mathbb{Z}h_2 + 2\mathbb{Z}h_3 + 2\mathbb{Z}h_4.$$

DYNKIN NAME $\tilde{F}_4^t$ KAC NAME $^2\tilde{E}_6$

The lattice $M^* \subset (H_\mathbb{R}^0)^*$.

$$M^* = \mathbb{Z}\alpha_1 + \mathbb{Z}\alpha_2 + \mathbb{Z}\alpha_3 + \mathbb{Z}\alpha_4.$$

The fundamental alcove $A \subset H_\mathbb{R}^0$.

$$A = \{h \in H_\mathbb{R}^0 \; ; \; \alpha_i(h) > 0 \text{ for } i=1,\ldots,4$$
$$2\alpha_1(h) + 3\alpha_2(h) + 2\alpha_3(h) + \alpha_4(h) < 1\}.$$

The fundamental alcove $A^* \subset (H_\mathbb{R}^0)^*$.

$$A^* = \{\lambda \in (H_\mathbb{R}^0)^* \; ; \; \lambda(h_i) > 0 \text{ for } i=1,\ldots,4$$
$$2\lambda(h_1) + 3\lambda(h_2) + 4\lambda(h_3) + 2\lambda(h_4) < 1\}.$$

The root system Φ in terms of the root system Φ^0 of L^0.

$$\Phi_{\text{Re},s} = \{\alpha + r\delta \; ; \; \alpha \in \Phi_s^0, \; r \in \mathbb{Z}\}$$
$$\Phi_{\text{Re},l} = \{\alpha + 2r\delta \; ; \; \alpha \in \Phi_l^0, \; r \in \mathbb{Z}\}$$
$$\Phi_{\text{Im}} = \{2k\delta \; ; \; k \in \mathbb{Z}, \; k \neq 0\} \quad \text{Multiplicity 4}$$
$$\cup \{(2k+1)\delta \; ; \; k \in \mathbb{Z}\} \quad \text{Multiplicity 2}.$$

The fundamental weights $\omega_i \in H_\mathbb{R}^*$ $\quad i=0,\ldots,l$ in terms of the fundamental weights $\bar{\omega}_i$, $i=1,\ldots,l$ of L^0.

$$\omega_0 = \gamma, \quad \omega_1 = \bar{\omega}_1 + 2\gamma, \quad \omega_2 = \bar{\omega}_2 + 3\gamma, \quad \omega_3 = \bar{\omega}_3 + 4\gamma, \quad \omega_4 = \bar{\omega}_4 + 2\gamma.$$

The standard invariant form on H.

$$\langle h_i, h_j \rangle = a_j c_j^{-1} A_{ij} \quad i,j = 0,\ldots l$$
$$\langle h_0, d \rangle = 1, \quad \langle h_i, d \rangle = 0 \quad i = 1,\ldots l$$
$$\langle d, d \rangle = 0.$$

The standard invariant form on H^*.

$$\langle \alpha_i, \alpha_j \rangle = a_i^{-1} c_i A_{ij} \quad i,j = 0,\ldots,l$$
$$\langle \alpha_0, \gamma \rangle = 1, \quad \langle \alpha_i, \gamma \rangle = 0 \quad i = 1,\ldots l$$
$$\langle \gamma, \gamma \rangle = 0.$$

DYNKIN NAME $\tilde{G}_2$ KAC NAME $\tilde{G}_2$

Dynkin diagram with labelling.

$$\underset{0}{\circ}\!\!-\!\!\underset{1}{\circ}\!\!\Rrightarrow\!\!\underset{2}{\circ}$$

Generalised Cartan matrix.

$$\begin{array}{c} \begin{array}{ccc} 0 & 1 & 2 \end{array} \\ \begin{array}{c} 0 \\ 1 \\ 2 \end{array}\!\!\left(\begin{array}{ccc} 2 & -1 & 0 \\ -1 & 2 & -1 \\ 0 & -3 & 2 \end{array}\right) \end{array}$$

The integers $a_0, a_1, \ldots, a_l$.

$$\underset{\circ}{\overset{1}{\circ}}\!\!-\!\!\underset{\circ}{\overset{2}{\circ}}\!\!\Rrightarrow\!\!\underset{\circ}{\overset{3}{\circ}}$$

The integers $c_0, c_1, \ldots, c_l$.

$$\underset{\circ}{\overset{1}{\circ}}\!\!-\!\!\underset{\circ}{\overset{2}{\circ}}\!\!\Rrightarrow\!\!\underset{\circ}{\overset{1}{\circ}}$$

The central element c.

$$c = h_0 + 2h_1 + h_2.$$

The basic imaginary root δ.

$$\delta = \alpha_0 + 2\alpha_1 + 3\alpha_2.$$

The element $h_\theta \in H_{\mathbb{R}}^0$.

$$h_\theta = 2h_1 + h_2.$$

The element $\theta \in \left(H_{\mathbb{R}}^0\right)^*$.

$$\theta = 2\alpha_1 + 3\alpha_2.$$

The Coxeter number. $h = 6.$

The dual Coxeter number. $h^\vee = 4.$

The Lie algebra L^0. $L^0 = G_2.$

The lattice $M \subset H_{\mathbb{R}}^0$.

$$M = \mathbb{Z}h_1 + \mathbb{Z}h_2.$$

DYNKIN NAME $\tilde{G}_2$ KAC NAME $\tilde{G}_2$

The lattice $M^* \subset (H_{\mathbb{R}}^0)^*$.

$$M^* = \mathbb{Z}\alpha_1 + 3\mathbb{Z}\alpha_2.$$

The fundamental alcove $A \subset H_{\mathbb{R}}^0$.

$$A = \{h \in H_{\mathbb{R}}^0 \; ; \; \alpha_i(h) > 0 \quad \text{for } i = 1, 2$$
$$2\alpha_1(h) + 3\alpha_2(h) < 1\}.$$

The fundamental alcove $A^* \subset (H_{\mathbb{R}}^0)^*$.

$$A^* = \{\lambda \in (H_{\mathbb{R}}^0)^* \; ; \; \lambda(h_i) > 0 \quad \text{for } i = 1, 2$$
$$2\lambda(h_1) + \lambda(h_2) < 1\}.$$

The root system Φ in terms of the root system Φ^0 of L^0.

$$\Phi_{\text{Re}} = \{\alpha + r\delta \; ; \; \alpha \in \Phi^0, r \in \mathbb{Z}\}$$
$$\Phi_{\text{Im}} = \{k\delta \; ; \; k \in \mathbb{Z}, k \neq 0\} \qquad \text{Multiplicity 2.}$$

The fundamental weights $\omega_i \in H_{\mathbb{R}}^*$ $i = 0, \ldots, l$ in terms of the fundamental weights $\bar{\omega}_i$, $i = 1, \ldots, l$ of L^0.

$$\omega_0 = \gamma, \quad \omega_1 = \bar{\omega}_1 + 2\gamma, \quad \omega_2 = \bar{\omega}_2 + \gamma.$$

The standard invariant form on H.

$$\langle h_i, h_j \rangle = a_j c_j^{-1} A_{ij} \qquad i, j = 0, \ldots, l$$
$$\langle h_0, d \rangle = 1, \quad \langle h_i, d \rangle = 0 \qquad i = 1, \ldots, l$$
$$\langle d, d \rangle = 0.$$

The standard invariant form on H^*

$$\langle \alpha_i, \alpha_j \rangle = a_i^{-1} c_i A_{ij} \qquad i, j = 0, \ldots, l$$
$$\langle \alpha_0, \gamma \rangle = 1, \quad \langle \alpha_i, \gamma \rangle = 0 \qquad i = 1, \ldots, l$$
$$\langle \gamma, \gamma \rangle = 0.$$

DYNKIN NAME $\tilde{G}_2^t$ KAC NAME $^3\tilde{D}_4$

Dynkin diagram with labelling.

$$\underset{0}{\circ} \text{---} \underset{1}{\circ} \Leftarrow \underset{2}{\circ}$$

Generalised Cartan matrix.

$$\begin{array}{c} \\ 0 \\ 1 \\ 2 \end{array} \begin{pmatrix} 0 & 1 & 2 \\ 2 & -1 & 0 \\ -1 & 2 & -3 \\ 0 & -1 & 2 \end{pmatrix}$$

The integers $a_0, a_1, \ldots, a_l$.

$$\underset{\circ}{1} \text{---} \underset{\circ}{2} \Leftarrow \underset{\circ}{1}$$

The integers $c_0, c_1, \ldots, c_l$.

$$\underset{\circ}{1} \text{---} \underset{\circ}{2} \Leftarrow \underset{\circ}{3}$$

The central element c.

$$c = h_0 + 2h_1 + 3h_2.$$

The basic imaginary root δ.

$$\delta = \alpha_0 + 2\alpha_1 + \alpha_2.$$

The element $h_\theta \in H_\mathbb{R}^0$.

$$h_\theta = 2h_1 + 3h_2.$$

The element $\theta \in (H_\mathbb{R}^0)^*$.

$$\theta = 2\alpha_1 + \alpha_2.$$

The Coxeter number. $h = 4$.

The dual Coxeter number. $h^v = 6$.

The Lie algebra L^0. $L^0 = G_2$.

The lattice $M \subset H_\mathbb{R}^0$.

$$M = \mathbb{Z}h_1 + 3\mathbb{Z}h_2.$$

DYNKIN NAME $\tilde{G}_2^t$ KAC NAME $^3\tilde{D}_4$

The lattice $M^* \subset (H_\mathbb{R}^0)^*$.

$$M^* = \mathbb{Z}\alpha_1 + \mathbb{Z}\alpha_2.$$

The fundamental alcove $A \subset H_\mathbb{R}^0$.

$$A = \{h \in H_\mathbb{R}^0 \; ; \; \alpha_i(h) > 0 \text{ for } i = 1, 2$$
$$2\alpha_1(h) + \alpha_2(h) < 1\}.$$

The fundamental alcove $A^* \subset (H_\mathbb{R}^0)^*$.

$$A^* = \{\lambda \in (H_\mathbb{R}^0)^* \; ; \; \lambda(h_i) > 0 \text{ for } i = 1, 2$$
$$2\lambda(h_1) + 3\lambda(h_2) < 1\}.$$

The root system Φ_0 in terms of the root system Φ^0 of L^0.

$$\Phi_{\text{Re},s} = \{\alpha + r\delta \; ; \; \alpha \in \Phi_s^0, \; r \in \mathbb{Z}\}$$
$$\Phi_{\text{Re},l} = \{\alpha + 3r\delta \; ; \; \alpha \in \Phi_l^0, \; r \in \mathbb{Z}\}$$
$$\Phi_{\text{Im}} = \{3k\delta \; ; \; k \in \mathbb{Z}, \; k \neq 0\} \quad \text{Multiplicity 2}$$
$$\cup \{(3k+1)\delta \; ; \; k \in \mathbb{Z}\} \quad \text{Multiplicity 1}$$
$$\cup \{(3k+2)\delta \; ; \; k \in \mathbb{Z}\} \quad \text{Multiplicity 1.}$$

The fundamental weights $\omega_i \in H_\mathbb{R}^*$ $i = 0, \ldots, l$ in terms of the fundamental weights $\bar{\omega}_i$ $i = 1, \ldots, l$ of L^0.

$$\omega_0 = \gamma, \quad \omega_1 = \bar{\omega}_1 + 2\gamma, \quad \omega_2 = \bar{\omega}_2 + 3\gamma.$$

The standard invariant form on H.

$$\langle h_i, h_j \rangle = a_j c_j^{-1} A_{ij} \quad i, j = 0, 1, 2$$
$$\langle h_0, d \rangle = 1, \quad \langle h_i, d \rangle = 0 \quad i = 1, 2$$
$$\langle d, d \rangle = 0.$$

The standard invariant form on H^*.

$$\langle \alpha_i, \alpha_j \rangle = a_i^{-1} c_i A_{ij} \quad i, j = 0, 1, 2$$
$$\langle \alpha_0, \gamma \rangle = 1, \quad \langle \alpha_i, \gamma \rangle = 0 \quad i = 1, 2$$
$$\langle \gamma, \gamma \rangle = 0.$$

Notation

Symbol	Meaning	Page of definition
$[xy]$	Lie product of elements	1
$[HK]$	Lie product of subspaces	1
$[A]$	the Lie algebra of an associative algebra A	2
$v \otimes v'$	tensor product	152
$v \wedge v'$	exterior product	271
$\langle , \rangle$	the Killing form on a finite dimensional Lie algebra	39
$\langle , \rangle$	the standard invariant form on a Kac–Moody algebra	367
$\langle , \rangle_t$	a bilinear form on the loop algebra	418
$\langle , \rangle_0$	a contravariant form	520
$\{ , \}$	a symmetric scalar product	121
$\succ$	partial order on weights	185
$[V : L(\mu)]$	the multiplicity of $L(\mu)$ in V	459
$a_0, a_1, \ldots, a_l$	vector associated with an affine Cartan matrix	386
$\operatorname{ad} x$	the adjoint map	7
$A = (A_{ij})$	a Cartan matrix	71
$A = (A_{ij})$	a generalised Cartan matrix (GCM)	319
A_J	a principal minor of A	344
A^0	the underlying Cartan matrix of an affine Cartan matrix A	394
A	the fundamental alcove	410

Notation

Symbol	Meaning	Page of definition
$\bar{A}$	the closure of the fundamental alcove	413
A^*	the fundamental alcove in the dual space	415
$\overline{A^*}$	the closure of the fundamental dual alcove	415
$\mathfrak{A}$	the set of alcoves	411
B	the subalgebra $H \oplus N$	177
c	the Casimir element	238
c	the canonical central element	391
$c_0, c_1, \ldots, c_l$	vector associated with an affine Cartan matrix	388
$c(\lambda)$	scalar action of generalised Casimir operator	487
ch V	the character of an L-module V	241
ch V	the character of a module in category $\mathcal{O}$	459
C	the fundamental chamber	112, 247, 378
$\bar{C}$	closure of the fundamental chamber	247, 378
$C(V)$	the Clifford algebra	282
$C(V)^+$	positive part of the Clifford algebra	283
$C(V)^-$	negative part of the Clifford algebra	283
$\mathbb{C}[t, t^{-1}]$	the algebra of Laurent polynomials	417
d	the scaling element	388
$d_1, \ldots, d_l$	degrees of the basic polynomial invariants	222
$d^0(\lambda)$	the Weyl dimension of an L^0-module	491
$D = (d_i)$	a diagonal matrix	110, 390
D_0	an endomorphism in the basic representation	516
e_α	a root vector	88
e_i	a fundamental root vector	96
e_i	a generator of $\tilde{L}(A)$ or $L(A)$	323, 332
e_λ	a characteristic function	242
$e(\lambda)$	the characteristic function e_λ	487
E_i	a generator of L^0	421
f_i	a root vector for $-\alpha_i$	96

Symbol	Meaning	Page of definition
f_i	a generator of $\tilde{L}(A)$ or $L(A)$	323, 332
F_i	a generator of L^0	421
$FL(X)$	the free Lie algebra on a set X	161
$\mathfrak{gl}_n(k)$	the general linear Lie algebra of degree n over k	5
G	the adjoint group	207
h	the Coxeter number of a simple Lie algebra	252
h	the Coxeter number of an affine algebra	485
h^v	the dual Coxeter number of an affine algebra	485
h_i	a fundamental coroot	88, 320
h_α	the coroot of the root α	89, 397
h'_α	the element of H corresponding to α in H^*	46
h_θ	the coroot of θ	405
$h_\alpha(n)$	an endomorphism of the basic module	513
$h(0)$	an endomorphism of the basic module	515
ht α	the height of a root α	62
H	a Cartan subalgebra of a Lie algebra	23
H	a Cartan subalgebra of a Kac–Moody algebra	334
H^*	the dual space of H	46
$H_\mathbb{Q}$	a rational vector space in H	56
$H_\mathbb{R}$	a real vector space in H	56
$H^*_\mathbb{R}$	the dual space of $H_\mathbb{R}$	57
H_i	a generator of L^0	421
H_i	a hyperplane	112
H_i^+	the positive side of hyperplane H_i	112
H_i^-	the negative side of hyperplane H_i	112
$\tilde{H}$	the diagonal subalgebra of $\tilde{L}$	324
I	the kernel of the map from $\tilde{L}(A)$ to $L(A)$	105, 331
I^+	the positive subspace of I	105, 479
I^-	the negative subspace of I	105, 479

Notation

Symbol	Meaning	Page of definition
J	an orbit	166
$J(\lambda)$	the maximal submodule of $M(\lambda)$	185, 455
$j(\tau)$	the modular j-function	533
K	a set of positive imaginary roots	380
K_λ	the kernel of the map from $\mathfrak{U}(L)$ to $M(\lambda)$	178
K_A	the set of vectors u with $Au \geq 0$	339
$\mathfrak{K}$	the generalised partition function	473
l	the rank of L	59
$l(w)$	the length of w	63
L	a Lie algebra	1
L^n	a power of the Lie algebra L	7
$L^{(n)}$	a power of the Lie algebra L	8
$L_{0,x}$	the null component of x in L	23
L_α	a root space of L	36, 333
$L(X, R)$	the Lie algebra with generators X and relations R	163
$L(A)$	the simple Lie algebra with Cartan matrix A	99
$L(A)$	the Kac–Moody algebra with GCM A	331
$L(A)'$	the derived subalgebra of the Kac–Moody algebra $L(A)$	335
$\tilde{L}(A)$	a Lie algebra associated with Cartan matrix A	99
$\tilde{L}(A)$	a Lie algebra associated with GCM A	323,
$L(A)^\sigma$	the fixed point subalgebra of σ on $L(A)$	166
$L(\lambda)$	the irreducible module with highest weight λ	186, 455, 525
$\tilde{L}_\alpha$	a root space of $\tilde{L}(A)$	328
L^0	the simple Lie algebra with Cartan matrix A^0	416
L_α	a reflecting hyperplane	246
$L_{\alpha,k}$	an affine hyperplane	409
$L_{\theta,1}$	a wall of the fundamental alcove	409
$L_0, L_1, \ldots, L_l$	the walls of the fundamental alcove	412

Symbol	Meaning	Page of definition
$\mathfrak{L}$	a set of affine hyperplanes	409
$\mathfrak{L}^*$	a set of affine hyperplanes in the dual space	414
$\mathfrak{L}(L^0)$	the loop algebra of L^0	417
$\tilde{\mathfrak{L}}(L^0)$	a central extension of the loop algebra	420
$\hat{\mathfrak{L}}(L^0)$	realisation of an untwisted Kac–Moody algebra	420
$\hat{\mathfrak{L}}(L^0)^\tau$	realisation of a twisted Kac–Moody algebra	432
m_λ	a highest weight vector in a Verma module	180, 452
m_α	multiplicity of a root α	454
M	a lattice	407
M^*	a lattice in the dual space	413
$M(\lambda)$	a Verma module	178, 452
$M^\perp$	the orthogonal subspace of a subspace M	40
$\mathfrak{M}$	Monster Lie algebra	535
n_i	an automorphism of $L(A)$	373
$n(w)$	the number of positive roots made negative by w	63
$N_{\alpha,\beta}$	structure constant	89
$\tilde{N}$	positive subalgebra of $\tilde{L}(A)$	103, 324
$\tilde{N}^-$	negative subalgebra of $\tilde{L}(A)$	103, 324
N	positive subalgebra of $L(A)$	107, 331
N^-	negative subalgebra of $L(A)$	107, 331
$N(H)$	normaliser of a subalgebra H	23
$\mathcal{O}$	Bernstein–Gelfand–Gelfand category of modules	452
$p(k)$	the number of partitions of k	507
$p_l(k)$	the number of partitions of k into l colours	508
$P(L)$	algebra of polynomial functions on L	208
$P(L)^G$	G-invariant polynomial functions on L	210

Notation

Symbol	Meaning	Page of definition
$P(H)^W$	W-invariant polynomial functions on H	211
$PSL_2(\mathbb{Z})$	the modular group	531
$\mathfrak{P}(\lambda)$	the number of partitions of λ into positive roots	182
Q	the root lattice	103, 328
Q^+	positive part of the root lattice	103, 328
Q^-	negative part of the root lattice	103, 328
Q^0	root lattice of L^0	404
$Q(x_1, \ldots, x_l)$	quadratic form	73
$\mathfrak{R}$	a ring of functions on H^*	242
s_i	fundamental reflection	63, 373
s_α	reflection	60
s_θ	reflection corresponding to root θ	405
$s_{\alpha,k}$	affine reflection	410
$\mathfrak{sl}_n(\mathbb{C})$	special linear Lie algebra of degree n over $\mathbb{C}$	52
supp α	support of a root α	378
Supp f	support of a function f	241
$S(L)$	symmetric algebra of L	201
$S(L)^G$	G-invariants in the symmetric algebra of L	223
$S(H)^W$	W-invariants in the symmetric algebra of H	223
t_x	a linear map on H	406
t_α	a linear map on H^*	413
$t(M)$	translation subgroup of the affine Weyl group	407
$t(M^*)$	translation subgroup of the affine Weyl group	413
$T(L)$	tensor algebra of L	152
$T(V)$	tensor algebra of V	324
T	a subalgebra of $L(A)$	500
T^-	the negative part of T	500
$\mathfrak{U}(L)$	universal enveloping algebra of L	153
$\mathfrak{U}(L)^+$	the ideal $L\mathfrak{U}(L)$ of $\mathfrak{U}(L)$	475
V^*	the dual module of V	306
w_0	longest element of the Weyl group	65

Notation

Symbol	Meaning	Page of definition
$(w_0)_J$	longest element of W_J	170
w_i	the weight of α_i	267
W	the Weyl group of a semisimple Lie algebra	60
W	the Weyl group of a Kac–Moody algebra	373
W^0	the Weyl group of L^0	394
W_J	a Weyl subgroup of W	170
W^σ	the group of σ-stable elements of W	169
X	the weight lattice	190, 466
X^+	dominant integral weights	190, 466
X^{++}	strictly dominant integral weights	469
$Y_\alpha(z)$	a vertex operator	514
$Z(L)$	the centre of the enveloping algebra	226
$\alpha_1, \ldots, \alpha_l$	fundamental roots of a semisimple Lie algebra	62
$\alpha_1, \ldots, \alpha_n$	fundamental roots of a Kac–Moody algebra	377
$\alpha^\vee$	dual root	150
γ	the fundamental weight ω_0 of an affine algebra	389
$\Gamma_\alpha(j)$	component in a vertex operator	515
δ	the basic imaginary root of an affine algebra	384
Δ	the Weyl denominator	245
Δ	the Kac denominator	469
Δ	the Dynkin diagram of a semisimple Lie algebra	80
$\Delta(A)$	the Dynkin diagram of a Kac–Moody algebra	353
$\Delta(\tau)$	Dedekind's delta function	533
$\phi(q)$	Euler's ϕ-function	491
Φ	the root system of a finite dimensional Lie algebra	36
Φ	the root system of a Kac–Moody algebra	377
Φ^+	set of positive roots	58, 377
Φ^-	set of negative roots	58, 377

Notation

Symbol	Meaning	Page of definition
$\Phi^\vee$	the dual root system of Φ	148
Φ_{Re}	the real roots in Φ	377
Φ_{Im}	the imaginary roots in Φ	377
$\Phi_{\text{Re},s}$	the short real roots in Φ	395
$\Phi_{\text{Re},l}$	the long real roots in Φ	395
$\Phi_{\text{Re},i}$	the intermediate real roots in Φ	395
Φ^0	the root system of L^0	394
Φ^0_s	the short roots in Φ^0	400
Φ^0_l	the long roots in Φ^0	400
χ_λ	a central character	226
$\chi^0(\lambda)$	an irreducible character of L^0	487
$\Lambda(V)$	the exterior algebra of V	271
$\Lambda^i(V)$	the i th exterior power of V	271
Π	fundamental roots of a semisimple Lie algebra	58
Π	fundamental roots of a Kac–Moody algebra	320, 334
Π	fundamental roots of a Borcherds algebra	524
$\Pi^\vee$	fundamental roots in $\Phi^\vee$	148
$\Pi^\vee$	fundamental coroots of a Kac–Moody algebra	320, 397
Π^0	fundamental roots in Φ^0	394
Π_{re}	real fundamental roots of a Borcherds algebra	525
Π_{im}	imaginary fundamental roots of a Borcherds algebra	527
ρ	sum of the fundamental weights	228
ρ	an element satisfying $\rho(h_i) = 1$	460
ρ	an element in Borcherds' character formula	526
θ	the element $\delta - a_0\alpha_0$	404
θ_i	an automorphism of L	108
θ_1	the highest root	251
θ_s	the highest short root	251
θ_0	an orbit representative	433
ω	an automorphism of $L(A)$	333
$\tilde{\omega}$	an automorphism of $\tilde{L}(A)$	323

Symbol	Meaning	Page of definition
$\omega_1, \ldots, \omega_l$	the fundamental weights of a simple Lie algebra	190
$\omega_0, \omega_1, \ldots, \omega_l$	the fundamental weights of an affine Kac–Moody algebra	494
Ω	generalised Casimir operator	461
Ω_0	an operator on a T-module in category $\mathcal{O}$	500

Bibliography

Bibliography of books on Lie algebras

Bäuerle, G. G. A. and de Kerf, E. A. *Lie Algebras, Part 1.* Studies in Mathematical Physics. 1. (1990). 394 pp. North-Holland.

Bäuerle, G. G. A., de Kerf, E. A. and ten Kroode, A. P. E. *Lie Algebras Part 2.* Studies in Mathematical Physics. 7. (1997). 554 pp. North-Holland.

Benkart, G. and Osborn, J. M. (Editors). *Lie Algebras and Related Topics.* Contemporary Mathematics. 110. (1990). 313 pp. American Mathematical Society.

Bergvelt, M. J. and ten Kroode, A. P. E. *Lectures on Kac–Moody Algebras.* CWI Syllabi. 30. (1992). 97 pp. Stichting Mathematisch Centrum, Amsterdam.

Bourbaki, N. *Groupes et Algèbres de Lie*, Chapitre 1. Actualités Sci. Indust. 1285. (1960), Paris, Hermann.

Groupes et Algèbres de Lie, Chapitres 2, 3. Actualités Sci. Indust. 1349. (1972). Paris, Hermann.

Groupes et Algèbres de Lie, Chapitres 4, 5, 6. Actualités Sci. Indust. 1337. (1968). Paris, Hermann.

Groupes et Algèbres de Lie, Chapitres 7, 8. Actualités Sci. Indust. 1364. (1975). Paris, Hermann.

Carter, R. W., Segal, G. B. and Macdonald, I. G. *Lectures on Lie Groups and Lie Algebras.* London Math. Soc. Student Texts. 32. (1995). 190 pp. Cambridge University Press.

Collingwood, D. H. and McGovern, W. M. *Nilpotent Orbits in Semisimple Lie Algebras.* (1993). 186 pp. Van Nostrand Reinhold.

Frenkel, I., Lepowsky, J. and Meurman, A. *Vertex Operator Algebras and the Monster.* Pure and Applied Mathematics. 134. (1988). 508 pp. Academic Press.

Fulton, W. and Harris, J. *Representation Theory.* Graduate Texts in Mathematics. 129. (1991). 551 pp. Springer.

Goddard, P. and Olive, D. (Editors). *Kac–Moody and Virasoro Algebras.* (1988). 586 pp. World Scientific Publishing.

Humphreys, J. E. *Introduction to Lie Algebras and Representation Theory.* Graduate Texts in Mathematics. 9. (1972). 169 pp. Springer.

Jacobson, N. *Lie Algebras.* Interscience Tracts in Pure and Applied Mathematics. 10. (1962). 331 pp. John Wiley.
  Exceptional Lie Algebras. (1971), 125 pp. Marcel Dekker.
Kac, V. *Infinite-Dimensional Lie Algebras.* Progress in Mathematics. 44. (1983). 245 pp. Birkhäuser.
  Infinite-Dimensional Lie Algebras. Third edition. (1990). 400 pp. Cambridge University Press.
  Vertex Algebras for Beginners. University lecture series. 10. (1997). 141 pp. American Mathematical Society.
Kaplansky, I. *Lie Algebras and Locally Compact Groups.* Chicago Lectures in Mathematics. (1995). 148 pp. University of Chicago Press.
Kass, S. N. (Editor). *Infinite-Dimensional Lie Algebras and their Applications.* (1988). 251 pp. World Scientific Publishing.
Kass, S. N., Moody, R. V., Patera, J. and Slansky, R. *Affine Lie Algebras, Weight Multiplicities and Branching Rules.* Vols. 1, 2. Los Alamos Series in Basic and Applied Sciences. 9. (1990). 893 pp. Berkeley, University of California Press.
Moody, R. V. and Pianzola, A. *Lie Algebras with Triangular Decomposition* (1995). 685 pp. John Wiley.
Sagle, A. A. and Walde, R. E. *Introduction to Lie Groups and Lie Algebras.* Pure and Applied Mathematics. 51. (1973). 361 pp. Academic Press.
Samelson, H. *Notes on Lie Algebras.* Second edition. Universitext. (1990). 162 pp. Springer.
Vinberg, E. B., Gorbatsevich, V. V. and Onishchik, A. L. *Structure of Lie Groups and Lie Algebras.* Current Problems in Mathematics. 41. (1990). 259 pp. Itogi Nauki i Tekhniki, Akad. Nauk. SSSR. Moscow (in Russian).
Wan, Z. X. *Lie Algebras.* International Series of Monographs in Pure and Applied Mathematics. 104. (1975). 231 pp. Pergamon Press.
  Introduction to Kac–Moody Algebra. (1991). 159 pp. World Scientific Publishing.
Winter, D. J. *Abstract Lie Algebras.* (1972). 150 pp. M.I.T. Press.

Bibliography

Bibliography of articles on Kac–Moody algebras

Allison, B., Berman, S., Gao, Y. and Pianzola, A. A characterisation of affine Kac–Moody Lie algebras. *Comm. Math. Phys.* **185**: 3 (1997), 671–688.

Altschüler, D. The critical representations of affine Lie algebras. *Modern Phys. Lett. A* **1**: 9–10 (1986), 557–564.

Ariki, S., Nakajima, T. and Yamada, H. Weight vectors of the basic $A_1^{(1)}$-module and the Littlewood–Richardson rule. *J. Phys. A* **28**: 13 (1995), 357–361.

Back-Valente, V., Bardy-Panse, N., Ben Messaoud, H. and Rousseau, G. Formes presque-déployées des algèbres de Kac–Moody: classification et racines relatives. *J. Algebra* **171**: 1 (1995), 43–96.

Bausch, J. and Rousseau, G. Algèbres de Kac–Moody affines. Automorphismes et formes réeles. Université de Nancy, Institut Elie Cartan. (1989).

Benkart, G. A Kac–Moody bibliography and some related references. Lie algebras and related topics. *CMS Conf. Proc.* **5** (1986), 111–135. American Mathematical Society.

Benkart, G. and Moody, R. Derivations, central extensions, and affine Lie algebras. *Algebras Groups Geom.* **3**: 4 (1986), 456–492.

Benkart, G., Kang, S. and Misra, K. Indefinite Kac–Moody algebras of classical type. *Adv. Math.* **105**: 1 (1994), 76–110.

Indefinite Kac–Moody algebras of special linear type. *Pacific J. Math.* **170**: 2 (1995), 379–404.

Weight multiplicity polynomials for affine Kac–Moody algebras of type $A_r^{(1)}$. *Compositio Math.* **104**: 2 (1996), 153–187.

Berman, S., Moody, R. and Wonenbuger, M. Certain matrices with null roots and finite Cartan matrices. *Indiana University Math. J.* **21** (1971–2), 1091–1099.

Berman, S. and Moody, R. Lie algebra multiplicities. *Proc. Amer. Math. Soc.* **76**: 2 (1979), 223–228.

Bernard, D. Towards generalised Macdonald's identities. Infinite-dimensional Lie algebras and groups. *Adv. Ser. Math. Phys.* **7** (1989), 467–482. World Scientific Publishing.

Bernard, D. and Thierry–Mieg, J. Level one representations of the simple affine Kac–Moody algebras in their homogeneous gradations. *Comm. Math. Phys.* **111**: 2 (1987), 181–246.

Bernstein, J., Gelfand, I. and Gelfand, S. A certain category of g-modules. *Funkcional Anal. i Prilozen.* **10**: 2 (1976), 1–8.

Billig, Y. Conjugacy theorem for the strictly hyperbolic Kac–Moody algebras. *Comm. Algebra* **23**: 3 (1995), 819–839.

Borcherds, R. Vertex algebras, Kac–Moody algebras, and the Monster. *Proc. Nat. Acad. Sci. USA* **83**: 10 (1986), 3068–3071.

Generalised Kac–Moody algebras. *J. Algebra* **115**: 2 (1988), 501–512.

The monster Lie algebra. *Adv. Math.* **83**: 1 (1990), 30–47.

Central extensions of generalised Kac–Moody algebras. *J. Algebra* **140**: 2 (1991), 330–335.

Introduction to the Monster Lie algebra. *London Math. Soc. Lecture Note Ser.* **165** (1992) 99–107. Cambridge University Press.

A characterisation of generalised Kac–Moody algebras. *J. Algebra* **174**: 3 (1995), 1073–1079.

Automorphic forms on $O_{s+2,2}(\mathbb{R})$ and infinite products. *Invent. Math.* **120**: 1 (1995), 161–213.

Capps, R. Representations of affine Kac–Moody algebras and the affine scalar product. *J. Math. Phys.* **31**: 8 (1990), 1853–1858.

Capps, R. and Lyons, M. Multiplicity formulae for a class of representations of affine Kac–Moody algebras. *Rev. Math. Phys.* **6**: 1 (1994), 97–114.

Carter, R. W. Representations of the Monster. *Textos de Matemática (Serie B)*, no. 33 (2002). 59 pp.

Casperson, D. Strictly imaginary roots of Kac–Moody algebras. *J. Algebra* **168**: 1 (1994), 90–122.

Chari, V. Integrable representations of affine Lie algebras. *Invent. Math.* **85**: 2 (1986), 317–335.

Chari, V. and Pressley, A. Integrable representations of twisted affine Lie algebras. *J. Algebra* **113**: 2 (1988), 438–464.

Coleman, A. J. Killing and the Coxeter transformation of Kac–Moody algebras. *Invent. Math.* **95**: 3 (1989), 447–477.

Coleman, A. J. and Howard, M. Root multiplicities for general Kac–Moody algebras. *C. R. Math. Rep. Acad. Sci. Canada* **11**: 1 (1989), 15–18.

Date, E., Jimbo, M., Kashiwara, M. and Miwa, T. Solitons, τ functions and Euclidean Lie algebras. *Progr. Math.* **37** (1983), 261–279. Birkhäuser.

Date, E., Jimbo, M., Kuniba, A., Miwa, T. and Okado, M. A new realization of the basic representation of $A_n^{(1)}$. *Lett. Math. Phys.* **17**: 1 (1989), 51–54.

Path space realization of the basic representation of $A_n^{(1)}$. *Adv. Ser. Math. Phys.* **7** (1989), 108–123. World Scientific Publishing.

Demazure, M. *Identités de Macdonald. Séminaire Bourbaki. Exp. 483.* Lecture Notes in Math. 567. (1977), 191–201. Springer.

Deodhar, V., Gabber, O. and Kac, V. Structure of some categories of representations of infinite dimensional Lie algebras. *Adv. Math.* **45**: 1 (1982), 92–116.

Deodhar, V. and Kumaresan, S. A finiteness theorem for affine Lie algebras. *J. Algebra* **103**: 2 (1986), 403–426.

Dolan, L. *Why Kac–Moody Subalgebras are Interesting in Physics*. Lectures in Appl. Math. 21. (1985), 307–324. American Mathematical Society.

    The beacon of Kac–Moody symmetry for physics. *Notices Amer. Math. Soc.* **42**: 12 (1995), 1489–1495.

Drinfel'd, V. and Sokolov, V. *Lie Algebras and Equations of Korteweg–de Vries Type*. Current Problems in Mathematics. 24. (1984), 81–180. Akad. Nauk SSSR, Moscow (in Russian).

Etinghof, P. and Kirillov, A. Jr. On the affine analogue of Jack and Macdonald polynomials. *Duke Math. J.* **78**: 2 (1995), 229–256.

Feigin, B. and Zelevinsky, A. *Representations of contragredient Lie Algebras and the Kac–Macdonald Identities*. Representations of Lie Groups and Lie Algebras, Akad. Kiadó, Budapest (1985), 25–77.

Feingold, A. A hyperbolic GCM Lie algebra and the Fibonacci numbers. *Proc. Amer. Math. Soc.* **80**: 3 (1980), 379–385.

    Tensor products of certain modules for the generalised Cartan matrix Lie algebra $A_1^{(1)}$. *Comm. Algebra* **9**: 12 (1981), 1323–1341.

    Constructions of vertex operator algebras. *Proc. Sympos. Pure Math.* **56** (1994), 317–336. American Mathematical Society.

Feingold, A. and Lepowsky, J. The Weyl–Kac character formula and power series identities. *Adv. Math.* **29**: 3 (1978), 271–309.

Feingold, A. and Frenkel, I. A hyperbolic Kac–Moody algebra and the theory of Siegel modular forms of genus 2. *Math. Ann.* **263**: 1 (1983), 87–144.

    Classical affine algebras. *Adv. Math.* **56**: 2 (1985), 117–172.

Feingold, A., Frenkel, I. and Ries, J. Representations of hyperbolic Kac–Moody algebras. *J. Algebra* **156**: 2 (1993), 433–453.

Frenkel, I. Spinor representations of affine Lie algebras. *Proc. Nat. Acad. Sci. USA* **77**: 11 (1980), 6303–6306.

    Two constructions of affine Lie algebra representations and boson–fermion correspondence in quantum field theory. *J. Funct. Anal.* **44**: 3 (1981), 259–327.

    Orbital theory for affine Lie algebras. *Invent. Math.* **77**: 2 (1984), 301–352.

    Representations of Kac–Moody Algebras and Dual Resonance Models. Lectures in Appl. Math. 21. (1985), 325–353. American Mathematical Society.

    Beyond affine Lie algebras. *Proc. Int. Cong. Math.* (1986), 821–839. American Mathematical Society.

Frenkel, I. and Kac, V. Basic representations of affine Lie algebras and dual resonance models. *Invent. Math.* **62**: 1 (1980/81), 23–66.

Frenkel, I. and Zhu, Y. Vertex operator algebras associated to representations of affine and Virasoro algebras. *Duke Math. J.* **66**: 1 (1992), 123–168.

Fuchs, J., Schellekens, B. and Schweigert, C. From Dynkin diagram symmetries to fixed point structures. *Comm. Math. Phys.* **180**: 1 (1996), 39–97.

Futornyi, V. On two constructions of representations of affine Lie algebras. Akad. Nauk Ukrain. SSR, Inst. Mat. Kiev (1987), 86–88 (in Russian).

Gabber, O. and Kac, V. On defining relations of certain infinite-dimensional Lie algebras. *Bull. Amer. Math. Soc.* **5**: 2 (1981), 185–189.

Garland, H. The arithmetic theory of loop algebras. *J. Algebra* **53**: 2 (1978), 480–551.

Garland H. and Lepowsky, J. Lie algebra homology and the Macdonald–Kac formulae. *Invent. Math.* **34**: 1 (1976), 37–76.

The Macdonald–Kac formulae as a consequence of the Euler–Poincaré principle, in *Contributions to Algebra*. (1977), pp. 165–173. Academic Press.

Gebert, R. Introduction to vertex algebras, Borcherds algebras and the Monster Lie algebra. *Internat. J. Modern Phys. A* **8**: 31 (1993), 5441–5503.

Goddard, P. and Olive, D. Kac–Moody and Virasoro algebras in relation to quantum physics. *Internat. J. Modern Phys. A* **1**: 2 (1986), 303–414.

The vertex operator construction for non-simply-laced Kac–Moody algebras, in *Topological and Geometrical Methods in Field Theory*. World Scientific Publishing. (1986), 29–57.

Gould, M. Tensor product decompositions for affine Kac–Moody algebras. *Rev. Math. Phys.* **6**: 6 (1994), 1269–1299.

Howe, R. A century of Lie theory. *Amer. Math. Soc. Centennial Publications, Vol. 2* (1988), 101–320.

Jimbo, M. and Miwa, T. Irreducible decomposition of fundamental modules for $A_l^{(1)}$ and $C_l^{(1)}$, and Hecke modular forms. *Adv. Stud. Pure Math.* **4** (1984), 97–119. North-Holland.

Joseph, A. Infinite dimensional Lie algebras in mathematics and physics, in *Group Theoretical Methods in Physics, Vol. 2* (1974), 582–665. Marseille.

Jurisich, E. Generalised Kac–Moody Lie algebras, free Lie algebras and the structure of the Monster Lie algebra. *J. Pure Appl. Algebra* **126** (1998), 233–266.

Kac, V. Simple graded Lie algebras of finite height. *Funkcional Anal. i Prilozen.* **1**: 4 (1967), 82–83 (in Russian).

Simple irreducible graded Lie algebras of finite growth. *Izv. Akad. Nauk SSSR Ser. Mat.* **32** (1968), 1323–1367 (in Russian).

Automorphisms of finite order of semisimple Lie algebras. *Funkcional Anal. i Prilozen* **3**: 3 (1969), 94–96 (in Russian).

Infinite dimensional algebras, Dedekind's η-function, classical Möbius function and the very strange formula. *Adv. in Math.* **30**: 2 (1978), 85–136.

Infinite root systems, representations of graphs and invariant theory. *Invent. Math.* **56**: 1 (1980), 57–92.

Some Remarks on Representations of Quivers and Infinite Root Systems. Lecture Notes in Math. 832. (1980), 311–327. Springer.

An elucidation of : Infinite dimensional algebras, Dedekind's η-function, classical Möbius function and the very strange formula. $E_8^{(1)}$ and the cube root of the modular invariant j. *Adv. Math.* **35**: 3 (1980), 264–273.

Infinite root systems, representations of graphs and invariant theory. II. *J. Algebra* **78**: 1 (1982), 141–162.

Laplace operators of infinite-dimensional Lie algebras and theta functions. *Proc. Nat. Acad. Sci. USA* **81**: 2 (1984), 645–647.

Modular invariance in mathematics and physics. *Amer. Math. Soc. Centennial Publications, Vol. II* (1992), 337–350.

Kac, V. and Kazhdan, D. Structure of representations with highest weight of infinite-dimensional Lie algebras. *Adv. Math.* **34**: 1 (1979), 97–108.

Kac, V., Kazhdan, D., Lepowsky, J. and Wilson, R. Realization of the basic representations of the Euclidean Lie algebras. *Adv. Math.* **42**: 1 (1981), 83–112.

Kac, V. and Peterson, D. Affine Lie algebras and Hecke modular forms. *Bull. Amer. Math. Soc.* **3**: 3 (1980), 1057–1061.

Spin and wedge representations of infinite-dimensional Lie algebras and groups. *Proc. Nat. Acad. Sci. USA* **78**: 6 (1981), 3308–3312.

Infinite flag varieties and conjugacy theorems. *Proc. Nat. Acad. Sci. USA* **80**: 6 (1983), 1778–1782.

Infinite-dimensional Lie algebras, theta functions and modular forms. *Adv. Math.* **53**: 2 (1984), 125–264.

Unitary structure in representations of infinite-dimensional groups and a convexity theorem. *Invent. Math.* **76**: 1 (1984), 1–14.

112 constructions of the basic representation of the loop group of E_8. *Symposium on Anomalies, Geometry, Topology*. (1985), 276–298. World Scientific Publishing.

Kac, V. and Wakimoto, M. Modular and conformal invariance constraints in representation theory of affine algebras. *Adv. in Math.* **70**: 2 (1988), 156–236.

Modular invariant representations of infinite-dimensional Lie algebras and superalgebras. *Proc. Nat. Acad. Sci. USA* **85**: 14 (1988), 4956–4960.

Kac, V. and Wang, S. On automorphisms of Kac–Moody algebras and groups. *Adv. Math.* **92**: 2 (1992), 129–195.

Kang, S. Kac–Moody Lie algebras, spectral sequences, and the Witt formula. *Trans. Amer. Math. Soc.* **339**: 2 (1993), 463–493.

Kang, S. Root multiplicities of Kac–Moody algebras. *Duke Math. J.* **74**: 3 (1994), 635–666.

Kang, S. and Melville, D. Rank 2 symmetric hyperbolic Kac–Moody algebras. *Nagoya Math. J.* **140** (1995), 41–75.

Kim, W. Some properties of Verma modules over affine Lie algebras. *Comm. Korean Math. Soc.* **10**: 4 (1995), 789–795.

Kobayashi, Z. Polynomial invariants of Euclidean Lie algebras. *Tsukuba J. Math.* **8**: 2 (1984), 255–259.

Kobayashi, Z. and Morita, J. Automorphisms of certain root lattices. *Tsukuba J. Math.* **7**: 2 (1983), 323–336.

Kroode ten, A. The homogeneous realization of $L(\Lambda_0)$ over $A_1^{(1)}$ and the Toda lattice, in *Infinite-Dimensional Lie Algebras and their Applications* (1988), 234–241. World Scientific Publishing.

Kroode ten, A. and Bergvelt, M. The homogeneous realization of the basic representation of $A_1^{(1)}$ and the Toda lattice. *Lett. Math. Phys.* **12**: 2 (1986), 139–147.

Kroode ten, F. and van de Leur, J. Level one representations of the twisted affine algebras $A_n^{(2)}$ and $D_n^{(2)}$. *Acta Appl. Math.* **27**: 3 (1992), 153–224.

Level one representations of the affine Lie algebra $B_n^{(1)}$. *Acta Appl. Math.* **31**: 1 (1993), 1–73.

Ku, J. Structure of the Verma module $M(-\rho)$ over Euclidean Lie algebras. *J. Algebra* **124**: 2 (1989), 367–387.

Relative version of Weyl–Kac character formula. *J. Algebra* **130**: 1 (1990), 191–197.

Kumar, S. A homology vanishing theorem for Kac–Moody algebras with coefficients in the category $\mathcal{O}$. *J. Algebra* **102**: 2 (1986), 444–462.

Demazure character formula in arbitrary Kac–Moody setting. *Invent. Math.* **89**: 2 (1987), 395–423.

Bernstein–Gel'fand–Gel'fand resolution for arbitrary Kac–Moody algebras. *Math. Ann.* **286**: 4 (1990), 709–729.

Toward proof of Lusztig's conjecture concerning negative level representations of affine Lie algebras. *J. Algebra* **164**: 2 (1994), 515–527.

Lepowsky, J. Macdonald-type identities. *Advances in Math.* **27**: 3 (1978), 230–234.

Generalised Verma modules, loop space cohomology and Macdonald-type identities. *Ann. Sci. Ecole Norm. Sup.* **12**: 2 (1979), 169–234.

Application of the numerator formula to k-rowed plane partitions. *Adv. Math.* **35**: 2 (1980), 179–194.

Euclidean Lie algebras and the modular function j. *Proc. Sympos. Pure Math.* 37 (1980), 567–570. American Mathematical Society.

Affine Lie algebras and combinatorial identities. Lecture Notes in Math. 933. (1982), 130–156. Springer.

Some constructions of the affine Lie algebra $A_1^{(1)}$. Lectures in Appl. Math. 21. (1985), 375–397. American Mathematical Society.

Calculus of twisted vertex operators. *Proc. Nat. Acad. Sci. USA* **82**: 24 (1985), 8295–8299.

Perspectives on vertex operators and the Monster. *Proc. Sympos. Pure Math.* 48. (1988), 181–197. American Mathematical Society.

Lepowsky, J. and Milne, S. Lie algebras and classical partition identities. *Proc. Nat. Acad. Sci. USA* **75**: 2 (1978), 578–579.

Lepowsky, J. and Moody, R. Hyperbolic Lie algebras and quasiregular cusps on Hilbert modular surfaces. *Math. Ann.* **245**: 1 (1979), 63–88.

Lepowsky, J. and Primc, M. Standard modules for type one affine Lie algebras. *Lecture Notes in Mathematics*. 1052. (1984), 194–251. Springer.

Lepowsky, J. and Wilson, R. Construction of the affine Lie algebra $A_1^{(1)}$. *Comm. Math. Phys.* **62**: 1 (1978), 43–53.

The Rogers–Ramanujan identities: Lie theoretic interpretation and proof. *Proc. Nat. Acad. Sci. USA* **78**: 2 (1981), 699–701.

A new family of algebras underlying the Rogers–Ramanujan identities and generalizations. *Proc. Nat. Acad. Sci. USA* **78**: 12 (1981), 7254–7258.

The structure of standard modules. I. Universal algebras and the Rogers–Ramanujan identities. *Invent. Math.* **77**: 2 (1984), 199–290.

The structure of standard modules. II. The case $A_1^{(1)}$, principal gradation. *Invent. Math.* **79**: 3 (1985), 417–442.

Z-algebras and the Rogers–Ramanujan identities. Vertex operators in mathematics and physics. Math. Sci. Res. Inst. Publ. 3. (1985), 97–142. Springer.

Li, W. Classification of generalised Cartan matrices of hyperbolic type. *Chinese Ann. Math. Ser. B* **9**: 1 (1988), 68–77.

Littelmann, P. A Littlewood–Richardson rule for symmetrisable Kac–Moody algebras. *Invent. Math.* **116** (1994), 329–346.

Paths and root operators in representation theory. *Ann. Math.* **142**: 3 (1995), 499–525.

The path model for representations of symmetrisable Kac–Moody algebras. *Proc. Int. Cong. Math.* (1995), 298–308. Birkhäuser.

Liu, L. Kostant's formula for Kac–Moody Lie algebras. *J. Algebra* **149**: 1 (1992), 155–178.

Macdonald, I. Affine root systems and Dedekind's η-function. *Invent. Math.* **15** (1972), 91–143.

Affine Lie algebras and modular forms. Lecture Notes in Math. 901. (1981), 258–276. Springer.

Mathieu, O. Formules de caractères pour les algèbres de Kac–Moody générales. *Astérisque* 159–160 (1988). 267 pp.

Meurman, A. and Primc, M. Vertex operator algebras and representations of affine Lie algebras. *Acta Appl. Math.* **44**: 1–2, (1996), 207–215.

Misra, K. Structure of certain standard modules for $A_n^{(1)}$ and the Rogers–Ramanujan identities. *J. Algebra* **88**: 1 (1984), 196–227.

Structure of some standard modules for $C_n^{(1)}$. *J. Algebra* **90**: 2 (1984), 385–409.

Mitzman, D. Integral bases for affine Lie algebras and their universal enveloping algebras. *Contemporary Mathematics.* **40**. (1985). American Mathematical Society.

Moody, R. Lie algebras associated with generalised Cartan matrices. *Bull. Amer. Math. Soc.* **73** (1967), 217–221.

A new class of Lie algebras. *J. Algebra* **10** (1968), 211–230.

Euclidean Lie algebras. *Canad. J. Math.* **21** (1969), 1432–1454.

Simple quotients of Euclidean Lie algebras. *Canad. J. Math.* **22** (1970), 839–846.

Macdonald identities and Euclidean Lie algebras. *Proc. Amer. Math. Soc.* **48** (1975), 43–52.

Root systems of hyperbolic type. *Adv. Math.* **33**: 2 (1979), 144–160.

Isomorphisms of upper Cartan-matrix Lie algebras. *Proc. London Math. Soc.* **40**: 3 (1980), 430–442.

Generalised root systems and characters. Contemp. Math. 45. (1985), 245–269. American Mathematical Society.

Moody, R. and Pianzola, A. Infinite-dimensional Lie algebras (a unifying overview). *Algebras Groups Geom.* **4**: 2 (1987), 165–213.

Moody, R. and Yokonuma, T. Root systems and Cartan matrices. *Canad. J. Math.* **34**: 1 (1982), 63–79.

Nakajima, H. Instantons and affine Lie algebras. *Nuclear Phys. B Proc. Suppl.* 46. (1996), 154–161.

Neidhardt, W. An algorithm for computing weight multiplicities in irreducible highest weight modules over Kac–Moody algebras. *Algebras Groups Geom.* **3**: 1 (1986), 19–26.

Olive, D. Kac–Moody algebras: an introduction for physicists. *Rend. Circ. Mat. Palermo* **9** (1986), 177–198.

The vertex operator construction for non-simply laced Kac–Moody algebras, in *String Theory.* (1987). 172–179. World Scientific Publishing.

Peng, L. and Xiao, J. A realization of affine Lie algebras of type $\tilde{A}_{n-1}$ via the derived categories of cyclic quivers. *CMS Conf. Proc.* **18** (1996), 539–554.

Rousseau, G. On forms of Kac–Moody algebras. *Proc. Symp. Pure Math.* **56** (1994), 393–399. Amer. Math. Soc.

Saçlioğlu, C. Dynkin diagrams for hyperbolic Kac–Moody algebras. *J. Phys. A* **22**: 18 (1989), 3753–3769.

Santhanam, T. Kostant's partition functions in Kac–Moody algebras, in *Particle Physics – Superstring Theory* (1988), 457–465. World Scientific Publishing.

Segal, G. *Loop Groups.* Lecture Notes in Math. 1111. (1985), 155–168. Springer.

Singer, P. *Serre Relations.* Contemp. Math. 110 (1990), 231–239. American Mathematical Society.

Soergel, W. Charakterformeln für Kipp-Moduln über Kac–Moody Algebren. *Represent. Theory* **1** (1997), 115–132.

Verdier, J.-L. Les représentations des algèbres de Lie affines: applications à quelques problèmes de physique. *Astérisque* **92–93** (1982), 365–377.

Wakimoto, M. Fock representations of the affine Lie algebra $A_1^{(1)}$. *Comm. Math. Phys.* **104**: 4 (1986), 605–609.

Wilson, R. *Euclidean Lie Algebras are Universal Central Extensions.* Lecture Notes in Math. 933. (1982), 210–213. Springer.

Yang, J. Kac–Moody algebras, the Monstrous Moonshine, Jacobi forms and infinite products. *Number Theory, Geometry and Related Topics.* Pyungsan Inst. Math. Sci. Seoul (1996), 13–82.

Index

abelian Lie algebra, 7, 46
adjoint group, 210
adjoint module, 7, 466
affine algebras – summary of properties, 392
affine Cartan matrix, 358
affine hyperplane, 409
affine Kac–Moody algebra, 386
affine list, 354
affine reflection, 410
affine type, 337, 342, 344
affine Weyl group, 404, 408
alcove, 410, 413, 415
arrow on Dynkin diagram, 80
associated graded algebra, 203
automorphism, 25, 323, 333, 373, 519

basic polynomial invariants, 220
basic representation, 508, 516
bijection between H, H^*, 46, 390
Borcherds algebra, 519
Borcherds' character formula, 525, 526
Borcherds' denominator formula, 526, 539

canonical central element, 391, 392
Cartan decomposition, 36, 45
Cartan matrix, 71
Cartan subalgebra, 23, 334
Casimir element, 238
category $\mathcal{O}$, 452
central character, 226, 235, 239
centre of universal enveloping algebra, 226
chain of roots, 50
chamber, 246
character of an L-module, 241
character of a module in $\mathcal{O}$, 454
character of a Verma module, 244, 255, 454
characteristic function, 242
Chevalley group, 120
Chevalley's theorem, 222
classification of Cartan matrices, 82

classification of Dynkin diagrams, 74
classification of simple Lie algebras, 118
classification of GCMs of affine type, 356
classification of GCMs of finite type, 352
classification of GCMs of indefinite type, 359
Clifford algebra, 282
closure of fundamental chamber, 247
cocycle, 418
complete reducibility theorem, 262
completely reducible module, 7
composition series of a Verma module, 257
conjugacy, 27
conjugacy of Cartan subalgebras, 34
contraction map, 298, 302
contravariant bilinear form, 520
coroot, 89, 397
Coxeter group, 66, 376
Coxeter number, 252, 485

Dedekind's Δ-function, 533
deletion condition, 64
derivation, 25
differential operator, 509
direct sum of Lie algebras, 43
dominant integral weights, 190
dominant maximal weights, 497, 499
dual Coxeter number, 485
dual module, 306
dual root, 150
dual root system, 148
Dynkin diagram, 72, 353
Dynkin name, 540

Eisenstein series, 533
Engel's theorem, 20
equivalent Cartan matrices, 81
equivalent GCMs, 336
equivalent representations, 5
Euler's ϕ-function, 491

629

Euler's identity, 494
existence theorem for simple Lie
  algebras, 117
exterior powers, 272
extraspecial pair of roots, 94

factor algebra, 3
filtered algebra, 202
finite list, 352
finite type, 336, 344, 350
fixed point subalgebra, 166, 172, 432, 445
free associative algebra, 98, 99
free Lie algebra, 161
fundamental alcove, 415
fundamental chamber, 112, 247, 249, 378
fundamental module, 267
fundamental modules for A_l, 273
fundamental modules for B_l, 276, 289
fundamental modules for C_l, 302
fundamental modules for D_l, 280, 292
fundamental modules for E_6, 307
fundamental modules for E_7, 308
fundamental modules for E_8, 310
fundamental modules for F_4, 314
fundamental modules for G_2, 316
fundamental modules for $L(\tilde{A}_1)$, 504
fundamental reflection, 62, 373
fundamental region, 413, 415, 532
fundamental root, 334, 377, 395
fundamental system of roots, 58, 61
fundamental weight, 190, 192, 494

Gauss' identity, 493, 494
general linear Lie algebra, 5
generalised Cartan matrix (GCM), 319
generalised Casimir operator, 461, 465
generalised eigenspace, 16
generalised partition function, 473
generators and relations for a Lie algebra, 99, 163, 482
graded algebra, 202
graph automorphism of a simple Lie
  algebra, 165
graph automorphism of an affine algebra, 426

Harish-Chandra homomorphism, 227
height of a root, 62
highest root, 251
highest short root, 251
highest weight vector, 452
homogeneous realisation, 517
homomorphism of Lie algebras, 4

ideal of a Lie algebra, 2
identification theorem for Kac–Moody
  algebras, 331

imaginary root, 377, 382, 384
indecomposable Cartan matrix, 83
indecomposable GCM, 336
indecomposable module, 7
indefinite type, 337, 344, 350
inner automorphism, 26
integrable module, 466
intermediate root, 395, 400
invariant bilinear form, 363, 520
invariant polynomial function, 210
irreducible module, 7
irreducible module for a semisimple Lie
  algebra, 199
irreducible module in $\mathcal{O}$, 455
isomorphism of Lie algebras, 4

j-function, 533
Jacobi identity, 1
Jacobian determinant, 28
Jacobian matrix, 28
Jordan canonical form, 14

Kac' character formula, 469, 472
Kac' denominator formula, 471
Kac–Moody algebra, 331
Kac name for a twisted affine algebra, 451, 540
Killing form, 39
Killing isomorphism, 223
Kostant's multiplicity formula, 260

lattice, 407, 414
Laurent polynomial, 417
length of element of Weyl group, 63
level of module, 494
Lie algebra, 1
Lie algebra of type A_l, 122, 543
Lie algebra of type B_l, 128, 545
Lie algebra of type C_l, 132, 547
Lie algebra of type D_l, 124, 549
Lie algebra of type E_6, 140, 551
Lie algebra of type E_7, 140, 553
Lie algebra of type E_8, 140, 555
Lie algebra of type F_4, 138, 557
Lie algebra of type G_2, 135, 559
Lie algebra of type $\tilde{A}_1$, 561
Lie algebra of type $\tilde{A}'_1 \cong {}^2\tilde{A}_2$, 563, 565
Lie algebra of type $\tilde{A}_l$, $l \geq 2$, 567
Lie algebra of type $\tilde{B}_l$, 570
Lie algebra of type $\tilde{B}^t_l \cong {}^2\tilde{A}_{2l-1}$, 573
Lie algebra of type $\tilde{C}_l$, 576
Lie algebra of type $\tilde{C}^t_l \cong {}^2\tilde{D}_{l+1}$, 579
Lie algebra of type $\tilde{C}'_l \cong {}^2\tilde{A}_{2l}$, 582, 585
Lie algebra of type $\tilde{D}_4$, 588
Lie algebra of type $\tilde{D}_l$, $l \geq 5$, 590

Index

Lie algebra of type $\tilde{E}_6$, 593
Lie algebra of type $\tilde{E}_7$, 596
Lie algebra of type $\tilde{E}_8$, 599
Lie algebra of type $\tilde{F}_4$, 602
Lie algebra of type $\tilde{F}_4^t \cong {}^2\tilde{E}_6$, 604
Lie algebra of type $\tilde{G}_2$, 606
Lie algebra of type $\tilde{G}_2^t \cong {}^3\tilde{D}_4$, 608
Lie monomial, 163
Lie word, 163
Lie's theorem, 11
locally nilpotent map, 107
long root, 145, 395, 397
longest element of Weyl group, 65
loop algebra, 418

Macdonald's identities, 487, 488
Macdonald's identity for $\tilde{A}_1$, 489
Macdonald's identity for $\tilde{A}_1'$, 490
Macdonald's ϕ-function identity, 491
Macdonald's twisted ϕ-function identity, 492
maximal weight, 497
minimal realisation, 320, 396
modular form, 532
modular group, 531
modular j-function, 533
module for a Lie algebra, 5
module for the basic representation, 516
Monster Lie algebra, 535, 539
multiplicity formula, 260, 474
multiplicity of a root, 334, 425, 440, 450, 454

nilpotent Lie algebra, 8
no-ghost theorem, 536
normaliser, 23
null component, 23

orbit, 166

partial order on weights, 185
partition function, 182, 507
Poincaré–Birkhoff–Witt (PBW) basis theorem, 155
polarisation, 215
polynomial functions, 208
positive definite quadratic form, 74
positive imaginary roots, 382, 527
positive system of roots, 58, 61
primitive vector, 478
principal minor, 344

quadratic form, 73
quintuple product identity, 490

rank, 35
real minimal realisation, 334
real root, 377, 394, 443

realisation, 319
realisation of twisted affine algebra, 432, 445
realisation of untwisted affine algebra, 421
reduced expression, 63
reflecting hyperplane, 246
reflection, 60
regular element, 23
representation of a Borcherds algebra, 524
representation of a Kac–Moody algebra, 452
representation of a Lie algebra, 5
representation of a nilpotent Lie algebra, 18
representation of a semisimple Lie algebra, 199
representation of a soluble Lie algebra, 13
residue, 418
Riemann surface, 531
root, 36, 334, 377
root lattice, 148, 328
root space, 36, 48, 334

scaling element, 388
semisimple Lie algebra, 9, 42, 85
short root, 145, 395, 396
simple Lie algebra, 9
soluble Lie algebra, 8
soluble radical, 9
special linear Lie algebra, 52
special pair of roots, 94
spin modules, 289, 292
spin representations, 281
standard invariant form, 367
standard list of Dynkin diagrams, 81
standard list of Cartan matrices, 82
Steinberg's multiplicity formula, 265
string of weights, 497, 504
structure constants, 89
subalgebra, 2
submodule, 7
support of a function, 241
symmetric algebra, 201
symmetric GCM, 345
symmetric tensor, 205
symmetrisable GCM, 346, 348
symmetrisation, 206

tensor algebra, 152, 324
total order on a vector space, 58
triangular decomposition, 104, 107, 326, 331
trichotomy theorem, 337, 350
triple product identity, 489
trivial simple Lie algebra, 9
twisted affine algebras, 429
twisted graph automorphisms, 429
twisted Harish-Chandra homomorphism, 228, 234

uniqueness theorem for simple Lie
    algebras, 95
universal Borcherds algebra, 521
universal enveloping algebra, 153
universal property of enveloping algebra, 153
universal property of free Lie algebra, 161
untwisted affine Cartan matrix, 417
upper half plane, 531

vacuum vector, 517
Verma module, 178, 182, 184, 452
vertex operator, 512

W-invariant polynomial function,
    211, 220
walls of a chamber, 247
weight, 19
weight lattice, 190
weight space, 19, 102, 184
weight space decomposition, 19
weight vector, 102
Weyl group, 60, 373
Weyl's character formula, 258
Weyl's denominator formula, 253
Weyl's dimension formula, 261, 267

Printed in the United States
By Bookmasters